ENVIRONMENTAL CHEMISTRY

Seventh Edition

STANLEY E. MANAHAN

COLEG SIR GAR	
Cypher	23.03.02
628.5	£30.99

LEWIS PUBLISHERS
Boca Raton London New York Washington, D.C.

KT-498-586

Cover image: Soybeans are a major source of protein for human and animal nutrition. Bacteria in root nodules on the soybean fix atmospheric nitrogen in a form that the plant can use, thus lowering the requirement for chemically fixed nitrogen added with fertilizer.

Library of Congress Cataloging-in-Publication Data

Manahan, Stanley E.
    Environmental chemistry / Stanley E. Manahan.—7th ed.
        p.   cm.
    Includes bibliographical references and index.
    ISBN 1-56670-492-8 (alk. paper)
    1. Environmental chemistry. I. Title.
TD193 .M36 1999
628.5′01′54—dc21   99-047521
                                                                                                        CIP

This book contains information obtained from authentic and highly regarded sources. Reprinted material is quoted with permission, and sources are indicated. A wide variety of references are listed. Reasonable efforts have been made to publish reliable data and information, but the author and the publisher cannot assume responsibility for the validity of all materials or for the consequences of their use.

Neither this book nor any part may be reproduced or transmitted in any form or by any means, electronic or mechanical, including photocopying, microfilming, and recording, or by any information storage or retrieval system, without prior permission in writing from the publisher.

The consent of CRC Press LLC does not extend to copying for general distribution, for promotion, for creating new works, or for resale. Specific permission must be obtained in writing from CRC Press LLC for such copying.

Direct all inquiries to CRC Press LLC, 2000 N.W. Corporate Blvd., Boca Raton, Florida 33431.

Trademark Notice: Product or corporate names may be trademarks or registered trademarks, and are used only for identification and explanation, without intent to infringe.

© 2000 by CRC Press LLC
Lewis Publishers is an imprint of CRC Press LLC

No claim to original U.S. Government works
International Standard Book Number 1-56670-492-8
Library of Congress Card Number 99-047521
Printed in the United States of America   2 3 4 5 6 7 8 9 0
Printed on acid-free paper

PREFACE TO THE SEVENTH EDITION

Environmental chemistry, Seventh Edition, continues much the same organizational structure, level, and emphasis that have been developed through preceding editions. In addition to providing updated material in the rapidly developing area of environmental chemistry, this edition emphasizes several major concepts that are proving essential to the practice of environmental chemistry at the beginning of the new millennium. These include the concept of the anthrosphere as a distinct sphere of the environment and the practice of industrial ecology, sometimes known as "green chemistry" as it applies to chemical science.

Chapter 1 serves as an introduction to environmental science, technology, and chemistry. Chapter 2 defines and discusses the anthrosphere, industrial ecosystems, and their relationship to environmental chemistry. Chapters 3 through 8 deal with aquatic chemistry.

Chapters 9 through 14 discuss atmospheric chemistry. Chapter 14 emphasizes the greatest success story of environment chemistry to date, the study of ozone-depleting chlorofluorocarbons which resulted in the first Nobel prize awarded in environmental chemistry. It also emphasizes the greenhouse effect, which may be the greatest of all threats to the global environment as we know it.

Chapters 15 and 16 deal with the geosphere, the latter chapter emphasizing soil and agricultural chemistry. Included in the discussion of agricultural chemistry is the important and controversial new area of of transgenic crops. Another area discussed is that of conservation tillage, which makes limited use of herbicides to grow crops with minimum soil disturbance.

Chapters 17 through 20 cover several aspects of industrial ecology and how it relates to material and energy resources, recycling, and hazardous waste.

Chapters 21 through 23 cover the biosphere. Chapter 21 is an overview of biochemistry with emphasis upon environmental aspects. Chapter 22 introduces and outlines the topic of toxicological chemistry. Chapter 23 discusses the toxicological chemistry of various classes of chemical substances.

Chapters 24 through 27 deal with environmental chemical analysis, including water, wastes, air, and xenobiotics in biological materials.

ACKNOWLEDGMENTS

The author is pleased to acknowledge the excellent support from the staff of CRC Press/Lewis Publishers for their outstanding efforts in bringing this book to print. Arline Massey, William Heyward, and others have been unfailingly helpful in producing this work.

Special thanks are due to Scott Martin, a graduate of the Environmental Chemistry program at the University of Missouri and now with the University of Kansas Department of Pharmaceutical Chemistry, and Kenny Garrison of the University of Missouri. Scott wrote the answer manual for the Seventh Edition of *Environmental Chemistry* and was extremely helpful in making needed corrections to the work. His outstanding research effort while at the University of Missouri has made a lasting contribution to the Environmental Chemistry program. Kenny Garrison, a natural teacher, kept the Environmental Chemistry program running during the period that the book was being revised, enabling the author to devote the time required to finish the Seventh Edition. The efforts of both of these fine young scientists are very much appreciated.

00014663

CONTENTS

1 ENVIRONMENTAL SCIENCE, TECHNOLOGY, AND CHEMISTRY

_____

1.1. WHAT IS ENVIRONMENTAL SCIENCE?

This book is about environmental chemistry. To understand that topic, it is important to have some appreciation of environmental science as a whole. **Environmental science** in its broadest sense is the science of the complex interactions that occur among the terrestrial, atmospheric, aquatic, living, and anthropological environments. It includes all the disciplines, such as chemistry, biology, ecology, sociology, and government, that affect or describe these interactions. For the purposes of this book, environmental science will be defined as _the study of the earth, air, water, and living environments, and the effects of technology thereon._ To a significant degree, environmental science has evolved from investigations of the ways by which, and places in which, living organisms carry out their life cycles. This is the discipline of **natural history**, which in recent times has evolved into **ecology**, the study of environmental factors that affect organisms and how organisms interact with these factors and with each other.[1]

For better or for worse, the environment in which all humans must live has been affected irrreversibly by technology. Therefore, technology is considered strongly in this book in terms of how it affects the environment and in the ways by which, applied intelligently by those knowledgeable of environmental science, it can serve, rather than damage, this Earth upon which all living beings depend for their welfare and existence.

The Environment

Air, water, earth, life, and technology are strongly interconnected as shown in Figure 1.1. Therefore, in a sense this figure summarizes and outlines the theme of the rest of this book.

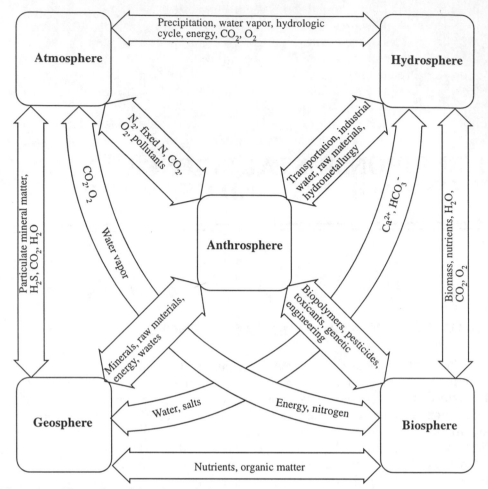

Figure 1.1. Illustration of the close relationships among the air, water, and earth environments with each other and with living systems, as well as the tie-in with technology (the anthrosphere).

Traditionally, environmental science has been divided among the study of the atmosphere, the hydrosphere, the geosphere, and the biosphere. The **atmosphere** is the thin layer of gases that cover Earth's surface. In addition to its role as a reservoir of gases, the atmosphere moderates Earth's temperature, absorbs energy and damaging ultraviolet radiation from the sun, transports energy away from equatorial regions, and serves as a pathway for vapor-phase movement of water in the hydrologic cycle. The **hydrosphere** contains Earth's water. Over 97% of Earth's water is in oceans, and most of the remaining fresh water is in the form of ice. Therefore, only a relatively small percentage of the total water on Earth is actually involved with terrestrial, atmospheric, and biological processes. Exclusive of seawater, the water that circulates through environmental processes and cycles occurs in the atmosphere, underground as groundwater, and as surface water in streams, rivers, lakes, ponds, and reservoirs. The **geosphere** consists of the solid earth, including soil, which supports most plant life. The part of the geosphere that is directly involved with environmental processes through contact with the atmosphere, the hydrosphere, and living things is the solid **lithosphere**. The lithosphere varies from 50 to 100 km in thickness. The most important part of it insofar as interactions with the other spheres of the environment are concerned is its thin outer skin composed largely of lighter silicate-based minerals and called the **crust**. All living entities on Earth compose the **biosphere**. Living organisms and the aspects of the environment pertaining directly to them are called **biotic**, and other portions of the environment are **abiotic**.

To a large extent, the strong interactions among living organisms and the various spheres of the abiotic environment are best described by cycles of matter that involve biological, chemical, and geological processes and phenomena. Such cycles are called **biogeochemical cycles**, and are discussed in more detail in Section 1.6 and elsewhere in this book.

1.2. ENVIRONMENTAL CHEMISTRY AND ENVIRONMENTAL BIOCHEMISTRY

Environmental chemistry encompasses many diverse topics. It may involve a study of Freon reactions in the stratosphere or an analysis of PCB deposits in ocean sediments. It also covers the chemistry and biochemistry of volatile and soluble organometallic compounds biosynthesized by anaerobic bacteria. Literally thousands of other examples of environmental chemical phenomena could be given. **Environmental chemistry** may be defined as *the study of the sources, reactions, transport, effects, and fates of chemical species in water, soil, air, and living environments, and the effects of technology thereon.*

Environmental chemistry is not a new discipline. Excellent work has been done in this field for the greater part of a century. Until about 1970, most of this work was done in academic departments or industrial groups other than those primarily concerned with chemistry. Much of it was performed by people whose basic education was not in chemistry. Thus, when pesticides were synthesized, biologists observed firsthand some of the less desirable consequences of their use. When detergents were formulated, sanitary engineers were startled to see sewage treatment plant aeration tanks vanish under meter-thick blankets of foam, while limnologists wondered why previously normal lakes suddenly became choked with stinking cyanobacteria. Despite these long standing environmental effects, and even more recent and serious problems, such as those from hazardous wastes, relatively few chemists have been exposed to material dealing with environmental chemistry as part of their education.

Environmental Chemistry and the Environmental Chemist

An encouraging trend is that in recent years many chemists have become deeply involved with the investigation of environmental problems. Academic chemistry departments have found that environmental chemistry courses appeal to students, and many graduate students are attracted to environmental chemistry research. Help-wanted ads have included significant numbers of openings for environmental chemists among those of the more traditional chemical subdisciplines. Industries have found that well-trained environmental chemists at least help avoid difficulties with regulatory agencies, and at best are instrumental in developing profitable pollution-control products and processes.

Some background in environmental chemistry should be part of the training of every chemistry student. The ecologically illiterate chemist can be a very dangerous species. Chemists must be aware of the possible effects their products and processes might have upon the environment. Furthermore, any serious attempt to solve environmental problems must involve the extensive use of chemicals and chemical processes.

There are some things that environmental chemistry is not. It is not just the same old chemistry with a different cover and title. Because it deals with natural systems, it is more complicated and difficult than "pure" chemistry. Students sometimes find this hard to grasp, and some traditionalist faculty find it impossible. Accustomed to the clear-cut concepts of relatively simple, well-defined, though often unrealistic systems, they may find environmental chemistry to be poorly delineated, vague, and confusing. More often than not, it is impossible to come up with a simple answer to an environmental chemistry problem. But, building on an ever-increasing body of knowledge, the environmental chemist can make educated guesses as to how environmental systems will behave.

Chemical Analysis in Environmental Chemistry

One of environmental chemistry's major challenges is the determination of the nature and quantity of specific pollutants in the environment. Thus, chemical analysis is a vital first step in environmental chemistry research. The difficulty of analyzing for many environmental pollutants can be awesome. Significant levels of air pollutants may consist of less than a microgram per cubic meter of air. For many water pollutants, one part per million by weight (essentially 1 milligram per liter) is a very high value. Environmentally significant levels of some pollutants may be only a few parts per trillion. Thus, it is obvious that the chemical analyses used to study some environmental systems require a very low limit of detection.

However, environmental chemistry is not the same as analytical chemistry, which is only one of the many subdisciplines that are involved in the study of the chemistry of the environment. Although a "brute-force" approach to environmental control, involving attempts to monitor each environmental niche for every possible pollutant, increases employment for chemists and raises sales of analytical instruments, it is a wasteful way to detect and solve environmental problems, degenerating into a mindless exercise in the collection of marginally useful numbers. Those responsible for environmental protection must be smarter than that. In order for chemistry to make a maximum contribution to the solution of environmental problems, the chemist must work toward an understanding of the nature, reactions, and transport of chemical species in the environment. Analytical chemistry is a fundamental and crucial part of that endeavor.

Environmental Biochemistry

The ultimate environmental concern is that of life itself. The discipline that deals specifically with the effects of environmental chemical species on life is

environmental biochemistry. A related area, **toxicological chemistry**, is *the chemistry of toxic substances with emphasis upon their interactions with biologic tissue and living organisms.*[2] Toxicological chemistry, which is discussed in detail in Chapters 22 and 23, deals with the chemical nature and reactions of toxic substances and involves their origins, uses, and chemical aspects of exposure, fates, and disposal.

1.3. WATER, AIR, EARTH, LIFE, AND TECHNOLOGY

In light of the above definitions, it is now possible to consider environmental chemistry from the viewpoint of the interactions among water, air, earth, life, and the anthrosphere outlined in Figure 1.1. These five environmental "spheres" and the interrelationships among them are summarized in this section. In addition, the chapters in which each of these topics is discussed in greater detail are designated here.

Water and the Hydrosphere

Water, with a deceptively simple chemical formula of H_2O, is a vitally important substance in all parts of the environment. Water covers about 70% of Earth's surface. It occurs in all spheres of the environment—in the oceans as a vast reservoir of saltwater, on land as surface water in lakes and rivers, underground as groundwater, in the atmosphere as water vapor, in the polar icecaps as solid ice, and in many segments of the anthrosphere such as in boilers or municipal water distribution systems. Water is an essential part of all living systems and is the medium from which life evolved and in which life exists.

Energy and matter are carried through various spheres of the environment by water. Water leaches soluble constituents from mineral matter and carries them to the ocean or leaves them as mineral deposits some distance from their sources. Water carries plant nutrients from soil into the bodies of plants by way of plant roots. Solar energy absorbed in the evaporation of ocean water is carried as latent heat and released inland. The accompanying release of latent heat provides a large fraction of the energy that is transported from equatorial regions toward Earth's poles and powers massive storms.

Water is obviously an important topic in environmental sciences. Its environmental chemistry is discussed in detail in Chapters 3-8.

Air and the Atmosphere

The atmosphere is a protective blanket which nurtures life on the Earth and protects it from the hostile environment of outer space. It is the source of carbon dioxide for plant photosynthesis and of oxygen for respiration. It provides the nitrogen that nitrogen-fixing bacteria and ammonia-manufacturing industrial plants use to produce chemically-bound nitrogen, an essential component of life molecules. As a basic part of the hydrologic cycle (Chapter 3, Figure 3.1), the atmosphere transports water from the oceans to land, thus acting as the condenser in a vast solar-powered still. The atmosphere serves a vital protective function, absorbing harmful ultraviolet radiation from the sun and stabilizing Earth's temperature.

Atmospheric science deals with the movement of air masses in the atmosphere, atmospheric heat balance, and atmospheric chemical composition and reactions. Atmospheric chemistry is covered in this book in Chapters 9–14.

Earth

The **geosphere**, or solid Earth, discussed in general in Chapter 15, is that part of the Earth upon which humans live and from which they extract most of their food, minerals, and fuels. The earth is divided into layers, including the solid, iron-rich inner core, molten outer core, mantle, and crust. Environmental science is most concerned with the **lithosphere**, which consists of the outer mantle and the **crust**. The latter is the earth's outer skin that is accessible to humans. It is extremely thin compared to the diameter of the earth, ranging from 5 to 40 km thick.

Geology is the science of the geosphere. As such, it pertains mostly to the solid mineral portions of Earth's crust. But it must also consider water, which is involved in weathering rocks and in producing mineral formations; the atmosphere and climate, which have profound effects on the geosphere and interchange matter and energy with it; and living systems, which largely exist on the geosphere and in turn have significant effects on it. Geological science uses chemistry in the form of geochemistry to explain the nature and behavior of geological materials, physics to explain their mechanical behavior, and biology to explain the mutual interactions between the geosphere and the biosphere.[3] Modern technology, for example the ability to move massive quantities of dirt and rock around, has a profound influence on the geosphere.

The most important part of the geosphere for life on earth is **soil** formed by the disintegrative weathering action of physical, geochemical, and biological processes on rock. It is the medium upon which plants grow, and virtually all terrestrial organisms depend upon it for their existence. The productivity of soil is strongly affected by environmental conditions and pollutants. Because of the importance of soil, all of Chapter 16 is devoted to it.

Life

Biology is the science of life. It is based on biologically synthesized chemical species, many of which exist as large molecules called *macromolecules*. As living beings, the ultimate concern of humans with their environment is the interaction of the environment with life. Therefore, biological science is a key component of environmental science and environmental chemistry

The role of life in environmental science is discussed in numerous parts of this book. For example, the crucial effects of microorganisms on aquatic chemistry are covered in Chapter 6, "Aquatic Microbial Biochemistry." Chapter 21, "Environmental Biochemistry," addresses biochemistry as it applies to the environment. The effects on living beings of toxic substances, many of which are environmental pollutants, are addressed in Chapter 22, "Toxicological Chemistry," and Chapter 23, "Toxicological Chemistry of Chemical Substances." Other chapters discuss aspects of the interaction of living systems with various parts of the environment.

The Anthrosphere and Technology

Technology refers to the ways in which humans do and make things with materials and energy. In the modern era, technology is to a large extent the product of engineering based on scientific principles. Science deals with the discovery, explanation, and development of theories pertaining to interrelated natural phenomena of energy, matter, time, and space. Based on the fundamental knowledge of science, engineering provides the plans and means to achieve specific practical objectives. Technology uses these plans to carry out the desired objectives.

It is essential to consider technology, engineering, and industrial activities in studying environmental science because of the enormous influence that they have on the environment. Humans will use technology to provide the food, shelter, and goods that they need for their well-being and survival. The challenge is to interweave technology with considerations of the environment and ecology such that the two are mutually advantageous rather than in opposition to each other.

Technology, properly applied, is an enormously positive influence for environmental protection. The most obvious such application is in air and water pollution control. As necessary as "end-of-pipe" measures are for the control of air and water pollution, it is much better to use technology in manufacturing processes to prevent the formation of pollutants. Technology is being used increasingly to develop highly efficient processes of energy conversion, renewable energy resource utilization, and conversion of raw materials to finished goods with minimum generation of hazardous waste by-products. In the transportation area, properly applied technology in areas such as high speed train transport can enormously increase the speed, energy efficiency, and safety of means for moving people and goods.

Until very recently, technological advances were made largely without heed to environmental impacts. Now, however, the greatest technological challenge is to reconcile technology with environmental consequences. The survival of humankind and of the planet that supports it now requires that the established two-way interaction between science and technology become a three-way relationship including environmental protection.

1.4. ECOLOGY AND THE BIOSPHERE

The Biosphere

The **biosphere** is the name given to that part of the environment consisting of organisms and living biological material. Virtually all of the biosphere is contained by the geosphere and hydrosphere in the very thin layer where these environmental spheres interface with the atmosphere. There are some specialized life forms at extreme depths in the ocean, but these are still relatively close to the atmospheric interface.

The biosphere strongly influences, and in turn is strongly influenced by, the other parts of the environment. It is believed that organisms were responsible for converting Earth's original reducing atmosphere to an oxygen-rich one, a process that also resulted in the formation of massive deposits of oxidized minerals, such as iron in deposits of Fe_2O_3. Photosynthetic organisms remove CO_2 from the atmosphere, thus preventing runaway greenhouse warming of Earth's surface. Organisms strongly influence bodies of water, producing biomass required for life in the water and mediating oxidation-reduction reactions in the water. Organisms are strongly involved with weathering processes that break down rocks in the geosphere and convert rock matter to soil. Lichens, consisting of symbiotic (mutually advantageous) combinations of algae and fungi, attach strongly to rocks; they secrete chemical species that slowly dissolve the rock surface and retain surface moisture that promotes rock weathering.

The biosphere is based upon plant photosynthesis, which fixes solar energy (hv) and carbon from atmospheric CO_2 in the form of high-energy biomass, represented as $\{CH_2O\}$:

$$CO_2 + H_2O \xrightarrow{hv} \{CH_2O\} + O_2(g) \tag{1.4.1}$$

In so doing, plants and algae function as autotrophic organisms, those that utilize solar or chemical energy to fix elements from simple, nonliving inorganic material into complex life molecules that compose living organisms. The opposite process, biodegradation, breaks down biomass either in the presence of oxygen (aerobic respiration),

$$\{CH_2O\} + O_2(g) \rightarrow CO_2 + H_2O \tag{1.4.2}$$

or absence of oxygen (anaerobic respiration):

$$2\{CH_2O\} \rightarrow CO_2(g) + CH_4(g) \tag{1.4.3}$$

Both aerobic and anaerobic biodegradation get rid of biomass and return carbon dioxide to the atmosphere. The latter reaction is the major source of atmospheric methane. Nondegraded remains of these processes constitute organic matter in aquatic sediments and in soils, which has an important influence on the characteristics of these solids. Carbon that was originally fixed photosynthetically forms the basis of all fossil fuels in the geosphere.

There is a strong interconnection between the biosphere and the anthrosphere. Humans depend upon the biosphere for food, fuel, and raw materials. Human influence on the biosphere continues to change it drastically. Fertilizers, pesticides, and cultivation practices have vastly increased yields of biomass, grains, and food. Destruction of habitat is resulting in the extinction of vast numbers of species, in some cases even before they are discovered. Bioengineering of organisms with recombinant DNA technology and older techniques of selection and hybridization are causing great changes in the characteristics of organisms and promise to result in even more striking alterations in the future. It is the responsibility of humankind to make such changes intelligently and to protect and nurture the biosphere.

Ecology

Ecology is the science that deals with the relationships between living organisms with their physical environment and with each other.[4] Ecology can be approached from the viewpoints of (1) the environment and the demands it places on the organisms in it or (2) organisms and how they adapt to their environmental conditions. An **ecosystem** consists of an assembly of mutually interacting organisms and their environment in which materials are interchanged in a largely cyclical manner. An ecosystem has physical, chemical, and biological components along with energy sources and pathways of energy and materials interchange. The environment in which a particular organism lives is called its **habitat.** The role of an organism in a habitat is called its **niche.**

For the study of ecology it is often convenient to divide the environment into four broad categories. The **terrestrial environment** is based on land and consists of **biomes,** such as grasslands, savannas, deserts, or one of several kinds of forests. The **freshwater environment** can be further subdivided between *standing-water habitats* (lakes, reservoirs) and *running-water habitats* (streams, rivers). The oceanic **marine environment** is characterized by saltwater and may be divided broadly into the shallow waters of the continental shelf composing the **neritic zone** and the deeper waters of the ocean that constitute the **oceanic region.** An environment in which two or more kinds of organisms exist together to their mutual benefit is termed a **symbiotic environment.**

A particularly important factor in describing ecosystems is that of **populations** consisting of numbers of a specific species occupying a specific habitat. Populations may be stable, or they may grow exponentially as a **population explosion.** A population explosion that is unchecked results in resource depletion, waste accumulation, and predation culminating in an abrupt decline called a **population crash. Behavior** in areas such as hierarchies, territoriality, social stress, and feeding patterns plays a strong role in determining the fates of populations.

Two major subdivisions of modern ecology are **ecosystem ecology,** which views ecosystems as large units, and **population ecology,** which attempts to explain ecosystem behavior from the properties of individual units. In practice, the two approaches are usually merged. **Descriptive ecology** describes the types and nature of organisms and their environment, emphasizing structures of ecosystems and communities, and dispersions and structures of populations. **Functional ecology** explains how things work in an ecosystem, including how populations respond to environmental alteration and how matter and energy move through ecosystems.

An understanding of ecology is essential in the management of modern industrialized societies in ways that are compatible with environmental preservation and enhancement. **Applied ecology** deals with predicting the impacts of technology and development and making recommendations such that these activities will have minimum adverse impact, or even positive impact, on ecosystems.

1.5. ENERGY AND CYCLES OF ENERGY

Biogeochemical cycles and virtually all other processes on Earth are driven by energy from the sun. The sun acts as a so-called blackbody radiator with an effective surface temperature of 5780 K (absolute temperature in which each unit is the same as a Celsius degree, but with zero taken at absolute zero).[5] It transmits energy to Earth as electromagnetic radiation (see below) with a maximum energy flux at about 500 nanometers, which is in the visible region of the spectrum. A 1-square-meter area perpendicular to the line of solar flux at the top of the atmosphere receives energy at a rate of 1,340 watts, sufficient, for example, to power an electric iron. This is called the **solar flux** (see Chapter 9, Figure 9.3).

Light and Electromagnetic Radiation

Electromagnetic radiation, particularly light, is of utmost importance in considering energy in environmental systems. Therefore, the following important points related to electromagnetic radiation should be noted:

- Energy can be carried through space at the speed of light (c), 3.00×10^8 meters per second (m/s) in a vacuum, by **electromagnetic radiation**, which includes visible light, ultraviolet radiation, infrared radiation, microwaves, radio waves, gamma rays, and X-rays.

- Electromagnetic radiation has a **wave character**. The waves move at the speed of light, c, and have characteristics of **wavelength** (λ), amplitude, and **frequency** (ν, Greek "nu") as illustrated below:

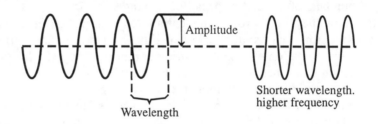

- The wavelength is the distance required for one complete cycle, and the frequency is the number of cycles per unit time. They are related by the following equation:

$$\nu \lambda = c$$

where ν is in units of cycles per second (s^{-1}, a unit called the **hertz**, Hz) and λ is in meters (m).

- In addition to behaving as a wave, electromagnetic radiation has characteristics of particles.

- The dual wave/particle nature of electromagnetic radiation is the basis of the **quantum theory** of electromagnetic radiation, which states that radiant energy may be absorbed or emitted only in discrete packets called **quanta** or **photons**. The energy, E, of each photon is given by

$$E = h\nu$$

where h is Planck's constant, 6.63×10^{-34} J-s (joule $\times$ second).

- From the preceding, it is seen that *the energy of a photon is higher when the frequency of the associated wave is higher* (and the wavelength shorter).

Energy Flow and Photosynthesis in Living Systems

Whereas materials are recycled through ecosystems, the flow of <u>useful</u> energy is essentially a one-way process. Incoming solar energy can be regarded as high-grade energy because it can cause useful reactions to occur, such as production of electricity in photovoltaic cells or photosynthesis in plants. As shown in Figure 1.2, solar energy captured by green plants energizes chlorophyll, which in turn powers metabolic processes that produce carbohydrates from water and carbon dioxide. These carbohydrates are repositories of stored chemical energy that can be converted to heat and work by metabolic reactions with oxygen in organisms. Ultimately, most of the energy is converted to low-grade heat, which is eventually reradiated away from Earth by infrared radiation.

Energy Utilization

During the last two centuries, the growing, enormous human impact on energy utilization has resulted in many of the environmental problems now facing humankind. This time period has seen a transition from the almost exclusive use of energy captured by photosynthesis and utilized as biomass (food to provide muscle power, wood for heat) to the use of fossil fuel petroleum, natural gas, and coal for about 90 percent, and nuclear energy for about 5 percent, of all energy employed commercially. Although fossil sources of energy have greatly exceeded the pessimistic estimates made during the "energy crisis" of the 1970s, they are limited and their pollution potential is high. Of particular importance is the fact that all fossil fuels produce carbon dioxide, a greenhouse gas. Therefore, it will be necessary to move toward the utilization of alternate renewable energy sources, including solar energy and biomass. The study of energy utilization is crucial in the environmental sciences, and it is discussed in greater detail in Chapter 18, "Industrial Ecology, Resources, and Energy."

1.6. MATTER AND CYCLES OF MATTER

Cycles of matter (Figure 1.3), often based on elemental cycles, are of utmost importance in the environment.[6] These cycles are summarized here and are discussed further in later chapters. Global geochemical cycles can be regarded from the viewpoint of various reservoirs, such as oceans, sediments, and the atmosphere, connected by conduits through which matter moves continuously. The movement of a specific kind of matter between two particular reservoirs may be reversible or irreversible. The fluxes of movement for specific kinds of matter vary greatly as do the contents of such matter in a specified reservoir. Cycles of matter would occur even in the absence of life on Earth but are strongly influenced by life forms, particularly plants and microorganisms. Organisms participate in **biogeochemical cycles**, which describe the circulation of matter, particularly plant and animal nutrients, through ecosystems. As part of the carbon cycle, atmospheric carbon in CO_2 is fixed as biomass; as part of the nitrogen cycle, atmospheric N_2 is fixed in organic matter. The reverse of these kinds of processes is **mineralization**, in which biologically bound elements are returned to inorganic states. Biogeochemical cycles are ultimately powered by solar energy, which is fine-tuned and directed by energy expended by organisms. In a sense, the solar-energy-powered hydrologic cycle (Figure 3.1) acts as an endless conveyer belt to move materials essential for life through ecosystems.

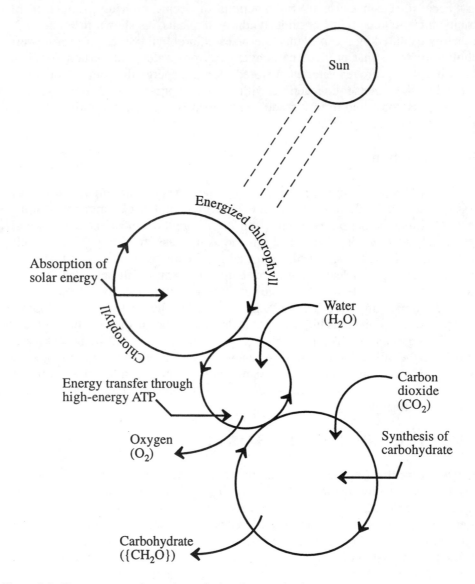

Figure 1.2. Energy conversion and transfer by photosynthesis.

Figure 1.3 shows a general cycle with all five spheres or reservoirs in which matter may be contained. Human activities now have such a strong influence on materials cycles that it is useful to refer to the "anthrosphere" along with the other environmental "spheres" as a reservoir of materials. Using Figure 1.3 as a model, it is possible to arrive at any of the known elemental cycles. Some of the numerous possibilities for materials exchange are summarized in Table 1.1.

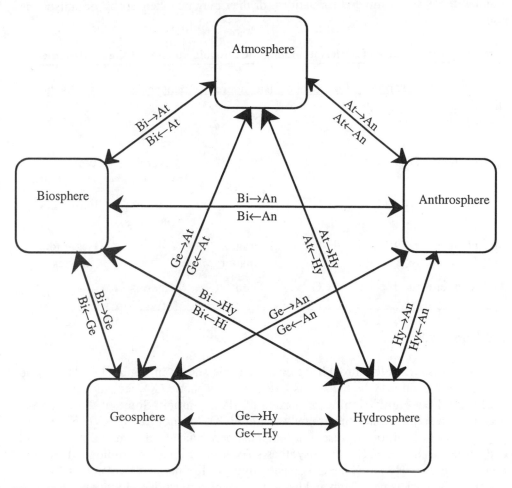

Figure 1.3. General cycle showing interchange of matter among the atmosphere, biosphere, anthrosphere, geosphere, and hydrosphere.

Endogenic and Exogenic Cycles

Materials cycles may be divided broadly between **endogenic cycles**, which predominantly involve subsurface rocks of various kinds, and **exogenic cycles**, which occur largely on Earth's surface and usually have an atmospheric component.[7] These two kinds of cycles are broadly outlined in Figure 1.4. In general, sediment and soil can be viewed as being shared between the two cycles and constitute the predominant interface between them.

Most biogeochemical cycles can be described as **elemental cycles** involving **nutrient elements** such as carbon, nitrogen, oxygen, phosphorus, and sulfur. Many are exogenic cycles in which the element in question spends part of the cycle in the atmosphere—O_2 for oxygen, N_2 for nitrogen, CO_2 for carbon. Others, notably the phosphorus cycle, do not have a gaseous component and are endogenic cycles. All sedimentary cycles involve **salt solutions** or **soil solutions** (see Section 16.2) that contain dissolved substances leached from weathered minerals; these substances may be deposited as mineral formations, or they may be taken up by organisms as nutrients.

Table 1.1. Interchange of Materials among the Possible Spheres of the Environment

From To	Atmosphere	Hydrosphere	Biosphere	Geosphere	Anthrosphere
Atmosphere	—	H_2O	O_2	H_2S, part-icles	SO_2, CO_2
Hydrosphere	H_2O	—	$\{CH_2O\}$	Mineral solutes	Water pollutants
Biosphere	O_2, CO_2	H_2O	—	Mineral nutrients	Fertilizers
Geosphere	H_2O	H_2O	Organic matter	—	Hazardous wastes
Anthrosphere	O_2, N_2	H_2O	Food	Minerals	—

Carbon Cycle

Carbon is circulated through the **carbon cycle** shown in Figure 1.5. This cycle shows that carbon may be present as gaseous atmospheric CO_2, constituting a relatively small but highly significant portion of global carbon. Some of the carbon is dissolved in surface water and groundwater as HCO_3^- or molecular $CO_2(aq)$. A very large amount of carbon is present in minerals, particularly calcium and magnesium carbonates such as $CaCO_3$. Photosynthesis fixes inorganic C as biological carbon, represented as $\{CH_2O\}$, which is a consituent of all life molecules. Another fraction of carbon is fixed as petroleum and natural gas, with a much larger amount as hydro-carbonaceous kerogen (the organic matter in oil shale), coal, and lignite, represented as C_xH_{2x}. Manufacturing processes are used to convert hydrocarbons to xenobiotic compounds with functional groups containing halogens, oxygen, nitrogen, phosphorus, or sulfur. Though a very small amount of total environmental carbon, these compounds are particularly significant because of their toxicological chemical effects.

An important aspect of the carbon cycle is that it is the cycle by which solar energy is transferred to biological systems and ultimately to the geosphere and anthrosphere as fossil carbon and fossil fuels. Organic, or biological, carbon, $\{CH_2O\}$, is contained in energy-rich molecules that can react biochemically with molecular oxygen, O_2, to regenerate carbon dioxide and produce energy. This can occur biochemically in an organism through aerobic respiration as shown in Equation 1.4.2, or it may occur as combustion, such as when wood or fossil fuels are burned.

Microorganisms are strongly involved in the carbon cycle, mediating crucial biochemical reactions discussed later in this section. Photosynthetic algae are the predominant carbon-fixing compounds in water; as they consume CO_2, the pH of the

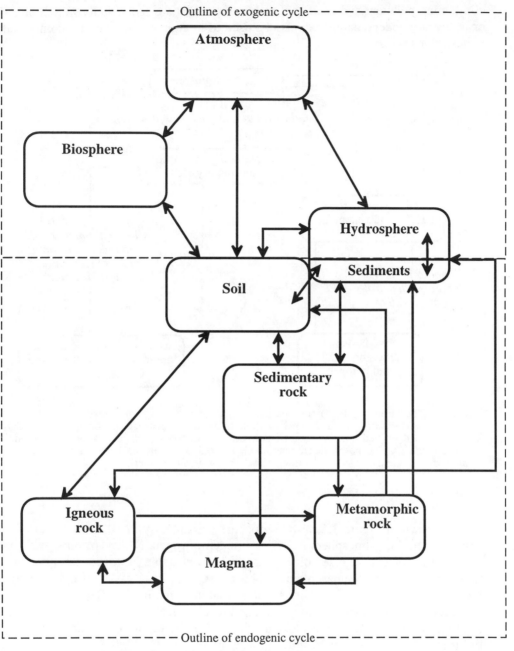

Figure 1.4. General outline of exogenic and endogenic cycles.

water is raised enabling precipitation of $CaCO_3$ and $CaCO_3 \cdot MgCO_3$. Organic carbon fixed by microorganisms is transformed by biogeochemical processes to fossil petroleum, kerogen, coal, and lignite. Microorganisms degrade organic carbon from biomass, petroleum, and xenobiotic sources, ultimately returning it to the atmosphere as CO_2. Hydrocarbons such as those in crude oil and some synthetic hydrocarbons are degraded by microorganisms. This is an important mechanism for eliminating pollutant hydrocarbons, such as those that are accidentally spilled on soil or in water. Biodegradation can also be used to treat carbon-containing compounds in hazardous wastes.

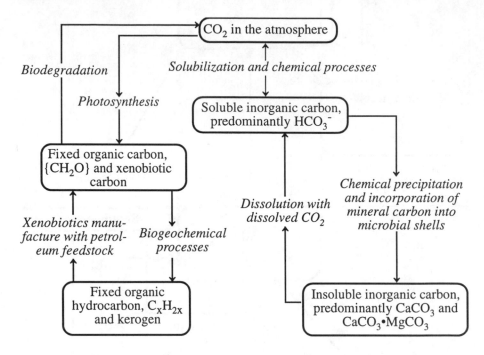

Figure 1.5. The carbon cycle. Mineral carbon is held in a reservoir of limestone, $CaCO_3$, from which it may be leached into a mineral solution as dissolved hydrogen carbonate ion, HCO_3^-, formed when dissolved $CO_2(aq)$ reacts with $CaCO_3$. In the atmosphere carbon is present as carbon dioxide, CO_2. Atmospheric carbon dioxide is fixed as organic matter by photosynthesis, and organic carbon is released as CO_2 by microbial decay of organic matter.

The Nitrogen Cycle

As shown in Figure 1.6, nitrogen occurs prominently in all the spheres of the environment. The atmosphere is 78% elemental nitrogen, N_2, by volume and comprises an inexhaustible reservoir of this essential element. Nitrogen, though constituting much less of biomass than carbon or oxygen, is an essential constituent of proteins. The N_2 molecule is very stable so that breaking it down into atoms that can be incorporated with inorganic and organic chemical forms of nitrogen is the limiting step in the nitrogen cycle. This does occur by highly energetic processes in lightning discharges that produce nitrogen oxides. Elemental nitrogen is also incorporated into chemically bound forms, or **fixed** by biochemical processes mediated by microorganisms. The biological nitrogen is mineralized to the inorganic form during the decay of biomass. Large quantities of nitrogen are fixed synthetically under high temperature and high pressure conditions according to the following overall reaction:

$$N_2 + 3H_2 \rightarrow 2NH_3 \tag{1.6.1}$$

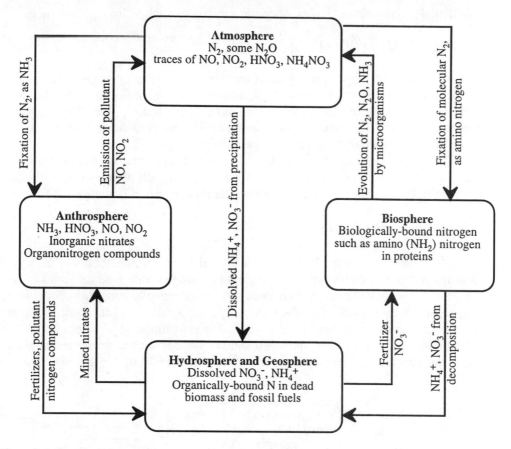

Figure 1.6. The nitrogen cycle.

The production of gaseous N_2 and N_2O by microorganisms and the evolution of these gases to the atmosphere completes the nitrogen cycle through a process called **denitrification**. The nitrogen cycle is discussed from the viewpoint of microbial processes in Section 6.11.

The Oxygen Cycle

The **oxygen cycle** is discussed in Chapter 9 and is illustrated in Figure 9.11. It involves the interchange of oxygen between the elemental form of gaseous O_2, contained in a huge reservoir in the atmosphere, and chemically bound O in CO_2, H_2O, and organic matter. It is strongly tied with other elemental cycles, particularly the carbon cycle. Elemental oxygen becomes chemically bound by various energy-yielding processes, particularly combustion and metabolic processes in organisms. It is released in photosynthesis. This element readily combines with and oxidizes other species such as carbon in aerobic respiration (Equation 1.4.2), or carbon and hydrogen in the combustion of fossil fuels such as methane:

$$CH_4 + 2O_2 \rightarrow CO_2 + 2H_2O \tag{1.6.2}$$

Elemental oxygen also oxidizes inorganic substances such as iron(II) in minerals:

$$4FeO + O_2 \rightarrow 2Fe_2O_3 \qquad (1.6.3)$$

A particularly important aspect of the oxygen cycle is stratospheric ozone, O_3. As discussed in Chapter 9, Section 9.9, a relatively small concentration of ozone in the stratosphere, more than 10 kilometers high in the atmosphere, filters out ultraviolet radiation in the wavelength range of 220-330 nm, thus protecting life on Earth from the highly damaging effects of this radiation.

The oxygen cycle is completed by the return of elemental O_2 to the atmosphere. The only significant way in which this is done is through photosynthesis mediated by plants. The overall reaction for photosynthesis is given in Equation 1.4.1.

The Phosphorus Cycle

The phosphorus cycle, Figure 1.7, is crucial because phosphorus is usually the limiting nutrient in ecosystems. There are no common stable gaseous forms of phosphorus, so the phosphorus cycle is endogenic. In the geosphere, phosphorus is held largely in poorly soluble minerals, such as hydroxyapatite a calcium salt, deposits of which constitute the major reservoir of environmental phosphate. Soluble phosphorus from phosphate minerals and other sources such as fertilizers is taken up by plants and incorporated into nucleic acids which make up the genetic material of

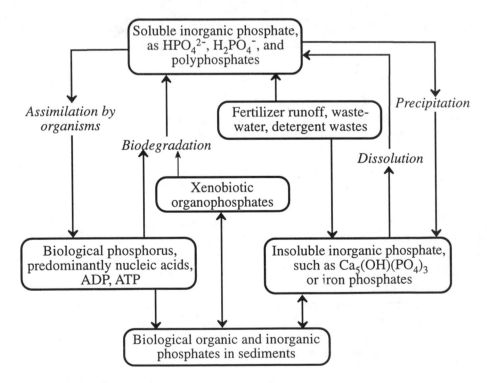

Figure 1.7. The phosphorus cycle.

of organisms. Mineralization of biomass by microbial decay returns phosphorus to the salt solution from which it may precipitate as mineral matter.

The anthrosphere is an important reservoir of phosphorus in the environment. Large quantities of phosphates are extracted from phosphate minerals for fertilizer, industrial chemicals, and food additives. Phosphorus is a constituent of some extremely toxic compounds, especially organophosphate insecticides and military poison nerve gases.

The Sulfur Cycle

The sulfur cycle, which is illustrated in Figure 1.8, is relatively complex in that it involves several gaseous species, poorly soluble minerals, and several species in solution. It is tied with the oxygen cycle in that sulfur combines with oxygen to form gaseous sulfur dioxide, SO_2, an atmospheric pollutant, and soluble sulfate ion, SO_4^{2-}.

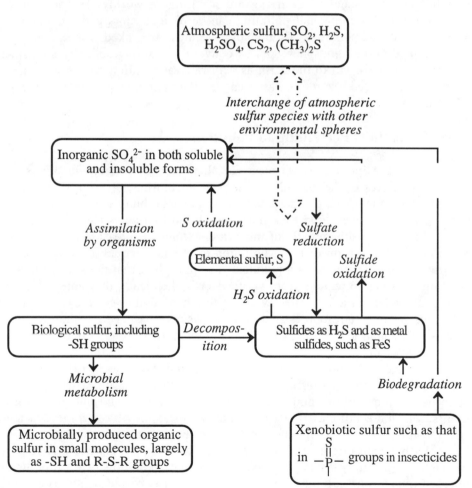

Figure 1.8. The sulfur cycle.

Among the significant species involved in the sulfur cycle are gaseous hydrogen sulfide, H_2S; mineral sulfides, such as PbS, sulfuric acid, H_2SO_4, the main constituent of acid rain; and biologically bound sulfur in sulfur-containing proteins.

Insofar as pollution is concerned, the most significant part of the sulfur cycle is the presence of pollutant SO_2 gas and H_2SO_4 in the atmosphere. The former is a somewhat toxic gaseous air pollutant evolved in the combustion of sulfur-containing fossil fuels. Sulfur dioxide is discussed further as an air pollutant in Chapter 11, and its toxicological chemistry is covered in Chapter 22. The major detrimental effect of sulfur dioxide in the atmosphere is its tendency to oxidize in the atmosphere to produce sulfuric acid. This species is responsible for acidic precipitation, "acid rain," discussed as a major atmospheric pollutant in Chapter 14.

1.7. HUMAN IMPACT AND POLLUTION

The demands of increasing population coupled with the desire of most people for a higher material standard of living are resulting in worldwide pollution on a massive scale. Environmental pollution can be divided among the categories of water, air, and land pollution. All three of these areas are linked. For example, some gases emitted to the atmosphere can be converted to strong acids by atmospheric chemical processes, fall to the earth as acid rain, and pollute water with acidity. Improperly discarded hazardous wastes can leach into groundwater that is eventually released as polluted water into streams.

Some Definitions Pertaining to Pollution

In some cases pollution is a clear-cut phenomenon, whereas in others it lies largely in the eyes of the beholder. Toxic organochlorine solvent residues leached into water supplies from a hazardous waste chemical dump are pollutants in anybody's view. However, loud rock music amplified to a high decibel level by the sometimes questionable miracle of modern electronics is pleasant to some people, and a very definite form of noise pollution to others. Frequently, time and place determine what may be called a pollutant. The phosphate that the sewage treatment plant operator has to remove from wastewater is chemically the same as the phosphate that the farmer a few miles away has to buy at high prices for fertilizer. Most pollutants are, in fact, resources gone to waste; as resources become more scarce and expensive, economic pressure will almost automatically force solutions to many pollution problems.

A reasonable definition of a **pollutant** is a substance present in greater than natural concentration as a result of human activity that has a net detrimental effect upon its environment or upon something of value in that environment. **Contaminants**, which are not classified as pollutants unless they have some detrimental effect, cause deviations from the normal composition of an environment.

Every pollutant originates from a **source**. The source is particularly important because it is generally the logical place to eliminate pollution. After a pollutant is released from a source, it may act upon a receptor. The **receptor** is anything that is affected by the pollutant. Humans whose eyes smart from oxidants in the atmosphere are receptors. Trout fingerlings that may die after exposure to dieldrin in water are also receptors. Eventually, if the pollutant is long-lived, it may be deposited in a **sink**, a long-time repository of the pollutant. Here it will remain for a long time, though not necessarily permanently. Thus, a limestone wall may be a sink for atmospheric sulfuric acid through the reaction,

$$CaCO_3 + H_2SO_4 \rightarrow CaSO_4 + H_2O + CO_2 \qquad (1.7.1)$$

which fixes the sulfate as part of the wall composition.

Pollution of Various Spheres of the Environment

Pollution of surface water and groundwater are discussed in some detail in Chapter 7, Particulate air pollutants are covered in Chapter 10, gaseous inorganic air pollutants in Chapter 11, and organic air pollutants and associated photochemical smog in Chapters 12 and 13. Some air pollutants, particularly those that may result in irreversible global warming or destruction of the protective stratospheric ozone layer, are of such a magnitude that they have the potential to threaten life on earth. These are discussed in Chapter 14, "The Endangered Global Atmosphere." The most serious kind of pollutant that is likely to contaminate the geosphere, particularly soil, consists of hazardous wastes. A simple definition of a **hazardous waste** is that it is a potentially dangerous substance that has been discarded, abandoned, neglected, released, or designated as a waste material, or is one that may interact with other substances to pose a threat. Hazardous wastes are addressed specifically in Chapters 19 and 20.

1.8. TECHNOLOGY: THE PROBLEMS IT POSES AND THE SOLUTIONS IT OFFERS

Modern technology has provided the means for massive alteration of the environment and pollution of the environment. However, technology intelligently applied with a strong environmental awareness also provides the means for dealing with problems of environmental pollution and degradation.

Some of the major ways in which modern technology has contributed to environmental alteration and pollution are the following:

- Agricultural practices that have resulted in intensive cultivation of land, drainage of wetlands, irrigation of arid lands, and application of herbicides and insecticides

- Manufacturing of huge quantities of industrial products that consumes vast amounts of raw materials and produces large quantities of air pollutants, water pollutants, and hazardous waste by-products

- Extraction and production of minerals and other raw materials with accompanying environmental disruption and pollution

- Energy production and utilization with environmental effects that include disruption of soil by strip mining, pollution of water by release of saltwater from petroleum production, and emission of air pollutants such as acid-rain-forming sulfur dioxide

• Modern transportation practices, particularly reliance on the automobile, that cause scarring of land surfaces from road construction, emission of air pollutants, and greatly increased demands for fossil fuel resources

Despite all of the problems that it raises, technology based on a firm foundation of environmental science can be very effectively applied to the solution of environmental problems. One important example of this is the redesign of basic manufacturing processes to minimize raw material consumption, energy use, and waste production. Consider a generalized manufacturing process shown in Figure 1.9. With proper design the environmental acceptability of such a process can be greatly enhanced. In some cases raw materials and energy sources can be chosen in ways that minimize environmental impact. If the process involves manufacture of a chemical, it may be possible to completely alter the reactions used so that the entire operation is more environmentally friendly. Raw materials and water may be recycled to the maximum extent possible. Best available technologies may be employed to minimize air, water, and solid waste emissions.

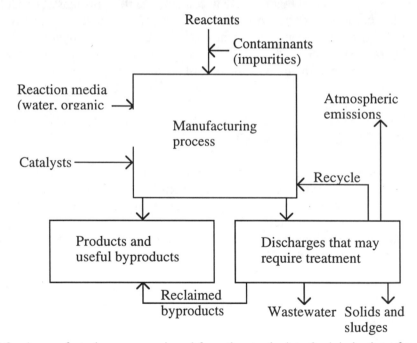

Figure 1.9. A manufacturing process viewed from the standpoint of minimization of environmental impact.

There are numerous ways in which technology can be applied to minimize environmental impact. Among these are the following:

• Use of state-of-the-art computerized control to achieve maximum energy efficiency, maximum utilization of raw materials, and minimum production of pollutant by-products

- Use of materials that minimize pollution problems, for example heat-resistant materials that enable use of high temperatures for efficient thermal processes
- Application of processes and materials that enable maximum materials recycling and minimum waste product production, for example, advanced membrane processes for wastewater treatment to enable water recycling
- Application of advanced biotechnologies such as in the biological treatment of wastes
- Use of best available catalysts for efficient synthesis
- Use of lasers for precision machining and processing to minimize waste production

The applications of modern technology to environmental improvement are addressed in several chapters of this book. Chapter 8, "Water Treatment," discusses technologically-based treatment of water. The technology of air pollution control is discussed in various sections of Chapters 10-13. Hazardous waste treatment is addressed specifically in Chapter 20.

LITERATURE CITED

1. Cunningham, William P., and Barbara Woodworth Saigo, *Environmental Science, a Global Concern*, 5th ed., Wm. C. Brown/McGraw-Hill, New York, 1998.

2. Manahan, Stanley E., *Toxicological Chemistry*, 2nd ed., Lewis Publishers/CRC Press, Boca Raton, FL, 1992.

3. Montgomery, Carla W., *Environmental Geology*, 5th ed., Wm. C. Brown/McGraw-Hill, New York, 1997.

4. Smith, Robert Leo, *Elements of Ecology*, 4th ed., Benjamin Cummings, Menlo Park, CA, 1998.

5. Graedel, T. E., and Paul J. Crutzen *Atmospheric Change, An Earth System Perspective*, W. H. Freeman and Company, New York, 1993.

6. Berner, Elizabeth K. and Robert A. Berner, *Global Environment: Water, Air, and Geochemical Cycles*, Prentice Hall, Englewood Cliffs, NJ, 1994.

7. "Geochemical Cycles," Chapter 23 in *Inorganic Geochemistry*, Gunter Faure, Macmillan Publishing Co., New York, pp. 500-525, 1991.

SUPPLEMENTARY REFERENCES

Alexander, David E. and Rhodes W. Fairbridge, Eds., *Encyclopedia of Environmental Science*, Kluwer Academic Publishers, Hingham, MA, 1999.

Andrews, J. E., *An Introduction to Environmental Chemistry*, Blackwell Science, Cambridge, MA, 1996.

Anderson, Terry L., and Donald R. Leal, *Free Market Environmentalism*, Westview, Boulder, CO, 1991.

Attilio, Bisio and Sharon G. Boots, *Energy Technology and the Environment*, John Wiley & Sons, Inc., New York, 1995.

Attilio, Bisio and Sharon G. Boots, *The Wiley Encyclopedia of Energy and the Environment*, John Wiley & Sons, Inc., New York, 1996.

Brown, Lester R., Christopher Flavin, and Hilary French, *State of the World 1998*, Worldwatch Publications, Washington, D.C., 1998.

Brown, Lester R., Gary Gardner, and Brian Halweil, *Beyond Malthus: Nineteen Dimensions of the Population Challenge*, Worldwatch Publications, Washington, DC, 1999.

Cahil, Lawrence B., *Environmental Audits*, 7th ed., Government Institutes, Rockville, MD, 1996.

Cairncross, Francis, *Costing the Earth*, Harvard Business School Press, Boston, 1992.

Crosby, Donald G., *Environmental Toxicology and Chemistry*, Oxford University Press, New York, 1998.

Costanza, Robert, Ed., *Ecological Economics*, Columbia University Press, New York, 1992.

Dooge, J. C. I., Ed., *An Agenda of Science for Environment and Development into the 21st Century*, Cambridge University Press, New York, 1992.

Dunnette, David A., and Robert J. O'Brien, *The Science of Global Change*, American Chemical Society, Washington, D.C., 1992.

Ehrlich, Paul R., and Anne H. Ehrlich, *Healing the Planet*, Addison-Wesley, Reading, MA, 1992.

Elsom, Derek, *Earth*, Macmillan, New York, 1992.

Encyclopedia of Environmental Analysis and Remediation, John Wiley & Sons, Inc., New York, 1998.

Hollander, Jack M., Ed., *The Energy-Environment Connection*, Island Press, Washington, D.C., 1992.

Marriott, Betty Bowers, *Environmental Impact Assessment: A Practical Guide*, McGraw-Hill, New York, 1997.

Meyers, Robert A., Ed., *Encyclopedia of Environmental Pollution and Cleanup*, John Wiley & Sons, Inc., New York, 1999.

Mungall, Constance, and Digby J. McLaren, Eds., *Planet Under Stress*, Oxford University Press, New York, 1991.

Real, Leslie A., and James H. Brown, Eds., *Foundations of Ecology*, University of Chicago Press, Chicago, 1991.

Silver, Brian L., *The Ascent of Science*, Oxford University Press, New York, 1998.

Rodes, Barbara K., and Rice Odell, *A Dictionary of Environmental Quotations*, Simon and Schuster, New York, NY, 1992.

Stokes, Kenneth M., *Man and the Biosphere*, Sharpe, Armonk, NY, 1992.

Sullivan, Thomas F. P., Ed., *Environmental Law Handbook*, 15th ed., Government Institutes, Rockville, MD, 1999.

White, Rodney R., *North, South, and the Environmental Crisis*, University of Toronto Press, Buffalo, NY, 1993.

Yen, Teh Fu, *Environmental Chemistry: Essentials for Engineering Practice*, Vol. 4A, Prentice Hall, Upper Saddle River, NJ, 1999.

QUESTIONS AND PROBLEMS

1. Under what circumstances does a contaminant become a pollutant?

2. Examine Figure 1.1.1 and abbreviate each "sphere" with its first two letters of At, Hy, An, Bi, Ge. Then place each of the following with the appropriate arrow, indicating the direction of its movement with a notation such as At → Hy: (a) Iron ore used for steel making, (b) waste heat from coal-fired electricity generation, (c) hay, (d) cotton, (e) water from the ocean as it enters the hydrologic cycle, (f) snow, (g) argon used as an inert gas shield for welding.

3. Explain how Figure 1.1 illustrates the definition of environmental chemistry given at the beginning of Section 1.2.

4. Explain how toxicological chemistry differs from environmental biochemistry.

5. Distinguish among geosphere, lithosphere, and crust of the Earth. Which science deals with these parts of the environment?

6. Define ecology and relate this science to Figure 1.1.

7. Although energy is not destroyed, why is it true that the flow of <u>useful</u> energy through an environmental system is essentially a one-way process?

8. Describe some ways in which use of energy has "resulted in many of the environmental problems now facing humankind."

9. Compare nuclear energy to fossil fuel energy sources and defend or refute the statement, "Nuclear energy, with modern, safe, and efficient reactors, is gaining increasing attention as a reliable, environmentally friendly energy source."

10. What is shown by the reaction below?

$$2\{CH_2O\} \rightarrow CO_2(g) + CH_4(g)$$

How is this process related to aerobic respiration?

11. Define cycles of matter and explain how the definition given relates to the definition of environmental chemistry.

12. What are the main features of the carbon cycle?

13. Describe the role of organisms in the nitrogen cycle.

14. Describe how the oxygen cycle is closely related to the carbon cycle.

15. In what important respect does the phosphorus cycle differ from cycles of other similar elements such as nitrogen and sulfur?

2 THE ANTHROSPHERE, INDUSTRIAL ECOSYSTEMS, AND ENVIRONMENTAL CHEMISTRY

2.1. THE ANTHROSPHERE

The **anthrosphere** may be defined as *that part of the environment made or modified by humans and used for their activities*. Of course, there are some ambiguities associated with that definition. Clearly, a factory building used for manufacture is part of the anthrosphere as is an ocean-going ship used to ship goods made in the factory. The ocean on which the ship moves belongs to the hydrosphere, but it is clearly used by humans. A pier constructed on the ocean shore and used to load the ship is part of the anthrosphere, but it is closely associated with the hydrosphere.

During most of its time on Earth, humankind made little impact on the planet, and its small, widely scattered anthrospheric artifacts—simple huts or tents for dwellings, narrow trails worn across the land for movement, clearings in forests to grow some food—rested lightly on the land with virtually no impact. However, with increasing effect as the industrial revolution developed, and especially during the last century, humans have built structures and modified the other environmental spheres, especially the geosphere, such that it is necessary to consider the anthrosphere as a separate area with pronounced, sometimes overwhelming influence on the environment as a whole.

Components of the Anthrosphere

As discussed later in this book, the various spheres of the environment are each divided into several subcategories. For example, the hydrosphere consists of oceans, streams, groundwater, ice in polar icecaps, and other components. The anthrosphere, too, consists of a number of different parts. These may be categorized by considering where humans live; how they move; how they make or provide the things or services they need or want; how they produce food, fiber, and wood; how they obtain, distribute, and use energy; how they communicate; how they extract and process nonrenewable minerals; and how they collect, treat, and dispose of wastes. With these factors in mind, it is possible to divide the anthrosphere into the following categories:

- Structures used for dwellings

- Structures used for manufacturing, commerce, education, and other activities

- Utilities, including water, fuel, and electricity distribution systems, and waste distribution systems, such as sewers

- Structures used for transportation, including roads, railroads, airports, and waterways constructed or modified for water transport

- Structures and other parts of the environment modified for food production, such as fields used for growing crops and water systems used to irrigate the fields

- Machines of various kinds, including automobiles, farm machinery, and airplanes

- Structures and devices used for communications, such as telephone lines or radio transmitter towers

- Structures, such as mines or oil wells, associated with extractive industries

From the list given above it is obvious that the anthrosphere is very complex with an enormous potential to affect the environment. Prior to addressing these environmental effects, several categories of the anthrosphere will be discussed in more detail.

2.2. TECHNOLOGY AND THE ANTHROSPHERE

Since the anthrosphere is the result of technology, it is appropriate to discuss technology at this point. **Technology** refers to the ways in which humans do and make things with materials and energy. In the modern era, technology is to a large extent the product of engineering based on scientific principles. Science deals with the discovery, explanation, and development of theories pertaining to interrelated natural phenomena of energy, matter, time, and space. Based on the fundamental knowledge of science, engineering provides the plans and means to achieve specific practical objectives. Technology uses these plans to carry out the desired objectives.

Technology has a long history and, indeed, goes back into prehistory to times when humans used primitive tools made from stone, wood, and bone. As humans settled in cities, human and material resources became concentrated and focused such that technology began to develop at an accelerating pace. Technological advances predating the Roman era include the development of **metallurgy**, beginning with native copper around 4000 B.C., domestication of the horse, discovery of the wheel, architecture to enable construction of substantial buildings, control of water for canals and irrigation, and writing for communication. The Greek and Roman eras saw the development of **machines**, including the windlass, pulley, inclined plane, screw, catapult for throwing missiles in warfare, and water screw for moving water. Later, the water wheel was developed for power, which was transmitted by wooden gears. Many technological innovations such as printing with wood blocks starting around 740 and gunpowder about a century later, originated in China.

The 1800s saw an explosion in technology. Among the major advances during this century were widespread use of steam power, steam-powered railroads, the telegraph, telephone, electricity as a power source, textiles, the use of iron and steel in building and bridge construction, cement, photography, and the invention of the internal combustion engine, which revolutionized transportation in the following century.

Since about 1900, advancing technology has been characterized by vastly increased uses of energy; greatly increased speed in manufacturing processes, information transfer, computation, transportation, and communication; automated control; a vast new variety of chemicals; new and improved materials for new applications; and, more recently, the widespread application of computers to manufacturing, communication, and transportation. In transportation, the development of passenger-carrying airplanes has affected an astounding change in the ways in which people get around and how high-priority freight is moved. Rapid advances in biotechnology now promise to revolutionize food production and medical care.

The technological advances of the present century are largely attributable to two factors. The first of these is the application of electronics, now based upon solid state devices, to technology in areas such as communications, sensors, and computers for manufacturing control. The second area largely responsible for modern technological innovations is based upon improved materials. For example, special strong alloys of aluminum were used in the construction of airliners before World War II and now these alloys are being supplanted by even more advanced composites. Synthetic materials with a significant impact on modern technology include plastics, fiber-reinforced materials, composites, and ceramics.

Until very recently, technological advances were made largely without heed to environmental impacts. Now, however, the greatest technological challenge is to reconcile technology with environmental consequences. The survival of humankind and of the planet that supports it now requires that the established two-way interaction between science and technology become a three-way relationship including environmental protection.

Engineering

Engineering uses fundamental knowledge acquired through science to provide the plans and means to achieve specific objectives in areas such as manufacturing, communication, and transportation. At one time engineering could be divided conveniently between military and civil engineering. With increasing sophistication, civil engineering evolved into even more specialized areas such as mechanical engineering, chemical engineering, electrical engineering, and environmental engineering. Other engineering specialties include aerospace engineering, agricultural engineering, biomedical engineering, CAD/CAM (computer-aided design and computer-aided manufacturing engineering), ceramic engineering, industrial engineering, materials engineering, metallurgical engineering, mining engineering, plastics engineering, and petroleum engineering. Some of the main categories of engineering are defined below:

- **Mechanical engineering**, which deals with machines and the manner in which they handle forces, motion, and power

- **Electrical engineering** dealing with the generation, transmission, and utilization of electrical energy

- **Electronics engineering** dealing with phenomena based on the behavior of electrons in vacuum tubes and other devices

- **Chemical engineering,**which uses the principles of chemical science, physics, and mathematics to design and operate processes that generate products and materials

The role of engineering in constructing and operating the various components of the anthrosphere is obvious. In the past, engineering was often applied without much if any consideration of environmental factors. As examples, huge machines designed by mechanical engineers were used to dig up and rearrange Earth's surface without regard for the environmental consequences, and chemical engineering was used to make a broad range of products without consideration of the wastes produced. Fortunately, that approach is changing rapidly. Examples of environmentally friendly engineering include machinery designed to minimize noise, much improved energy efficiency in machines, and the uses of earth-moving equipment for environmentally beneficial purposes, such as restoration of strip-mined lands and construction of wetlands. Efficient generation, distribution, and utilization of electrical energy based on the principles of electrical engineering constitute one of the most promising avenues of endeavor leading to environmental improvement. Automated factories developed through applications of electronic engineering can turn out goods with the lowest possible consumption of energy and materials while minimizing air and water pollutants and production of hazardous wastes. Chemical factories can be engineered to maximize the most efficient utilization of energy and materials while minimizing waste production.

2.3. INFRASTRUCTURE

The **infrastructure** is the utilities, facilities, and systems used in common by members of a society and upon which the society depends for its normal function. The infrastructure includes both physical components—roads, bridges, and pipe-lines—and the instruction—laws, regulations, and operational procedures—under which the physical infrastructure operates. Parts of the infrastructure may be publicly owned, such as the U.S. Interstate Highway system and some European railroads, or privately owned, as is the case with virtually all railroads in the U.S. Some of the major components of the infrastructure of a modern society are the following:[1]

- Transportation systems, including railroads, highways, and air transport systems

- Energy generating and distribution systems

- Buildings

- Telecommunications systems

- Water supply and distribution systems

- Waste treatment and disposal systems, including those for municipal wastewater, municipal solid refuse, and industrial wastes

In general, the infrastructure refers to the facilities that large segments of a population must use in common in order for a society to function. In a sense, the infrastructure is analogous to the operating system of a computer. A computer operating system determines how individual applications operate and the manner in which they distribute and store the documents, spreadsheets, and illustrations created by the applications. Similarly, the infrastructure is used to move raw materials and power to factories and to distribute and store their output. An outdated, cumbersome computer operating system with a tendency to crash is detrimental to the efficient operation of a computer. In a similar fashion, an outdated, cumbersome, broken-down infrastructure causes society to operate in a very inefficient manner and is subject to catastrophic failure.

For a society to be successful, it is of the utmost importance to maintain a modern, viable infrastructure. Such an infrastructure is consistent with environmental protection. Properly designed utilities and other infrastructural elements, such as water supply systems and wastewater treatment systems, minimize pollution and environmental damage.

Components of the infrastructure are subject to deterioration. To a large extent this is due to natural aging processes. Fortunately, many of these processes can be slowed or even reversed. Corrosion of steel structures, such as bridges, is a big problem for infrastructures; however, use of corrosion-resistant materials and maintenance with corrosion-resistant coatings can virtually stop this deterioration process. The infrastructure is subject to human insult, such as vandalism, misuse, and neglect. Often the problem begins with the design and basic concept of a particular component of the infrastructure. For example, many river dikes destroyed by flooding should never have been built because they attempt to thwart to an impossible extent the natural tendency of rivers to flood periodically.

Technology plays a major role in building and maintaining a successful infrastructure. Many of the most notable technological advances applied to the infrastructure were made from 150 to 100 years ago. By 1900 railroads, electric utilities, telephones, and steel building skeletons had been developed. The net effect of most of these technological innovations was to enable humankind to "conquer" or at least temporarily subdue Nature. The telephone and telegraph helped to overcome isolation, high speed rail transport and later air transport conquered distance, and dams were used to control rivers and water flow.

The development of new and improved materials is having a significant influence on the infrastructure. From about 1970 to 1985 the strength of steel commonly used in construction nearly doubled. During the latter 1900s significant advances were made in the properties of structural concrete. Superplasticizers enabled mixing cement with less water, resulting in a much less porous, stronger concrete product. Polymeric and metallic fibers used in concrete made it much stronger. For dams and other applications in which a material stronger than earth but not as strong as conventional concrete is required, roller-compacted concrete consisting of a mixture of cement with silt or clay has been found to be useful. The silt or clay used is obtained on site with the result that both construction costs and times are lowered.

The major challenge in designing and operating the infrastructure in the future will be to use it to work with the environment and to enhance environmental quality to the benefit of humankind. Obvious examples of environmentally friendly infrastructures are state-of-the-art sewage treatment systems, high-speed rail systems that can replace inefficient highway transport, and stack gas emission control systems in power plants. More subtle approaches with a tremendous potential for making the infrastructure more environmentally friendly include employment of workers at computer terminals in their homes so that they do not need to commute, instantaneous electronic mail that avoids the necessity of moving letters physically, and solar electric-powered installations to operate remote signals and relay stations, which avoids having to run electric power lines to them.

Whereas advances in technology and the invention of new machines and devices enabled rapid advances in the development of the infrastructure during the 1800s and early 1900s, it may be anticipated that advances in electronics and computers will have a comparable effect in the future. One of the areas in which the influence of modern electronics and computers is most visible is in telecommunications. Dial telephones and mechanical relays were perfectly satisfactory in their time, but have been made totally obsolete by innovations in electronics, computer control, and fiber-optics. Air transport controlled by a truly modern, state-of-the-art computerized control system (which, unfortunately, is not yet fully installed in the U.S.) could enable present airports to handle many more airplanes safely and efficiently, thus reducing the need for airport construction. Sensors for monitoring strain, temperature, movement, and other parameters can be imbedded in the structural members of bridges and other structures. Information from these sensors can be processed by computer to warn of failure and to aid in proper maintenance. Many similar examples could be cited.

Although the payoff is relatively long term, intelligent investment in infrastructure pays very high returns. In addition to the traditional rewards in economics and convenience, properly designed additions and modifications to the infrastructure can pay large returns in environmental improvement as well.

2.4. DWELLINGS

The dwellings of humans have an enormous influence on their well-being and on the surrounding environment. In relatively affluent societies the quality of living space has improved dramatically during the last century. Homes have become much more spacious per occupant and largely immune to the extremes of weather conditions. Such homes are equipped with a huge array of devices, such as indoor plumbing, climate control, communications equipment, and entertainment centers. The comfort factor for occupants has increased enormously.

The construction and use of modern homes and the other buildings in which people spend most of their time place tremendous strains on their environmental support systems and cause a great deal of environmental damage. Typically, as part of the siting and construction of new homes, shopping centers, and other buildings, the landscape is rearranged drastically at the whims of developers. Topsoil is removed, low places are filled in, and hills are cut down in an attempt to make the surrounding environment conform to a particular landscape scheme. The construction of modern buildings consumes large amounts of resources such as concrete, steel, plastic, and glass, as well as the energy required to make synthetic building materials. The operation of a modern building requires additional large amounts of energy, and of materials such as water. It has been pointed out that all too often the design and operation of modern homes and other buildings takes place "out of the context" of the surroundings and the people who must work in and occupy the buildings.[2]

There is a large potential to design, construct, and operate homes and other buildings in a manner consistent with environmental preservation and improvement. One obvious way in which this can be done is by careful selection of the kinds of materials used in buildings. Use of renewable materials such as wood, and non-fabricated materials such as quarried stone, can save large amounts of energy and minimize environmental impact. In some parts of the world sun-dried adobe blocks made from soil are practical building materials that require little energy to fabricate. Recycling of building materials and of whole buildings can save large amounts of materials and minimize environmental damage. At a low level, stone, brick, and concrete can be used as fill material upon which new structures may be constructed. Bricks are often recyclable, and recycled bricks often make useful and quaint materials for walls and patios. Given careful demolition practices, wood can often be recycled. Buildings can be designed with recycling in mind. This means using architectural design conducive to adding stories and annexes and to rearranging existing space. Utilities may be placed in readily accessible passageways rather than being imbedded in structural components in order to facilitate later changes and additions.

Technological advances can be used to make buildings much more environmentally friendly. Advanced window design that incorporates multiple panes and infrared-blocking glass can significantly reduce energy consumption. Modern insulation materials are highly effective. Advanced heating and air conditioning systems operate with a high degree of efficiency. Automated and computerized control of building utilities, particularly those used for cooling and heating, can significantly reduce energy consumption by regulating temperatures and lighting to the desired levels at specific locations and times in the building.

Advances in making buildings airtight and extremely well insulated can lead to problems with indoor air quality. Carpets, paints, paneling, and other manufactured components of buildings give off organic vapors such as formaldehyde, solvents, and monomers used to make plastics and fabrics. In a poorly insulated building that is not very airtight, such indoor air pollutants cause few if any problems for the building occupants. However, extremely airtight buildings can accumulate harmful levels of indoor air pollutants. Therefore, building design and operation to minimize accumulation of toxic indoor air pollutants is receiving a much higher priority.

2.5. TRANSPORTATION

Few aspects of modern industrialized society have had as much influence on the environment as developments in transportation. These effects have been both direct and indirect. The direct effects are those resulting from the construction and use of transportation systems. The most obvious example of this is the tremendous effects that the widespread use of automobiles, trucks, and buses have had upon the environment. Entire landscapes have been entirely rearranged to construct highways, interchanges, and parking lots. Emissions from the internal combustion engines used in automobiles are the major source of air pollution in many urban areas.

The indirect environmental effects of the widespread use of automobiles are enormous. The automobile has made possible the "urban sprawl" that is characteristic of residential and commercial patterns of development in the U.S., and in many other industrialized countries as well. Huge new suburban housing tracts and the commercial developments, streets, and parking lots constructed to support them continue to consume productive farmland at a frightening rate. The paving of vast areas of watershed and alteration of runoff patterns have contributed to flooding and water pollution. Discarded, worn-out automobiles have caused significant waste disposal problems. Vast enterprises of manufacturing, mining, and petroleum production and refining required to support the "automobile habit" have been very damaging to the environment.

On the positive side, however, applications of advanced engineering and technology to transportation can be of tremendous benefit to the environment. Modern rail and subway transportation systems, concentrated in urban areas and carefully connected to airports for longer distance travel, can enable the movement of people rapidly, conveniently, and safely with minimum environmental damage. Although pitifully few in number in respect to the need for them, examples of such systems are emerging in progressive cities, showing the way to environmentally friendly transportation systems of the future.

A new development that is just beginning to reshape the way humans move, where they live, and how they live, is the growth of a **telecommuter society** composed of workers who do their work at home and "commute" through their computers, modems, FAX machines, and the Internet connected by way of high-speed telephone communication lines. These new technologies, along with several other developments in modern society, have made such a work pattern possible and desirable. An increasing fraction of the work force deals with information in their jobs. In principle, information can be handled just as well from a home office as it can from a centralized location, which is often an hour or more commuting distance from the worker's dwelling. Within approximately the next 10 years, it is estimated that almost 20% of the U.S. work force, a total of around 30 million people, may be working out of their homes.

2.6. COMMUNICATIONS

It has become an overworked cliché that we live in an information age. Nevertheless, the means to acquire, store, and communicate information are expanding at an incredible pace. This phenomenon is having a tremendous effect upon society and has the potential to have numerous effects upon the environment.

The major areas to consider with respect to information are its acquisition, recording, computing, storing, displaying, and communicating. Consider, for example, the detection of a pollutant in a major river. Data pertaining to the nature and concentration of the pollutant may be obtained with a combination gas chromatograph and mass spectrometer. Computation by digital computer is employed to determine the identity and concentration of the pollutant. The data can be stored on a magnetic disk, displayed on a video screen, and communicated instantaneously all over the world by satellite and fiber-optic cable.

All the aspects of information and communication listed above have been tremendously augmented by recent technological advances. Perhaps the greatest such advance has been that of silicon integrated circuits. Optical memory consisting of information recorded and read by microscopic beams of laser light has enabled the storage of astounding quantities of information on a single compact disk. The use of optical fibers to transmit information digitally by light has resulted in a comparable advance in the communication of information.

The central characteristic of communication in the modern age is the combination of telecommunications with computers called **telematics**.[3] Automatic teller machines use telematics to make cash available to users at locations far from the user's bank. Information used for banking, for business transactions, and in the media depends upon telematics.

There exists a tremendous potential for good in the applications of the "information revolution" to environmental improvement. An important advantage is the ability to acquire, analyze, and communicate information about the environment. For example, such a capability enables detection of perturbations in environmental systems, analysis of the data to determine the nature and severity of the pollution problems causing such perturbations, and rapid communication of the findings to all interested parties.

2.7. FOOD AND AGRICULTURE

The most basic human need is the need for food. Without adequate supplies of food, the most pristine and beautiful environment becomes a hostile place for human life. The industry that provides food is **agriculture**, an enterprise concerned primarily with growing crops and livestock.

The environmental impact of agriculture is enormous. One of the most rapid and profound changes in the environment that has ever taken place was the conversion of vast areas of the North American continent from forests and grasslands to cropland. Throughout most of the continental United States, this conversion took place predominantly during the 1800s. The effects of it were enormous. Huge acreages of forest lands that had been undisturbed since the last Ice Age were suddenly deprived of stabilizing tree cover and subjected to water erosion. Prairie lands put to the plow were destabilized and subjected to extremes of heat, drought, and wind that caused topsoil to blow away, culminating in the Dust Bowl of the 1930s.

In recent decades, valuable farmland has faced a new threat posed by the urbanization of rural areas. Prime agricultural land has been turned into subdivisions and paved over to create parking lots and streets. Increasing urban sprawl has led to the need for more highways. In a vicious continuing circle, the availability of new highway systems has enabled even more development. The ultimate result of this pattern of development has been the removal of once productive farmland from agricultural use.

On a positive note, agriculture has been a sector in which environmental improvement has seen some notable advances during the last 50 to 75 years. This has occurred largely under the umbrella of soil conservation. The need for soil conservation became particularly obvious during the Dust Bowl years of the 1930s, when it appeared that much of the agricultural production capacity of the U.S. would be swept away from drought-stricken soil by erosive winds. In those times and areas in which wind erosion was not a problem, water erosion took its toll. Ambitious programs of soil conservation have largely alleviated these problems. Wind erosion has been minimized by practices such as low-tillage agriculture, strip cropping in which crops are grown in strips alternating with strips of summer-fallowed crop stubble, and reconversion of marginal cultivated land to pasture. The application of low-tillage agriculture and the installation of terraces and grass waterways have greatly reduced water erosion.

Food production and consumption are closely linked with industrialization and the growth of technology. It is an interesting observation that those countries that develop high population densities prior to major industrial development experience two major changes that strongly impact food production and consumption:[4]

1. Cropland is lost as a result of industrialization; if the industrialization is rapid, increases in grain crop productivity cannot compensate fast enough for the loss of cropland to prevent a significant fall in production.

2. As industrialization raises incomes, the consumption of livestock products increases, such that demand for grain to produce more meat, milk, and eggs rises significantly.

To date, the only three countries that have experienced rapid industrialization after achieving a high population density are Japan, Taiwan, and South Korea. In each case, starting as countries that were largely self-sufficient in grain supplies, they lost 20-30 percent of their grain production and became heavy grain importers over an approximately 30-year period. The effects of these changes on global grain supplies and prices was small because of the limited population of these countries — the largest, Japan, had a population of only about 100 million. Since approximately 1990, however, China has been experiencing economic growth at a rate of about 10% per year. With a population of 1.2 billion people, China's economic activity has an enormous effect on global markets. It may be anticipated that this economic growth, coupled with a projected population increase of more than 400 million people during the next 30 years, will result in a demand for grain and other food supplies that will cause disruptive food shortages and dramatic price increases.

In addition to the destruction of farmland to build factories, roads, housing, and other parts of the infrastructure associated with industrialization, there are other factors that tend to decrease grain production as economic activity increases. One of the major factors is air pollution, which can lower grain yields significantly. Water pollution can seriously curtail fish harvests. Intensive agriculture uses large quantities of water for irrigation. If groundwater is used for irrigation, aquifers may become rapidly depleted.

The discussion above points out several factors that are involved in supplying food to a growing world population. There are numerous complex interactions among the industrial, societal, and agricultural sectors. Changes in one inevitably result in changes in the others.

2.8. MANUFACTURING

Once a device or product is designed and developed through the applications of engineering (see Section 2.2), it must be made—synthesized or manufactured. This may consist of the synthesis of a chemical from raw materials, casting of metal or plastic parts, assembly of parts into a device or product, or any of the other things that go into producing a product that is needed in the marketplace.

Manufacturing activities have a tremendous influence on the environment. Energy, petroleum to make petrochemicals, and ores to make metals must be dug from, pumped from, or grown on the ground to provide essential raw materials. The potential for environmental pollution from mining, petroleum production, and intensive cultivation of soil is enormous. Huge land-disrupting factories and roads must be built to transport raw materials and manufactured products. The manufacture of goods carries with it the potential to cause significant air and water pollution and production of hazardous wastes. The earlier in the design and development process that environmental considerations are taken into account, the more "environmentally friendly" a manufacturing process will be.

Three relatively new developments that have revolutionized manufacturing and that continue to do so are automation, robotics, and computers. These topics are discussed briefly below.

Automation

Automation uses automatic devices to perform repetitive tasks such as assembly line operations. The greatest application of automation is in manufacturing and assembly. Automation employs mechanical and electrical devices integrated into systems to replace or extend human physical and mental activities. Primitive forms of automation were known in ancient times, with devices such as floats used to control water levels in Roman plumbing systems. A key component of an automated system is the **control system**, which regulates the response of components of a system as a function of conditions, particularly those of time or location.

The simplest level of automation is **mechanization**, in which a machine is designed to increase the strength, speed, or precision of human activities. A backhoe for dirt excavation is an example of mechanization. **Open-loop, multifunctional** devices perform tasks according to preset instructions, but without any feedback regarding whether or how the task is done. **Closed-loop, multifunctional** devices use process feedback information to adjust the process on a continuous basis. The highest level of automation is **artificial intelligence** in which information is combined with simulated reasoning to arrive at a solution to a new problem or perturbation that may arise in the process.

Not all of the effects of automation on society and on the environment are necessarily good. One obvious problem is increased unemployment and attendant social unrest resulting from displaced workers. Another is the ability that automation provides to enormously increase the output of consumer goods at more affordable prices. This capability greatly increases demands for raw materials and energy, putting additional strain on the environment. To attempt to address such concerns by cutting back on automation is unrealistic, so societies must learn to live with it and to use it in beneficial ways. There are many beneficial applications of technology. Automated processes can result in much more efficient utilization of energy and materials for production, transportation, and other human needs. A prime example is the greatly increased gasoline mileage achieved during the last approximately 20 years by the application of computerized, automated control of automobile engines. Automation in manufacturing and chemical synthesis is used to produce maximum product from minimum raw material. Production of air and water pollutants and of hazardous wastes can be minimized by the application of automated processes. By replacing workers in dangerous locations, automation can contribute significantly to worker health and well-being.

Robotics

Robotics refers to the use of machines to simulate human movements and activities. **Robots** are machines that perform such functions using computer-driven mechanical components to grip, move, reorient, and manipulate objects. Modern robots are characterized by intricately related mechanical, electronic, and computational systems. A robot can perform a variety of functions according to pre-programmed instructions that can be changed according to human direction or in response to changed circumstances.

There are a variety of mechanical mechanisms associated with robots. These are servomechanisms in which low-energy signals from electronic control devices are used to direct the actions of a relatively large and powerful mechanical system. Robot arms may bend relative to each other through the actions of flexible joints. Specialized end effectors are attached to the ends of robot arms to accomplish specific functions. The most common such device is a gripper used like a hand to grasp objects.

Sensory devices and systems are crucial in robotics to sense position, direction, speed, and other factors required to control the functions of the robot. Sensors may be used to respond to sound, light, and temperature. One of the more sophisticated types of sensors involves a form of vision. Images captured by a video camera can be processed by computer to provide information required by the robot to perform its assigned task.

Robots interact strongly with their environment. In addition to sensing their surroundings, robots must be able to respond to it in desired ways. In so doing, robots rely on sophisticated computer control. Rapid developments in computer hardware, power, and software continue to increase the ability of robots to interact with their environment. Commonly, instructions to robots are provided by computer software programmed by humans. It is now possible in many cases to lead a robot through its desired motions and have it "learn" the sequence by computer.

Robots are now used for numerous applications. The main applications are for moving materials and objects, and performing operations in manufacturing, assembly, and inspection. A promising use of robots is in surroundings that are hazardous to humans. For example, robots can be used to perform tasks in the presence of hazardous substances that would threaten human health and safety.

Computers

The explosive growth of digital computer hardware and software is one of the most interesting and arguably the most influential phenomenon of our time. Computers have found applications throughout manufacturing. The most important of these are outlined here.

Computer-aided design (CAD) is employed to convert an idea to a manufactured product. Whereas innumerable sketches, engineering drawings, and physical mockups used to be required to bring this transition about, computer graphics are now used. Thus computers can be used to provide a realistic visual picture of a product, to analyze its characteristics and performance, and to redesign it based upon the results of computer analysis. The capabilities of computers in this respect are enormous. As an example, Boeing's large, extremely complex 777 passenger airliner, which entered commercial service in 1995, was designed by computer and brought to production without construction of a full-scale mockup.

Closely linked to CAD is **computer-aided manufacturing (CAM)** which employs computers to plan and control manufacturing operations, for quality control, and to manage entire manufacturing plants. The CAD/CAM combination continues to totally change manufacturing operations to the extent that it may be called a "new industrial revolution." Computerized control of manufacturing, production, and distribution has greatly increased the efficiency of these activities, enabling economic growth with price stability.

The application of computers has had a profound influence on environmental concerns. One example is the improved accuracy of weather forecasting that has resulted from sophisticated and powerful computer programs and hardware. Related to this are the uses of weather satellites, which could not be placed in orbit or operated without computers. Satellites operated by computer control are used to monitor pollutants and map their patterns of dispersion. Computers are widely used in modeling to mimic complex ecosystems, climate, and other environmentally relevant systems.

Computers and their networks are susceptible to mischief and sabotage by outsiders. The exploits of "computer hackers" in breaking into government and private sector computers have been well documented. Important information has been stolen and the operation of computers has been seriously disrupted by hackers with malicious intent. Most computer operations are connected with others through the Internet, enabling communication with employees at remote locations and instant contact with suppliers and customers. The problem of deliberate disruption is potentially so great that companies throughout the world have spent billions of dollars on outside experts in computer security. The U.S. Federal Bureau of Investigation (FBI) now trains its new agents in cyberspace crime, and maintains special computer crime squads in New York, Washington, and San Francisco.

Numerous kinds of protection are available for computer installations. Such protection comes in the form of both software and hardware. Special encryption software can be used to put computer messages in code that is hard to break. Hardware and software barriers to unauthorized corporate computer access, "firewalls," continue to become more sophisticated and effective.

2.9. EFFECTS OF THE ANTHROSPHERE ON EARTH

The effects of the anthrosphere on Earth have been many and profound. Persistent and potentially harmful products of human activities have been widely dispersed and concentrated in specific locations in the anthrosphere as well as other spheres of the environment as the result of human activities. Among the most troublesome of these are toxic heavy metals and organochlorine compounds. Such materials have accumulated in the anthrosphere in painted and coated surfaces, such as organotin-containing paints used to prevent biofouling on boats; under and adjacent to airport runways; under and along highway paving; buried in old factory sites; in landfills; and in materials dredged from waterways and harbors that are sometimes used as landfill on which buildings, airport runways and other structures have been placed. In many cases productive topsoil used to grow food has been contaminated with discarded industrial wastes, phosphate fertilizers, and dried sewage sludge.

Some of the portions of the anthrosphere that may be severely contaminated by human activities are shown in Figure 2.1. In some cases the contamination has been so pervasive and persistent that the effects will remain for centuries. Some of the most vexsome environmental and waste problems are due to contamination of various parts of the anthrosphere by persistent and toxic waste materials.

Potentially harmful wastes and pollutants of anthrospheric origin have found their way into water, air, soil, and living organisms. For example, chlorofluoro-carbons (Freons) have been released to the atmosphere in such quantities and are so stable that they are now constituents of "normal" atmospheric air and pose a threat to the protective ozone layer in the stratosphere. Lake sediments, stream beds, and deltas deposited by flowing rivers are contaminated with heavy metals and refractory organic compounds of anthrospheric origin. The most troubling repository of wastes in the hydrosphere is groundwater. Some organisms have accumulated high enough levels of persistent organic compounds or heavy metals to do harm to themselves or to humans that use them as a food source.

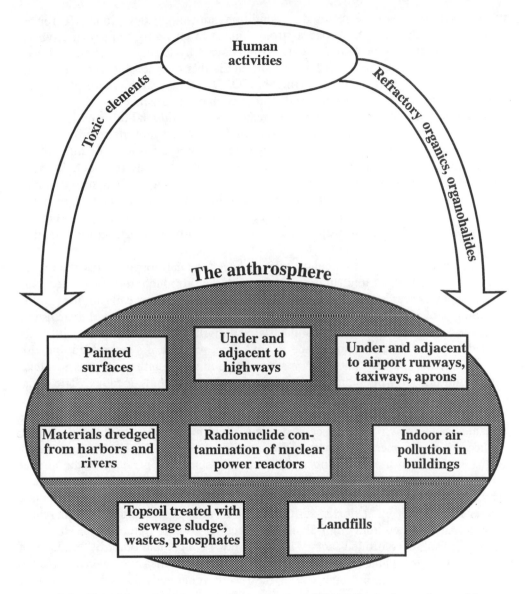

Figure 2.1. The anthrosphere is a repository of many of the pollutant by-products of human activities.

2.10. INTEGRATION OF THE ANTHROSPHERE INTO THE TOTAL ENVIRONMENT

Over the eons of Earth's existence, natural processes free from sudden, catastrophic disturbances (such as those that have occurred from massive asteroid impacts, for example) have resulted in a finely tuned balance among the systems composing Earth's natural environment. Fortuitously, these conditions—adequate water, moderate temperatures, an atmosphere that serves as a shield against damaging solar radiation—have resulted in conditions amenable to various life forms. Indeed, these life forms have had a strong impact in changing their own environments. According to the **Gaia hypothesis** advanced by the British chemist James Lovelock, organisms on Earth have modified Earth's climate and other environmental conditions, such as by regulating the CO_2/O_2 balance in the atmosphere, in a manner conducive to the existence and reproduction of the organisms.

To a degree, the early anthrosphere created by pre-industrial humans integrated well with the other spheres of the environment and caused minimal environmental degradation. That this was so resulted less from any noble instincts of humankind toward nature than it did from the lack of power to alter the environment. In those cases where humans had the capability of modifying or damaging their surroundings, such as by burning forests to provide cropland, the effects on the natural environment could be profound and very damaging. In general, though, preindustrial humans integrated their anthrosphere, such as it was, with the natural environment as a whole.

The relatively harmonious relationship between the anthrosphere and the rest of the environment began to change markedly with the introduction of machines, particularly power sources, beginning with the steam engine, that greatly multiplied the capabilities of humans to alter their surroundings. As humans developed their use of machines and other attributes of industrialized civilization, they did so with little consideration of the environment and in a way that was out of synchronization with the other environmental spheres. A massive environmental imbalance has resulted, the magnitude of which has been realized only in recent decades. The most commonly cited manifestation of this imbalance has been pollution of air or water.

Because of the detrimental effects of human activities undertaken without due consideration of environmental consequences, significant efforts have been made to reduce the environmental impacts of these activities. Figure 2.2 shows three stages of the evolution of the anthrosphere from an unintegrated appendage to the natural environment to a system more attuned to its surroundings. The first approach to dealing with the pollutants and wastes produced by industrial activities—particulate matter from power plant stacks, sulfur dioxide from copper smelters, and mercury-contaminated wastes from chlor-alkali manufacture—was to ignore them. However, as smoke from uncontrolled factory furnaces, raw sewage, and other by-products of human activities became more troublesome, "end-of-pipe" measures were adopted to prevent the release of pollutants after they were generated. Such measures have included electrostatic precipitators and flue gas desulfurization to remove particulate matter and sulfur dioxide from flue gas; physical processes used in primary sewage treatment; microbial processes used for secondary sewage treatment; and physical, chemical, and biological processes for advanced (tertiary) sewage treatment. Such treatment measures are often very sophisticated and effective. Another kind of end-of-pipe treatment is the disposal of wastes in a supposedly safe place. In some cases, such as municipal solid wastes, radioactive materials, hazardous chemicals, power plant ash, and contaminated soil, disposal of sequestered wastes in a secure location is practiced as a direct treatment process. In other cases, including flue-gas desulfurization sludge, sewage sludge, and sludge from chemical treatment of industrial wastewater, disposal is practiced as an adjunct to other end-of-pipe measures. Waste disposal practices later found to be inadequate have spawned an entirely separate end-of-pipe treatment called **remediation** in which discarded wastes are dug up, sometimes subjected to additional treatment, and then placed in a more secure disposal site.

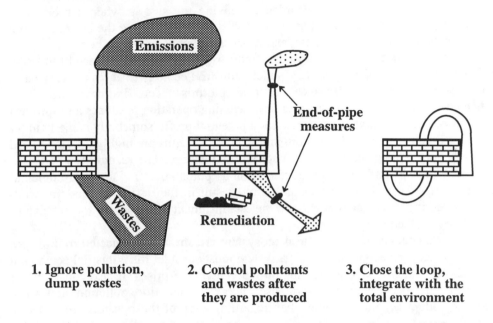

Figure 2.2. Steps in evolution of the anthrosphere to a more environmentally compatible form.

Although sometimes unavoidable, the production of pollutants followed by measures taken to control or remediate them to reduce the quantities and potential harmfulness of wastes are not very desirable. Such measures do not usually eliminate wastes and may, in fact, transfer a waste problem from one part of the environment to another. An example of this is the removal of air pollutants from stack gas and their disposal on land, where they have the potential to cause groundwater pollution. Clearly, it is now unacceptable to ignore pollution and to dump wastes, and the control of pollutants and wastes after they are produced is not a good permanent solution to waste problems. Therefore, it has become accepted practice to "close the loop" on industrial processes, recycling materials as much as possible and allowing only benign waste products to be released to the environment. Such an approach is the basis of industrial ecology discussed in the following section.

2.11. THE ANTHROSPHERE AND INDUSTRIAL ECOLOGY

Industrial ecology is an approach based upon systems engineering and ecological principles that integrates the production and consumption aspects of the design, production, use, and termination (decommissioning) of products and services in a manner that minimizes environmental impact. Industrial ecology functions within groups of enterprises that utilize each others' materials and by-products such that waste materials are reduced to the absolute minimum. Such a system is an **industrial ecosystem**, which is analogous to a natural ecosystem. In a manner analogous to natural ecosystems, industrial ecosystems utilize energy and process materials through a process of **industrial metabolism**. In such systems, products, effluents, and wastes are not regarded as leaving the system when a product or service is sold to a consumer, but are regarded as remaining in the system until a complete cycle of manufacture, use, and disposal is completed.

The first clear delineation of modern industrial ecology can be traced to a 1989 article by Frosch and Gallopoulos.[5] In fact, industrial ecology in at least a very basic form has been practiced ever since industrial enterprises were first developed. That is because whenever a manufacturing or processing operation produces a by-product that can be used by another enterprise for a potential profit, somebody is likely to try to do so. Potentially, at least, modern industrial ecosystems are highly developed and very efficient in their utilization of materials and energy. The recognition that such systems can exist and that they have enormous potential to reduce environmental pollution in a cost-effective manner should result in the design of modern, well-coordinated industrial ecosystems, and the establishment of economic and regulatory incentives for their establishment.

The components of an industrial ecosystem are strongly connected by linkages of time, space, and economics. Such a system must consider the industrial ecosystem as a whole, rather than concentrating on individual enterprises. It also considers the total impact of the system on material use, energy consumption, pollution, and waste and by-product generation rather than regarding each of these aspects in isolation from the rest.

Industrial Ecosystems

A group of interrelated firms functioning together in the practice of industrial ecology constitutes an industrial ecosystem (see Figure 2.3). As noted above, such a system functions in a manner analogous to natural ecosystems. Each constituent of the system consumes energy and materials, and each produces a product or service. A well-developed industrial ecosystem is characterized by a very high level of exchange of materials among its various segments. It is often based upon a firm that is a primary producer of materials or energy. To the greatest extent possible, materials are kept within the industrial ecosystem and few wastes (ideally none) are produced that require disposal.

A functioning system of industrial ecology has at least five main constituents. The first of these is a primary materials producer, which generates the materials used throughout the rest of the system. Also required is a source of energy that keeps the system operating. The energy enters the system as high-grade energy that is used as efficiently as possible throughout the system before it is dissipated as waste heat. Another segment of the system is a materials processing and manufacturer sector, usually consisting of a number of firms. There is a large and diverse consumer sector. And finally there is a well-developed waste processing sector. It is this last kind of enterprise that distinguishes a functional industrial ecosystem from conventional industrial systems.

A key to the success of a sustainable industrial ecosystem is the symbiotic relationships that occur among various constituents of the system. Such relationships are analogous to symbiosis among organisms in a natural ecosystem. In a natural ecosystem the essential interdependencies among various biological species constituting the system have developed through long periods of evolution. Similarly, mutually advantageous relationships evolve naturally in functional industrial ecosystems.

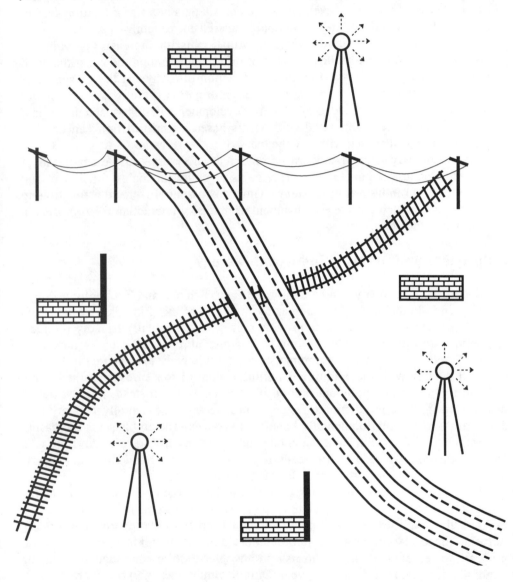

Figure 2.3. An industrial ecosystem consists of a variety of enterprises linked together by transportation and communications systems and processing materials in a manner that maximizes the efficient utilization of materials and minimizes wastes.

An essential key to the success of an industrial ecosystem is the facile exchange of materials and information in the system. Because of the central importance of materials exchange, industrial ecosystems tend to be clustered in relatively small geographic areas. Furthermore, they are often centered around a transportation system, such as a navigable river, a railroad line, or an interstate highway network. For high-value goods that can be shipped economically by air, a major airport is essential, and the constituents of the industrial ecosystem may be separated by great distances. Usually, several, or even all, of the constituents of a transportation system mentioned above are utilized in a functional industrial ecosystem.

In the modern "information age," information is an increasingly valuable commodity. Therefore, the facile exchange of information among the constituents of an industrial ecosystem is essential for it to function properly. At the present time industrial ecosystems are developing in which information is the major commodity involved. Such systems are aided by modern developments in high-speed data transmission and computers. Industrial ecosystems based on information can be geographically highly dispersed with a scope that is truly worldwide.

Despite the advances of modern communication systems, the oldest form of communication, direct human contact, remains essential for the establishment and functioning of an industrial ecosystem. Thus it is essential to have a functioning social system conducive to the establishment of mutually interacting relationships in an industrial ecosystem.

Industrial Applications of Industrial Ecology

According to a survey conducted by the Rand Science and Technology Policy Institute, industrial concerns are increasing their use of the principles of industrial ecology to curtail pollution and to provide more environmentally friendly products and services.[6] A detailed report dealing with industrial priorities in environmentally related research has shown that companies have a relatively low interest in funding research dealing with the traditional pollution control measures of remediation, effluent treatment, and pollutant monitoring and analysis.[7] Instead, interest lies in research to increase production efficiency, produce environmentally friendly products, and provide environmentally beneficial services. The emphasis is increasing on research that deals with environmental concerns at an early stage of development, and "end-of-pipe" measures are receiving less emphasis. Basically, this means increasing emphasis upon the principles of industrial ecology.

There are a number of promising examples of how large companies are using the principles of industrial ecology to reduce environmental impact. Some of the best news is that such efforts can actually increase profitability. Xerox has greatly reduced disposal problems for photocopiers and components with a vigorous program to recycle parts and refurbish photocopiers, while simultaneously saving hundreds of millions of dollars per year. DuPont emphasizes market opportunities in the environmental area, such as the enzymatic production from cornstarch of intermediates required to make polyesters. The company also advises purchasers of hazardous chemicals on ways to reduce inventories of such chemicals, and in some cases even ways to produce the materials on site to reduce the transportation of hazardous materials. Intel, which is a world leader in the rapidly changing semi-conductor industry, completely retools its product line every two years. In so doing, the concern attempts to avoid triggering environmental permit requirements by designing processes that keep emissions low. It has been a leader in programs to conserve and recycle water, especially at a new facility in Arizona, where water is always a critical issue. Monsanto, arguably the world's leading concern in transgenic crops, is developing crops that are selectively resistant to Monsanto herbicides and that produce their own insecticides using genetic material transferred from bacteria that produce a naturally occurring insecticide.

Scenario Creation to Avoid Environmental Problems

A major concern with rapidly developing new technologies, of which transgenic biotechnology is a prime example, is the emergence of problems, often related to public concern, that were unforeseen in the development of the technologies. In order to avoid such problems, increasing use is being made of **scenario creation**, popularly known as story building, to visualize problems and take remedial action before they become unmanageable. Such an approach has been employed by members of the World Business Council for Sustainable Development to consider potential problems with newly emerging biotechnologies.[8] In a remark attributed to Patricia Solaro of Germany's Hoechst A. G., the job of a panel on scenario creation is "to think the unthinkable and speak the unspeakable, not to say what we think will or should happen." That statement is supported by the fact that scenario creation was developed by the Rand Corporation in the 1950s under sponsorship by U. S. Federal Government to explore scenarios under which nuclear warfare might break out.

Several scenarios have been considered related to biotechnology. One of these is that biotechnology is entirely positive, leading to an abundance of products that lengthen life and improve its quality. Even with this positive scenario, problems could develop, such as those resulting from a much increased population of the elderly made possible by improved drugs and nutrition. Another possible course could result from some relatively small, unforeseen event such as human illness attributed to consumption of food from transgenic plants. The bad publicity resulting from such an event could cause a cascade of opposition that would result in onerous regulation that would seriously cripple efforts in biotechnology. Such an event would be an example of chaos theory, which holds that complex systems can be altered drastically by small perturbations which cause a catastrophic ripple effect in the system. A third scenario is that potential producers and users of biotechnology conclude that the benefits are not worth the effort, costs, and risks, so the technology simply fades away. Such an outcome is similar to that which has occurred with the nuclear power industry in the U. S. and most of Europe. (It could be argued, as well, that the course of the nuclear power industry illustrates chaos theory, with the Chernobyl nuclear power plant disaster the event that triggered a cascade of events leading to its demise.) Obviously, the development of computers and powerful computer programs have been extremely useful in scenario creation. In the late 1990s scenario creation was applied to possible problems created by the "Y2K bug" in which older computer programs had difficulty recognizing the year 2000.

2.12. ENVIRONMENTAL CHEMISTRY

In Chapter 1, environmental chemistry was defined as *the study of the sources, reactions, transport, effects, and fates of chemical species in water, soil, air, and living environments and the effects of technology thereon.* This definition is illustrated for a typical environmental pollutant in Figure 2.4. Pollutant sulfur dioxide is generated in the combustion of sulfur in coal, transported to the atmosphere with flue gas, and oxidized by chemical and photochemical processes to sulfuric acid. The sulfuric acid, in turn, falls as acidic precipitation, where it may have detrimental effects such as toxic effects on trees and other plants. Eventually the sulfuric acid is carried by stream runoff to a lake or ocean where its ultimate fate is to be stored in solution in the water or precipitated as solid sulfates.

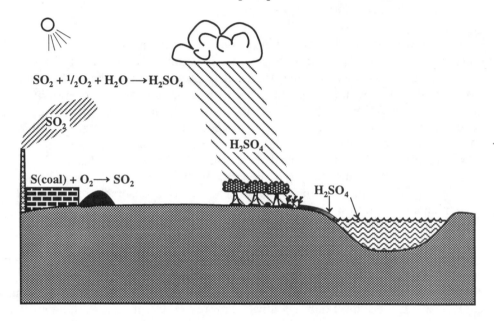

$$SO_2 + \tfrac{1}{2}O_2 + H_2O \rightarrow H_2SO_4$$

SO_2

H_2SO_4

$$S(coal) + O_2 \rightarrow SO_2$$

H_2SO_4

H_2SO_4

Figure 2.4. Illustration of the definition of environmental chemistry by the example of pollutant sulfuric acid formed by the oxidation of sulfur dioxide generated during the combustion of sulfur-containing coal.

Some idea of the complexity of environmental chemistry as a discipline may be realized by examining Figure 2.5, which shows the interchange of chemical species among various environmental spheres. Throughout an environmental system there are variations in temperature, mixing, intensity of solar radiation, input of materials, and various other factors that strongly influence chemical conditions and behavior. Because of its complexity, environmental chemistry must be approached with simplified models.

Potentially, environmental chemistry and industrial ecology have many strong connections. The design of an integrated system of industrial ecology must consider the principles and processes of environmental chemistry. Environmental chemistry must be considered in the extraction of materials from the geosphere and other environmental spheres to provide the materials required by industrial systems in a manner consistent with minimum environmental impact. The facilities and processes of an industrial ecology system can be sited and operated for minimal adverse environmental impact if environmental chemistry is considered in their planning and operation. Environmental chemistry clearly points the way to minimize the environmental impacts of the emissions and by-products of industrial systems, and is very helpful in reaching the ultimate goal of a system of industrial ecology, which is to reduce these emissions and by-products to zero.

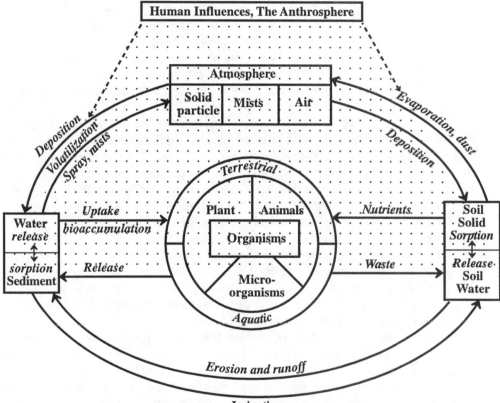

Figure 2.5. Interchange of environmental chemical species among the atmosphere, hydrosphere, geosphere, and biosphere. Human activities (the anthrosphere) have a strong influence on the various processes shown.

Environmental chemistry can be divided into several categories. The first of these addressed in this book is **aquatic chemistry**, the branch of environmental chemistry that deals with chemical phenomena in water. To a large extent, aquatic chemistry addresses chemical phenomena in so-called "natural waters," consisting of water in streams, lakes, oceans, underground aquifers, and other places where the water is rather freely exposed to the atmosphere, soil, rock, and living systems. Aquatic chemistry is introduced in Chapter 3, "Fundamentals of Aquatic Chemistry," which discusses some of the fundamental phenomena that apply to chemical species dissolved in water, particularly acid-base reactions and complexation. Oxidation-reduction phenomena in water are addressed in Chapter 4, "Oxidation-Reduction." Many important aquatic chemical interactions occur between species dissolved in water and those in gaseous, solid, and immiscible liquid phases, the subject of Chapter 5, "Phase Interactions." One thing that clearly distinguishes the chemistry of natural waters from water isolated, contained, and purified by humans is the strong influence of microorganisms on aquatic chemistry, the topic of Chapter 6, "Aquatic Microbial Biochemistry." Water Pollution is discussed in Chapter 7, and Water Treatment in Chapter 8.

As the name implies, **atmospheric chemistry** deals with chemical phenomena in the atmosphere. To understand these processes it is first necessary to have a basic knowledge of the structure and composition of the atmosphere as given in Chapter 9, "The Atmosphere and Atmospheric Chemistry." This chapter also introduces the unique and important concepts of atmospheric chemistry, particularly photochemistry. Photochemical reactions occur when electromagnetic radiation from the sun energizes gas molecules forming reactive species that initiate chain reactions that largely determine key atmospheric chemical phenomena. Particles in the atmosphere are discussed in Chapter 10, and gaseous inorganic air pollutants, such as carbon monoxide, nitrogen oxides, and sulfur oxides, in Chapter 11. Organic species in the atmosphere result in some important pollution phenomena that also influence inorganic species. Organic air pollutants are the topic of Chapter 12, and the most important effect of organic air pollution, photochemical smog, is explained in Chapter 13. Some potentially catastrophic atmospheric pollution effects are outlined in Chapter 14, "The Endangered Global Atmosphere."

The **geosphere** is the topic of two chapters. The first of these, Chapter 15, "The Geosphere and Geochemistry," outlines the physical nature and chemical characteristics of the geosphere and introduces some basic geochemistry. Soil is a uniquely important part of the geosphere that is essential to life on earth. Soil chemistry is covered in Chapter 16, "Soil Environmental Chemistry."

Human activities have such a profound effect on the environment that it is convenient to invoke a fifth sphere of the environment called the "anthrosphere." Much of the influence of human activity on the environment is addressed in Chapters 1-16, particularly as it relates to water and air pollution. Specific environmental aspects of the anthrosphere addressed from the perspective of industrial ecology are covered in Chapter 17, "Principles of Industrial Ecology," and Chapter 18, "Industrial Ecology, Resources, and Energy." A unique environmental problem arising from anthrospheric activities is that of hazardous wastes, which is discussed in Chapter 19, "Nature, Sources, and Environmental Chemistry of Hazardous Wastes," and Chapter 20, "Industrial Ecology for Waste Minimization, Utilization, and Treatment."

The **biosphere** related to environmental chemistry is mentioned in various contexts throughout the book as it relates to environmental chemical processes in water and soil and it is the main topic of discussion of three chapters. The first of these, Chapter 21, "Environmental Biochemistry," covers general aspects of biochemical phenomena in the environment. The second, Chapter 22, "Toxicological Chemistry," deals specifically with the chemistry and environmental chemistry of toxic substances. The third of the biologically oriented chapters is Chapter 23, "Toxicological Chemistry of Chemical Substances," which covers the toxicological chemistry of specific chemical substances and classes of substances.

Analytical chemistry is uniquely important in environmental chemistry. As it applies to environmental chemistry, it is summarized in Chapter 24, "Chemical Analysis of Water and Wastewater;" Chapter 25, "Chemical Analysis of Wastes and Solids;" Chapter 26, "Air and Gas Analysis;" and Chapter 27, "Chemical Analysis of Biological Materials and Xenobiotics.

LITERATURE CITED

1. Ausubel, Jesse H. and Robert Herman, Eds., *Cities and Their Vital Systems: Infrastructure Past, Present, and Future*, National Academy Press, Washington, D.C., 1988.

2. Lennsen, Nicholas and David M. Roodman, "Making Better Buildings," Chapter 6 in *State of the World 1995*, Lester R. Brown, Ed., W. W. Norton and Company, New York, 1995.

3. Gille, Dean, "Combining Communications and Computing: Telematics Infrastructure," Chapter 10 in *Cities and their Vital Systems: Infrastructure Past, Present, and Future*, Jesse H. Ausubel and Robert Herman, Eds., National Academy Press, Washington, D.C., 1988, pp. 233-257.

4. Brown, Lester R., *Who Will Feed China?*, W. W. Norton and Company, New York, 1995.

5. Frosch, Robert A. and Nicholas E. Gallopoulos, "Strategies for Manufacturing," *Scientific American*, **261**, 94-102 (1989).

6. Johnson, Jeff, "The Environment and the Bottom Line," *Chemical and Engineering News*, June 21, 1999, pp. 25-26.

7. "Technology Forces at Work," Rand Science and Technology Policy Institute, Washington, D.C., 1999.

8. Feder, Barnaby J., "Plotting Corporate Futures: Biotechnology Examines What Could go Wrong," *New York Times*, June 24, 1999, p. C1.

SUPPLEMENTARY REFERENCES

Baird, Colin, *Environmental Chemistry*, 2nd ed., W. H. Freeman & Co., San Francisco, 1998.

Bedworth, David D., Mark R. Henderson, and Philip M. Wolfe, *Computer Integrated Design and Manufacturing*, McGraw-Hill, New York, 1991.

Berlow, Lawrence, *The Reference Guide to Famous Engineering Landmarks of the World: Bridges, Tunnels, Dams, Roads, and Other Structures*, Oryx Press, Phoenix, 1997.

Billington, David P., *The Innovators: The Engineering Pioneers Who Made America Modern*, John Wiley & Sons, New York, 1996.

Garratt, James, *Design and Technology* , 2nd ed., Cambridge University Press, New York, 1996.

Garrison, Ervan G., *A History of Engineering and Technology: Artful Methods*, CRC Press, Boca Raton, FL, 1998.

Hastings, Ashley J., CAD at Work: Making the Most of Computer-Aided Design, McGraw-Hill, New York, 1997.

Hawker, Darryl W., D. W. Connell, Micha Warne, and Peter D. Vowles, *Basic Concepts of Environmental Chemistry*, CRC Press, Lewis Publishers, Boca Raton, FL, 1997.

Holtzapple, Mark T. and W. Dan Reece, *Foundations of Engineering*, McGraw-Hill, New York, 1999.

Kortenkamp, David, R. Peter Bonasso, and Robin Murphy, Eds., *Artificial Intelligence and Mobile Robots: Case Studies of Successful Robot Systems*, MIT Press, Cambridge, MA, 1998.

Macalady, Donald L., Ed., *Perspectives in Environmental Chemistry*, Oxford University Press, New York, 1998.

McGrayne, Sharon Bertsch, *Blue Genes and Polyester Plants: 365 More Surprising Scientific Facts, Breakthroughs, and Discoveries*, John Wiley & Sons, New York, 1997.

McMahon, Chris, and Jimmie Browne, *CADCAM: Principles, Practice, and Manufacturing Management*, 2nd ed., Addison-Wesley, Reading, MA, 1998.

Petroski, Henry, *Invention by Design: How Engineers Get from Thought to Thing*, Harvard University Press, Cambridge, MA, 1996.

Pond, Robert J., *Introduction to Engineering Technology*, Prentice Hall, Upper Saddle River, NJ, 1999.

Sandler, Ben-Zion, *Robotics: Designing the Mechanisms for Automated Machinery*, Academic Press, San Diego, 1999.

Spiro, Thomas G. and William M. Stigliani, *Chemistry of the Environment*, Prentice Hall, Upper Saddle River, NJ, 1995.

QUESTIONS AND PROBLEMS

1. Much of The Netherlands consists of land reclaimed from the sea that is actually below sea level as the result of dredging and dike construction. Discuss how this may relate to the anthrosphere and the other spheres of the environment.

2. With a knowledge of the chemical behavior of iron and copper, explain why copper was used as a metal long before iron, even though iron has some superior qualities for a number of applications.

3. How does engineering relate to basic science and to technology?

4. Suggest ways in which an inadequate infrastructure of a city may contribute to environmental degradation.

5. In what sense are automobiles not part of the infrastructure whereas trains are?

6. Discuss how the application of computers can make an existing infrastructure run more efficiently.

7. Although synthetic materials require relatively more energy and nonrenewable resources for their fabrication, how may it be argued that they are often the best choice from an environmental viewpoint for construction of buildings?

8. What is a telecommuter society and what are its favorable environmental characteristics?

9. What are the major areas to consider with respect to information?

10. What was the greatest threat to farmland in the U.S. during the 1930s, and what was done to alleviate that threat? What is currently the greatest threat, and what can be done to alleviate it?

11. What typically happens with regard to food production and demand in a country that acquires a high population density and then becomes industrialized?

12. What is the distinction between automation and robotics?

13. What is the function of a sensory device such as a thermocouple on a robot?

14. What is the CAD/CAM combination?

15. What are some of the parts of the anthrosphere that may be severely contaminated by human activities?

16. What largely caused or marked the change between the "relatively harmonious relationship between the anthrosphere and the rest of the environment" that characterized most of human existence on Earth, and the current situation in which the anthrosphere is a highly perturbing, potentially damaging influence?

17. What are three major stages in the evolution of industry with respect to how it relates to the environment?

18. How is an industrial facility based on the principles of industrial ecology similar to a natural ecological system?

19. Describe, with an example, if possible, what is meant by "end of pipe" measures for pollution control. Why are such measures sometimes necessary? Why are they relatively less desirable? What are the alternatives?

20. Discuss how at least one kind of air pollutant might become a water pollutant.

21. Suggest one or two examples of how technology, properly applied, can be "environmentally friendly."

3 FUNDAMENTALS OF AQUATIC CHEMISTRY

3.1. WATER QUALITY AND QUANTITY

Throughout history, the quality and quantity of water available to humans have been vital factors in determining their well-being. Whole civilizations have disappeared because of water shortages resulting from changes in climate. Even in temperate climates, fluctuations in precipitation cause problems. Devastating droughts in Africa during the 1980s resulted in catastrophic crop failures and starvation. In 1997 an unprecedented flood struck huge sections of North Dakota inundating most of the City of Grand Forks, and in 1998 floods produced by torrential rains from Hurricane Mitch killed thousands in Central America. In 1999 heat killed a number of people in the central and eastern United States and a drought devastated crops and water supplies. In September 1999 Hurricane Floyd produced a 500-year flood in eastern North Carolina.

Waterborne diseases such as cholera and typhoid killed millions of people in the past and still cause great misery in less developed countries. Ambitious programs of dam and dike construction have reduced flood damage, but they have had a number of undesirable side effects in some areas, such as inundation of farmland by reservoirs and failure of unsafe dams. Globally, problems with quantity and quality of water supply remain and in some respects are becoming more serious. These problems include increased water use due to population growth, contamination of drinking water by improperly discarded hazardous wastes (see Chapter 19), and destruction of wildlife by water pollution.

Aquatic chemistry, the subject of this chapter, must consider water in rivers, lakes, estuaries, oceans, and underground, as well as the phenomena that determine the distribution and circulation of chemical species in natural waters. Its study requires some understanding of the sources, transport, characteristics, and composition of water. The chemical reactions that occur in water and the chemical species found in it are strongly influenced by the environment in which the water is found. The chemistry of water exposed to the atmosphere is quite different from that of water at the bottom of a lake. Microorganisms play an essential role in determining the chemical composition of water. Thus, in discussing water chemistry, it is necessary to consider the many general factors that influence this chemistry.

The study of water is known as **hydrology** and is divided into a number of subcategories. **Limnology** is the branch of the science dealing with the characteristics of fresh water including biological properties, as well as chemical and physical properties. **Oceanography** is the science of the ocean and its physical and chemical characteristics. The chemistry and biology of the Earth's vast oceans are unique because of the ocean's high salt content, great depth, and other factors.

Sources and Uses of Water: The Hydrologic Cycle

The world's water supply is found in the five parts of the **hydrologic cycle** (Figure 3.1). About 97% of Earth's water is found in the oceans. Another fraction is present as water vapor in the atmosphere (clouds). Some water is contained in the solid state as ice and snow in snowpacks, glaciers, and the polar ice caps. Surface water is found in lakes, streams, and reservoirs. Groundwater is located in aquifers underground.

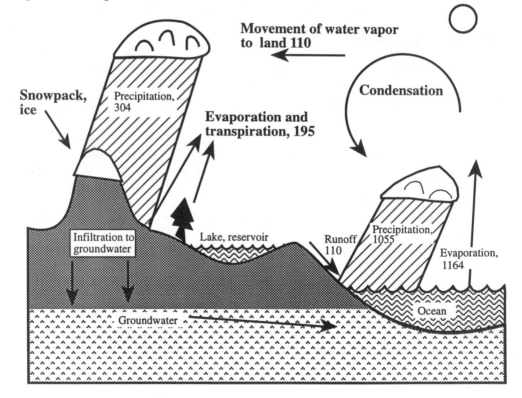

Figure 3.1. The hydrologic cycle, quantities of water in trillions of liters per day.

There is a strong connection between the *hydrosphere,* where water is found, and the *lithosphere*, which is that part of the geosphere accessible to water. Human activities affect both. For example, disturbance of land by conversion of grasslands or forests to agricultural land or intensification of agricultural production may reduce vegetation cover, decreasing **transpiration** (loss of water vapor by plants) and affecting the microclimate. The result is increased rain runoff, erosion, and accumulation of silt in bodies of water. The nutrient cycles may be accelerated, leading to nutrient enrichment of surface waters. This, in turn, can profoundly affect the chemical and biological characteristics of bodies of water.

The water that humans use is primarily fresh surface water and groundwater, the sources of which may differ from each other significantly. In arid regions, a small fraction of the water supply comes from the ocean, a source that is likely to become more important as the world's supply of fresh water dwindles relative to demand. Saline or brackish groundwaters may also be utilized in some areas.

In the continental United States, an average of approximately 1.48×10^{13} liters of water fall as precipitation each day, which translates to 76 cm per year. Of that amount, approximately 1.02×10^{13} liters per day, or 53 cm per year, are lost by evaporation and transpiration. Thus, the water theoretically available for use is approximately 4.6×10^{12} liters per day, or only 23 centimeters per year. At present, the U.S. uses 1.6×10^{12} liters per day, or 8 centimeters of the average annual precipitation. This amounts to an almost tenfold increase from a usage of 1.66×10^{11} liters per day at the turn of the century. Even more striking is the per capita increase from about 40 liters per day in 1900 to around 600 liters per day now. Much of this increase is accounted for by high agricultural and industrial use, which each account for approximately 46% of total consumption. Municipal use consumes the remaining 8%.

Since about 1980, however, water use in the U.S. has shown an encouraging trend with total consumption down by about 9% during a time in which population grew 16%, according to figures compiled by the U.S. Geological Survey.[1] This trend, which is illustrated in Figure 3.2, has been attributed to the success of efforts to conserve water, especially in the industrial (including power generation) and agricultural sectors. Conservation and recycling have accounted for much of the decreased use in the industrial sector. Irrigation water has been used much more efficiently by replacing spray irrigators, which lose large quantities of water to the action of wind and to evaporation, with irrigation systems that apply water directly to soil. Trickle irrigation systems that apply just the amount of water needed directly to plant roots are especially efficient.

A major problem with water supply is its nonuniform distribution with location and time. As shown in Figure 3.3, precipitation falls unevenly in the continental U.S. This causes difficulties because people in areas with low precipitation often consume more water than people in regions with more rainfall. Rapid population growth in the more arid southwestern states of the U.S. during the last four decades has further aggravated the problem. Water shortages are becoming more acute in this region which contains six of the nation's eleven largest cities (Los Angeles, Houston, Dallas, San Diego, Phoenix, and San Antonio). Other problem areas include Florida, where overdevelopment of coastal areas threatens Lake Okeechobee; the Northeast, plagued by deteriorating water systems; and the High Plains, ranging from the Texas panhandle to Nebraska, where irrigation demands on the Oglala aquifer are dropping the water table steadily with no hope of recharge. These problems are minor, however, in comparison to those in some parts of Africa where water shortages are contributing to real famine conditions.

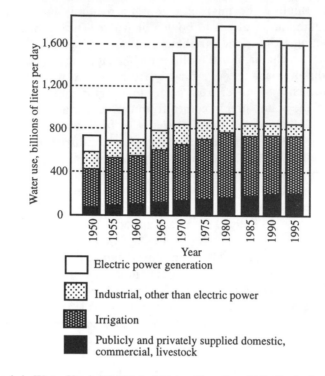

Figure 3.2. Trends in Water Use in The United States (Data from U.S. Geological Survey).

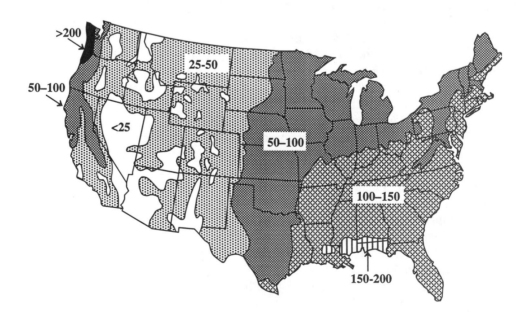

Figure 3.3. Distribution of precipitation in the continental U.S., showing average annual rainfall in centimeters.

3.2. THE PROPERTIES OF WATER, A UNIQUE SUBSTANCE

Water has a number of unique properties that are essential to life. Some of the special characteristics of water include its polar character, tendency to form hydrogen bonds, and ability to hydrate metal ions. These properties are listed in Table 3.1.

Table 3.1. Important Properties of Water

Property	Effects and Significance
Excellent solvent	Transport of nutrients and waste products, making biological processes possible in an aqueous medium
Highest dielectric constant of any common liquid	High solubility of ionic substances and their ionization in solution
Higher surface tension than any other liquid	Controlling factor in physiology; governs drop and surface phenomena
Transparent to visible and longer-wavelength fraction of ultraviolet light	Colorless, allowing light required for photosynthesis to reach considerable depths in bodies of water
Maximum density as a liquid at 4°C	Ice floats; vertical circulation restricted in stratified bodies of water
Higher heat of evaporation than any other material	Determines transfer of heat and water molecules between the atmosphere and bodies of water
Higher latent heat of fusion than any other liquid except ammonia	Temperature stabilized at the freezing point of water
Higher heat capacity than any other liquid except ammonia	Stabilization of temperatures of organisms and geographical regions

The Water Molecule

Water's properties can best be understood by considering the structure and bonding of the water molecule:

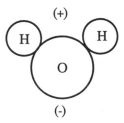

The water molecule is made up of two hydrogen atoms bonded to an oxygen atom. The three atoms are not in a straight line; instead, as shown above, they form an angle of 105°. Because of water's bent structure and the fact that the oxygen atom attracts the negative electrons more strongly than do the hydrogen atoms, the water molecule behaves like a *dipole* having opposite electrical charges at either end. The water dipole may be attracted to either positively or negatively charged ions. For example, when NaCl dissolves in water as positive Na^+ ions and negative Cl^- ions, the positive sodium ions are surrounded by water molecules with their negative ends pointed at the ions, and the chloride ions are surrounded by water molecules with their positive ends pointing at the negative ions, as shown in Figure 3.4. This kind of attraction for ions is the reason why water dissolves many ionic compounds and salts that do not dissolve in other liquids.

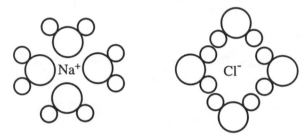

Figure 3.4. Polar water molecules surrounding Na^+ ion (left) and Cl^- ion (right).

A second important characteristic of the water molecule is its ability to form **hydrogen bonds**. Hydrogen bonds are a special type of bond that can form between the hydrogen in one water molecule and the oxygen in another water molecule. This bonding takes place because the oxygen has a partial negative charge and the hydrogen a partial positive charge. Hydrogen bonds, shown in Figure 3.5 as dashed lines, hold the water molecules together in large groups.

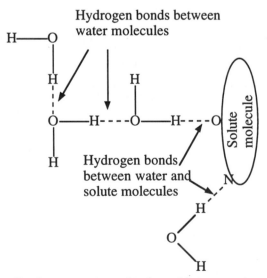

Figure 3.5. Hydrogen bonding between water molecules and between water molecules and a solute molecule in solution.

Hydrogen bonds also help to hold some solute molecules or ions in solution. This happens when hydrogen bonds form between the water molecules and hydrogen, nitrogen, or oxygen atoms on the solute molecule (see Figure 3.5). Hydrogen bonding also aids in retaining extremely small particles called *colloidal particles* in suspension in water (see Section 5.4).

Water is an excellent solvent for many materials; thus it is the basic transport medium for nutrients and waste products in life processes. The extremely high dielectric constant of water relative to other liquids has a profound effect upon its solvent properties in that most ionic materials are dissociated in water. With the exception of liquid ammonia, water has the highest heat capacity of any liquid or solid, $1 \text{ cal} \times \text{g}^{-1} \times \text{deg}^{-1}$. Because of this high heat capacity, a relatively large amount of heat is required to change appreciably the temperature of a mass of water; hence, a body of water can have a stabilizing effect upon the temperature of nearby geographic regions. In addition, this property prevents sudden large changes of temperature in large bodies of water and thereby protects aquatic organisms from the shock of abrupt temperature variations. The extremely high heat of vaporization of water, 585 cal/g at 20°C, likewise stabilizes the temperature of bodies of water and the surrounding geographic regions. It also influences the transfer of heat and water vapor between bodies of water and the atmosphere. Water has its maximum density at 4°C, a temperature above its freezing point. The fortunate consequence of this fact is that ice floats, so that few large bodies of water ever freeze solid. Furthermore, the pattern of vertical circulation of water in lakes, a determining factor in their chemistry and biology, is governed largely by the unique temperature-density relationship of water.

3.3. THE CHARACTERISTICS OF BODIES OF WATER

The physical condition of a body of water strongly influences the chemical and biological processes that occur in water. **Surface water** occurs primarily in streams, lakes, and reservoirs. **Wetlands** are flooded areas in which the water is shallow enough to enable growth of bottom-rooted plants. **Estuaries** are arms of the ocean into which streams flow. The mixing of fresh and salt water gives estuaries unique chemical and biological properties. Estuaries are the breeding grounds of much marine life, which makes their preservation very important.

Water's unique temperature-density relationship results in the formation of distinct layers within nonflowing bodies of water, as shown in Figure 3.6. During the summer a surface layer (**epilimnion**) is heated by solar radiation and, because of its lower density, floats upon the bottom layer, or **hypolimnion**. This phenomenon is called **thermal stratification.** When an appreciable temperature difference exists between the two layers, they do not mix but behave independently and have very different chemical and biological properties. The epilimnion, which is exposed to light, may have a heavy growth of algae. As a result of exposure to the atmosphere and (during daylight hours) because of the photosynthetic activity of algae, the epilimnion contains relatively higher levels of dissolved oxygen and generally is aerobic. In the hypolimnion, bacterial action on biodegradable organic material may cause the water to become anaerobic (lacking dissolved oxygen). As a consequence, chemical species in a relatively reduced form tend to predominate in the hypolimnion.

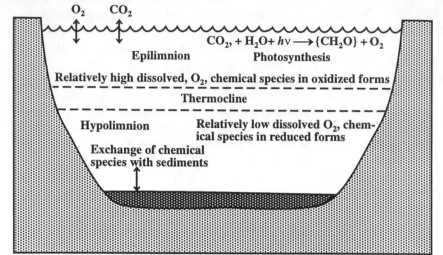

Figure 3.6. Stratification of a lake.

The shear-plane, or layer between epilimnion and hypolimnion, is called the **thermocline**. During the autumn, when the epilimnion cools, a point is reached at which the temperatures of the epilimnion and hypolimnion are equal. This disappearance of thermal stratification causes the entire body of water to behave as a hydrological unit, and the resultant mixing is known as **overturn**. An overturn also generally occurs in the spring. During the overturn, the chemical and physical characteristics of the body of water become much more uniform, and a number of chemical, physical, and biological changes may result. Biological activity may increase from the mixing of nutrients. Changes in water composition during overturn may cause disruption in water-treatment processes.

3.4. AQUATIC LIFE

The living organisms (**biota**) in an aquatic ecosystem may be classified as either autotrophic or heterotrophic. **Autotrophic** organisms utilize solar or chemical energy to fix elements from simple, nonliving inorganic material into complex life molecules that compose living organisms. Algae are the most important autotrophic aquatic organisms because they are **producers** that utilize solar energy to generate biomass from CO_2 and other simple inorganic species.

Heterotrophic organisms utilize the organic substances produced by autotrophic organisms as energy sources and as the raw materials for the synthesis of their own biomass. **Decomposers** (or **reducers**) are a subclass of the heterotrophic organisms and consist of chiefly bacteria and fungi, which ultimately break down material of biological origin to the simple compounds originally fixed by the autotrophic organisms.

The ability of a body of water to produce living material is known as its **productivity.** Productivity results from a combination of physical and chemical factors. High productivity requires an adequate supply of carbon (CO_2), nitrogen (nitrate), phosphorus (orthophosphate), and trace elements such as iron. Water of low productivity generally is desirable for water supply or for swimming. Relatively high productivity is required for the support of fish and to serve as the basis of the food chain in an aquatic ecosystem. Excessive productivity results in decay of the biomass produced, consumption of dissolved oxygen, and odor production, a condition called **eutrophication**.

Life forms higher than algae and bacteria—fish, for example—comprise a comparatively small fraction of the biomass in most aquatic systems. The influence of these higher life forms upon aquatic chemistry is minimal. However, aquatic life is strongly influenced by the physical and chemical properties of the body of water in which it lives. *Temperature, transparency,* and *turbulence* are the three main physical properties affecting aquatic life. Very low water temperatures result in very slow biological processes, whereas very high temperatures are fatal to most organisms. The transparency of water is particularly important in determining the growth of algae. Turbulence is an important factor in mixing processes and transport of nutrients and waste products in water. Some small organisms (**plankton**) depend upon water currents for their own mobility.

Dissolved oxygen (DO) frequently is the key substance in determining the extent and kinds of life in a body of water. Oxygen deficiency is fatal to many aquatic animals such as fish. The presence of oxygen can be equally fatal to many kinds of anaerobic bacteria. **Biochemical oxygen demand, BOD,** discussed as a water pollutant in Section 7.9, refers to the amount of oxygen utilized when the organic matter in a given volume of water is degraded biologically.

Carbon dioxide is produced by respiratory processes in water and sediments and can also enter water from the atmosphere. Carbon dioxide is required for the photosynthetic production of biomass by algae and in some cases is a limiting factor. High levels of carbon dioxide produced by the degradation of organic matter in water can cause excessive algal growth and productivity.

The salinity of water also determines the kinds of life forms present. Irrigation waters may pick up harmful levels of salt. Marine life obviously requires or tolerates salt water, whereas many freshwater organisms are intolerant of salt.

3.5. INTRODUCTION TO AQUATIC CHEMISTRY

To understand water pollution, it is first necessary to have an appreciation of chemical phenomena that occur in water. The remaining sections of this chapter discuss aquatic acid-base and complexation phenomena. Oxidation-reduction reactions and equilibria are discussed in Chapter 4, and details of solubility calculations and interactions between liquid water and other phases are given in Chapter 5. The main categories of aquatic chemical phenomena are illustrated in Figure 3.7.

Aquatic environmental chemical phenomena involve processes familiar to chemists, including acid-base, solubility, oxidation-reduction, and complexation reactions. Although most aquatic chemical phenomena are discussed here from the thermodynamic (equilibrium) viewpoint, it is important to keep in mind that kinetics—rates of reactions—are very important in aquatic chemistry. Biological processes play a key role in aquatic chemistry. For example, algae undergoing photosynthesis can raise the pH of water by removing aqueous CO_2, thereby converting an HCO_3^- ion to a CO_3^{2-} ion; this ion in turn reacts with Ca^{2+} in water to precipitate $CaCO_3$.

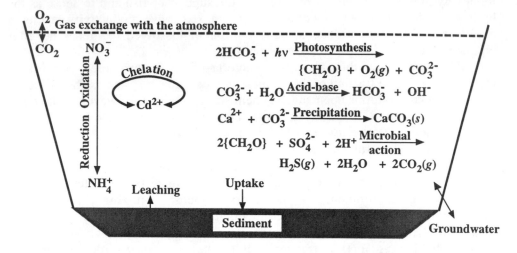

Figure 3.7. Major aquatic chemical processes.

Compared to the carefully controlled conditions of the laboratory, it is much more difficult to describe chemical phenomena in natural water systems. Such systems are very complex and a description of their chemistry must take many variables into consideration. In addition to water, these systems contain mineral phases, gas phases, and organisms. As open, dynamic systems, they have variable inputs and outputs of energy and mass. Therefore, except under unusual circumstances, a true equilibrium condition is not obtained, although an approximately steady-state aquatic system frequently exists. Most metals found in natural waters do not exist as simple hydrated cations in the water, and oxyanions often are found as polynuclear species, rather than as simple monomers. The nature of chemical species in water containing bacteria or algae is strongly influenced by the action of these organisms. Thus, an exact description of the chemistry of a natural water system based upon acid-base, solubility, and complexation equilibrium constants, redox potential, pH, and other chemical parameters is not possible. Therefore, the systems must be described by simplified **models**, often based around equilibrium chemical concepts. Though not exact, nor entirely realistic, such models can yield useful generalizations and insights pertaining to the nature of aquatic chemical processes, and provide guidelines for the description and measurement of natural water systems. Though greatly simplified, such models are very helpful in visualizing the conditions that determine chemical species and their reactions in natural waters and wastewaters.

3.6. GASES IN WATER

Dissolved gases—O_2 for fish and CO_2 for photosynthetic algae—are crucial to the welfare of living species in water. Some gases in water can also cause problems, such as the death of fish from bubbles of nitrogen formed in the blood caused by exposure to water supersaturated with N_2. Volcanic carbon dioxide evolved from the waters of Lake Nyos in the African country of Cameroon asphyxiated 1,700 people in 1986.

The solubilities of gases in water are calculated with **Henry's Law**, which states that *the solubility of a gas in a liquid is proportional to the partial pressure of that gas in contact with the liquid.* These calculations are discussed in some detail in Chapter 5.

Oxygen in Water

Without an appreciable level of dissolved oxygen, many kinds of aquatic organisms cannot exist in water. Dissolved oxygen is consumed by the degradation of organic matter in water. Many fish kills are caused not from the direct toxicity of pollutants but from a deficiency of oxygen because of its consumption in the biodegradation of pollutants.

Most elemental oxygen comes from the atmosphere, which is 20.95% oxygen by volume of dry air. Therefore, the ability of a body of water to reoxygenate itself by contact with the atmosphere is an important characteristic. Oxygen is produced by the photosynthetic action of algae, but this process is really not an efficient means of oxygenating water because some of the oxygen formed by photosynthesis during the daylight hours is lost at night when the algae consume oxygen as part of their metabolic processes. When the algae die, the degradation of their biomass also consumes oxygen.

The solubility of oxygen in water depends upon water temperature, the partial pressure of oxygen in the atmosphere, and the salt content of the water. It is important to distinguish between oxygen *solubility*, which is the maximum dissolved oxygen concentration at equilibrium, and dissolved oxygen *concentration*, which is generally not the equilibrium concentration and is limited by the rate at which oxygen dissolves. The calculation of oxygen solubility as a function of partial pressure is discussed in Section 5.3, where it is shown that the concentration of oxygen in water at 25°C in equilibrium with air at atmospheric pressure is only 8.32 mg/L. Thus, water in equilibrium with air cannot contain a high level of dissolved oxygen compared to many other solute species. If oxygen-consuming processes are occurring in the water, the dissolved oxygen level may rapidly approach zero unless some efficient mechanism for the reaeration of water is operative, such as turbulent flow in a shallow stream or air pumped into the aeration tank of an activated sludge secondary waste treatment facility (see Chapter 8). The problem becomes largely one of kinetics, in which there is a limit to the rate at which oxygen is transferred across the air-water interface. This rate depends upon turbulence, air bubble size, temperature, and other factors.

If organic matter of biological origin is represented by the formula $\{CH_2O\}$, the consumption of oxygen in water by the degradation of organic matter may be expressed by the following biochemical reaction:

$$\{CH_2O\} + O_2 \rightarrow CO_2 + H_2O \qquad (3.6.1)$$

The weight of organic material required to consume the 8.3 mg of O_2 in a liter of water in equilibrium with the atmosphere at 25°C is given by a simple stoichiometric calculation based on Equation 3.6.1, which yields a value of 7.8 mg of $\{CH_2O\}$. Thus, the microorganism-mediated degradation of only 7 or 8 mg of organic material can completely consume the oxygen in one liter of water initially saturated with air at 25°C. The depletion of oxygen to levels below those that will sustain aerobic organisms requires the degradation of even less organic matter at higher temperatures (where the solubility of oxygen is less) or in water not initially saturated with atmospheric oxygen. Furthermore, there are no common aquatic chemical reactions that replenish dissolved oxygen; except for oxygen provided by photosynthesis, it must come from the atmosphere.

The temperature effect on the solubility of gases in water is especially important in the case of oxygen. The solubility of oxygen in water decreases from 14.74 mg/L at 0°C to 7.03 mg/L at 35°C. At higher temperatures, the decreased solubility of oxygen, combined with the increased respiration rate of aquatic organisms, frequently causes a condition in which a higher demand for oxygen accompanied by lower solubility of the gas in water results in severe oxygen depletion.

3.7. WATER ACIDITY AND CARBON DIOXIDE IN WATER

Acid-base phenomena in water involve loss and acceptance of H^+ ion. Many species act as **acids** in water by releasing H^+ ion, others act as **bases** by accepting H^+, and the water molecule itself does both. An important species in the acid-base chemistry of water is bicarbonate ion, HCO_3^-, which may act as either an acid or a base:

$$HCO_3^- \; \xleftarrow{\rightarrow} \; CO_3^{2-} + H^+ \qquad\qquad (3.7.1)$$

$$HCO_3^- + H^+ \; \xleftarrow{\rightarrow} \; CO_2(aq) + H_2O \qquad\qquad (3.7.2)$$

Acidity as applied to natural water and wastewater is the capacity of the water to neutralize OH^-; it is analogous to alkalinity, the capacity to neutralize H^+, which is discussed in the next section. Although virtually all water has some alkalinity acidic water is not frequently encountered, except in cases of severe pollution. Acidity generally results from the presence of weak acids, particularly CO_2, but sometimes includes others such as $H_2PO_4^-$, H_2S, proteins, and fatty acids. Acidic metal ions, particularly Fe^{3+}, may also contribute to acidity.

From the pollution standpoint, strong acids are the most important contributors to acidity. The term *free mineral acid* is applied to strong acids such as H_2SO_4 and HCl in water. Acid mine water is a common water pollutant that contains an appreciable concentration of free mineral acid. Whereas total acidity is determined by titration with base to the phenolphthalein endpoint (pH 8.2), free mineral acid is determined by titration with base to the methyl orange endpoint (pH 4.3).

The acidic character of some hydrated metal ions may contribute to acidity:

$$Al(H_2O)_6^{3+} \; \xleftarrow{\rightarrow} \; Al(H_2O)_5OH^{2+} + H^+ \qquad\qquad (3.7.3)$$

Some industrial wastes, such as spent steel pickling liquor, contain acidic metal ions and often some excess strong acid. The acidity of such wastes must be measured in calculating the amount of lime or other chemicals required to neutralize the acid.

Carbon Dioxide in Water

The most important weak acid in water is carbon dioxide, CO_2. Because of the presence of carbon dioxide in air and its production from microbial decay of organic matter, dissolved CO_2 is present in virtually all natural waters and waste-waters. Rainfall from even an absolutely unpolluted atmosphere is slightly acidic due to the presence of dissolved CO_2. Carbon dioxide, and its ionization products, bicarbonate ion (HCO_3^-), and carbonate ion (CO_3^{2-}) have an extremely important influence upon the chemistry of water. Many minerals are deposited as salts of the carbonate ion. Algae in water utilize dissolved CO_2 in the synthesis of biomass. The equilibrium of dissolved CO_2 with gaseous carbon dioxide in the atmosphere,

$$CO_2(water) \longleftrightarrow CO_2(atmosphere) \tag{3.7.4}$$

and equilibrium of CO_3^{2-} ion between aquatic solution and solid carbonate minerals,

$$MCO_3(slightly\ soluble\ carbonate\ salt) \longleftrightarrow M^{2+} + CO_3^{2-} \tag{3.7.5}$$

have a strong buffering effect upon the pH of water.

Carbon dioxide is only about 0.037% by volume of normal dry air. As a consequence of the low level of atmospheric CO_2, water totally lacking in alkalinity (capacity to neutralize H^+, see Section 3.8) in equilibrium with the atmosphere contains only a very low level of carbon dioxide. However, the formation of HCO_3^- and CO_3^{2-} greatly increases the solubility of carbon dioxide. High concentrations of free carbon dioxide in water may adversely affect respiration and gas exchange of aquatic animals. It may even cause death and should not exceed levels of 25 mg/L in water.

A large share of the carbon dioxide found in water is a product of the breakdown of organic matter by bacteria. Even algae, which utilize CO_2 in photo-synthesis, produce it through their metabolic processes in the absence of light. As water seeps through layers of decaying organic matter while infiltrating the ground, it may dissolve a great deal of CO_2 produced by the respiration of organisms in the soil. Later, as water goes through limestone formations, it dissolves calcium carbonate because of the presence of the dissolved CO_2:

$$CaCO_3(s) + CO_2(aq) + H_2O \longleftrightarrow Ca^{2+} + 2HCO_3^- \tag{3.7.6}$$

This process is the one by which limestone caves are formed. The implications of the above reaction for aquatic chemistry are discussed in greater detail in Section 3.9.

The concentration of gaseous CO_2 in the atmosphere varies with location and season; it is increasing by about one part per million (ppm) by volume per year. For purposes of calculation here, the concentration of atmospheric CO_2 will be taken as 350 ppm (0.0350%) in dry air. At 25°C, water in equilibrium with unpolluted air containing 350 ppm carbon dioxide has a $CO_2(aq)$ concentration of 1.146×10^{-5} M (see Henry's law calculation of gas solubility in Section 5.3), and this value will be used for subsequent calculations.

Although CO_2 in water is often represented as H_2CO_3, the equilibrium constant for the reaction

$$CO_2(aq) + H_2O \longleftrightarrow H_2CO_3 \qquad (3.7.7)$$

is only around 2×10^{-3} at 25°C, so just a small fraction of the dissolved carbon dioxide is actually present as H_2CO_3. In this text, nonionized carbon dioxide in water will be designated simply as CO_2, which in subsequent discussions will stand for the total of dissolved molecular CO_2 and undissociated H_2CO_3.

The CO_2-HCO_3^--CO_3^{2-} system in water may be described by the equations,

$$CO_2 + H_2O \longleftrightarrow HCO_3^- + H^+ \qquad (3.7.8)$$

$$K_{a1} = \frac{[H^+][HCO_3^-]}{[CO_2]} = 4.45 \times 10^{-7} \quad pK_{a1} = 6.35 \qquad (3.7.9)$$

$$HCO_3^- \longleftrightarrow CO_3^{2-} + H^+ \qquad (3.7.10)$$

$$K_{a2} = \frac{[H^+][CO_3^{2-}]}{[HCO_3^-]} = 4.69 \times 10^{-11} \quad pK_{a2} = 10.33 \qquad (3.7.11)$$

where $pK_a = -\log K_a$. The predominant species formed by CO_2 dissolved in water depends upon pH. This is best shown by a **distribution of species diagram** with pH as a master variable as illustrated in Figure 3.8. Such a diagram shows the major

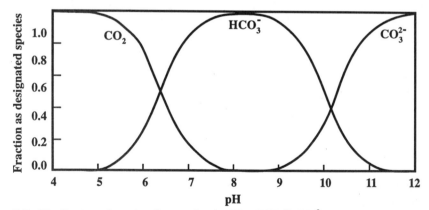

Figure 3.8. Distribution of species diagram for the CO_2–HCO_3^-–CO_3^{2-} system in water.

species present in solution as a function of pH. For CO_2 in aqueous solution, the diagram is a series of plots of the fractions present as CO_2, HCO_3^-, and CO_3^{2-} as a function of pH. These fractions, designated as α_x, are given by the following expressions:

$$\alpha_{CO_2} = \frac{[CO_2]}{[CO_2] + [HCO_3^-] + [CO_3^{2-}]} \tag{3.7.12}$$

$$\alpha_{HCO_3^-} = \frac{[HCO_3^-]}{[CO_2] + [HCO_3^-] + [CO_3^{2-}]} \tag{3.7.13}$$

$$\alpha_{CO_3^{2-}} = \frac{[CO_3^{2-}]}{[CO_2] + [HCO_3^-] + [CO_3^{2-}]} \tag{3.7.14}$$

Substitution of the expressions for K_{a1} and K_{a2} into the α expressions gives the fractions of species as a function of acid dissociation constants and hydrogen ion concentration:

$$\alpha_{CO_2} = \frac{[H^+]^2}{[H^+]^2 + K_{a1}[H^+] + K_{a1}K_{a2}} \tag{3.7.15}$$

$$\alpha_{HCO_3^-} = \frac{K_{a1}[H^+]}{[H^+]^2 + K_{a1}[H^+] + K_{a1}K_{a2}} \tag{3.7.16}$$

$$\alpha_{CO_3^{2-}} = \frac{K_{a1}K_{a2}}{[H^+]^2 + K_{a1}[H^+] + K_{a1}K_{a2}} \tag{3.7.17}$$

Calculations from these expressions show the following:

- For pH significantly below pK_{a1}, α_{CO_2} is essentially 1
- When pH = pK_{a1}, $\alpha_{CO_2} = \alpha_{HCO_3^-}$
- When pH = $1/2(pK_{a1} + pK_{a2})$, $\alpha_{HCO_3^-}$ is at its maximum value of 0.98
- When pH = pK_{a2}, $\alpha_{HCO_3^-} = \alpha_{CO_3^{2-}}$
- For pH significantly above pK_{a2}, $\alpha_{CO_3^{2-}}$ is essentially 1.

The distribution of species diagram in Figure 3.8 shows that hydrogen carbonate (bicarbonate) ion (HCO_3^-) is the predominant species in the pH range found in most waters, with CO_2 predominating in more acidic waters.

As mentioned above, the value of $[CO_2(aq)]$ in water at 25°C in equilibrium with air that is 350 ppm CO_2 is 1.146×10^{-5} M. The carbon dioxide dissociates partially in water to produce equal concentrations of H^+ and HCO_3^-:

$$CO_2 + H_2O \longleftrightarrow HCO_3^- + H^+ \tag{3.7.18}$$

The concentrations of H^+ and HCO_3^- are calculated from K_{a1}:

$$K_{a1} = \frac{[H^+][HCO_3^-]}{[CO_2]} = \frac{[H^+]^2}{1.146 \times 10^{-5}} = 4.45 \times 10^{-7} \tag{3.7.19}$$

$$[H^+] = [HCO_3^-] = (1.146 \times 10^{-5} \times 4.45 \times 10^{-7})^{1/2} = 2.25 \times 10^{-6}$$

$$pH = 5.65$$

This calculation explains why pure water that has equilibrated with the unpolluted atmosphere is slightly acidic with a pH somewhat less than 7.

3.8. ALKALINITY

The capacity of water to accept H^+ ions (protons) is called **alkalinity**. Alkalinity is important in water treatment and in the chemistry and biology of natural waters. Frequently, the alkalinity of water must be known to calculate the quantities of chemicals to be added in treating the water. Highly alkaline water often has a high pH and generally contains elevated levels of dissolved solids. These characteristics may be detrimental for water to be used in boilers, food processing, and municipal water systems. Alkalinity serves as a pH buffer and reservoir for inorganic carbon, thus helping to determine the ability of a water to support algal growth and other aquatic life, so it can be used as a measure of water fertility. Generally, the basic species responsible for alkalinity in water are bicarbonate ion, carbonate ion, and hydroxide ion:

$$HCO_3^- + H^+ \rightarrow CO_2 + H_2O \tag{3.8.1}$$

$$CO_3^{2-} + H^+ \rightarrow HCO_3^- \tag{3.8.2}$$

$$OH^- + H^+ \rightarrow H_2O \tag{3.8.3}$$

Other, usually minor, contributors to alkalinity are ammonia and the conjugate bases of phosphoric, silicic, boric, and organic acids.

At pH values below 7, $[H^+]$ in water detracts significantly from alkalinity, and its concentration must be subtracted in computing the total alkalinity. Therefore, the following equation is the complete equation for alkalinity in a medium where the only contributors to it are HCO_3^-, CO_3^{2-}, and OH^-:

$$[alk] = [HCO_3^-] + 2[CO_3^{2-}] + [OH]^- - [H^+] \tag{3.8.4}$$

Alkalinity generally is expressed as *phenolphthalein alkalinity*, corresponding to titration with acid to the pH at which HCO_3^- is the predominant carbonate species (pH 8.3), or *total alkalinity*, corresponding to titration with acid to the methyl orange endpoint (pH 4.3), where both bicarbonate and carbonate species have been converted to CO_2.

It is important to distinguish between high *basicity*, manifested by an elevated pH, and high *alkalinity*, the capacity to accept H^+. Whereas pH is an *intensity* factor, alkalinity is a *capacity* factor. This may be illustrated by comparing a solution of 1.00×10^{-3} M NaOH with a solution of 0.100 M HCO_3^-. The sodium hydroxide solution is quite basic, with a pH of 11, but a liter of it will neutralize only 1.00×10^{-3} mole of acid. The pH of the sodium bicarbonate solution is 8.34, much lower than that of the NaOH. However, a liter of the sodium bicarbonate solution will neutralize 0.100 mole of acid; therefore, its alkalinity is 100 times that of the more basic NaOH solution.

In engineering terms, alkalinity frequently is expressed in units of mg/L of $CaCO_3$, based upon the following acid-neutralizing reaction:

$$CaCO_3 + 2H^+ \rightarrow Ca^{2+} + CO_2 + H_2O \qquad (3.8.5)$$

The equivalent weight of calcium carbonate is one-half its formula weight. Expressing alkalinity in terms of mg/L of $CaCO_3$ can, however, lead to confusion, and the preferable notation for the chemist is equivalents/L, the number of moles of H^+ neutralized by the alkalinity in a liter of solution.

Contributors to Alkalinity at Different pH Values

Natural water typically has an alkalinity, designated here as "[alk]," of 1.00×10^{-3} equivalents per liter (eq/L), meaning that the alkaline solutes in 1 liter of the water will neutralize 1.00×10^{-3} moles of acid. The contributions made by different species to alkalinity depend upon pH. This is shown here by calculation of the relative contributions to alkalinity of HCO_3^-, CO_3^{2-}, and OH^- at pH 7.00 and pH 10.00. First, for water at pH 7.00, $[OH^-]$ is too low to make any significant contribution to the alkalinity. Furthermore, as shown by Figure 3.8, at pH 7.00 $[HCO_3^-] \gg [CO_3^{2-}]$. Therefore, the alkalinity is due to HCO_3^- and $[HCO_3^-] = 1.00 \times 10^{-3}$ M. Substitution into the expression for K_{a1} shows that at pH 7.00 and $[HCO_3^-] = 1.00 \times 10^{-3}$ M, the value of $[CO_2(aq)]$ is 2.25×10^{-4} M, somewhat higher than the value that arises from water in equilibrium with atmospheric air, but readily reached due to the presence of carbon dioxide from bacterial decay in water and sediments.

Consider next the case of water with the same alkalinity, 1.00×10^{-3} eq/L that has a pH of 10.00. At this higher pH both OH^- and CO_3^{2-} are present at significant concentrations compared to HCO_3^- and the following may be calculated:

$$[alk] = [HCO_3^-] + 2[CO_3^{2-}] + [OH^-] = 1.00 \times 10^{-3} \qquad (3.8.6)$$

The concentration of CO_3^{2-} is multiplied by 2 because each CO_3^{2-} ion can neutralize 2 H^+ ions. The other two equations that must be solved to get the concentrations of HCO_3, CO_3^{2-}, and OH^- are

$$[OH^-] = \frac{K_w}{[H^+]} = \frac{1.00 \times 10^{-14}}{1.00 \times 10^{-10}} = 1.00 \times 10^{-4} \qquad (3.8.7)$$

and

$$[CO_3^{2-}] = \frac{K_{a2}[HCO_3^-]}{[H^+]} \qquad (3.8.8)$$

Solving these three equations gives $[HCO_3^-] = 4.64 \times 10^{-4}$ M and $[CO_3^{2-}] = 2.18 \times 10^{-4}$ M, so the contributions to the alkalinity of this solution are the following:

$$\begin{array}{r} 4.64 \times 10^{-4} \text{ eq/L from } HCO_3^- \\ 2 \times 2.18 \times 10^{-4} = 4.36 \times 10^{-4} \text{ eq/L from } CO_3^{2-} \\ \underline{1.00 \times 10^{-4} \text{ eq/L from } OH^-} \\ alk = 1.00 \times 10^{-3} \text{ eq/L} \end{array}$$

Dissolved Inorganic Carbon and Alkalinity

The values given above can be used to show that at the same alkalinity value the concentration of total dissolved inorganic carbon, [C],

$$[C] = [CO_2] + [HCO_3^-] + [CO_3^{2-}] \qquad (3.8.9)$$

varies with pH. At pH 7.00,

$$[C]_{pH\ 7} = 2.25 \times 10^{-4} + 1.00 \times 10^{-3} + 0 = 1.22 \times 10^{-3} \qquad (3.8.10)$$

whereas at pH 10.00,

$$[C]_{pH\ 10} = 0 + 4.64 \times 10^{-4} + 2.18 \times 10^{-4} = 6.82 \times 10^{-4} \qquad (3.8.11)$$

The calculation above shows that the dissolved inorganic carbon concentration at pH 10.00 is only about half that at pH 7.00. This is because at pH 10 major contributions to alkalinity are made by CO_3^{2-} ion, each of which has twice the alkalinity of each HCO_3^- ion, and by OH^-, which does not contain any carbon. The lower inorganic carbon concentration at pH 10 shows that the aquatic system can donate dissolved inorganic carbon for use in photosynthesis with a change in pH but none in alkalinity. This pH-dependent difference in dissolved inorganic carbon concentration represents a significant potential source of carbon for algae growing in water which fix carbon by the overall reactions

$$CO_2 + H_2O + h\nu \rightarrow \{CH_2O\} + O_2 \qquad (3.8.12)$$

and

$$HCO_3^- + H_2O + h\nu \rightarrow \{CH_2O\} + OH^- + O_2 \qquad (3.8.13)$$

As dissolved inorganic carbon is used up to synthesize biomass, $\{CH_2O\}$, the water becomes more basic. The amount of inorganic carbon that can be consumed before the water becomes too basic to allow algal reproduction is proportional to the alkalinity. In going from pH 7.00 to pH 10.00, the amount of inorganic carbon consumed from 1.00 L of water having an alkalinity of 1.00×10^{-3} eq/L is

$$[C]_{pH\ 7} \times 1 \text{ L} - [C]_{pH\ 10} \times 1 \text{ L} =$$

$$1.22 \times 10^{-3} \text{ mol} - 6.82 \times 10^{-4} \text{ mol} = 5.4 \times 10^{-4} \text{ mol} \qquad (3.8.14)$$

This translates to an increase of 5.4×10^{-4} mol/L of biomass. Since the formula mass of $\{CH_2O\}$ is 30, the weight of biomass produced amounts to 16 mg/L.

Assuming no input of additional CO_2, at higher alkalinity more biomass is produced for the same change in pH, whereas at lower alkalinity less is produced. Because of this effect, biologists use alkalinity as a measure of water fertility.

Influence of Alkalinity on CO_2 Solubility

The increased solubility of carbon dioxide in water with an elevated alkalinity can be illustrated by comparing its solubility in pure water (alkalinity 0) to its solubility in water initially containing 1.00×10^{-3} M NaOH (alkalinity 1.00×10^{-3} eq/L). The number of moles of CO_2 that will dissolve in a liter of pure water from the atmosphere containing 350 ppm carbon dioxide is

$$\text{Solubility} = [CO_2(aq)] + [HCO_3^-] \qquad (3.8.15)$$

Substituting values calculated in Section 3.7 gives

$$\text{Solubility} = 1.146 \times 10^{-5} + 2.25 \times 10^{-6} = 1.371 \times 10^{-5} \, M$$

The solubility of CO_2 in water, initially 1.00×10^{-3} M in NaOH, is about 100-fold higher because of uptake of CO_2 by the reaction

$$CO_2(ag) + OH^- \xleftarrow{\longrightarrow} HCO_3^- \qquad (3.8.16)$$

so that

$$\text{Solubility} = [CO_2(aq)] + [HCO_3^-]$$
$$= 1.146 \times 10^{-5} + 1.00 \times 10^{-3} = 1.01 \times 10^{-3} \, M \quad (3.8.17)$$

3.9. CALCIUM AND OTHER METALS IN WATER

Metal ions in water, commonly denoted M^{n+}, exist in numerous forms. A bare metal ion, Ca^{2+} for example, cannot exist as a separate entity in water. In order to secure the highest stability of their outer electron shells, metal ions in water are bonded, or *coordinated*, to other species. These may be water molecules or other stronger bases (electron-donor partners) that might be present. Therefore, metal ions in water solution are present in forms such as the **hydrated** metal cation $M(H_2O)_x^{n+}$. Metal ions in aqueous solution seek to reach a state of maximum stability through chemical reactions including acid-base,

$$Fe(H_2O)_6^{3+} \xleftarrow{\longrightarrow} FeOH(H_2O)_5^{2+} + H^+ \qquad (3.9.1)$$

precipitation,

$$Fe(H_2O)_6^{3+} \xleftarrow{\longrightarrow} Fe(OH)_3(s) + 3H_2O + 3H^+ \qquad (3.9.2)$$

and oxidation-reduction reactions:

$$Fe(H_2O)_6^{2+} \xleftarrow{\longrightarrow} Fe(OH)_3(s) + 3H_2O + e^- + 3H^+ \qquad (3.9.3)$$

These all provide means through which metal ions in water are transformed to more stable forms. Because of reactions such as these and the formation of dimeric species, such as $Fe_2(OH)_2^{4+}$, the concentration of simple hydrated $Fe(H_2O)_6^{3+}$ ion in water is vanishingly small; the same holds true for many other hydrated metal ions dissolved in water.

Hydrated Metal Ions as Acids

Hydrated metal ions, particularly those with a charge of +3 or more, tend to lose H^+ ion from the water molecules bound to them in aqueous solution, and fit the definition of Brönsted acids, according to which acids are H^+ donors and bases are H^+ acceptors. The acidity of a metal ion increases with charge and decreases with increasing radius. As shown by the reaction,

$$Fe(H_2O)_6^{3+} \longleftrightarrow Fe(H_2O)_5OH^{2+} + H^+ \tag{3.9.4}$$

hydrated iron(III) ion is an acid, a relatively strong one with a K_{a1} of 8.9×10^{-4}, so that solutions of iron(III) tend to have low pH values. Hydrated trivalent metal ions, such as iron(III), generally are minus at least one hydrogen ion at neutral pH values or above. For tetravalent metal ions, the completely protonated forms, $M(H_2O)_x^{4+}$, are rare even at very low pH values. Commonly, O^{2-} is coordinated to tetravalent metal ions; an example is the vanadium(IV) species, VO^{2+}. Generally, divalent metal ions do not lose a hydrogen ion at pH values below 6, whereas monovalent metal ions such as Na^+ do not act as acids at all, and exist in water solution as simple hydrated ions.

The tendency of hydrated metal ions to behave as acids may have a profound effect upon the aquatic environment. A good example is *acid mine water* (see Chapter 7), which derives part of its acidic character from the acidic nature of hydrated iron(III):

$$Fe(H_2O)_6^{3+} \longleftrightarrow Fe(OH)_3(s) + 3H^+ + 3H_2O \tag{3.9.5}$$

Hydroxide, OH^-, bonded to a metal ion, may function as a bridging group to join two or more metals together as shown by the following dehydration-dimerization process:

$$2Fe(H_2O)_5OH^{2+} \rightarrow (H_2O)_4Fe \underset{\underset{H}{O}}{\overset{\overset{H}{O}}{<}}> Fe(H_2O)_4^{4+} + 2H_2O \tag{3.9.6}$$

Among the metals other than iron(III) forming polymeric species with OH^- as a bridging group are Al(III), Be(II), Bi(III), Ce(IV), Co(III), Cu(II), Ga(III), Mo(V), Pb(II), Sc(II), Sn(IV), and U(VI). Additional hydrogen ions may be lost from water molecules bonded to the dimers, furnishing OH^- groups for further bonding and leading to the formation of polymeric hydrolytic species. If the process continues, colloidal hydroxy polymers are formed and, finally, precipitates are produced. This process is thought to be the general one by which hydrated iron(III) oxide, $Fe_2O_3 \cdot x(H_2O)$, (also called ferric hydroxide, $Fe(OH)_3$), is precipitated from solutions containing iron(III).

Calcium in Water

Of the cations found in most fresh-water systems, calcium generally has the highest concentration. The chemistry of calcium, although complicated enough, is simpler than that of the transition metal ions found in water. Calcium is a key element in many geochemical processes, and minerals constitute the primary sources of calcium ion in waters. Among the primary contributing minerals are gypsum, $CaSO_4 \cdot 2H_2O$; anhydrite, $CaSO_4$; dolomite, $CaMg(CO_3)_2$; and calcite and aragonite, which are different mineral forms of $CaCO_3$.

Calcium ion, along with magnesium and sometimes iron(II) ion, accounts for **water hardness**. The most common manifestation of water hardness is the curdy precipitate formed by soap in hard water. *Temporary hardness* is due to the presence of calcium and bicarbonate ions in water and may be eliminated by boiling the water:

$$Ca^{2+} + 2HCO_3^- \longleftrightarrow CaCO_3(s) + CO_2(g) + H_2O \qquad (3.9.7)$$

Increased temperature may force this reaction to the right by evolving CO_2 gas, and a white precipitate of calcium carbonate may form in boiling water having temporary hardness.

Water containing a high level of carbon dioxide readily dissolves calcium from its carbonate minerals:

$$CaCO_3(s) + CO_2(aq) + H_2O \longleftrightarrow Ca^{2+} + 2HCO_3^- \qquad (3.9.8)$$

When this reaction is reversed and CO_2 is lost from the water, calcium carbonate deposits are formed. The concentration of CO_2 in water determines the extent of dissolution of calcium carbonate. The carbon dioxide that water may gain by equilibration with the atmosphere is not sufficient to account for the levels of calcium dissolved in natural waters, especially groundwaters. Rather, the respiration of microorganisms degrading organic matter in water, sediments, and soil,

$$\{CH_2O\} + O_2 \rightarrow CO_2 + H_2O \qquad (3.9.9)$$

accounts for the very high levels of CO_2 and HCO_3^- observed in water and is very important in aquatic chemical processes and geochemical transformations.

Dissolved Carbon Dioxide and Calcium Carbonate Minerals

The equilibrium between dissolved carbon dioxide and calcium carbonate minerals is important in determining several natural water chemistry parameters such as alkalinity, pH, and dissolved calcium concentration (Figure 3.9). For fresh water, the typical figures quoted for the concentrations of both HCO_3^- and Ca^{2+} are

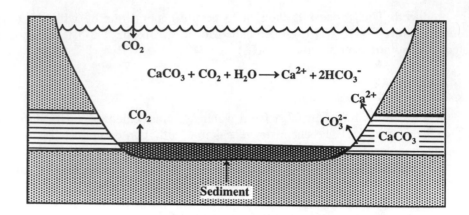

Figure 3.9. Carbon dioxide-calcium carbonate equilibria.

1.00×10^{-3} M. It may be shown that these are reasonable values when the water is in equilibrium with limestone, $CaCO_3$, and with atmospheric CO_2. The concentration of CO_2 in water in equilibrium with air has already been calculated as 1.146×10^{-5} M. The other constants needed to calculate $[HCO_3^-]$ and $[Ca^{2+}]$ are the acid dissociation constant for CO_2:

$$K_{a1} = \frac{[H^+][HCO_3^-]}{[CO_2]} = 4.45 \times 10^{-7} \tag{3.9.10}$$

the acid dissociation constant of HCO_3^-:

$$K_{a2} = \frac{[H^+][CO_3^{2-}]}{[HCO_3^-]} = 4.69 \times 10^{-11} \tag{3.9.11}$$

and the solubility product of calcium carbonate (calcite):

$$K_{sp} = [Ca^{2+}][CO_3^{2-}] = 4.47 \times 10^{-9} \tag{3.9.12}$$

The reaction between calcium carbonate and dissolved CO_2 is

$$CaCO_3(s) + CO_2(aq) + H_2O \xleftarrow{\quad} \rightarrow Ca^{2+} + 2HCO_3^- \tag{3.9.13}$$

for which the equilibrium expression is the following:

$$K' = \frac{[Ca^{2+}][HCO_3^-]^2}{[CO_2]} = \frac{K_{sp}K_{a1}}{K_{a2}} = 4.24 \times 10^{-5} \tag{3.9.14}$$

The stoichiometry of Reaction 3.9.13 gives a bicarbonate ion concentration that is twice that of calcium. Substitution of the value of CO_2 concentration into the expression for K' yields values of 4.99×10^{-4} M for $[Ca^{2+}]$ and 9.98×10^{-4} for $[HCO_3^-]$. Substitution into the expression for K_{sp} yields 8.96×10^{-6} M for $[CO_3^{2-}]$. When known concentrations are substituted into the product $K_{a1}K_{a2}$,

$$K_{a1}K_{a2} = \frac{[H^+]^2[CO_3^{2-}]}{[CO_2]} = 2.09 \times 10^{-17} \tag{3.9.15}$$

a value of 5.17×10^{-9} M is obtained for $[H^+]$ (pH 8.29). The alkalinity is essentially equal to $[HCO_3^-]$, which is much higher than $[CO_3^{2-}]$ or $[OH^-]$.

To summarize, for water in equilibrium with solid calcium carbonate and atmospheric CO_2, the following concentrations are calculated:

$$[CO_2] = 1.146 \times 10^{-5} \text{ M} \qquad [Ca^{2+}] = 4.99 \times 10^{-4} \text{ M}$$

$$[HCO_3^-] = 9.98 \times 10^{-4} \text{ M} \qquad [H^+] = 5.17 \times 10^{-9} \text{ M}$$

$$[CO_3^{2-}] = 8.96 \times 10^{-6} \text{ M} \qquad \text{pH} = 8.29$$

Factors such as nonequilibrium conditions, high CO_2 concentrations in bottom regions, and increased pH due to algal uptake of CO_2 cause deviations from these values. Nevertheless, they are close to the values found in a large number of natural water bodies.

3.10. COMPLEXATION AND CHELATION

The properties of metals dissolved in water depend largely upon the nature of metal species dissolved in the water. Therefore, **speciation** of metals plays a crucial role in their environmental chemistry in natural waters and wastewaters. In addition to the hydrated metal ions, for example, $Fe(H_2O)_6^{3+}$ and hydroxy species such as $FeOH(H_2O)_5^{2+}$ discussed in the preceding section, metals may exist in water reversibly bound to inorganic anions or to organic compounds as **metal complexes**. For example, a cyanide ion can bond to dissolved iron(II):

$$Fe(H_2O)_6^{2+} + CN^- \longleftrightarrow FeCN(H_2O)_5^+ + H_2O \tag{3.10.1}$$

Additional cyanide ions may bond to the iron to form $Fe(CN)_2$, $Fe(CN)_3^-$, $Fe(CN)_4^{2-}$, $Fe(CN)_5^{3-}$, and $Fe(CN)_6^{4-}$, where the water molecules still bound to the iron(II) are omitted for simplicity. This phenomenon is called **complexation**; the species that binds with the metal ion, CN^- in the example above, is called a **ligand**, and the product in which the ligand is bound with the metal ion is a **complex**, **complex ion**, or **coordination compound**. A special case of complexation in which a ligand bonds in two or more places to a metal ion is called **chelation**. In addition to being present as metal complexes, metals may occur in water as **organometallic** compounds containing carbon-to-metal bonds. The solubilities, transport properties, and biological effects of such species are often vastly different from those of the metal ions themselves. Subsequent sections of this chapter consider metal species with an emphasis upon metal complexation, especially chelation, in which particularly strong metal complexes are formed.

In the example above, the cyanide ion is a **unidentate ligand**, which means that it possesses only one site that bonds to a metal ion. Complexes of unidentate ligands are of relatively little importance in solution in natural waters. Of considerably more importance are complexes with **chelating agents.** A chelating agent has more than one atom that may be bonded to a central metal ion at one time to form a ring structure. Thus, pyrophosphate ion, $P_2O_7^{4-}$, bonds to two sites on a calcium ion to form a chelate:

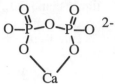

In general, since a chelating agent may bond to a metal ion in more than one place simultaneously (Figure 3.10), chelates are more stable than complexes involv-

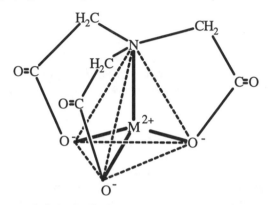

Figure 3.10. Nitrilotriacetate chelate of a divalent metal ion in a tetrahedral configuration.

ing unidentate ligands. Stability tends to increase with the number of chelating sites available on the ligand. Structures of metal chelates take a number of different forms, all characterized by rings in various configurations. The structure of a tetrahedrally coordinated chelate of nitrilotriacetate ion is shown in Figure 3.10.

The ligands found in natural waters and wastewaters contain a variety of functional groups which can donate the electrons required to bond the ligand to a metal ion.[2] Among the most common of these groups are:

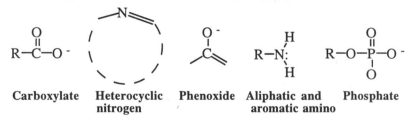

| Carboxylate | Heterocyclic nitrogen | Phenoxide | Aliphatic and aromatic amino | Phosphate |

These ligands complex most metal ions found in unpolluted waters and biological systems (Mg^{2+}, Ca^{2+}, Mn^{2+}, Fe^{2+}, Fe^{3+}, Cu^{2+}, Zn^{2+}, VO^{2+}). They also bind to contaminant metal ions such as Co^{2+}, Ni^{2+}, Sr^{2+}, Cd^{2+}, and Ba^{2+}.

Complexation may have a number of effects, including reactions of both ligands and metals. Among the ligand reactions are oxidation-reduction, decarboxylation, and hydrolysis. Complexation may cause changes in oxidation state of the metal and may result in a metal becoming solubilized from an insoluble compound. The formation of insoluble complex compounds removes metal ions from solution.

Complex compounds of metals such as iron (in hemoglobin) and magnesium (in chlorophyll) are vital to life processes. Naturally occurring chelating agents, such as humic substances and amino acids, are found in water and soil. The high concentration of chloride ion in seawater results in the formation of some chloro complexes. Synthetic chelating agents such as sodium tripolyphosphate, sodium ethylenediaminetetraacetate (EDTA), sodium nitrilotriacetate (NTA), and sodium citrate are produced in large quantities for use in metal-plating baths, industrial water treatment, detergent formulations, and food preparation. Small quantities of these compounds enter aquatic systems through waste discharges.

Occurrence and Importance of Chelating Agents in Water

Chelating agents are common potential water pollutants. These substances can occur in sewage effluent and industrial wastewater such as metal plating wastewater. Chelates formed by the strong chelating agent ethylenediaminetetraacetate (EDTA, structure illustrated at the beginning of Section 3.13) have been shown to greatly increase the migration rates of radioactive ^{60}Co from pits and trenches used by the Oak Ridge National Laboratory in Oak Ridge, Tennessee, for disposal of intermediate-level radioactive waste[3]. EDTA was used as a cleaning and solubilizing agent for the decontamination of hot cells, equipment, and reactor components. Analysis of water from sample wells in the disposal pits showed EDTA concentrations of 3.4×10^{-7} M. The presence of EDTA 12-15 years after its burial attests to its low rate of biodegradation. In addition to cobalt, EDTA strongly chelates radioactive plutonium and radioisotopes of Am^{3+}, Cm^{3+}, and Th^{4+}. Such chelates with negative charges are much less strongly sorbed by mineral matter and are vastly more mobile than the unchelated metal ions.

Contrary to the above findings, only very low concentrations of chelatable radioactive plutonium were observed in groundwater near the Idaho Chemical Processing Plant's low-level waste disposal well.[4] No plutonium was observed in wells at any significant distance from the disposal well. The waste processing procedure used was designed to destroy any chelating agents in the waste prior to disposal, and no chelating agents were found in the water pumped from the test wells.

The fate of radionuclide metal chelates that have been discarded in soil is obviously important. If some mechanism exists to destroy the chelating agents, the radioactive metals will be much less mobile. Although EDTA is only poorly biodegradable, nitrilotriacetate, NTA, is degraded by the action of *Chlatobacter heintzii* bacteria. In addition to uncomplexed NTA, these bacteria have been shown to degrade NTA that is chelated to metals, including cobalt, iron, zinc, aluminum, copper, and nickel.[5]

Complexing agents in wastewater are of concern primarily because of their ability to solubilize heavy metals from plumbing, and from deposits containing heavy metals. Complexation may increase the leaching of heavy metals from waste disposal sites and reduce the efficiency with which heavy metals are removed with sludge in conventional biological waste treatment. Removal of chelated iron is difficult with conventional municipal water treatment processes. Iron(III) and perhaps several other essential micronutrient metal ions are kept in solution by chelation in algal cultures. The availability of chelating agents may be a factor in determining algal growth. The yellow-brown color of some natural waters is due to naturally-occurring chelates of iron.

3.11. BONDING AND STRUCTURE OF METAL COMPLEXES

This section discusses some of the fundamentals helpful in understanding complexation in water. A complex consists of a central metal atom to which neutral or negatively charged ligands possessing electron-donor properties are bonded. The resulting complex may be neutral or may have a positive or negative charge. The ligands are said to be contained within the **coordination sphere** of the central metal atom. Depending upon the type of bonding involved, the ligands within the coordination sphere are held in a definite structural pattern. However, in solution, ligands of many complexes exchange rapidly between solution and the coordination sphere of the central metal ion.

The **coordination number** of a metal atom, or ion, is the number of ligand electron-donor groups that are bonded to it. The most common coordination numbers are 2, 4, and 6. Polynuclear complexes contain two or more metal atoms joined together through bridging ligands, frequently OH, as shown for iron(III) in Reaction 3.9.6.

Selectivity and Specificity in Chelation

Although chelating agents are never entirely specific for a particular metal ion, some complicated chelating agents of biological origin approach almost complete specificity for certain metal ions. One example of such a chelating agent is ferrichrome, synthesized by and extracted from fungi, which forms extremely stable chelates with iron(III). It has been observed that cyanobacteria of the *Anabaena* species secrete appreciable quantities of iron-selective hydroxamate chelating agents during periods of heavy algal bloom.[6] These photosynthetic organisms readily take up iron chelated by hydroxamate-chelated iron, whereas some competing green algae, such as *Scenedesmus,* do not. Thus, the chelating agent serves a dual function of promoting the growth of certain cyanobacteria while suppressing the growth of competing species, allowing the cyanobacteria to exist as the predominant species

3.12. CALCULATIONS OF SPECIES CONCENTRATIONS

The stability of complex ions in solution is expressed in terms of **formation constants**. These can be **stepwise formation constants** (K expressions) representing the bonding of individual ligands to a metal ion, or **overall formation constants** (β expressions) representing the binding of two or more ligands to a metal ion. These concepts are illustrated for complexes of zinc ion with ammonia by the following:

$$Zn^{2+} + NH_3 \longleftrightarrow ZnNH_3^{2+} \qquad (3.12.1)$$

$$K_1 = \frac{[ZnNH_3^{2+}]}{[Zn^{2+}][NH_3]} = 3.9 \times 10^2 \text{ (Stepwise formation constant)} \qquad (3.12.2)$$

$$ZnNH_3^{2+} + NH_3 \longleftrightarrow Zn(NH_3)_2^{2+} \qquad\qquad (3.12.3)$$

$$K_2 = \frac{[Zn(NH_3)_2^{2+}]}{[ZnNH_3^{2+}][NH_3]} = 2.1 \times 10^2 \qquad\qquad (3.12.4)$$

$$Zn^{2+} + 2NH_3 \longleftrightarrow Zn(NH_3)_2^{2+} \qquad\qquad (3.12.5)$$

$$\beta_2 = \frac{[Zn(NH_3)_2^{2+}]}{[Zn^{2+}][NH_3]^2} = K_1K_2 = 8.2 \times 10^4 \text{ (Overall formation constant)} \quad (3.12.6)$$

(For $Zn(NH_3)_3^{2+}$, $\beta_3 = K_1K_2K_3$ and for $Zn(NH_3)_4^{2+}$, $\beta_4 = K_1K_2K_3K_4$.)

The following sections show some calculations involving chelated metal ions in aquatic systems. Because of their complexity, the details of these calculations may be beyond the needs of some readers, who may choose to simply consider the results. In addition to the complexation itself, consideration must be given to competition of H^+ for ligands, competition among metal ions for ligands, competition among different ligands for metal ions, and precipitation of metal ions by various precipitants. Not the least of the problems involved in such calculations is the lack of accurately known values of equilibrium constants to be used under the conditions being considered, a factor which can yield questionable results from even the most elegant computerized calculations. Furthermore, kinetic factors are often quite important. Such calculations, however, can be quite useful to provide an overall view of aquatic systems in which complexation is important, and as general guidelines to determine areas in which more data should be obtained.

3.13. COMPLEXATION BY DEPROTONATED LIGANDS

In most circumstances, metal ions and hydrogen ions compete for ligands, making the calculation of species concentrations more complicated. Before going into such calculations, however, it is instructive to look at an example in which the ligand has lost all ionizable hydrogen. At pH values of 11 or above, EDTA is essentially all in the completely ionized tetranegative form, Y^{4-}, illustrated below:

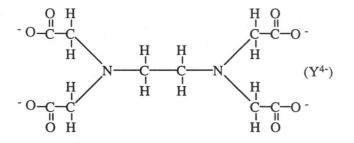

Consider a wastewater with an alkaline pH of 11 containing copper(II) at a total level of 5.0 mg/L and excess uncomplexed EDTA at a level of 200 mg/L (expressed as the disodium salt, $Na_2H_2C_{10}H_{12}O_8N_2 \cdot 2H_2O$, formula weight 372). At this pH uncomplexed EDTA is present as ionized Y^{4-}. The questions to be asked are: Will most of the copper be present as the EDTA complex? If so, what will be the equilibrium concentration of the hydrated copper(II) ion, Cu^{2+}? To answer the former question it is first necessary to calculate the molar concentration of uncomplexed excess EDTA, Y^{4-}. Since disodium EDTA with a formula weight of 372 is present at 200 mg/L (ppm), the total molar concentration of EDTA as Y^{4-} is 5.4×10^{-4} M. The formation constant K_1 of the copper-EDTA complex CuY^{2-} is

$$K_1 = \frac{[CuY^{2-}]}{[Cu^{2+}][Y^{4-}]} = 6.3 \times 10^{18} \qquad (3.13.1)$$

The ratio of complexed copper to uncomplexed copper is

$$\frac{[CuY^{2-}]}{[Cu^{2+}]} = [Y^{4-}]K_1 = 5.4 \times 10^{-4} \times 6.3 \times 10^{18} = 3.3 \times 10^{15} \qquad (3.13.2)$$

and, therefore, essentially all of the copper is present as the complex ion. The molar concentration of total copper(II) in a solution containing 5.0 mg/L copper(II) is 7.9×10^{-5} M, which in this case is essentially all in the form of the EDTA complex. The very low concentration of uncomplexed, hydrated copper(II) ion is given by

$$[Cu^{2+}] = \frac{[CuY^{2-}]}{K_1[Y^{4-}]} = \frac{7.9 \times 10^{-5}}{6.3 \times 10^{18} \times 5.4 \times 10^{-4}} = 2.3 \times 10^{-20} \text{ M} \qquad (3.13.3)$$

It is seen that in the medium described, the concentration of hydrated copper(II) ion is extremely low compared to total copper(II) ion. Any phenomenon in solution that depends upon the concentration of the hydrated copper(II) ion (such as a physiological effect or an electrode response) would be very different in the medium described, as compared to the effect observed if all of the copper at a level of 5.0 mg/L were present as Cu^{2+} in a more acidic solution and in the absence of complexing agent. The phenomenon of reducing the concentration of hydrated metal ion to very low values through the action of strong chelating agents is one of the most important effects of complexation in natural aquatic systems.

3.14. COMPLEXATION BY PROTONATED LIGANDS

Generally, complexing agents, particularly chelating compounds, are conjugate bases of Brönsted acids; for example, glycinate anion, $H_2NCH_2CO_2^-$, is the conjugate base of glycine, $^+H_3NCH_2CO_2^-$. Therefore, in many cases hydrogen ion competes with metal ions for a ligand, so that the strength of chelation depends upon pH. In the nearly neutral pH range usually encountered in natural waters, most organic ligands are present in a conjugated acid form.

In order to understand the competition between hydrogen ion and metal ion for a ligand, it is useful to know the distribution of ligand species as a function of pH.

Consider nitrilotriacetic acid, commonly designated H_3T, as an example. The trisodium salt of this compound, (NTA) is used as a detergent phosphate substitute and is a strong chelating agent. Biological processes are required for NTA degradation, and under some conditions it persists for long times in water. Given the ability of NTA to solubilize and transport heavy metal ions, this material is of considerable environmental concern.

Nitrilotriacetic acid, H_3T, loses hydrogen ion in three steps to form the nitrilotriacetate anion, T^{3-}, the structural formula of which is

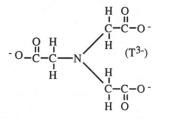

The T^{3-} species may coordinate through three $-CO_2^-$ groups and through the nitrogen atom, as shown in Figure 3.10. Note the similarity of the NTA structure to that of EDTA, discussed in Section 3.13. The stepwise ionization of H_3T is described by the following equilibria:

$$H_3T \leftarrow\rightarrow H^+ + H_2T^- \tag{3.14.1}$$

$$K_{a1} = \frac{[H^+][H_2T^-]}{[H_3T]} = 2.18 \times 10^{-2} \qquad pK_{a1} = 1.66 \tag{3.14.2}$$

$$H_2T^- \leftarrow\rightarrow H^+ + HT^{2-} \tag{3.14.3}$$

$$K_{a2} = \frac{[H^+][HT^{2-}]}{[H_2T^-]} = 1.12 \times 10^{-3} \qquad pK_{a2} = 2.95 \tag{3.14.4}$$

$$HT^{2-} \leftarrow\rightarrow H^+ + T^{3-} \tag{3.14.5}$$

$$K_{a3} = \frac{[H^+][T^{3-}]}{[HT^{2-}]} = 5.25 \times 10^{-11} \qquad pK_{a3} = 10.28 \tag{3.14.6}$$

These expressions show that uncomplexed NTA may exist in solution as any one of the four species, H_3T, H_2T^-, HT^{2-}, or T^{3-}, depending upon the pH of the solution. As was shown for the $CO_2/HCO_3^-/CO_3^{2-}$ system in Section 3.7 and Figure 3.8, fractions of NTA species can be illustrated graphically by a diagram of the distribution-of-species with pH as a master (independent) variable. The key points used to plot such a diagram for NTA are given in Table 3.2, and the plot of fractions of species (α values) as a function of pH is shown in Figure 3.11. Examination of the plot shows that the complexing anion T^{3-} is the predominant species only at relatively high pH values, much higher than usually would be encountered in natural waters. The HT^{2-} species has an extremely wide range of predominance, however, spanning the entire normal pH range of ordinary fresh waters.

Table 3.2. Fractions of NTA Species at Selected pH Values

pH value	α_{H_3T}	$\alpha_{H_2T^-}$	$\alpha_{HT^{2-}}$	$\alpha_{T^{3-}}$
pH below 1.00	1.00	0.00	0.00	0.00
pH = pK_{a1}	0.49	0.49	0.02	0.00
pH = $\frac{1}{2}(pK_{a1} + pK_{a2})$	0.16	0.68	0.16	0.00
pH = pK_{a2}	0.02	0.49	0.49	0.00
pH = $\frac{1}{2}(pK_{a2} + pK_{a3})$	0.00	0.00	1.00	0.00
pH = pK_{a3}	0.00	0.00	0.50	0.50
pH above 12	0.00	0.00	0.00	1.00

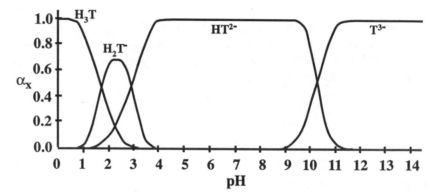

Figure 3.11. Plot of fraction of species α_x as a function of pH for NTA species in water.

3.15. SOLUBILIZATION OF LEAD ION FROM SOLIDS BY NTA

A major concern regarding the widespread introduction of strong chelating agents such as NTA into aquatic ecosystems from sources such as detergents or electroplating wastes is that of possible solubilization of toxic heavy metals from solids through the action of chelating agents. Experimentation is required to determine whether this may be a problem, but calculations are helpful in predicting probable effects. The extent of solubilization of heavy metals depends upon a number of factors, including the stability of the metal chelates, the concentration of the complexing agent in the water, pH, and the nature of the insoluble metal deposit. Several example calculations are given here.

Consider first the solubilization of lead from solid $Pb(OH)_2$ by NTA at pH 8.00. As illustrated in Figure 3.11, essentially all uncomplexed NTA is present as HT^{2-} ion at pH 8.00. Therefore, the solubilization reaction is

$$Pb(OH)_2(s) + HT^{2-} \longleftrightarrow PbT^- + OH^- + H_2O \qquad (3.15.1)$$

which may be obtained by adding the following reactions:

$$Pb(OH)_2(s) \longleftrightarrow Pb^{2+} + 2OH^- \tag{3.15.2}$$

$$K_{sp} = [Pb^{2+}][OH^-]^2 = 1.61 \times 10^{-20} \tag{3.15.3}$$

$$HT^{2-} \longleftrightarrow H^+ + T^{3-} \tag{3.15.4}$$

$$K_{a3} = \frac{[H^+][T^{3-}]}{[HT^{2-}]} = 5.25 \times 10^{-11} \tag{3.15.5}$$

$$Pb^{2+} + T^{3-} \longleftrightarrow PbT^- \tag{3.15.6}$$

$$K_f = \frac{[PbT^-]}{[Pb^{2+}][T^{3-}]} = 2.45 \times 10^{11} \tag{3.15.7}$$

$$H^+ + OH^- \longleftrightarrow H_2O \tag{3.15.8}$$

$$\frac{1}{K_w} = \frac{1}{[H^+][OH^-]} = \frac{1}{1.00 \times 10^{-14}} \tag{3.15.9}$$

$$Pb(OH)_2(s) + HT^{2-} \longleftrightarrow PbT^- + OH^- + H_2O \tag{3.15.1}$$

$$K = \frac{[PbT^-][OH^-]}{[HT^{2-}]} = \frac{K_{sp}K_{a3}K_f}{K_w} = 2.07 \times 10^{-5} \tag{3.15.10}$$

Assume that a sample of water contains 25 mg/L of $N(CH_2CO_2Na)_3$, the trisodium NTA salt, formula weight 257. The total concentration of both complexed and uncomplexed NTA is 9.7×10^{-5} mmol/mL. Assuming a system in which NTA at pH 8.00 is in equilibrium with solid $Pb(OH)_2$, the NTA may be primarily in the uncomplexed form, HT^{2-}, or in the lead complex, PbT^-. The predominant species may be determined by calculating the $[PbT^-]/[HT^{2-}]$ ratio from the expression for K, noting that at pH 8.00, $[OH^-] = 1.00 \times 10^{-6}$ M:

$$\frac{[PbT^-]}{[HT^{2-}]} = \frac{K}{[OH^-]} = \frac{2.07 \times 10^{-5}}{1.00 \times 10^{-6}} = 20.7 \tag{3.15.11}$$

Since $[PbT^-]/[HT^{2-}]$ is approximately 20 to 1, most of the NTA in solution is present as the lead chelate. The molar concentration of PbT^- is just slightly less than the 9.7×10^{-5} mmols/mL total NTA present. The atomic weight of lead is 207, so the concentration of lead in solution is approximately 20 mg/L. This reaction is pH-dependent such that the fraction of NTA chelated decreases with increasing pH.

Reaction of NTA with Metal Carbonate

Carbonates are common forms of heavy metal ion solids. Solid lead carbonate, $PbCO_3$, is stable within the pH region and alkalinity conditions often found in natural waters and wastewaters. An example similar to the one in the preceding section may be worked, assuming that equilibrium is established with $PbCO_3$ rather than with solid $Pb(OH)_2$. In this example it is assumed that 25 mg/L of trisodium

NTA is in equilibrium with $PbCO_3$ at pH 7.00 and a calculation is made to determine whether the lead will be complexed appreciably by the NTA. The carbonate ion, CO_3^{2-}, reacts with H^+ to form HCO_3^-. As discussed in Section 3.7, the acid-base equilibrium reactions for the $CO_2/HCO_3^-/CO_3^{2-}$ system are

$$CO_2 + H_2O \longleftrightarrow HCO_3^- + H^+ \tag{3.3.6}$$

$$K'_{a1} = \frac{[H^+][HCO_3^-]}{[CO_2]} = 4.45 \times 10^{-7} \quad pK'_{a1} = 6.35 \tag{3.3.7}$$

$$HCO_3^- \longleftrightarrow CO_3^{2-} + H^+ \tag{3.3.8}$$

$$K'_{a2} = \frac{[H^+][CO_3^{2-}]}{[HCO_3^-]} = 4.69 \times 10^{-11} \quad pK'_{a2} = 10.33 \tag{3.3.9}$$

where the acid dissociation constants of the carbonate species are designated as K'_a to distinguish them from the acid dissociation constants of NTA. Figure 3.8 shows that within a pH range of about 7 to 10 the predominant carbonic species is HCO_3^-; therefore, the CO_3^{2-} released by the reaction of NTA with $PbCO_3$ will go into solution as HCO_3^- :

$$PbCO_3(s) + HT^{2-} \longleftrightarrow PbT^- + HCO_3^- \tag{3.15.12}$$

This reaction and its equilibrium constant are obtained as follows:

$$PbCO_3(s) \longleftrightarrow Pb^{2+} + CO_3^{2-} \tag{3.15.13}$$

$$K_{sp} = [Pb^{2+}][CO_3^{2-}] = 1.48 \times 10^{-13} \tag{3.15.14}$$

$$Pb^{2+} + T^{3-} \longleftrightarrow PbT^- \tag{3.15.6}$$

$$K_f = \frac{[PbT^-]}{[Pb^{2+}][T^{3-}]} = 2.45 \times 10^{11} \tag{3.15.7}$$

$$HT^{2-} \longleftrightarrow H^+ + T^{3-} \tag{3.15.4}$$

$$K_{a3} = \frac{[H^+][T^{3-}]}{[HT^{2-}]} = 5.25 \times 10^{-11} \tag{3.15.5}$$

$$CO_3^{2-} + H^+ \longleftrightarrow HCO_3^- \tag{3.15.15}$$

$$\frac{1}{K'_{a2}} = \frac{[HCO_3^-]}{[CO_3^{2-}][H^+]} = \frac{1}{4.69 \times 10^{-11}} \tag{3.15.16}$$

$$\overline{PbCO_3(s) + HT^{2-} \longleftrightarrow PbT^- + HCO_3^-} \tag{3.15.12}$$

$$K = \frac{[PbT^-][HCO_3^-]}{[HT^{2-}]} = \frac{K_{sp}K_{a3}K_f}{K'_{a2}} = 4.06 \times 10^{-2} \tag{3.15.17}$$

From the expression for K, Equation 3.15.17, it may be seen that the degree to which $PbCO_3$ is solubilized as PbT^- depends upon the concentration of HCO_3^-. Although this concentration will vary appreciably, the figure commonly used to describe natural waters is a bicarbonate ion concentration of 1.00×10^{-3}, as shown in Section 3.9. Using this value the following may be calculated:

$$\frac{[PbT^-]}{[HT^{2-}]} = \frac{K}{[HCO_3^-]} = \frac{4.06 \times 10^{-2}}{1.00 \times 10^{-3}} = 40.6 \qquad (3.15.18)$$

Thus, under the given conditions, most of the NTA in equilibrium with solid $PbCO_3$ would be present as the lead complex. As in the previous example, at a trisodium NTA level of 25 mg/L, the concentration of soluble lead(II) would be approximately 20 mg/L. At relatively higher concentrations of HCO_3^-, the tendency to solubilize lead would be diminished, whereas at lower concentrations of HCO_3^-, NTA would be more effective in solubilizing lead.

Effect of Calcium Ion upon the Reaction of Chelating Agents with Slightly Soluble Salts

Chelatable calcium ion, Ca^{2+}, which is generally present in natural waters and wastewaters, competes for the chelating agent with a metal in a slightly soluble salt, such as $PbCO_3$. At pH 7.00, the reaction between calcium ion and NTA is

$$Ca^{2+} + HT^{2-} \longleftrightarrow CaT^- + H^+ \qquad (3.15.19)$$

described by the following equilibrium expression:

$$K' = \frac{[CaT^-][H^+]}{[Ca^{2+}][HT^{2-}]} = 1.48 \times 10^8 \times 5.25 \times 10^{-11} = 7.75 \times 10^{-3} \qquad (3.15.20)$$

The value of K' is the product of the formation constant of CaT^-, (1.48×10^8), and K_{a3} of NTA, 5.25×10^{-11}. The fraction of NTA bound as CaT^- depends upon the concentration of Ca^{2+} and the pH. Typically, $[Ca^{2+}]$ in water is 1.00×10^{-3} M. Assuming this value and pH 7.00, the ratio of NTA present in solution as the calcium complex to that present as HT^{2-} is:

$$\frac{[CaT^-]}{[HT^{2-}]} = \frac{[Ca^{2+}]}{[H^+]} K' = \frac{1.00 \times 10^{-3}}{1.00 \times 10^{-7}} \times 7.75 \times 10^{-3} \qquad (3.15.21)$$

$$\frac{[CaT^-]}{[HT^{2-}]} = 77.5$$

Therefore, most of the NTA in equilibrium with 1.00×10^{-3} M Ca^{2+} would be present as the calcium complex, CaT^-, which would react with lead carbonate as follows:

$$PbCO_3(s) + CaT^- + H^+ \longleftrightarrow Ca^{2+} + HCO_3^- + PbT^- \tag{3.15.22}$$

$$K'' = \frac{[Ca^{2+}][HCO_3^-][PbT^-]}{[CaT^-][H^+]} \tag{3.15.23}$$

Reaction 3.15.22 may be obtained by subtracting Reaction 3.15.19 from Reaction 3.15.12, and its equilibrium constant may be obtained by dividing the equilibrium constant of Reaction 3.15.19 into that of Reaction 3.15.12:

$$PbCO_3(s) + HT^{2-} \longleftrightarrow PbT^- + HCO_3^- \tag{3.15.12}$$

$$K = \frac{[PbT^-][HCO_3^-]}{[HT^{2-}]} = \frac{K_s K_{a3} K_f}{K'_{a2}} = 4.06 \times 10^{-2} \tag{3.15.17}$$

$$-(Ca^{2+} + HT^{2-} \longleftrightarrow CaT^- + H^+) \tag{3.15.19}$$

$$K' = \frac{[CaT^-][H^+]}{[Ca^{2+}][HT^{2-}]} = 7.75 \times 10^{-3} \tag{3.15.20}$$

$$PbCO_3(s) + CaT^- + H^+ \longleftrightarrow Ca^{2+} + HCO_3^- + PbT^- \tag{3.15.22}$$

$$K'' = \frac{K}{K'} = \frac{4.06 \times 10^{-2}}{7.75 \times 10^{-3}} = 5.24 \tag{3.15.24}$$

Having obtained the value of K", it is now possible to determine the distribution of NTA between PbT^- and CaT^-. Thus, for water containing NTA chelated to calcium at pH 7.00, a concentration of HCO_3^- of 1.00×10^{-3}, a concentration of Ca^{2+} of 1.00×10^{-3}, and in equilibrium with solid $PbCO_3$, the distribution of NTA between the lead complex and the calcium complex is:

$$\frac{[PbT^-]}{[CaT^-]} = \frac{[H^+]K''}{[Ca^{2+}][HCO_3^-]} = \frac{1.00 \times 10^{-7} \times 5.24}{1.00 \times 10^{-3} \times 1.00 \times 10^{-3}} = 0.524$$

It may be seen that only about 1/3 of the NTA would be present as the lead chelate, whereas under the identical conditions, but in the absence of Ca^{2+}, approximately all of the NTA in equilibrium with solid $PbCO_3$ was chelated to NTA. Since the fraction of NTA present as the lead chelate is directly proportional to the solubilization of $PbCO_3$, differences in calcium concentration will affect the degree to which NTA solubilizes lead from lead carbonate.

3.16. POLYPHOSPHATES IN WATER

Phosphorus occurs as many oxoanions, anionic forms in combination with oxygen. Some of these are strong complexing agents. Since about 1930, salts of polymeric phosphorus oxoanions have been used for water treatment, for water softening, and as detergent builders. When used for water treatment, polyphosphates "sequester" calcium ion in a soluble or suspended form. The effect is to reduce the equilibrium concentration of calcium ion and prevent the precipitation of calcium carbonate in installations such as water pipes and boilers. Furthermore, when water is softened properly with polyphosphates, calcium does not form precipitates with soaps or interact detrimentally with detergents.

The simplest form of phosphate is orthophosphate, PO_4^{3-}:

The orthophosphate ion possesses three sites for attachment of H^+. Orthophosphoric acid, H_3PO_4, has a pK_{a1} of 2.17, a pK_{a2} of 7.31, and a pK_{a3} of 12.36. Because the third hydrogen ion is so difficult to remove from orthophosphate, as evidenced by the very high value of pK_{a3}, very basic conditions are required for PO_4^{3-} to be present at significant levels in water. It is possible for orthophosphate in natural waters to originate from the hydrolysis of polymeric phosphate species.

Pyrophosphate ion, $P_2O_7^{4-}$, is the first of a series of unbranched chain polyphosphates produced by the condensation of orthophosphate:

$$2PO_4^{3-} + H_2O \longleftrightarrow P_2O_7^{4-} + 2OH^- \qquad (3.16.1)$$

A long series of linear polyphosphates may be formed, the second of which is triphosphate ion, $P_3O_{10}^{5-}$. These species consist of PO_4 tetrahedra with adjacent tetrahedra sharing a common oxygen atom at one corner. The structural formulas of the acidic forms, $H_4P_2O_7$ and $H_5P_3O_{10}$, are:

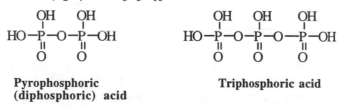

Pyrophosphoric   **Triphosphoric acid**
(diphosphoric) acid

It is easy to visualize the longer chains composing the higher linear polyphosphates. **Vitreous sodium phosphates** are mixtures consisting of linear phosphate chains with from 4 to approximately 18 phosphorus atoms each. Those with intermediate chain lengths comprise the majority of the species present.

The acid-base behavior of the linear-chain polyphosphoric acids may be explained in terms of their structure by comparing them to orthophosphoric acid. Pyrophosphoric acid, $H_4P_2O_7$, has four ionizable hydrogens. The value of pK_{a1} is quite small (relatively strong acid), whereas pK_{a2} is 2.64, pK_{a3} is 6.76, and pK_{a4} is 9.42. In the case of triphosphoric acid, $H_5P_3O_{10}$, the first two pK_a values are small, pK_{a3} is 2.30, pK_{a4} is 6.50, and pK_{a5} is 9.24. When linear polyphosphoric acids are titrated with base, the titration curve has an inflection at a pH of approximately 4.5 and another inflection at a pH close to 9.5. To understand these phenomena, consider the following ionization of triphosphoric acid:

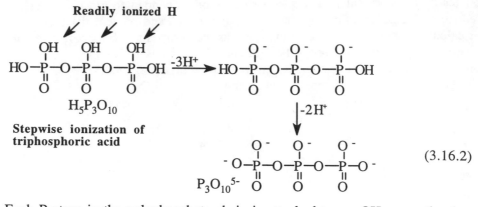

$$(3.16.2)$$

Each P atom in the polyphosphate chain is attached to an -OH group that has one readily ionizable hydrogen that is readily removed in titrating to the first equivalence point. The end phosphorus atoms have two OH groups each. One of the OH groups on an end phosphorus atom has a readily ionizable hydrogen, whereas the other loses its hydrogen much less readily. Therefore, one mole of triphosphoric acid, $H_5P_3O_{10}$, loses three moles of hydrogen ion at a relatively low pH (below 4.5), leaving the $H_2P_3O_{10}^{3-}$ species with two ionizable hydrogens. At intermediate pH values (below 9.5), an additional two moles of "end hydrogens" are lost to form the $P_3O_{10}^{5-}$ species. Titration of a linear-chain polyphosphoric acid up to pH 4.5 yields the number of moles of phosphorus atoms per mole of acid, and titration from pH 4.5 to pH 9.5 yields the number of end phosphorus atoms. Orthophosphoric acid, H_3PO_4, differs from the linear chain polyphosphoric acids in that it has a third ionizable hydrogen which, as noted earlier, is removed in only extremely basic media.

Hydrolysis of Polyphosphates

All of the polymeric phosphates hydrolyze to simpler products in water. The rate of hydrolysis depends upon a number of factors, including pH, and the ultimate product is always some form of orthophosphate. The simplest hydrolytic reaction of a polyphosphate is that of pyrophosphoric acid to produce orthophosphoric acid:

$$H_4P_2O_7 + H_2O \rightarrow 2H_3PO_4 \qquad (3.16.3)$$

Researchers have found evidence that algae and other microorganisms catalyze the hydrolysis of polyphosphates. Even in the absence of biological activity, polyphosphates hydrolyze chemically at a significant rate in water. Therefore, there is much less concern about the possibility of polyphosphates binding to heavy metal ions and transporting them than is the case with organic chelating agents such as NTA or EDTA, which must depend upon microbial degradation for their decomposition.

Complexation by Polyphosphates

In general, chain phosphates are good complexing agents and even form complexes with alkali-metal ions. Ring phosphates form much weaker complexes than do chain species. The different chelating abilities of chain and ring phosphates are due to structural hindrance of bonding by the ring polyphosphates.

3.17. COMPLEXATION BY HUMIC SUBSTANCES

The most important class of complexing agents that occur naturally are the **humic substances**.[7] These are degradation-resistant materials formed during the decomposition of vegetation that occur as deposits in soil, marsh sediments, peat, coal, lignite, or in almost any location where large quantities of vegetation have decayed. They are commonly classified on the basis of solubility. If a material containing humic substances is extracted with strong base, and the resulting solution is acidified, the products are (a) a nonextractable plant residue called **humin**; (b) a material that precipitates from the acidified extract, called **humic acid**; and (c) an organic material that remains in the acidified solution, called **fulvic acid**. Because of their acid-base, sorptive, and complexing properties, both the soluble and insoluble humic substances have a strong effect upon the properties of water. In general, fulvic acid dissolves in water and exerts its effects as the soluble species. Humin and humic acid remain insoluble and affect water quality through exchange of species, such as cations or organic materials, with water.

Humic substances are high-molecular-weight, polyelectrolytic macromolecules. Molecular weights range from a few hundred for fulvic acid to tens of thousands for the humic acid and humin fractions. These substances contain a carbon skeleton with a high degree of aromatic character and with a large percentage of the molecular weight incorporated in functional groups, most of which contain oxygen. The elementary composition of most humic substances is within the following ranges: C, 45-55%; O, 30-45%, H, 3-6%; N, 1-5%; and S, 0-1%. The terms *humin, humic acid,* and *fulvic acid* do not refer to single compounds but to a wide range of compounds of generally similar origin with many properties in common. Humic substances have been known since before 1800, but their structural and chemical characteristics are still being explained.

Some feeling for the nature of humic substances may be obtained by considering the structure of a hypothetical molecule of fulvic acid shown below:

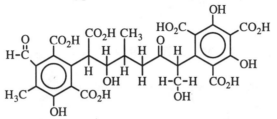

This structure is typical of the type of compound composing fulvic acid. The compound has a formula weight of 666, and its chemical formula may be represented by $C_{20}H_{15}(CO_2H)_6(OH)_5(CO)_2$. As shown in the hypothetical compound, the functional groups that may be present in fulvic acid are carboxyl, phenolic hydroxyl, alcoholic hydroxyl, and carbonyl. The functional groups vary with the particular acid sample. Approximate ranges in units of milliequivalents per gram of acid are: total acidity, 12-14; carboxyl, 8-9; phenolic hydroxyl, 3-6; alcoholic hydroxyl, 3-5; and carbonyl, 1-3. In addition, some methoxyl groups, -OCH_3, may be encountered at low levels.

The binding of metal ions by humic substances is one of the most important environmental qualities of humic substances. This binding can occur as chelation between a carboxyl group and a phenolic hydroxyl group, as chelation between two carboxyl groups, or as complexation with a carboxyl group (see below):

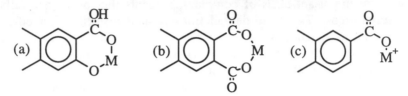

Figure 3.1. Binding of a metal ion, M^{2+}, by humic substances (a) by chelation between carboxyl and phenolic hydroxyl, (b) by chelation between two carboxyl groups, and (c) by complexation with a carboxyl group.

Iron and aluminum are very strongly bound to humic substances, whereas magnesium is rather weakly bound. Other common ions, such as Ni^{2+}, Pb^{2+}, Ca^{2+}, and Zn^{2+}, are intermediate in their binding to humic substances.

The role played by soluble fulvic acid complexes of metals in natural waters is not well known. They probably keep some of the biologically important transition-metal ions in solution, and are particularly involved in iron solubilization and transport. Yellow fulvic acid-type compounds called **Gelbstoffe** and frequently encountered along with soluble iron, are associated with color in water.

Insoluble humins and humic acids, effectively exchange cations with water and may accumulate large quantities of metals. Lignite coal, which is largely a humic acid material, tends to remove some metal ions from water.

Special attention has been given to humic substances since about 1970, following the discovery of **trihalomethanes** (THMs, such as chloroform and dibromochloromethane) in water supplies. It is now generally believed that these suspected carcinogens can be formed in the presence of humic substances during the disinfection of raw municipal drinking water by chlorination (see Chapter 8). The humic substances produce THMs by reaction with chlorine. The formation of THMs can be reduced by removing as much of the humic material as possible prior to chlorination.

3.18. COMPLEXATION AND REDOX PROCESSES

Complexation may have a strong effect upon oxidation-reduction equilibria by shifting reactions, such as that for the oxidation of lead,

$$Pb \longleftrightarrow Pb^{2+} + 2e^- \qquad (3.18.1)$$

strongly to the right by binding to the product ion, thus cutting its concentration down to very low levels. Perhaps more important is the fact that upon oxidation,

$$M + \tfrac{1}{2}O_2 \longleftrightarrow MO \qquad\qquad (3.18.2)$$

many metals form self-protective coatings of oxides, carbonates, or other insoluble species which prevent further chemical reaction. Copper and aluminum roofing and structural iron are examples of materials which are thus self-protecting. A chelating agent in contact with such metals can result in continual dissolution of the protective coating so that the exposed metal corrodes readily. For example, chelating agents in wastewater may increase the corrosion of metal plumbing, thus adding heavy metals to effluents. Solutions of chelating agents employed to clean metal surfaces in metal plating operations have a similar effect.

LITERATURE CITED

1. Stevens, William K., "Expectation Aside, Water Use in U.S. Is Showing Decline," *New York Times*, November 10, 1998, p. 1.

2. Martell, A. E., "Principles of Complex Formation," in *Organic Compounds in Aquatic Environments*, S. D. Faust and J. V. Hunter, Eds., Marcell Dekker, Inc., New York, 1971, pp. 262–392.

3. Means, J. L., D. A. Crerar, and J. O. Duguid, "Migration of Radioactive Wastes: Radionuclide Mobilization by Complexing Agents," *Science*, **200**, 1978, pp. 1477-81.

4. Cleveland, K. J. M. and T. F. Rees, "Characterization of Plutonium in Ground Water near the Idaho Chemical Processing Plant," *Environmental Science and Technology*, **16**, 1982, pp. 437–439.

5. Bolton, Harvey, Jr., Don C. Girvin, Andrew E. Plymale, Scott D. Harvey, and Darla J. Workman, "Degradation of Metal-Nitrilotriacetate Complexes by *Chelatobacter heintzii*," *Environmental Science and Technology*, **30**, 931-938 (1996).

6. Murphy, T. P., D. R. S. Lean, and C. Nalewajko, "Blue–Green Algae: Their Excretion of Iron–Selective Chelators Enables them to Dominate other Algae," *Science*, **192**, 1976, pp. 900-2.

7. Manahan, Stanley E., "Humic Substances and the Fates of Hazardous Waste Chemicals," Chapter 6 in *Influence of Aquatic Humic Substances on Fate and Treatment of Pollutants*, Advances in Chemistry Series **219**, American Chemical Society, Washington, D.C., 1989, pp. 83-92

SUPPLEMENTARY REFERENCES

Appelo, C.A.J., and D. Postma, *Geochemistry, Groundwater, and Pollution*, A. A. Balkema Publishers, Rotterdam, 1993.

Baker, Lawrence A., Ed., *Environmental Chemistry of Lakes and Reservoirs*, American Chemical Society, Washington, D.C., 1994.

Brownlow, Arthur H., *Geochemistry*, 2nd ed., Prentice Hall, Upper Saddle River, NJ, 1996.

Butler, James N., *Ionic Equilibrium: Solubility and pH Calculations*, John Wiley & Sons, New York, 1998.

Drever, James I., *The Geochemistry of Natural Waters: Surface and Groundwater Environments*, 3rd ed., Prentice Hall, Upper Saddle River, NJ, 1997.

Faure, Gunter, *Principles and Applications of Geochemistry: A Comprehensive Textbook for Geology Students*, 2nd ed., Prentice Hall, Upper Saddle River, NJ, 1998.

Gilbert, Janine, Dan L. Danielopol, and Jack Stanford, Eds., *Groundwater Ecology*, Academic Press, Orlando, FL, 1994.

Hem, J. D., *Study and Interpretation of the Chemical Characteristics of Natural Water*, 2nd ed., U.S. Geological Survey Paper **1473**, Washington, D.C., 1970.

Hessen, D. O. and L. J. Tranvik, Eds., *Aquatic Humic Substances: Ecology and Biogeochemistry*, Springer Verlag, Berlin, 1998.

Howard, Alan G., *Aquatic Environmental Chemistry*, Oxford University Press, Oxford, UK, 1998.

Kegley, Susan E. and Joy Andrews, *The Chemistry of Water*, University Science Books , Mill Valley, CA, 1997.

Knapp, Brian, *Air and Water Chemistry*, Atlantic Europe, Henley-on-Thames, UK, 1998).

Langmuir, Donald, *Aqueous Environmental Geochemistry*, Prentice Hall, Upper Saddle River, NJ, 1997.

Matshullat, Jorg, Heinz Jurgen Tobschall, and Hans-Jurgen Voigt, Eds., *Geochemistry and the Environment: Relevant Processes in the Atmosphere*, Springer, Berlin, 1997.

Oceanography in the Next Decade, National Academy Press, Washington, D.C. 1992.

Patrick, Ruth, *Surface Water Quality*, Princeton University Press, Princeton, NJ, 1992.

Sigg, Laura and Werner Stumm, *Aquatische Chemie: Eine Einfuehrung in die Chimie Waessriger Loesungen und Natuerlicher Gewaesser*, Tuebner, Stuttgart, Germany, 1994.

Snoeyink, Vernon L., and David Jenkins, *Water Chemistry*, John Wiley and Sons, Inc., New York, 1980.

Stumm, Werner and James J. Morgan, *Aquatic Chemistry: Chemical Equilibria and Rates in Natural Waters*, 3rd ed., John Wiley and Sons, Inc., New York, 1995.

Stumm, Werner, *Chemistry of the Solid-Water Interface: Processes at the Mineral-Water and Particle-Water Interface in Natural Systems*, John Wiley and Sons, Inc., New York, 1992.

Stumm, Werner, Ed., *Aquatic Chemical Kinetics: Reaction Rates of Processes in Natural Waters*, John Wiley and Sons, Inc., New York, 1990.

Suffet, I. H., and Patrick MacCarthy, Eds., *Aquatic Humic Substances: Influence on Fate and Treatment of Pollutants*, Advances in Chemistry Series **219**, American Chemical Society, Washington, D.C., 1989.

U. S. Environmental Protection Agency, *Ground Water Handbook*, 2nd ed., Government Institutes, Rockville, MD, 1992.

van der Leeden, Frits, Fred L. Troise, and David K. Todd, Eds., *The Water Encyclopedia*, Water Information Center, Plainview, NY, 1991.

QUESTIONS AND PROBLEMS

1. Alkalinity is determined by titration with standard acid. The alkalinity is often expressed as mg/L of $CaCO_3$. If V_p mL of acid of normality N are required to titrate V_s mL of sample to the phenolphthalein endpoint, what is the formula for the phenolphthalein alkalinity as mg/L of $CaCO_3$?

2. Exactly 100 pounds of cane sugar (dextrose), $C_{12}H_{22}O_{11}$, were accidentally discharged into a small stream saturated with oxygen from the air at 25°C. How many liters of this water could be contaminated to the extent of removing all the dissolved oxygen by biodegradation?

3. Water with an alkalinity of 2.00×10^{-3} equivalents/liter has a pH of 7.00. Calculate $[CO_2]$, $[HCO_3^-]$, $[CO_3^{2-}]$, and $[OH^-]$.

4. Through the photosynthetic activity of algae, the pH of the water in Problem 3 was changed to 10.00. Calculate all the preceding concentrations and the weight of biomass, $\{CH_2O\}$, produced. Assume no input of atmospheric CO_2.

5. Calcium chloride is quite soluble, whereas the solubility product of calcium fluoride, CaF_2, is only 3.9×10^{-11}. A waste stream of 1.00×10^{-3} M HCl is injected into a formation of limestone, $CaCO_3$, where it comes into equilibrium. Give the chemical reaction that occurs and calculate the hardness and alkalinity of the water at equilibrium. Do the same for a waste stream of 1.00×10^{-3} M HF.

6. For a solution having 1.00×10^{-3} equivalents/liter total alkalinity (contributions from HCO_3^-, CO_3^{2-}, and OH^-) at $[H^+] = 4.69 \times 10^{-11}$, what is the percentage contribution to alkalinity from CO_3^{2-}?

7. A wastewater disposal well for carrying various wastes at different times is drilled into a formation of limestone ($CaCO_3$), and the wastewater has time to come to complete equilibrium with the calcium carbonate before leaving the formation through an underground aquifer. Of the following components in the wastewater, the one that would not cause an increase in alkalinity due either to

the component itself or to its reaction with limestone, is (a) NaOH, (b) CO_2, (c) HF, (d) HCl, (e) all of the preceding would cause an increase in alkalinity.

8. Calculate the ratio $[PbT^-]/[HT^{2-}]$ for NTA in equilibrium with $PbCO_3$ in a medium having $[HCO_3^-] = 3.00 \times 10^{-3}$ M.

9. If the medium in Problem 8 contained excess calcium such that the concentration of uncomplexed calcium, $[Ca^{2+}]$, were 5.00×10^{-3} M, what would be the ratio $[PbT^-]/[CaT^-]$ at pH 7?

10. A wastewater stream containing 1.00×10^{-3} M disodium NTA, Na_2HT, as the only solute is injected into a limestone ($CaCO_3$) formation through a waste disposal well. After going through this aquifer for some distance and reaching equilibrium, the water is sampled through a sampling well. What is the reaction between NTA species and $CaCO_3$? What is the equilibrium constant for the reaction? What are the equilibrium concentrations of CaT^-, HCO_3^-, and HT^{2-}? (The appropriate constants may be looked up in this chapter.)

11. If the wastewater stream in Problem 10 were 0.100 M in NTA and contained other solutes that exerted a buffering action such that the final pH were 9.00, what would be the equilibrium value of HT^{2-} concentration in moles/liter?

12. Exactly 1.00×10^{-3} mole of $CaCl_2$, 0.100 mole of NaOH, and 0.100 mole of Na_3T were mixed and diluted to 1.00 liter. What was the concentration of Ca^{2+} in the resulting mixture?

13. How does chelation influence corrosion?

14. The following ligand has more than one site for binding to a metal ion. How many such sites does it have?

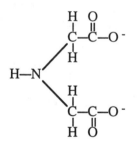

15. If a solution containing initially 25 mg/L trisodium NTA is allowed to come to equilibrium with solid $PbCO_3$ at pH 8.50 in a medium that contains 1.76×10^{-3} M HCO_3^- at equilibrium, what is the value of the ratio of the concentration of NTA bound with lead to the concentration of unbound NTA, $[PbT^-]/[HT^{2-}]$?

16. After a low concentration of NTA has equilibrated with $PbCO_3$ at pH 7.00, in a medium having $[HCO_3^-] = 7.50 \times 10^{-4}$ M, what is the ratio $[PbT^-]/[HT^{2-}]$?

17. What detrimental effect may dissolved chelating agents have upon conventional biological waste treatment?

18. Why is chelating agent usually added to artificial algal growth media?

19. What common complex compound of magnesium is essential to certain life processes?

20. What is always the ultimate product of polyphosphate hydrolysis?

21. A solution containing initially 1.00×10^{-5} M CaT^- is brought to equilibrium with solid $PbCO_3$. At equilibrium, pH = 7.00, $[Ca^{2+}] = 1.50 \times 10^{-3}$ M, and $[HCO_3^-] = 1.10 \times 10^{-3}$ M. At equilibrium, what is the fraction of total NTA in solution as PbT^-?

22. What is the fraction of NTA present after HT^{2-} has been brought to equilibrium with solid $PbCO_3$ at pH 7.00, in a medium in which $[HCO_3^-] = 1.25 \times 10^{-3}$ M?

23. Describe ways in which measures taken to alleviate water supply and flooding problems might actually aggravate such problems.

24. The study of water is known as _____, _____, _____ is the branch of the science dealing with the characteristics of fresh water, and the science that deals with about 97% of all Earth's water is called _____.

25. Consider the hydrologic cycle in Figure 3.1. List or discuss the kinds or classes of environmental chemistry that might apply to each major part of this cycle.

26. Consider the unique and important properties of water. What molecular or bonding characteristics of the water molecules are largely responsible for these properties. List or describe one of each of the following unique properties of water related to (a) thermal characteristics, (b) transmission of light, (c) surface tension, (d) solvent properties.

27. Discuss how thermal stratification of a body of water may affect its chemistry.

28. Relate aquatic life to aquatic chemistry. In so doing, consider the following: autotrophic organisms, producers, heterotrophic organisms, decomposers, eutrophication, dissolved oxygen, biochemical oxygen demand.

4 OXIDATION-REDUCTION

4.1. THE SIGNIFICANCE OF OXIDATION-REDUCTION

Oxidation-reduction (**redox**) reactions are those involving changes of oxidation states of reactants. Such reactions are easiest to visualize as the transfer of electrons from one species to another. For example, soluble cadmium ion, Cd^{2+}, is removed from wastewater by reaction with metallic iron. The overall reaction is

$$Cd^{2+} + Fe \rightarrow Cd + Fe^{2+} \qquad (4.1.1)$$

This reaction is the sum of two **half-reactions**, a reduction half-reaction in which cadmium ion accepts two electrons and is reduced,

$$Cd^{2+} + 2e^- \rightarrow Cd \qquad (4.1.2)$$

and an oxidation half-reaction in which elemental iron is oxidized:

$$Fe \rightarrow Fe^{2+} + 2e^- \qquad (4.1.3)$$

When these two half-reactions are added algebraically, the electrons cancel on both sides and the result is the overall reaction given in Equation 4.1.1.

Oxidation-reduction phenomena are highly significant in the environmental chemistry of natural waters and wastewaters. In a lake, for example, the reduction of oxygen (O_2) by organic matter (represented by $\{CH_2O\}$),

$$\{CH_2O\} + O_2 \rightarrow CO_2 + H_2O \qquad (4.1.4)$$

results in oxygen depletion which can be fatal to fish. The rate at which sewage is oxidized is crucial to the operation of a waste treatment plant. Reduction of insoluble iron(III) to soluble iron(II),

$$Fe(OH)_3(s) + 3H^+ + e^- \rightarrow Fe^{2+} + 3H_2O \qquad (4.1.5)$$

in a reservoir contaminates the water with dissolved iron, which is hard to remove in the water treatment plant. Oxidation of NH_4^+ to NO_3^- in water,

$$NH_4^+ + 2O_2 \rightarrow NO_3^- + 2H^+ + H_2O \qquad (4.1.6)$$

converts ammonium nitrogen to nitrate, a form more assimilable by algae in the water. Many other examples can be cited of the ways in which the types, rates, and equilibria of redox reactions largely determine the nature of important solute species in water.

This chapter discusses redox processes and equilibria in water. In so doing, it emphasizes the concept of pE, analogous to pH and defined as the negative log of electron activity. Low pE values are indicative of reducing conditions and high pE values reflect oxidizing conditions.

Two important points should be stressed regarding redox reactions in natural waters and wastewaters. First, as is discussed in Chapter 6, "Aquatic Microbial Biochemistry," many of the most important redox reactions are catalyzed by micro-organisms. Bacteria are the catalysts by which molecular oxygen reacts with organic matter, iron(III) is reduced to iron(II), and ammonia is oxidized to nitrate ion.

The second important point regarding redox reactions in the hydrosphere is their close relationship to acid-base reactions. Whereas the activity of the H^+ ion is used to express the extent to which water is acidic or basic, the activity of the electron, e^-, is used to express the degree to which an aquatic medium is oxidizing or reducing. Water with a high hydrogen ion activity, such as runoff from "acid rain", is *acidic*. By analogy, water with a high *electron* activity, such as that in the anaerobic digester of a sewage treatment plant, is said to be *reducing*. Water with a low H^+ ion activity (high concentration of OH^-)—such as landfill leachate contaminated with waste sodium hydroxide—is *basic*, whereas water with a low electron activity—highly chlorinated water, for example—is said to be *oxidizing*. Actually, neither free electrons nor free H^+ ions as such are found dissolved in aquatic solution; they are always strongly associated with solvent or solute species. However, the concept of electron activity, like that of hydrogen ion activity, remains a very useful one to the aquatic chemist.

Many species in water undergo exchange of both electrons and H^+ ions. For example, acid mine water contains the hydrated iron(III) ion, $Fe(H_2O)_6^{3+}$, which readily loses H^+ ion

$$Fe(H_2O)_6^{3+} \leftarrow \rightarrow Fe(H_2O)_5OH^{2+} + H^+ \qquad (4.1.7)$$

to contribute acidity to the medium. The same ion accepts an electron

$$Fe(H_2O)_6^{3+} + e^- \leftarrow \rightarrow Fe(H_2O)_6^{2+} \qquad (4.1.8)$$

to give iron(II).

Generally, the transfer of electrons in a redox reaction is accompanied by H^+ ion transfer, and there is a close relationship between redox and acid-base processes. For example, if iron(II) loses an electron at pH 7, three hydrogen ions are also lost to form highly insoluble iron(II) hydroxide,

$$Fe(H_2O)_6^{2+} \longleftrightarrow e^- + Fe(OH)_3(s) + 3H_2O + 3H^+ \qquad (4.1.9)$$

an insoluble, gelatinous solid.

The stratified body of water shown in Figure 4.1 can be used to illustrate redox phenomena and relationships in an aquatic system. The anaerobic sediment layer is so reducing that carbon can be reduced to its lowest possible oxidation state, -4 in CH_4. If the lake becomes anaerobic, the hypolimnion may contain elements in their reduced states: NH_4^+ for nitrogen, H_2S for sulfur, and soluble $Fe(H_2O)_6^{2+}$ for iron. Saturation with atmospheric oxygen makes the surface layer a relatively oxidizing medium. If allowed to reach thermodynamic equilibrium, it is characterized by the more oxidized forms of the elements present: CO_2 for carbon, NO_3^- for nitrogen, iron as insoluble $Fe(OH)_3$, and sulfur as SO_4^{2-}. Substantial changes in the distribution of chemical species in water resulting from redox reactions are vitally important to aquatic organisms and have tremendous influence on water quality.

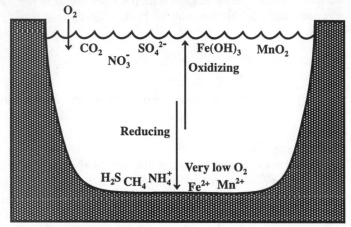

Figure 4.1. Predominance of various chemical species in a stratified body of water that has a high oxygen concentration (oxidizing, high pE) near the surface and a low oxygen concentration (reducing, low pE) near the bottom.

It should be pointed out that the systems presented in this chapter are assumed to be at equilibrium, a state almost never achieved in any real natural water or wastewater system. Most real aquatic systems are dynamic systems that may approach a steady-state, rather than true equilibrium. Nevertheless, the picture of a system at equilibrium is very useful in visualizing trends in natural water and wastewater systems, yet the model is still simple enough to comprehend. It is important to realize the limitations of such a model, however, especially in trying to make measurements of the redox status of water.

4.2. THE ELECTRON AND REDOX REACTIONS

In order to explain redox processes in natural waters it is necessary to have an understanding of redox reactions. In a formal sense such reactions can be viewed as the transfer of electrons between species. This section considers such reactions in a simple system. All redox reactions involve changes in the oxidation states of some of the species that take part in the reaction. Consider, for example, a solution containing iron(II) and iron(III) that is sufficiently acidic to prevent precipitation of solid $Fe(OH)_3$; such a medium might be acid mine water or a steel pickling liquor waste. Suppose that the solution is treated with elemental hydrogen gas over a suitable catalyst to bring about the reduction of iron(III) to iron(II). The overall reaction can be represented as

$$2Fe^{3+} + H_2 \leftarrow\rightarrow 2Fe^{2+} + 2H^+ \qquad (4.2.1)$$

The reaction is written with a double arrow, indicating that it is *reversible* and could proceed in either direction; for normal concentrations of reaction participants, this reaction goes to the right. As the reaction goes to the right, the hydrogen is *oxidized* as it changes from an *oxidation state* (number) of 0 in elemental H_2 to a higher oxidation number of +1 in H^+. The oxidation state of iron goes from +3 in Fe^{3+} to +2 in Fe^{2+}; the oxidation number of iron decreases, which means that it is *reduced*.

All redox reactions such as this one can be broken down into a reduction *half-reaction*, in this case

$$2Fe^{3+} + 2e^- \leftarrow\rightarrow 2Fe^{2+} \qquad (4.2.2)$$

(for one electron, $Fe^{3+} + e^- \leftarrow\rightarrow Fe^{3+}$) and an oxidation half-reaction, in this case

$$H_2 \leftarrow\rightarrow 2H^+ + 2e^- \qquad (4.2.3)$$

Note that adding these two half-reactions together gives the overall reaction. *The addition of an oxidation half-reaction and a reduction half-reaction, each expressed for the same number of electrons so that the electrons cancel on both sides of the arrows, gives a whole redox reaction.*

The equilibrium of a redox reaction, that is, the degree to which the reaction as written tends to lie to the right or left, can be deduced from information about its constituent half-reactions. To visualize this, assume that the two half-reactions can be separated into two half-cells of an electrochemical cell as shown for Reaction 4.2.1 in Figure 4.2.

If the initial activities of H^+, Fe^{2+}, and Fe^{3+} were of the order of 1 (concentrations of 1 M), and if the pressure of H_2 were 1 atm, H_2 would be oxidized to H^+ in the left half-cell, Fe^{3+} would be reduced to Fe^{2+} in the right half-cell, and ions would migrate through the salt bridge to maintain electroneutrality in both half-cells. The net reaction occurring would be Reaction 4.2.1.

If a voltmeter were inserted in the circuit between the two electrodes, no significant current could flow and the two half-reactions could not take place. However, the voltage registered by the voltmeter would be a measure of the relative tendencies of the two half-reactions to occur. In the left half-cell the oxidation half-reaction,

$$H_2 \leftarrow\rightarrow 2H^+ + 2e^- \qquad (4.2.3)$$

will tend to go to the right, releasing electrons to the platinum electrode in the half-cell and giving that electrode a relatively negative (-) potential. In the right half-cell the reduction half-reaction,

$$Fe^{3+} + e^- \longleftrightarrow Fe^{2+} \tag{4.2.2}$$

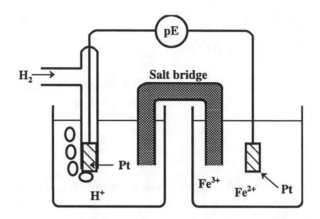

Figure 4.2. Electrochemical cell in which the reaction $2Fe^{3+} + H_2 \longleftrightarrow 2Fe^{2+} + 2H^+$ can be carried out in two half-cells.

will tend to go to the right, taking electrons from the platinum electrode in the half-cell and giving that electrode a relatively positive (+) potential. The difference in these potentials is a measure of the "driving force" of the overall reaction. If each of the reaction participants were at unit activity, the potential difference would be 0.77 volts.

The left electrode shown in Figure 4.2 is the standard electrode against which all other electrode potentials are compared. It is called the **standard hydrogen electrode, SHE**. This electrode has been assigned a potential value of exactly 0 volts by convention, and its half-reaction is written as the following:

$$2H^+ + 2e^- \longleftrightarrow H_2 \qquad E^0 = 0.00 \text{ volts} \tag{4.2.4}$$

The measured potential of the right-hand electrode in Figure 4.2.1 versus the standard hydrogen electrode is called the **electrode potential, E.** If the Fe^{2+} and Fe^{3+} ions in solution are both at unit activity, the potential is the **standard electrode potential** (according to IUPAC convention, the **standard reduction potential**), E^0. The standard electrode potential for the Fe^{3+}/Fe^{2+} couple is 0.77 volts expressed conventionally as follows:

$$Fe^{3+} + e^- \longleftrightarrow Fe^{2+} \qquad E^0 = +0.77 \text{ volts} \tag{4.2.5}$$

4.3. ELECTRON ACTIVITY AND pE

In this book, for the most part, pE and pE^0 are used instead of E and E^0 to more clearly illustrate redox equilibria in aquatic systems over many orders of magnitude of electron activity in a manner analogous to pH. Numerically, pE and pE^0 are simply the following:

$$pE = \frac{\frac{E}{2.303RT}}{F} = \frac{E}{0.0591} \text{ (at 25°C)} \tag{4.3.1}$$

$$pE^0 = \frac{\frac{E^0}{2.303RT}}{F} = \frac{E^0}{0.0591} \text{ (at 25°C)} \tag{4.3.2}$$

where R is the molar gas constant, T is the absolute temperature, and F is the Faraday constant. The "pE concept" is explained below.

Just as pH is defined as

$$pH = -\log(a_{H^+}) \tag{4.3.3}$$

where a_{H^+} is the activity of hydrogen ion in solution, pE is defined as

$$pE = -\log(a_{e^-}) \tag{4.3.4}$$

where a_{e^-} is the activity of the electron in solution. Since hydrogen ion concentration may vary over many orders of magnitude, pH is a convenient way of expressing a_{H^+} in terms of manageable numbers. Similarly, electron activities in water may vary over more than 20 orders of magnitude so that it is convenient to express a_{e^-} as pE.

Values of pE are defined in terms of the following half-reaction for which pE^0 is defined as exactly zero:*

$$2H^+(aq) + 2e^- \longleftrightarrow H_2(g) \qquad E^0 = +0.00 \text{ volts} \qquad pE^0 = 0.00 \tag{4.3.5}$$

Whereas it is relatively easy to visualize the activities of ions in terms of concentration, it is harder to visualize the activity of the electron, and therefore pE, in similar terms. For example, at 25°C in pure water, a medium of zero ionic strength, the hydrogen ion concentration is 1.0×10^{-7} M, the hydrogen-ion *activity* is 1.0×10^{-7}, and the pH is 7.0. The electron activity, however, must be defined in terms of Equation 4.3.5. When $H^+(aq)$ at unit activity is in equilibrium with hydrogen gas at 1 atmosphere pressure (and likewise at unit activity), the activity of the electron in the medium is exactly 1.00 and the pE is 0.0. If the electron activity were increased

---

* Thermodynamically, the free energy change for this reaction is defined as exactly zero when all reaction participants are at unit activity. For ionic solutes, activity — the effective concentration in a sense — approaches concentration at low concentrations and low ionic strengths. The activity of a gas is equal to its partial pressure. Furthermore, the free energy, G, decreases for spontaneous processes occurring at constant temperature and pressure. Processes for which the free energy change, ΔG, is zero have no tendency toward spontaneous change and are in a state of equilibrium. Reaction 4.2.4 is the one upon which free energies of formation of all ions in aqueous solution is based. It also forms the basis for defining free energy changes for oxidation-reduction processes in water.

by a factor of 10 (as would be the case if $H^+(aq)$ at an activity of 0.100 were in equilibrium with H_2 at an activity of 1.00), the electron activity would be 10 and the pE value would be -1.0.

4.4. THE NERNST EQUATION

The **Nernst equation** is used to account for the effect of different activities upon electrode potential. Referring to Figure 4.2, if the Fe^{3+} ion concentration is increased relative to the Fe^{2+} ion concentration, it is readily visualized that the potential and the pE of the right electrode will become more positive because the higher concentration of electron-deficient Fe^{3+} ions clustered around it tends to draw electrons from the electrode. Decreased Fe^{3+} ion or increased Fe^{2+} ion concentration has the opposite effect. Such concentration effects upon E and pE are expressed by the **Nernst equation**. As applied to the half-reaction

$$Fe^{3+} + e^- \longleftrightarrow Fe^{2+} \qquad E^0 = +0.77 \text{ volts} \qquad pE^0 = 13.2 \qquad (4.4.1)$$

the Nernst equation is

$$E = E^0 + \frac{2.303RT}{nF}\log\frac{[Fe^{3+}]}{[Fe^{2+}]} = E^0 + \frac{0.0591}{n}\log\frac{[Fe^{3+}]}{[Fe^{2+}]} \qquad (4.4.2)$$

where n is the number of electrons involved in the half-reaction (1 in this case), and the activities of Fe^{3+} and Fe^{2+} ions have been taken as their concentrations (a simplification valid for more dilute solutions, which will be made throughout this chapter). Considering that

$$pE = \frac{E}{\frac{2.303RT}{F}} \quad \text{and} \quad pE^0 = \frac{E^0}{\frac{2.303RT}{F}}$$

the Nernst equation can be expressed in terms of pE and pE^0

$$pE = pE^0 + \frac{1}{n}\log\frac{[Fe^{3+}]}{[Fe^{2+}]} \quad \text{(In this case n = 1)} \qquad (4.4.3)$$

The Nernst equation in this form is quite simple and offers some advantages in calculating redox relationships.

If, for example, the value of $[Fe^{3+}]$ is 2.35×10^{-3} M and $[Fe^{2+}] = 7.85 \times 10^{-5}$ M, the value of pE is

$$pE = 13.2 + \log\frac{2.35 \times 10^{-3}}{7.85 \times 10^{-5}} = 14.7 \qquad (4.4.4)$$

As the concentration of Fe^{3+} increases relative to the concentration of Fe^{2+}, the value of pE becomes higher (more positive) and as the concentration of Fe^{2+} increases relative to the concentration of Fe^{3+}, the value of pE becomes lower (more negative).

4.5. REACTION TENDENCY: WHOLE REACTION FROM HALF-REACTIONS

This section discusses how half-reactions can be combined to give whole reactions and how the pE^0 values of the half-reactions can be used to predict the directions in which reactions will go. The half-reactions discussed here are the following:

$$Hg^{2+} + 2e^- \longleftrightarrow Hg \qquad\qquad pE^0 = 13.35 \qquad\qquad (4.5.1)$$

$$Fe^{3+} + e^- \longleftrightarrow Fe^{2+} \qquad\qquad pE^0 = 13.2 \qquad\qquad (4.5.2)$$

$$Cu^{2+} + 2e^- \longleftrightarrow Cu \qquad\qquad pE^0 = 5.71 \qquad\qquad (4.5.3)$$

$$2H^+ + 2e^- \longleftrightarrow H_2 \qquad\qquad pE^0 = 0.00 \qquad\qquad (4.5.4)$$

$$Pb^{2+} + 2e^- \longleftrightarrow Pb \qquad\qquad pE^0 = -2.13 \qquad\qquad (4.5.5)$$

Such half-reactions and their pE^0 values can be used to explain observations such as the following: A solution of Cu^{2+} flows through a lead pipe and the lead acquires a layer of copper metal through the reaction

$$Cu^{2+} + Pb \rightarrow Cu + Pb^{2+} \qquad\qquad (4.5.6)$$

This reaction occurs because the copper(II) ion has a greater tendency to acquire electrons than the lead ion has to retain them. This reaction, which can be carried out in two separate half-cells as shown in Figure 4.3, can be obtained by subtracting the lead half-reaction, Equation 4.5.5, from the copper half-reaction, Equation 4.5.3:

$$Cu^{2+} + 2e^- \longleftrightarrow Cu \qquad\qquad pE^0 = 5.71 \qquad\qquad (4.5.3)$$

$$-(Pb^{2+} + 2e^- \longleftrightarrow Pb \qquad\qquad pE^0 = -2.13) \qquad\qquad (4.5.5)$$

$$\overline{Cu^{2+} + Pb \longleftrightarrow Cu + Pb^{2+} \qquad\qquad pE^0 = 7.84 \qquad\qquad (4.5.7)}$$

The positive values of pE^0 for this reaction, 7.84, indicate that the reaction tends to go to the right as written. This occurs when lead metal directly contacts a solution of copper(II) ion. Therefore, if a waste solution containing copper(II) ion, a relatively innocuous pollutant, comes into contact with lead in plumbing, toxic lead may go into solution.

In principle, half-reactions may be allowed to occur in separate electrochemical half-cells, as could occur for Reaction 4.5.7 in the cell shown in Figure 4.3 if the meter (pE) were bypassed by an electrical conductor; they are therefore called **cell reactions.**

If the activities of Cu^{2+} and Pb^{2+} are not unity, the direction of the reaction and value of pE are deduced from the Nernst equation. For Reaction 4.5.7 the Nernst equation is

$$pE = pE^0 + \frac{1}{n} \log \frac{[Cu^{2+}]}{[Pb^{2+}]} = 7.84 + \frac{1}{2} \log \frac{[Cu^{2+}]}{[Pb^{2+}]} \qquad (4.5.8)$$

By combining the appropriate half-reactions it can be shown that copper metal will not cause hydrogen gas to be evolved from solutions of strong acid (hydrogen ion has less attraction for electrons than does copper(II) ion), whereas lead metal, in contrast, will displace hydrogen gas from acidic solutions.

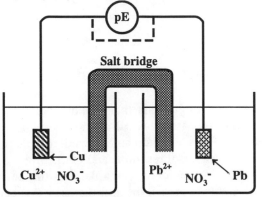

Figure 4.3. Cell for the measurement of pE between a lead half-cell and a copper half-cell. In this configuration "pE" has a very high resistance and current cannot flow.

4.6. THE NERNST EQUATION AND CHEMICAL EQUILIBRIUM

Refer again to Figure 4.3. Imagine that instead of the cell being set up to measure the potential between the copper and lead electrodes, the voltmeter, V, was removed and the electrodes directly connected with a wire so that the current might flow between them. The reaction

$$Cu^{2+} + Pb \longleftrightarrow Cu + Pb^{2+} \qquad pE^0 = 7.84 \qquad (4.5.7)$$

will occur until the concentration of lead ion beomes so high, and that of copper ion so low, that the reaction stops. The system is at equilibrium and, since current no longer flows, pE is exactly zero. The equilibrium constant K for the reaction is given by the expression

$$K = \frac{[Pb^{2+}]}{[Cu^{2+}]} \qquad (4.6.1)$$

The equilibrium constant can be calculated from the Nernst Equation, noting that under equilibrium conditions pE is zero and $[Cu^{2+}]$ and $[Pb^{2+}]$ are at equilibrium concentrations:

$$pE = pE^0 + \frac{1}{n} \log \frac{[Cu^{2+}]}{[Pb^{2+}]} = pE^0 + \frac{1}{n} \log \frac{1}{K}$$

$$pE = 0.00 = 7.84 - \frac{1}{2} \log \frac{[Pb^{2+}]}{[Cu^{2+}]} = 7.84 - \frac{1}{2} \log K \qquad (4.6.2)$$

Note that the reaction products are placed over reactants in the log term, and a minus sign is placed in front to put the equilibrium constant in the correct form (a purely mathematical operation). The value of log K obtained from solving the above equation is 15.7.

The equilibrium constant for a redox reaction involving n electrons is given in terms of pE simply by

$$\log K = n(pE^0) \qquad (4.6.3)$$

4.7. THE RELATIONSHIP OF pE TO FREE ENERGY

Aquatic systems and the organisms that inhabit them — just like the steam engine or students hoping to pass physical chemistry — must obey the laws of thermodynamics. Bacteria, fungi, and human beings derive their energy from acting as mediators (catalysts) of chemical reactions and extracting a certain percentage of useful energy from them. In predicting or explaining the behavior of an aquatic system, it is helpful to be able to predict the useful energy that can be extracted from chemical reactions in the system, such as microbially mediated oxidation of organic matter to CO_2 and water, or the fermentation of organic matter to methane by anaerobic bacteria in the absence of oxygen. Such information may be obtained by knowing the free-energy change, ΔG, for the redox reaction; ΔG, in turn, may be obtained from pE for the reaction. The free-energy change for a redox reaction involving n electrons at an absolute temperature of T is given by

$$\Delta G = -2.303nRT(pE) \qquad (4.7.1)$$

where R is the gas constant. When all reaction participants are in their standard states (pure liquids, pure solids, solutes at an activity of 1.00) ΔG is the standard free energy change, ΔG^0, given by

$$\Delta G^0 = -2.303nRT(pE^0) \qquad (4.7.2)$$

4.8. REACTIONS IN TERMS OF ONE ELECTRON-MOLE

For comparing free energy changes between different redox reactions, it is most meaningful to consider the reactions in terms of the transfer of exactly 1 mole of electrons. This concept may be understood by considering two typical and important redox reactions that occur in aquatic systems — nitrification

$$NH_4^+ + 2O_2 \longleftrightarrow NO_3^- + 2H^+ + H_2O \qquad pE^0 = 5.85 \qquad (4.8.1)$$

and oxidation of iron(II) to iron(III):

$$4Fe^{2+} + O_2 + 10H_2O \longleftrightarrow 4Fe(OH)_3(s) + 8H^+ \qquad pE^0 = 7.6 \qquad (4.8.2)$$

What do reactions written in this way really mean? If any thermodynamic calculations are to be made involving the reactions, Reaction 4.8.1 means that one mole of ammonium ion reacts with two moles of oxygen to yield one mole of nitrate ion, two moles of hydrogen ion, and one mole of water. Reaction 4.8.2 is taken to mean that four moles of iron(II) ion react with one mole of oxygen and ten moles of water to produce four moles of Fe(OH)$_3$ and eight moles of hydrogen ions. The free-energy changes calculated for these quantities of reaction participants do not enable meaningful comparisons of their free energy changes. Such comparisons may be made, though, on the common basis of the transfer of one mole of electrons, writing each reaction in terms of one electron-mole. The advantage of this approach is illustrated by considering Reaction 4.8.1, which involves an eight-electron change, and Reaction 4.8.2, which involves a four-electron change. Rewriting Equation 4.8.1 for one electron-mole yields

$$ 1/8NH_4^+ + 1/4O_2 \longleftrightarrow 1/8NO_3^- + 1/4H^+ + 1/8H_2O \quad pE^0 = 5.85 \quad (4.8.3) $$

whereas Reaction 4.8.2, when rewritten for one electron-mole rather than four, yields:

$$ Fe^{2+} + 1/4O_2 + 5/2H_2O \longleftrightarrow Fe(OH)_3(s) + 2H^+ \quad pE^0 = 7.6 \quad (4.8.4) $$

From Equation 4.7.2, the standard free energy change for a reaction is

$$ \Delta G^0 = -2.303nRT(pE^0) \tag{4.7.2} $$

which, for a one electron-mole reaction is simply

$$ \Delta G^0 = -2.303RT(pE^0) \tag{4.7.2} $$

Therefore, for reactions written in terms of one electron-mole, a comparison of pE^0 values provides a direct comparison of ΔG^0 values.

As shown in Equation 4.6.3 for a redox reaction involving n electrons, pE^0 is related to the equilibrium constant by

$$ \log K = n\,(pE^0) \tag{4.8.5} $$

which for a one electron-mole reaction becomes simply

$$ \log K = pE^0 \tag{4.8.6} $$

Reaction 4.8.3, the nitrification reaction written in terms of one electron-mole, has a pE^0 value of +5.85. The equilibrium-constant expression for this reaction is,

$$ K = \frac{[NO_3^-]^{1/8}[H^+]^{1/4}}{[NH_4^+]^{1/8}P_{O_2}^{1/4}} \tag{4.8.7} $$

a computationally cumbersome form for which log K is given very simply as the following:

$$\log K = pE^0 = 5.85 \text{ or } K = 7.08 \times 10^5 \tag{4.8.8}$$

Table 4.1 is a compilation of pE^0 values for redox reactions that are especially important in aquatic systems. Most of these values are calculated from thermo-dynamic data rather than from direct potentiometric measurements in an electro-chemical cell, as shown in Figure 4.2. Most electrode systems that might be devised do not give potential responses corresponding to the Nernst equation; that is, they do not behave *reversibly*. It is true that one may place a platinum electrode and a reference electrode in water and measure a potential. This potential, referred to the standard hydrogen electrode, is the so-called E_H **value**. Furthermore, the measured potential will be more positive (more oxidizing) in an oxidizing medium, such as the aerobic surface layers of a lake, than in a reducing medium, such as the anaerobic bottom regions of a body of water. However, attaching any quantitative significance to the E_H value measured directly with an electrode is a very dubious practice. Acid mine waters containing relatively high levels of sulfuric acid and dissolved iron give reasonably accurate E_H values by direct measurement, but most aquatic systems do not yield meaningful values of E_H.

4.9. THE LIMITS OF pE IN WATER

There are pH-dependent limits to the pE values at which water is thermodynam-ically stable. Water may be both oxidized

$$2H_2O \xleftarrow{\;\;} \rightarrow O_2 + 4H^+ + 4e^- \tag{4.9.1}$$

or it may be reduced:

$$2H_2O + 2e^- \xleftarrow{\;\;} \rightarrow H_2 + 2OH^- \tag{4.9.2}$$

These two reactions determine the limits of pE in water. On the oxidizing side (relatively more positive pE values), the pE value is limited by the oxidation of water, Half-reaction 4.9.1. The evolution of hydrogen, Half-reaction 4.9.2, limits the pE value on the reducing side.

The condition under which oxygen from the oxidation of water has a pressure of 1.00 atm can be regarded as the oxidizing limit of water whereas a hydrogen pressure of 1.00 atmosphere may be regarded as the reducing limit of water. These are **boundary conditions** that enable calculation of the stability boundaries of water. Writing the reverse of Reaction 4.9.1 for one electron and setting $P_{O_2} = 1.00$ yields:

$$1/4O_2 + H^+ + e^- \xleftarrow{\;\;} \rightarrow 1/2H_2O \quad pE^0 = 20.75 \text{ (from Table 4.1)} \tag{4.9.3}$$

Thus, Equation 4.9.5 defines the pH-dependent oxidizing limit of water. At a specified pH, pE values more positive than the one given by Equation 4.9.5 cannot exist at equilibrium in water in contact with the atmosphere.

Table 4.1. pE^0 Values of Redox Reactions Important in Natural Waters (at 25°C)

Reaction	pE^0	$pE^0(W)$[1]
(1) $1/4O_2(g) + H^+(W) + e^- \leftarrow\rightarrow 1/2H_2O$	+20.75	+13.75
(2) $1/5NO_3^- + 6/5H^+ + e^- \leftarrow\rightarrow 1/10N_2 + 3/5H_2O$	+21.05	+12.65
(3) $1/2MnO_2 + 1/2HCO_3^-(10^{-3}) + 3/2H^+(W) + e^- \leftarrow\rightarrow$ $1/2MnCO_3(s) + H_2O$	—	$+8.5$[2]
(4) $1/2NO_3^- + H^+(W) + e^- \leftarrow\rightarrow 1/2NO_2^- + 1/2H_2O$	+14.15	+7.15
(5) $1/8NO_3^- + 5/4H^+(W) + e^- \leftarrow\rightarrow 1/8NH_4^+ + 3/8H_2O$	+14.90	+6.15
(6) $1/6NO_2^- + 4/3H^+(W) + e^- \leftarrow\rightarrow$ $1/6NH_4^+ + 1/3H_2O$	+15.14	+5.82
(7) $1/2CH_3OH + H^+(W) + e^- \leftarrow\rightarrow$ $1/2CH_4(g) + 1/2H_2O$	+9.88	+2.88
(8) $1/4CH_2O + H^+(W) + e^- \leftarrow\rightarrow 1/4CH_4(g) + 1/4H_2O$	+6.94	-0.06
(9) $FeOOH(g) + HCO_3^-(10^{-3}) + 2H^+(W) + e^- \leftarrow\rightarrow$ $FeCO_3(s) + 2H_2O$	—	-1.67[2]
(10) $1/2CH_2O + H^+(W) + e^- \leftarrow\rightarrow + 1/2CH_3OH$	+3.99	-3.01
(11) $1/6SO_4^{2-} + 4/3H^+(W) + e^- \leftarrow\rightarrow 1/6S(s) + 2/3H_2O$	+6.03	-3.30
(12) $1/8SO_4^{2-} + 5/4H^+(W) + e^- \leftarrow\rightarrow$ $1/8H_2S(g) + 1/2H_2O$	+5.75	-3.50
(13) $1/8SO_4^{2-} + 5/4H^+(W) + e^- \leftarrow\rightarrow$ $1/8H_2S(g) + 1/2H_2O$	+4.13	-3.75
(14) $1/2S + H^+(W) + e^- \leftarrow\rightarrow 1/2H_2S$	+2.89	-4.11
(15) $1/8CO_2 + H^+ + e^- \leftarrow\rightarrow 1/8CH_4 + 1/4H_2O$	+2.87	-4.13
(16) $1/6N_2 + 4/3H^+(W) + e^- \leftarrow\rightarrow 1/3NH_4^+$	+4.68	-4.65
(17) $H^+(W) + e^- \leftarrow\rightarrow 1/2H_2(g)$	0.00	-7.00
(18) $1/4CO_2(g) + H^+(W) + e^- \leftarrow\rightarrow 1/4CH_2O + 1/4H_2O$	-1.20	-8.20

[1] (W) indicates $a_{H^+} = 1.00 \times 10^{-7}$ M and $pE^0(W)$ is a pE^0 at $a_{H^+} = 1.00 \times 10^{-7}$ M.

[2] These data correspond to $a_{HCO_3^-} = 1.00 \times 10^{-3}$ M rather than unity and so are not exactly $pE^0(W)$; they represent typical aquatic conditions more exactly than pE^0 values do.

Source: Stumm, Werner and James J. Morgan, *Aquatic Chemistry*, John Wiley and Sons, New York, 1970, p. 318. Reproduced by permission of John Wiley & Sons, Inc.

$$pE = pE^0 + \log(P_{O_2}^{1/4}[H^+]) \qquad (4.9.4)$$

$$pE = 20.75 - pH \qquad (4.9.5)$$

The pE-pH relationship for the reducing limit of water taken at P_{H_2} = exactly 1 atm is given by the following derivation:

$$H^+ + e^- \longleftrightarrow 1/2 H_2 \qquad\qquad pE^0 = 0.00 \qquad (4.9.6)$$

$$pE = pE^0 + \log [H^+] \qquad (4.9.7)$$

$$pE = -pH \qquad (4.9.8)$$

For neutral water (pH = 7.00), substitution into Equations 4.9.8 and 4.9.5 yields - 7.00 to 13.75 for the pE range of water. The pE-pH boundaries of stability for water are shown by the dashed lines in Figure 4.4 of Section 4.11.

The decomposition of water is very slow in the absence of a suitable catalyst. Therefore, water may have temporary nonequilibrium pE values more negative than the reducing limit or more positive than the oxidizing limit. A common example of the latter is a solution of chlorine in water.

4.10. pE VALUES IN NATURAL WATER SYSTEMS

Although it is not generally possible to obtain accurate pE values by direct potentiometric measurements in natural aquatic systems, in principle, pE values may be calculated from the species present in water at equilibrium. An obviously significant pE value is that of neutral water in thermodynamic equilibrium with the atmosphere. In water under these conditions, P_{O_2} = 0.21 atm and $[H^+]$ = 1.00 × 10^{-7} M. Substitution into Equation 4.9.4 yields:

$$pE = 20.75 + \log \{(0.21)^{1/4} \times 1.00 \times 10^{-7}\} = 13.8 \qquad (4.10.1)$$

According to this calculation, a pE value of around +13 is to be expected for water in equilibrium with the atmosphere, that is, an aerobic water. At the other extreme, consider anaerobic water in which methane and CO_2 are being produced by microorganisms. Assume $P_{CO_2} = P_{CH_4}$ and that pH = 7.00. The relevant half-reaction is

$$1/8 CO_2 + H^+ + e^- \longleftrightarrow 1/8 CH_4 + 1/4 H_2O \qquad (4.10.2)$$

for which the Nernst equation is

$$pE = 2.87 + \log \frac{P_{CO_2}^{1/8}[H^+]}{P_{CH_4}^{1/8}[H^+]} = 2.87 + \log[H^+] = 2.87 - 7.00 = -4.13 \qquad (4.10.3)$$

Note that the pE value of - 4.13 does not exceed the reducing limit of water at pH 7.00, which from Equation 4.9.8 is -7.00. It is of interest to calculate the pressure of oxygen in neutral water at this low pE value of - 4.13. Substitution into Equation 4.9.4 yields

$$- 4.13 = 20.75 + \log(P_{O_2}^{1/4} \times 1.00 \times 10^{-7}) \tag{4.10.4}$$

from which the pressure of oxygen is calculated to be 3.0×10^{-72} atm. This impossibly low figure for the pressure of oxygen means that equilibrium with respect to oxygen partial pressure is not achieved under these conditions. Certainly, under any condition approaching equilibrium between comparable levels of CO_2 and CH_4, the partial pressure of oxygen must be extremely low.

4.11. pE-pH DIAGRAMS

The examples cited so far have shown the close relationships between pE and pH in water. This relationship may be expressed graphically in the form of a **pE-pH diagram.** Such diagrams show the regions of stability and the boundary lines for various species in water. Because of the numerous species that may be formed, such diagrams may become extremely complicated. For example, if a metal is being considered, several different oxidation states of the metal, hydroxy complexes, and different forms of the solid metal oxide or hydroxide may exist in different regions described by the pE-pH diagram. Most waters contain carbonate, and many contain sulfates and sulfides, so that various metal carbonates, sulfates, and sulfides may predominate in different regions of the diagram. In order to illustrate the principles involved, however, a simplified pE-pH diagram is considered here. The reader is referred to more advanced works on geochemistry and aquatic chemistry for more complicated (and more realistic) pE-pH diagrams[1,2].

A pE-pH diagram for iron may be constructed assuming a maximum concentration of iron in solution, in this case 1.0×10^{-5} M. The following equilibria will be considered:

$$Fe^{3+} + e^- \longleftrightarrow Fe^{2+} \qquad\qquad pE^0 = + 13.2 \tag{4.11.1}$$

$$Fe(OH)_2(s) + 2H^+ \longleftrightarrow Fe^{2+} + 2H_2O \tag{4.11.2}$$

$$K_{sp} = \frac{[Fe^{2+}]}{[H^+]^2} = 8.0 \times 10^{12} \tag{4.11.3}$$

$$Fe(OH)_3(s) + 3H^+ \longleftrightarrow Fe^{3+} + 3H_2O \tag{4.11.4}$$

$$K_{sp}' = \frac{[Fe^{3+}]}{[H^+]^3} = 9.1 \times 10^3 \tag{4.11.5}$$

(The constants K_{sp} and K_{sp}' are derived from the solubility products of $Fe(OH)_2$ and $Fe(OH)_3$, respectively, and are expressed in terms of $[H^+]$ to facilitate the calculations.) Note that the formation of species such as $Fe(OH)^{2+}$, $Fe(OH)_2^+$, and solid $FeCO_3$ or FeS, all of which might be of significance in a natural water system, is not considered.

In constructing the pE-pH diagram, several boundaries must be considered. The first two of these are the oxidizing and reducing limits of water (see Section 4.9). At the high pE end, the stability limit of water is defined by Equation 4.9.5 derived previously:

$$pE = 20.75 - pH \qquad (4.9.5)$$

The low pE limit is defined by Equation 4.9.8:

$$pE = -pH \qquad (4.9.8)$$

The pE-pH diagram constructed for the iron system must fall between the boundaries defined by these two equations.

Below pH 3, Fe^{3+} may exist in equilibrium with Fe^{2+}. The boundary line that separates these two species, where $[Fe^{3+}] = [Fe^{2+}]$, is given by the following calculation:

$$pE = 13.2 + \log\frac{[Fe^{3+}]}{[Fe^{2+}]} \qquad (4.11.6)$$

$$[Fe^{3+}] = [Fe^{2+}] \qquad (4.11.7)$$

$$pE = 13.2 \text{ (independent of pH)} \qquad (4.11.8)$$

At pE exceeding 13.2, as the pH increases from very low values, $Fe(OH)_3$ precipitates from a solution of Fe^{3+}. The pH at which precipitation occurs depends, of course, upon the concentration of Fe^{3+}. In this example, a maximum soluble iron concentration of 1.00×10^{-5} M has been chosen so that at the $Fe^{3+}/Fe(OH)_3$ boundary, $[Fe^{3+}] = 1.00 \times 10^{-5}$ M. Substitution in Equation 4.11.5 yields:

$$[H^+]^3 = \frac{[Fe^{3+}]}{K_{sp'}} = \frac{1.00 \times 10^{-5}}{9.1 \times 10^3} \qquad (4.11.9)$$

$$pH = 2.99 \qquad (4.11.10)$$

In a similar manner, the boundary between Fe^{2+} and solid $Fe(OH)_2$ may be defined, assuming $[Fe^{2+}] = 1.00 \times 10^{-5}$ M (the maximum soluble iron concentration specified at the beginning of this exercise) at the boundary:

$$[H^+]^2 = \frac{[Fe^{2+}]}{K_{sp}} = \frac{1.00 \times 10^{-5}}{8.0 \times 10^{12}} \text{ (from Equation 4.11.3)} \qquad (4.11.11)$$

$$pH = 8.95 \qquad (4.11.12)$$

Throughout a wide pE-pH range, Fe^{2+} is the predominant soluble iron species in equilibrium with the solid hydrated iron(III) oxide, $Fe(OH)_3$. The boundary between these two species depends upon both pE and pH. Substituting Equation 4.11.5 into Equation 4.11.6 yields:

$$pE = 13.2 + \log\frac{K_{sp}'[H^+]^3}{[Fe^{2+}]} \qquad (4.11.13)$$

$$pE = 13.2 + \log 9.1 \times 10^3 - \log 1.00 \times 10^{-5} + 3 \times \log [H^+]$$

$$pE = 22.2 - 3\,pH \qquad (4.11.14)$$

The boundary between the solid phases $Fe(OH)_2$ and $Fe(OH)_3$ likewise depends upon both pE and pH, but it does not depend upon an assumed value for total soluble iron. The required relationship is derived from substituting both Equation 4.11.3 and Equation 4.11.5 into Equation 4.11.6:

$$pE = 13.2 + \log\frac{K_{sp}'[H^+]^3}{[Fe^{2+}]} \qquad (4.11.13)$$

$$pE = 13.2 + \log \frac{K_{sp}'[H^+]^3}{K_{sp}[H^+]^2} \qquad (4.11.15)$$

$$pE = 13.2 + \log \frac{9.1 \times 10^3}{8.0 \times 10^{12}} + \log [H^+]$$

$$pE = 4.3 - pH \qquad (4.11.16)$$

All of the equations needed to prepare the pE-pH diagram for iron in water have now been derived. To summarize, the equations are (4.9.5), O_2-H_2O boundary; (4.9.8), H_2–H_2O boundary; (4.11.8), Fe^{3+}-Fe^{2+} boundary; (4.11.10), Fe^{3+}-$Fe(OH)_3$ boundary; (4.11.12), Fe^{2+}-$Fe(OH)_2$ boundary; (4.11.14), Fe^{2+}-$Fe(OH)_3$ boundary; and (4.11.16), $Fe(OH)_2$-$Fe(OH)_3$ boundary.

The pE-pH diagram for the iron system in water is shown in Figure 4.4. In this system, at a relatively high hydrogen ion activity and high electron activity (an acidic reducing medium), iron(II) ion, Fe^{2+}, is the predominant iron species; some groundwaters contain appreciable levels of iron(II) under these conditions. (In most natural water systems the solubility range of Fe^{2+} is very narrow because of the precipitation of FeS or $FeCO_3$.) At a very high hydrogen ion activity and low electron activity (an acidic oxidizing medium), Fe^{3+} ion predominates. In an oxidizing medium at lower acidity, solid $Fe(OH)_3$ is the primary iron species present. Finally, in a basic reducing medium, with low hydrogen ion activity and high electron activity, solid $Fe(OH)_2$ is stable.

Note that within the pH regions normally encountered in a natural aquatic system (approximately pH 5 to 9) $Fe(OH)_3$ or Fe^{2+} are the predominant stable iron species. In fact, it is observed that in waters containing dissolved oxygen at any appreciable level (a relatively high pE), hydrated iron(III) oxide ($Fe(OH)_3$ is essentially the only inorganic iron species found. Such waters contain a high level of suspended iron, but any truly soluble iron must be in the form of a complex (see Chapter 3).

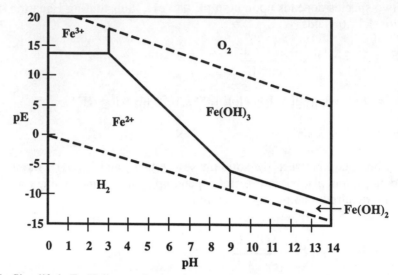

Figure 4.4. Simplifed pE-pH diagram for iron in water. The maximum soluble iron concentration is 1.00×10^{-5} M.

   In highly anaerobic, low pE water, appreciable levels of Fe^{2+} may be present. When such water is exposed to atmospheric oxygen, the pE rises and $Fe(OH)_3$ precipitates. The resulting deposits of hydrated iron(III) oxide can stain laundry and bathroom fixtures with a refractory red/brown stain. This phenomenon also explains why red iron oxide deposits are found near pumps and springs that bring deep, anaerobic water to the surface. In shallow wells, where the water may become aerobic, solid $Fe(OH)_3$ may precipitate on the well walls, clogging the aquifer outlet. This usually occurs through bacterially-mediated reactions, which are discussed in Chapter 6.

   One species not yet considered is elemental iron. For the half-reaction

$$Fe^{2+} + 2e^- \longleftrightarrow Fe \qquad\qquad pE^0 = -7.45 \qquad\qquad (4.11.17)$$

the Nernst equation gives pE as a function of $[Fe^{2+}]$

$$pE = -7.45 + \tfrac{1}{2}\log[Fe^{2+}] \qquad\qquad (4.11.18)$$

For iron metal in equilibrium with 1.00×10^{-5} M Fe^{2+}, the following pE value is obtained:

$$pE = -7.45 + \tfrac{1}{2}\log 1.00 \times 10^{-5} = -9.95 \qquad\qquad (4.11.19)$$

Examination of Figure 4.4 shows that the pE values for elemental iron in contact with Fe^{2+} is below the reducing limit of water. This shows that iron metal in contact with water is thermodynamically unstable with respect to reducing water and going into solution as Fe^{2+}, a factor that contributes to the tendency of iron to undergo corrosion.

4.12. CORROSION

One of the most damaging redox phenomena is **corrosion**, defined as the destructive alteration of metal through interactions with its surroundings. In addition to its multibillion dollar annual costs due to destruction of equipment and structures, corrosion introduces metals into water systems and destroys pollution control equipment and waste disposal pipes; it is aggravated by water and air pollutants and some kinds of hazardous wastes (see corrosive wastes in Chapter 19, Section 19.6).

Thermodynamically, all commonly-used metals are unstable relative to their environments. Elemental metals tend to undergo chemical changes to produce the more stable forms of ions, salts, oxides, and hydroxides. Fortunately, the rates of corrosion are normally slow, so that metals exposed to air and water may endure for long periods of time. However, protective measures are necessary. Sometimes these measures fail; for example, witness the gaping holes in automobile bodies exposed to salt used to control road ice.

Corrosion normally occurs when an electrochemical cell is set up on a metal surface. The area corroded is the anode, where the following oxidation reaction occurs, illustrated for the formation of a divalent metal ion from a metal, M:

$$M \rightarrow M^{2+} + 2e^- \tag{4.12.1}$$

Several cathodic reactions are possible. One of the most common of these is the reduction of H^+ ion:

$$2H^+ + 2e^- \rightarrow H_2 \tag{4.12.2}$$

Oxygen may also be involved in cathodic reactions, including reduction to hydroxide, reduction to water, and reduction to hydrogen peroxide:

$$O_2 + 2H_2O + 4e^- \rightarrow 4OH^- \tag{4.12.3}$$

$$O_2 + 4H^+ + 4e^- \rightarrow 2H_2O \tag{4.12.4}$$

$$O_2 + 2H_2O + 2e^- \rightarrow 2OH^- + H_2O_2 \tag{4.12.5}$$

Oxygen may either accelerate corrosion processes by participating in reactions such as these, or retard them by forming protective oxide films. As discussed in Chapter 6, bacteria are often involved with corrosion.

LITERATURE CITED

1. Stumm, Werner and James J. Morgan, *Aquatic Chemistry: Chemical Equilibria and Rates in Natural Waters*, 3rd ed., John Wiley and Sons, Inc., New York, 1995.

2. Garrels, R. M., and C. M. Christ, *Solutions, Minerals, and Equilibria*, Harper and Row, New York, 1965.

SUPPLEMENTARY REFERENCES

Baas Becking, L. G. M., I. R. Kaplan, and D. Moore, "Limits of the Natural Environment in Terms of pH and Oxidation-Reduction Potentials in Natural Waters," *J. Geol.*, **68**, 243-284 (1960).

Bates, Roger G., "The Modern Meaning of pH," *Crit. Rev. Anal. Chem.*, **10**, 247-278 (1981).

Brubaker, G. R. and P. Phipps, Eds., *Corrosion Chemistry*, American Chemical Society, 1979.

Dowdy, R. H., *Chemistry in the Soil Environment*, ASA Special Publication **40**, American Society of Agronomy, Madison, WI, 1981.

"Electrochemical Phenomena," Chapter 6 in *The Chemistry of Soils*, Garrison Sposito, Oxford University Press, New York, 1989, pp. 106-126.

"Element Fixation in Soil," Chapter 3 in *The Soil Chemistry of Hazardous Materials*, James Dragun, Hazardous Materials Control Research Insitute, Silver Spring, MD, 1988, pp. 75-152.

Lingane, James J., *Electroanalytical Chemistry*, Wiley-Interscience, New York, 1958.

Pankow, James F., *Aquatic Chemistry Concepts*, Lewis Publishers/CRC Press, Boca Raton, FL, 1991.

Pankow, James F., *Aquatic Chemistry Problems*, Titan Press-OR, Portland, OR, 1992.

Stumm, Werner, *Redox Potential as an Environmental Parameter: Conceptual Significance and Operational Limitation*, Third International Conference on Water Pollution Research, (Munich, Germany), Water Pollution Control Federation, Washington, D.C., 1966.

Rowell, D. L., "Oxidation and Reduction," in *The Chemistry of Soil Processes*, D. J. Greenland and M. H. B. Hayes, Eds., Wiley, Chichester, U.K., 1981.

QUESTIONS AND PROBLEMS

1. The acid-base reaction for the dissociation of acetic acid is

   $$HOAc + H_2O \rightarrow H_3O^+ + OAc^-$$

   with $K_a=1.75 \times 10^{-5}$. Break this reaction down into two half-reactions involving H^+ ion. Break down the redox reaction

   $$Fe^{2+} + H^+ \rightarrow Fe^{3+} + 1/2H_2$$

   into two half-reactions involving the electron. Discuss the analogies between the acid-base and redox processes.

2. Assuming a bicarbonate ion concentration $[HCO_3^-]$ of $1.00 \times 10^{-3}\,M$ and a value of 3.5×10^{-11} for the solubility product of $FeCO_3$, what would you expect to be the stable iron species at pH 9.5 and pE -8.0, as shown in Figure 4.4?

3. Assuming that the partial pressure of oxygen in water is that of atmospheric O_2, 0.21 atm, rather than the 1.00 atm assumed in deriving Equation 4.9.5, derive an equation describing the oxidizing pE limit of water as a function of pH.

4. Plot log P_{O_2} as a function of pE at pH 7.00.

5. Calculate the pressure of oxygen for a system in equilibrium in which $[NH_4^+] = [NO_3^-]$ at pH 7.00.

6. Calculate the values of $[Fe^{3+}]$, pE, and pH at the point in Figure 4.4 where Fe^{2+} is at a concentration of 1.00×10^{-5} M, $Fe(OH)_2$, and $Fe(OH)_3$ are all in equilibrium.

7. What is the pE value in a solution in equilibrium with air (21% O_2 by volume) at pH 6.00?

8. What is the pE value at the point on the Fe^{2+}–$Fe(OH)_3$ boundary line (see Figure 4.4) in a solution with a soluble iron concentration of 1.00×10^{-4} M at pH 6.00?

9. What is the pE value in an acid mine water sample having $[Fe^{3+}] = 7.03 \times 10^{-3}$ M and $[Fe^{2+}] = 3.71 \times 10^{-4}$ M?

10. At pH 6.00 and pE 2.58, what is the concentration of Fe^{2+} in equilibrium with $Fe(OH)_2$?

11. What is the calculated value of the partial pressure of O_2 in acid mine water of pH 2.00, in which $[Fe^{3+}] = [Fe^{2+}]$?

12. What is the major advantage of expressing redox reactions and half-reactions in terms of exactly one electron-mole?

13. Why are pE values that are determined by reading the potential of a platinum electrode versus a reference electrode generally not very meaningful?

14. What determines the oxidizing and reducing limits, respectively, for the thermodynamic stability of water?

15. How would you expect pE to vary with depth in a stratified lake?

16. Upon what half-reaction is the rigorous definition of pE based?

5 PHASE INTERACTIONS

5.1. CHEMICAL INTERACTIONS INVOLVING SOLIDS, GASES, AND WATER

Homogeneous chemical reactions occurring entirely in aqueous solution are rather rare in natural waters and wastewaters. Instead, most significant chemical and biochemical phenomena in water involve interactions between species in water and another phase. Some of these important interactions are illustrated in Figure 5.1. Several examples of phase interactions in water illustrated by the figure are the fol-

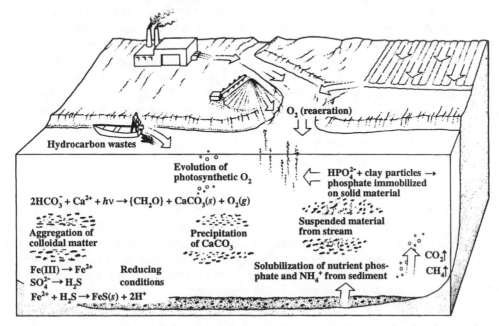

Figure 5.1. Most important environmental chemical processes in water involve interactions between water and another phase.

lowing: production of solid biomass through the photosynthetic activity of algae occurs within a suspended algal cell and involves exchange of dissolved solids and gases between the surrounding water and the cell. Similar exchanges occur when bacteria degrade organic matter (often in the form of small particles) in water. Chemical reactions occur that produce solids or gases in water. Iron and many important trace-level elements are transported through aquatic systems as colloidal chemical compounds or are sorbed to solid particles. Pollutant hydrocarbons and some pesticides may be present on the water surface as an immiscible liquid film. Sediment can be washed physically into a body of water.

This chapter discusses the importance of interactions among different phases in aquatic chemical processes. In a general sense, in addition to water, these phases may be divided between *sediments* (bulk solids) and *suspended colloidal material*. The ways in which sediments are formed and the significance of sediments as repositories and sources of aquatic solutes are discussed. Mentioned in earlier chapters, solubilities of solids and gases (Henry's law) are covered here in some detail.

Much of this chapter deals with the behavior of colloidal material, which consists of very fine particles of solids, gases, or immiscible liquids suspended in water. Colloidal material is involved with many significant aquatic chemical phenomena. It is very reactive because of its high surface-area-to-volume ratio.

5.2. IMPORTANCE AND FORMATION OF SEDIMENTS

Sediments are the layers of relatively finely divided matter covering the bottoms of rivers, streams, lakes, reservoirs, bays, estuaries, and oceans. Sediments typically consist of mixtures of fine-, medium-, and coarse-grained minerals, including clay, silt, and sand, mixed with organic matter. They may vary in composition from pure mineral matter to predominantly organic matter. Sediments are repositories of a variety of biological, chemical, and pollutant detritus in bodies of water. Of particular concern is the transfer of chemical species from sediments into aquatic food chains via organisms that spend significant parts of their life cycles in contact with or living in sediments. Among the sediment-dwelling organisms are various kinds of shellfish (shrimp, crayfish, crab, clams) and a variety of worms, insects, amphipods, bivalves, and other smaller organisms that are of particular concern because they are located near the bottom of the food chain.

Although the classic picture of pollutant transfer from sediments to organisms invokes an intermediate stage in water solution, it is now believed that direct transfer from sediments to organisms occurs to a large extent. This is probably particularly important for poorly-water-soluble organophilic pollutants, such as organohalide pesticides. The portion of substances held in sediments that is probably most available to organisms is that contained in **pore water**, contained in microscopic pores within the sediment mass. Pore water is commonly extracted from sediments for measurements of toxicity to aquatic test organisms.

Formation of Sediments

Physical, chemical, and biological processes may all result in the deposition of sediments in the bottom regions of bodies of water. Sedimentary material may be simply carried into a body of water by erosion or through sloughing (caving in) of the shore. Thus, clay, sand, organic matter, and other materials may be washed into a lake and settle out as layers of sediment.

Sediments may be formed by simple precipitation reactions, several of which are discussed below. When a phosphate-rich wastewater enters a body of water containing a high concentration of calcium ion, the following reaction occurs to produce solid hydroxyapatite:

$$5Ca^{2+} + H_2O + 3HPO_4^{2-} \rightarrow Ca_5OH(PO_4)_3(s) + 4H^+ \tag{5.2.1}$$

Calcium carbonate sediment may form when water rich in carbon dioxide and containing a high level of calcium as temporary hardness (see Section 3.5) loses carbon dioxide to the atmosphere,

$$Ca^{2+} + 2HCO_3^- \rightarrow CaCO_3(s) + CO_2(g) + H_2O \tag{5.2.2}$$

or when the pH is raised by a photosynthetic reaction:

$$Ca^{2+} + 2HCO_3^- + h\nu \rightarrow \{CH_2O\} + CaCO_3(s) + O_2(g) \tag{5.2.3}$$

Oxidation of reduced forms of an element can result in its transformation to an insoluble species, such as occurs when iron(II) is oxidized to iron(III) to produce a precipitate of insoluble iron(III) hydroxide:

$$4Fe^{2+} + 10H_2O + O_2 \rightarrow 4Fe(OH)_3(s) + 8H^+ \tag{5.2.4}$$

A decrease in pH can result in the production of an insoluble humic acid sediment from base-soluble organic humic substances in solution (see Section 3.17).

Biological activity is responsible for the formation of some aquatic sediments. Some bacterial species produce large quantities of iron(III) oxide (see Section 6.14) as part of their energy-extracting mediation of the oxidation of iron(II) to iron(III). In anaerobic bottom regions of bodies of water, some bacteria use sulfate ion as an electron receptor,

$$SO_4^{2-} \rightarrow H_2S \tag{5.2.5}$$

whereas other bacteria reduce iron(III) to iron(II):

$$Fe(OH)_3(s) \rightarrow Fe^{2+} \tag{5.2.6}$$

The net result is a precipitation reaction producing a black layer of iron(II) sulfide sediment:

$$Fe^{2+} + H_2S \rightarrow FeS(s) + 2H^+ \tag{5.2.7}$$

This frequently occurs during the winter, alternating with the production of calcium carbonate by-product from photosynthesis (Reaction 5.2.3) during the summer. Under such conditions, a layered bottom sediment is produced composed of alternate layers of black FeS and white $CaCO_3$ as shown in Figure 5.2.

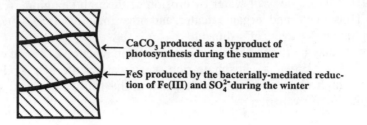

CaCO₃ produced as a byproduct of photosynthesis during the summer

FeS produced by the bacterially-mediated reduction of Fe(III) and SO₄²⁻ during the winter

Figure 5.2. Alternate layers of FeS and $CaCO_3$ in a lake sediment. This phenomenon has been observed in Lake Zürich in Switzerland.

The preceding are only a few examples of reactions that result in the formation of bottom sediments in bodies of water. Eventually these sediments may become covered and form sedimentary minerals.

5.3. SOLUBILITIES

The formation and stabilities of nonaqueous phases in water are strongly dependent upon solubilities. Calculations of the solubilities of solids and gases are addressed in this section.

Solubilities of Solids

Generally, the solubility of a solid in water is of concern when the solid is slightly soluble, often having such a low solubility that it is called "insoluble." In section 3.11 the solubility of lead carbonate was considered. This salt can introduce toxic lead ion into water by reactions such as

$$PbCO_3(s) \xleftarrow{}{\rightarrow} Pb^{2+} + CO_3^{2-} \tag{3.10.13}$$

A relatively straightforward calculation of the solubility of an ionic solid can be performed on barium sulfate,[1] which dissolves according to the reaction

$$BaSO_4(s) \xleftarrow{}{\rightarrow} Ba^{2+} + SO_4^{2-} \tag{5.3.1}$$

for which the equilibrium constant is the following:

$$K_{sp} = [Ba^{2+}][SO_4^{2-}] = 1.23 \times 10^{-10} \tag{5.3.2}$$

An equilibrium constant in this form that expresses the solubility of a solid that forms ions in water is a **solubility product** and is designated K_{sp}. In the simplest cases a solubility product can be used alone to calculate the solubility of a slightly soluble salt in water. The solubility (S, moles per liter) of barium sulfate is calculated as follows:

$$[Ba^{2+}] = [SO_4^{2-}] = S \tag{5.3.3}$$

$$[Ba^{2+}][SO_4^{2-}] = S \times S = K_{sp} = 1.23 \times 10^{-10} \tag{5.3.4}$$

$$S = (K_{sp})^{1/2} = (1.23 \times 10^{-10})^{1/2} = 1.11 \times 10^{-5} \tag{5.3.5}$$

Even such a simple calculation may be complicated by variations in activity coefficients resulting from differences in ionic strength.

Intrinsic solubilities account for the fact that a significant portion of the solubility of an ionic solid is due to the dissolution of the neutral form of the salt and must be added to the solubility calculated from K_{sp} to obtain the total solubility, as illustrated below for the calculation of the solubility of calcium sulfate. When calcium sulfate dissolves in water the two major reactions are

$$CaSO_4(s) \longleftrightarrow CaSO_4(aq) \tag{5.3.6}$$

$$[CaSO_4(aq)] = 5.0 \times 10^{-3} \, M \quad (25°C) \tag{5.3.7}$$
$$\text{(Intrinsic solubility of } CaSO_4)$$

$$CaSO_4(s) \longleftrightarrow Ca^{2+} + SO_4^{2-} \tag{5.3.8}$$

$$[Ca^{2+}][SO_4^{2-}] = K_{sp} = 2.6 \times 10^{-5} \quad (25°C) \tag{5.3.9}$$

and the total solubility of $CaSO_4$ is calculated as follows:

$$S = [Ca^{2+}] + [CaSO_4(aq)] \tag{5.3.10}$$

Contribution to solubility   Contribution to solubility
from solubility product   from intrinsic solubilty

$$S = (K_{sp})^{1/2} + [CaSO_4(aq)] = (2.6 \times 10^{-5})^{1/2} + 5.0 \times 10^{-3}$$
$$= 5.1 \times 10^{-3} + 5.0 \times 10^{-3} = 1.01 \times 10^{-2} \, M \tag{5.3.11}$$

It is seen that, in this case, the intrinsic solubility accounts for half of the solubility of the salt.

In Section 3.15 it was seen that solubilities of ionic solids can be very much affected by reactions of cations and anions. It was shown that the solubility of $PbCO_3$ is increased by the chelation of lead ion by NTA,

$$Pb^{2+} + T^{3-} \longleftrightarrow PbT^- \tag{3.15.6}$$

increased by reaction of carbonate ion with H^+,

$$H^+ + CO_3^{2-} \longleftrightarrow HCO_3^- \tag{5.3.12}$$

and decreased by the presence of carbonate ion from water alkalinity:

$$CO_3^{2-}(\text{from dissociation of } HCO_3^-) + Pb^{2+} \longleftrightarrow PbCO_3(s) \qquad (5.3.13)$$

These examples illustrate that reactions of both cations and anions must often be considered in calculating the solubilities of ionic solids.

Solubilities of Gases

The solubilities of gases in water are described by Henry's Law which states that *at constant temperature the solubility of a gas in a liquid is proportional to the partial pressure of the gas in contact with the liquid.* For a gas, "X," this law applies to equilibria of the type

$$X(g) \longleftrightarrow X(aq) \qquad (5.3.14)$$

and does not account for additional reactions of the gas species in water such as,

$$NH_3 + H_2O \longleftrightarrow NH_4^+ + OH^- \qquad (5.3.15)$$

$$SO_2 + HCO_3^- \text{ (From water alkalinity)} \longleftrightarrow CO_2 + HSO_3^- \qquad (5.3.16)$$

which may result in much higher solubilities than predicted by Henry's law alone. Mathematically, Henry's Law is expressed as

$$[X(aq)] = K\,P_X \qquad (5.3.17)$$

where $[X(aq)]$ is the aqueous concentration of the gas, P_X is the partial pressure of the gas, and K is the Henry's Law constant applicable to a particular gas at a specified temperature. For gas concentrations in units of moles per liter and gas pressures in atmospheres, the units of K are $mol \times L^{-1} \times atm^{-1}$. Some values of K for dissolved gases that are significant in water are given in Table 5.1.

Table 5.1. Henry's Law Constants for Some Gases in Water at 25°C.

Gas	K, $mol \times L^{-1} \times atm^{-1}$
O_2	1.28×10^{-3}
CO_2	3.38×10^{-2}
H_2	7.90×10^{-4}
CH_4	1.34×10^{-3}
N_2	6.48×10^{-4}
NO	2.0×10^{-4}

In calculating the solubility of a gas in water, a correction must be made for the partial pressure of water by subtracting it from the total pressure of the gas. At 25°C the partial pressure of water is 0.0313 atm; values at other temperatures are readily obtained from standard handbooks. The concentration of oxygen in water saturated with air at 1.00 atm and 25°C may be calculated as an example of a simple gas solubility calculation. Considering that dry air is 20.95% by volume oxygen, factoring in the partial pressure of water gives the following:

$$P_{O_2} = (1.0000 \text{ atm} - 0.0313 \text{ atm}) \times 0.2095 = 0.2029 \text{ atm} \qquad (5.3.18)$$

$$\begin{aligned}[O_2(aq)] = K \times P_{O_2} &= 1.28 \times 10^{-3} \text{ mol} \times L^{-1} \times \text{atm}^{-1} \times 0.2029 \text{ atm} \\ &= 2.60 \times 10^{-4} \text{ mol} \times L^{-1} \end{aligned} \qquad (5.3.19)$$

Since the molecular weight of oxygen is 32, the concentration of dissolved oxygen in water in equilibrium with air under the conditions given above is 8.32 mg/L, or 8.32 parts per million (ppm).

The solubilities of gases decrease with increasing temperature. Account is taken of this factor with the **Clausius-Clapeyron** equation,

$$\log \frac{C_2}{C_1} = \frac{\Delta H}{2.303R} \left[\frac{1}{T_1} - \frac{1}{T_2} \right] \qquad (5.3.20)$$

where C_1 and C_2 denote the gas concentration in water at absolute temperatures of T_1 and T_2, respectively; ΔH is the heat of solution; and R is the gas constant. The value of R is $1.987 \text{ cal} \times \text{deg}^{-1} \times \text{mol}^{-1}$, which gives ΔH in units of cal/mol.

5.4. COLLOIDAL PARTICLES IN WATER

Many minerals, some organic pollutants, proteinaceous materials, some algae, and some bacteria are suspended in water as very small particles. Such particles, which have some characteristics of both species in solution and larger particles in suspension, which range in diameter from about 0.001 micrometer (μm) to about 1 μm, and which scatter white light as a light blue hue observed at right angles to the incident light, are classified as **colloidal particles**. The characteristic light-scattering phenomenon of colloids results from their being the same order of size as the wavelength of light and is called the **Tyndall effect**. The unique properties and behavior of colloidal particles are strongly influenced by their physical-chemical characteristics, including high specific area, high interfacial energy, and high surface/charge density ratio.

Occurrence of Colloids in Water

Colloids composed of a variety of organic substances (including humic substances), inorganic materials (especially clays), and pollutants occur in natural water and wastewater.[2] These substances have a number of effects, including effects on organisms and pollutant transport. The characterization of colloidal materials in water is obviously very important, and a variety of means are used to isolate and characterize these materials. The two most widely used methods are filtration and centrifugation, although other techniques including voltammetry, gels, and field-flow fractionation can be used.[3]

Kinds of Colloidal Particles

Colloids may be classified as *hydrophilic colloids*, *hydrophobic colloids*, or *association colloids*. These three classes are briefly summarized below.

Hydrophilic colloids generally consist of macromolecules, such as proteins and synthetic polymers, that are characterized by strong interaction with water resulting in spontaneous formation of colloids when they are placed in water. In a sense, hydrophilic colloids are solutions of very large molecules or ions. Suspensions of hydrophilic colloids are less affected by the addition of salts to water than are suspensions of hydrophobic colloids.

Hydrophobic colloids interact to a lesser extent with water and are stable because of their positive or negative electrical charges as shown in Figure 5.3. The charged surface of the colloidal particle and the **counter-ions** that surround it compose an **electrical double layer**, which causes the particles to repel each other.

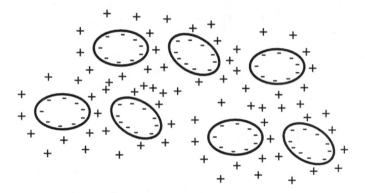

Figure 5.3. Representation of negatively charged hydrophobic colloidal particles surrounded in solution by positively charged counter-ions, forming an electrical double layer. (Colloidal particles suspended in water may have either a negative or positive charge.)

Hydrophobic colloids are usually caused to settle from suspension by the addition of salts. Examples of hydrophobic colloids are clay particles, petroleum droplets, and very small gold particles.

Association colloids consist of special aggregates of ions and molecules called **micelles**. To understand how this occurs, consider sodium stearate, a typical soap with the structural formula shown below:

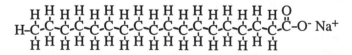

Represented as 〰〰〰〰〰Ⓞ

The stearate ion has both a hydrophilic $-CO_2^-$ head and a long organophilic tail, $CH_3(CH_2)_{16}^-$. As a result, stearate anions in water tend to form clusters consisting of as many as 100 anions clustered together with their hydrocarbon "tails" on the inside of a spherical colloidal particle and their ionic "heads" on the surface in contact with water and with Na^+ counterions. This results in the formation of **micelles** as illustrated in Figure 5.4. Micelles can be visualized as droplets of oil about 3-4 nanometers (nm) in diameter and covered with ions or polar groups. According to this model, micelles form when a certain concentration of surfactant species, typically around 1×10^{-3}, is reached. The concentration at which this occurs is called the **critical micelle concentration**.

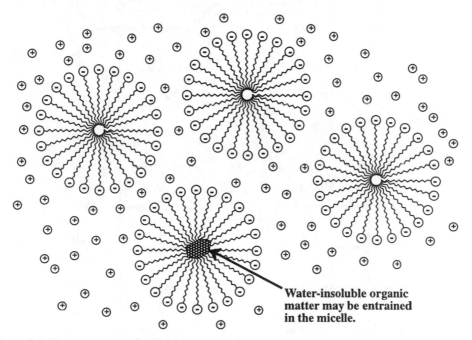

Water-insoluble organic matter may be entrained in the micelle.

Figure 5.4. Representation of colloidal soap micelle particles.

Colloid Stability

The stability of colloids is a prime consideration in determining their behavior. It is involved in important aquatic chemical phenomena including the formation of sediments, dispersion and agglomeration of bacterial cells, and dispersion and removal of pollutants (such as crude oil from an oil spill).

Discussed above, the two main phenomena contributing to the stabilization of colloids are **hydration** and **surface charge**. The layer of water on the surface of hydrated colloidal particles prevents contact, which would result in the formation of larger units. A surface charge on colloidal particles may prevent aggregation, since like-charged particles repel each other. The surface charge is frequently pH-dependent; around pH 7 most colloidal particles in natural waters are negatively charged. Negatively charged aquatic colloids include algal cells, bacterial cells, proteins, and colloidal petroleum droplets.

One of the three major ways in which a particle may acquire a surface charge is by **chemical reaction at the particle surface**. This phenomenon, which frequently involves hydrogen ion and is pH-dependent, is typical of hydroxides and oxides and is illustrated for manganese dioxide, MnO_2, in Figure 5.5.

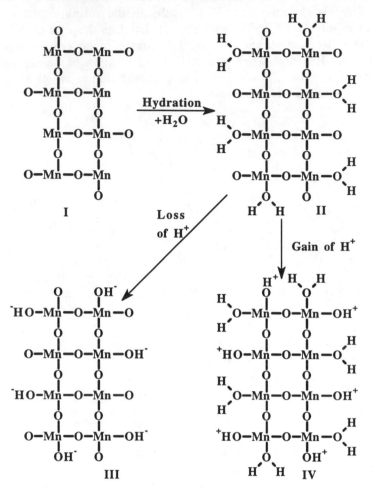

Figure 5.5. Acquisition of surface charge by colloidal MnO_2 in water. Anhydrous MnO_2 (I) has two O atoms per Mn atom. Suspended in water as a colloid, it binds to water molecules to form hydrated MnO_2 (II). Loss of H^+ from the bound H_2O yields a negatively charged colloidal particle (III). Gain of H^+ by surface O atoms yields a positively charged particle (IV). The former process (loss of H^+ ion) predominates for metal oxides.

As an illustration of pH-dependent charge on colloidal particle surfaces, consider the effects of pH on the surface charge of hydrated manganese oxide, represented by the chemical formula $MnO_2(H_2O)(s)$. In a relatively acidic medium, the reaction

$$MnO_2(H_2O)(s) + H^+ \rightarrow MnO_2(H_3O)^+(s) \qquad (5.4.1)$$

may occur on the surface giving the particle a net positive charge. In a more basic medium, hydrogen ion may be lost from the hydrated oxide surface to yield negatively charged particles:

$$MnO_2(H_2O)(s) \rightarrow MnO_2(OH)^-(s) + H^+ \qquad (5.4.2)$$

At some intermediate pH value, called the **zero point of charge (ZPC)**, colloidal particles of a given hydroxide will have a net charge of zero, which favors aggregation of particles and precipitation of a bulk solid:

$$\text{Number of } MnO_2(H_3O)^+ \text{ sites} = \text{Number of } MnO_2(OH)^- \text{ sites} \qquad (5.4.3)$$

Individual cells of microorganisms that behave as colloidal particles have a charge that is pH-dependent. The charge is acquired through the loss and gain of H^+ ion by carboxyl and amino groups on the cell surface:

$$^+H_3N(+ \text{ cell})CO_2H \quad ^+H_3N(\text{neutral cell})CO_2^- \quad H_2N(- \text{ cell})CO_2^-$$

$$\text{low pH} \qquad\qquad \text{intermediate pH} \qquad\qquad \text{high pH}$$

Ion absorption is a second way in which colloidal particles become charged. This phenomenon involves attachment of ions onto the colloidal particle surface by means other than conventional covalent bonding, including hydrogen bonding and London (Van der Waal) interactions.

Ion replacement is a third way in which a colloidal particle may gain a net charge; for example, replacement of some of the Si(IV) with Al(III) in the basic SiO_2 chemical unit in the crystalline lattice of some clay minerals as shown in Equation 5.4.4,

$$[SiO_2] + Al(III) \rightarrow [AlO_2^-] + Si(IV) \qquad (5.4.4)$$

yields sites with a net negative charge. Similarly, replacement of Al(III) by a divalent metal ion such as Mg(II) in the clay crystalline lattice produces a net negative charge.

5.5. THE COLLOIDAL PROPERTIES OF CLAYS

Clays constitute the most important class of common minerals occurring as colloidal matter in water. The composition and properties of clays are discussed in some detail in Section 15.7 (as solid terrestrial minerals) and are briefly summarized here. **Clays** consist largely of hydrated aluminum and silicon oxides and are **secondary minerals**, which are formed by weathering and other processes acting on primary rocks (see Sections 15.2 and 15.8). The general formulas of some common clays are given below:

- Kaolinite: $Al_2(OH)_4Si_2O_5$
- Montmorillonite: $Al_2(OH)_2Si_4O_{10}$
- Nontronite: $Fe_2(OH)_2Si_4O_{10}$
- Hydrous mica: $KAl_2(OH)_2(AlSi_3)O_{10}$

Iron and manganese are commonly associated with clay minerals. The most common clay minerals are illites, montmorillonites, chlorites, and kaolinites. These clay minerals are distinguished from each other by general chemical formula, structure, and chemical and physical properties.

Clays are characterized by layered structures consisting of sheets of silicon oxide alternating with sheets of aluminum oxide. Units of two or three sheets make up **unit layers**. Some clays, particularly the montmorillonites, may absorb large quantities of water between unit layers, a process accompanied by swelling of the clay.

As described in Section 5.4, clay minerals may attain a net negative charge by ion replacement, in which Si(IV) and Al(III) ions are replaced by metal ions of similar size but lesser charge. This negative charge must be compensated by association of cations with the clay layer surfaces. Since these cations need not fit specific sites in the crystalline lattice of the clay, they may be relatively large ions, such as K^+, Na^+, or NH_4^+. These cations are called **exchangeable cations** and are exchangeable for other cations in water. The amount of exchangeable cations, expressed as milliequivalents (of monovalent cations) per 100 g of dry clay, is called the **cation-exchange capacity, CEC**, of the clay and is a very important characteristic of colloids and sediments that have cation-exchange capabilities.

Because of their structure and high surface area per unit weight, clays have a strong tendency to sorb chemical species from water. Thus, clays play a role in the transport and reactions of biological wastes, organic chemicals, gases, and other pollutant species in water. However, clay minerals also may effectively immobilize dissolved chemicals in water and so exert a purifying action. Some microbial processes occur at clay particle surfaces and, in some cases, sorption of organics by clay inhibits biodegradation. Thus, clay may play a role in the microbial degradation or nondegradation of organic wastes.

5.6. AGGREGATION OF PARTICLES

The processes by which particles aggregate and precipitate from colloidal suspension are quite important in the aquatic environment. For example, the settling of biomass during biological waste treatment depends upon the aggregation of bacterial cells. Other processes involving the aggregation of colloidal particles are the formation of bottom sediments and the clarification of turbid water for domestic or industrial use. Particle aggregation is complicated and may be divided into the two general classes of *coagulation* and *flocculation*. These are discussed below.

Colloidal particles are prevented from aggregating by the electrostatic repulsion of the electrical double layers (adsorbed-ion layer and counter-ion layer). **Coagulation** involves the reduction of this electrostatic repulsion such that colloidal particles of identical materials may aggregate. **Flocculation** uses **bridging compounds,** which form chemically bonded links between colloidal particles and enmesh the particles in relatively large masses called **floc networks.**

Hydrophobic colloids are often readily coagulated by the addition of small quantities of salts that contribute ions to solution. Such colloids are stabilized by electrostatic repulsion. Therefore, the simple explanation of coagulation by ions in solution is that the ions reduce the electrostatic repulsion between particles to such an extent that the particles aggregate. Because of the double layer of electrical charge surrounding a charged particle, this aggregation mechanism is sometimes called **double-layer compression.** It is particularly noticeable in estuaries where sediment-laden fresh water flows into the sea, and is largely responsible for deltas formed where large rivers enter oceans.

The binding of positive ions to the surface of an initially negatively charged colloid can result in precipitation followed by colloid restabilization as shown in Figure 5.6. This kind of behavior is explained by an initial neutralization of the negative surface charge on the particles by sorption of positive ions, allowing coagulation to occur. As more of the source of positive ions is added, their sorption results in the formation of positive colloidal particles.

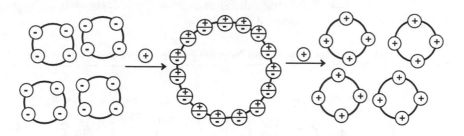

Figure 5.6. Aggregation of negatively charged colloidal particles by reaction with positive ions, followed by restabilization as a positively charged colloid.

Flocculation of Colloids by Polyelectrolytes

Polyelectrolytes of both natural and synthetic origin may cause colloids to flocculate. Polyelectrolytes are polymers with a high formula weight that normally contain ionizable functional groups. Typical examples of synthetic polyelectrolytes are shown in Table 5.2.

It can be seen from Table 5.2 that anionic polyelectrolytes have negatively charged functional groups, such as $-SO_3^-$ and $-CO_2^-$. Cationic polyelectrolytes have positively charged functional groups, normally H^+ bonded to N. Nonionic polymers that serve as flocculants normally do not have charged functional groups.

Somewhat paradoxically, *anionic* polyelectrolytes may flocculate *negatively charged* colloidal particles. The mechanism by which this occurs involves bridging between the colloidal particles by way of the polyelectrolyte anions. Strong chemical bonding has to be involved, since both the particles and the polyelectrolytes are negatively charged. However, the process does occur and is particularly important in biological systems, for example, in the cohesion of tissue cells, clumping of bacterial cells, and antibody-antigen reactions.

The flocculation process induced by anionic polyelectrolytes is greatly facilitated by the presence of a low concentration of a metal ion capable of binding with the functional groups on the polyelectrolyte. The positively charged metal ion serves to form a bridge between the negatively charged anionic polyelectrolytes and negatively charged functional groups on the colloidal particle surface.

Flocculation of Bacteria by Polymeric Materials

The aggregation and settling of microorganism cells is a very important process in aquatic systems and is essential to the function of biological waste treatment systems. In biological waste treatment processes, such as the activated sludge process (Chapter 8), microorganisms utilize carbonaceous solutes in the water to produce biomass. The primary objective of biological waste treatment is the removal of carbonaceous material and, consequently, its oxygen demand. Part of the carbon is evolved from the water as CO_2, produced by the energy-yielding metabolic processes of the bacteria. However, a significant fraction of the carbon is removed as **bacterial floc**, consisting of aggregated bacterial cells that have settled from the water. The formation of this floc is obviously an important phenomenon in biological waste treatment. Polymeric substances, including polyelectrolytes, that are formed by the bacteria induce bacterial flocculation.

Table 5.2. Synthetic Polyelectrolytes and Neutral Polymers Used as Flocculants

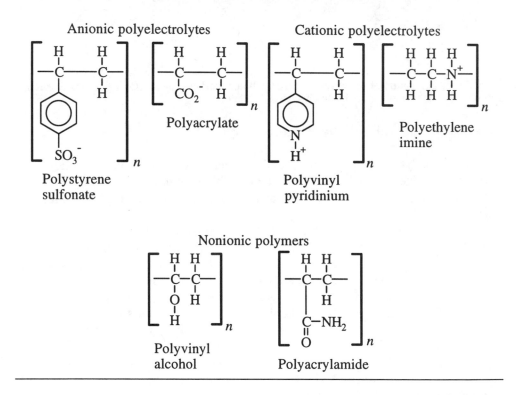

Within the pH range of normal natural waters (pH 5-9), bacterial cells are negatively charged. The ZPC of most bacteria is within the pH range 2-3. However, even at the ZPC, stable bacterial suspensions may exist. Therefore, surface charge is not necessarily required to maintain bacterial cells in suspension in water, and it is likely that bacterial cells remain in suspension because of the hydrophilic character of their surfaces. As a consequence, some sort of chemical interaction involving bridging species must be involved in bacterial flocculation.

5.7. SURFACE SORPTION BY SOLIDS

Many of the properties and effects of solids in contact with water have to do with the sorption of solutes by solid surfaces. Surfaces in finely divided solids tend to have excess surface energy because of an imbalance of chemical forces among surface atoms, ions, and molecules. Surface energy level may be lowered by a reduction in surface area. Normally this reduction is accomplished by aggregation of particles or by sorption of solute species.

Some kinds of surface interactions can be illustrated with metal oxide surfaces binding with metal ions in water. (Such a surface, its reaction with water, and its subsequent acquisition of a charge by loss or gain of H^+ ion were shown in Figure 5.5 for MnO_2.) Other inorganic solids, such as clays, probably behave much like solid metal oxides. Soluble metal ions, such as Cd^{2+}, Cu^{2+}, Pb^{2+}, or Zn^{2+}, may be bound with metal oxides such as $MnO_2 \cdot xH_2O$ by nonspecific ion exchange adsorption, complexation with surface –OH groups, coprecipitation in solid solution with the metal oxide, or as a discrete oxide or hydroxide of the sorbed metal.[4] Sorption of metal ions, Mt^{z+}, by complexation to the surface is illustrated by the reaction

$$M–OH + Mt^{z+} \longleftrightarrow M\text{-}OMt^{z-1} + H^+ \qquad (5.7.1)$$

and chelation by the following process:

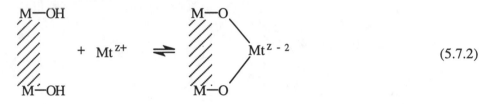

$$\qquad (5.7.2)$$

A metal ion complexed with a ligand, L, may bond by displacement of either H^+ or OH^-:

$$M–OH + MtL^{z+} \longleftrightarrow M\text{-}OMtL^{(z-1)} + H^+ \qquad (5.7.3)$$

$$M–OH + MtL^{z+} \longleftrightarrow M\text{-}(MtL)^{(z+1)} + OH^- \qquad (5.7.4)$$

Furthermore, in the presence of a ligand, dissociation of the complex and sorption of the metal complex and ligand must be considered as shown by the scheme below in which "(sorbed)" represents sorbed species and "(aq)" represents dissolved species:

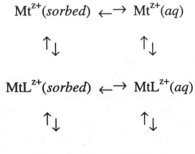

Some hydrated metal oxides, such as manganese(IV) oxide and iron(III) oxide, are especially effective in sorbing various species from aquatic solution. The sorption ability is especially pronounced for relatively fresh metal hydroxides or hydrated oxides such as colloidal MnO_2. This oxide usually is produced in natural waters by the oxidation of Mn(II) present in natural waters placed there by the bacterially-mediated reduction of manganese oxides in anaerobic bottom sediments. Colloidal hydrated manganese(II) oxide can also be produced by the reduction of manganese(VII), which often is deliberately added to water as an oxidant in the form of permanganate salts to diminish taste and odor or to oxidize iron(II).

Freshly precipitated MnO_2 may have a surface area as large as several hundred square meters per gram. The hydrated oxide acquires a charge by loss and gain of H^+ ion and has a ZPC in an acidic pH range between 2.8 and 4.5. Since the pH of most normal natural waters exceeds 4.5, hydrous MnO_2 colloids are usually negatively charged.

The sorption of anions by solid surfaces is harder to explain than the sorption of cations. Phosphates may be sorbed on hydroxylated surfaces by displacement of hydroxides (ion exchange):

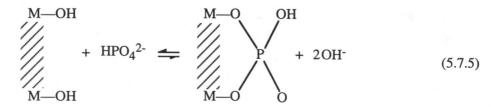

$$(5.7.5)$$

The degree of anion sorption varies. As with phosphate, sulfate may be sorbed by chemical bonding, usually at a pH less than 7. Chloride and nitrate are sorbed by electrostatic attraction, such as occurs with positively charged colloidal particles in soil at a low pH. More specific bonding mechanisms may be involved in the sorption of fluoride, molybdate, selenate, selenite, arsenate, and arsenite anions.

5.8. ION EXCHANGE WITH BOTTOM SEDIMENTS

Bottom sediments are important sources of inorganic and organic matter in streams, fresh-water impoundments, estuaries, and oceans. It is incorrect to consider bottom sediments simply as wet soil. Normal soils are in contact with the atmosphere and are aerobic, whereas the environment around bottom sediments is usually anaerobic, so sediments are subjected to reducing conditions. Bottom sediments undergo continuous leaching, whereas soils do not. The level of organic matter in sediments is generally higher than that in soils.

One of the most important characteristics of bottom sediments is their ability to exchange cations with the surrounding aquatic medium. **Cation-exchange capacity** (CEC) measures the capacity of a solid, such as a sediment, to sorb cations. It varies with pH and with salt concentration. Another parameter, **exchangeable cation status** (ECS), refers to the amounts of specific ions bonded to a given amount of sediment. Generally, both CEC and ECS are expressed as milliequivalents per 100 g of solid.

Because of the generally anaerobic nature of bottom sediments, special care must be exercised in their collection and treatment. Particularly, contact with atmospheric oxygen rapidly oxidizes exchangeable Fe^{2+} and Mn^{2+} to nonexchangeable oxides containing the metals in higher oxidation states as Fe_2O_3 and MnO_2. Therefore, sediment samples must be sealed and frozen as soon as possible after they are collected.

A common method for the determination of CEC consists of: (1) treating the sediment with a solution of an ammonium salt so that all exchangeable sites are occupied by NH_4^+ ion; (2) displacing the ammonium ion with a solution of NaCl; and (3) determining the quantity of displaced ammonium ion. The CEC values may then be expressed as the number of milliequivalents of ammonium ion exchanged per 100 g of dried sample. Note that the sample must be dried *after* exchange.

The basic method for the determination of ECS consists of stripping all of the exchangeable metal cations from the sediment sample with ammonium acetate. Metal cations, including Fe^{2+}, Mn^{2+}, Zn^{2+}, Cu^{2+}, Ni^{2+}, Na^+, K^+, Ca^{2+}, and Mg^{2+}, are then determined in the leachate. Exchangeable hydrogen ion is very difficult to determine by direct methods. It is generally assumed that the total cation exchange capacity minus the sum of all exchangeable cations except hydrogen ion is equal to the exchangeable hydrogen ion.

Freshwater sediments typically have CEC values of 20-30 milliequivalents/100 g. The ECS values for individual cations typically range from less than 1 to 10-20 milliequivalents/100 g. Sediments are important repositories of metal ions that may be exchanged with surrounding waters. Furthermore, because of their capacity to sorb and release hydrogen ions, sediments have an important buffering effect in some waters.

Trace-Level Metals in Suspended Matter and Sediments

Sediments and suspended particles are important repositories for trace amounts of metals such as chromium, cadmium, copper, molybdenum, nickel, cobalt, and manganese. These metals may be present as discrete compounds, ions held by cation-exchanging clays, bound to hydrated oxides of iron or manganese, or chelated by insoluble humic substances. The form of the metals depends upon pE. Examples of specific trace-metal-containing compounds that may be stable in natural waters under oxidizing and reducing conditions are given in Table 5.3. Solubilization of metals from sedimentary or suspended matter is often a function of the complexing agents present. These include amino acids, such as histidine, tyrosine, or cysteine; citrate ion; and, in the presence of seawater, chloride ion. Suspended particles containing trace elements may be in the submicrometer size range. Although less available than metals in true solution, metals held by very small particles are more accessible than those in sediments. Among the factors involved in metal availability are the identity of the metal, its chemical form (type of binding, oxidation state), the nature of the suspended material, the type of organism taking up the metal, and the physical and chemical conditions in the water. The pattern of trace-metal occurrence in suspended matter in relatively unpolluted water tends to correlate well with that of the parent minerals from which the suspended solids originated; anomalies appear in polluted waters where industrial sources add to the metal content of the stream.

Table 5.3. Inorganic Trace Metal Compounds That May be Stable under Oxidizing and Reducing Conditions.

Metal	Discrete compound that may be present	
	Oxidizing conditions	Reducing conditions
Cadmium	$CdCO_3$	CdS
Copper	$Cu_2(OH)_2CO_3$	CuS
Iron	$Fe_2O_3 \cdot x(H_2O)$	FeS, FeS_2
Mercury	HgO	HgS
Manganese	$MnO_2 \cdot x(H_2O)$	MnS, $MnCO_3$
Nickel	$Ni(OH)_2$, $NiCO_3$	NiS
Lead	$2PbCO_3 \cdot Pb(OH)_2$, $PbCO_3$	PbS
Zinc	$ZnCO_3$, $ZnSiO_3$	ZnS

The toxicities of heavy metals in sediments and their availability to organisms are very important in determining the environmental effects of heavy metals in aquatic systems. Many sediments are anaerobic, so that microbial reduction of sulfate to sulfide leads to a preponderance of metal sulfides in sediments. The very low solubilities of sulfides tend to limit bioavailability of metals in anaerobic sediments. However, exposure of such sediments to air, and subsequent oxidation of sulfide to sulfate, can release significant amounts of heavy metals. Dredging operations can expose anaerobic sediments to air, leading to oxidation of sulfides and release of metals such as lead, mercury, cadmium, zinc, and copper.[5]

Phosphorus Exchange with Bottom Sediments

Phosphorus is one of the key elements in aquatic chemistry and is thought to be the limiting nutrient in the growth of algae under many conditions. Exchange with sediments plays a role in making phosphorus available for algae and contributes, therefore, to eutrophication. Sedimentary phosphorus may be classified into the following types:

- **Phosphate minerals**, particularly hydroxyapatite, $Ca_5OH(PO_4)_3$

- **Nonoccluded phosphorus**, such as orthophosphate ion bound to the surface of SiO_2 or $CaCO_3$. Such phosphorus is generally more soluble and more available than occluded phosphorus (below).

- **Occluded phosphorus** consisting of orthophosphate ions contained within the matrix structures of amorphous hydrated oxides of iron and aluminum and amorphous aluminosilicates. Such phosphorus is not as readily available as nonoccluded phosphorus.

- **Organic phosphorus** incorporated within aquatic biomass, usually of algal or bacterial origin.

In some waters receiving heavy loads of domestic or industrial wastes, inorganic polyphosphates (from detergents, for example) may be present in sediments. Runoff from fields where liquid polyphosphate fertilizers have been used might possibly provide polyphosphates sorbed on sediments.

Organic Compounds on Sediments and Suspended Matter

Many organic compounds interact with suspended material and sediments in bodies of water.[6] Colloids can play a significant role in the transport of organic pollutants in surface waters, through treatment processes, and even to a limited extent in groundwater. Settling of suspended material containing sorbed organic matter carries organic compounds into the sediment of a stream or lake. For example, this phenomenon is largely responsible for the presence of herbicides in sediments containing contaminated soil particles eroded from crop land. Some organics are carried into sediments by the remains of organisms or by fecal pellets from zooplankton that have accumulated organic contaminants.

Suspended particulate matter affects the mobility of organic compounds sorbed to particles. Furthermore, sorbed organic matter undergoes chemical degradation and biodegradation at different rates and by different pathways compared to organic matter in solution. There is, of course, a vast variety of organic compounds that get into water. As one would expect, they react with sediments in different ways, the type and strength of binding varying with the type of compound. An indication of the variable nature of the binding of organic compounds to sediments is provided by evidence that release of the compounds from sediments to water often occurs in two stages, the first rapid and the second slow.[7]

The most common types of sediments considered for their organic binding abilities are clays, organic (humic) substances, and complexes between clay and humic substances. Both clays and humic substances act as cation exchangers. Therefore, these materials sorb cationic organic compounds through ion exchange. This is a relatively strong sorption mechanism, greatly reducing the mobility and biological activity of the organic compound. When sorbed by clays, cationic organic compounds are generally held between the layers of the clay mineral structure where their biological activity is essentially zero.

Since most sediments lack strong anion exchange sites, negatively charged organics are not held strongly at all. Thus, these compounds are relatively mobile and biodegradable in water despite the presence of solids.

The degree of sorption of organic compounds is generally inversely proportional to their water solubility. The more water-insoluble compounds tend to be taken up strongly by lipophilic ("fat-loving") solid materials, such as humic substances (see Section 3.17). Compounds having a relatively high vapor pressure can be lost from water or solids by evaporation. When this happens, photochemical processes (see Chapter 9) can play an important role in their degradation.

The herbicide 2,4-D (2,4-dichlorophenoxyacetic acid) has been studied extensively in regard to sorption reactions. Most of these studies have dealt with pure clay minerals, however, whereas soils and sediments are likely to have a strong clay-fulvic acid complex component. The sorption of 2,4-D by such a complex can be described using an equation of the Freundlich isotherm type,

$$X = K C^n \qquad (5.8.1)$$

where X is the amount sorbed per unit weight of solid, C is the concentration of 2,4-D in water solution at equilibrium, and n and K are constants. These values are determined by plotting log X versus log C. If a Freundlich-type equation is obeyed, the plot will be linear with a slope of n and an intercept of log K. In a study of the sorption of 2,4-D on an organoclay complex,[8] n was found to be 0.76 and log K was 0.815 at 5°C; at 25°C, n was 0.83 and log K was 0.716.

Sorption of comparatively nonvolatile hydrocarbons by sediments removes these materials from contact with aquatic organisms but also greatly retards their bio-degradation. Aquatic plants produce some of the hydrocarbons that are found in sed-iments. Photosynthetic organisms, for example, produce quantities of n-heptadecane. Pollutant hydrocarbons in sediments are indicated by a smooth, chain-length distri-bution of n-alkanes and thus can be distinguished from hydrocarbons generated pho-tosynthetically in the water. An analysis of sediments in Lake Zug, Switzerland, for example, has shown a predominance of pollutant petroleum hydrocarbons near densely populated areas.

The sorption of neutral species like petroleum obviously cannot be explained by ion-exchange processes. It probably involves phenomena such as Van der Waals forces (a term sometimes invoked when the true nature of an attractive force is not understood, but generally regarded as consisting of induced dipole-dipole interaction involving a neutral molecule), hydrogen bonding, charge-transfer complexation, and hydrophobic interactions.

In some cases pollutant compounds become covalently bound as "bound residues" to humic substances in soil.[9] It is plausible that the same occurs in sediments formed from soil washed into a body of water or directly onto sediments. It is very difficult to remove such residues from humic substances thermally, biochemically, or by exposure to acid or base (hydrolysis). The binding is thought to occur through the action of enzymes from some organisms. These enzymes are extracellular enzymes (those acting outside the cell) that act as oxidoreductases, which catalyze oxidation-reduction reactions. Such enzymes are capable of causing polymerization of aromatic compounds, as illustrated below for the coupling of pollutant 2,4-dichlorophenol to an aryl ring on a humic substance molecule:

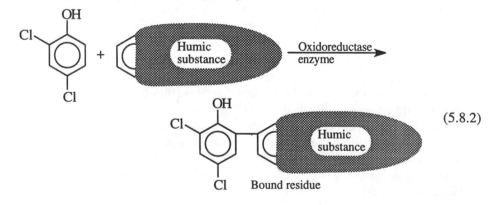

$$(5.8.2)$$

Obviously, uptake of organic matter by suspended and sedimentary material in water is an important phenomenon. Were it not for this phenomenon, it is likely that pesticides in water would be much more toxic. Biodegradation is generally slowed down appreciably, however, by sorption of a substance to a solid. In certain intensively farmed areas, there is a very high accumulation of pesticides in the sediments of streams, lakes, and reservoirs. The sorption of pesticides by solids and the resulting influence on their biodegradation is an important consideration in the licensing of new pesticides.

The transfer of surface water to groundwater often results in sorption of some water contaminants by soil and mineral material. To take advantage of this purification effect, some municipal water supplies are drawn from beneath the surface of natural or artificial river banks as a first step in water treatment. The movement of water from waste landfills to aquifers is also an important process (see Chapter 19) in which pollutants in the landfill leachate may be sorbed by solid material through which the water passes.

The sorption of dilute solutions of halogenated and aryl hydrocarbons by soil and sand has been studied under simulated water infiltration conditions.[10] The relationship between the sorption equilibria observed may be expressed by the formula

$$S = K_p C \tag{5.8.2}$$

where S and C are the concentrations of hydrocarbons in the solid and liquid phases, respectively, and K_p is the partition coefficient. It was found that the two most important factors in estimating the sorption of nonpolar organic compounds were: (1) the fraction of organic carbon, f_{oc}, in the solid sorbents; and (2) the 1-octanol/water partition coefficient, K_{ow}, of the organic compound. (The K_{ow} value is a measure of the tendency of a solute to dissolve from water into immiscible 1-octanol. This long-chain alcohol mimics lipid (fat) tissue, and K_{ow} is used to indicate a tendency toward bioaccumulation of solutes in water.) The K_p of individual compounds was determined using the following empirical relationship:

$$\text{Log } K_p = 0.72 \log K_{ow} + \log f_{oc} + 0.49 \tag{5.8.3}$$

The sorption was found to be reversible on the solids studied, which included natural aquifer material, river sediment, soil, sand, and sewage sludge. The organic compounds studied included methylbenzene compounds containing from 1 to 4 chlorine atoms, tetrachloroethylene, n-butylbenzene, benzene, acetophenone, tetrachloroethane, naphthalene, parathion, β-BHC, DDT (the latter three compounds are insecticides), pyrene, and tetracene.

5.9. SORPTION OF GASES—GASES IN INTERSTITIAL WATER

Interstitial water or *pore water* consisting of water held by sediments is an important reservoir for gases in natural water systems. Generally, the gas concentrations in interstitial waters are different from those in the overlying water. The results of the analyses of gases in interstitial water in some sediments taken from Chesapeake Bay[11] are given in Table 5.4. Examination of this table shows that CH_4 could not be detected at the sediment surface, which is because the equilibrium concentration of methane in air is very low, and it is biodegradable under aerobic conditions. However, of the gases analyzed, by far the highest concentration at a depth of one meter was that of methane. The methane is produced by the anaerobic fermentation of biodegradable organic matter, $\{CH_2O\}$, (see Section 6.6):

$$2\{CH_2O\} \rightarrow CH_4(g) + CO_2(g) \tag{5.9.1}$$

The concentrations of argon and nitrogen are much lower at a depth of one meter than they are at the sediment surface. This finding may be explained by the stripping action of the fermentation-produced methane rising to the sediment surface.

Table 5.4. Gases in Interstitial Waters from Chesapeake Bay Sediments.

Gas	Depth	Gas concentration, mL/L
N_2	surface	13.5
N_2	1.00 m	2.4
Ar	surface	0.35
Ar	1.00 m	0.12
CH_4	surface	0.00
CH_4	1.00 m	140

LITERATURE CITED

1. Manahan, Stanley E., *Quantitative Chemical Analysis*, Brooks/Cole Publishing Co., Pacific Grove, CA, 1986.

2. Ranville, James F. and Schmiermund, Ronald L., "An Overview of Environmental Colloids," *Perspectives in Environmental Chemistry*, 25-56 (1998).

3. Lead, J. R., W. Davison, J. Hamilton-Taylor, and J. Buffle, "Characterizing Colloidal Material in Natural Waters, *Aquatic Geochemistry*, **3**, 213-232 (1997).

4. Martinez, Carmen Enid and Murray B. McBride, "Solubility of Cd^{2+}, Cu^{2+}, Pb^{2+}, or Z n^{2+} i n Aged Coprecipitates w i t h Amorphous Iron Hydroxides," *Environmental Science and Technology*, **32**, 743-748 (1998).

5. Christensen, E. R., "Metals, Acid-volatile Sulfides, Organics, and Particle Distributions of Contaminated Sediments," *Water Science and Technology*, **37**, 149-156 (1998).

6. Lick, Wilbert, Zenitha Chroneer, and Venkatrao Rapaka, "Modeling the Dynamics of the Sorption of Hydrophobic Organic Chemicals to Suspended Sediments," *Water, Air, and Soil Pollutution*, **99**, 225-235 (1997).

7. Cornelissen, Gerard, Paul C. M. van Noort, and Harrie A. J. Govers, "Mechanism of Slow Desorption of Organic Compounds from Sediments: A Study Using Model Sorbents," *Environmental Science and Technology*, **32**, 3124-3131 (1998).

8. Means, J. C., and R. Wijayratne, "Role of Natural Colloids in the Transport of Hydrophobic Pollutants," *Science*, **215**, 968–970 (1982).

9. Bollag, Jean-Marc, "Decontaminating Soil with Enzymes," *Environmental Science and Technology*, **26**, 1876-81 (1992).

10. Schwarzenbach, R. P., and J. Westall, "Transport of Nonpolar Organic Compounds from Surface Water to Groundwater. Laboratory Sorption Studies," *Environmental Science and Technology* **15**, 1360-7 (1982).

11. Reeburgh, W. S., "Determination of Gases in Sediments," *Environmental Science and Technology*, **2**, 140-1 (1968).

SUPPLEMENTARY REFERENCES

Allen, Herbert E.; Ed., *Metal Contaminated Aquatic Sediments*, Ann Arbor Press, Chelsea, MI, 1995.

Beckett, Ronald, Ed., *Surface and Colloid Chemistry in Natural Waters and Water Treatment*, Plenum, New York, 1990.

Chamley, Hervé, *Sedimentology*, Springer-Verlag, New York, 1990.

Evans, R. Douglas, Joe Wisniewski, and Jan R. Wisniewski, Eds., *The Interactions between Sediments and Water*, Kluwer, Dordrecht, Netherlands, 1997.

Golterman, H. L., Ed., *Sediment-Water Interaction 6*, Kluwer, Dordrecht, The Netherlands, 1996.

Gustafsson, Orjan and Philip M. Gschwend, "Aquatic Colloids: Concepts, Definitions, and Current Challenges., *Limnology and Oceanography*, **42**, 519-528 (1997).

Hart, Barry T., Ronald Beckett, and Deirdre Murphy, "Role of Colloids in Cycling Contaminants in Rivers," *Science Reviews*, Northwood, U.K., 1997.

Hunter, Robert J., *Foundations of Colloid Science*, Clarendon (Oxford University Press), New York, 1992.

Jones, Malcolm N. and Nicholas D. Bryan, "Colloidal Properties of Humic Substances, *Advances in Colloid and Interfacial Science*, **78**, 1-48 (1998).

Mudroch, Alena, Jose M. Mudroch, and Paul Mudroch, Eds., *Manual of Physico-Chemical Analysis of Aquatic Sediments*, CRC Press, Inc., Boca Raton, FL, 1997.

Nowell, Lisa H., Peter D. Dileanis, and Paul D. Capel, *Pesticides in Bed Sediments and Aquatic Biota in Streams*, Ann Arbor Press, Chelsea, MI, 1996.

Stumm, Werner, Laura Sigg, and Barbara Sulzberger, *Chemistry of the Solid-Water Interface*, John Wiley and Sons, Inc., New York, 1992.

Stumm, Werner and James J. Morgan, *Aquatic Chemistry: Chemical Equilibria and Rates in Natural Waters*, 3rd ed., John Wiley and Sons, Inc., New York, 1995.

QUESTIONS AND PROBLEMS

1. A sediment sample was taken from a lignite strip-mine pit containing highly alkaline (pH 10) water. Cations were displaced from the sediment by treatment with HCl. A total analysis of cations in the leachate yielded, on the basis of millimoles per 100 g of dry sediment, 150 mmol of Na^+, 5 mmol of K^+, 20 mmol of Mg^{2+}, and 75 mmol of Ca^{2+}. What is the cation exchange capacity of the sediment in milliequivalents per 100 g of dry sediment?

2. What is the value of $[O_2(aq)]$ for water saturated with a mixture of 50% O_2, 50% N_2 by volume at 25°C and a total pressure of 1.00 atm?

3. Of the following, the least likely mode of transport of iron(III) in a normal stream is: (a) bound to suspended humic material, (b) bound to clay particles by cation exchange processes, (c) as suspended Fe_2O_3, (d) as soluble Fe^{3+} ion, (e) bound to colloidal clay-humic substance complexes.

4. How does freshly precipitated colloidal iron(III) hydroxide interact with many divalent metal ions in solution?

5. What stabilizes colloids made of bacterial cells in water?

6. The solubility of oxygen in water is 14.74 mg/L at 0°C and 7.03 mg/L at 35°C. Estimate the solubility at 50°C.

7. What is thought to be the mechanism by which bacterial cells aggregate?

8. What is a good method for the production of freshly precipitated MnO_2?

9. A sediment sample was equilibrated with a solution of NH_4^+ ion, and the NH_4^+ was later displaced by Na^+ for analysis. A total of 33.8 milliequivalents of NH_4^+ were bound to the sediment and later displaced by Na^+. After drying, the sediment weighed 87.2 g. What was its CEC in milliequivalents/100 g?

10. A sediment sample with a CEC of 67.4 milliequivalents/100 g was found to contain the following exchangeable cations in milliequivalents/100 g: Ca^{2+}, 21.3; Mg^{2+}, 5.2; Na^+, 4.4; K^+, 0.7. The quantity of hydrogen ion, H^+, was not measured directly. What was the ECS of H^+ in milliequivalents/100 g?

11. What is the meaning of *zero point of charge* as applied to colloids? Is the surface of a colloidal particle totally without charged groups at the ZPC?

12. The concentration of methane in an interstitial water sample was found to be 150 mL/L at STP. Assuming that the methane was produced by the fermentation of organic matter, {CH_2O}, what weight of organic matter was required to produce the methane in a liter of the interstitial water?

13. What is the difference between CEC and ECS?

14. Match the sedimentary mineral on the left with its conditions of formation on the right:

    (a) $FeS(s)$ (1) May be formed when anaerobic water is exposed to O_2.
    (b) $Ca_5OH(PO_4)_3$ (2) May be formed when aerobic water becomes anaerobic.

(c) $Fe(OH)_3$
(d) $CaCO_3$

(3) Photosynthesis by-product.

(4) May be formed when wastewater containing a particular kind of contaminant flows into a body of very hard water.

15. In terms of their potential for reactions with species in solution, how might metal atoms, M, on the surface of a metal oxide, MO, be described?

16. Air is 20.95% oxygen by volume. If air at 1.0000 atm pressure is bubbled through water at 25°C, what is the partial pressure of O_2 in the water?

17. The volume percentage of CO_2 in a mixture of that gas with N_2 was determined by bubbling the mixture at 1.00 atm and 25°C through a solution of 0.0100 M $NaHCO_3$ and measuring the pH. If the equilibrium pH was 6.50, what was the volume percentage of CO_2?

18. For what purpose is a polymer with the following general formula used?

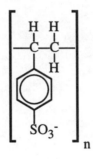

6 AQUATIC MICROBIAL BIOCHEMISTRY

6.1. AQUATIC BIOCHEMICAL PROCESSES

Microorganisms—**bacteria, fungi, protozoa**, and **algae**—are living catalysts that enable a vast number of chemical processes to occur in water and soil. A majority of the important chemical reactions that take place in water, particularly those involving organic matter and oxidation-reduction processes, occur through bacterial intermediaries. Algae are the primary producers of biological organic matter (biomass) in water. Microorganisms are responsible for the formation of many sediment and mineral deposits; they also play the dominant role in secondary waste treatment. Some of the effects of microorganisms on the chemistry of water in nature are illustrated in Figure 6.1.

Pathogenic microorganisms must be eliminated from water purified for domestic use. In the past, major epidemics of typhoid, cholera, and other water-borne diseases resulted from pathogenic microorganisms in water supplies. Even today, constant vigilance is required to ensure that water for domestic use is free of pathogens.

Although they are not involved in aquatic chemical transformations, which constitute most of this chapter, special mention should be made of viruses in water. Viruses cannot grow by themselves, but reproduce in the cells of host organisms. They are only about 1/30-1/20 the size of bacterial cells, and they cause a number of diseases, such as polio, viral hepatitis, and perhaps cancer. It is thought that many of these diseases are waterborne.

Because of their small size (0.025-0.100 μm) and biological characteristics, viruses are difficult to isolate and culture. They often survive municipal water treatment, including chlorination. Thus, although viruses have no effect upon the overall environmental chemistry of water, they are an important consideration in the treatment and use of water.

Microorganisms are divided into the two broad categories of **prokaryotes** and **eukaryotes**; the latter have well-defined cell nuclei enclosed by a nuclear membrane, whereas the former lack a nuclear membrane and the nuclear genetic material is more diffuse in the cell. Other differences between these two classes of organisms include location of cell respiration, means of photosynthesis, means of motility, and reproductive processes. All classes of microorganisms produce **spores**, metabolically inactive bodies that form and survive under adverse conditions in a "resting" state until conditions favorable for growth occur.

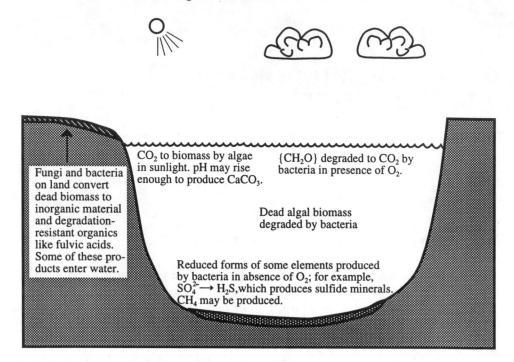

Figure 6.1. Effects of microorganisms on the chemistry of water in nature.

Fungi, protozoa, and bacteria (with the exception of photosynthetic bacteria and protozoa) are classified as **reducers**, which break down chemical compounds to more simple species and thereby extract the energy needed for their growth and metabolism. Algae are classified as **producers** because they utilize light energy and store it as chemical energy. In the absence of sunlight, however, algae utilize chemical energy for their metabolic needs. In a sense, therefore, bacteria, protozoa, and fungi may be looked upon as environmental catalysts, whereas algae function as aquatic solar fuel cells.

All microorganisms can be put into one of the four following classifications based on the sources of energy and carbon that they utilize: chemoheterotrophs, chemoautotrophs, photoheterotrophs, and photoautotrophs. These classifications are based upon (1) the energy source and (2) the carbon source utilized by the organism. **Chemotrophs** use chemical energy derived from oxidation-reduction reactions of simple inorganic chemical species for their energy needs. **Phototrophs** utilize light energy from photosynthesis. **Heterotrophs** obtain their carbon from other organisms; **autotrophs** use carbon dioxide and ionic carbonates for the C that they require. Figure 6.2 summarizes the classifications into which microorganisms may be placed with these definitions.

Energy source → / Carbon sources ↓	Chemical	Photochemical (light)
Organic matter	**Chemoheterotrophs** All fungi and protozoans, most bacteria. Chemoheterotrophs use organic sources for both energy and carbon.	**Photoheterotrophs** A few specialized bacteria that use photoenergy, but are dependent on organic matter for a carbon source
Inorganic carbon (CO_2, HCO_3^-)	**Chemoautotrophs** Use CO_2 for biomass and oxidize substances such as H_2 (*Pseudomonas*), NH_4^+ (*Nitrosomonas*), S (*Thiobacillus*) for energy	**Photoautotrophs** Algae, cyanobacteria ("blue-green algae"), photosynthetic bacteria that use light energy to convert CO_2 (HCO_3^-) to biomass by photosynthesis

Figure 6.2. Classification of microorganisms among chemoheterotrophs, chemoautotrophs, photoheterotrophs, and photoautotrophs.

Microorganisms at Interfaces

Aquatic microorganisms tend to grow at interfaces. Many such microorganisms grow on solids that are suspended in water or are present in sediments. Large populations of aquatic bacteria typically reside on the surface of water at the air-water interface.[1] In addition to being in contact with air that aerobic microorganisms need for their metabolic processes, this interface also accumulates food in the form of lipids (oils, fats), polysaccharides, and proteins. Bacteria at this interface are generally different from those in the body of water and may have a hydrophobic cell character. When surface bubbles burst, bacteria at the air-water interface can be incorporated into aerosol water droplets and carried by wind. This is a matter of some concern with respect to sewage treatment plants as a possible vector for spreading disease-causing microorganisms.

6.2. ALGAE

For the purposes of discussion here, **algae** may be considered as generally microscopic organisms that subsist on inorganic nutrients and produce organic matter from carbon dioxide by photosynthesis.[2] In addition to single cells, algae grow as filaments, sheets, and colonies. Some algae, particularly the marine kelps, are huge multicellular organisms. The study of algae is called **phycology**.

The four main classes of unicellular algae of importance in environmental chemistry are the following:

- **Chrysophyta**, which contain pigments that give these organisms a yellow-green or golden-brown color. Chrysophyta are found in both freshwater and marine systems. They store food as carbohydrate or oil. The most well-known of these algae are **diatoms**, characterized by silica-containing cell walls.

- **Chlorophyta**, commonly known as green algae, are responsible for most of the primary productivity in fresh waters.

- **Pyrrophyta**, commonly known as dinoflagellates, are motile with structures that enable them to move about in water. (In some cases the distinction between algae and single-celled "animal" protozoa is blurred; see the example of *Pfiesteria* discussed in Section 6.4.) Pyrrophyta occur in both marine and freshwater environments. "Blooms" of *Gymnodinium* and *Gonyaulax* species release toxins that cause harmful "red tides."

- **Euglenophyta** likewise exhibit characteristics of both plants and animals. Though capable of photosynthesis, these algae are not exclusively photo-autotrophic (see Figure 6.2), and they utilize biomass from other sources for at least part of their carbon needs

The general nutrient requirements of algae are carbon (obtained from CO_2 or HCO_3^-), nitrogen (generally as NO_3^-), phosphorus (as some form of orthophosphate), sulfur (as SO_4^{2-}), and trace elements including sodium, potassium, calcium, magnesium, iron, cobalt, and molybdenum.

In a highly simplified form, the production of organic matter by algal photosynthesis is described by the reaction

$$CO_2 + H_2O \xrightarrow{h\nu} \{CH_2O\} + O_2(g) \tag{6.2.1}$$

where $\{CH_2O\}$ represents a unit of carbohydrate and $h\nu$ stands for the energy of a quantum of light. Fogg[3] has represented the overall formula of the algae *Chlorella* as $C_{5.7}H_{9.8}O_{2.3}NP_{0.06}$. Using Fogg's formula for algal biomass exclusive of the phosphorus, the overall reaction for photosynthesis is:

$$5.7CO_2 + 3.4H_2O + NH_3 \xrightarrow{h\nu} C_{5.7}H_{9.8}O_{2.3}N + 6.25O_2(g) \tag{6.2.2}$$

In the absence of light, algae metabolize organic matter in the same manner as do nonphotosynthetic organisms. Thus, algae may satisfy their metabolic demands by utilizing chemical energy from the degradation of stored starches or oils, or from the consumption of algal protoplasm itself. In the absence of photosynthesis, the metabolic process consumes oxygen, so during the hours of darkness an aquatic system with a heavy growth of algae may become depleted in oxygen.

Symbiotic relationships of algae with other organisms are common. There are even reports of unicellular green algae growing inside hairs on polar bears, which are hollow for purposes of insulation; the sight of a green polar bear is alleged to have driven more than one arctic explorer to the brink of madness. The most common symbiotic relationship involving algae is that of **lichen** in which algae coexist with fungi; both kinds of organisms are woven into the same thallus (tubular vegetative unit). The fungus provides moisture and nutrients required by the algae, which generates food photosynthetically. Lichen are involved in weathering processes of rocks.

The main role of algae in aquatic systems is the production of biomass. This occurs through photosynthesis, which fixes carbon dioxide and inorganic carbon from dissolved carbonate species as organic matter, thus providing the basis of the food chain for the other organisms in the system. Unless it occurs to an excessive extent, leading to accumulation of biomass that exhausts dissolved oxygen when it decays (eutrophication), the production of biomass is beneficial to the other organisms in the aquatic system. Under some conditions, the growth of algae can produce metabolites that are responsible for odor and even toxicity in water.[4]

6.3. FUNGI

Fungi are nonphotosynthetic, often filamentous, organisms exhibiting a wide range of morphology (structure).[5] Some fungi are as simple as the microscopic unicellular yeasts, whereas other fungi form large, intricate toadstools. The microscopic filamentous structures of fungi generally are much larger than bacteria, and usually are 5-10 μm in width. Fungi are aerobic (oxygen-requiring) organisms and generally can thrive in more acidic media than can bacteria. They are also more tolerant of higher concentrations of heavy metal ions than are bacteria.

Perhaps the most important function of fungi in the environment is the breakdown of cellulose in wood and other plant materials. To accomplish this, fungal cells secrete an extracellular enzyme (exoenzyme), *cellulase*, that hydrolyzes insoluble cellulose to soluble carbohydrates that can be absorbed by the fungal cell.

Fungi do not grow well in water. However, they play an important role in determining the composition of natural waters and wastewaters because of the large amount of their decomposition products that enter water. An example of such a product is humic material, which interacts with hydrogen ions and metals (see Section 3.17).

6.4. PROTOZOA

Protozoa are microscopic animals consisting of single eukaryotic cells. The numerous kinds of protozoa are classified on the bases of morphology (physical structure), means of locomotion (flagella, cilia, pseudopodia), presence or absence of chloroplasts, presence or absence of shells, ability to form cysts (consisting of a reduced-size cell encapsulated in a relatively thick skin that can be carried in the air or by animals in the absence of water), and ability to form spores. Protozoa occur in a wide variety of shapes and their movement in the field of a microscope is especially fascinating to watch. Some protozoa contain chloroplasts and are photosynthetic.

Protozoa play a relatively small role in environmental biochemical processes, but are nevertheless significant in the aquatic and soil environment for the following reasons:

- Several devastating human diseases, including malaria, sleeping sickness, and some kinds of dysentery, are caused by protozoa that are parasitic to the human body.

- Parasitic protozoa can cause debilitating, even fatal, diseases in livestock and wildlife.

- Vast limestone ($CaCO_3$) deposits have been formed by the deposition of shells from the *foramifera* group of protozoa.

- Protozoa are active in the oxidation of degradable biomass, particularly in sewage treatment.

- Protozoa may affect bacteria active in degrading biodegradable substances by "grazing" on bacterial cells.

Though they are single-celled, protozoa have a fascinating variety of structures that enable them to function. The protozoal cell membrane is protected and supported by a relatively thick pellicle, or by a mineral shell that may act as an exoskeleton. Food is ingested through a structure called a cytosome from which it is concentrated in a cytopharynx or oral groove, then digested by enzymatic action in a food vacuole. Residue from food digestion is expelled through a cytopyge and soluble metabolic products, such as urea or ammonia, are eliminated by a contractile vacuole, which also expels water from the cell interior.

One of the most troublesome aquatic protozoans in recent times is *Pfiesteria piscicida*, a single-celled organism that is reputed to have more than 20 life stages, including flagellated, amoeboid, and encysted forms, some of which are capable of photosynthesis and some of which are capable of living as parasites on fish.[6] In certain amoeboid or dinoflagellate stages, which are induced to form by substances in fish excreta, these organisms secrete a neurotoxin that incapacitates fish, enabling the *Pfiesteria* to attach to the fish and cause often fatal lesions. Large outbreaks of *Pfiesteria* occurred in North Carolina, and in the Pocomoke River of Maryland and the Rappahannock River of Virginia in the mid-late 1990s. In addition to killing fish, these microorganisms have caused symptoms in exposed humans, particularly a condition manifested by short-term memory loss. These "blooms" of *Pfiesteria* have been attributed to excessive enrichment of water with nitrogen and phosphorus, particularly from sewage and from runoff of swine-raising operations, leading to excessive algal growth and eutrophication.

6.5. BACTERIA

Bacteria are single-celled prokaryotic microorganisms that may be shaped as rods (**bacillus**), spheres (**coccus**), or spirals (**vibrios, spirilla, spirochetes**). Bacteria cells may occur individually or grow as groups ranging from two to millions of individual cells. Most bacteria fall into the size range of 0.5-3.0 micrometers. However, considering all species, a size range of 0.3-50 μm is observed. Characteristics of most bacteria include a semirigid cell wall, motility with flagella for those capable of movement, unicellular nature (although clusters of cloned bacterial cells are common), and multiplication by binary fission in which each of two daughter cells is genetically identical to the parent cell. Like other microorganisms, bacteria produce spores.

The metabolic activity of bacteria is greatly influenced by their small size. Their surface-to-volume ratio is extremely large, so that the inside of a bacterial cell is highly accessible to a chemical substance in the surrounding medium. Thus, for the same reason that a finely divided catalyst is more efficient than a more coarsely divided one, bacteria may bring about very rapid chemical reactions compared to those mediated by larger organisms. Bacteria excrete exoenzymes that break down solid food material to soluble components which can penetrate bacterial cell walls, where the digestion process is completed.

Although individual bacteria cells cannot be seen by the naked eye, bacterial colonies arising from individual cells are readily visible. A common method of counting individual bacterial cells in water consists of spreading a measured volume of an appropriately diluted water sample on a plate of agar gel containing bacterial nutrients. Wherever a viable bacterial cell adheres to the plate, a bacterial colony consisting of many cells will grow. These visible colonies are counted and related to the number of cells present initially. Because bacteria cells may already be present in groups, and because individual cells may not live to form colonies or even have the ability to form colonies on a plate, plate counts tend to grossly underestimate the number of viable bacteria.

Autotrophic and Heterotrophic Bacteria

Bacteria may be divided into two main categories, autotrophic and heterotrophic. **Autotrophic bacteria** are not dependent upon organic matter for growth and thrive in a completely inorganic medium; they use carbon dioxide or other carbonate species as a carbon source. A number of sources of energy may be used, depending upon the species of bacteria; however, a biologically mediated chemical reaction always supplies the energy.

An example of autotrophic bacteria is *Gallionella*. In the presence of oxygen, these bacteria are grown in a medium consisting of NH_4Cl, phosphates, mineral salts, CO_2 (as a carbon source), and solid FeS (as an energy source). It is believed that the following is the energy-yielding reaction for this species:

$$4FeS(s) + 9O_2 + 10H_2O \rightarrow 4Fe(OH)_3(s) + 4SO_4^{2-} + 8H^+ \qquad (6.5.1)$$

Starting with the simplest inorganic materials, autotrophic bacteria must synthesize all of the complicated proteins, enzymes, and other materials needed for their life processes. It follows, therefore, that the biochemistry of autotrophic bacteria is quite complicated. Because of their consumption and production of a wide range of minerals, autotrophic bacteria are involved in many geochemical transformations.

Heterotrophic bacteria depend upon organic compounds, both for their energy and for the carbon required to build their biomass. They are much more common in occurrence than autotrophic bacteria. Heterotrophic bacteria are the microorganisms primarily responsible for the breakdown of pollutant organic matter in water, and of organic wastes in biological waste-treatment processes.

Aerobic and Anaerobic Bacteria

Another classification system for bacteria depends upon their requirement for molecular oxygen. **Aerobic bacteria** require oxygen as an electron receptor:

$$O_2 + 4H^+ + 4e^- \rightarrow 2H_2O \tag{6.5.2}$$

Anaerobic bacteria function only in the complete absence of molecular oxygen. Frequently, molecular oxygen is quite toxic to anaerobic bacteria.

A third class of bacteria, **facultative bacteria,** utilize free oxygen when it is available and use other substances as electron receptors (oxidants) when molecular oxygen is not available. Common oxygen substitutes in water are nitrate ion (see Section 6.11) and sulfate ion (see Section 6.12).

6.6. THE PROKARYOTIC BACTERIAL CELL

Figure 6.3 illustrates a generic prokaryotic bacterial cell. Bacterial cells are enclosed in a **cell wall**, which holds the contents of the bacterial cell and determines the shape of the cell. The cell wall in many bacteria is frequently surrounded by a **slime layer** (capsule). This layer protects the bacteria and helps the bacterial cells to adhere to surfaces.

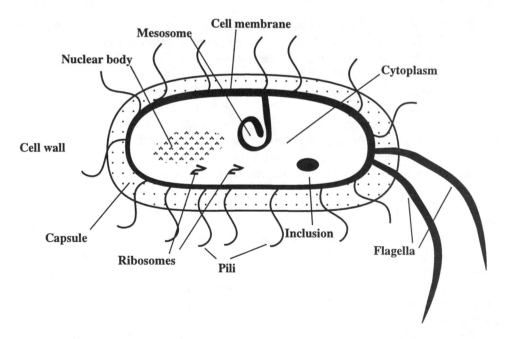

Figure 6.3. Generic prokaryotic bacterial cell illustrating major cell features.

The **cell membrane** or **cytoplasmic membrane** composed of protein and phospholipid occurs as a thin layer only about 7 nanometers in thickness on the inner surface of the cell wall enclosing the cellular cytoplasm. The cytoplasmic membrane is of crucial importance to cell function in that it controls the nature and quantity of materials transported into and out of the cell. It is also very susceptible to damage from some toxic substances.

Folds in the cytoplasmic membrane called **mesosomes** serve several functions. One of these is to increase the surface area of the membrane to enhance transport of materials through it. Another function is to act as a site for division of the cell during reproduction. Bacterial DNA is separated at the mesosome during cell division.

Hairlike **pili** on the surface of a bacterial cell enable the cell to stick to surfaces. Specialized **sex pili** enable nucleic acid transfer between bacterial cells during an exchange of genetic material. Somewhat similar to pili — but larger, more complex, and fewer in number — are **flagella**, moveable appendages that cause bacterial cells to move by their whipping action. Bacteria with flagella are termed **motile**.

Bacterial cells are filled with an aqueous solution and suspension containing proteins, lipids, carbohydrates, nucleic acids, ions, and other materials. Collectively, these materials are referred to as **cytoplasm**, the medium in which the cell's metabolic processes are carried out. The major consituents of cytoplasm are the following:

- **Nuclear body** consisting of a single DNA macromolecule that controls metabolic processes and reproduction.

- **Inclusions** of reserve food material consisting of fats, carbohydrates, and even elemental sulfur.

- **Ribosomes**, which are sites of protein synthesis and which contain protein and RNA.

6.7. KINETICS OF BACTERIAL GROWTH

The population size of bacteria and unicellular algae as a function of time in a growth culture is illustrated by Figure 6.4, which shows a **population curve** for a bacterial culture. Such a culture is started by inoculating a rich nutrient medium with a small number of bacterial cells. The population curve consists of four regions. The first region is characterized by little bacterial reproduction and is called the **lag phase**. The lag phase occurs because the bacteria must become acclimated to the new medium. Following the lag phase comes a period of very rapid bacterial growth. This is the **log phase**, or exponential phase, during which the population doubles over a regular time interval called the **generation time**. This behavior can be described by a mathematical model in which growth rate is proportional to the number of individuals present and there are no limiting factors such as death or lack of food:

$$\frac{dN}{dt} = kN \qquad (6.7.1)$$

This equation can be integrated to give

$$\ln\frac{N}{N_0} = kt \ \text{ or } \ N = N_0 e^{kt} \qquad (6.7.2)$$

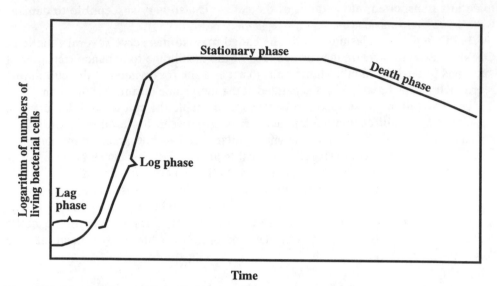

Figure 6.4. Population curve for a bacterial culture.

where N is the population at time t and N_0 is the population at time t = 0. Thus, another way of describing population growth during the log phase is to say that the logarithm of bacterial population increases linearly with time. The generation time, or doubling time, is $(\ln 2)/k$, analogous to the half-life of radioactive decay. Fast growth during the log phase can cause very rapid microbial transformations of chemical species in water.

The log phase terminates and the **stationary phase** begins when a limiting factor is encountered. Typical factors limiting growth are depletion of an essential nutrient, build-up of toxic material, and exhaustion of oxygen. During the stationary phase, the number of viable cells remains virtually constant. After the stationary phase, the bacteria begin to die faster than they reproduce, and the population enters the **death phase**.

6.8. BACTERIAL METABOLISM

Bacteria obtain the energy and raw materials needed for their metabolic processes and reproduction by mediating chemical reactions. Nature provides a large number of such reactions, and bacterial species have evolved that utilize many of these. As a consequence of their participation in such reactions, bacteria are involved in many biogeochemical processes in water and soil. Bacteria are essential partici- pants in many important elemental cycles in nature, including those of nitrogen, carbon, and sulfur. They are responsible for the formation of many mineral deposits, including some of iron and manganese. On a smaller scale, some of these deposits form through bacterial action in natural water systems and even in pipes used to transport water.

Bacterial metabolism addresses the biochemical processes by which chemical species are modified in bacterial cells. It is basically a means of deriving energy and cellular material from nutrient substances. Figure 6.5 summarizes the essential features of bacterial metabolism. The two major divisions of bacterial metabolism are **catabolism**, energy-yielding degradative metabolism which breaks macromolecules down to their small monomeric constituents, and **anabolism**, synthetic metabolism in which small molecules are assembled into large ones.

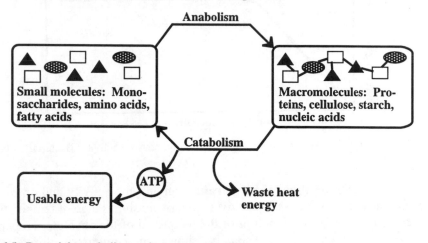

Figure 6.5. Bacterial metabolism and energy production.

A key distinction among bacteria has to do with the terminal electron acceptor in the electron transport chain involved in the process by which bacteria gain energy by oxidizing food materials. If the terminal electron acceptor is molecular O_2, the process is **aerobic respiration**. If it is another reducible species, commonly including SO_4^{2-}, NO_3^-, HCO_3^-, or iron(III), the process is called **anaerobic respiration**. As examples, *Desulfovibrio* bacteria convert SO_4^{2-} to H_2S, *Methanobacterium* reduce HCO_3^- to CH_4, and assorted bacteria reduce NO_3^- to NO_2^-, N_2O, N_2, or NH_4^+.

Factors Affecting Bacterial Metabolism

Bacterial metabolic reactions are mediated by enzymes, biochemical catalysts endogenous to living organisms that are discussed in detail in Chapter 21. Enzymatic processes in bacteria are essentially the same as those in other organisms. At this point, however, it is useful to review several factors that influence bacterial enzyme activity and, therefore, bacterial growth.

Figure 6.6 illustrates the effect of **substrate concentration** on enzyme activity, where a substrate is a substance upon which an enzyme acts. It is seen that enzyme activity increases in a linear fashion up to a value that represents saturation of the enzyme activity. Beyond this concentration, increasing substrate levels do not result in increased enzyme activity. This kind of behavior is reflected in bacterial activity which increases with available nutrients up to a saturation value. Superimposed on this plot in a bacterial system is increased bacterial population which, in effect, increases the amount of available enzyme.

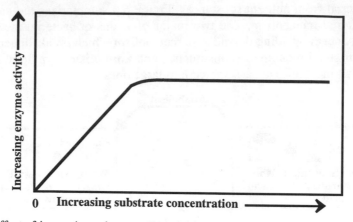

Figure 6.6. Effect of increasing substrate concentration on enzyme activity. Bacterial metabolism parallels such a plot.

Figure 6.7 shows the effect of **temperature** on enzyme activity and on bacterial growth and metabolism. It is seen that over a relatively short range of temperature, a plot of enzyme activity as a function of the reciprocal of the absolute temperature, $1/T$, is linear (an Arrhenius plot). The curve shows a maximum growth rate with an optimum temperature that is skewed toward the high temperature end of the curve, and exhibits an abrupt dropoff beyond the temperature maximum. This occurs because enzymes are destroyed by being denatured at temperatures not far above the optimum. Bacteria show different temperature optima. **Psychrophilic bacteria** are bacteria having temperature optima below approximately 20°C. The temperature optima of **mesophilic bacteria** lie between 20°C and 45°C. Bacteria having temperature optima above 45°C are called **thermophilic bacteria**. The temperature range for optimum growth of bacteria is remarkably wide, with some bacteria being able to grow at 0°C, and some thermophilic bacteria existing in boiling hot water.

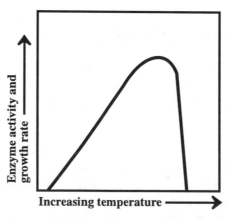

Figure 6.7. Enzyme activity as a function of temperature. A plot of bacterial growth *vs.* temperature has the same shape.

Figure 6.8 is a plot of pH *vs.* bacterial enzyme activity. Although the optimum pH will vary somewhat, enzymes typically have a pH optimum around neutrality.

Enzymes tend to become denatured at pH extremes. For some bacteria, such as those that generate sulfuric acid by the oxidation of sulfide or that produce organic acids by fermentation of organic matter, the pH optimum may be quite acidic, illustrating the ability of bacteria to adapt to very extreme environments.

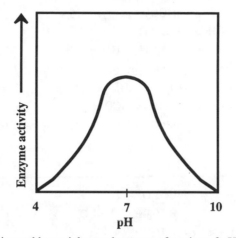

Figure 6.8. Enzyme activity and bacterial growth rate as a function of pH.

Microbial Oxidation and Reduction

The metabolic processes by which bacteria obtain their energy involve mediation of oxidation-reduction reactions. The most environmentally important oxidation-reduction reactions occurring in water and soil through the action of bacteria are summarized in Table 6.1. Much of the remainder of this chapter is devoted to a discussion of important redox reactions mediated by bacteria, particularly those summarizcd in Table 6.1.

6.9. MICROBIAL TRANSFORMATIONS OF CARBON

Carbon is an essential life element and composes a high percentage of the dry weight of microorganisms. For most microorganisms, the bulk of net energy-yielding or energy-consuming metabolic processes involve changes in the oxidation state of carbon. These chemical transformations of carbon have important environmental implications. For example, when algae and other plants fix CO_2 as carbohydrate, represented as $\{CH_2O\}$,

$$CO_2 + H_2O \xrightarrow{h\nu} \{CH_2O\} + O_2(g) \tag{6.2.1}$$

carbon changes from the +4 to the 0 oxidation state. Energy from sunlight is stored as chemical energy in organic compounds. However, when the algae die, bacterial decomposition occurs through aerobic respiration in the reverse of the biochemical process represented by the above reaction for photosynthesis, energy is released, and oxygen is consumed.

Table 6.1. Principal Microbially Mediated Oxidation and Reduction Reactions

Oxidation	$pE^0(w)$[1]
(1) $\quad 1/4\{CH_2O\} + 1/4H_2O \longleftrightarrow 1/4CO_2 + H^+(w) + e^-$	-8.20
(1a) $1/2HCOO^- \longleftrightarrow 1/2CO(g) + 1/2H^+(w) + e^-$	-8.73
(1b) $1/2\{CH_2O\} + 1/2H_2O \longleftrightarrow 1/2HCOO^- + 3/2H^+(w) + e^-$	-7.68
(1c) $1/2CH_3OH \longleftrightarrow 1/2\{CH_2O\} + H^+(w) + e^-$	-3.01
(1d) $1/2CH_4(g) + 1/2H_2O \longleftrightarrow 1/2CH_3OH + H^+(w) + e^-$	-2.88
(2) $\quad 1/8HS^- + 1/2H_2O \longleftrightarrow 1/8SO_4^{2-} + 9/8H^+(w) + e^-$	-3.75
(3) $\quad 1/8NH_4^+ + 3/8H_2O \longleftrightarrow 1/8NO_3^- + 5/4H^+(w) + e^-$	+6.16
(4)[1] $FeCO_3(s) + 2H_2O \longleftrightarrow FeOOH(s) + HCO_3^-(10^{-3}) + 2H^+(w) + e^-$	-1.67
(5)[1] $1/2MnCO_3(s) + H_2O \longleftrightarrow 1/2MnO_2 + 1/2HCO_3^-(10^{-3}) + 3/2H^+(w) + e^-$	-8.5

Reduction	
(A) $\quad 1/4O_2(g) + H^+(w) + e^- \longleftrightarrow 1/2H_2O$	+13.75
(B) $\quad 1/5NO_3^- + 6/5H^+(w) + e^- \longleftrightarrow 1/10N_2 + 3/5H_2O$	+12.65
(C) $\quad 1/8NO_3^- + 5/4H^+(w) + e^- \longleftrightarrow 1/8NH_4^+ + 3/8H_2O$	+6.15
(D) $\quad 1/2\{CH_2O\} + H^+(w) + e^- \longleftrightarrow 1/2CH_3OH$	-3.01
(E) $\quad 1/8SO_4^{2-} + 9/8H^+(w) + e^- \longleftrightarrow 1/8HS^- + 1/2H_2O$	-3.75
(F) $\quad 1/8CO_2(g) + H^+(w) + e^- \longleftrightarrow 1/8CH_4(g) + 1/4H_2O$	-4.13
(G) $\quad 1/6N_2 + 4/3H^+(w) + e^- \longleftrightarrow 1/3NH_4^+$	-4.68

Sequence of Microbial Mediation

Model 1: Excess of organic material (water initially contains O_2, NO_3^-, SO_4^{2-}, and HCO_3^-). Examples: Hypolimnion of a eutrophic lake, sediments, sewage treatment plant digester.

	Combination	$pE^0(w)$[2]	$\Delta G^0(w)$, kcal
Aerobic respiration	(1) + (A)	21.95	-29.9
Denitrification	(1) + (B)	20.85	-28.4
Nitrate reduction	(1) + (C)	14.36	-19.6
Fermentation[3]	(1b) + (D)	4.67	-6.4
Sulfate reduction	(1) + (E)	4.45	-5.9
Methane fermentataion	(1) + (F)	4.07	-5.6
N-fixation	(1) + (G)	3.52	-4.8

Model 2: Excess O_2 (water initially contains organic matter SH^-, NH_4^+, and possibly Fe(II) and Mn(II)). Examples: aerobic waste treatment, self-purification in streams, epilimnion of lake.

Table 6.1 (cont.)

	Combination	$pE^0(w)^2$	$\Delta G^0(w)$, kcal
Aerobic respiration	(A) + (1)	21.95	-29.9
Sulfide oxidation	(A) + (2)	17.50	-23.8
Nitrification	(A) + (3)	7.59	-10.3
Iron(II) oxidation[4]	(A) + (4)	15.42	21.0
Manganese(II) oxidation[4]	(A) + (5)	5.75	-7.2

[1] These pE^0 values are at H^+ ion activity of 1.00×10^{-7}; $H^+(w)$ designates water in which $[H^+] = 1.00 \times 10^{-7}$. pE^0 values for half-reactions (1)-(5) are given for reduction, although the reaction is written as an oxidation.

[2] pE^0 values = $logK(w)$ for a reaction written for a one-electron transfer. The term $K(w)$ is the equilibrium constant for the reaction in which the activity of the hydrogen ion has been set at 1.00×10^{-7} and incorporated into the equililbrium constant.

[3] Fermentation is interpreted as an organic redox reaction where one organic substance is reduced by oxidizing another organic substance (for example, alcoholic fermentation; the products are metastable thermodynamically with respect to CO_2 and CH_4).

[4] The data for $pE^0(w)$ or $\Delta G^0(w)$ of these reactions correspond to an activity of HCO_3^- ion of 1.00×10^{-3} rather than unity.

Source: Stumm, Werner, and James J. Morgan, *Aquatic Chemistry*, Wiley-Interscience, New York, 1970, pp. 336-337. Reproduced with permission of John Wiley & Sons, Inc.

In the presence of oxygen, the principal energy-yielding reaction of bacteria is the oxidation of organic matter. Since it is generally more meaningful to compare reactions on the basis of the reaction of one electron-mole, the aerobic degradation of organic matter is conveniently written as

$$1/4\{CH_2O\} + 1/4O_2(g) \rightarrow 1/4CO_2 + 1/4H_2O \tag{6.9.1}$$

for which the free-energy change is -29.9 kcal (see aerobic respiration, Table 6.1). From this general type of reaction, bacteria and other microorganisms extract the energy needed to carry out their metabolic processes, to synthesize new cell material, for reproduction, and for locomotion.

Partial microbial decomposition of organic matter is a major step in the production of peat, lignite, coal, oil shale, and petroleum. Under reducing conditions, particularly below water, the oxygen content of the original plant material (approximate empirical formula, $\{CH_2O\}$) is lowered, leaving materials with relatively higher carbon contents.

Methane-Forming Bacteria

The production of methane in anoxic (oxygenless) sediments is favored by high organic levels and low nitrate and sulfate levels. Methane production plays a key role in local and global carbon cycles as the final step in the anaerobic decomposition of organic matter. This process is the source of about 80% of the methane entering the atmosphere.

The carbon from microbially produced methane can come from either the reduction of CO_2 or the fermentation of organic matter, particularly acetate. The anoxic production of methane can be represented in the following simplified manner. When carbon dioxide acts as an electron receptor in the absence of oxygen, methane gas is produced:

$$1/8 CO_2 + H^+ + e^- \rightarrow 1/8 CH_4 + 1/4 H_2O \qquad (6.9.2)$$

This reaction is mediated by methane-forming bacteria. When organic matter is degraded microbially, the half-reaction for one electron-mole of $\{CH_2O\}$ is

$$1/4\{CH_2O\} + 1/4 H_2O \rightarrow 1/4 CO_2 + H^+ + e^- \qquad (6.9.3)$$

Adding half-reactions 6.9.2 and 6.9.3 yields the overall reaction for the anaerobic degradation of organic matter by methane-forming bacteria, which involves a free-energy change of -5.55 kcal per electron-mole:

$$1/4\{CH_2O\} \rightarrow 1/8 CH_4 + 1/8 CO_2 \qquad (6.9.4)$$

This reaction, in reality a series of complicated processes, is a **fermentation reaction**, defined as a redox process in which both the oxidizing agent and reducing agent are organic substances. It may be seen that only about one-fifth as much free energy is obtained from one electron-mole of methane formation as from a one electron-mole reaction involving complete oxidation of one electron-mole of the organic matter, Reaction 6.9.1.

There are four main categories of methane-producing bacteria. These bacteria, differentiated largely by morphology, are *Methanobacterium*, *Methanobacillus*, *Methanococcus*, and *Methanosarcina*. The methane-forming bacteria are *obligately anaerobic*; that is, they cannot tolerate the presence of molecular oxygen. The necessity of avoiding any exposure to oxygen makes the laboratory culture of these bacteria very difficult.

Methane formation is a valuable process responsible for the degradation of large quantities of organic wastes, both in biological waste-treatment processes (see Chapter 8) and in nature. Methane production is used in biological waste treatment plants to further degrade excess sludge from the activated sludge process. In the bottom regions of natural waters, methane-forming bacteria degrade organic matter in the absence of oxygen. This eliminates organic matter which would otherwise require oxygen for its biodegradation. If this organic matter were transported to aerobic water containing dissolved O_2, it would exert a biological oxygen demand (BOD). Methane production is a very efficient means for the removal of BOD. The reaction,

$$CH_4 + 2O_2 \rightarrow CO_2 + 2H_2O \qquad (6.9.5)$$

shows that 1 mole of methane requires 2 moles of oxygen for its oxidation to CO_2. Therefore, the production of 1 mole of methane and its subsequent evolution from water are equivalent to the removal of 2 moles of oxygen demand. In a sense, therefore, the removal of 16 grams (1 mole) of methane is equivalent to the addition of 64 grams (2 moles) of available oxygen to the water.

In favorable cases, methane fuel can be produced cost-effectively as a renewable resource from anaerobic digestion of organic wastes. Some installations use cattle feedlot wastes. Methane is routinely generated by the action of anaerobic bacteria and is used for heat and engine fuel at sewage treatment plants (see Chapter 8). Methane produced underground in municipal landfills is being tapped by some municipalities; however, methane seeping into basements of buildings constructed on landfill containing garbage has caused serious explosions and fires.

Bacterial Utilization of Hydrocarbons

Methane is oxidized under aerobic conditions by a number of strains of bacteria. One of these, *Methanomonas*, is a highly specialized organism that cannot use any material other than methane as an energy source. Methanol, formaldehyde, and formic acid are intermediates in the microbial oxidation of methane to carbon dioxide. As discussed in Section 6.10, several types of bacteria can degrade higher hydrocarbons and use them as energy and carbon sources.

Microbial Utilization of Carbon Monoxide

Carbon monoxide is removed from the atmosphere by contact with soil. It has been found that carbon monoxide is removed rapidly from air in contact with soil. Since neither sterilized soil nor green plants grown under sterile conditions show any capacity to remove carbon monoxide from air, this ability must be due to microorganisms in the soil. Fungi capable of CO metabolism include some commonly-occurring strains of the ubiquitous *Penicillium* and *Aspergillus*. It is also possible that some bacteria are involved in CO removal. Whereas some microorganisms metabolize CO, other aquatic and terrestrial organisms produce this gas.

6.10. BIODEGRADATION OF ORGANIC MATTER

The biodegradation of organic matter in the aquatic and terrestrial environments is a crucial environmental process. Some organic pollutants are biocidal; for example, effective fungicides must be antimicrobial in action. Therefore, in addition to killing harmful fungi, fungicides frequently harm beneficial saprophytic fungi (fungi that decompose dead organic matter) and bacteria. Herbicides, which are designed for plant control, and insecticides, which are used to control insects, generally do not have any detrimental effect upon microorganisms.

The biodegradation of organic matter by microorganisms occurs by way of a number of stepwise, microbially catalyzed reactions. These reactions will be discussed individually with examples.

Oxidation

Oxidation occurs by the action of oxygenase enzymes (see Chapter 21 for a discussion of biochemical terms). The microbially catalyzed conversion of aldrin to dieldrin is an example of epoxide formation, a major step in many oxidation mechanisms. **Epoxidation** consists of adding an oxygen atom between two C atoms in an unsaturated system as shown below:

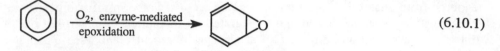

$$(6.10.1)$$

a particularly important means of metabolic attack upon aromatic rings that abound in many xenobiotic compounds.

Microbial Oxidation of Hydrocarbons

The degradation of hydrocarbons by microbial oxidation is an important environmental process because it is the primary means by which petroleum wastes are eliminated from water and soil. Bacteria capable of degrading hydrocarbons include *Micrococcus, Pseudomonas, Mycobacterium,* and *Nocardia.*

The most common initial step in the microbial oxidation of alkanes involves conversion of a terminal $-CH_3$ group to a $-CO_2$ group. More rarely, the initial enzymatic attack involves the addition of an oxygen atom to a nonterminal carbon, forming a ketone. After formation of a carboxylic acid from the alkane, further oxidation normally occurs by a process illustrated by the following reaction, a β–oxidation:

$$CH_3CH_2CH_2CH_2CO_2H + 3O_2 \rightarrow CH_3CH_2CO_2H + 2CO_2 + 2H_2O \qquad (6.10.2)$$

Since 1904, it has been known that the oxidation of fatty acids involves oxidation of the β–carbon atom, followed by removal of two-carbon fragments. A complicated cycle with a number of steps is involved. The residue at the end of each cycle is an organic acid with two fewer carbon atoms than its precursor at the beginning of the cycle.

Hydrocarbons vary significantly in their biodegradability, and microorganisms show a strong preference for straight-chain hydrocarbons. A major reason for this preference is that branching inhibits β–oxidation at the site of the branch. The presence of a quaternary carbon (below) particularly inhibits alkane degradation.

Despite their chemical stability, aromatic (aryl) rings are susceptible to microbial oxidation. The overall process leading to ring cleavage is

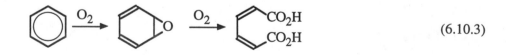

$$(6.10.3)$$

in which cleavage is preceded by addition of –OH to adjacent carbon atoms. Among the microorganisms that attack aromatic rings is the fungus *Cunninghamella elegans*.[7] It metabolizes a wide range of hydrocarbons including: C_3-C_{32} alkanes; alkenes; and aryls, including toluene, naphthalene, anthracene, biphenyl, and phenanthrene. A study of the metabolism of naphthalene by this organism led to the isolation of the following metabolites (the percentage yields are given in parentheses):

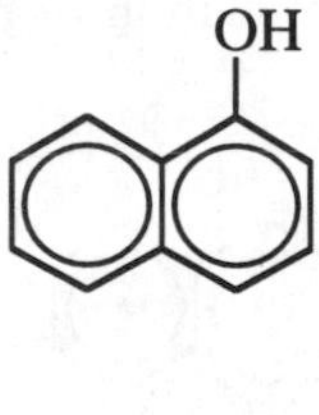

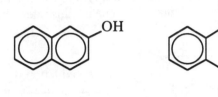

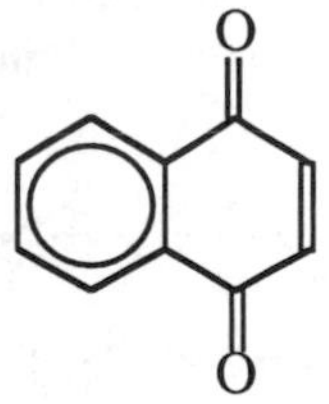

1-Naphthol (67.9%)   2-Naphthol (6.3%)   1,4-Naphthoquinone (2.8%)

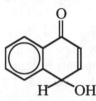

4-Hydroxy-1-tetralone (16.7%) 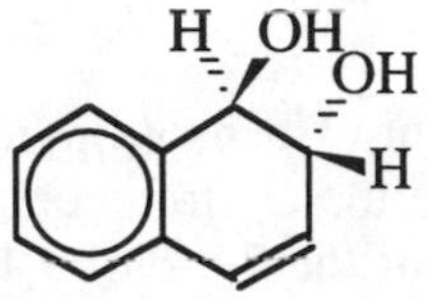   *Trans* -1,2-dihydroxy-
                                    1,2-dihydronaphthalene (5.3%)

The initial attack of oxygen on naphthalene produces 1,2-naphthalene oxide (below), which reacts to form the other products shown above.

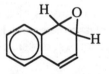

The biodegradation of petroleum is essential to the elimination of oil spills (of the order of a million metric tons per year). This oil is degraded by both marine bacteria and filamentous fungi. In some cases, the rate of degradation is limited by available nitrate and phosphate.

The physical form of crude oil makes a large difference in its degradability. Degradation in water occurs at the water-oil interface. Therefore, thick layers of crude oil prevent contact with bacterial enzymes and O_2. Apparently, bacteria synthesize an emulsifier that keeps the oil dispersed in the water as a fine colloid and therefore accessible to the bacterial cells.

Hydroxylation often accompanies microbial oxidation. It is the attachment of −OH groups to hydrocarbon chains or rings. In the biodegradation of foreign compounds, hydroxylation often follows epoxidation as shown by the following rearrangement reaction for benzene epoxide:

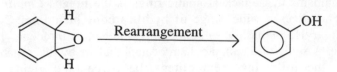

Hydroxylation can consist of the addition of more than one hydroxide group. An example of epoxidation and hydroxylation is the metabolic production of the 7,8-diol-9,10-epoxide of benzo(a)pyrene as illustrated below:

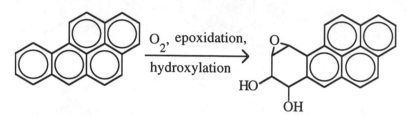

The metabolism of benzo(a)pyrene according to the above reaction is particularly important from the environmental and toxicological viewpoints because one of the stereoisomers of the 7,8-diol-9,10-epoxide binds strongly to cellular DNA, which can cause cancer in an organism.

Other Biochemical Processes

Hydrolysis, which involves the addition of H_2O to a molecule accompanied by cleavage of the molecule into two products, is a major step in microbial degradation of many pollutant compounds, especially pesticidal esters, amides, and organophosphate esters. The types of enzymes that bring about hydrolysis are **hydrolase enzymes**; those that enable the hydrolysis of esters are called **esterases**, whereas those that hydrolyze amides are **amidases**. At least one species of *Pseudomonas* hydrolyzes malathion in a type of hydrolysis reaction typical of those by which pesticides are degraded:

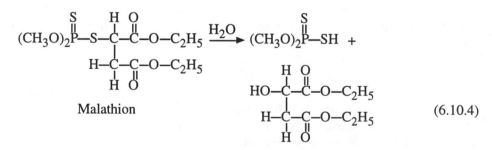

(6.10.4)

Reductions are carried out by **reductase enzymes**; for example, nitroreductase enzyme catalyzes the reduction of the nitro group. Table 6.2 gives the major kinds of functional groups reduced by microorganisms.

Dehalogenation reactions of organohalide compounds involve the bacterially-mediated replacement of a covalently-bound halogen atom (F, Cl, Br, I) with –OH, and are discussed in more detail in Section 6.13.

Table 6.2. Functional Groups that Undergo Microbial Reduction

Reactant	Process	Product
O ‖ R—C—H	Aldehyde reduction	H | R—C—OH | H
O ‖ R—C—R'	Ketone reduction	OH | R—C—R' | H
O ‖ R—S—R'	Sulfoxide reduction	R—S—R'
R—SS—R'	Disulfide reduction	R—SH, R'—SH
R⟍ H C=C H⟋ R'	Alkene reduction	H H | | R—C—C—R' | | H H

R—NO$_2$	Nitro group reduction	R—NO, R—NH$_2$,
		          H
		R—N⟍
		       OH

Ring cleavage is a crucial step in the ultimate degradation of organic compounds having aryl rings. Normally, ring cleavage follows the addition of –OH groups (hydroxylation).

Many environmentally significant organic compounds contain alkyl groups, such as the methyl (–CH$_3$) group, attached to atoms of O, N, and S. An important step in the microbial metabolism of many of these compounds is **dealkylation**, replacement of alkyl groups by H as shown in Figure 6.9. Examples of these kinds of reactions include O-dealkylation of methoxychlor insecticides, N-dealkylation of carbaryl insecticide, and S-dealkylation of dimethyl sulfide. Alkyl groups removed by dealkylation usually are attached to oxygen, sulfur, or nitrogen atoms; those attached to carbon are normally not removed directly by microbial processes.

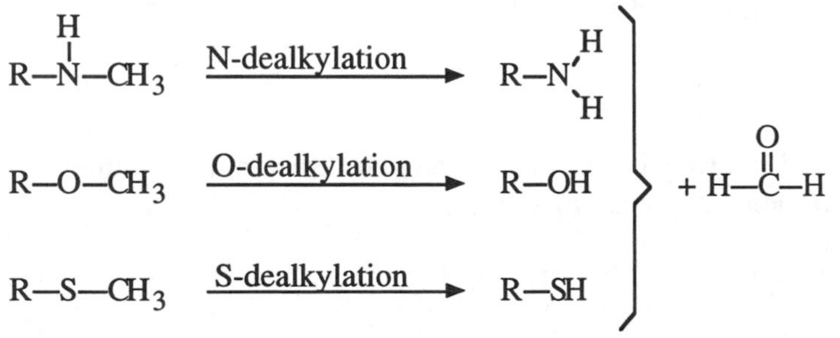

Figure 6.9. Metabolic dealkylation reactions shown for the removal of CH$_3$ from N, O, and S atoms in organic compounds.

6.11. MICROBIAL TRANSFORMATIONS OF NITROGEN

Some of the most important microorganism-mediated chemical reactions in aquatic and soil environments are those involving nitrogen compounds. They are summarized in the **nitrogen cycle** shown in Figure 6.10. This cycle describes the dynamic processes through which nitrogen is interchanged among the atmosphere, organic matter, and inorganic compounds. It is one of nature's most vital dynamic processes.

Among the biochemical transformations in the nitrogen cycle are nitrogen fixation, whereby molecular nitrogen is fixed as organic nitrogen; nitrification, the process of oxidizing ammonia to nitrate; nitrate reduction, the process by which nitrogen in nitrate ion is reduced to form compounds having nitrogen in a lower oxidation state; and denitrification, the reduction of nitrate and nitrite to N_2, with a resultant net loss of nitrogen gas to the atmosphere. Each of these important chemical processes will be discussed separately.

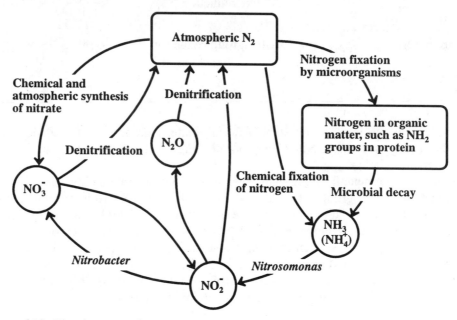

Figure 6.10. The nitrogen cycle.

Nitrogen Fixation

The overall microbial process for **nitrogen fixation**, the binding of atmospheric nitrogen in a chemically combined form,

$$3\{CH_2O\} + 2N_2 + 3H_2O + 4H^+ \rightarrow 3\,CO_2 + 4NH_4^+ \qquad (6.11.1)$$

is actually quite complicated and not completely understood. Biological nitrogen fixation is a key biochemical process in the environment and is essential for plant growth in the absence of synthetic fertilizers.

Only a few species of aquatic microorganisms have the ability to fix atmospheric nitrogen. Among the aquatic bacteria having this capability are photosynthetic bacteria, *Azotobacter*, several species of *Clostridium*, and cyanobacteria, formerly called blue-green algae. In most natural fresh water systems, however, the fraction of nitrogen fixed by organisms in the water relative to that originating from the decay of organic material, fertilizer runoff and other external sources, is quite low.

The best-known and most important form of nitrogen-fixing bacteria is *Rhizobium*, which enjoys a symbiotic (mutually advantageous) relationship with leguminous plants such as clover or alfalfa. The *Rhizobium* bacteria are found in root nodules, special structures attached to the roots of legumes (see Figure 16.2). The nodules develop as a result of the bacteria "irritating" the root hairs of the developing legume plants. The nodules are connected directly to the vascular (circulatory) system of the plant, enabling the bacteria to derive photosynthetically-produced energy directly from the plant. Thus, the plant provides the energy required to break the strong triple bonds in the dinitrogen molecule, converting the nitrogen to a reduced form which is directly assimilated by the plant. When the legumes die and decay, NH_4^+ ion is released and is converted by microorganisms to nitrate ion which is assimilable by other plants. Some of the ammonium ion and nitrate released may be carried into natural water systems.

Some nonlegume angiosperms fix nitrogen through the action of actinomycetes bacteria contained in root nodules. Shrubs and trees in the nitrogen-fixing category are abundant in fields, forests, and wetlands throughout the world. Their rate of nitrogen fixation is comparable to that of legumes.

Free-living bacteria associated with some grasses are stimulated by the grasses to fix nitrogen. One such bacterium is *Spirillum lipoferum*. In tropical surroundings, the amount of reduced nitrogen fixed by such bacteria can amount to the order of 100 kg per hectare per year.

Because of the cost of energy required to fix nitrogen synthetically, efforts are underway to increase the efficiency of natural means of nitrogen fixation. One approach uses recombinant DNA methodologies in attempts to transfer the nitrogen-fixing capabilities of nitrogen-fixing bacteria directly to plant cells. Though a fascinating possibility, this transfer has not yet been achieved on a practical basis. The other approach uses more conventional plant breeding and biological techniques in attempts to increase the range and effectiveness of the symbiotic relationship existing between some plants and nitrogen-fixing bacteria.

One matter of concern is that successful efforts to increase nitrogen fixation may upset the global nitrogen balance. Total annual global fixation of nitrogen is now more than 50% higher than the pre-industrial level of 150 million metric tons estimated for 1850. Potential accumulation of excess fixed nitrogen is the subject of some concern because of aquatic nitrate pollution and microbial production of N_2O gas. Some atmospheric scientists fear that excess N_2O gas may be involved in depletion of the protective atmospheric ozone layer (see Chapter 14).

Nitrification

Nitrification, the conversion of N(-III) to N(V), is a very common and extremely important process in water and in soil. Aquatic nitrogen in thermodynamic equilibrium with air is in the +5 oxidation state as NO_3^-, whereas in most biological compounds, nitrogen is present as N(-III), such as $-NH_2$ in amino acids. The equilibrium constant of the overall nitrification reaction, written for one electron-mole,

$$1/4 O_2 + 1/8 NH_4^+ \rightarrow 1/8 NO_3^- + 1/4 H^+ + 1/8 H_2O \qquad (6.11.2)$$

is $10^{7.59}$ (Table 6.1), showing that the reaction is highly favored from a thermo-dynamic viewpoint.

Nitrification is especially important in nature because nitrogen is absorbed by plants primarily as nitrate. When fertilizers are applied in the form of ammonium salts or anhydrous ammonia, a microbial transformation to nitrate enables maximum assimilation of nitrogen by the plants.

The nitrification conversion of ammoniacal nitrogen to nitrate ion takes place if extensive aeration is allowed to occur in the activated sludge sewage-treatment process (see Chapter 8). As the sewage sludge settles out in the settler, the bacteria in the sludge carry out denitrification while using this nitrate as an oxygen source (Reaction 6.11.8), producing N_2. The bubbles of nitrogen gas cause the sludge to rise, so that it does not settle properly. This can hinder the proper treatment of sewage through carryover of sludge into effluent water.

In nature, nitrification is catalyzed by two groups of bacteria, *Nitrosomonas* and *Nitrobacter*. *Nitrosomonas* bacteria bring about the transition of ammonia to nitrite,

$$NH_3 + 3/2 O_2 \rightarrow H^+ + NO_2^- + H_2O \qquad (6.11.3)$$

whereas *Nitrobacter* mediates the oxidation of nitrite to nitrate:

$$NO_2^- + 1/2 O_2 \rightarrow NO_3^- \qquad (6.11.4)$$

Both of these highly specialized types of bacteria are *obligate aerobes*; that is, they function only in the presence of molecular O_2. These bacteria are also *chemo-lithotrophic*, meaning that they can utilize oxidizable inorganic materials as electron donors in oxidation reactions to yield needed energy for metabolic processes.

For the aerobic conversion of one electron-mole of ammoniacal nitrogen to nitrite ion at pH 7.00,

$$1/4 O_2 + 1/6 NH_4^+ \rightarrow 1/6 NO_2^- + 1/3 H^+ + 1/6 H_2O \qquad (6.11.5)$$

the free-energy change is -10.8 kcal. The free-energy change for the aerobic oxidation of one electron-mole of nitrite ion to nitrate ion,

$$1/4 O_2 + 1/2 NO_2^- \rightarrow 1/2 NO_3^- \qquad (6.11.6)$$

is -9.0 kcal. Both steps of the nitrification process involve an appreciable yield of free energy. It is interesting to note that the free-energy yield per electron-mole is approximately the same for the conversion of NH_4^+ to NO_2^- as it is for the conversion of NO_2^- to NO_3^-, about 10 kcal/electron-mole.

Nitrate Reduction

As a general term, **nitrate reduction** refers to microbial processes by which nitrogen in chemical compounds is reduced to lower oxidation states. In the absence of free oxygen, nitrate may be used by some bacteria as an alternate electron receptor. The most complete possible reduction of nitrogen in nitrate ion involves the acceptance of 8 electrons by the nitrogen atom, with the consequent conversion of nitrate to ammonia (+V to -III oxidation state). Nitrogen is an essential component of protein, and any organism that utilizes nitrogen from nitrate for the synthesis of protein must first reduce the nitrogen to the -III oxidation state (ammoniacal form). However, incorporation of nitrogen into protein generally is a relatively minor use of the nitrate undergoing microbially mediated reactions and is more properly termed nitrate assimilation.

Nitrate ion functioning as an electron receptor usually produces NO_2^-:

$$1/2NO_3^- + 1/4\{CH_2O\} \rightarrow 1/2NO_2^- + 1/4H_2O + 1/4CO_2 \qquad (6.11.7)$$

The free-energy yield per electron-mole is only about 2/3 of the yield when oxygen is the oxidant; however, nitrate ion is a good electron receptor in the absence of O_2. One of the factors limiting the use of nitrate ion in this function is its relatively low concentration in most waters. Furthermore, nitrite, NO_2^-, is relatively toxic and tends to inhibit the growth of many bacteria after building up to a certain level. Sodium nitrate is has been used as a "first-aid" treatment in sewage lagoons that have become oxygen-deficient. It provides an emergency source of oxygen to reestablish normal bacterial growth.

Nitrate ion can be an effective oxidizing agent for a number of species in water that are oxidized by the action of microorganisms. One place in which this process is of interest is in biological sewage treatment. Nitrate has been shown to act as a microbial oxidizing agent for the conversion of iron(II) to iron(III) under conditions corresponding to the biological treatment of sewage.[8]

$$2NO_3^- + 10Fe^{2+} + 24H_2O \rightarrow N_2 + 10Fe(OH)_3 + 18H^+ \qquad (6.11.8)$$

Denitrification

An important special case of nitrate reduction is **denitrification**, in which the reduced nitrogen product is a nitrogen-containing gas, usually N_2. At pH 7.00, the free-energy change per electron-mole of reaction,

$$1/5NO_3^- + 1/4\{CH_2O\} + 1/5H^+ \rightarrow 1/10N_2 + 1/4CO_2 + 7/20H_2O \qquad (6.11.9)$$

is -2.84 kcal. The free-energy yield per mole of nitrate reduced to N_2 (5 electron-moles) is lower than that for the reduction of the same quantity of nitrate to nitrite. More important, however, the reduction of a nitrate ion to N_2 gas consumes 5 electrons, compared to only 2 electrons for the reduction of NO_3^- to NO_2^-.

Denitrification is an important process in nature. It is the mechanism by which fixed nitrogen is returned to the atmosphere. Denitrification is also used in advanced water treatment for the removal of nutrient nitrogen (see Chapter 8). Because nitrogen gas is a nontoxic volatile substance that does not inhibit microbial growth, and since nitrate ion is a very efficient electron acceptor, denitrification allows the extensive growth of bacteria under anaerobic conditions.

Loss of nitrogen to the atmosphere may also occur through the formation of N_2O and NO by bacterial action on nitrate and nitrite catalyzed by the action of several types of bacteria. Production of N_2O relative to N_2 is enhanced during denitrification in soils by increased concentrations of NO_3^-, NO_2^-, and O_2.

Competitive Oxidation of Organic Matter by Nitrate Ion and Other Oxidizing Agents

The successive oxidation of organic matter by dissolved O_2, NO_3^-, and SO_4^{2-} brings about an interesting sequence of nitrate-ion levels in sediments and hypolimnion waters initially containing O_2 but lacking a mechanism for reaeration.[9] This is shown in Figure 6.11, where concentrations of dissolved O_2, NO_3^-, and SO_4^{2-} are

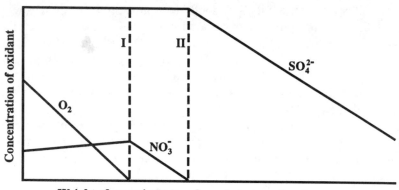

Figure 6.11. Oxidation of organic matter by O_2, NO_3^-, and SO_4^{2-}.

plotted as a function of total organic matter metabolized. This behavior can be explained by the following sequence of biochemical processes:

$$O_2 + \text{organic matter} \rightarrow \text{products} \tag{6.11.10}$$

$$NO_3^- + \text{organic matter} \rightarrow \text{products} \tag{6.11.11}$$

$$SO_4^{2-} + \text{organic matter} \rightarrow \text{products} \tag{6.11.12}$$

So long as some O_2 is present, some nitrate may be produced from organic matter. After exhaustion of molecular oxygen, nitrate is the favored oxidizing agent, and its concentration falls from a maximum value (I) to zero (II). Sulfate, which is usually present in a large excess over the other two oxidants, then becomes the favored electron receptor, enabling biodegradation of organic matter to continue.

6.12. MICROBIAL TRANSFORMATIONS OF PHOSPHORUS AND SULFUR

Phosphorus Compounds

Biodegradation of phosphorus compounds is important in the environment for two reasons. The first of these is that it provides a source of algal nutrient orthophosphate from the hydrolysis of polyphosphates (see Section 3.16). Secondly, biodegradation deactivates highly toxic organophosphate compounds, such as the organophosphate insecticides.

The organophosphorus compounds of greatest environmental concern tend to be sulfur-containing **phosphorothionate** and **phosphorodithioate** ester insecticides with the general formulas illustrated in Figure 6.12, where R and R' represents a hydro-carbon substituted hydrocarbon moieties. These are used because they exhibit higher ratios of insect:mammal toxicity than do their nonsulfur analogs. The metabolic conversion of P=S to P=O (oxidative desulfuration, such as in the conversion of parathion to paraoxon) in organisms is responsible for the insecticidal activity and mammalian toxicity of phosphorothionate and phosphorodithioate insecticides. The biodegradation of these compounds is an important environmental chemical process. Fortunately, unlike the organohalide insecticides that they largely displaced, the organophosphates readily undergo biodegradation and do not bioaccumulate.

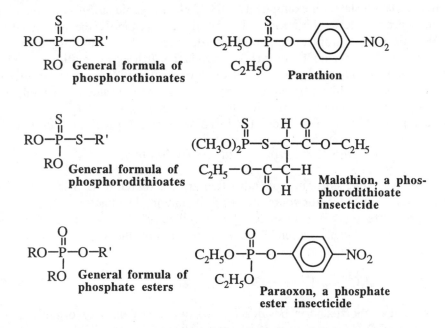

Figure 6.12. Phosphorothionate, phosphorodithioate, and phosphate ester insecticides.

Hydrolysis is an important step in the biodegradation of phosphorothionate, phosphorodithioate, and phosphate ester insecticides as shown by the following general reactions where R is an alkyl group, Ar is a substituent group that is frequently aromatic, and X is either S or O:

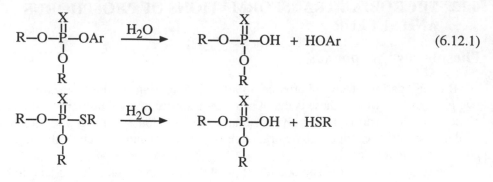

$$R-O-\overset{\overset{\displaystyle X}{\|}}{\underset{\underset{\displaystyle R}{|}}{P}}-OAr \xrightarrow{\ H_2O\ } R-O-\overset{\overset{\displaystyle X}{\|}}{\underset{\underset{\displaystyle R}{|}}{P}}-OH\ +\ HOAr \qquad (6.12.1)$$

$$R-O-\overset{\overset{\displaystyle X}{\|}}{\underset{\underset{\displaystyle R}{|}}{P}}-SR \xrightarrow{\ H_2O\ } R-O-\overset{\overset{\displaystyle X}{\|}}{\underset{\underset{\displaystyle R}{|}}{P}}-OH\ +\ HSR$$

Sulfur Compounds

Sulfur compounds are very common in water. Sulfate ion, SO_4^{2-}, is found in varying concentrations in practically all natural waters. Organic sulfur compounds, both those of natural origin and pollutant species, are very common in natural aquatic systems, and the degradation of these compounds is an important microbial process. Sometimes the degradation products, such as odiferous and toxic H_2S, cause serious problems with water quality.

There is a strong analogy between sulfur in the environment and nitrogen in the environment. Sulfur in living material is present primarily in its most reduced state, for example, as the hydrosulfide group, -SH. Nitrogen in living material is present in the (-III) oxidation state, for example, as $-NH_2$. When organic sulfur compounds are decomposed by bacteria, the initial sulfur product is generally the reduced form, H_2S. When organic nitrogen compounds are decomposed by microorganisms, the reduced form of nitrogen, NH_3 or NH_4^+, is produced. Just as some microorganisms can produce elemental nitrogen from nitrogen compounds, some bacteria produce and store elemental sulfur from sulfur compounds. In the presence of oxygen, some bacteria convert reduced forms of sulfur to the oxidized form in SO_4^{2-} ion, whereas other bacteria catalyze the oxidation of reduced nitrogen compounds to nitrate ion.

Oxidation of H₂S and Reduction of Sulfate by Bacteria

Although organic sulfur compounds often are the source of H_2S in water, they are not required as the sulfur source for H_2S formation. The bacteria *Desulfovibrio* can reduce sulfate ion to H_2S. In so doing, they utilize sulfate as an electron acceptor in the oxidation of organic matter. The overall reaction for the microbially-mediated oxidation of biomass with sulfate is,

$$SO_4^{2-} + 2\{CH_2O\} + 2H^+ \rightarrow H_2S + 2CO_2 + 2H_2O \qquad (6.12.3)$$

and it requires other bacteria besides *Desulfovibrio* to oxidize organic matter completely to CO_2. The oxidation of organic matter by *Desulfovibrio* generally terminates with acetic acid, and accumulation of acetic acid is evident in bottom waters. Because of the high concentration of sulfate ion in seawater, bacterially-mediated formation of H_2S causes pollution problems in some coastal areas and is a major source of atmospheric sulfur. In waters where sulfide formation occurs, the sediment is often black in color due to the formation of FeS.

Bacterially-mediated reduction of sulfur in calcium sulfate deposits produces elemental sulfur interspersed in the pores of the limestone product. The highly generalized chemical reaction for this process is

$$2CaSO_4 + 3\{CH_2O\} \xrightarrow{\text{Bacteria}} 2CaCO_3 + 2S + CO_2 + 3H_2O \quad (6.12.4)$$

although the stoichiometric amount of free sulfur is never found in these deposits due to the formation of volatile H_2S, which escapes.

Whereas some bacteria can reduce sulfate ion to H_2S, others can oxidize hydrogen sulfide to higher oxidation states. The purple sulfur bacteria and green sulfur bacteria derive energy for their metabolic processes through the oxidation of H_2S. These bacteria utilize CO_2 as a carbon source and are strictly anaerobic. The aerobic colorless sulfur bacteria may use molecular oxygen to oxidize H_2S,

$$2H_2S + O_2 \rightarrow 2S + 2H_2O \qquad (6.12.5)$$

elemental sulfur,

$$2S + 2H_2O + 3O_2 \rightarrow 4H^+ + 2SO_4^{2-} \qquad (6.12.6)$$

or thiosulfate ion:

$$S_2O_3^{2-} + H_2O + 2O_2 \rightarrow 2H^+ + 2SO_4^{2-} \qquad (6.12.7)$$

Oxidation of sulfur in a low oxidation state to sulfate ion produces sulfuric acid, a strong acid. One of the colorless sulfur bacteria, *Thiobacillus thiooxidans* is tolerant of 1 normal acid solutions, a remarkable acid tolerance. When elemental sulfur is added to excessively alkaline soils, the acidity is increased because of a microorganism-mediated reaction (6.12.6), which produces sulfuric acid. Elemental sulfur may be deposited as granules in the cells of purple sulfur bacteria and colorless sulfur bacteria. Such processes are important sources of elemental sulfur deposits.

Microorganism-Mediated Degradation of Organic Sulfur Compounds

Sulfur occurs in many types of biological compounds. As a consequence, organic sulfur compounds of natural and pollutant origin are very common in water. The degradation of these compounds is an important microbial process having a strong effect upon water quality.

Among some of the common sulfur-containing functional groups found in aquatic organic compounds are hydrosulfide (–SH), disulfide (–SS–), sulfide (–S–), sulfoxide ($-\overset{\overset{\displaystyle O}{\|}}{S}-$), sulfonic acid (–SO_2OH), thioketone ($-\overset{\overset{\displaystyle S}{\|}}{C}-$), and thiazole (a heterocyclic sulfur group). Protein contains some amino acids with sulfur functional groups—cysteine, cystine, and methionine—whose breakdown is important in natural waters. The amino acids are readily degraded by bacteria and fungi.

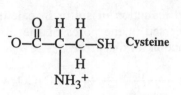

The biodegradation of sulfur-containing amino acids can result in production of volatile organic sulfur compounds such as methane thiol, CH_3SH, and dimethyl disulfide, CH_3SSCH_3. These compounds have strong, unpleasant odors. Their formation, in addition to that of H_2S, accounts for much of the odor associated with the biodegradation of sulfur-containing organic compounds.

Hydrogen sulfide is formed from a large variety of organic compounds through the action of a number of different kinds of microorganisms. A typical sulfur-cleavage reaction producing H_2S is the conversion of cysteine to pyruvic acid through the action of cysteine desulfhydrase enzyme in bacteria:

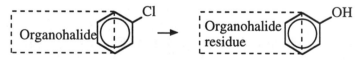

$$H_3C-\overset{\overset{O}{\|}}{C}-\overset{\overset{O}{\|}}{C}-OH \ + \ H_2S \ + \ NH_3 \qquad (6.12.8)$$

Because of the numerous forms in which organic sulfur may exist, a variety of sulfur products and biochemical reaction paths must be associated with the biodegradation of organic sulfur compounds.

6.13. MICROBIAL TRANSFORMATIONS OF HALOGENS AND ORGANOHALIDES

Dehalogenation reactions involving the replacement of a halogen atom, for example,

represent a major pathway for the biodegradation of organohalide hydrocarbons. In some cases, organohalide compounds serve as sole carbon sources, sole energy sources, or electron acceptors for anaerobic bacteria.[10] Microorganisms need not utilize a particular organohalide compound as a sole carbon source in order to cause its degradation. This is due to the phenomenon of **cometabolism**, which results from a lack of specificity in the microbial degradation processes. Thus, bacterial degradation of small amounts of an organohalide compound may occur while the microorganism involved is metabolizing much larger quantities of another substance.

Organohalide compounds can undergo biodegradation anaerobically as shown by the example of 1,1,2,2-tetrachloroethane.[11] Microbially mediated dichloroelimination from this compound can produce one of three possible isomers of dichloroethylene.

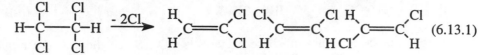

Successive hydrogenolysis reactions can produce vinyl chloride and ethene (ethylene).

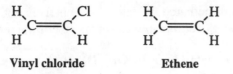

Vinyl chloride   **Ethene**

Successive hydrogenolysis reactions of 1,1,2,2-tetrachloroethane can produce ethane derivatives with 3, 2, 1, and 0 chlorine atoms.

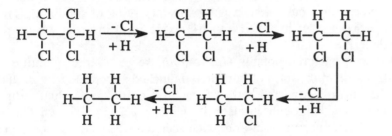

(6.13.2)

Bioconversion of DDT to replace Cl with H yields DDD:

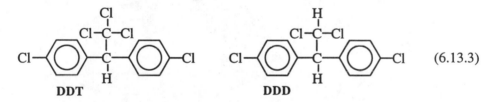

DDT   **DDD**

(6.13.3)

The latter compound is more toxic to some insects than DDT and has even been manufactured as a pesticide. The same situation applies to microbially mediated conversion of aldrin to dieldrin:

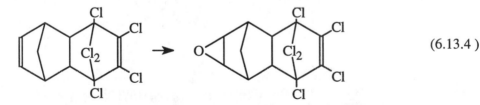

(6.13.4)

6.14. MICROBIAL TRANSFORMATIONS OF METALS AND METALLOIDS

Some bacteria, including *Ferrobacillus*, *Gallionella*, and some forms of *Sphaerotilus*, utilize iron compounds in obtaining energy for their metabolic needs. These bacteria catalyze the oxidation of iron(II) to iron(III) by molecular oxygen:

$$4Fe(II) + 4H^+ + O_2 \rightarrow 4Fe(III) + 2H_2O \tag{6.14.1}$$

The carbon source for some of these bacteria is CO_2. Since they do not require organic matter for carbon, and because they derive energy from the oxidation of inorganic matter, these bacteria may thrive in environments where organic matter is absent.

The microorganism-mediated oxidation of iron(II) is not a particularly efficient means of obtaining energy for metabolic processes. For the reaction

$$FeCO_3(s) + \tfrac{1}{4}O_2 + \tfrac{3}{2}H_2O \rightarrow Fe(OH)_3(s) + CO_2 \tag{6.14.2}$$

the change in free energy is approximately 10 kcal/electron-mole. Approximately 220 g of iron(II) must be oxidized to produce 1.0 g of cell carbon. The calculation assumes CO_2 as a carbon source and a biological efficiency of 5%. The production of only 1.0 g of cell carbon would produce approximately 430 g of solid $Fe(OH)_3$. It follows that large deposits of hydrated iron(III) oxide form in areas where iron-oxidizing bacteria thrive.

Some of the iron bacteria, notably *Gallionella*, secrete large quantities of hydrated iron(III) oxide in the form of intricately branched structures. The bacterial cell grows at the end of a twisted stalk of the iron oxide. Individual cells of *Gallionella*, photographed through an electron microscope, have shown that the stalks consist of a number of strands of iron oxide secreted from one side of the cell (Figure 6.13).

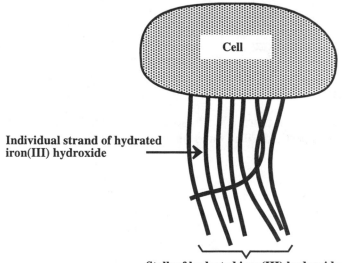

Figure 6.13. Sketch of a cell of *Gallionella* showing iron(III) oxide secretion.

At nearly neutral pH values, bacteria deriving energy by mediating the air oxidation of iron(II) must compete with direct chemical oxidation of iron(II) by O_2. The latter process is relatively rapid at pH 7. As a consequence, these bacteria tend to grow in a narrow layer in the region between the oxygen source and the source of iron(II). Therefore, iron bacteria are sometimes called *gradient organisms*, and they grow at intermediate pE values.

Bacteria are strongly involved in the oceanic manganese cycle. Manganese nodules, a potentially important source of manganese, copper, nickel, and cobalt that occur on ocean floors, yield different species of bacteria which enzymatically mediate both the oxidation and reduction of manganese.

Acid Mine Waters

One consequence of bacterial action on metal compounds is acid mine drainage, one of the most common and damaging problems in the aquatic environment. Many waters flowing from coal mines and draining from the "gob piles" left over from coal processing and washing are practically sterile due to high acidity.

Acid mine water results from the presence of sulfuric acid produced by the oxidation of pyrite, FeS_2. Microorganisms are closely involved in the overall process, which consists of several reactions. The first of these reactions is the oxidation of pyrite:

$$2FeS_2(s) + 2H_2O + 7O_2 \rightarrow 4H^+ + 4SO_4^{2-} + 2Fe^{2+} \tag{6.14.3}$$

The next step is the oxidation of iron(II) ion to iron(III) ion,

$$4Fe^{2+} + O_2 + 4H^+ \rightarrow 4Fe^{3+} + 2H_2O \tag{6.14.4}$$

a process that occurs very slowly at the low pH values found in acid mine waters. Below pH 3.5, the iron oxidation is catalyzed by the iron bacterium *Thiobacillus ferrooxidans*, and in the pH range 3.5-4.5 it may be catalyzed by a variety of *Metallogenium*, a filamentous iron bacterium. Other bacteria that may be involved in acid mine water formation are *Thiobacillus thiooxidans* and *Ferrobacillus ferro-oxidans*. The Fe^{3+} ion further dissolves pyrite,

$$FeS_2(s) + 14Fe^{3+} + 8H_2O \rightarrow 15Fe^{2+} + 2SO_4^{2-} + 16H^+ \tag{6.14.5}$$

which in conjunction with Reaction 6.14.4 constitutes a cycle for the dissolution of pyrite. $Fe(H_2O)_6^{3+}$ is an acidic ion and at pH values much above 3, the iron(III) precipitates as the hydrated iron(III) oxide:

$$Fe^{3+} + 3H_2O \xleftarrow{\hspace{0.5em}}\rightarrow Fe(OH)_3(s) + 3H^+ \tag{6.14.6}$$

The beds of streams afflicted with acid mine drainage often are covered with "yellowboy," an unsightly deposit of amorphous, semigelatinous $Fe(OH)_3$. The most damaging component of acid mine water, however, is sulfuric acid. It is directly toxic and has other undesirable effects.

In past years, the prevention and cure of acid mine water has been one of the major challenges facing the environmental chemist. One approach to eliminating excess acidity involves the use of carbonate rocks. When acid mine water is treated with limestone, the following reaction occurs:

$$CaCO_3(s) + 2H^+ + SO_4^{2-} \rightarrow Ca^{2+} + SO_4^{2-} + H_2O + CO_2(g) \qquad (6.14.7)$$

Unfortunately, because iron(III) is generally present, $Fe(OH)_3$ precipitates as the pH is raised (Reaction 6.11.6). The hydrated iron(III) oxide product covers the particles of carbonate rock with a relatively impermeable layer. This armoring effect prevents further neutralization of the acid.

Microbial Transitions of Selenium

Directly below sulfur in the periodic table, selenium is subject to bacterial oxidation and reduction. These transitions are important because selenium is a crucial element in nutrition, particularly of livestock. Diseases related to either selenium excesses or deficiency have been reported in at least half of the states of the U.S. and in 20 other countries, including the major livestock-producing countries. Livestock in New Zealand, in particular, suffer from selenium deficiency.

Microorganisms are closely involved with the selenium cycle, and microbial reduction of oxidized forms of selenium has been known for some time. Reductive processes under anaerobic conditions can reduce both SeO_3^{2-} and SeO_4^{2-} ions to elemental selenium, which can accumulate as a sink for selenium in anoxic sediments. Some bacteria such as selected strains of *Thiobacillus* and *Leptothrix* can oxidize elemental selenium to selenite, SeO_3^{2-}, thus remobilizing this element from deposits of Se(0).[12]

6.15. MICROBIAL CORROSION

Corrosion is a redox phenomenon and was discussed in Section 4.12. Much corrosion is bacterial in nature.[13,14] Bacteria involved with corrosion set up their own electrochemical cells in which a portion of the surface of the metal being corroded forms the anode of the cell and is oxidized. Structures called tubercles form in which bacteria pit and corrode metals as shown in Figure 6.14.

It is beyond the scope of this book to discuss corrosion in detail. However, its significance and effects should be kept in mind by the environmental chemist.

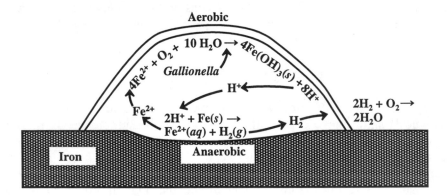

Figure 6.14. Tubercle in which the bacterially-mediated corrosion of iron occurs through the action of *Gallionella*.

LITERATURE CITED

1. Schäfer, Anke, Hauke Harms, and Alexander J. B. Zehnder, "Bacterial Accumulation at the Air-Water Interface," *Environmental Science and Technology* **32**, 3704-3712 (1998).

2. Van Den Hoek, C., D. G. Mann, and Hans Martin Jahns, *Algae: An Introduction to Phycology* , Cambridge University Press, Cambridge, UK, 1995.

3. Fogg, G. E., *The Metabolism of Algae*, John Wiley and Sons, Inc., New York, 1953.

4. Rashash, Diana M. C., Robert C. Hoehn, Andrea M. Dietrich, and Thomas J. Grizzard, *Identification and Control of Odorous Algal Metabolites*, American Water Works Association, Denver, CO, 1997.

5. Alexopoulos, C. J., C. W. Mims, and Meredith Blackwell, *Introductory Mycology*, 4th ed., John Wiley & Sons, New York, 1995.

6. Hileman, Betty, *"Pfisteria* Health Concerns Realized," *Chemical and Engineering News*, October 13, 1999, pp. 14-15."

7. Cerniglia, C. E., and D. T. Gibson, "Metabolism of Naphthalene by *Cunninghamella elegans*," *Applied and Environmental Microbiology*, **34**, 363-70 (1977).

8. Nielsen, Jeppe L. and Per H. Nielsen, "Microbial Nitrate-Dependent Oxidation of Ferrous Iron in Activated Sludge, *Environmental Science and Technology*, **32**, 3556-3561 (1998).

9. Bender, M. L., K. A. Fanning, P. H. Froehlich, and V. Maynard, "Interstitial Nitrate Profiles and Oxidation of Sedimentary Organic Matter in Eastern Equatorial Atlantic," *Science*, **198**, 605-8 (1977).

10. Braus-Stromeyer, Susanna A., Alasdair M. Cook, and Thomas Lesinger, "Biotransformation of Chloromethane to Methanethiol," *Environmental Science and Technology* **27**, 1577-1579 (1993).

11. Lorah, Michelle M. and Lisa D. Olsen, "Degradation of 1,1,2,2-Tetrachloroethane in a Freshwater Tidal Wetland: Field and Laboratory Evidence," *Environmental Science and Technology*, **33**, 227-234 (1999).

12. Dowdle, Phillip R. and Ronald S. Oremland, "Microbial Oxidation of Elemental Selenium in Soil Cultures and Bacterial Cultures," *Environmental Science and Technology*, **32**, 3749-3755 (1998).

13. Licina, G. J., "Detection and Control of Microbiologically Influenced Corrosion," *Official Proceedings of the 57th International Water Conference*, 632-641 (1996).

14. Little, Brenda J., Richard I. Ray, and Patricia A. Wagner, "Tame Microbiologically Influenced Corrosion," *Chem. Eng. Prog.*, **94**(9), 51-60 (1998)

SUPPLEMENTARY REFERENCES

Bitton, Gabriel, *Wastewater Microbiology*, Wiley-Liss, New York, NY, 1999.

Butcher, Samuel S., Ed., *Global Biogeochemical Cycles*, Academic Press, San Diego, CA, 1992.

Csuros, Maria and Csaba Csuros, *Microbiological Examination of Water and Wastewater*, CRC Press/Lewis Publishers, Boca Raton, FL, 1999.

Cullimore, D. Roy, *Practical Manual for Groundwater Microbiology*, CRC Press/Lewis Publishers, Boca Raton, FL, 1991,

Deacon, J. W., *Introduction to Modern Mycology*, Blackwell Science Inc., Cambridge, MA, 1997.

Fenchel, Tom, Gary King, and T. H. Blackburn, *Bacterial Biogeochemistry*, Academic Press, San Diego, CA, 1998.

Geldreich, Edwin E., *Microbial Quality of Water Supply in Distribution*, CRC Press/Lewis Publishers, Boca Raton, FL, 1996.

Howard, Alan D., Ed., *Algal Modelling: Processes and Management*, Kluwer Academic Publishing, 1999.

Lee, Robert Edward, *Phycology*, 3rd ed., Cambridge University Press, New York, 1999.

Madigan, Michael T., John M. Martinko, and Jack Parker, *Brock Biology of Microorganisms*, 9th edition, Prentice Hall, Upper Saddle River, NJ, 1999.

McKinney, R. E., *Microbiology for Sanitary Engineers*, McGraw-Hill Book Company, New York, 1962.

Mitchell, R., *Water Pollution Microbiology*, Vol. 1, Wiley-Interscience, New York, 1970.

Mitchell, R., *Water Pollution Microbiology*, Vol. 2, Wiley-Interscience, New York, 1978.

Postgate, John, *Microbes and Man*, 4th ed., Cambridge University Press, New York, 1999.

Spellman, Frank R., *Microbiology for Water/Wastewater Operators*, Technomic Publishing Co., Lancaster, PA, 1997.

Stevenson, R. Jan, Max L. Bothwell, and Rex L. Lowe, *Algal Ecology: Freshwater Benthic Ecosystems*, Academic Press, San Diego, CA, 1996

Sutton, Brian, Ed., *A Century of Mycology*, Cambridge University Press, New York, 1996.

Talaro, Kathleen Park and Arthur Talaro, *Foundations in Microbiology: Basic Principles*, 3rd ed., WCB/McGraw-Hill, Boston , 1999.

QUESTIONS AND PROBLEMS

1. As $CH_3CH_2CH_2CH_2CO_2H$ biodegrades in several steps to carbon dioxide and water, various chemical species are observed. What stable chemical species would be observed as a result of the first step of this degradation process?

2. Which of the following statements is true regarding the production of methane in water: (a) it occurs in the presence of oxygen, (b) it consumes oxygen, (c) it removes biological oxygen demand from the water, (d) it is accomplished by aerobic bacteria, (e) it produces more energy per electron-mole than does aerobic respiration.

3. At the time zero, the cell count of a bacterial species mediating aerobic respiration of wastes was 1×10^6 cells per liter. At 30 minutes it was 2×10^6; at 60 minutes it was 4×10^6; at 90 minutes, 7×10^6; at 120 minutes, 10×10^6; and at 150 minutes,13×10^6. From these data, which of the following logical conclusions would you draw? (a) The culture was entering the log phase at the end of the 150-minute period, (b) the culture was in the log phase throughout the 150-minute period, (c) the culture was leaving the log phase at the end of the 150-minute period, (d) the culture was in the lag phase throughout the 150-minute period, (e) the culture was in the death phase throughout the 150-minute period.

4. What may be said about the biodegradability of a hydrocarbon containing the following structure?

5. Suppose that the anaerobic fermentation of organic matter, {CH_2O}, in water yields 15.0 L of CH_4 (at standard temperature and pressure). How many grams of oxygen would be consumed by the aerobic respiration of the same quantity of {CH_2O}? (Recall the significance of 22.4 L in chemical reaction of gases.)

6. What weight of $FeCO_3(s)$, using Reaction (A) + (4) in Table 6.1, gives the same free energy yield as 1.00 g of organic matter, using Reaction (A) + (1), when oxidized by oxygen at pH 7.00?

7. How many bacteria would be produced after 10 hours by one bacterial cell, assuming exponential growth with a generation time of 20 minutes?

8. Referring to Reaction 6.11.2, calculate the concentration of ammonium ion in equilibrium with oxygen in the atmosphere and 1.00×10^{-5} M NO_3^- at pH 7.00.

9. When a bacterial nutrient medium is inoculated with bacteria grown in a markedly different medium, the lag phase (Fig. 6.4) often is quite long, even if the bacteria eventually grow well in the new medium. Can you explain this behavior?

10. Most plants assimilate nitrogen as nitrate ion. However, ammonia (NH_3) is a popular and economical fertilizer. What essential role do bacteria play when ammonia is used as a fertilizer? Do you think any problems might occur when using ammonia in a waterlogged soil lacking oxygen?

11. Why is the growth rate of bacteria as a function of temperature (Fig. 6.7) not a symmetrical curve?

12. Discuss the analogies between bacteria and a finely divided chemical catalyst.

13. Would you expect autotrophic bacteria to be more complex physiologically and biochemically than heterotrophic bacteria? Why?

14. Wastewater containing 8 mg/L O_2 (atomic weight O = 16), 1.00×10^{-3} M NO_3^-, and 1.00×10^{-2} M soluble organic matter, $\{CH_2O\}$, is stored isolated from the atmosphere in a container richly seeded with a variety of bacteria. Assume that denitrification is one of the processes which will occur during storage. After the bacteria have had a chance to do their work, which of the following statements will be true? (a) No $\{CH_2O\}$ will remain, (b) some O_2 will remain, (c) some NO_3^- will remain, (d) denitrification will have consumed more of the organic matter than aerobic respiration, (e) the composition of the water will remain unchanged.

15. Of the four classes of microorganisms—algae, fungi, bacteria, and virus—which has the least influence on water chemistry?

16. Figure 6.3 shows the main structural features of a bacterial cell. Which of these do you think might cause the most trouble in water-treatment processes such as filtration or ion exchange, where the maintenance of a clean, unfouled surface is critical? Explain.

17. A bacterium capable of degrading 2,4-D herbicide was found to have its maximum growth rate at 32°C. Its growth rate at 12°C was only 10% of the maximum. Do you think there is another temperature at which the growth rate would also be 10% of the maximum? If you believe this to be the case, of the following temperatures, choose the one at which it is most plausible for the bacterium to also have a growth rate of 10% of the maximum: 52°C, 37°C, 8°C, 20°C.

18. The day after a heavy rain washed a great deal of cattle feedlot waste into a farm pond, the following counts of bacteria were obtained:

Time	Thousands of viable cells per mL
6:00 a.m.	0.10
7:00 a.m.	0.11
8:00 a.m.	0.13
9:00 a.m.	0.16
10:00 a.m.	0.20
11:00 a.m.	0.40
12:00 Noon	0.80
1:00 p.m.	1.60
2:00 p.m.	3.20

To which portion of the bacterial growth curve, Figure 6.3, does this time span correspond?

19. Addition of which two half-reactions in Table 6.1 is responsible for: (a) elimination of an algal nutrient in secondary sewage effluent using methanol as a carbon source, (b) a process responsible for a bad-smelling pollutant when bacteria grow in the absence of oxygen, (c) A process that converts a common form of commercial fertilizer to a form that most crop plants can absorb, (d) a process responsible for the elimination of organic matter from wastewater in the aeration tank of an activated sludge sewage-treatment plant, (e) a characteristic process that occurs in the anaerobic digester of a sewage treatment plant.

20. What is the surface area in square meters of 1.00 gram of spherical bacterial cells, 1.00 μm in diameter, having density of 1.00 g/cm^3?

21. What is the purpose of exoenzymes in bacteria?

22. Match each species of bacteria listed in the left column with its function on the right.

(a) *Spirillum lipoferum*	(1) Reduces sulfate to H_2S
(b) *Rhizobium*	(2) Catalyzes oxidation of Fe^{2+} to Fe^{3+}
(c) *Thiobacillus ferrooxidans*	(3) Fixes nitrogen in grasses
(d) *Desulfovibrio*	(4) On legume roots

23. What factors favor the production of methane in anoxic surroundings?

7 WATER POLLUTION

7.1. NATURE AND TYPES OF WATER POLLUTANTS

Throughout history, the quality of drinking water has been a factor in determining human welfare. Fecal pollution of drinking water has frequently caused waterborne diseases that have decimated the populations of whole cities. Unwholesome water polluted by natural sources has caused great hardship for people forced to drink it or use it for irrigation.

Although there are still occasional epidemics of bacterial and viral diseases caused by infectious agents carried in drinking water, waterborne diseases have in general been well controlled, and drinking water in technologically advanced countries is now remarkably free of the disease-causing agents that were very common water contaminants only a few decades earlier.

Currently, waterborne toxic chemicals pose the greatest threat to the safety of water supplies in industrialized nations. This is particularly true of groundwater in the U.S., which exceeds in volume the flow of all U.S. rivers, lakes, and streams. In some areas, the quality of groundwater is subject to a number of chemical threats. There are many possible sources of chemical contamination. These include wastes from industrial chemical production, metal plating operations, and pesticide runoff from agricultural lands. Some specific pollutants include industrial chemicals such as chlorinated hydrocarbons; heavy metals, including cadmium, lead, and mercury; saline water; bacteria, particularly coliforms; and general municipal and industrial wastes.

Since World War II there has been a tremendous growth in the manufacture and use of synthetic chemicals. Many of the chemicals have contaminated water supplies. Two examples are insecticide and herbicide runoff from agricultural land, and industrial discharge into surface waters. Also, there is a threat to groundwater from waste chemical dumps and landfills, storage lagoons, treating ponds, and other facilities. These threats are discussed in more detail in Chapter 19.

It is clear that water pollution should be a concern of every citizen. Understanding the sources, interactions, and effects of water pollutants is essential for controlling pollutants in an environmentally safe and economically acceptable manner. Above all, an understanding of water pollution and its control depends upon a basic knowledge of aquatic environmental chemistry. That is why this text covers the principles of aquatic chemistry prior to discussing pollution. Water pollution may be studied much more effectively with a sound background in the fundamental properties of water, aquatic microbial reactions, sediment-water interactions, and other factors involved with the reactions, transport, and effects of these pollutants.

Water pollutants can be divided among some general categories, as summarized in Table 7.1. Most of these categories of pollutants, and several subcategories, are discussed in this chapter. An enormous amount of material is published on this subject each year, and it is impossible to cover it all in one chapter. In order to be up to date on this subject the reader may want to survey journals and books dealing with water pollution, such as those listed in the Supplementary References section at the end of this chapter.

Table 7.1. General Types of Water Pollutants

Class of pollutant	Significance
Trace Elements	Health, aquatic biota, toxicity
Heavy metals	Health, aquatic biota, toxicity
Organically-bound metals	Metal transport
Radionuclides	Toxicity
Inorganic pollutants	Toxicity, aquatic biota
Asbestos	Human health
Algal nutrients	Eutrophication
Acidity, alkalinity, salinity (in excess)	Water quality, aquatic life
Trace organic pollutants	Toxicity
Polychlorinated biphenyls	Possible biological effects
Pesticides	Toxicity, aquatic biota, wildlife
Petroleum wastes	Effect on wildlife, esthetics
Sewage, human and animal wastes	Water quality, oxygen levels
Biochemical oxygen demand	Water quality, oxygen levels
Pathogens	Health effects
Detergents	Eutrophication, wildlife, esthetics
Chemical carcinogens	Incidence of cancer
Sediments	Water quality, aquatic biota, wildlife
Taste, odor, and color	Esthetics

7.2. ELEMENTAL POLLUTANTS

Trace element is a term that refers to those elements that occur at very low levels of a few parts per million or less in a given system. The term **trace substance** is a more general one applied to both elements and chemical compounds.

Table 7.2 summarizes the more important trace elements encountered in natural waters. Some of these are recognized as nutrients required for animal and plant life, including some that are essential at low levels but toxic at higher levels. This is typical behavior for many substances in the aquatic environment, a point that must be

Table 7.2. Important Trace Elements in Natural Waters

Element	Sources	Effects and Significance
Arsenic	Mining byproduct, chemical waste	Toxic[1], possibly carcinogenic
Beryllium	Coal, industrial wastes	Toxic
Boron	Coal, detergents, wastes	Toxic
Chromium	Metal plating	Essential as Cr(III), toxic as Cr(VI)
Copper	Metal plating, mining, industrial waste	Essential trace element, toxic to plants and algae at higher levels
Fluorine (F⁻)	Natural geological sources, wastes, water additive	Prevents tooth decay at around 1 mg/L, toxic at higher levels
Iodine (I⁻)	Industrial wastes, natural brines, seawater intrusion	Prevents goiter
Iron	Industrial wastes, corrosion, acid mine water, microbial action	Essential nutrient, damages fixtures by staining
Lead	Industrial waste, mining, fuels	Toxic, harmful to wildlife
Manganese	Industrial wastes, acid mine water, microbial action	Toxic to plants, damages fixtures by staining
Mercury	Industrial waste, mining, coal	Toxic, mobilized as methyl mercury compounds by anaerobic bacteria
Molybdenum	Industrial wastes, natural sources	Essential to plants, toxic to animals
Selenium	Natural sources, coal	Essential at lower levels, toxic at higher levels
Zinc	Industrial waste, metal plating, plumbing	Essential element, toxic to plants at higher levels

[1] Toxicities of these elements are discussed in Chapter 23.

kept in mind in judging whether a particular element is beneficial or detrimental. Some of these elements, such as lead or mercury, have such toxicological and environmental significance that they are discussed in detail in separate sections.

Some of the **heavy metals** are among the most harmful of the elemental pollutants and are of particular concern because of their toxicities to humans. These elements are in general the transition metals, and some of the representative elements, such as lead and tin, in the lower right-hand corner of the periodic table. Heavy metals include essential elements like iron as well as toxic metals like cadmium and mercury. Most of them have a tremendous affinity for sulfur and disrupt enzyme function by forming bonds with sulfur groups in enzymes. Protein carboxylic acid ($-CO_2H$) and amino ($-NH_2$) groups are also chemically bound by heavy metals. Cadmium, copper, lead, and mercury ions bind to cell membranes, hindering transport processes through the cell wall. Heavy metals may also precipitate phosphate biocompounds or catalyze their decomposition. The biochemical effects of metals are discussed in Chapter 23.

Some of the **metalloids**, elements on the borderline between metals and nonmetals, are significant water pollutants. Arsenic, selenium, and antimony are of particular interest.

Inorganic chemicals manufacture has the potential to contaminate water with trace elements. Among the industries regulated for potential trace element pollution of water are those producing chlor-alkali, hydrofluoric acid, sodium dichromate (sulfate process and chloride ilmenite process), aluminum fluoride, chrome pigments, copper sulfate, nickel sulfate, sodium bisulfate, sodium hydrosulfate, sodium bisulfite, titanium dioxide, and hydrogen cyanide.

7.3. HEAVY METALS

Cadmium

Pollutant **cadmium** in water may arise from industrial discharges and mining wastes. Cadmium is widely used in metal plating. Chemically, cadmium is very similar to zinc, and these two metals frequently undergo geochemical processes together. Both metals are found in water in the +2 oxidation state.

The effects of acute cadmium poisoning in humans are very serious. Among them are high blood pressure, kidney damage, destruction of testicular tissue, and destruction of red blood cells. It is believed that much of the physiological action of cadmium arises from its chemical similarity to zinc. Specifically, cadmium may replace zinc in some enzymes, thereby altering the stereostructure of the enzyme and impairing its catalytic activity. Disease symptoms ultimately result.

Cadmium and zinc are common water and sediment pollutants in harbors surrounded by industrial installations. Concentrations of more than 100 ppm dry weight sediment have been found in harbor sediments. Typically, during periods of calm in the summer when the water stagnates, the anaerobic bottom layer of harbor water has a low soluble Cd concentration because microbial reduction of sulfate produces sulfide,

$$2\{CH_2O\} + SO_4^{2-} + H^+ \rightarrow 2CO_2 + HS^- + 2H_2O \tag{7.3.1}$$

which precipitates cadmium as insoluble cadmium sulfide:

$$CdCl^+(chloro\ complex\ in\ seawater) + HS^- \rightarrow CdS(s) + H^+ + Cl^- \qquad (7.3.2)$$

Mixing of bay water from outside the harbor and harbor water by high winds during the winter results in desorption of cadmium from harbor sediments by aerobic bay water. This dissolved cadmium is carried out into the bay where it is absorbed by suspended solid materials, which then become incorporated with the bay sediments. This is an example of the sort of complicated interaction of hydraulic, chemical solution-solid, and microbiological factors involved in the transport and distribution of a pollutant in an aquatic system.

Lead

Inorganic **lead** arising from a number of industrial and mining sources occurs in water in the +2 oxidation state. Lead from leaded gasoline used to be a major source of atmospheric and terrestrial lead, much of which eventually entered natural water systems. In addition to pollutant sources, lead-bearing limestone and galena (PbS) contribute lead to natural waters in some locations.

Despite greatly increased total use of lead by industry, evidence from hair samples and other sources indicates that body burdens of this toxic metal have decreased during recent decades. This may be the result of less lead used in plumbing and other products that come in contact with food or drink.

Acute lead poisoning in humans causes severe dysfunction in the kidneys, reproductive system, liver, and the brain and central nervous system. Sickness or death results. Lead poisoning from environmental exposure is thought to have caused mental retardation in many children. Mild lead poisoning causes anemia. The victim may have headaches and sore muscles, and may feel generally fatigued and irritable.

Except in isolated cases, lead is probably not a major problem in drinking water, although the potential exists in cases where old lead pipe is still in use. Lead used to be a constituent of solder and some pipe-joint formulations, so that household water does have some contact with lead. Water that has stood in household plumbing for some time may accumulate significant levels of lead (along with zinc, cadmium, and copper) and should be drained for a while before use.

Mercury

Because of its toxicity, mobilization as methylated forms by anaerobic bacteria, and other pollution factors, **mercury** generates a great deal of concern as a heavy-metal pollutant. Mercury is found as a trace component of many minerals, with continental rocks containing an average of around 80 parts per billion, or slightly less, of this element. Cinnabar, red mercuric sulfide, is the chief commercial mercury ore. Fossil fuel coal and lignite contain mercury, often at levels of 100 parts per billion or even higher, a matter of some concern with increased use of these fuels for energy resources.

Metallic mercury is used as an electrode in the electrolytic generation of chlorine gas, in laboratory vacuum apparatus, and in other applications. Significant quantities of inorganic mercury(I) and mercury(II) compounds are used annually. Organic mercury compounds used to be widely applied as pesticides, particularly fungicides. These mercury compounds include aryl mercurials such as phenyl mercuric dimethyldithiocarbamate

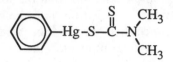

(formerly used in paper mills as a slimicide and as a mold retardant for paper), and alkyl-mercurials such as ethylmercuric chloride, C_2H_5HgCl, that was used as a seed fungicide. Because of their resistance to degradation and their mobility, the alkyl mercury compounds are generally considered to be more of an environmental threat than either the aryl or inorganic compounds.

Mercury enters the environment from a large number of miscellaneous sources related to human use of the element. These include discarded laboratory chemicals, batteries, broken thermometers, amalgam tooth fillings, and formerly lawn fungicides and pharmaceutical products. Taken individually, each of these sources may not contribute much of the toxic metal, but the total effect can be substantial. Sewage effluent sometimes contains up to 10 times the level of mercury found in typical natural waters.

The toxicity of mercury was tragically illustrated in the Minamata Bay area of Japan during the period 1953-1960. A total of 111 cases of mercury poisoning and 43 deaths were reported among people who had consumed seafood from the bay that had been contaminated with mercury waste from a chemical plant that drained into Minamata Bay. Congenital defects were observed in 19 babies whose mothers had consumed seafood contaminated with mercury. The level of metal in the contaminated seafood was 5-20 parts per million.

Among the toxicological effects of mercury are neurological damage, including irritability, paralysis, blindness, or insanity; chromosome breakage; and birth defects. The milder symptoms of mercury poisoning such as depression and irritability have a psychopathological character. Because of the resemblance of these symptoms to common human behavior, mild mercury poisoning may escape detection. Some forms of mercury are relatively nontoxic and were formerly used as medicines, for example, in the treatment of syphilis. Other forms of mercury, particularly organic compounds, are highly toxic.

Because there are few major natural sources of mercury, and since most inorganic compounds of this element are relatively insoluble, it was assumed for some time that mercury was not a serious water pollutant. However, in 1970, alarming mercury levels were discovered in fish in Lake Saint Clair located between Michigan and Ontario, Canada. A subsequent survey by the U.S. Federal Water Quality Administration revealed a number of other waters contaminated with mercury. It was found that several chemical plants, particularly caustic chemical manufacturing operations, were each releasing up to 14 or more kilograms of mercury in wastewaters each day.

The unexpectedly high concentrations of mercury found in water and in fish tissues result from the formation of soluble monomethylmercury ion, CH_3Hg^+, and volatile dimethylmercury, $(CH_3)_2Hg$, by anaerobic bacteria in sediments. Mercury from these compounds becomes concentrated in fish lipid (fat) tissue and the concentration factor from water to fish may exceed 10^3. The methylating agent by which inorganic mercury is converted to methylmercury compounds is methylcobalamin, a vitamin B_{12} analog:

$$HgCl_2 \xrightarrow{\text{Methylcobalamin}} CH_3HgCl + Cl^- \qquad (7.3.3)$$

It is believed that the bacteria that synthesize methane produce methylcobalamin as an intermediate in the synthesis. Thus, waters and sediments in which anaerobic decay is occurring provide the conditions under which methylmercury production occurs. In neutral or alkaline waters, the formation of dimethylmercury, $(CH_3)_2Hg$, is favored. This volatile compound can escape to the atmosphere.

7.4. METALLOIDS

The most significant water pollutant metalloid element is arsenic, a toxic element that has been the chemical villain of more than a few murder plots. Acute arsenic poisoning can result from the ingestion of more than about 100 mg of the element. Chronic poisoning occurs with the ingestion of small amounts of arsenic over a long period of time. There is some evidence that this element is also carcinogenic.

Arsenic occurs in the Earth's crust at an average level of 2-5 ppm. The combustion of fossil fuels, particularly coal, introduces large quantities of arsenic into the environment, much of it reaching natural waters. Arsenic occurs with phosphate minerals and enters into the environment along with some phosphorus compounds. Some formerly-used pesticides, particularly those from before World War II, contain highly toxic arsenic compounds. The most common of these are lead arsenate, $Pb_3(AsO_4)_2$; sodium arsenite, Na_3AsO_3; and Paris Green, $Cu_3(AsO_3)_2$. Another major source of arsenic is mine tailings. Arsenic produced as a by-product of copper, gold, and lead refining exceeds the commercial demand for arsenic, and it accumulates as waste material.

Like mercury, arsenic may be converted to more mobile and toxic methyl derivatives by bacteria, according to the following reactions:

$$H_3AsO_4 + 2H^+ + 2e^- \rightarrow H_3AsO_3 + H_2O \qquad (7.4.1)$$

$$H_3AsO_3 \xrightarrow{\text{Methylcobalamin}} \underset{\text{Methylarsinic acid}}{CH_3AsO(OH)_2} \qquad (7.4.2)$$

$$CH_3AsO(OH)_2 \xrightarrow{\text{Methylcobalamin}} \underset{\text{(Dimethylarsinic acid)}}{(CH_3)_2AsO(OH)} \qquad (7.4.3)$$

$$(CH_3)_2AsO(OH) + 4H^+ + 4e^- \rightarrow (CH_3)_2AsH + 2H_2O \qquad (7.4.4)$$

7.5. ORGANICALLY BOUND METALS AND METALLOIDS

An appreciation of the strong influence of complexation and chelation on heavy metals' behavior in natural waters and wastewaters may be gained by reading Sections 3.10-3.17, which deal with that subject. Methylmercury formation is discussed in Section 7.3. Both topics involve the combination of metals and organic entities in water. It must be stressed that the interaction of metals with organic compounds is of utmost importance in determining the role played by the metal in an aquatic system.

There are two major types of metal-organic interactions to be considered in an aquatic system. The first of these is complexation, usually chelation when organic ligands are involved. A reasonable definition of complexation by organics applicable to natural water and wastewater systems is a system in which a species is present that reversibly dissociates to a metal ion and an organic complexing species as a function of hydrogen ion concentration:

$$ML + 2H^+ \longleftrightarrow M^{2+} + H_2L \tag{7.5.1}$$

In this equation, M^{2+} is a metal ion and H_2L is the acidic form of a complexing—frequently chelating—ligand, L^{2-}, illustrated here as a compound that has two ionizable hydrogens.

Organometallic compounds, on the other hand, contain metals bound to organic entities by way of a carbon atom and do not dissociate reversibly at lower pH or greater dilution. Furthermore, the organic component, and sometimes the particular oxidation state of the metal involved, may not be stable apart from the organometallic compound. A simple way to classify organometallic compounds for the purpose of discussing their toxicology is the following:

1. Those in which the organic group is an alkyl group such as ethyl in tetraethyllead, $Pb(C_2H_5)_4$:

2. **Carbonyls**, some of which are quite volatile and toxic, having carbon monoxide bonded to metals:

$$:C\equiv O:$$

(In the preceding Lewis formula of CO each dash, –, represents a pair of bonding electrons, and each pair of dots, :, represents an unshared pair of electrons.)

3. Those in which the organic group is a π electron donor, such as ethylene or benzene.

Combinations exist of the three general types of compounds outlined above, the most prominent of which are arene carbonyl species in which a metal atom is bonded to both an aryl entity such as benzene and to several carbon monoxide molecules.

A large number of compounds exist that have at least one bond between the metal and a C atom on an organic group, as well as other covalent or ionic bonds between the metal and atoms other than carbon. Because they have at least one metal-carbon bond, as well as properties, uses, and toxicological effects typical of organometallic compounds, it is useful to consider such compounds along with organometallic compounds. Examples are monomethylmercury chloride, CH_3HgCl, in which the organometallic CH_3Hg^+ ion is ionically bonded to the chloride anion. Another example is phenyldichloroarsine, $C_6H_5AsCl_2$, in which a phenyl group is covalently bonded to arsenic through an As-C bond, and two Cl atoms are also covalently bonded to arsenic.

A number of compounds exist that consist of organic groups bonded to a metal atom through atoms other than carbon. Although they do not meet the strict definition thereof, such compounds can be classified as organometallics for the discussion of their toxicology and aspects of their chemistry. An example of such a compound is isopropyl titanate, $Ti(i\text{-}OC_3H_7)_4$, also called titanium isopropylate,

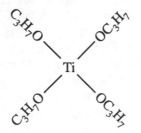

a colorless liquid melting at 14.8°C and boiling at 104°C. Its behavior is more that of an organometallic compound than that of an inorganic compound, and by virtue of its titanium content it is not properly classified as an organic compound. The term "organometal" is sometimes applied to such a compound, which for environmental considerations may be regarded as an organometallic compound.

The interaction of trace metals with organic compounds in natural waters is too vast an area to cover in detail in this chapter; however, it may be noted that metal-organic interactions may involve organic species of both pollutant (such as EDTA) and natural (such as fulvic acids) origin. These interactions are influenced by, and sometimes play a role in, redox equilibria; formation and dissolution of precipitates; colloid formation and stability; acid-base reactions; and microorganism-mediated reactions in water. Metal-organic interactions may increase or decrease the toxicity of metals in aquatic ecosystems, and they have a strong influence on the growth of algae in water.

Organotin Compounds

Of all the metals, tin has the greatest number of organometallic compounds in commercial use, with global production on the order of 40,000 metric tons per year.

In addition to synthetic organotin compounds, methylated tin species can be produced biologically in the environment. Figure 7.1 gives some examples of the many known organotin compounds.

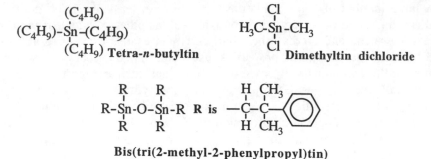

Figure 7.1. Examples of organotin compounds.

Major industrial uses of organotin compounds include applications of tin compounds in fungicides, acaricides, disinfectants, antifouling paints, stabilizers to lessen the effects of heat and light in PVC plastics, catalysts, and precursors for the formation of films of SnO_2 on glass. Tributyl tin chloride and related tributyl tin (TBT) compounds have bactericidal, fungicidal, and insecticidal properties and are of particular environmental significance because of their use as industrial biocides. In addition to tributyl tin chloride, other tributyl tin compounds used as biocides include the hydroxide, the naphthenate, bis(tributyltin) oxide, and tris(tributyl-stannyl) phosphate. TBT has been widely used in boat and ship hull coatings to prevent the growth of fouling organisms. Other applications include preservation of wood, leather, paper, and textiles. Antifungal TBT compounds have been used as slimicides in cooling tower water.

Obviously the many applications of organotin compounds for a variety of uses poses a significant potential for environmental pollution. Because of their applications near or in contact with bodies of water, organotin compounds are potentially significant water pollutants and have been linked to endocrine disruption in shellfish, oysters, and snails. Because of such concerns, several countries, including the U.S., England, and France, prohibited TBT application on vessels smaller than 25 meters in length during the 1980s. In response to concerns over water pollution, in 1998 the International Maritime Organization agreed to ban organotin antifouling paints on all ships by 2003.

7.6. INORGANIC SPECIES

Some important inorganic water pollutants were mentioned in Sections 7.2-7.4 as part of the discussion of pollutant trace elements. Inorganic pollutants that contribute acidity, alkalinity, or salinity to water are considered separately in this chapter. Still another class is that of algal nutrients. This leaves unclassified, however, some important inorganic pollutant species, of which cyanide ion, CN^-, is probably the most important. Others include ammonia, carbon dioxide, hydrogen sulfide, nitrite, and sulfite.

Cyanide

Cyanide, a deadly poisonous substance, exists in water as HCN, a weak acid, K_a of 6×10^{-10}. The cyanide ion has a strong affinity for many metal ions, forming relatively less-toxic ferrocyanide, $Fe(CN)_6^{4-}$, with iron(II), for example. Volatile HCN is very toxic and has been used in gas chamber executions in the U.S.

Cyanide is widely used in industry, especially for metal cleaning and electroplating. It is also one of the main gas and coke scrubber effluent pollutants from gas works and coke ovens. Cyanide is widely used in certain mineral-processing operations. In August of 1995 approximately 2.7 billion liters of cyanide-contaminated water mixed with red clay from mine tailings spilled into the Essequibo River from a breached dam at a gold mining operation in the South American country of Guyana. The water contained cyanide at a level of approximately 25 parts per million, which killed all the fish in the small Omai Creek leading from the breached dam to the Essequibo River. No human fatalities or confirmed health effects were reported, however.

Ammonia and Other Inorganic Pollutants

Excessive levels of ammoniacal nitrogen cause water-quality problems. **Ammonia** is the initial product of the decay of nitrogenous organic wastes, and its presence frequently indicates the presence of such wastes. It is a normal constituent of low-pE groundwaters and is sometimes added to drinking water, where it reacts with chlorine to provide residual chlorine (see Section 8.11). Since the pK_a of ammonium ion, NH_4^+, is 9.26, most ammonia in water is present as NH_4^+ rather than as NH_3.

Hydrogen sulfide, H_2S, is a product of the anaerobic decay of organic matter containing sulfur. It is also produced in the anaerobic reduction of sulfate by microorganisms (see Chapter 6) and is evolved as a gaseous pollutant from geothermal waters. Wastes from chemical plants, paper mills, textile mills, and tanneries may also contain H_2S. Its presence is easily detected by its characteristic rotten-egg odor. In water, H_2S is a weak diprotic acid with pK_{a1} of 6.99 and pK_{a2} of 12.92; S^{2-} is not present in normal natural waters. The sulfide ion has tremendous affinity for many heavy metals, and precipitation of metallic sulfides often accompanies production of H_2S.

Free **carbon dioxide**, CO_2, is frequently present in water at high levels due to decay of organic matter. It is also added to softened water during water treatment as part of a recarbonation process (see Chapter 8). Excessive carbon dioxide levels may make water more corrosive and may be harmful to aquatic life.

Nitrite ion, NO_2^-, occurs in water as an intermediate oxidation state of nitrogen over a relatively narrow pE range. Nitrite is added to some industrial process water as a corrosion inhibitor. However, it rarely occurs in drinking water at levels over 0.1 mg/L.

Sulfite ion, SO_3^{2-}, is found in some industrial wastewaters. Sodium sulfite is commonly added to boiler feedwaters as an oxygen scavenger:

$$2SO_3^{2-} + O_2 \rightarrow 2SO_4^{2-} \tag{7.6.1}$$

Since pK_{a1} of sulfurous acid is 1.76 and pK_{a2} is 7.20, sulfite exists as either HSO_3^- or SO_3^{2-} in natural waters, depending upon pH. It may be noted that hydrazine, N_2H_4, also functions as an oxygen scavenger:

$$N_2H_4 + O_2 \rightarrow 2H_2O + N_2(g) \tag{7.6.2}$$

Asbestos in Water

The toxicity of inhaled asbestos is well established. The fibers scar lung tissue and cancer eventually develops, often 20 or 30 years after exposure. It is not known for sure whether asbestos is toxic in drinking water. This has been a matter of considerable concern because of the dumping of taconite (iron ore tailings) containing asbestos-like fibers into Lake Superior. The fibers have been found in drinking waters of cities around the lake. After having dumped the tailings into Lake Superior since 1952, the Reserve Mining Company at Silver Bay on Lake Superior solved the problem in 1980 by constructing a 6-square-mile containment basin inland from the lake. This $370-million facility keeps the taconite tailings covered with a 3-meter layer of water to prevent escape of fiber dust.

7.7. ALGAL NUTRIENTS AND EUTROPHICATION

The term **eutrophication**, derived from the Greek word meaning "well-nourished," describes a condition of lakes or reservoirs involving excess algal growth. Although some algal productivity is necessary to support the food chain in an aquatic ecosystem, excess growth under eutrophic conditions may eventually lead to severe deterioration of the body of water. The first step in eutrophication of a body of water is an input of plant nutrients (Table 7.3) from watershed runoff or sewage. The nutrient-rich body of water then produces a great deal of plant biomass by photosynthesis, along with a smaller amount of animal biomass. Dead biomass accumulates in the bottom of the lake, where it partially decays, recycling nutrient carbon dioxide, phosphorus, nitrogen, and potassium. If the lake is not too deep, bottom-rooted plants begin to grow, accelerating the accumulation of solid material in the basin. Eventually a marsh is formed, which finally fills in to produce a meadow or forest.

Eutrophication is often a natural phenomenon; for instance, it is basically responsible for the formation of huge deposits of coal and peat. However, human activity can greatly accelerate the process. To understand why this is so, refer to Table 7.3, which shows the chemical elements needed for plant growth. Most of these are present at levels more than sufficient to support plant life in the average lake or reservoir. Hydrogen and oxygen come from the water itself. Carbon is provided by CO_2 from the atmosphere or from decaying vegetation. Sulfate, magnesium, and calcium are normally present in abundance from mineral strata in contact with the water. The micronutrients are required at only very low levels (for example, approximately 40 ppb for copper). Therefore, the nutrients most likely to be limiting are the "fertilizer" elements: nitrogen, phosphorus, and potassium. These are all present in sewage and are, of course, found in runoff from heavily fertilized fields. They are also constituents of various kinds of industrial wastes. Each of these elements can also come from natural sources—phosphorus and potassium from mineral formations, and nitrogen fixed by bacteria, cyanobacteria, or discharge of lightning in the atmosphere.

Table 7.3. Essential Plant Nutrients: Sources and Functions

Nutrient	Source	Function
Macronutrients		
Carbon (CO_2)	Atmosphere, decay	Biomass constituent
Hydrogen	Water	Biomass constituent
Oxygen	Water	Biomass constituent
Nitrogen (NO_3^-)	Decay, pollutants, atmosphere (from nitrogen-fixing organisms)	Protein constituent
Phosphorus (phosphate)	Decay, minerals, pollutants	DNA/RNA constituent
Potassium	Minerals, pollutants	Metabolic function
Sulfur (sulfate)	Minerals	Proteins, enzymes
Magnesium	Minerals	Metabolic function
Calcium	Minerals	Metabolic function
Micronutrients		
B, Cl, Co, Cu, Fe, Mo, Mn, Na, Si, V, Zn	Minerals, pollutants	Metabolic function and/or constituent of enzymes

In most cases, the single plant nutrient most likely to be limiting is phosphorus, and it is generally named as the culprit in excessive eutrophication. Household detergents are a common source of phosphate in wastewater, and eutrophication control has concentrated upon eliminating phosphates from detergents, removing phosphate at the sewage treatment plant, and preventing phosphate-laden sewage effluents (treated or untreated) from entering bodies of water. (See Chapter 3 for additional details regarding phosphates and detergent phosphate substitutes in water.)

In some cases, nitrogen or even carbon may be limiting nutrients. This is particularly true of nitrogen in seawater.

The whole eutrophication picture is a complex one, and continued research is needed to solve the problem. It is indeed ironic that in a food-poor world, nutrient-rich wastes from overfertilized fields or from sewage are causing excessive plant growth in many lakes and reservoirs. This illustrates a point that in many cases pollutants are resources (in this case, plant nutrients) gone to waste.

7.8. ACIDITY, ALKALINITY, AND SALINITY

Aquatic biota are sensitive to extremes of pH. Largely because of osmotic effects, they cannot live in a medium having a salinity to which they are not adapted. Thus, a fresh-water fish soon succumbs in the ocean, and sea fish normally cannot live in fresh water. Excess salinity soon kills plants not adapted to it. There are, of course, ranges in salinity and pH in which organisms live. As shown in Figure 7.2, these ranges frequently may be represented by a reasonably symmetrical curve, along the fringes of which an organism may live without really thriving. These curves do not generally exhibit a sharp cutoff at one end or the other, as does the high-temperature end of the curve representing the growth of bacteria as a function of temperature (Figure 6.7).

The most common source of **pollutant acid** in water is acid mine drainage. The sulfuric acid in such drainage arises from the microbial oxidation of pyrite or other sulfide minerals as described in Chapter 6. The values of pH encountered in acid-polluted water may fall below 3, a condition deadly to most forms of aquatic life except the culprit bacteria mediating the pyrite and iron(II) oxidation, which thrive under very low pH conditions. Industrial wastes frequently have the potential to contribute strong acid to water. Sulfuric acid produced by the air oxidation of pollutant sulfur dioxide (see Chapter 11) enters natural waters as acidic rainfall. In cases where the water does not have contact with a basic mineral, such as limestone, the water pH may become dangerously low. This condition occurs in some Canadian lakes, for example.

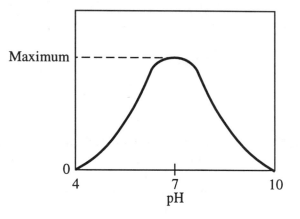

Figure 7.2. A generalized plot of the growth of an aquatic organism as a function of pH.

Excess **alkalinity,** and frequently accompanying high pH, generally are not introduced directly into water from anthropogenic sources. However, in many geographic areas, the soil and mineral strata are alkaline and impart a high alkalinity to water. Human activity can aggravate the situation—for example, by exposure of alkaline overburden from strip mining to surface water or groundwater. Excess alkalinity in water is manifested by a characteristic fringe of white salts at the edges of a body of water or on the banks of a stream.

Water **salinity** may be increased by a number of human activities. Water passing through a municipal water system inevitably picks up salt from a number of processes; for example, recharging water softeners with sodium chloride is a major contributor to salinity in municipal wastewater. Salts can leach from spoil piles. One of the major environmental constraints on the production of shale oil, for example, is the high percentage of leachable sodium sulfate in piles of spent shale. Careful control of these wastes is necessary to prevent further saline pollution of water in areas where salinity is already a problem. Irrigation adds a great deal of salt to water, a phenomenon responsible for the Salton Sea in California, and is a source of conflict between the United States and Mexico over saline contamination of the Rio Grande and Colorado rivers. Irrigation and intensive agricultural production have caused saline seeps in some of the Western states. These occur when water seeps into a slight depression in tilled, sometimes irrigated, fertilized land, carrying salts (particularly sodium, magnesium, and calcium sulfates) along with it. The water evaporates in the dry summer heat, leaving behind a salt-laden area which no longer supports much plant growth. With time, these areas spread, destroying the productivity of crop land.

7.9. OXYGEN, OXIDANTS, AND REDUCTANTS

Oxygen is a vitally important species in water (see Chapter 2). In water, oxygen is consumed rapidly by the oxidation of organic matter, $\{CH_2O\}$:

$$\{CH_2O\} + O_2 \xrightarrow{\text{Microorganisms}} CO_2 + H_2O \qquad (7.9.1)$$

Unless the water is reaerated efficiently, as by turbulent flow in a shallow stream, it rapidly loses oxygen and will not support higher forms of aquatic life.

In addition to the microorganism-mediated oxidation of organic matter, oxygen in water may be consumed by the biooxidation of nitrogenous material,

$$NH_4^+ + 2O_2 \rightarrow 2H^+ + NO_3^- + H_2O \qquad (7.9.2)$$

and by the chemical or biochemical oxidation of chemical reducing agents:

$$4Fe^{2+} + O_2 + 10H_2O \rightarrow 4Fe(OH)_3(s) + 8H^+ \qquad (7.9.3)$$

$$2SO_3^{2-} + O_2 \rightarrow 2SO_4^{2-} \qquad (7.9.4)$$

All these processes contribute to the deoxygenation of water.

The degree of oxygen consumption by microbially-mediated oxidation of contaminants in water is called the **biochemical oxygen demand** (or biological oxygen demand), **BOD**. This parameter is commonly measured by determining the quantity of oxygen utilized by suitable aquatic microorganisms during a five-day period. Despite the somewhat arbitrary five-day period, this test remains a respectable measure of the short-term oxygen demand exerted by a pollutant.[1]

The addition of oxidizable pollutants to streams produces a typical oxygen sag curve as shown in Figure 7.3. Initially, a well-aerated, unpolluted stream is relatively free of oxidizable material; the oxygen level is high; and the bacterial population is relatively low. With the addition of oxidizable pollutant, the oxygen level drops because reaeration cannot keep up with oxygen consumption. In the decomposition zone, the bacterial population rises. The septic zone is characterized by a high bacterial population and very low oxygen levels. The septic zone terminates when the oxidizable pollutant is exhausted, and then the recovery zone begins. In the recovery zone, the bacterial population decreases and the dissolved oxygen level increases until the water regains its original condition.

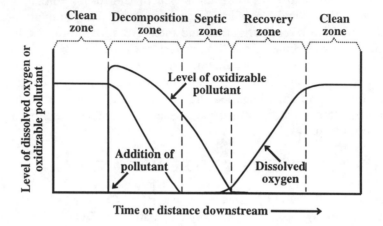

Figure 7.3. Oxygen sag curve resulting from the addition of oxidizable pollutant material to a stream.

Although BOD is a reasonably realistic measure of water quality insofar as oxygen is concerned, the test for determining it is time-consuming and cumbersome to perform. Total organic carbon (TOC), is frequently measured by catalytically oxidizing carbon in the water and measuring the CO_2 that is evolved. It has become popular because TOC is readily determined instrumentally.

7.10. ORGANIC POLLUTANTS

Sewage

As shown in Table 7.4, sewage from domestic, commercial, food-processing, and industrial sources contains a wide variety of pollutants, including organic pollutants. Some of these pollutants, particularly oxygen-demanding substances (see Section 7.9)—oil, grease, and solids—are removed by primary and secondary sewage-treatment processes. Others, such as salts, heavy metals, and refractory (degradation-resistant) organics, are not efficiently removed.

Disposal of inadequately treated sewage can cause severe problems. For example, offshore disposal of sewage, once commonly practiced by coastal cities, results in the formation of beds of sewage residues. Municipal sewage typically contains about 0.1% solids, even after treatment, and these settle out in the ocean in a typical pattern, illustrated in Figure 7.4. The warm sewage water rises in the cold hypolimnion and is carried in one direction or another by tides or currents. It does not

Table 7.4. Some of the Primary Constituents of Sewage from a City Sewage System

Constituent	Potential sources	Effects in water
Oxygen-demanding substances	Mostly organic materials, particularly human feces	Consume dissolved oxygen
Refractory organics	Industrial wastes, household products	Toxic to aquatic life
Viruses	Human wastes	Cause disease (possibly cancer); major deterrent to sewage recycle through water systems
Detergents	Household detergents	Esthetics, prevent grease and oil removal, toxic to aquatic life
Phosphates	Detergents	Algal nutrients
Grease and oil	Cooking, food processing, industrial wastes	Esthetics, harmful to some aquatic life
Salts	Human wastes, water softeners, industrial wastes	Increase water salinity
Heavy metals	Industrial wastes, chemical laboratories	Toxicity
Chelating agents	Some detergents, industrial wastes	Heavy metal ion solubilization and transport
Solids	All sources	Esthetics, harmful to aquatic life

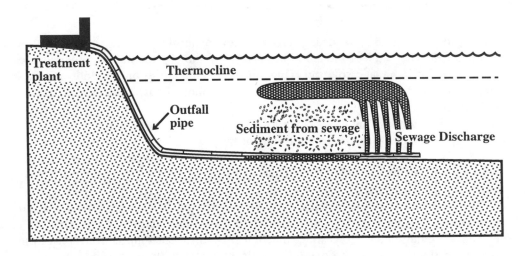

Figure 7.4. Settling of solids from an ocean-floor sewage effluent discharge.

rise above the thermocline; instead, it spreads out as a cloud from which the solids rain down on the ocean floor. Aggregation of sewage colloids is aided by dissolved salts in seawater (see Chapter 5), thus promoting the formation of sludge-containing sediment.

Another major disposal problem with sewage is the sludge produced as a product of the sewage treatment process (see Chapter 8). This sludge contains organic material which continues to degrade slowly; refractory organics; and heavy metals. The amounts of sludge produced are truly staggering. For example, the city of Chicago produces about 3 million tons of sludge each year. A major consideration in the safe disposal of such amounts of sludge is the presence of potentially dangerous components such as heavy metals.

Careful control of sewage sources is needed to minimize sewage pollution problems. Particularly, heavy metals and refractory organic compounds need to be controlled at the source to enable use of sewage, or treated sewage effluents, for irrigation, recycling to the water system, or groundwater recharge.

Soaps, detergents, and associated chemicals are potential sources of organic pollutants. These pollutants are discussed briefly here.

Soaps, Detergents, and Detergent Builders

Soaps

Soaps are salts of higher fatty acids, such as sodium stearate, $C_{17}H_{35}COO^-Na^+$. The cleaning action of soap results largely from its emulsifying power and its ability to lower the surface tension of water. This concept may be understood by considering the dual nature of the soap anion. An examination of its structure shows that the stearate ion consists of an ionic carboxyl "head" and a long hydrocarbon "tail":

In the presence of oils, fats, and other water-insoluble organic materials, the tendency is for the "tail" of the anion to dissolve in the organic matter, whereas the "head" remains in aquatic solution. Thus, the soap emulsifies, or suspends, organic material in water. In the process, the anions form colloidal soap micelles, as shown in Figure 5.4

The primary disadvantage of soap as a cleaning agent comes from its reaction with divalent cations to form insoluble salts of fatty acids:

$$2C_{17}H_{35}COO^-Na^+ + Ca^{2+} \rightarrow Ca(C_{17}H_{35}CO_2)_2(s) + 2Na^+ \tag{7.10.1}$$

These insoluble solids, usually salts of magnesium or calcium, are not at all effective as cleaning agents. In addition, the insoluble "curds" form unsightly deposits on clothing and in washing machines. If sufficient soap is used, all of the divalent cations may be removed by their reaction with soap, and the water containing excess soap will have good cleaning qualities. This is the approach commonly used when soap is employed with unsoftened water in the bathtub or wash basin, where the insoluble calcium and magnesium salts can be tolerated. However, in applications such as washing clothing, the water must be softened by the removal of calcium and magnesium or their complexation by substances such as polyphosphates (see Section 3.16).

Although the formation of insoluble calcium and magnesium salts has resulted in the essential elimination of soap as a cleaning agent for clothing, dishes, and most other materials, it has distinct advantages from the environmental standpoint. As soon as soap gets into sewage or an aquatic system, it generally precipitates as calcium and magnesium salts. Hence, any effects that soap might have in solution are eliminated. With eventual biodegradation, the soap is completely eliminated from the environment. Therefore, aside from the occasional formation of unsightly scum, soap does not cause any substantial pollution problems.

Detergents

Synthetic **detergents** have good cleaning properties and do not form insoluble salts with "hardness ions" such as calcium and magnesium. Such synthetic detergents have the additional advantage of being the salts of relatively strong acids and, therefore, they do not precipitate out of acidic waters as insoluble acids, an undesirable characteristic of soaps. The potential of detergents to contaminate water is high because of their heavy use throughout the consumer, institutional, and industrial markets. It has been projected that by 2004, about 3.0 billion pounds of detergent surfactants will be consumed in the U.S. household market alone, with slightly more consumed in Europe.[2] Most of this material, along with the other ingredients associated with detergent formulations, is discared with wastewater.

The key ingredient of detergents is the **surfactant** or surface-active agent, which acts in effect to make water "wetter" and a better cleaning agent. Surfactants concentrate at interfaces of water with gases (air), solids (dirt), and immiscible liquids (oil). They do so because of their **amphiphilic structure**, meaning that one part of the molecule is a polar or ionic group (head) with a strong affinity for water, and the other part is a hydrocarbon group (tail) with an aversion to water. This kind of structure is illustrated below for the structure of alkyl benzene sulfonate (ABS) surfactant:

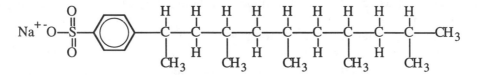

Until the early 1960s, ABS was the most common surfactant used in detergent formulations. However, it suffered the distinct disadvantage of being only very slowly biodegradable because of its branched-chain structure (see Section 6.10). The most objectionable manifestation of the nonbiodegradable detergents, insofar as the average citizen was concerned, was the "head" of foam that began to appear in glasses of drinking water in areas where sewage was recycled through the domestic water supply. Sewage-plant operators were disturbed by spectacular beds of foam which appeared near sewage outflows and in sewage treatment plants. Occasionally, the entire aeration tank of an activated sludge plant would be smothered by a blanket of foam. Among the other undesirable effects of persistent detergents upon waste-treatment processes were lowered surface tension of water; deflocculation of colloids; flotation of solids; emulsification of grease and oil; and destruction of useful bacteria. Consequently, ABS was replaced by a biodegradable surfactant known as linear alkyl sulfonate LAS.

LAS, α–benzenesulfonate, has the general structure illustrated at the top of the next page where the benzene ring may be attached at any point on the alkyl chain except at the ends. LAS is more biodegradable than ABS because the alkyl portion of LAS is not branched and does not contain the tertiary carbon which is so detrimental to biodegradability. Since LAS has replaced ABS in detergents, the prob-

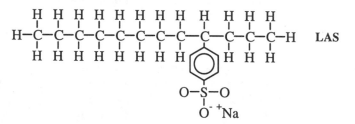

lems arising from the surface-active agent in the detergents (such as toxicity to fish fingerlings) have greatly diminished and the levels of surface-active agents found in water have decreased markedly.

Most of the environmental problems currently attributed to detergents do not arise from the surface-active agents, which basically improve the wetting qualities of water. The **builders** added to detergents continued to cause environmental problems for a longer time, however. Builders bind to hardness ions, making the detergent solution alkaline and greatly improving the action of the detergent surfactant. A commercial solid detergent contains only 10-30% surfactant. In addition, some detergents still contain polyphosphates added to complex calcium and to function as builders. Other ingredients include ion exchangers, alkalies (sodium carbonate), anti-corrosive sodium silicates, amide foam stabilizers, soil-suspending carboxymethyl-cellulose, bleaches, fabric softeners, enzymes, optical brighteners, fragrances, dyes, and diluent sodium sulfate. Of these materials, the polyphosphates have caused the most concern as environmental pollutants, although these problems have largely been resolved.

Increasing demands on the performance of detergents have led to a growing use of enzymes in detergent formulations destined for both domestic and commercial applications. To a degree, enzymes can take the place of chlorine and phosphates, both of which can have detrimental environmental consequences. Lipases and cellulases are the most useful enzymes for detergent applications.

Biorefractory Organic Pollutants

Millions of tons of organic compounds are manufactured globally each year. Significant quantities of several thousand such compounds appear as water pollutants. Most of these compounds, particularly the less biodegradable ones, are substances to which living organisms have not been exposed until recent years. Frequently, their effects upon organisms are not known, particularly for long-term exposures at very low levels. The potential of synthetic organics for causing genetic damage, cancer, or other ill effects is uncomfortably high. On the positive side, organic pesticides enable a level of agricultural productivity without which millions would starve. Synthetic organic chemicals are increasingly taking the place of natural products in short supply. Thus it is that organic chemicals are essential to the operation of a modern society. Because of their potential danger, however, acquisition of knowledge about their environmental chemistry must have a high priority.

Biorefractory organics are the organic compounds of most concern in wastewater, particularly when they are found in sources of drinking water. These are poorly biodegradable substances, prominent among which are aryl or chlorinated hydrocarbons. Included in the list of biorefractory organic industrial wastes are benzene, bornyl alcohol, bromobenzene, bromochlorobenzene, butylbenzene, camphor chloroethyl ether, chloroform, chloromethylethyl ether, chloronitrobenzene, chloropyridine, dibromobenzene, dichlorobenzene, dichloroethyl ether, dinitrotoluene, ethylbenzene, ethylene dichloride, 2-ethylhexanol, isocyanic acid, isopropylbenzene, methylbiphenyl, methyl chloride, nitrobenzene, styrene, tetrachloroethylene, trichloroethane, toluene, and 1,2-dimethoxybenzene. Many of these compounds have been found in drinking water, and some are known to cause taste and odor problems in water. Biorefractory compounds are not completely removed by biological treatment, and water contaminated with these compounds must be treated by physical and chemical means, including air stripping, solvent extraction, ozonation, and carbon adsorption.

Methyl *tert*-butyl ether (MTBE),

$$\begin{array}{c} \text{CH}_3 \\ | \\ \text{H}_3\text{C}-\text{O}-\overset{|}{\underset{|}{\text{C}}}-\text{CH}_3 \\ | \\ \text{CH}_3 \end{array}$$

is now showing up as a low-level water pollutant in the U.S. This compound is added to gasoline as an octane booster and to decrease emissions of automotive exhaust air pollutants. A detailed study of the occurrence of MTBE in Donner Lake (California) showed significant levels of this pollutant, which spiked upward dramatically over the July 4 holiday.[3] They were attributed largely to emissions of unburned fuel from recreational motorboats and personal watercraft having two-cycle engines that discharge their exhausts directly to the water. In 1999 the U.S. Environmental Protection Agency proposed phasing out the use of MTBE in gasoline, largely because of its potential to pollute water.

Naturally Occurring Chlorinated and Brominated Compounds

Although halogenated organic compounds in water, such as those discussed as pesticides in Section 7.11, are normally considered to be from anthropogenic sources, approximately 2400 such compounds have been identified from natural sources. These are produced largely by marine species, especially some kinds of red algae, probably as chemical defense agents. Some marine microorganisms, worms, sponges, and tunicates are also known to produce organochlorine and organobromine compounds. An interesting observation has been made of the possible bio-accumulation of a class of compounds with the formula $C_{10}H_6N_2Br_4Cl_2$ in several species of sea birds from the Pacific ocean region.[4] Although the structural formula of the compound could not be determined with certainty, mass spectral data indicate that it is 1,1'-dimethyl-tetrabromodichloro-2,2'-bipyrrole (below):

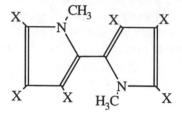

X is Br (4) and Cl (2)

7.11. PESTICIDES IN WATER

The introduction of DDT during World War II marked the beginning of a period of very rapid growth in pesticide use. Pesticides are employed for many different purposes. Chemicals used in the control of invertebrates include **insecticides**, **molluscicides** for the control of snails and slugs, and **nematicides** for the control of microscopic roundworms. Vertebrates are controlled by **rodenticides** which kill rodents, **avicides** used to repel birds, and **piscicides** used in fish control. **Herbicides** are used to kill plants. Plant **growth regulators**, **defoliants**, and **plant desiccants** are used for various purposes in the cultivation of plants. **Fungicides** are used against fungi, **bactericides** against bacteria, **slimicides** against slime-causing organisms in water, and **algicides** against algae. As of the mid-1990s, U,S. agriculture used about 365 million kg of pesticides per year, whereas about 900 million kg of insecticides were used in nonagricultural applications including forestry, landscaping, gardening, food distribution, and home pest control. Insecticide production has remained about level during the last three or four decades. However, insecticides and fungicides are the most important pesticides with respect to human exposure in food because they are applied shortly before or even after harvesting. Herbicide production has increased as chemicals have increasingly replaced cultivation of land in the control of weeds and now accounts for the majority of agricultural pesticides. The potential exists for large quantities of pesticides to enter water either directly, in applications such as mosquito control or indirectly, primarily from drainage of agricultural lands.

Natural Product Insecticides, Pyrethrins, and Pyrethroids

Several significant classes of insecticides are derived from plants. These include **nicotine** from tobacco, **rotenone** extracted from certain legume roots, and **pyrethrins** (see structural formulas in Figure 7.5). Because of the ways that they are applied and their biodegradabilities, these substances are unlikely to be significant water pollutants.

Pyrethrins and their synthetic analogs represent both the oldest and newest of insecticides. Extracts of dried chrysanthemum or pyrethrum flowers, which contain pyrethrin I and related compounds, have been known for their insecticidal properties for a long time, and may have even been used as botanical insecticides in China almost 2000 years ago. The most important commerical sources of insecticidal pyrethrins are chrysanthemum varieties grown in Kenya. Pyrethrins have several advantages as insecticides, including facile enzymatic degradation, which makes them relatively safe for mammals; ability to rapidly paralyze ("knock down") flying insects; and good biodegradability characteristics.

Synthetic analogs of the pyrethrins, **pyrethroids**, have been widely produced as insecticides during recent years. The first of these was allethrin, and another common example is fenvalerate (see structures in Figure 7.5).

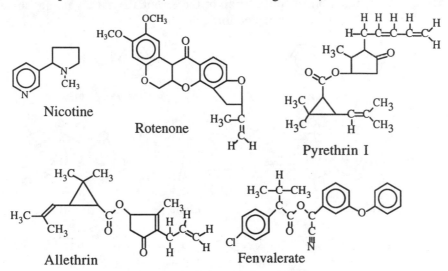

Figure 7.5. Common botanical insecticides and synthetic analogs of the pyrethrins.

DDT and Organochlorine Insecticides

Chlorinated hydrocarbon or organochlorine insecticides are hydrocarbon compounds in which various numbers of hydrogen atoms have been replaced by Cl atoms. The structural formulas of several chlorinated hydrocarbon insecticides are shown in Figure 7.6. It can be seen that the structural formulas of many of these insecticides are very similar; dieldrin and endrin are stereoisomers. The most commonly used insecticides in the 1960s, these compounds have been largely phased out of general use because of their toxicities, and particularly their accumulation and persistence in food chains. They are discussed briefly here, largely because of their historical interest, and because their residues in soils and sediments still contribute to water pollution.

Of the organochlorine insecticides, the most notable has been **DDT** (dichlorodiphenyltrichloroethane or 1,1,1-trichloro-2,2-bis(4-chlorophenyl)ethane), which was used in massive quantities following World War II. It has a low acute toxicity to mammals, although there is some evidence that it might be carcinogenic. It is a very persistent insecticide and accumulates in food chains. It has been banned in the U.S. since 1972. For some time, **methoxychlor** was a popular DDT substitute, reasonably biodegradable, and with a low toxicity to mammals. Structurally similar **chlordane**, **aldrin**, **dieldrin/endrin**, and **heptachlor**, all now banned for application in the U.S., share common characteristics of high persistence and suspicions of potential carcinogenicity. **Toxaphene** is a mixture of up to 177 individual compounds produced by chlorination of camphene, a terpene isolated from pine trees, to give a material that contains about 68% Cl and has an empirical formula of $C_{10}H_{10}Cl_8$. This compound had the widest use of any agricultural insecticide, particularly on cotton. It was employed to augment other insecticides, especially DDT, and in later years methyl parathion. A mixture of five isomers, 1,2,3,4,5,6-hexachlorocyclohexane has been widely produced for insecticidal use. Only the gamma isomer is effective as an insecticide, whereas the other isomers give the product a musty odor and tend to undergo bioaccumulation. A formulation of the essentially pure gamma isomer has been marketed as the insecticide called **lindane**.

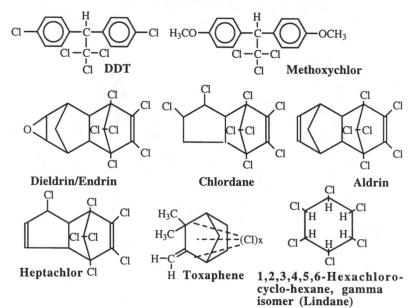

Figure 7.6. Common organochlorine insecticides.

Organophosphate Insecticides

Organophosphate insecticides are insecticidal organic compounds that contain phosphorus, some of which are organic esters of orthophosphoric acid, such as paraoxon:

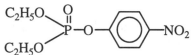

More commonly, insecticidal phosphorus compounds are phosphorothionate compounds, such as parathion or chlorpyrifos,

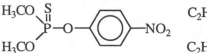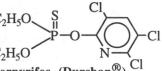

which have an =S group rather than an =O group bonded to P.

The toxicities of organophosphate insecticides vary a great deal. For example, as little as 120 mg of parathion has been known to kill an adult human, and a dose of 2 mg has killed a child. Most accidental poisonings have occurred by absorption through the skin. Since its use began, several hundred people have been killed by parathion. In contrast, **malathion** shows how differences in structural formula can cause pronounced differences in the properties of organophosphate pesticides. Malathion has two carboxyester linkages which are hydrolyzable by carboxylase enzymes to relatively nontoxic products as shown by the following reaction:

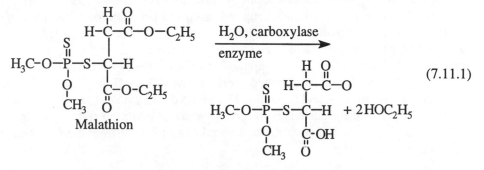

(7.11.1)

The enzymes that accomplish malathion hydrolysis are possessed by mammals, but not by insects, so mammals can detoxify malathion and insects cannot. The result is that malathion has selective insecticidal activity. For example, although malathion is a very effective insecticide, its LD_{50} (dose required to kill 50% of test subjects) for adult male rats is about 100 times that of parathion, reflecting the much lower mammalian toxicity of malathion compared to some of the more toxic organophosphate insecticides, such as parathion.

Unlike the organohalide compounds they largely displaced, the organophosphates readily undergo biodegradation and do not bioaccumulate. Because of their high biodegradability and restricted use, organophosphates are of comparatively little significance as water pollutants.

Carbamates

Pesticidal organic derivatives of carbamic acid, for which the formula is shown in Figure 7.7, are known collectively as **carbamates**. Carbamate pesticides have been widely used because some are more biodegradable than the formerly popular organochlorine insecticides, and have lower dermal toxicities than most common organophosphate pesticides.

Carbaryl has been widely used as an insecticide on lawns or gardens. It has a low toxicity to mammals. **Carbofuran** has a high water solubility and acts as a plant systemic insecticide. As a plant systemic insecticide, it is taken up by the roots and leaves of plants so that insects are poisoned by the plant material on which they feed. **Pirimicarb** has been widely used in agriculture as a systemic aphicide. Unlike many carbamates, it is rather persistent, with a strong tendency to bind to soil.

The toxic effects of carbamates to animals are due to the fact that these compounds inhibit acetylcholinesterase. Unlike some of the organophosphate insecticides, they do so without the need for undergoing a prior biotransformation and are therefore classified as direct inhibitors. Their inhibition of acetylcholinesterase is relatively reversible. Loss of acetylcholinesterase inhibition activity may result from hydrolysis of the carbamate ester, which can occur metabolically.

Herbicides

Herbicides are applied over millions of acres of farmland worldwide and are widespread water pollutants as a result of this intensive use. A 1994 report by the private Environmental Working Group indicated the presence of herbicides in 121 Midwestern U.S. drinking water supplies.[5] The herbicides named were atrazine, simazine, cyanazine, metolachlor, and alachlor, of which the first three are the most widely used. These substances are applied to control weeds on corn and soybeans, and the communities most affected were in the "Corn Belt" states of Kansas, Nebraska, Iowa, Illinois, and Missouri. The group doing the study applied the EPA's most strict standard for pesticides in food to water to come up with an estimate of approximately 3.5 million people at additional risk of cancer from these pesticides in drinking water.

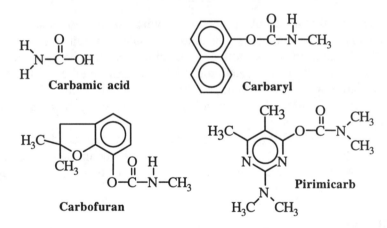

Figure 7.7 Carbamic acid and three insecticidal carbamates.

Bipyridilium Compounds

As shown by the structures in Figure 7.8, a bipyridilium compound contains 2 pyridine rings per molecule. The two important pesticidal compounds of this type are the herbicides **diquat** and **paraquat**, the structural formulas of which are illustrated below:

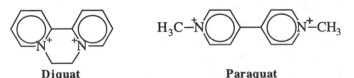

Figure 7.8. The two major bipyridilium herbicides (cation forms).

Other members of this class of herbicides include chlormequat, morfamquat, and difenzoquat. Applied directly to plant tissue, these compounds rapidly destroy plant cells and give the plant a frostbitten appearance. However, they bind tenaciously to soil, especially the clay mineral fraction, which results in rapid loss of herbicidal activity so that sprayed fields can be planted within a day or two of herbicide application.

Paraquat, which was registered for use in 1965, has been one of the most used of the bipyridilium herbicides. Highly toxic, it is reputed to have "been responsible for hundreds of human deaths."[6] Exposure to fatal or dangerous levels of paraquat can occur by all pathways, including inhalation of spray, skin contact, ingestion, and even suicidal hypodermic injections. Despite these possibilities and its widespread application, paraquat is used safely without ill effects when proper procedures are followed.

Because of its widespread use as a herbicide, the possibility exists of substantial paraquat contamination of food. Drinking water contamination by paraquat has also been observed.

Herbicidal Heterocyclic Nitrogen Compounds

A number of important herbicides contain three heterocyclic nitrogen atoms in ring structures and are therefore called **triazines**. Triazine herbicides inhibit photosynthesis. Selectivity is gained by the inability of target plants to metabolize and detoxify the herbicide. The most long established and common example of this class is atrazine, widely used on corn, and a widespread water pollutant in corn-growing regions. Another member of this class is metribuzin, which is widely used on soybeans, sugarcane, and wheat.

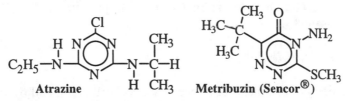

Atrazine   Metribuzin (Sencor®)

Chlorophenoxy Herbicides

The chlorophenoxy herbicides, including 2,4-D and 2,4,5-trichlorophenoxyacetic acid (2,4,5-T) shown below, were manufactured on a large scale for weed and brush control and as military defoliants. At one time the latter was of particular concern because of contaminant TCDD (see below) present as a manufacturing byproduct.

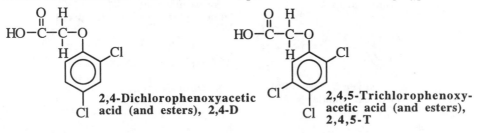

2,4-Dichlorophenoxyacetic acid (and esters), 2,4-D

2,4,5-Trichlorophenoxy-acetic acid (and esters), 2,4,5-T

Substituted Amide Herbicides

A diverse group of herbicides consists of substituted amides. Prominent among these are propanil, applied to control weeds in rice fields, and alachlor, marketed as Lasso® and widely applied to fields to kill germinating grass and broad-leaved weed seedlings:

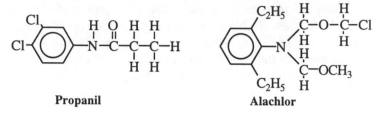

Propanil   **Alachlor**

Nitroaniline Herbicides

Nitroaniline herbicides are characterized by the presence of NO_2 and a substituted $-NH_2$ group on a benzene ring as shown for trifluralin:

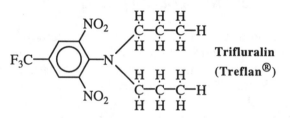

Trifluralin
(Treflan®)

This class of herbicides is widely represented in agricultural applications and includes benefin (Balan®), oryzalin (Surflan®), pendimethalin (Prowl®), and fluchoralin (Basalin®).

Miscellaneous Herbicides

A wide variety of chemicals have been used as herbicides, and have been potential water pollutants. One such compound is R-mecoprop. Other types of herbicides include substituted ureas, carbamates, and thiocarbamates:

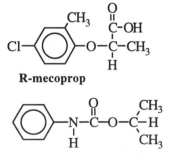

R-mecoprop

3-(4-Chlorophenyl)-1,1-
dimethylurea (substituted urea)

Isopropyl carbanilate
(a carbamate)

Sodium N-methyl-
dithiocarbamate

Until about 1960, arsenic trioxide and other inorganic arsenic compounds (see Section 7.4) were employed to kill weeds. Because of the incredibly high rates of application of up to several hundred kilograms per acre, and because arsenic is non-biodegradable, the potential still exists for arsenic pollution of surface water and groundwater from fields formerly dosed with inorganic arsenic. Organic arsenicals, such as cacodylic acid,

Hydroxydimethylarsine oxide, cacodylic acid

have also been widely applied to kill weeds.

Byproducts of Pesticide Manufacture

A number of water pollution and health problems have been associated with the manufacture of organochlorine pesticides. For example, degradation-resistant hexachlorobenzene,

is used as a raw material for the synthesis of other pesticides and has often been found in water.

The most notorious byproducts of pesticide manufacture are **polychlorinated dibenzodioxins**. From 1 to 8 Cl atoms may be substituted for H atoms on dibenzo-p-dioxin (Figure 7.9), giving a total of 75 possible chlorinated derivatives. Commonly referred to as "dioxins," these species have a high environmental and toxicological significance. Of the dioxins, the most notable pollutant and hazardous waste compound is **2,3,7,8-tetrachlorodibenzo-p-dioxin** (**TCDD**), often referred to simply as "**dioxin**." This compound, which is one of the most toxic of all synthetic substances to some animals, was produced as a low-level contaminant in the manufacture of some aryl, oxygen-containing organohalide compounds such as chlorophenoxy herbicides (mentioned previously in this section) synthesized by processes used until the 1960s.

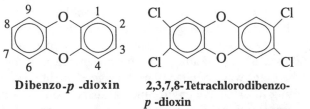

Dibenzo-p-dioxin   2,3,7,8-Tetrachlorodibenzo-p-dioxin

Figure 7.9.   Dibenzo-p-dioxin and 2,3,7,8-tetrachlorodibenzo-p-dioxin (TCDD), often called simply "dioxin." In the structure of dibenzo-p-dioxin, each number refers to a numbered carbon atom to which an H atom is bound, and the names of derivatives are based upon the carbon atoms where another group has been substituted for the H atoms, as is seen by the structure and name of 2,3,7,8-tetrachlorodibenzo-p-dioxin.

TCDD has a very low vapor pressure of only 1.7×10^{-6} mm Hg at 25°C, a high melting point of 305°C, and a water solubility of only 0.2 μg/L. It is stable thermally up to about 700°C, has a high degree of chemical stability, and is poorly biodegradable. It is very toxic to some animals, with an LD_{50} of only about 0.6 μg/kg body mass in male guinea pigs. (The type and degree of its toxicity to humans is largely unknown; it is known to cause a severe skin condition called chloracne). Because of its properties, TCDD is a stable, persistent environmental pollutant and hazardous waste constituent of considerable concern. It has been identified in some municipal incineration emissions, in which it is believed to form when chlorine from the combustion of organochlorine compounds reacts with carbon in the incinerator.

TCDD has been a highly publicized environmental pollutant from improper waste disposal. The most notable case of TCDD contamination resulted from the spraying of waste oil mixed with TCDD on roads and horse arenas in Missouri in the early 1970s. The oil was used to try to keep dust down in these areas. The extent of contamination was revealed by studies conducted in late 1982 and early 1983. As a result, the U.S. EPA bought out the entire TCDD-contaminated town of Times Beach, Missouri, in March 1983, at a cost of $33 million. Subsequently, permission was granted by the courts to incinerate the soil at Times Beach, as well as TCDD-contaminated soil from other areas at a total estimated cost of about $80 million. TCDD has been released in a number of industrial accidents, the most massive of which exposed several tens of thousands of people to a cloud of chemical emissions spread over an approximately 3-square-mile area at the Givaudan-La Roche Icmesa manufacturing plant near Seveso, Italy, in 1976. On an encouraging note from a toxicological perspective, no abnormal occurrences of major malformations were found in a study of 15,291 children born in the area within 6 years after the release.[7]

One of the greater environmental disasters ever to result from pesticide manufacture involved the production of Kepone, structural formula

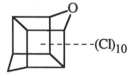

This pesticide has been used for the control of banana-root borer, tobacco wireworm, ants, and cockroaches. Kepone exhibits acute, delayed, and cumulative toxicity in birds, rodents, and humans, and it causes cancer in rodents. It was manufactured in Hopewell, Virginia, during the mid-1970s. During this time, workers were exposed to Kepone and are alleged to have suffered health problems as a result. The plant was connected to the Hopewell sewage system, and frequent infiltration of Kepone wastes caused the Hopewell sewage treatment plant to become inoperative at times. As much as 53,000 kg of Kepone may have been dumped into the sewage system during the years the plant was operated. The sewage effluent was discharged to the James River, resulting in extensive environmental dispersion and toxicity to aquatic organisms. Decontamination of the river would have required dredging and detoxification of 135 million cubic meters of river sediment at a prohibitively high cost of several billion dollars.

7.12. POLYCHLORINATED BIPHENYLS

First discovered as environmental pollutants in 1966, **polychlorinated biphenyls (PCB** compounds) have been found throughout the world in water, sediments, bird tissue, and fish tissue. These compounds constitute an important class of special wastes. They are made by substituting from 1 to 10 Cl atoms onto the biphenyl aryl structure as shown on the left in Figure 7.10. This substitution can produce 209 different compounds (congeners), of which one example is shown on the right in Figure 7.10.

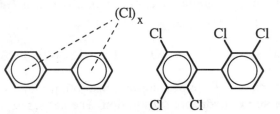

Figure 7.10. General formula of polychlorinated biphenyls (left, where X may range from 1 to 10) and a specific 5-chlorine congener (right).

Polychlorinated biphenyls have very high chemical, thermal, and biological stability; low vapor pressure; and high dielectric constants. These properties have led to the use of PCBs as coolant-insulation fluids in transformers and capacitors; for the impregnation of cotton and asbestos; as plasticizers; and as additives to some epoxy paints. The same properties that made extraordinarily stable PCBs so useful also contributed to their widespread dispersion and accumulation in the environment. By regulations issued in the U.S. under the authority of the Toxic Substances Control Act passed in 1976, the manufacture of PCBs was discontinued in the U.S., and their uses and disposal were strictly controlled.

Several chemical formulations have been developed to substitute for PCBs in electrical applications. Disposal of PCBs from discarded electrical equipment and other sources have caused problems, particularly since PCBs can survive ordinary incineration by escaping as vapors through the smokestack. However, they can be destroyed by special incineration processes.

PCBs are especially prominent pollutants in the sediments of the Hudson River as a result of waste discharges from two capacitor manufacturing plants which operated about 60 km upstream from the southernmost dam on the river from 1950 to 1976. The river sediments downstream from the plants exhibit PCB levels of about 10 ppm, 1-2 orders of magnitude higher than levels commonly encountered in river and estuary sediments.

Biodegradation of PCBs

The biodegradation of PCBs in New York's Hudson River provides an interesting example of microbial degradation of environmental chemicals. As a result of the dumping of PCBs in the Hudson River mentioned above, these virtually insoluble, dense, hydrophobic materials accumulated in the river's sediment, causing serious concern about their effects on water quality as a result of their bioaccumulation in fish. Methods of removal, such as dredging, were deemed prohibitively expensive and likely to cause severe contamination and disposal problems. Although it was well known that aerobic bacteria could degrade PCBs with only one or two Cl atom constituents, most of the PCB congeners discharged to the sediments had multiple chlorine atom constituents, specifically an average of 3.5 Cl atoms per PCB molecule at the time the PCBs were discharged. However, investigations during the late 1980s revealed that the PCBs in the sediments had been largely converted to mono- and dichloro-substituted forms. This conversion had to be due to the action of anaerobic bacteria, a fact later confirmed with laboratory cultures.[8] Such bacteria do not use PCBs as a carbon source, but rather as electron acceptors in the overall process,

$$\{CH_2O\} + H_2O + 2Cl\text{-}PCB \rightarrow CO_2 + 2H^+ + 2Cl^- + 2H\text{-}PCB \qquad (7.12.1)$$

where Cl-PCB represents a site of chlorine substitution on a PCB molecule and H-PCB represents a site of hydrogen substitution. The net result of this process is the replacement of Cl by H on the more highly chlorine-substituted PCB molecules. The circumstances under which this took place in the sediments included anaerobic conditions and very long residence times. These are ideal for the growth of anaerobic bacteria, which tend to carry out their metabolic processes slowly with relatively low efficiency compared to aerobic organisms, and in complex synergistic relationships with other anaerobic bacteria.

The mono- and dichloro substituted PCBs are degraded by aerobic bacteria. Aerobic processes oxidize the molecules and cleave the aryl rings as shown in Figure 7.11. Ultimately, the PCBs are mineralized by conversion to inorganic chloride, carbon dioxide, and water.

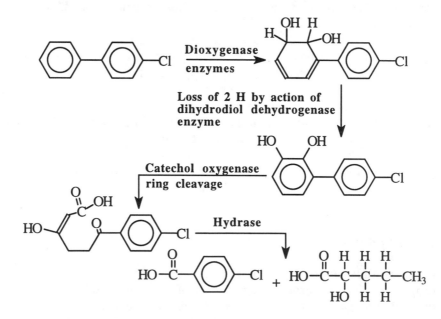

Figure 7.11. Enzymatic processes involved in the initial attack of aerobic bacteria on mono- and dichloro substituted PCBs.

Ideally, in the case of PCBs in sediments, about half of the work of remediation by biodegradation—conversion of highly substituted PCBs to molecules with one or two chlorines—is done by the slow, steady action of anaerobic bacteria without human intervention. Since the PCB products tend to stay in anaerobic surroundings, some assistance is required to provide oxygen to finish the biodegradation aerobically by introducing aerobic bacteria acclimated to PCB biodegradation, along with the oxygen and nutrients required for their growth.[9]

Polybrominated Biphenyls

Polybrominated biphenyl (PBB) is a chemical fire retardant which was accidently mistaken for magnesium oxide and mixed with cattle feed distributed in Michigan in 1973. As a result, over 30,000 cattle, approximately 6000 hogs, 1500 sheep, 1.5 million chickens, 18,000 pounds of cheese, 2700 pounds of butter, 34,000 pounds of dry milk products, and 5 million eggs had to be destroyed. Farm families eating PBB-contaminated foods had ingested the substance, and it was detected in the blood of many Michigan residents, causing a great deal of anxiety to people who may have been exposed to the substance. Persons who had ingested PBB tended to have less disease resistance and increased incidence of rashes, liver ailments, and headaches. The economic cost of the Michigan PBB incident was very high, with some estimates exceeding $100 million.

7.13. RADIONUCLIDES IN THE AQUATIC ENVIRONMENT

The massive production of **radionuclides** (radioactive isotopes) by weapons and nuclear reactors since World War II has been accompanied by increasing concern about the effects of radioactivity upon health and the environment. Radionuclides are produced as fission products of heavy nuclei of such elements as uranium or plutonium. They are also produced by the reaction of neutrons with stable nuclei. These phenomena are illustrated in Figure 7.12 and specific examples are given in Table 7.5. Radionuclides are formed in large quantities as waste products in nuclear power generation. Their ultimate disposal is a problem that has caused much controversy regarding the widespread use of nuclear power. Artificially produced radionuclides are also widely used in industrial and medical applications, particularly as "tracers." With so many possible sources of radionuclides, it is impossible to entirely eliminate radioactive contamination of aquatic systems. Furthermore, radionuclides may enter aquatic systems from natural sources. Therefore, the transport, reactions, and biological concentration of radionuclides in aquatic ecosystems are of great importance to the environmental chemist.

Radionuclides differ from other nuclei in that they emit **ionizing radiation** — alpha particles, beta particles, and gamma rays. The most massive of these emissions is the **alpha particle**, a helium nucleus of atomic mass 4, consisting of two neutrons and two protons. The symbol for an alpha particle is $^4_2\alpha$. An example of alpha production is found in the radioactive decay of uranium-238:

$$^{238}_{92}U \rightarrow \ ^{234}_{90}Th \ + \ ^4_2\alpha \qquad\qquad (7.13.1)$$

This transformation occurs when a uranium nucleus, atomic number 92 and atomic mass 238, loses an alpha particle, atomic number 2 and atomic mass 4, to yield a thorium nucleus, atomic number 90 and atomic mass 234.

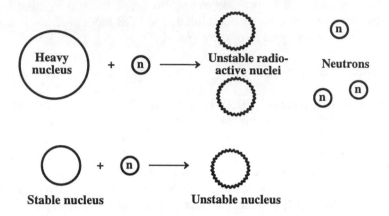

Figure 7.12. A heavy nucleus, such as that of ^{235}U, may absorb a neutron and break up (undergo fission), yielding lighter radioactive nuclei. A stable nucleus may absorb a neutron to produce a radioactive nucleus.

Beta radiation consists of either highly energetic, negative electrons, which are designated $_{-1}^{0}\beta$, or positive electrons, called positrons, and designated $_{-1}^{0}\beta$. A typical beta emitter, chlorine-38, may be produced by irradiating chlorine with neutrons. The chlorine-37 nucleus, natural abundance 24.5%, absorbs a neutron to produce chlorine-38 and gamma radiation:

$$_{17}^{37}Cl + _{0}^{1}n \longrightarrow _{17}^{38}Cl + \gamma \tag{7.13.2}$$

The chlorine-38 nucleus is radioactive and loses a negative **beta particle** to become an argon-38 nucleus:

$$_{17}^{38}Cl \longrightarrow _{18}^{38}Ar + _{-1}^{0}\beta \tag{7.13.3}$$

Since the negative beta particle has essentially no mass and a -1 charge, the stable product isotope, argon-38, has the same mass and a charge 1 greater than chlorine-38.

Gamma rays are electromagnetic radiation similar to X-rays, though more energetic. Since the energy of gamma radiation is often a well-defined property of the emitting nucleus, it may be used in some cases for the qualitative and quantitative analysis of radionuclides.

The primary effect of alpha particles, beta particles, and gamma rays upon materials is the production of ions; therefore, they are called **ionizing radiation**. Due to their large size, alpha particles do not penetrate matter deeply, but cause an enormous amount of ionization along their short path of penetration. Therefore, alpha particles present little hazard outside the body, but are very dangerous when

Table 7.5. Radionuclides in Water

Radionuclide	Half-life	Nuclear reaction, description, source
Naturally occurring and from cosmic reactions		
Carbon-14	5730 y[1]	$^{14}N(n,p)^{14}C$,[2] thermal neutrons from cosmic or nuclear-weapon sources reacting with N_2
Silicon-32	~300 y	$^{40}Ar(p,x)^{32}Si$, nuclear spallation (splitting of the nucleus) of atmospheric argon by cosmic-ray protons
Potassium-40	~1.4×10^9 y	0.0119% of natural potassium including potassium in the body
Naturally occurring from ^{238}U series		
Radium-226	1620 y	Diffusion from sediments, atmosphere
Lead-210	21 y	$^{226}Ra \rightarrow 6$ steps $\rightarrow ^{210}Pb$
Thorium-230	75,200 y	$^{238}U \rightarrow 3$ steps $\rightarrow ^{230}Th$ produced *in situ*
Thorium-234	24 d	$^{238}U \rightarrow ^{234}Th$ produced *in situ*

From reactor and weapons fission[3]

Strontium-90 (28 y)   Iodine-131 (8 d)   Cesium-137 (30 y)

Barium-140 (13 d) > Zirconium-95 (65 d) >Cerium-141 (33d) > Strontium-89 (51 d)

> Ruthenium-103 (40 d) > Krypton-85 (10.3 y)

From nonfission sources		
Cobalt-60	5.25 y	From nonfission neutron reactions in reactors
Manganese-54	310 d	From nonfission neutron reactions in reactors
Iron-55	2.7 y	$^{56}Fe(n,2n)^{55}Fe$, from high-energy neutrons acting on iron in weapons hardware
Plutonium-239	24,300 y	$^{238}U(n,\gamma)^{239}Pu$, neutron capture by uranium

[1] Abbreviations: y, years; d, days

[2] This notation shows the isotope nitrogen-14 reacting with a neutron, n, giving off a proton, p, and forming the carbon-14 isotope; other nuclear reactions may be deduced from this notation where x represents nuclear fragments from spallation.

[3] The first three fission-product radioisotopes listed below as products of reactor and weapons fission are of most significance because of their high yields and biological activity. The other fission products are listed in generally decreasing order of yield.

ingested. Although beta particles are more penetrating than alpha particles, they produce much less ionization per unit path length. Gamma rays are much more penetrating than particulate radiation, but cause much less ionization. Their degree of penetration is proportional to their energy.

The **decay** of a specific radionuclide follows first-order kinetics; that is, the number of nuclei disintegrating in a short time interval is directly proportional to the number of radioactive nuclei present. The rate of decay, $-dN/dt$, is given by the equation,

$$\text{Decay rate} = -\frac{dN}{dt} = \lambda N \qquad (7.13.4)$$

where N is the number of radioactive nuclei present and λ is the rate constant, which has units of reciprocal time. Since the exact number of disintegrations per second is difficult to determine in the laboratory, radioactive decay is often described in terms of the **activity**, A, which is proportional to the absolute rate of decay. The first-order decay equation may be expressed in terms of A,

$$A = A_0 e^{-\lambda t} \qquad (7.13.5)$$

where A is the activity at time t; A_0 is the activity when t is zero; and e is the natural logarithm base. The **half-life**, $t_{1/2}$, is generally used instead of λ to characterize a radionuclide:

$$t_{1/2} = \frac{0.693}{\lambda} \qquad (7.13.6)$$

As the term implies, a half-life is the period of time during which half of a given number of atoms of a specific kind of radionuclide decay. Ten half-lives are required for the loss of 99.9% of the activity of a radionuclide.

Radiation damages living organisms by initiating harmful chemical reactions in tissues. For example, bonds are broken in the macromolecules that carry out life processes. In cases of acute radiation poisoning, bone marrow which produces red blood cells is destroyed and the concentration of red blood cells is diminished. Radiation-induced genetic damage is of great concern. Such damage may not become apparent until many years after exposure. As humans have learned more about the effects of ionizing radiation, the dosage level considered to be safe has steadily diminished. For example, the United States Nuclear Regulatory Commission has dropped the maximum permissible concentration of some radioisotopes to levels of less than one ten-thousandth of those considered safe in the early 1950s. Although it is possible that even the slightest exposure to ionizing radiation entails some damage, some radiation is unavoidably received from natural sources. For the majority of the population, exposure to natural radiation exceeds that from artificial sources.

The study of the ecological and health effects of radionuclides involves consideration of many factors. Among these are the type and energy of radiation emitter and the half-life of the source. In addition, the degree to which the particular element is absorbed by living species and the chemical interactions and transport of the element in aquatic ecosystems are important factors. Radionuclides having very short half-lives may be hazardous when produced but decay too rapidly to affect the environment into which they are introduced. Radionuclides with very long half-lives may be quite persistent in the environment but of such low activity that little environmental damage is caused. Therefore, in general, radionuclides with intermediate half-lives are the most dangerous. They persist long enough to enter living systems while still retaining a high activity. Because they may be incorporated within living tissue, radionuclides of "life elements" are particularly dangerous. Much concern has been expressed over strontium-90, a common waste product of nuclear testing. This element is interchangeable with calcium in bone. Strontium-90 fallout drops onto pasture and crop land and is ingested by cattle. Eventually, it enters the bodies of infants and children by way of cow's milk.

Some radionuclides found in water, primarily radium and potassium-40, originate from natural sources, particularly leaching from minerals. Others come from pollutant sources, primarily nuclear power plants and testing of nuclear weapons. The levels of radionuclides found in water typically are measured in units of picocuries/liter, where a curie is 3.7×10^{10} disintegrations per second, and a picocurie is 1×10^{-12} that amount, or 3.7×10^{-2} disintegrations per second. (2.2 disintegrations per minute).

The radionuclide of most concern in drinking water is **radium**, Ra. Areas in the United States where significant radium contamination of water has been observed include the uranium-producing regions of the western U.S., Iowa, Illinois, Wisconsin, Missouri, Minnesota, Florida, North Carolina, Virginia, and the New England states.

The maximum contaminant level (MCL) for total radium (^{226}Ra plus ^{228}Ra) in drinking water is specified by the U.S. Environmental Protection Agency as 5 pCi/L (picocuries per liter). Perhaps as many as several hundred municipal water supplies in the U.S. exceed this level and require additional treatment to remove radium. Fortunately, conventional water softening processes, which are designed to take out excessive levels of calcium, are relatively efficient in removing radium from water.

As the use of nuclear power has increased, the possible contamination of water by fission-product radioisotopes has become more of a cause for concern. (If nations continue to refrain from testing nuclear weapons above ground, it is hoped that radioisotopes from this source will contribute only minor amounts of radioactivity to water.) Table 7.5 summarizes the major natural and artificial radionuclides likely to be encountered in water.

Transuranic elements are of growing concern in the oceanic environment. These alpha emitters are long-lived and highly toxic. As their production increases, so does the risk of environmental contamination. Included among these elements are various isotopes of neptunium, plutonium, americium, and curium. Specific isotopes, with half-lives in years given in parentheses, are: Np-237 (2.14×10^6); Pu-236 (2.85); Pu-238 (87.8); Pu-239 (2.44×10^4); Pu-240 (6.54×10^3); Pu-241 (15); Pu-242 (3.87×10^5); Am-241 (433); Am-243 (7.37×10^6); Cm-242 (0.22); and Cm-244 (17.9).

LITERATURE CITED

1. Clesceri, Lenore S., Arnold E. Greenberg, and Andrew D. Eaton, *Standard Methods for the Examination of Water and Wastewater*, 20th ed., American Public Health Association, Washington, D.C., 1998.

2. Morse, Paige Marie, "Soaps and Detergents," *Chemical and Engineering News*, February 1, 1999, pp. 35-48.

3. Reuter, John E., Brant C. Allen, Robert C. Richards, James F. Pankow, Charles R. Goldman, Roger L. Scholl, and J. Scott Seyfried, "Concentrations, Sources, and Fate of the Gasoline Oxygenate Methyl *tert*-Butyl Ether (MTBE) in a Multiple-Use Lake," *Environmental Science and Technology*, **32**, 3666-3672 (1998).

4. Tittlemier, Sheryl A., Mary Simon, Walter M. Jarman, John E. Elliott, and Ross J. Norstrom, "Identification of a Novel $C_{10}H_6N_2Br_4Cl_2$ Heterocyclic Compound in Seabird Eggs. A Bioaccumulating Marine Natural Product?," *Environmental Science and Technology*, **33**, 26-33 (1999).

5. "Drinking Water in Midwest Has Pesticides, Report Says," *New York Times*, Oct. 18, 1994, p. C19.

6. Gosselin, Robert E., Roger P. Smith, and Harold C. Hodge, "Paraquat," in *Clinical Toxicology of Commercial Products*, 5th ed., Williams and Wilkins, Baltimore/London, 1984, pp. III-328–III-336.

7. "Dioxin is Found Not to Increase Birth Defects," *New York Times*, March 18, 1988, p. 12.

8. Rhee, G.-Yull, Roger C. Sokol, Brian Bush, and Charlotte M. Bethoney "Long-Term Study of the Anaerobic Dechlorination of Arochlor 1254 with and without Biphenyl Enrichment," *Environmental Science and Technology*, **27**, 714-719 (1993).

9. Liou, Raycharn, James H. Johnson, and John P. Tharakan, "Anaerobic Dechlorination and Aerobic Degradation of PCBs in Soil Columns and Slurries," *Hazardous Industrial Wastes*, **29**, 414-423 (1997)

SUPPLEMENTARY REFERENCES

Allen, Herbert E., A. Wayne Garrison, and George W. Luther, *Metals in Surface Waters*, CRC Press/Lewis Publishers, Boca Raton, FL, 1998.

Barbash, Jack E., Elizabeth A. Resek, and Robert J. Gilliom, Eds., *Pesticides in Ground Water: Distribution, Trends, and Governing Factors*, Vol. 2, Ann Arbor Press, Chelsea, MI, 1997.

Dolan, Edward F., *Our Poisoned Waters*, Cobblehill Books, New York, NY, 1997.

Eckenfelder, W. Wesley, *Industrial Water Pollution Control*, 3rd ed., McGraw-Hill, Boston, 1999.

Gustafson, David, *Pesticides in Drinking Water*, Van Nostrand Reinhold, New York, 1993.

Larson, Steven J., Paul D. Capel, Michael S. Majewski, Eds., *Pesticides in Ground Water: Distribution, Trends, and Governing Factors*, Vol. 3, CRC Press, Boca Raton, FL, 1997.

Laws, Edward A., *Aquatic Pollution: An Introductory Text*, 2nd ed., Wiley, New York, 1993.

Malins, Donald C. and Gary K. Ostrander, Eds., *Aquatic Toxicology: Molecular, Biochemical, and Cellular Perspectives*, CRC Press/Lewis Publishers, Boca Raton, FL, 1994.

Mansour, Mohammed, Ed., *Fate and Prediction of Environmental Chemicals in Soils, Plants, and Aquatic Systems*, Lewis Publishers/CRC Press, Boca Raton, FL, 1993.

Mason, C. F., *Biology of Freshwater Pollution*, 3rd ed., Longman, Harlow, Essex, England, 1996.

Meyer, M. T. and E. M. Thurman, Eds., *Herbicide Metabolites in Surface Water and Groundwater*, American Chemical Society, Washington, D.C., 1996.

Misra, S. G., D. Prasad, and H. S. Gaur, *Environmental Pollution, Water*, Venus Publishing House, New Delhi, 1992.

National Research Council, *Setting Priorities for Drinking Water*, National Academy Press, Washington, D.C., 1999.

Ney, Ronald E., *Fate and Transport of Organic Chemicals in the Environment*, 2nd ed., Government Institutes, Rockville, MD, 1995.

Novotny, Vladimir and Harvey Olem, *Water Quality: Prevention, Identification, and Management of Diffuse Pollution*, Van Nostrand Reinhold, New York, 1994.

Ongley, Edwin D., *Control of Water Pollution from Agriculture*, Food and Agriculture Organization of the United Nations, Rome, 1996.

Pollution Prevention Committee, *Controlling Industrial Laundry Discharges to Wastewater*, The Water Environment Federation, Alexandria, VA, 1999.

Patrick, Ruth, *Rivers of the United States: Pollution and Environmental Management*, John Wiley & Sons, New York, 1999.

Schmitz, Richard J., *Introduction to Water Pollution Biology*, Gulf Publications, Houston, TX, 1996.

Spellman, Frank R., *The Science of Water: Concepts and Applications*, Technomic Publishing Co, Lancaster, PA 1998.

Spellman, Frank R. and Nancy E. Whiting, *Water Pollution Control Technology: Concepts and Applications*, Government Institutes, Rockville, MD, 1999.

Taylor, E. W., Ed., *Aquatic Toxicology*, Cambridge University Press, New York, 1995.

Thornton, Jeffrey A., Ed., *Assessment and Control of Nonpoint Source Pollution of Aquatic Ecosystems: A Practical Approach*, Parthenon Publication Group, New York, 1998.

Trudgill, T., Des E. Walling, and Bruce W. Webb, *Water Quality: Processes and Policy*, Wiley, New York, 1999.

Turkington, Carol A., *Protect Yourself from Contaminated Food and Drink*, Ballantine Books, New York, 1999.

Ware, George W., Ed., *Reviews of Environmental Contamination and Toxicology*, Springer-Verlag, New York, published annually.

Wickramanayake, Godage B. and Robert E. Hinchee, Eds., *Bioremediation and Phytoremediation: Chlorinated and Recalcitrant Compounds*, Batelle Press, Columbus, OH, 1998.

QUESTIONS AND PROBLEMS

1. Which of the following statements is true regarding chromium in water: (a) chromium(III) is suspected of being carcinogenic, (b) chromium(III) is less likely to be found in a soluble form than chromium(VI), (c) the toxicity of chromium(III) in electroplating wastewaters is decreased by oxidation to chromium(VI), (d) chromium is not an essential trace element, (e) chromium is known to form methylated species analogous to methylmercury compounds.

2. What do mercury and arsenic have in common in regard to their interactions with bacteria in sediments?

3. What are some characteristics of radionuclides that make them especially hazardous to humans?

4. To what class do pesticides containing the following group belong?

5. Consider the following compound:

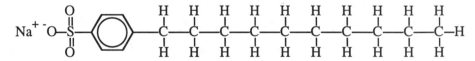

   Which of the following characteristics is not possessed by the compound: (a) one end of the molecule is hydrophilic and the other end in hydrophobic, (b) surface-active qualities, (c) the ability to lower surface tension of water, (d) good biodegradability, (e) tendency to cause foaming in sewage treatment plants.

6. A certain pesticide is fatal to fish fingerlings at a level of 0.50 parts per million in water. A leaking metal can containing 5.00 kg of the pesticide was dumped into a stream with a flow of 10.0 liters per second moving at 1 kilometer per hour. The container leaks pesticide at a constant rate of 5 mg/sec. For what distance (in km) downstream is the water contaminated by fatal levels of the pesticide by the time the container is empty?

7. Give a reason that Na_3PO_4 would not function well as a detergent builder, whereas $Na_3P_3O_{10}$ is satisfactory, though it is a source of pollutant phosphate.

8. Of the compounds $CH_3(CH_2)_{10}CO_2H$, $(CH_3)_3C(CH_2)_2CO_2H$, $CH_3(CH_2)_{10}CH_3$, and ϕ-$(CH_2)_{10}CH_3$ (where ϕ represents a benzene ring), which is the most readily biodegradable?

9. A pesticide sprayer got stuck while trying to ford a stream flowing at a rate of 136 liters per second. Pesticide leaked into the stream for exactly 1 hour and at a rate that contaminated the stream at a uniform 0.25 ppm of methoxychlor. How much pesticide was lost from the sprayer during this time?

10. A sample of water contaminated by the accidental discharge of a radionuclide used for medicinal purposes showed an activity of 12,436 counts per second at the time of sampling and 8,966 cps exactly 30 days later. What is the half-life of the radionuclide?

11. What are the two reasons that soap is environmentally less harmful than ABS surfactant used in detergents?

12. What is the exact chemical formula of the specific compound designated as PCB?

13. Match each compound designated by a letter with the description corresponding to it designated by a number.

(a) CdS (b) $(CH_3)_2AsH$ (c) Cl_x (d) Cl_x

(1) Pollutant released to a U.S. stream by a poorly controlled manufacturing process.

(2) Insoluble form of a toxic trace element likely to be found in anaerobic sediments.

(3) Common environmental pollutant formerly used as a transformer coolant.

(4) Chemical species thought to be produced by bacterial action.

14. A radioisotope has a nuclear half-life of 24 hours and a biological half-life of 16 hours (half of the element is eliminated from the body in 16 hours). A person accidentally swallowed sufficient quantities of this isotope to give an initial "whole body" count rate of 1000 counts per minute. What was the count rate after 16 hours?

15. What is the primary detrimental effect upon organisms of salinity in water arising from dissolved NaCl and Na_2SO_4?

16. Give a specific example of each of the following general classes of water pollutants: (a) trace elements, (b) metal-organic combinations, (c) pesticides

17. A polluted water sample is suspected of being contaminated with one of the following: soap, ABS surfactant, or LAS surfactant. The sample has a very low BOD relative to its TOC. Which is the contaminant?

8 WATER TREATMENT

8.1. WATER TREATMENT AND WATER USE

The treatment of water may be divided into three major categories:

- Purification for domestic use

- Treatment for specialized industrial applications

- Treatment of wastewater to make it acceptable for release or reuse

The type and degree of treatment are strongly dependent upon the source and intended use of the water. Water for domestic use must be thoroughly disinfected to eliminate disease-causing microorganisms, but may contain appreciable levels of dissolved calcium and magnesium (hardness). Water to be used in boilers may contain bacteria but must be quite soft to prevent scale formation. Wastewater being discharged into a large river may require less rigorous treatment than water to be reused in an arid region. As world demand for limited water resources grows, more sophisticated and extensive means will have to be employed to treat water.

Most physical and chemical processes used to treat water involve similar phenomena, regardless of their application to the three main categories of water treatment listed above. Therefore, after introductions to water treatment for municipal use, industrial use, and disposal, each major kind of treatment process is discussed as it applies to all of these applications.

8.2. MUNICIPAL WATER TREATMENT

The modern water treatment plant is often called upon to perform wonders with the water fed to it. The clear, safe, even tasteful water that comes from a faucet may have started as a murky liquid pumped from a polluted river laden with mud and swarming with bacteria. Or, its source may have been well water, much too hard for domestic use and containing high levels of stain-producing dissolved iron and manganese. The water treatment plant operator's job is to make sure that the water plant product presents no hazards to the consumer.

A schematic diagram of a typical municipal water treatment plant is shown in Figure 8.1. This particular facility treats water containing excessive hardness and a high level of iron. The raw water taken from wells first goes to an aerator. Contact of the water with air removes volatile solutes such as hydrogen sulfide, carbon dioxide, methane, and volatile odorous substances such as methane thiol (CH_3SH) and bacterial metabolites. Contact with oxygen also aids iron removal by oxidizing soluble iron(II) to insoluble iron(III). The addition of lime as CaO or $Ca(OH)_2$ after aeration raises the pH and results in the formation of precipitates containing the hardness ions Ca^{2+} and Mg^{2+}. These precipitates settle from the water in a primary basin. Much of the solid material remains in suspension and requires the addition of coagulants (such as iron(III) and aluminum sulfates, which form gelatinous metal hydroxides) to settle the colloidal particles. Activated silica or synthetic polyelectrolytes may also be added to stimulate coagulation or flocculation. The settling occurs in a secondary basin after the addition of carbon dioxide to lower the pH. Sludge from both the primary and secondary basins is pumped to a sludge lagoon. The water is finally chlorinated, filtered, and pumped to the city water mains.

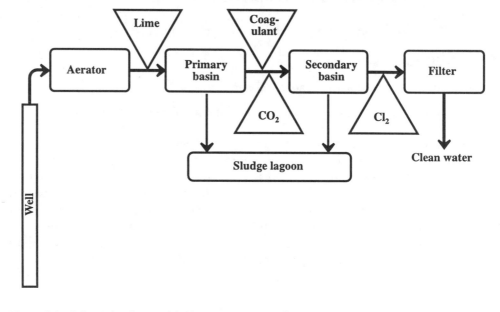

Figure 8.1. Schematic of a municipal water treatment plant.

8.3. TREATMENT OF WATER FOR INDUSTRIAL USE

Water is widely used in various process applications in industry. Other major industrial uses are boiler feedwater and cooling water. The kind and degree of treatment of water in these applications depends upon the end use. As examples, cooling water may require only minimal treatment, removal of corrosive substances and scale-forming solutes is essential for boiler feedwater, and water used in food processing must be free of pathogens and toxic substances. Improper treatment of water for industrial use can cause problems such as corrosion, scale formation, reduced heat transfer in heat exchangers, reduced water flow, and product contamination. These effects may cause reduced equipment performance or equipment failure, increased energy costs due to inefficient heat utilization or cooling, increased costs for pumping water, and product deterioration. Obviously, the effective treatment of water at minimum cost for industrial use is a very important area of water treatment.

Numerous factors must be taken into consideration in designing and operating an industrial water treatment facility. These include the following:

- Water requirement

- Quantity and quality of available water sources

- Sequential use of water (successive uses for applications requiring progressively lower water quality)

- Water recycle

- Discharge standards

The various specific processes employed to treat water for industrial use are discussed in later sections of this chapter. **External treatment**, usually applied to the plant's entire water supply, uses processes such as aeration, filtration, and clarification to remove material that may cause problems from water. Such substances include suspended or dissolved solids, hardness, and dissolved gases. Following this basic treatment, the water may be divided into different streams, some to be used without further treatment, and the rest to be treated for specific applications.

Internal treatment is designed to modify the properties of water for specific applications. Examples of internal treatment include the following:

- Reaction of dissolved oxygen with hydrazine or sulfite

- Addition of chelating agents to react with dissolved Ca^{2+} and prevent formation of calcium deposits

- Addition of precipitants, such as phosphate used for calcium removal

- Treatment with dispersants to inhibit scale

- Addition of inhibitors to prevent corrosion

- Adjustment of pH

- Disinfection for food processing uses or to prevent bacterial growth in cooling water

8.4. SEWAGE TREATMENT

Typical municipal sewage contains oxygen-demanding materials, sediments, grease, oil, scum, pathogenic bacteria, viruses, salts, algal nutrients, pesticides, refractory organic compounds, heavy metals, and an astonishing variety of flotsam ranging from children's socks to sponges. It is the job of the waste treatment plant to remove as much of this material as possible.

Several characteristics are used to describe sewage. These include turbidity (international turbidity units), suspended solids (ppm), total dissolved solids (ppm), acidity (H^+ ion concentration or pH), and dissolved oxygen (in ppm O_2). Biochemical oxygen demand is used as a measure of oxygen-demanding substances.

Current processes for the treatment of wastewater may be divided into three main categories of primary treatment, secondary treatment, and tertiary treatment, each of which is discussed separately. Also discussed are total wastewater treatment systems, based largely upon physical and chemical processes

Waste from a municipal water system is normally treated in a **publicly owned treatment works, POTW**. In the United States these systems are allowed to discharge only effluents that have attained a certain level of treatment, as mandated by Federal law.

Primary Waste Treatment

Primary treatment of wastewater consists of the removal of insoluble matter such as grit, grease, and scum from water. The first step in primary treatment normally is screening. Screening removes or reduces the size of trash and large solids that get into the sewage system. These solids are collected on screens and scraped off for subsequent disposal. Most screens are cleaned with power rakes. Comminuting devices shred and grind solids in the sewage. Particle size may be reduced to the extent that the particles can be returned to the sewage flow.

Grit in wastewater consists of such materials as sand and coffee grounds which do not biodegrade well and generally have a high settling velocity. **Grit removal** is practiced to prevent its accumulation in other parts of the treatment system, to reduce clogging of pipes and other parts, and to protect moving parts from abrasion and wear. Grit normally is allowed to settle in a tank under conditions of low flow velocity, and it is then scraped mechanically from the bottom of the tank.

Primary sedimentation removes both settleable and floatable solids. During primary sedimentation there is a tendency for flocculent particles to aggregate for better settling, a process that may be aided by the addition of chemicals. The material that floats in the primary settling basin is known collectively as grease. In addition to fatty substances, the grease consists of oils, waxes, free fatty acids, and insoluble soaps containing calcium and magnesium. Normally, some of the grease settles with the sludge and some floats to the surface, where it may be removed by a skimming device.

Secondary Waste Treatment by Biological Processes

The most obvious harmful effect of biodegradable organic matter in wastewater is BOD, consisting of a biochemical oxygen demand for dissolved oxygen by microorganism-mediated degradation of the organic matter. **Secondary wastewater treatment** is designed to remove BOD, usually by taking advantage of the same kind of biological processes that would otherwise consume oxygen in water receiving the wastewater. Secondary treatment by biological processes takes many forms but consists basically of the action of microorganisms provided with added oxygen degrading organic material in solution or in suspension until the BOD of the waste has been reduced to acceptable levels. The waste is oxidized biologically under conditions controlled for optimum bacterial growth, and at a site where this growth does not influence the environment.

One of the simplest biological waste treatment processes is the **trickling filter** (Fig. 8.2) in which wastewater is sprayed over rocks or other solid support material covered with microorganisms. The structure of the trickling filter is such that contact of the wastewater with air is allowed and degradation of organic matter occurs by the action of the microorganisms.

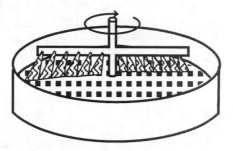

Figure 8.2. Trickling filter for secondary waste treatment.

Rotating biological reactors (**contactors**), another type of treatment system, consist of groups of large plastic discs mounted close together on a rotating shaft. The device is positioned such that at any particular instant half of each disc is immersed in wastewater and half exposed to air. The shaft rotates constantly, so that the submerged portion of the discs is always changing. The discs, usually made of high-density polyethylene or polystyrene, accumulate thin layers of attached biomass, which degrades organic matter in the sewage. Oxygen is absorbed by the biomass and by the layer of wastewater adhering to it during the time that the biomass is exposed to air.

Both trickling filters and rotating biological reactors are examples of fixed-film biological (FFB) or attached growth processes. The greatest advantage of these processes is their low energy consumption. The energy consumption is minimal because it is not necessary to pump air or oxygen into the water, as is the case with the popular activated sludge process described below. The trickling filter has long been a standard means of wastewater treatment, and a number of wastewater treatment plants use trickling filters at present.

The **activated sludge process**, Figure 8.3, is probably the most versatile and effective of all wastewater treatment processes. Microorganisms in the aeration tank convert organic material in wastewater to microbial biomass and CO_2. Organic nitrogen is converted to ammonium ion or nitrate. Organic phosphorus is converted to orthophosphate. The microbial cell matter formed as part of the waste degradation processes is normally kept in the aeration tank until the microorganisms are past the log phase of growth (Section 6.3), at which point the cells flocculate relatively well to form settleable solids. These solids settle out in a settler and a fraction of them is discarded. Part of the solids, the return sludge, is recycled to the head of the aeration tank and comes into contact with fresh sewage. The combination of a high concentration of "hungry" cells in the return sludge and a rich food source in the influent sewage provides optimum conditions for the rapid degradation of organic matter.

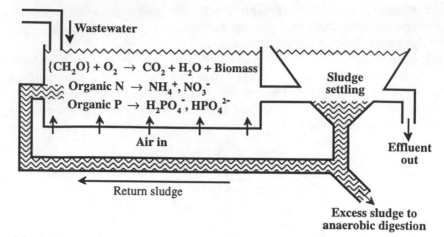

Figure 8.3. Activated sludge process.

The degradation of organic matter that occurs in an activated sludge facility also occurs in streams and other aquatic environments. However, in general, when a degradable waste is put into a stream, it encounters only a relatively small population of microorganisms capable of carrying out the degradation process. Thus, several days may be required for the buildup of a sufficient population of organisms to degrade the waste. In the activated sludge process, continual recycling of active organisms provides the optimum conditions for waste degradation, and a waste may be degraded within the very few hours that it is present in the aeration tank.

The activated sludge process provides two pathways for the removal of BOD, as illustrated schematically in Figure 8.4. BOD may be removed by (1) oxidation of organic matter to provide energy for the metabolic processes of the microorganisms, and (2) synthesis, incorporation of the organic matter into cell mass. In the first pathway, carbon is removed in the gaseous form as CO_2. The second pathway provides for removal of carbon as a solid in biomass. That portion of the carbon converted to CO_2 is vented to the atmosphere and does not present a disposal problem. The disposal of waste sludge, however, is a problem, primarily because it is only about 1% solids and contains many undesirable components. Normally, partial water removal is accomplished by drying on sand filters, vacuum filtration, or centrifugation. The dewatered sludge may be incinerated or used as landfill. To a certain extent, sewage

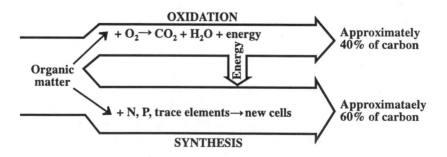

Figure 8.4. Pathways for the removal of BOD in biological wastewater treatment.

sludge may be digested in the absence of oxygen by methane-producing anaerobic bacteria to produce methane and carbon dioxide,

$$2\{CH_2O\} \rightarrow CH_4 + CO_2 \qquad\qquad (8.4.1)$$

a process that reduces both the volatile-matter content and the volume of the sludge by about 60%. A carefully designed plant may produce enough methane to provide for all of its power needs.

One of the most desirable means of sludge disposal is to use it to fertilize and condition soil. However, care has to be taken that excessive levels of heavy metals are not applied to the soil as sludge contaminants. Problems with various kinds of sludges resulting from water treatment are discussed further in Section 8.10.

Activated sludge wastewater treatment is the most common example of an aerobic suspended culture process. Many factors must be considered in the design and operation of an activated sludge wastewater treatment system.[1] These include parameters involved with the process modeling and kinetics. The microbiology of the system must be considered. In addition to BOD removal, phosphorus and nitrogen removal must also be taken into account. Oxygen transfer and solids separation are important. Industrial wastes and the fates and effects of industrial chemicals (xenobiotics) must also be considered.

Nitrification (the microbially mediated conversion of ammonium nitrogen to nitrate; see Section 6.11) is a significant process that occurs during biological waste treatment. Ammonium ion is normally the first inorganic nitrogen species produced in the biodegradation of nitrogenous organic compounds. It is oxidized, under the appropriate conditions, first to nitrite by *Nitrosomonas* bacteria, then to nitrate by *Nitrobacter*:

$$2NH_4^+ + 3O_2 \rightarrow 4H^+ + 2NO_2^- + 2H_2O \qquad\qquad (8.4.2)$$

$$2NO_2^- + O_2 \rightarrow 2NO_3^- \qquad\qquad (8.4.3)$$

These reactions occur in the aeration tank of the activated sludge plant and are favored in general by long retention times, low organic loadings, large amounts of suspended solids, and high temperatures. Nitrification can reduce sludge settling efficiency because the denitrification reaction

$$4NO_3^- + 5\{CH_2O\} + 4H^+ \rightarrow 2N_2(g) + 5CO_2(g) + 7H_2O \qquad\qquad (8.4.4)$$

occurring in the oxygen-deficient settler causes bubbles of N_2 to form on the sludge floc (aggregated sludge particles), making it so buoyant that it floats to the top. This prevents settling of the sludge and increases the organic load in the receiving waters. Under the appropriate conditions, however, advantage can be taken of this phenomenon to remove nutrient nitrogen from water (see Section 8.9).

Tertiary Waste Treatment

Unpleasant as the thought may be, many people drink used water—water that has been discharged from a municipal sewage treatment plant or from some industrial process. This raises serious questions about the presence of pathogenic organisms or toxic substances in such water. Because of high population density and heavy industrial development, the problem is especially acute in Europe where some municipalities process 50% or more of their water from "used" sources. Obviously, there is a great need to treat wastewater in a manner that makes it amenable to reuse. This requires treatment beyond the secondary processes.

Tertiary waste treatment (sometimes called **advanced waste treatment**) is a term used to describe a variety of processes performed on the effluent from secondary waste treatment.[2] The contaminants removed by tertiary waste treatment fall into the general categories of (1) suspended solids, (2) dissolved organic compounds, and (3) dissolved inorganic materials, including the important class of algal nutrients. Each of these categories presents its own problems with regard to water quality. Suspended solids are primarily responsible for residual biological oxygen demand in secondary sewage effluent waters. The dissolved organics are the most hazardous from the standpoint of potential toxicity. The major problem with dissolved inorganic materials is that presented by algal nutrients, primarily nitrates and phosphates. In addition, potentially hazardous toxic metals may be found among the dissolved inorganics.

In addition to these chemical contaminants, secondary sewage effluent often contains a number of disease-causing microorganisms, requiring disinfection in cases where humans may later come into contact with the water. Among the bacteria that may be found in secondary sewage effluent are organisms causing tuberculosis, dysenteric bacteria (*Bacillus dysenteriae, Shigella dysenteriae, Shigella paradysenteriae, Proteus vulgaris*), cholera bacteria (*Vibrio cholerae*), bacteria causing mud fever (*Leptospira icterohemorrhagiae*), and bacteria causing typhoid fever (*Salmonella typhosa, Salmonella paratyphi*). In addition, viruses causing diarrhea, eye infections, infectious hepatitis, and polio may be encountered. Ingestion of sewage still causes disease, even in more developed nations.

Physical-Chemical Treatment of Municipal Wastewater

Complete physical-chemical wastewater treatment systems offer both advantages and disadvantages relative to biological treatment systems. The capital costs of physical-chemical facilities can be less than those of biological treatment facilities, and they usually require less land. They are better able to cope with toxic materials and overloads. However, they require careful operator control and consume relatively large amounts of energy.

Basically, a physical-chemical treatment process involves:

- Removal of scum and solid objects

- Clarification, generally with addition of a coagulant, and frequently with the addition of other chemicals (such as lime for phosphorus removal)

- Filtration to remove filterable solids

- Activated carbon adsorption

- Disinfection

The basic steps of a complete physical-chemical wastewater treatment facility are shown in Figure 8.5.

During the early 1970s, it appeared likely that physical-chemical treatment would largely replace biological treatment. However, higher chemical and energy costs since then have slowed the development of physical-chemical facilities.

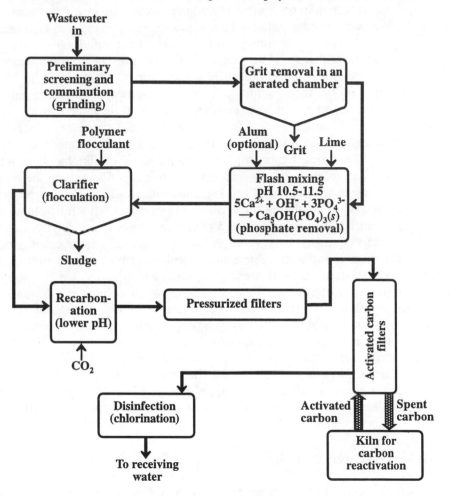

Figure 8.5. Major components of a complete physical-chemical treatment facility for municipal wastewater.

8.5. INDUSTRIAL WASTEWATER TREATMENT

Before treatment, industrial wastewater should be characterized fully and the biodegradability of wastewater constituents determined. The options available for the treatment of wastewater are summarized briefly in this section and discussed in greater detail in later sections.

One of two major ways of removing organic wastes is biological treatment by an activated sludge or related process (see Section 8.4 and Figure 8.3). It may be necessary to acclimate microorganisms to the degradation of constituents that are not normally biodegradable. Consideration needs to be given to possible hazards of biotreatment sludges, such as those containing excessive levels of heavy metal ions. The other major process for the removal of organics from wastewater is sorption by activated carbon (see Section 8.8), usually in columns of granular activated carbon. Activated carbon and biological treatment can be combined with the use of powdered activated carbon in the activated sludge process. The powdered activated carbon sorbs some constituents that may be toxic to microorganisms and is collected with the sludge. A major consideration with the use of activated carbon to treat wastewater is the hazard that spent activated carbon may present from the wastes it retains. These hazards may include those of toxicity or reactivity, such as those posed by wastes from the manufacture of explosives sorbed to activated carbon. Regeneration of the carbon is expensive and can be hazardous in some cases.

Wastewater can be treated by a variety of chemical processes, including acid/base neutralization, precipitation, and oxidation/reduction. Sometimess these steps must precede biological treatment; for example, acidic or alkaline wastewater must be neutralized in order for microorganisms to thrive in it. Cyanide in the wastewater may be oxidized with chlorine and organics with ozone, hydrogen peroxide promoted with ultraviolet radiation, or dissolved oxygen at high temperatures and pressures. Heavy metals may be precipitated with base, carbonate, or sulfide.

Wastewater can be treated by several physical processes. In some cases, simple density separation and sedimentation can be used to remove water-immiscible liquids and solids. Filtration is frequently required, and flotation by gas bubbles generated on particle surfaces may be useful. Wastewater solutes can be concentrated by evaporation, distillation, and membrane processes, including reverse osmosis, hyperfiltration, and ultrafiltration. Organic constituents can be removed by solvent extraction, air stripping, or steam stripping.

Synthetic resins are useful for removing some pollutant solutes from wastewater. Organophilic resins have proven useful for the removal of alcohols; aldehydes; ketones; hydrocarbons; chlorinated alkanes, alkenes, and aryl compounds; esters, including phthalate esters; and pesticides. Cation exchange resins are effective for the removal of heavy metals.

8.6. REMOVAL OF SOLIDS

Relatively large solid particles are removed from water by simple **settling** and **filtration**. A special type of filtration procedure known as **microstraining** is especially effective in the removal of the very small particles. These filters are woven from stainless steel wire so fine that it is barely visible. This enables preparation of filters with openings only 60-70 μm across. These openings may be reduced to 5-15 μm by partial clogging with small particles, such as bacterial cells. The cost of this treatment is likely to be substantially lower than the costs of competing processes. High flow rates at low back pressures are normally achieved.

The removal of colloidal solids from water usually requires **coagulation**. Salts of aluminum and iron are the coagulants most often used in water treatment. Of these, alum or filter alum is most commonly used. This substance is a hydrated aluminum sulfate, $Al_2(SO_4)_3 \cdot 18H_2O$. When this salt is added to water, the aluminum ion hydrolyzes by reactions that consume alkalinity in the water, such as:

$$Al(H_2O)_6^{3+} + 3HCO_3^- \rightarrow Al(OH)_3(s) + 3CO_2 + 6H_2O \qquad (8.6.1)$$

The gelatinous hydroxide thus formed carries suspended material with it as it settles. Furthermore, it is likely that positively charged hydroxyl-bridged dimers such as

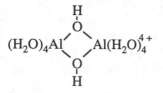

and higher polymers are formed which interact specifically with colloidal particles, bringing about coagulation. Sodium silicate partially neutralized by acid aids coagulation, particularly when used with alum. Metal ions in coagulants also react with virus proteins and destroy viruses in water.

Anhydrous iron(III) sulfate added to water forms iron(III) hydroxide in a reaction analogous to Reaction 8.6.1. An advantage of iron(III) sulfate is that it works over a wide pH range of approximately 4-11. Hydrated iron(II) sulfate, or copperas, $FeSO_4 \cdot 7H_2O$, is also commonly used as a coagulant. It forms a gelatinous precipitate of hydrated iron(III) oxide; in order to function, it must be oxidized to iron(III) by dissolved oxygen in the water at a pH higher than 8.5, or by chlorine, which can oxidize iron(II) at lower pH values.

Natural and synthetic polyelectrolytes are used in flocculating particles. Among the natural compounds so used are starch and cellulose derivatives, proteinaceous materials, and gums composed of polysaccharides. More recently, selected synthetic polymers, including neutral polymers and both anionic and cationic polyelectrolytes, that are effective flocculants have come into use.

Coagulation-filtration is a much more effective procedure than filtration alone for the removal of suspended material from water. As the term implies, the process consists of the addition of coagulants that aggregate the particles into larger size particles, followed by filtration. Either alum or lime, often with added polyelectrolytes, is most commonly employed for coagulation .

The filtration step of coagulation-filtration is usually performed on a medium such as sand or anthracite coal. Often, to reduce clogging, several media with progressively smaller interstitial spaces are used. One example is the **rapid sand filter**, which consists of a layer of sand supported by layers of gravel particles, the particles becoming progressively larger with increasing depth. The substance that actually filters the water is coagulated material that collects in the sand. As more material is removed, the buildup of coagulated material eventually clogs the filter and must be removed by back-flushing.

An important class of solids that must be removed from wastewater consists of suspended solids in secondary sewage effluent that arise primarily from sludge that was not removed in the settling process. These solids account for a large part of the BOD in the effluent and may interfere with other aspects of tertiary waste treatment, such as by clogging membranes in reverse osmosis water treatment processes. The quantity of material involved may be rather high. Processes designed to remove suspended solids often will remove 10-20 mg/L of organic material from secondary sewage effluent. In addition, a small amount of the inorganic material is removed.

8.7. REMOVAL OF CALCIUM AND OTHER METALS

Calcium and magnesium salts, which generally are present in water as bicarbonates or sulfates, cause water hardness. One of the most common manifestations of water hardness is the insoluble "curd" formed by the reaction of soap with calcium or magnesium ions. The formation of these insoluble soap salts is discussed in Section 7.10. Although ions that cause water hardness do not form insoluble products with detergents, they do adversely affect detergent performance. Therefore, calcium and magnesium must be complexed or removed from water in order for detergents to function properly.

Another problem caused by hard water is the formation of mineral deposits. For example, when water containing calcium and bicarbonate ions is heated, insoluble calcium carbonate is formed:

$$Ca^{2+} + 2HCO_3^- \rightarrow CaCO_3(s) + CO_2(g) + H_2O \qquad (8.7.1)$$

This product coats the surfaces of hot water systems, clogging pipes and reducing heating efficiency. Dissolved salts such as calcium and magnesium bicarbonates and sulfates can be especially damaging in boiler feedwater. Clearly, the removal of water hardness is essential for many uses of water.

Several processes are used for softening water. On a large scale, such as in community water-softening operations, the lime-soda process is used. This process involves the treatment of water with lime, $Ca(OH)_2$, and soda ash, Na_2CO_3. Calcium is precipitated as $CaCO_3$ and magnesium as $Mg(OH)_2$. When the calcium is present primarily as "bicarbonate hardness," it can be removed by the addition of $Ca(OH)_2$ alone:

$$Ca^{2+} + 2HCO_3^- + Ca(OH)_2 \rightarrow 2CaCO_3(s) + 2H_2O \qquad (8.7.2)$$

When bicarbonate ion is not present at substantial levels, a source of CO_3^{2-} must be provided at a high enough pH to prevent conversion of most of the carbonate to bicarbonate. These conditions are obtained by the addition of Na_2CO_3. For example, calcium present as the chloride can be removed from water by the addition of soda ash:

$$Ca^{2+} + 2Cl^- + 2Na^+ + CO_3^{2-} \rightarrow CaCO_3(s) + 2Cl^- + 2Na^+ \qquad (8.7.3)$$

Note that the removal of bicarbonate hardness results in a net removal of soluble salts from solution, whereas removal of nonbicarbonate hardness involves the addition of at least as many equivalents of ionic material as are removed.

The precipitation of magnesium as the hydroxide requires a higher pH than the precipitation of calcium as the carbonate:

$$Mg^{2+} + 2OH^- \rightarrow Mg(OH)_2(s) \qquad (8.7.4)$$

The high pH required may be provided by the basic carbonate ion from soda ash:

$$CO_3^{2-} + H_2O \rightarrow HCO_3^- + OH^- \qquad (8.7.5)$$

Some large-scale lime-soda softening plants make use of the precipitated calcium carbonate product as a source of additional lime. The calcium carbonate is first heated to at least 825°C to produce quicklime, CaO:

$$CaCO_3 + heat \rightarrow CaO + CO_2(g) \qquad (8.7.6)$$

The quicklime is then slaked with water to produce calcium hydroxide:

$$CaO + H_2O \rightarrow Ca(OH)_2 \qquad (8.7.7)$$

The water softened by lime-soda softening plants usually suffers from two defects. First, because of super-saturation effects, some $CaCO_3$ and $Mg(OH)_2$ usually remain in solution. If not removed, these compounds will precipitate at a later time and cause harmful deposits or undesirable cloudiness in water. The second problem results from the use of highly basic sodium carbonate, which gives the product water an excessively high pH, up to pH 11. To overcome these problems, the water is recarbonated by bubbling CO_2 into it. The carbon dioxide converts the slightly soluble calcium carbonate and magnesium hydroxide to their soluble bicarbonate forms:

$$CaCO_3(s) + CO_2 + H_2O \rightarrow Ca^{2+} + 2HCO_3^- \qquad (8.7.8)$$

$$Mg(OH)_2(s) + 2CO_2 \rightarrow Mg^{2+} + 2HCO_3^- \qquad (8.7.9)$$

The CO_2 also neutralizes excess hydroxide ion:

$$OH^- + CO_2 \rightarrow HCO_3^- \qquad (8.7.10)$$

The pH generally is brought within the range 7.5-8.5 by recarbonation. The source of CO_2 used in the recarbonation process may be from the combustion of carbonaceous fuel. Scrubbed stack gas from a power plant frequently is utilized. Water adjusted to a pH, alkalinity, and Ca^{2+} concentration very close to $CaCO_3$ saturation is labeled *chemically stabilized*. It neither precipitates $CaCO_3$ in water mains, which can clog the pipes, nor dissolves protective $CaCO_3$ coatings from the pipe surfaces. Water with Ca^{2+} concentration much below $CaCO_3$ saturation is called an *aggressive* water.

Calcium may be removed from water very efficiently by the addition of orthophosphate:

$$5Ca^{2+} + 3PO_4^{3-} + OH^- \rightarrow Ca_5OH(PO_4)_3(s) \qquad (8.7.11)$$

It should be pointed out that the chemical formation of a slightly soluble product for the removal of undesired solutes such as hardness ions, phosphate, iron, and manganese must be followed by sedimentation in a suitable apparatus. Frequently, coagulants must be added, and filtration employed for complete removal of these sediments.

Water may be purified by ion exchange, the reversible transfer of ions between aquatic solution and a solid material capable of bonding ions. The removal of NaCl from solution by two ion exchange reactions is a good illustration of this process. First the water is passed over a solid cation exchanger in the hydrogen form, represented by $H^{+-}\{Cat(s)\}$:

$$H^{+-}\{Cat(s)\} + Na^+ + Cl^- \rightarrow Na^{+-}\{Cat(s)\} + H^+ + Cl^- \qquad (8.7.12)$$

Next, the water is passed over an anion exchanger in the hydroxide ion form, represented by $OH^{-+}\{An(s)\}$:

$$OH^{-+}\{An(s)\} + H^+ + Cl^- \rightarrow Cl^{-+}\{An(s)\} + H_2O \qquad (8.7.13)$$

Thus, the cations in solution are replaced by hydrogen ion and the anions by hydroxide ion, yielding water as the product.

The softening of water by ion exchange does not require the removal of all ionic solutes, just those cations responsible for water hardness. Generally, therefore, only a cation exchanger is necessary. Furthermore, the sodium rather than the hydrogen form of the cation exchanger is used, and the divalent cations are replaced by sodium ion. Sodium ion at low concentrations is harmless in water to be used for most purposes, and sodium chloride is a cheap and convenient substance with which to recharge the cation exchangers.

A number of materials have ion-exchanging properties. Among the minerals especially noted for their ion exchange properties are the aluminum silicate minerals, or **zeolites**. An example of a zeolite which has been used commercially in water softening is glauconite, $K_2(MgFe)_2Al_6(Si_4O_{10})_3(OH)_{12}$. Synthetic zeolites have been prepared by drying and crushing the white gel produced by mixing solutions of sodium silicate and sodium aluminate.

The discovery in the mid-1930s of synthetic ion exchange resins composed of organic polymers with attached functional groups marked the beginning of modern ion exchange technology. Structural formulas of typical synthetic ion exchangers are shown in Figures 8.6 and 8.7. The cation exchanger shown in Figure 8.6 is called a **strongly acidic cation exchanger** because the parent $-SO_3^-H^+$ group is a strong acid. When the functional group binding the cation is the $-CO_2^-$ group, the exchange resin is called a **weakly acidic cation exchanger**, because the $-CO_2H$ group is a weak acid. Figure 8.7 shows a **strongly basic anion exchanger** in which the functional group is a quaternary ammonium group, $-N^+(CH_3)_3$. In the hydroxide form, $-N^+(CH_3)_3OH^-$, the hydroxide ion is readily released, so the exchanger is classified as **strongly basic**.

The water-softening capability of a cation exchanger is shown in Figure 8.6, where sodium ion on the exchanger is exchanged for calcium ion in solution. The same reaction occurs with magnesium ion. Water softening by cation exchange is now a widely used, effective, and economical process. In many areas having a low water flow, however, it is not likely that home water softening by ion exchange may be used universally without some deterioration of water quality arising from the contamination of wastewater by sodium chloride. Such contamination results from

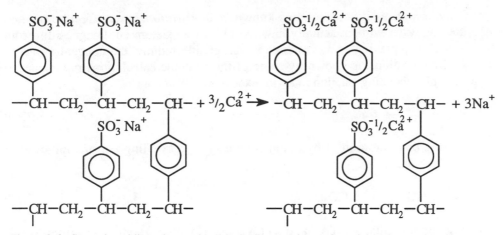

Figure 8.6. Strongly acidic cation exchanger. Sodium exchange for calcium in water is shown.

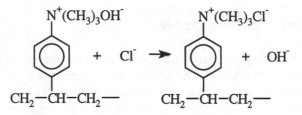

Figure 8.7. Strongly basic anion exchanger. Chloride exchange for hydroxide ion is shown.

the periodic need to regenerate a water softener with sodium chloride in order to displace calcium and magnesium ions from the resin and replace these hardness ions with sodium ions:

$$Ca^{2+} {}^{-}\{Cat(s)\}_2 + 2Na^+ + 2Cl^- \rightarrow 2Na^+ {}^{-}\{Cat(s)\} + Ca^{2+} + 2Cl^- \qquad (8.7.14)$$

During the regeneration process, a large excess of sodium chloride must be used — several pounds for a home water softener. Appreciable amounts of dissolved sodium chloride can be introduced into sewage by this route.

Strongly acidic cation exchangers are used for the removal of water hardness. Weakly acidic cation exchangers having the $-CO_2H$ group as a functional group are useful for removing alkalinity. Alkalinity generally is manifested by bicarbonate ion, a species that is a sufficiently strong base to neutralize the acid of a weak acid cation exchanger:

$$2R\text{-}CO_2H + Ca^{2+} + 2HCO_3^- \rightarrow [R\text{-}CO_2^-]_2Ca^{2+} + 2H_2O + 2CO_2 \qquad (8.7.15)$$

However, weak bases such as sulfate ion or chloride ion are not strong enough to remove hydrogen ion from the carboxylic acid exchanger. An additional advantage of these exchangers is that they may be regenerated almost stoichiometrically with dilute strong acids, thus avoiding the potential pollution problem caused by the use of excess sodium chloride to regenerate strongly acidic cation changers.

Chelation or, as it is sometimes known, *sequestration*, is an effective method of softening water without actually having to remove calcium and magnesium from solution. A complexing agent is added which greatly reduces the concentrations of free hydrated cations, as shown by some of the example calculations in Chapter 3. For example, chelating calcium ion with excess EDTA anion (Y^{4-}),

$$Ca^{2+} + Y^{4-} \rightarrow CaY^{2-} \qquad (8.7.16)$$

reduces the concentration of hydrated calcium ion, preventing the precipitation of calcium carbonate:

$$Ca^{2+} + CO_3^{2-} \rightarrow CaCO_3(s) \qquad (8.7.17)$$

Polyphosphate salts, EDTA, and NTA (see Chapter 3) are chelating agents commonly used for water softening. Polysilicates are used to complex iron.

Removal of Iron and Manganese

Soluble iron and manganese are found in many groundwaters because of reducing conditions which favor the soluble +2 oxidation state of these metals (see Chapter 4). Iron is the more commonly encountered of the two metals. In groundwater, the level of iron seldom exceeds 10 mg/L, and that of manganese is rarely higher than 2 mg/L. The basic method for removing both of these metals depends upon oxidation to higher insoluble oxidation states. The oxidation is generally accomplished by aeration. The rate of oxidation is pH-dependent in both cases, with a high pH favoring more rapid oxidation. The oxidation of soluble $Mn(II)$ to insoluble MnO_2 is a complicated process. It appears to be catalyzed by solid MnO_2, which is known to adsorb $Mn(II)$. This adsorbed $Mn(II)$ is slowly oxidized on the MnO_2 surface.

Chlorine and potassium permanganate are sometimes employed as oxidizing agents for iron and manganese. There is some evidence that organic chelating agents with reducing properties hold iron(II) in a soluble form in water. In such cases, chlorine is effective because it destroys the organic compounds and enables the oxidation of iron(II).

In water with a high level of carbonate, $FeCO_3$ and $MnCO_3$ may be precipitated directly by raising the pH above 8.5 by the addition of sodium carbonate or lime. This approach is less popular than oxidation, however.

Relatively high levels of insoluble iron(III) and manganese(IV) frequently are found in water as colloidal material which is difficult to remove. These metals may be associated with humic colloids or "peptizing" organic material that binds to colloidal metal oxides, stabilizing the colloid.

Heavy metals such as copper, cadmium, mercury, and lead are found in wastewaters from a number of industrial processes. Because of the toxicity of many heavy metals, their concentrations must be reduced to very low levels prior to release of the wastewater. A number of approaches are used in heavy metals removal.

Lime treatment, discussed earlier in this section for calcium removal, precipitates heavy metals as insoluble hydroxides, basic salts, or coprecipitated with calcium carbonate or iron(III) hydroxide. This process does not completely remove mercury, cadmium, or lead, so their removal is aided by addition of sulfide (most heavy metals are sulfide-seekers):

$$Cd^{2+} + S^{2-} \rightarrow CdS(s) \tag{8.7.18}$$

Heavy chlorination is frequently necessary to break down metal-solubilizing ligands (see Chapter 3). Lime precipitation does not normally permit recovery of metals and is sometimes undesirable from the economic viewpoint.

Electrodeposition (reduction of metal ions to metal by electrons at an electrode), *reverse osmosis* (see Section 8.9), and *ion exchange* are frequently employed for metal removal. Solvent extraction using organic-soluble chelating substances is also effective in removing many metals. **Cementation**, a process by which a metal deposits by reaction of its ion with a more readily oxidized metal, may be employed:

$$Cu^{2+} + Fe \text{ (iron scrap)} \rightarrow Fe^{2+} + Cu \tag{8.7.19}$$

Activated carbon adsorption effectively removes some metals from water at the part per million level. Sometimes a chelating agent is sorbed to the charcoal to increase metal removal.

Even when not specifically designed for the removal of heavy metals, most waste treatment processes remove appreciable quantities of the more troublesome heavy metals encountered in wastewater. Biological waste treatment effectively removes metals from water. These metals accumulate in the sludge from biological treatment, so sludge disposal must be given careful consideration.

Various physical-chemical treatment processes effectively remove heavy metals from wastewaters. One such treatment is lime precipitation followed by activated-carbon filtration. Activated-carbon filtration may also be preceded by treatment with iron(III) chloride to form an iron(III) hydroxide floc, which is an effective heavy metals scavenger. Similarly, alum, which forms aluminum hydroxide, may be added prior to activated-carbon filtration.

The form of the heavy metal has a strong effect upon the efficiency of metal removal. For instance, chromium(VI) is normally more difficult to remove than chromium(III). Chelation may prevent metal removal by solubilizing metals (see Chapter 3).

In the past, removal of heavy metals has been largely a fringe benefit of wastewater treatment processes. Currently, however, more consideration is being given to design and operating parameters that specifically enhance heavy-metals removal as part of wastewater treatment.

8.8. REMOVAL OF DISSOLVED ORGANICS

Very low levels of exotic organic compounds in drinking water are suspected of contributing to cancer and other maladies. Water disinfection processes, which by their nature involve chemically rather severe conditions, particularly of oxidation, have a tendency to produce **disinfection by-products**. Some of these are chlorinated organic compounds produced by chlorination of organics in water, especially humic substances. Removal of organics to very low levels prior to chlorination has been found to be effective in preventing trihalomethane formation. Another major class of disinfection by-products consists of organooxygen compounds such as aldehydes, carboxylic acids, and oxoacids.

A variety of organic compounds survive, or are produced by, secondary waste-water treatment and should be considered as factors in discharge or reuse of the treated water. Almost half of these are humic substances (see Section 3.17) with a molecular-weight range of 1000-5000. Among the remainder are found ether-extractable materials, carbohydrates, proteins, detergents, tannins, and lignins. The humic compounds, because of their high molecular weight and anionic character, influence some of the physical and chemical aspects of waste treatment. The ether-extractables contain many of the compounds that are resistant to biodegradation and are of particular concern regarding potential toxicity, carcinogenicity, and muta-genicity. In the ether extract are found many fatty acids, hydrocarbons of the *n*-alkane class, naphthalene, diphenylmethane, diphenyl, methylnaphthalene, isopro-pylbenzene, dodecylbenzene, phenol, phthalates, and triethylphosphate.

The standard method for the removal of dissolved organic material is adsorption on activated carbon, a product that is produced from a variety of carbonaceous materials including wood, pulp-mill char, peat, and lignite.[3] The carbon is produced by charring the raw material anaerobically below 600°C, followed by an activation step consisting of partial oxidation. Carbon dioxide may be employed as an oxidizing agent at 600-700°C.

$$CO_2 + C \rightarrow 2CO \qquad \qquad (8.8.1)$$

or the carbon may be oxidized by water at 800-900°C:

$$H_2O + C \rightarrow H_2 + CO \qquad \qquad (8.8.2)$$

These processes develop porosity, increase the surface area, and leave the C atoms in arrangements that have affinities for organic compounds.

Activated carbon comes in two general types: granulated activated carbon, consisting of particles 0.1-1 mm in diameter, and powdered activated carbon, in which most of the particles are 50-100 μm in diameter.

The exact mechanism by which activated carbon holds organic materials is not known. However, one reason for the effectiveness of this material as an adsorbent is its tremendous surface area. A solid cubic foot of carbon particles may have a combined pore and surface area of approximately 10 square miles!

Although interest is increasing in the use of powdered activated carbon for water treatment, currently granular carbon is more widely used. It may be employed in a fixed bed, through which water flows downward. Accumulation of particulate matter requires periodic backwashing. An expanded bed in which particles are kept slightly separated by water flowing upward may be used with less chance of clogging.

Economics require regeneration of the carbon, which is accomplished by heating it to 950°C in a steam-air atmosphere. This process oxidizes adsorbed organics and regenerates the carbon surface, with an approximately 10% loss of carbon.

Removal of organics may also be accomplished by adsorbent synthetic polymers. Such polymers as Amberlite XAD-4 have hydrophobic surfaces and strongly attract relatively insoluble organic compounds, such as chlorinated pesticides. The porosity of these polymers is up to 50% by volume, and the surface area may be as high as 850 m^2/g. They are readily regenerated by solvents such as isopropanol and acetone. Under appropriate operating conditions, these polymers remove virtually all nonionic organic solutes; for example, phenol at 250 mg/L is reduced to less than 0.1 mg/L by appropriate treatment with Amberlite XAD-4.

Oxidation of dissolved organics holds some promise for their removal. Ozone, hydrogen peroxide, molecular oxygen (with or without catalysts), chlorine and its derivatives, permanganate, or ferrate (iron(VI)) can be used. Electrochemical oxidation may be possible in some cases. High-energy electron beams produced by high-voltage electron accelerators also have the potential to destroy organic compounds.

Removal of Herbicides

Because of their widespread application and persistence, herbicides have proven to be particularly troublesome in drinking water sources. Their levels vary with season related to times that they are applied to control weeds. The more soluble ones, such as chlorophenoxy esters, are most likely to enter drinking water sources. One of the most troublesome is atrazine, which is often manifested by its metabolite desethylatrazine. Activated carbon treatment is the best means of removing herbicides and their metabolites from drinking water sources.[4] A problem with activated carbon is that of **preloading**, in which natural organic matter in the water loads up the carbon and hinders uptake of pollutant organics such as herbicides. Pretreatment to remove such organic matter, such as flocculation and precipitation of humic substances, can significantly increase the efficacy of activated carbon for the removal of herbicides and other organics.

8.9. REMOVAL OF DISSOLVED INORGANICS

In order for complete water recycling to be feasible, inorganic-solute removal is essential. The effluent from secondary waste treatment generally contains 300-400 mg/L more dissolved inorganic material than does the municipal water supply. It is obvious, therefore, that 100% water recycling without removal of inorganics would cause the accumulation of an intolerable level of dissolved material. Even when water is not destined for immediate reuse, the removal of the inorganic nutrients phosphorus and nitrogen is highly desirable to reduce eutrophication downstream. In some cases the removal of toxic trace metals is needed.

One of the most obvious methods for removing inorganics from water is distillation. However, the energy required for distillation is generally quite high, so that distillation is not generally economically feasible. Furthermore, volatile materials such as ammonia and odorous compounds are carried over to a large extent in the distillation process unless special preventative measures are taken. Freezing produces a very pure water, but is considered uneconomical with present technology. This leaves membrane processes as the most cost-effective means of removing inorganic materials from water. Membrane processes considered most promising for bulk removal of inorganics from water are electrodialysis, ion exchange, and reverse osmosis. (Other membrane processes used in water purification are nanofiltration, ultrafiltration,[5] microfiltration, and dialysis.)

Electrodialysis

Electrodialysis consists of applying a direct current across a body of water separated into vertical layers by membranes alternately permeable to cations and anions.[6] Cations migrate toward the cathode and anions toward the anode. Cations and anions both enter one layer of water, and both leave the adjacent layer. Thus, layers of water enriched in salts alternate with those from which salts have been removed. The water in the brine-enriched layers is recirculated to a certain extent to prevent excessive accumulation of brine. The principles involved in electrodialysis treatment are shown in Figure 8.8.

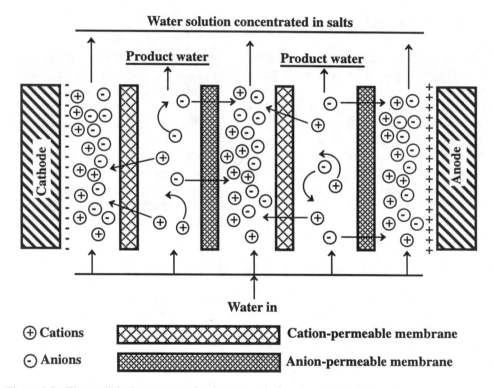

Figure 8.8. Electrodialysis apparatus for the removal of ionic material from water.

Fouling caused by various materials can cause problems with reverse osmosis treatment of water. Although the relatively small ions constituting the salts dissolved in wastewater readily pass through the membranes, large organic ions (proteins, for example) and charged colloids migrate to the membrane surfaces, often fouling or plugging the membranes and reducing efficiency. In addition, growth of microorganisms on the membranes can cause fouling.

Experience with pilot plants indicates that electrodialysis has the potential to be a practical and economical method to remove up to 50% of the dissolved inorganics from secondary sewage effluent after pretreatment to eliminate fouling substances.

Such a level of efficiency would permit repeated recycling of water without dissolved inorganic materials reaching unacceptably high levels.

Ion Exchange

The ion exchange method for softening water is described in detail in Section 8.7. The ion exchange process used for removal of inorganics consists of passing the water successively over a solid cation exchanger and a solid anion exchanger, which replace cations and anions by hydrogen ion and hydroxide ion, respectively, so that each equivalent of salt is replaced by a mole of water. For the hypothetical ionic salt MX, the reactions are the following where $^-\{Cat(s)\}$ represents the solid cation exchanger and $^+\{An(s)\}$ represents the solid anion exchanger.:

$$H^+{}^-\{Cat(s)\} + M^+ + X^- \rightarrow M^+{}^-\{Cat(s)\} + H^+ + X^- \qquad (8.9.1)$$

$$OH^-{}^+\{An(s)\} + H^+ + X^- \rightarrow X^-{}^+\{An(s)\} + H_2O \qquad (8.9.2)$$

The cation exchanger is regenerated with strong acid and the anion exchanger with strong base.

Demineralization by ion exchange generally produces water of a very high quality. Unfortunately, some organic compounds in wastewater foul ion exchangers, and microbial growth on the exchangers can diminish their efficiency. In addition, regeneration of the resins is expensive, and the concentrated wastes from regeneration require disposal in a manner that will not damage the environment.

Reverse Osmosis

Reverse osmosis, Figure 8.9, is a very useful and well-developed technique for the purification of water.[7] Basically, it consists of forcing pure water through a semipermeable membrane that allows the passage of water but not of other material. This process, which is not simply sieve separation or ultrafiltration, depends on the preferential sorption of water on the surface of a porous cellulose acetate or polyamide membrane. Pure water from the sorbed layer is forced through pores in the membrane under pressure. If the thickness of the sorbed water layer is d, the pore diameter for optimum separation should be 2d. The optimum pore diameter depends upon the thickness of the sorbed pure water layer and may be several times the diameters of the solute and solvent molecules.

Phosphorus Removal

Advanced waste treatment normally requires removal of phosphorus to reduce algal growth. Algae may grow at $PO_4{}^{3-}$ levels as low as 0.05 mg/L. Growth inhibition requires levels well below 0.5 mg/L. Since municipal wastes typically contain approximately 25 mg/L of phosphate (as orthophosphates, polyphosphates, and insoluble phosphates), the efficiency of phosphate removal must be quite high to prevent algal growth. This removal may occur in the sewage treatment process (1) in the primary settler; (2) in the aeration chamber of the activated sludge unit; or (3) after secondary waste treatment.

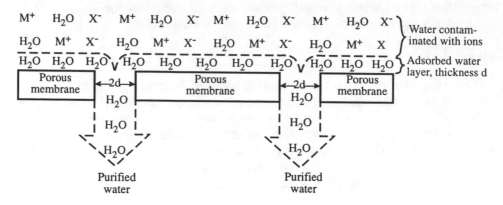

Figure 8.9. Solute removal from water by reverse osmosis.

Activated sludge treatment removes about 20% of the phosphorus from sewage. Thus, an appreciable fraction of largely biological phosphorus is removed with the sludge. Detergents and other sources contribute significant amounts of phosphorus to domestic sewage and considerable phosphate ion remains in the effluent. However, some wastes, such as carbohydrate wastes from sugar refineries, are so deficient in phosphorus that supplementation of the waste with inorganic phosphorus is required for proper growth of the microorganisms degrading the wastes.

Under some sewage plant operating conditions, much greater than normal phosphorus removal has been observed. In such plants, characterized by high dissolved oxygen and high pH levels in the aeration tank, removal of 60-90% of the phosphorus has been attained, yielding two or three times the normal level of phosphorus in the sludge. In a conventionally operated aeration tank of an activated sludge plant, the CO_2 level is relatively high because of release of the gas by the degradation of organic material. A high CO_2 level results in a relatively low pH, due to the presence of carbonic acid. The aeration rate is generally not maintained at a very high level because oxygen is transferred relatively more efficiently from air when the dissolved oxygen levels in water are relatively low. Therefore, the aeration rate normally is not high enough to sweep out sufficient dissolved carbon dioxide to bring its concentration down to low levels. Thus, the pH generally is low enough that phosphate is maintained primarily in the form of the $H_2PO_4^-$ ion. However, at a higher rate of aeration in a relatively hard water, the CO_2 is swept out, the pH rises, and reactions such as the following occur:

$$5Ca^{2+} + 3HPO_4^{2-} + H_2O \rightarrow Ca_5OH(PO_4)_3(s) + 4H^+ \tag{8.9.3}$$

The precipitated hydroxyapatite or other form of calcium phosphate is incorporated in the sludge floc. Reaction 8.9.3 is strongly hydrogen ion-dependent, and an increase in the hydrogen ion concentration drives the equilibrium back to the left. Thus, under anaerobic conditions when the sludge medium becomes more acidic due to higher CO_2 levels, the calcium returns to solution.

Chemically, phosphate is most commonly removed by precipitation. Some common precipitants and their products are shown in Table 8.1. Precipitation processes are capable of at least 90-95% phosphorus removal at reasonable cost.

Table 8.1. Chemical Precipitants for Phosphate and Their Products

Precipitant(s)	Products
$Ca(OH)_2$	$Ca_5OH(PO_4)_3$ (hydroxyapatite)
$Ca(OH)_2 + NaF$	$Ca_5F(PO_4)_3$ (fluorapatite)
$Al_2(SO_4)_3$	$AlPO_4$
$FeCl_3$	$FePO_4$
$MgSO_4$	$MgNH_4PO_4$

Lime, $Ca(OH)_2$, is the chemical most commonly used for phosphorus removal:

$$5Ca(OH)_2 + 3HPO_4^{2-} \rightarrow Ca_5OH(PO_4)_3(s) + 3H_2O + 6OH^- \qquad (8.9.4)$$

Lime has the advantages of low cost and ease of regeneration. The efficiency with which phosphorus is removed by lime is not as high as would be predicted by the low solubility of hydroxyapatite, $Ca_5OH(PO_4)_3$. Some of the possible reasons for this are slow precipitation of $Ca_5OH(PO_4)_3$, formation of nonsettling colloids; precipitation of calcium as $CaCO_3$ in certain pH ranges, and the fact that phosphate may be present as condensed phosphates (polyphosphates) which form soluble complexes with calcium ion.

Phosphate can be removed from solution by adsorption on some solids, particularly activated alumina, Al_2O_3. Removals of up to 99.9% of orthophosphate have been achieved with this method.

Nitrogen Removal

Next to phosphorus, nitrogen is the algal nutrient most commonly removed as part of advanced wastewater treatment. The techniques most often used for nitrogen removal are summarized in Table 8.2. Nitrogen in municipal wastewater generally is present as organic nitrogen or ammonia. Ammonia is the primary nitrogen product produced by most biological waste treatment processes. This is because it is expensive to aerate sewage sufficiently to oxidize the ammonia to nitrate through the action of nitrifying bacteria. If the activated sludge process is operated under conditions such that the nitrogen is maintained in the form of ammonia, the latter may be stripped in the form of NH_3 gas from the water by air. For ammonia stripping to work, the ammoniacal nitrogen must be converted to volatile NH_3 gas, which requires a pH substantially higher than the pK_a of the NH_4^+ ion. In practice, the pH is raised to approximately 11.5 by the addition of lime (which also serves to remove phosphate). The ammonia is stripped from the water by air.

Nitrification followed by denitrification is arguably the most effective technique for the removal of nitrogen from wastewater. The first step is an essentially complete conversion of ammonia and organic nitrogen to nitrate under strongly aerobic conditions, achieved by more extensive than normal aeration of the sewage:

$$NH_4^+ + 2O_2 \text{ (Nitrifying bacteria)} \rightarrow NO_3^- + 2H^+ + H_2O \qquad (8.9.5)$$

Table 8.2. Common Processes for the Removal of Nitrogen from Wastewater[1]

Process	Principles and conditions
Air stripping ammonia	Ammonium ion is the initial product of biodegradation of nitrogen waste. It is removed by raising the pH to approximately 11 with lime, and stripping ammonia gas from the water by air in a stripping tower. Scaling, icing, and air pollution are the main disadvantages.
Ammonium ion exchange	Clinoptilolite, a natural zeolite, selectively removes ammonium ion by ion exchange: $Na^+\{^-clinoptilolite\}$ $+ NH_4^+ \rightarrow NH_4^+\{^-clinoptilolite\} + Na^+$. The ion exchanger is regenerated with sodium or calcium salts.
Biosynthesis	The production of biomass in the sewage treatment system and its subsequent removal from the sewage effluent result in a net loss of nitrogen from the system.
Nitrification-denitrification	This approach involves the conversion of ammoniacal nitrogen to nitrate by bacteria under aerobic conditions, $$2NH_4^+ + 3O_2 \xrightarrow{Nitrosomonas} 4H^+ + 2NO_2^- + 2H_2O$$ $$2NO_2^- + O_2 \xrightarrow{Nitrobacter} 2NO_3^-$$ followed by production of elemental nitrogen (denitrification): $$4NO_3^- + 5\{CH_2O\} + 4H^+ \xrightarrow[bacteria]{Denitrifying} 2N_2(g) + 5CO_2(g) + 7H_2O$$ Typically, denitrification is carried out in an anaerobic column with added methanol as a food source (microbial reducing agent).
Chlorination	Reaction of ammonium ion and hypochlorite (from chlorine) results in denitrification by chemical reactions: $$NH_4^+ + HOCl \rightarrow NH_2Cl + H_2O + H^+$$ $$2NH_2Cl + HOCl \rightarrow N_2(g) + 3H^+ + 3Cl^- H_2O$$

The second step is the reduction of nitrate to nitrogen gas. This reaction is also bacterially catalyzed and requires a carbon source and a reducing agent such as methanol, CH_3OH.[8]

$$6NO_3^- + 5CH_3OH + 6H^+ \text{ (Denitrifying bacteria)} \rightarrow 3N_2(g) + 5CO_2 + 13H_2O \qquad (8.9.6)$$

The denitrification process may be carried out either in a tank or on a carbon column. In pilot plant operation, conversions of 95% of the ammonia to nitrate and 86% of the nitrate to nitrogen have been achieved.

8.10. SLUDGE

Perhaps the most pressing water treatment problem at this time has to do with sludge collected or produced during water treatment. Finding a safe place to put the sludge or a use for it has proven troublesome, and the problem is aggravated by the growing numbers of water treatment systems.

Some sludge is present in wastewater prior to treatment and may be collected from it. Such sludge includes human wastes, garbage grindings, organic wastes and inorganic silt and grit from storm water runoff, and organic and inorganic wastes from commercial and industrial sources. There are two major kinds of sludge generated in a waste treatment plant. The first of these is organic sludge from activated sludge, trickling filter, or rotating biological reactors. The second is inorganic sludge from the addition of chemicals, such as in phosphorus removal (see Section 8.9).

Most commonly, sewage sludge is subjected to anaerobic digestion in a digester designed to allow bacterial action to occur in the absence of air. This reduces the mass and volume of sludge and ideally results in the formation of a stabilized humus. Disease agents are also destroyed in the process.

Following digestion, sludge is generally conditioned and thickened to concentrate and stabilize it and make it more dewaterable. Relatively inexpensive processes, such as gravity thickening, may be employed to get the moisture content down to about 95%. Sludge may be further conditioned chemically by the addition of iron or aluminum salts, lime, or polymers.

Sludge dewatering is employed to convert the sludge from an essentially liquid material to a damp solid containing not more than about 85% water. This may be accomplished on sludge drying beds consisting of layers of sand and gravel. Mechanical devices may also be employed, including vacuum filtration, centrifugation, and filter presses. Heat may be used to aid the drying process.

Ultimately, disposal of the sludge is required. Two of the main alternatives for sludge disposal are land spreading and incineration.

Rich in nutrients, waste sewage sludge contains around 5% N, 3% P, and 0.5% K on a dry-weight basis and can be used to fertilize and condition soil. The humic material in the sludge improves the physical properties and cation-exchange capacity of the soil. Possible accumulation of heavy metals is of some concern insofar as the use of sludge on cropland is concerned. Sewage sludge is an efficient heavy metals scavenger and may contain elevated levels of zinc, copper, nickel, and cadmium These and other metals tend to remain immobilized in soil by chelation with organic matter, adsorption on clay minerals, and precipitation as insoluble compounds such as oxides or carbonates. However, increased application of sludge on cropland has caused distinctly elevated levels of zinc and cadmium in both leaves and grain of corn. Therefore, caution has been advised in heavy or prolonged application of sewage sludge to soil. Prior control of heavy metal contamination from industrial sources has greatly reduced the heavy metal content of sludge and enabled it to be used more extensively on soil.

An increasing problem in sewage treatment arises from sludge sidestreams. These consist of water removed from sludge by various treatment processes. Sewage treatment processes can be divided into mainstream treatment processes (primary clarification, trickling filter, activated sludge, and rotating biological reactor) and sidestream processes. During sidestream treatment, sludge is dewatered, degraded, and disinfected by a variety of processes, including gravity thickening, dissolved air flotation, anaerobic digestion, aerobic digestion, vacuum filtration, centrifugation, belt-filter press filtration, sand-drying-bed treatment, sludge-lagoon settling, wet air oxidation, pressure filtration, and Purifax treatment. Each of these produces a liquid by-product sidestream which is circulated back to the mainstream. These add to the biochemical oxygen demand and suspended solids of the mainstream.

A variety of chemical sludges are produced by various water treatment and industrial processes. Among the most abundant of such sludges is alum sludge produced by the hydrolysis of Al(III) salts used in the treatment of water, which creates gelatinous aluminum hydroxide:

$$Al^{3+} + 3OH^-(aq) \rightarrow Al(OH)_3(s) \tag{8.10.1}$$

Alum sludges normally are 98% or more water and are very difficult to dewater.

Both iron(II) and iron(III) compounds are used for the removal of impurities from wastewater by precipitation of $Fe(OH)_3$. The sludge contains $Fe(OH)_3$ in the form of soft, fluffy precipitates that are difficult to dewater beyond 10 or 12% solids.

The addition of either lime, $Ca(OH)_2$, or quicklime, CaO, to water is used to raise the pH to about 11.5 and cause the precipitation of $CaCO_3$, along with metal hydroxides and phosphates. Calcium carbonate is readily recovered from lime sludges and can be recalcined to produce CaO, which can be recycled through the system.

Metal hydroxide sludges are produced in the removal of metals such as lead, chromium, nickel, and zinc from wastewater by raising the pH to such a level that the corresponding hydroxides or hydrated metal oxides are precipitated. The disposal of these sludges is a substantial problem because of their toxic heavy metal content. Reclamation of the metals is an attractive alternative for these sludges.

Pathogenic (disease-causing) microorganisms may persist in the sludge left from the treatment of sewage. Many of these organisms present potential health hazards, and there is risk of public exposure when the sludge is applied to soil. Therefore, it is necessary both to be aware of pathogenic microorganisms in municipal wastewater treatment sludge and to find a means of reducing the hazards caused by their presence.

The most significant organisms in municipal sewage sludge include (1) indicators of fecal pollution, including fecal and total coliform; (2) pathogenic bacteria, including *Salmonellae* and *Shigellae*; (3) enteric (intestinal) viruses, including enterovirus and poliovirus; and (4) parasites, such as *Entamoeba histolytica* and *Ascaris lumbricoides*.

Several ways are recommended to significantly reduce levels of pathogens in sewage sludge. Aerobic digestion involves aerobic agitation of the sludge for periods of 40 to 60 days (longer times are employed with low sludge temperatures). Air drying involves draining and/or drying of the liquid sludge for at least three months in a layer 20-25 cm thick. This operation may be performed on underdrained sand beds or in basins. Anaerobic digestion involves maintenance of the sludge in an anaerobic state for periods of time ranging from 60 days at 20°C to 15 days at temperatures exceeding 35°C. Composting involves mixing dewatered sludge cake with bulking agents subject to decay, such as wood chips or shredded municipal refuse, and allowing the action of bacteria to promote decay at temperatures ranging up to 45-65°C. The higher temperatures tend to kill pathogenic bacteria. Finally, pathogenic organisms may be destroyed by lime stabilization in which sufficient lime is added to raise the pH of the sludge to 12 or higher.

8.11. WATER DISINFECTION

Chlorine is the most commonly used disinfectant employed for killing bacteria in water. When chlorine is added to water, it rapidly hydrolyzes according to the reaction

$$Cl_2 + H_2O \rightarrow H^+ + Cl^- + HOCl \qquad (8.11.1)$$

which has the following equilibrium constant:

$$K = \frac{[H^+][Cl^-][HOCl]}{[Cl_2]} = 4.5 \times 10^{-4} \qquad (8.11.2)$$

Hypochlorous acid, HOCl, is a weak acid that dissociates according to the reaction,

$$HOCl \xleftarrow{\longrightarrow} H^+ + OCl^- \qquad (8.11.3)$$

with an ionization constant of 2.7×10^{-8}. From the above it can be calculated that the concentration of elemental Cl_2 is negligible at equilibrium above pH 3 when chlorine is added to water at levels below 1.0 g/L.

Sometimes, hypochlorite salts are substituted for chlorine gas as a disinfectant. Calcium hypochlorite, $Ca(OCl)_2$, is commonly used. The hypochlorites are safer to handle than gaseous chlorine.

The two chemical species formed by chlorine in water, HOCl and OCl⁻, are known as **free available chlorine**. Free available chlorine is very effective in killing bacteria. In the presence of ammonia, monochloramine, dichloramine, and trichloramine are formed:

$$NH_4^+ + HOCl \rightarrow NH_2Cl \text{ (monochloramine)} + H_2O + H^+ \qquad (8.11.4)$$

$$NH_2Cl + HOCl \rightarrow NHCl_2 \text{ (dichloramine)} + H_2O \qquad (8.11.5)$$

$$NHCl_2 + HOCl \rightarrow NCl_3 \text{ (trichloramine)} + H_2O \qquad (8.11.6)$$

The chloramines are called **combined available chlorine**. Chlorination practice frequently provides for formation of combined available chlorine which, although a weaker disinfectant than free available chlorine, is more readily retained as a disinfectant throughout the water distribution system. Too much ammonia in water is considered undesirable because it exerts excess demand for chlorine.

At sufficiently high Cl:N molar ratios in water containing ammonia, some HOCl and OCl⁻ remain unreacted in solution, and a small quantity of NCl_3 is formed. The ratio at which this occurs is called the **breakpoint**. Chlorination beyond the break-point ensures disinfection. It has the additional advantage of destroying the more common materials that cause odor and taste in water.

At moderate levels of NH_3-N (approximately 20 mg/L), when the pH is between 5.0 and 8.0, chlorination with a minimum 8:1 weight ratio of Cl to NH_3-nitrogen produces efficient denitrification:

$$NH_4^+ + HOCl \rightarrow NH_2Cl + H_2O + H^+ \qquad (8.11.4)$$

$$2NH_2Cl + HOCl \rightarrow N_2(g) + 3H^+ + 3Cl^- + H_2O \qquad (8.11.7)$$

This reaction is used to remove pollutant ammonia from wastewater. However, problems can arise from chlorination of organic wastes. Typical of such by-products is chloroform, produced by the chlorination of humic substances in water.

Chlorine is used to treat water other than drinking water. It is employed to disinfect effluent from sewage treatment plants, as an additive to the water in electric power plant cooling towers, and to control microorganisms in food processing.

Chlorine Dioxide

Chlorine dioxide, ClO_2, is an effective water disinfectant that is of particular interest because, in the absence of impurity Cl_2, it does not produce impurity trihalomethanes in water treatment. In acidic and neutral water, respectively, the two half-reactions for ClO_2 acting as an oxidant are the following:

$$ClO_2 + 4H^+ + 5e^- \leftarrow\rightarrow Cl^- + 2H_2O \qquad (8.11.8)$$

$$ClO_2 + e^- \leftarrow\rightarrow ClO_2^- \qquad (8.11.9)$$

In the neutral pH range, chlorine dioxide in water remains largely as molecular ClO_2 until it contacts a reducing agent with which to react. Chlorine dioxide is a gas that is violently reactive with organic matter and explosive when exposed to light. For these reasons, it is not shipped, but is generated on-site by processes such as the reaction of chlorine gas with solid sodium hypochlorite:

$$2NaClO_2(s) + Cl_2(g) \leftarrow\rightarrow 2ClO_2(g) + 2NaCl(s) \qquad (8.11.10)$$

A high content of elemental chlorine in the product may require its purification to prevent unwanted side-reactions from Cl_2.

As a water disinfectant, chlorine dioxide does not chlorinate or oxidize ammonia or other nitrogen-containing compounds. Some concern has been raised over possible health effects of its main degradation byproducts, ClO_2^- and ClO_3^-.

Ozone

Ozone is sometimes used as a disinfectant in place of chlorine, particularly in Europe. Figure 8.10 shows the main components of an ozone water treatment system. Basically, air is filtered, cooled, dried, and pressurized, then subjected to an electrical discharge of approximately 20,000 volts. The ozone produced is then pumped into a contact chamber where water contacts the ozone for 10-15 minutes. Concern over possible production of toxic organochlorine compounds by water chlorination processes has increased interest in ozonation. Furthermore, ozone is more destructive to viruses than is chlorine. Unfortunately, the solubility of ozone in water is relatively low, which limits its disinfective power.

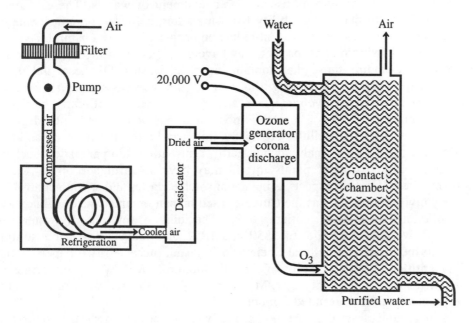

Figure 8.10. A schematic diagram of a typical ozone water-treatment system.

A major consideration with ozone is the rate at which it decomposes spontaneously in water, according to the overall reaction,

$$2O_3 \rightarrow 3O_2(g) \tag{8.11.11}$$

Because of the decomposition of ozone in water, some chlorine must be added to maintain disinfectant throughout the water distribution system.

Iron(VI) in the form of ferrate ion, FeO_4^{2-}, is a strong oxidizing agent with excellent disinfectant properties. It has the additional advantage of removing heavy metals, viruses, and phosphate. It may well find limited application for disinfection in the future.

8.12. NATURAL WATER PURIFICATION PROCESSES

Virtually all of the materials that waste treatment processes are designed to eliminate may be absorbed by soil or degraded in soil. In fact, most of these materials are essential for soil fertility. Wastewater may provide the water that is essential to plant growth, in addition to the nutrients—phosphorus, nitrogen and potassium—usually provided by fertilizers. Wastewater also contains essential trace elements and vitamins. Stretching the point a bit, the degradation of organic wastes provides the CO_2 essential for photosynthetic production of plant biomass.

Soil may be viewed as a natural filter for wastes. Most organic matter is readily degraded in soil and, in principle, soil constitutes an excellent primary, secondary, and tertiary treatment system for water. Soil has physical, chemical, and biological characteristics that can enable wastewater detoxification, biodegradation, chemical decomposition, and physical and chemical fixation. A number of soil characteristics are important in determining its use for land treatment of wastes. These characteristics include physical form, ability to retain water, aeration, organic content, acid-base characteristics, and oxidation-reduction behavior. Soil is a natural medium for a number of living organisms that may have an effect upon biodegradation of wastewaters, including those that contain industrial wastes. Of these, the most important are bacteria, including those from the genera *Agrobacterium*, *Arthrobacteri*, *Bacillus*, *Flavobacterium*, and *Pseudomonas*. Actinomycetes and fungi are important in decay of vegetable matter and may be involved in biodegradation of wastes. Other unicellular organisms that may be present in or on soil are protozoa and algae. Soil animals, such as earthworms, affect soil parameters such as soil texture. The growth of plants in soil may have an influence on its waste treatment potential in such aspects as uptake of soluble wastes and erosion control.

Early civilizations, such as the Chinese, used human organic wastes to increase soil fertility, and the practice continues today. The ability of soil to purify water was noted well over a century ago. In 1850 and 1852, J. Thomas Way, a consulting chemist to the Royal Agricultural Society in England, presented two papers to the Society entitled "Power of Soils to Absorb Manure." Mr. Way's experiments showed that soil is an ion exchanger. Much practical and theoretical information on the ion exchange process resulted from his work.

If soil treatment systems are not properly designed and operated, odor can become an overpowering problem. The author of this book is reminded of driving into a small town, recalled from some years before as a very pleasant place, and being assaulted with a virtually intolerable odor. The disgruntled residents pointed to a large spray irrigation system on a field in the distance—unfortunately upwind— spraying liquified pig manure as part of an experimental feedlot waste treatment operation. The experiment was not deemed a success and was discontinued by the investigators, presumably before they met with violence from the local residents.

Industrial Wastewater Treatment by Soil

Wastes that are amenable to land treatment are biodegradable organic substances, particularly those contained in municipal sewage and in wastewater from some industrial operations, such as food processing. However, through acclimation over a long period of time, soil bacterial cultures may develop that are effective in degrading normally recalcitrant compounds that occur in industrial wastewater. Acclimated microorganisms are found particularly at contaminated sites, such as those where soil has been exposed to crude oil for many years.

A variety of enzyme activities are exhibited by microorganisms in soil that enable them to degrade synthetic substances. Even sterilized soil may show enzyme activity due to extracellular enzymes secreted by microorganisms in soil. Some of these enzymes are hydrolase enzymes (see Chapter 21), such as those that catalyze the hydrolysis of organophosphate compounds as shown by the reaction,

$$R-O-\overset{\overset{\displaystyle X}{\|}}{\underset{\underset{\displaystyle R}{|}}{\underset{O}{P}}}-O-Ar \xrightarrow[\substack{\text{Phosphatase} \\ \text{enzyme}}]{H_2O} R-O-\overset{\overset{\displaystyle X}{\|}}{\underset{\underset{\displaystyle R}{|}}{\underset{O}{P}}}-OH \ + HOAr \qquad (8.12.1)$$

where R is an alkyl group, Ar is a substituent group that is frequently aryl, and X is either S or O. Another example of a reaction catalyzed by soil enzymes is the oxidation of phenolic compounds by diphenol oxidase:

$$\text{(diphenol)} + \{O\} \xrightarrow[\text{oxidase}]{\text{Diphenol}} \text{(quinone)} + \ H_2O \qquad (8.12.2)$$

Land treatment is most used for petroleum refining wastes and is applicable to the treatment of fuels and wastes from leaking underground storage tanks. It can also be applied to biodegradable organic chemical wastes, including some organohalide compounds. Land treatment is not suitable for the treatment of wastes containing acids, bases, toxic inorganic compounds, salts, heavy metals, and organic compounds that are excessively soluble, volatile, or flammable.

8.13. WATER REUSE AND RECYCLING

Water reuse and recycling are becoming much more common as demands for water exceed supply. **Unplanned reuse** occurs as the result of waste effluents entering receiving waters or groundwater and subsequently being taken into a water distribution system. A typical example of unplanned water reuse occurs in London, which withdraws water from the Thames River that may have been through other water systems at least once, and which uses groundwater sources unintentionally recharged with sewage effluents from a number of municipalities. **Planned reuse** utilizes wastewater treatment systems deliberately designed to bring water up to standards required for subsequent applications. The term **direct reuse** refers to water that has retained its identity from a previous application; reuse of water that has lost its identity is termed **indirect reuse**. The distinction also needs to be made between recycling and reuse. **Recycling** occurs internally before water is ever discharged. An example is condensation of steam in a steam power plant followed by return of the steam to boilers. **Reuse** occurs, for example, when water discharged by one user is taken as a water source by another user.

Reuse of water continues to grow because of two major factors. The first of these is lack of supply of water. The second is that widespread deployment of modern water treatment processes significantly enhances the quality of water available for reuse. These two factors come into play in semi-arid regions in countries with advanced technological bases. For example, Israel, which is dependent upon irrigation for essentially all its agriculture, reuses about 2/3 of the country's sewage effluent for irrigation, whereas the U.S., where water is relatively more available, uses only about 2–3% of its water for this purpose.

Since drinking water and water used for food processing require the highest quality of all large applications, intentional reuse for potable water is relatively less desirable, though widely practiced unintentionally or out of necessity. This leaves three applications with the greatest potential for reuse:

- Irrigation for cropland, golf courses, and other applications requiring water for plant and grass growth. This is the largest potential application for reused water and one that can take advantage of plant nutrients, particularly nitrogen and phosphorus, in water.

- Cooling and process water in industrial applications. For some industrial applications, relatively low quality water can be used and secondary sewage effluent is a suitable source.

- Groundwater recharge. Groundwater can be recharged with reused water either by direct injection into an aquifer or by applying the water to land, followed by percolation into the aquifer. The latter, especially, takes advantage of biodegradation and chemical sorption processes to further purify the water.

It is inevitable that water recycling and reuse will continue to grow. This trend will increase the demand for water treatment, both qualitatively and quantitatively. In addition, it will require more careful consideration of the original uses of water to minimize water deterioration and enhance its suitability for reuse.

LITERATURE CITED

1. Cowan, Robert M., Timothy G. Ellis, Matthew J. Alagappan, Gunaseelan Alagappan, and Keeyong Park, "Treatment Systems. Activated Sludge and Other Aerobic Suspended Culture Processes," *Water Environment Research*, **68**, 451-469 (1996).

2. Sebastian, Joseph, "Tertiary Treatment of Sewage for Reuse," *Chemical Engineering World*, **32** 55-57 (1997).

3. Kuo, Jih-Fen, James F. Stahl, Ching-Lin Chen, and Paul V. Bohlier,, "Dual Role of Activated Carbon Process for Water Reuse, *Water Environment Research*, **70**, 161-170 (1998).

4. Edell, Asa and Gregory M. Morrison, "Critical Evaluation of Pesticides in the Aquatic Environment and their Removal from Drinking Water" *Sweden Vatten*, **53**, 355-364 (1997).

5. Tchobanoglous, George, Jeannie Darby, Keith Bourgeous, John McArdle, Paul Genest, and Michael Tylla, "Ultrafiltration as an Advanced Tertiary Treatment Process for Municipal Wastewater," *Desalination*, **119**, 315-322 (1998).

6. Electrodialysis as an Alternative for Reverse Osmosis in an Integrated Membrane System. Van Der Hoek, J. P., D. O. Rijnbende, C. J. A. Lokin, P. A. C. Bonne, M. T. Loonen, and J. A. M. H. Hofman, *Desalination*, **117**, 159-172 (1998).

7. Thiemann, H., and H. Weiler, "One Year of Operational Experience with the Largest River Water Reverse Osmosis Plant in Germany," *VGB Kraftwerkstech*, **76**, 1017-1022 (1996).

8. Koch, G. and H. Siegrist, "Denitrification with Methanol in Tertiary Filtration, *Water Research*, **31**, 3029-3038 (1997).

SUPPLEMENTARY REFERENCES

Adin, Avner and Takashi Asano, "The Role of Physical Chemical Treatment in Wastewater Reclamation and Reuse," *Water Science and Technology*, **37**, 79-80 (1998).

American Water Works Association, *Reverse Osmosis and Nanofiltration*, American Water Works Association, Denver, CO, 1998.

Ash, Michael and Irene Ash, *Handbook of Water Treatment Chemicals*, Gower, Aldershot, England , 1996.

Balaban, Miriam, Ed., *Desalination and Water Re-use*, Hemisphere, New York, 1991.

Bitton, Gabriel, *Wastewater Microbiology*, Wiley-Liss, New York, 1999.

Casey, T. J. and J. T. Casey, *Unit Treatment Processes in Water and Wastewater Engineering*, John Wiley & Sons, New York, 1997.

Connell, Gerald F., *The Chlorination/Chloramination Handbook*, American Water Works Association, Denver, CO, 1996.

Design of Municipal Wastewater Treatment Plants-MOP 8, 4th ed., Water Environment Federation, Alexandria, VA, 1998.

Droste, Ronald, *Theory and Practice of Water and Wastewater Treatment*, John Wiley & Sons, New York, 1996.

Faust, Samuel D. and Osman M. Aly, Eds., *Chemistry of Water Treatment*, 2nd ed., American Water Works Association, Denver, CO, 1997.

Freeman, Harry M., Ed., *Standard Handbook of Hazardous Waste Treatment and Disposal*, 2nd ed., McGraw-Hill, New York, 1998.

Gallagher, Lynn M. and Leonard A. Miller, *Clean Water Handbook*, 2nd ed., Government Institutes, Rockville, MD, 1996.

Gates, Donald J., *The Chlorine Dioxide Handbook*, American Water Works Association, Denver, CO, 1998.

Geldreich, Edwin, *Microbial Quality of Water Supply in Distribution Systems*, CRC Press/Lewis Publishers, Boca Raton, FL, 1996.

Hahn, Hermann H., Erhard Hoffmann, and Ÿdegaard Hallvard, *Chemical Water and Wastewater Treatment V: Proceedings of the 7th Gothenburg Symposium 1998*, Springer, New York, 1998.

Kurbiel, J., Ed., *Advanced Wastewater Treatment and Reclamation*, Pergamon, London, 1991.

Langlais, Bruno, David A. Recknow, and Deborah R. Brink, Eds., *Ozone in Water Treatment: Application and Engineering*, Lewis Publishers, CRC Press, Boca Raton, FL, 1991.

Mathie, Alton J., *Chemical Treatment for Cooling Water*, Prentice Hall, Upper Saddle River, NJ, 1999.

Mays, Larry W., *Water Distribution Systems Handbook*, McGraw-Hill, New York, 1999.

Minear, Roger A. and Gary L. Amy, *Disinfection By-Products in Water Treatment: The Chemistry of Their Formation and Control*, CRC Press/Lewis Publishers, Boca Raton, FL, 1996.

Montgomery, James M., Consulting Engineers, *Water Treatment Principles and Design*, John Wiley & Sons, Inc., New York, 1985.

Norman, Terry and Gary Banuelos, *Phytoremediation of Contaminated Soil and Water*, CRC Press/Lewis Publishers, Boca Raton, FL, 1999.

Nyer, Evan K., *Groundwater Treatment Technology*, 2nd ed., van Nostrand Reinhold, New York, 1993.

Patrick David R. and George C. White, *Handbook of Chlorination and Alternative Disinfectants*, 3rd ed., Nostrand Reinhold, New York, 1993.

Polevoy, Savely, *Water Science and Engineering*, Blackie Academic & Professional, London, 1996.

Rice, Rip G., *Ozone Drinking Water Treatment Handbook*, CRC Press/Lewis Publishers, Boca Raton, FL, 1999.

Roques, Henri, *Chemical Water Treatment: Principles and Practice*, VCH, New York, 1996.

Scholze, R. J., Ed., *Biotechnology for Degradation of Toxic Chemicals in Hazardous Wastes*, Noyes Publications, Park Ridge, NJ, 1988.

Singer, Philip C., *Formation and Control of Disinfection By-Products in Drinking Water*, American Water Works Association, Denver, CO, 1999.

Speitel, Gerald E., *Advanced Oxidation and Biodegradation Processes for the Destruction of TOC and DBP Precursors*, AWWA Research Foundation, Denver, CO 1999.

Spellman, Frank R. and Nancy E. Whiting, *Water Pollution Control Technology: Concepts and Applications*, Government Institutes, Rockville, MD, 1999.

Steiner, V. "UV Irradiation in Drinking Water and Wastewater Treatment for Disinfection," *Wasser Rohrbau*, **49**, 22-31 (1998).

Stevenson, David G., *Water Treatment Unit Processes*, Imperial College Press, London, 1997.

Tomar, Mamta, *Laboratory Manual for the Quality Assessment of Water and Wastewater*, CRC Press, Boca Raton, FL, 1999.

Türkman, Aysen, and Orhan Uslu, Eds., *New Developments in Industrial Wastewater Treatment*, Kluwer, Norwell, MA, 1991.

Wachinski, James E. Etzel, Anthony M., *Environmental Ion Exchange: Principles and Design*, CRC Press/Lewis Publishers, Boca Raton, FL, 1997.

Wase, John and Christopher Forster, Eds., *Biosorbents for Metal Ions*, Taylor & Francis, London, 1997.

White, George C., *Handbook of Chlorination and Alternative Disinfectants*, John Wiley & Sons, New York, 1999.

QUESTIONS AND PROBLEMS

1. Consider the equilibrium reactions and expressions covered in Chapter 3. How many moles of NTA should be added to 1000 liters of water having a pH of 9 and containing CO_3^{2-} at 1.00×10^{-4} M to prevent precipitation of $CaCO_3$? Assume a total calcium level of 40 mg/L.

2. What is the purpose of the return sludge step in the activated sludge process?

3. What are the two processes by which the activated sludge process removes soluble carbonaceous material from sewage?

4. Why might hard water be desirable as a medium if phosphorus is to be removed by an activated sludge plant operated under conditions of high aeration?

5. How does reverse osmosis differ from a simple sieve separation or ultrafiltration process?

6. How many liters of methanol would be required daily to remove the nitrogen from a 200,000-L/day sewage treatment plant producing an effluent containing 50 mg/L of nitrogen? Assume that the nitrogen has been converted to NO_3^- in the plant. The denitrifying reaction is Reaction 8.9.6.

7. Discuss some of the advantages of physical-chemical treatment of sewage as opposed to biological wastewater treatment. What are some disadvantages?

8. Why is recarbonation necessary when water is softened by the lime-soda process?

9. Assume that a waste contains 300 mg/L of biodegradable $\{CH_2O\}$ and is processed through a 200,000-L/day sewage-treatment plant which converts 40% of the waste to CO_2 and H_2O. Calculate the volume of air (at 25°, 1 atm) required for this conversion. Assume that the O_2 is transferred to the water with 20% efficiency.

10. If all of the $\{CH_2O\}$ in the plant described in Question 9 could be converted to methane by anaerobic digestion, how many liters of methane (STP) could be produced daily?

11. Assuming that aeration of water does not result in the precipitation of calcium carbonate, of the following, which one that would not be removed by aeration: hydrogen sulfide, carbon dioxide, volatile odorous bacterial metabolites, alkalinity, iron?

12. In which of the following water supplies would moderately high water hardness be most detrimental: municipal water; irrigation water; boiler feedwater; drinking water (in regard to potential toxicity).

13. Which solute in water is commonly removed by the addition of sulfite or hydrazine?

14. A wastewater containing dissolved Cu^{2+} ion is to be treated to remove copper. Which of the following processes would *not* remove copper in an insoluble form; lime precipitation; cementation; treatment with NTA; ion exchange; reaction with metallic Fe.

15. Match each water contaminant in the left column with its preferred method of removal in the right column.

    (a) Mn^{2+} (1) Activated carbon
    (b) Ca^{2+} and HCO_3^- (2) Raise pH by addition of Na_2CO_3
    (c) Trihalomethane compounds   (3) Addition of lime
    (d) Mg^{2+} (4) Oxidation

16. A cementation reaction employs iron to remove Cd^{2+} present at a level of 350 mg/L from a wastewater stream. Given that the atomic weight of Cd is 112.4 and that of Fe is 55.8, how many kg of Fe are consumed in removing all the Cd from 4.50×10^6 liters of water?

17. Consider municipal drinking water from two different kinds of sources, one a flowing, well-aerated stream with a heavy load of particulate matter, and the other an anaerobic groundwater. Describe possible differences in the water treatment strategies for these two sources of water.

18. In treating water for industrial use, consideration is often given to "sequential use of the water." What is meant by this term? Give some plausible examples of sequential use of water.

19. Active biomass is used in the secondary treatment of municipal wastewater. Describe three ways of supporting a growth of the biomass, contacting it with wastewater, and exposing it to air.

20. Using appropriate chemical reactions for illustration, show how calcium present as the dissolved HCO_3^- salt in water is easier to remove than other forms of hardness, such as dissolved $CaCl_2$.

9 THE ATMOSPHERE AND ATMOSPHERIC CHEMISTRY

9.1. THE ATMOSPHERE AND ATMOSPHERIC CHEMISTRY

The **atmosphere** consists of the thin layer of mixed gases covering the earth's surface. Exclusive of water, atmospheric air is 78.1% (by volume) nitrogen, 21.0% oxygen, 0.9% argon, and 0.03% carbon dioxide. Normally, air contains 1-3% water vapor by volume. In addition, air contains a large variety of trace level gases at levels below 0.002%, including neon, helium, methane, krypton, nitrous oxide, hydrogen, xenon, sulfur dioxide, ozone, nitrogen dioxide, ammonia, and carbon monoxide.

The atmosphere is divided into several layers on the basis of temperature. Of these, the most significant are the troposphere extending in altitude from the earth's surface to approximately 11 kilometers (km), and the stratosphere from about 11 km to approximately 50 km. The temperature of the troposphere ranges from an average of 15°C at sea level to an average of -56°C at its upper boundary. The average temperature of the stratosphere increases from -56°C at its boundary with the troposphere to –2°C at its upper boundary. The reason for this increase is absorption of solar ultraviolet energy by ozone (O_3) in the stratosphere.

Various aspects of the environmental chemistry of the atmosphere are discussed in Chapters 9–14. The most significant feature of atmospheric chemistry is the occurrence of **photochemical reactions** resulting from the absorption by molecules of light photons, designated $h\nu$. (The energy, E, of a photon of visible or ultraviolet light is given by the equation, $E = h\nu$, where h is Planck's constant and ν is the frequency of light, which is inversely proportional to its wavelength. Ultraviolet radiation has a higher frequency than visible light and is, therefore, more energetic and more likely to break chemical bonds in molecules that absorb it.) One of the most significant photochemical reactions is the one responsible for the presence of ozone in the stratosphere (see above), which is initiated when O_2 absorbs highly energetic ultraviolet radiation in the wavelength ranges of 135-176 nanometers (nm) and 240-260 nm in the stratosphere:

$$O_2 + h\nu \rightarrow O + O \tag{2.3.1}$$

The oxygen atoms produced by the photochemical dissociation of O_2 react with oxygen molecules to produce ozone, O_3,

$$O + O_2 + M \rightarrow O_3 + M \tag{2.3.2}$$

where M is a third body, such as a molecule of N_2, which absorbs excess energy from the reaction. The ozone that is formed is very effective in absorbing ultraviolet radiation in the 220-330 nm wavelength range, which causes the temperature increase observed in the stratosphere. The ozone serves as a very valuable filter to remove ultraviolet radiation from the sun's rays. If this radiation reached the earth's surface, it would cause skin cancer and other damage to living organisms.

Gaseous Oxides in the Atmosphere

Oxides of carbon, sulfur, and nitrogen are important constituents of the atmosphere and are pollutants at higher levels. Of these, carbon dioxide, CO_2, is the most abundant. It is a natural atmospheric constituent, and it is required for plant growth. However, the level of carbon dioxide in the atmosphere, now at about 360 parts per million (ppm) by volume, is increasing by about 1 ppm per year. As discussed in Chapter 14, this increase.in atmospheric CO_2 may well cause general atmospheric warming—the "greenhouse effect," with potentially very serious consequences for the global atmosphere and for life on earth. Though not a global threat, carbon monoxide, CO, can be a serious health threat because it prevents blood from transporting oxygen to body tissues.

The two most serious nitrogen oxide air pollutants are nitric oxide, NO, and nitrogen dioxide, NO_2, collectively denoted as "NO_x." These tend to enter the atmosphere as NO, and photochemical processes in the atmosphere can convert NO to NO_2. Further reactions can result in the formation of corrosive nitrate salts or nitric acid, HNO_3. Nitrogen dioxide is particularly significant in atmospheric chemistry because of its photochemical dissociation by light with a wavelength less than 430 nm to produce highly reactive O atoms. This is the first step in the formation of photochemical smog (see below). Sulfur dioxide, SO_2, is a reaction product of the combustion of sulfur-containing fuels such as high-sulfur coal. Part of this sulfur dioxide is converted in the atmosphere to sulfuric acid, H_2SO_4, normally the predominant contributor to acid precipitation.

Hydrocarbons and Photochemical Smog

The most abundant hydrocarbon in the atmosphere is methane, CH_4, released from underground sources as natural gas and produced by the fermentation of organic matter. Methane is one of the least reactive atmospheric hydrocarbons and is produced by diffuse sources, so that its participation in the formation of pollutant photochemical reaction products is minimal. The most significant atmospheric pollutant hydrocarbons are the reactive ones produced as automobile exhaust emissions. In the presence of NO, under conditions of temperature inversion (see

Chapter 11), low humidity, and sunlight, these hydrocarbons produce undesirable **photochemical smog** manifested by the presence of visibility-obscuring particulate matter, oxidants such as ozone, and noxious organic species such as aldehydes.

Particulate Matter

Particles ranging from aggregates of a few molecules to pieces of dust readily visible to the naked eye are commonly found in the atmosphere and are discussed in detail in Chapter 10. Some atmospheric particles, such as sea salt formed by the evaporation of water from droplets of sea spray, are natural and even beneficial atmospheric constituents. Very small particles called **condensation nuclei** serve as bodies for atmospheric water vapor to condense upon and are essential for the formation of rain drops. Colloidal-sized particles in the atmosphere are called **aerosols**. Those formed by grinding up bulk matter are known as **dispersion aerosols**, whereas particles formed from chemical reactions of gases are **condensation aerosols**; the latter tend to be smaller. Smaller particles are in general the most harmful because they have a greater tendency to scatter light and are the most respirable (tendency to be inhaled into the lungs).

Much of the mineral particulate matter in a polluted atmosphere is in the form of oxides and other compounds produced during the combustion of high-ash fossil fuel. Smaller particles of **fly ash** enter furnace flues and are efficiently collected in a properly equipped stack system. However, some fly ash escapes through the stack and enters the atmosphere. Unfortunately, the fly ash thus released tends to consist of smaller particles that do the most damage to human health, plants, and visibility.

9.2. IMPORTANCE OF THE ATMOSPHERE

The atmosphere is a protective blanket which nurtures life on the Earth and protects it from the hostile environment of outer space. The atmosphere is the source of carbon dioxide for plant photosynthesis and of oxygen for respiration. It provides the nitrogen that nitrogen-fixing bacteria and ammonia-manufacturing plants use to produce chemically-bound nitrogen, an essential component of life molecules. As a basic part of the hydrologic cycle (Figure 3.1) the atmosphere transports water from the oceans to land, thus acting as the condenser in a vast solar-powered still. Unfortunately, the atmosphere also has been used as a dumping ground for many pollutant materials—ranging from sulfur dioxide to refrigerant Freon—a practice which causes damage to vegetation and materials, shortens human life, and alters the characteristics of the atmosphere itself.

In its essential role as a protective shield, the atmosphere absorbs most of the cosmic rays from outer space and protects organisms from their effects. It also absorbs most of the electromagnetic radiation from the sun, allowing transmission of significant amounts of radiation only in the regions of 300-2500 nm (near-ultraviolet, visible, and near-infrared radiation) and 0.01-40 m (radio waves). By absorbing electromagnetic radiation below 300 nm, the atmosphere filters out damaging ultraviolet radiation that would otherwise be very harmful to living organisms. Furthermore, because it reabsorbs much of the infrared radiation by which absorbed solar energy is re-emitted to space, the atmosphere stabilizes the earth's temperature, preventing the tremendous temperature extremes that occur on planets and moons lacking substantial atmospheres.

9.3. PHYSICAL CHARACTERISTICS OF THE ATMOSPHERE

Atmospheric science deals with the movement of air masses in the atmosphere, atmospheric heat balance, and atmospheric chemical composition and reactions. In order to understand atmospheric chemistry and air pollution, it is important to have an overall appreciation of the atmosphere, its composition, and physical characteristics as discussed in the first parts of this chapter.

Atmospheric Composition

Dry air within several kilometers of ground level consists of two **major components**

- Nitrogen, 78.08 % (by volume)
- Oxygen, 20.95 %

two **minor components**

- Argon, 0.934 %
- Carbon dioxide, 0.036 %

in addition to argon, four more **noble gases**,

- Neon, 1.818 x 10-3 %
- Helium, 5.24 x 10-4 %
- Krypton, 1.14 x 10-4 %
- Xenon, 8.7 x 10-6 %

and **trace gases** as given in Table 9.1. Atmospheric air may contain 0.1–5% water by volume, with a normal range of 1–3%.

Variation of Pressure and Density with Altitude

As anyone who has exercised at high altitudes well knows, the density of the atmosphere decreases sharply with increasing altitude as a consequence of the gas laws and gravity. More than 99% of the total mass of the atmosphere is found within approximately 30 km (about 20 miles) of the Earth's surface. Such an altitude is miniscule compared to the Earth's diameter, so it is not an exaggeration to characterize the atmosphere as a "tissue-thin" protective layer. Although the total mass of the global atmosphere is huge, approximately 5.14×10^{15} metric tons, it is still only about one millionth of the Earth's total mass.

The fact that atmospheric pressure decreases as an approximately exponential function of altitude largely determines the characteristics of the atmosphere. Ideally, in the absence of mixing and at a constant absolute temperature, T, the pressure at any given height, P_h, is given in the exponential form,

$$P_h = P_0 e^{-Mgh/RT}$$

(9.3.1)

Table 9.1. Atmospheric Trace Gases in Dry Air Near Ground Level

Gas or species	Volume percent[1]	Major sources	Process for removal from the atmosphere
CH_4	1.6×10^{-4}	Biogenic[2]	Photochemical[3]
CO	$\sim 1.2 \times 10^{-5}$	Photochemical, anthropogenic[4]	Photochemical
N_2O	3×10^{-5}	Biogenic	Photochemical
NO_x[5]	$10^{-10}\text{-}10^{-6}$	Photochemical, lightning, anthropogenic	Photochemical
HNO_3	$10^{-9}\text{-}10^{-7}$	Photochemical	Washed out by precipitation
NH_3	$10^{-8}\text{-}10^{-7}$	Biogenic	Photochemical, washed out by precipitation
H_2	5×10^{-5}	Biogenic, photochemical	Photochemical
H_2O_2	$10^{-8}\text{-}10^{-6}$	Photochemical	Washed out by precipitation
$HO\cdot$[6]	$10^{-13}\text{-}10^{-10}$	Photochemical	Photochemical
$HO_2\cdot$[6]	$10^{-11}\text{-}10^{-9}$	Photochemical	Photochemical
H_2CO	$10^{-8}\text{-}10^{-7}$	Photochemical	Photochemical
CS_2	$10^{-9}\text{-}10^{-8}$	Anthropogenic, biogenic	Photochemical
OCS	10^{-8}	Anthropogenic, biogenic, photochemical	Photochemical
SO_2	$\sim 2 \times 10^{-8}$	Anthropogenic, photochemical, volcanic	Photochemical
I_2	0-trace	—	—
CCl_2F_2[7]	2.8×10^{-5}	Anthropogenic	Photochemical
H_3CCCl_3[8]	$\sim 1 \times 10^{-8}$	Anthropogenic	Photochemical

[1] Levels in the absence of gross pollution
[2] From biological sources
[3] Reactions induced by the absorption of light energy as described later in this chapter and in Chapters 10-14
[4] Sources arising from human activities
[5] Sum of NO, NO_2, and NO_3, of which NO_3 is a major reactive species in the atmosphere at night
[6] Reactive free radical species with one unpaired electron, described later in Chapters 12 and 13; these are transient species whose concentrations become much lower at night
[7] A chlorofluorocarbon, Freon F-12
[8] Methyl chloroform

where P_0 is the pressure at zero altitude (sea level); M is the average molar mass of air (28.97 g/mole in the troposphere); g is the acceleration of gravity (981 cm x sec^{-2} at sea level); h is the altitude in cm; and R is the gas constant (8.314 x 10^7 erg x deg^{-1} x mole^{-1}). These units are given in the cgs (centimeter-gram-sec) system for consistency; altitude can be converted to meters or kilometers as appropriate.

The factor RT/Mg is defined as the **scale height,** which represents the increase in altitude by which the pressure drops by e^{-1}. At an average sea-level temperature of 288 K, the scale height is 8×10^5 cm or 8 km; at an altitude of 8 km, the pressure is only about 39% of that at sea-level.

Conversion of Equation 9.2.1 to the logarithmic (base 10) form and expression of h in km yields

$$\text{Log } P_h = \text{Log } P_0 - \frac{Mgh \times 10^5}{2.303RT} \tag{9.3.2}$$

and taking the pressure at sea level to be exactly 1 atm gives the following expression:

$$\text{Log } P_h = -\frac{Mgh \times 10^5}{2.303\ RT} \tag{9.3.3}$$

Plots of P_h and temperature *versus* altitude are shown in Figure 9.1. The plot of P_h is nonlinear because of variations arising from nonlinear variations in temperature with altitude that are discussed later in this section and from the mixing of air masses.

The characteristics of the atmosphere vary widely with altitude, time (season), location (latitude), and even solar activity. Extremes of pressure and temperature are illustrated in Figure 9.1. At very high altitudes, normally reactive species such as atomic oxygen, O, persist for long periods of time. That occurs because the pressure is very low at these altitudes such that the distance traveled by a reactive species

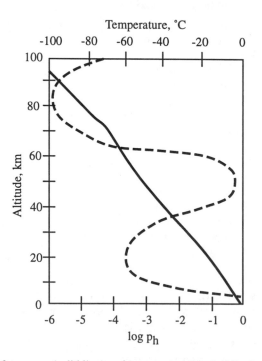

Figure 9.1. Variation of pressure (solid line) and temperature (dashed line) with altitude.

before it collides with a potential reactant — its **mean free path** — is quite high. A particle with a mean free path of 1 x 10^{-6} cm at sea level has a mean free path greater than 1 x 10^6 cm at an altitude of 500 km, where the pressure is lower by many orders of magnitude.

Stratification of the Atmosphere

As shown in Figure 9.2, the atmosphere is stratified on the basis of the temperature/density relationships resulting from interactions between physical and photochemical (light-induced chemical phenomena) processes in air.

The lowest layer of the atmosphere extending from sea level to an altitude of 10-16 km is the **troposphere**, characterized by a generally homogeneous composition of major gases other than water and decreasing temperature with increasing altitude from the heat-radiating surface of the earth. The upper limit of the troposphere, which has a temperature minimum of about -56°C, varies in altitude by a kilometer or more with atmospheric temperature, underlying terrestrial surface, and time. The homogeneous composition of the troposphere results from constant mixing by circulating air masses. However, the water vapor content of the troposphere is extremely variable because of cloud formation, precipitation, and evaporation of water from terrestrial water bodies.

The very cold temperature of the **tropopause** layer at the top of the troposphere serves as a barrier that causes water vapor to condense to ice so that it cannot reach altitudes at which it would photodissociate through the action of intense high-energy ultraviolet radiation. If this happened, the hydrogen produced would escape the earth's atmosphere and be lost. (Much of the hydrogen and helium gases originally present in the earth's atmosphere were lost by this process.)

The atmospheric layer directly above the troposphere is the **stratosphere**, in which the temperature rises to a maximum of about -2°C with increasing altitude. This phenomenon is due to the presence of ozone, O_3, which may reach a level of around 10 ppm by volume in the mid-range of the stratosphere. The heating effect is caused by the absorption of ultraviolet radiation energy by ozone, a phenomenon discussed later in this chapter.

The absence of high levels of radiation-absorbing species in the **mesosphere** immediately above the stratosphere results in a further temperature decrease to about −92°C at an altitude around 85 km. The upper regions of the mesosphere and higher define a region called the exosphere from which molecules and ions can completely escape the atmosphere. Extending to the far outer reaches of the atmosphere is the **thermosphere**, in which the highly rarified gas reaches temperatures as high as 1200°C by the absorption of very energetic radiation of wavelengths less than approximately 200 nm by gas species in this region.

9.4. ENERGY TRANSFER IN THE ATMOSPHERE

The physical and chemical characteristics of the atmosphere and the critical heat balance of the earth are determined by energy and mass transfer processes in the atmosphere. Energy transfer phenomena are addressed in this section and mass transfer in Section 9.4.

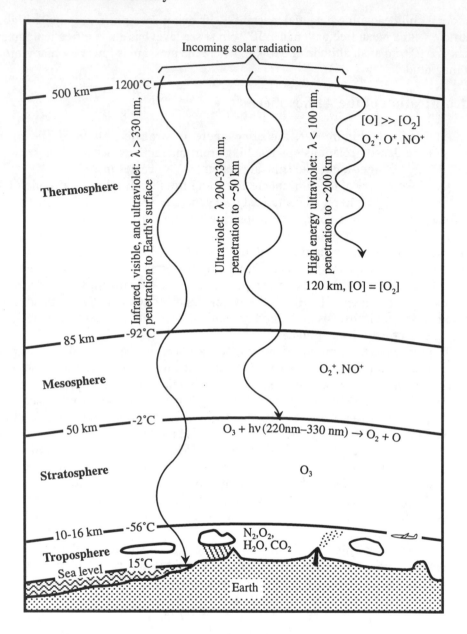

Figure 9.2. Major regions of the atmosphere (not to scale).

Incoming solar energy is largely in the visible region of the spectrum. The shorter wavelength blue solar light is scattered relatively more strongly by molecules and particles in the upper atmosphere, which is why the sky is blue as it is viewed by scattered light. Similarly, light that has been transmitted through scattering atmospheres appears red, particularly around sunset and sunrise, and under circumstances in which the atmosphere contains a high level of particles. The solar energy flux reaching the atmosphere is huge, amounting to 1.34×10^3 watts per square meter (19.2 kcal per minute per square meter) perpendicular to the line of solar flux at the top of the atmosphere, as illustrated in Figure 9.3. This value is the **solar constant**, and may be termed **insolation**, which stands for "incoming solar radiation." If all this energy reached the earth's surface and was retained, the planet would have vaporized long ago. As it is, the complex factors involved in maintaining the Earth's heat balance within very narrow limits are crucial to retaining conditions of climate that will support present levels of life on earth. The great changes of climate that resulted in ice ages during some periods, or tropical conditions during others, were caused by variations of only a few degrees in average temperature. Marked climate changes within recorded history have been caused by much smaller average temperature changes. The mechanisms by which the earth's average temperature is retained within its present narrow range are complex and not completely understood, but the main features are explained here.

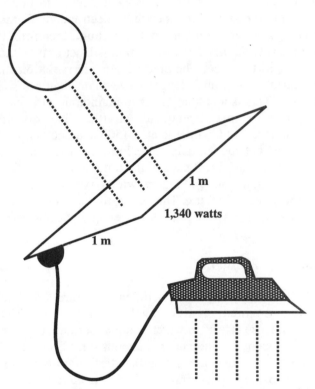

Figure 9.3. The solar flux at the distance of the Earth from the sun is 1.34×10^3 watts/m^2.

About half of the solar radiation entering the atmosphere reaches the earth's surface either directly or after scattering by clouds, atmospheric gases, or particles. The remaining half of the radiation is either reflected directly back or absorbed in the atmosphere, and its energy radiated back into space at a later time as infrared radiation. Most of the solar energy reaching the surface is absorbed and it must be returned to space in order to maintain heat balance. In addition, a very small amount of energy (less than 1% of that received from the sun) reaches the earth's surface by convection and conduction processes from the earth's hot mantle, and this, too, must be lost.

Energy transport, which is crucial to eventual reradiation of energy from the earth, is accomplished by three major mechanisms. These are conduction, convection, and radiation. **Conduction** of energy occurs through the interaction of adjacent atoms or molecules without the bulk movement of matter and is a relatively slow means of transferring energy in the atmosphere. **Convection** involves the movement of whole masses of air, which may be either relatively warm or cold. It is the mechanism by which abrupt temperature variations occur when large masses of air move across an area. As well as carrying **sensible heat** due to the kinetic energy of molecules, convection carries **latent heat** in the form of water vapor which releases heat as it condenses. An appreciable fraction of the earth's surface heat is transported to clouds in the atmosphere by conduction and convection before being lost ultimately by radiation.

Radiation of energy in earth's atmosphere occurs through electromagnetic radiation in the infrared region of the spectrum. Electromagnetic radiation is the only way in which energy is transmitted through a vacuum; therefore, it is the means by which all of the energy that must be lost from the planet to maintain its heat balance is ultimately returned to space. The electromagnetic radiation that carries energy away from the earth is of a much longer wavelength than the sunlight that brings energy to the earth. This is a crucial factor in maintaining the earth's heat balance, and one susceptible to upset by human activities. The maximum intensity of incoming radiation occurs at 0.5 micrometers (500 nanometers) in the visible region, with essentially none outside the range of 0.2 µm to 3 µm. This range encompasses the whole visible region and small parts of the ultraviolet and infrared adjacent to it. Outgoing radiation is in the infrared region, with maximum intensity at about 10 µm, primarily between 2 µm and 40 µm. Thus the earth loses energy by electromagnetic radiation of a much longer wavelength (lower energy per photon) than the radiation by which it receives energy.

Earth's Radiation Budget

The earth's radiation budget is illustrated in Figure 9.4. The average surface temperature is maintained at a relatively comfortable 15°C because of an atmospheric "greenhouse effect" in which water vapor and, to a lesser extent, carbon dioxide reabsorb much of the outgoing radiation and reradiate about half of it back to the surface. Were this not the case, the surface temperature would average around - 18°C. Most of the absorption of infrared radiation is done by water molecules in the atmosphere. Absorption is weak in the regions 7-8.5 µm and 11-14 µm, and nonexistent between 8.5 µm and 11 µm, leaving a "hole" in the infrared absorption spectrum through which radiation may escape. Carbon dioxide, though present at a much lower concentration than water vapor, absorbs strongly between 12 µm and 16.3 µm, and plays a key role in maintaining the heat balance. There is concern that an increase in the carbon dioxide level in the atmosphere could prevent sufficient energy loss to cause a perceptible and damaging increase in the earth's temperature. This phenomenon, discussed in more detail in Section 9.11 and Chapter 14, is popularly known as the **greenhouse effect** and may occur from elevated CO_2 levels caused by increased use of fossil fuels and the destruction of massive quantities of forests.

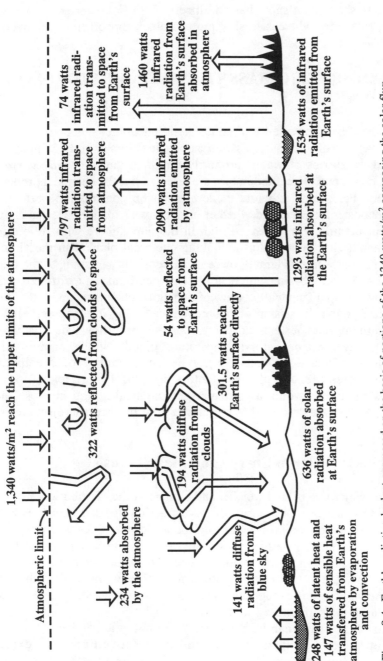

1,340 watts/m² reach the upper limits of the atmosphere

Atmospheric limit

234 watts absorbed by the atmosphere

322 watts reflected from clouds to space

194 watts diffuse radiation from clouds

141 watts diffuse radiation from blue sky

248 watts of latent heat and 147 watts of sensible heat transferred from Earth's atmosphere by evaporation and convection

636 watts of solar radiation absorbed at Earth's surface

301.5 watts reach Earth's surface directly

54 watts reflected to space from Earth's surface

1293 watts infrared radiation absorbed at the Earth's surface

797 watts infrared radiation trans- mitted to space from atmosphere

2090 watts infrared radiation emitted by atmosphere

74 watts infrared radi- ation trans- mitted to space from Earth's surface

1460 watts infrared radiation from Earth's surface absorbed in atmosphere

1534 watts of infrared radiation emitted from Earth's surface

Figure 9.4. Earth's radiation budget expressed on the basis of portions of the 1340 watts/m² composing the solar flux.

An important aspect of solar radiation that reaches earth's surface is the percentage reflected from the surface, described as **albedo**. Albedo is important in determining earth's heat balance in that absorbed radiation heats the surface, and reflected radiation does not. Albedo varies spectacularly with the surface. At the two extremes, freshly fallen snow has an albedo of 90 % because it reflects 9/10 of incoming radiation, whereas freshly plowed black topsoil has an albedo of only about 2.5 %.

9.5. ATMOSPHERIC MASS TRANSFER, METEOROLOGY, AND WEATHER

Meteorology is the science of atmospheric phenomena, encompassing the study of the movement of air masses as well as physical forces in the atmosphere—heat, wind, and transitions of water, primarily liquid to vapor or *vice versa*. Meteorological phenomena affect, and in turn are affected by, the chemical properties of the atmosphere. For example, before modern emission controls took effect, meteorological phenomena determined whether or not power plant stack gas heavily laced with sulfur dioxide was dispersed high in the atmosphere with little direct effect upon human health, or settled as a choking chemical blanket in the vicinity of the power plant. Los Angeles largely owes its susceptibility to smog to the meteorology of the Los Angeles basin, which holds hydrocarbons and nitrogen oxides long enough to cook up an unpleasant brew of damaging chemicals under the intense rays of the sun (see the discussion of photochemical smog in Chapter 13). Short-term variations in the state of the atmosphere constitute **weather**. The weather is defined in terms of seven major factors: temperature, clouds, winds, humidity, horizontal visibility (as affected by fog, etc.), type and quantity of precipitation, and atmospheric pressure. All of these factors are closely interrelated. Longer-term variations and trends within a particular geographical region in those factors that compose weather are described as **climate**, a term defined and discussed in Section 9.6.

Atmospheric Water in Energy and Mass Transfer

The driving force behind weather and climate is the distribution and ultimate re-radiation to space of solar energy. A large fraction of solar energy is converted to latent heat by evaporation of water into the atmosphere. As water condenses from atmospheric air, large quantities of heat are released. This is a particularly significant means for transferring energy from the ocean to land. Solar energy falling on the ocean is converted to latent heat by the evaporation of water, then the water vapor moves inland where it condenses. The latent heat released when the water condenses warms the surrounding land mass.

Atmospheric water can be present as vapor, liquid, or ice. The water vapor content of air can be expressed as **humidity**. **Relative humidity**, expressed as a percentage, describes the amount of water vapor in the air as a ratio of the maximum amount that the air can hold at that temperature. Air with a given relative humidity can undergo any of several processes to reach the saturation point at which water vapor condenses in the form of rain or snow. For this condensation to happen, air must be cooled below a temperature called the **dew point**, and **condensation nuclei** must be present. These nuclei are hygroscopic substances such as salts, sulfuric acid droplets, and some organic materials, including bacterial cells. Air pollution in some forms is an important source of condensation nuclei.

The liquid water in the atmosphere is present largely in **clouds**. Clouds normally form when rising, adiabatically cooling air can no longer hold water in the vapor form and the water forms very small aerosol droplets. Clouds may be classified in three major forms. Cirrus clouds occur at great altitudes and have a thin feathery appearance. Cumulus clouds are detached masses with a flat base and frequently a "bumpy" upper structure. Stratus clouds occur in large sheets and may cover all of the sky visible from a given point as overcast. Clouds are important absorbers and reflectors of radiation (heat). Their formation is affected by the products of human activities, especially particulate matter pollution and emission of deliquescent gases such as SO_2 and HCl.

The formation of precipitation from the very small droplets of water that compose clouds is a complicated and important process. Cloud droplets normally take somewhat longer than a minute to form by condensation. They average about 0.04 mm across and do not exceed 0.2 mm in diameter. Raindrops range from 0.5–4 mm in diameter. Condensation processes do not form particles large enough to fall as precipitation (rain, snow, sleet, or hail). The small condensation droplets must collide and coalesce to form precipitation-size particles. When droplets reach a threshold diameter of about 0.04 mm, they grow more rapidly by coalescence with other particles than by condensation of water vapor.

Air Masses

Distinct air masses are a major feature of the troposphere. These air masses are uniform and horizontally homogeneous. Their temperature and water-vapor content are particularly uniform. These characteristics are determined by the nature of the surface over which a large air mass forms. Polar continental air masses form over cold land regions; polar maritime air masses form over polar oceans. Air masses originating in the tropics may be similarly classified as tropical continental air masses or tropical maritime air masses. The movement of air masses and the conditions in them may have important effects upon pollutant reactions, effects, and dispersal.

Solar energy received by earth is largely redistributed by the movement of huge masses of air with different pressures, temperatures, and moisture contents separated by boundaries called **fronts**. Horizontally moving air is called **wind**, whereas vertically moving air is referred to as an **air current**. Atmospheric air moves constantly, with behavior and effects that reflect the laws governing the behavior of gases. First of all, gases will move horizontally and/or vertically from regions of *high atmopheric pressure* to those of *low atmospheric pressure*. Furthermore, expansion of gases causes cooling, whereas compression causes warming. A mass of warm air tends to move from earth's surface to higher altitudes where the pressure is lower; in so doing, it expands *adiabatically* (that is, without exchanging energy with its surroundings) and becomes cooler. If there is no condensation of moisture from the air, the cooling effect is about 10°C per 1000 meters of altitude, a figure known as the **dry adiabatic lapse rate**. A cold mass of air at a higher altitude does the opposite; it sinks and becomes warmer at about 10°C/1000 m. Often, however, when there is sufficient moisture in rising air, water condenses from it, releasing latent heat. This partially counteracts the cooling effect of the expanding air, giving a **moist adiabatic lapse rate** of about 6°C/1000 m. Parcels of air do not rise and fall, or even move horizontally in a completely uniform way, but exhibit eddies, currents, and various degrees of turbulence.

As noted above, *wind* is air moving horizontally, whereas *air currents* are created by air moving up or down. Wind occurs because of differences in air pressure from high pressure regions to low pressure areas. Air currents are largely **convection currents** formed by differential heating of air masses. Air that is over a solar heated land mass is warmed, becomes less dense and therefore rises and is replaced by cooler and more dense air. Wind and air currents are strongly involved with air pollution phenomena. Wind carries and disperses air pollutants. In some cases the absence of wind can enable pollutants to collect in a region and undergo processes that lead to even more (secondary) pollutants. Prevailing wind direction is an important factor in determining the areas most affected by an air pollution source. Wind is an important renewable energy resource (see Chapter 18). Furthermore, wind plays an important role in the propagation of life by dispersing spores, seeds, and organisms, such as spiders.

Topographical Effects

Topography, the surface configuration and relief features of the earth's surface may strongly affect winds and air currents. Differential heating and cooling of land surfaces and bodies of water can result in **local convective winds**, including land breezes and sea breezes at different times of the day along the seashore, as well as breezes associated with large bodies of water inland. Mountain topography causes complex and variable localized winds. The masses of air in mountain valleys heat up during the day causing upslope winds, and cool off at night causing downslope winds. Upslope winds flow over ridge tops in mountainous regions. The blocking of wind and of masses of air by mountain formations some distance inland from seashores can trap bodies of air, particularly when temperature inversion conditions occur (see Section 9.5).

Movement of Air Masses

Basically, weather is the result of the interactive effects of (1) redistribution of solar energy, (2) horizontal and vertical movement of air masses with varying moisture contents, and (3) evaporation and condensation of water, accompanied by uptake and release of heat. To see how these factors determine weather—and ultimately climate—on a global scale, first consider the cycle illustrated in Figure 9.5. This figure shows solar energy being absorbed by a body of water and causing some water to evaporate. The warm, moist mass of air thus produced moves from a region of high pressure to one of low pressure, and cools by expansion as it rises in what is called a **convection column**. As the air cools, water condenses from it and energy is released; this is a major pathway by which energy is transferred from the earth's surface to high in the atmosphere. As a result of condensation of water and loss of energy, the air is converted from warm, moist air to cool, dry air. Furthermore, the movement of the parcel of air to high altitudes results in a degree of "crowding" of air molecules and creates a zone of relatively high pressure high in the troposphere at the top of the convection column. This air mass, in turn, moves from the upper-level region of high pressure to one of low pressure; in so doing, it subsides, thus creating an upper-level low pressure zone, and becomes warm, dry air in the process. The pileup of this air at the surface creates a surface high pressure zone where the cycle described above began. The warm, dry air in this surface high pressure zone again picks up moisture, and the cycle begins again.

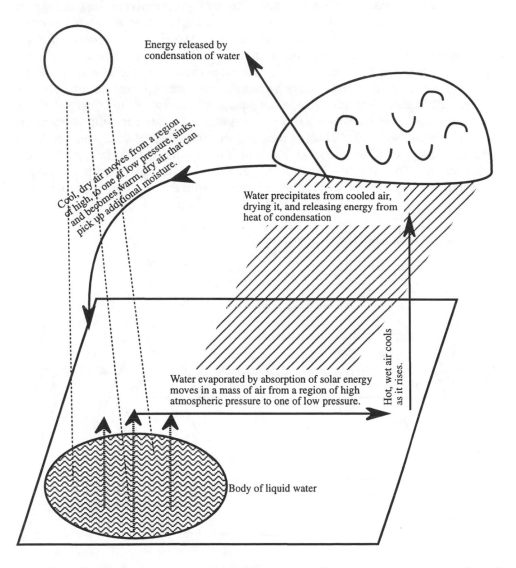

Figure 9.5. Circulation patterns involved with movement of air masses and water; uptake and release of solar energy as latent heat in water vapor.

Global Weather

The factors discussed above that determine and describe the movement of air masses are involved in the massive movement of air, moisture, and energy that occurs globally. The central feature of global weather is the redistribution of solar energy that falls unequally on earth at different latitudes (relative distances from the equator and poles). Consider Figure 9.6. Sunlight, and the energy flux from it, is most intense at the equator because, averaged over the seasons, solar radiation comes in perpendicular to earth's surface at the Equator. With increasing distance from the equator (higher latitudes) the angle is increasingly oblique and more of the energy-absorbing atmosphere must be traversed, so that progressively less energy is received per unit area of earth's surface. The net result is that equatorial regions receive a much greater share of solar radiation, progressively less is received farther from the Equator, and the poles receive a comparatively miniscule amount. The excess heat energy in the equatorial regions causes the air to rise. The air ceases to rise when it reaches the stratosphere because in the stratosphere the air becomes warmer with higher elevation. As the hot equatorial air rises in the troposphere, it cools by expansion and loss of water, then sinks again. The air circulation patterns in which this occurs are called **Hadley cells**. As shown in Figure 9.6, there are three major groupings of these cells, which result in very distinct climatic regions on earth's surface. The air in the Hadley cells does not move straight north and south, but is deflected by earth's rotation and by contact with the rotating earth; this is the

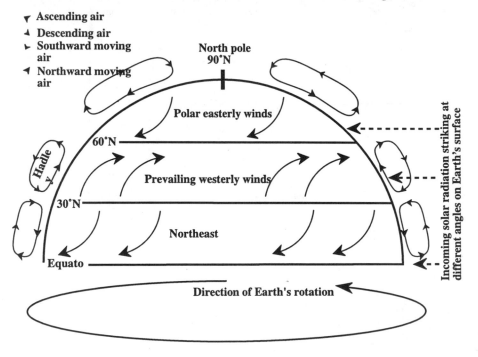

Figure 9.6. Global circulation of air in the northern hemisphere.

Coriolis effect, which results in spiral-shaped air circulation patterns called cyclonic or anticyclonic, depending upon the direction of rotation. These give rise to different directions of prevailing winds, depending on latitude. The boundaries between the massive bodies of circulating air shift markedly over time and season, resulting in significant weather instability.

The movement of air in Hadley cells combined with other atmospheric phenomena results in the development of massive **jet streams** that are, in a sense, shifting rivers of air that may be several kilometers deep and several tens of km wide. Jet streams move through discontinuities in the tropopause (see Section 9.2), generally from west to east at velocities around 200 km/hr (well over 100 mph); in so doing, they redistribute huge amounts of air and have a strong influence on weather patterns.

The air and wind circulation patterns described above shift massive amounts of energy over long distances on earth. If it weren't for this effect, the equatorial regions would be unbearably hot, and the regions closer to the poles intolerably cold. About half of the heat that is redistributed is carried as sensible heat by air circulation, almost 1/3 is carried by water vapor as latent heat, and the remaining approximately 20 % by ocean currents.

Weather Fronts and Storms

As noted earlier, the interface between two masses of air that differ in temperature, density, and water content is called a **front**. A mass of cold air moving such that it displaces one of warm air is a **cold front**, and a mass of warm air displacing one of cold air is a **warm front**. Since cold air is more dense than warm air, the air in a cold mass of air along a cold front pushes under warmer air. This causes the warm, moist air to rises such that water condense from it. The condensation of water releases energy, so the air rises further. The net effect can be formation of massive cloud formations (thunderheads) that may reach stratospheric levels. These spectacular thunderheads may produce heavy rainfall and even hail, and sometimes violent storms with strong winds, including tornadoes. Warm fronts cause somewhat similar effects as warm, moist air pushes over colder air. However, the front is usually much broader, and the weather effects milder, typically resulting in widespread drizzle rather than intense rainstorms.

Swirling **cyclonic storms**, such as typhoons, hurricanes, and tornadoes, are created in low pressure areas by rising masses of warm, moist air. As such air cools, water vapor condenses, and the latent heat released warms the air more, sustaining and intensifying its movement upward in the atmosphere. Air rising from surface level creates a low pressure zone into which surrounding air moves. The movement of the incoming air assumes a spiral pattern, thus causing a cyclonic storm.

9.6. INVERSIONS AND AIR POLLUTION

The complicated movement of air across the earth's surface is a crucial factor in the creation and dispersal of air pollution phenomena. When air movement ceases, stagnation can occur with a resultant buildup of atmospheric pollutants in localized regions. Although the temperature of air relatively near the earth's surface normally decreases with increasing altitude, certain atmospheric conditions can result in the opposite condition—increasing temperature with increasing altitude. Such conditions are characterized by high atmospheric stability and are known as **temperature inversions**. Because they limit the vertical circulation of air, temperature inversions result in air stagnation and the trapping of air pollutants in localized areas.

Inversions can occur in several ways. In a sense, the whole atmosphere is inverted by the warm stratosphere, which floats atop the troposphere, with relatively little mixing. An inversion can form from the collision of a warm air mass (warm front) with a cold air mass (cold front). The warm air mass overrides the cold air mass in the frontal area, producing the inversion. **Radiation inversions** are likely to form in still air at night when the Earth is no longer receiving solar radiation. The air closest to the earth cools faster than the air higher in the atmosphere, which remains warm, thus less dense. Furthermore, cooler surface air tends to flow into valleys at night, where it is overlain by warmer, less dense air. **Subsidence inversions**, often accompanied by radiation inversions, can become very widespread. These inversions can form in the vicinity of a surface high-pressure area when high-level air subsides to take the place of surface air blowing out of the high-pressure zone. The subsiding air is warmed as it compresses and can remain as a warm layer several hundred meters above ground level. A **marine inversion** is produced during the summer months when cool air laden with moisture from the ocean blows onshore and under warm, dry inland air.

As noted above, inversions contribute significantly to the effects of air pollution because, as shown in Figure 9.7, they prevent mixing of air pollutants, thus keeping the pollutants in one area. This not only prevents the pollutants from escaping, but also acts like a container in which additional pollutants accumulate. Furthermore, in the case of secondary pollutants formed by atmospheric chemical processes, such as photochemical smog (see Chapter 13), the pollutants may be kept together such that they react with each other and with sunlight to produce even more noxious products.

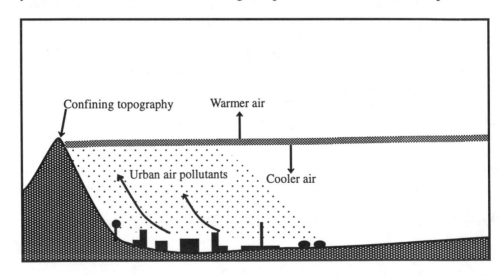

Figure 9.7. Illustration of pollutants trapped in a temperature inversion.

9.7. GLOBAL CLIMATE AND MICROCLIMATE

Perhaps the single most important influence on Earth's environment is **climate**, consisting of long-term weather patterns over large geographical areas. As a general rule, climatic conditions are characteristic of a particular region. This does not mean that climate remains the same throughout the year, of course, because it varies with season. One important example of such variation is the **monsoon**, seasonal variations in wind patterns between oceans and continents. The climates of Africa and the Indian subcontinent are particularly influenced by monsoons. In the latter, for example, summer heating of the Indian land mass causes air to rise, thereby creating a low pressure area that attracts warm, moist air from the ocean. This air rises on the slopes of the Himalayan mountains, which also block the flow of colder air from the north; moisture from the air condenses; and monsoon rains carrying enormous amounts of precipitation fall. Thus, from May until into August, summer monsoon rains fall in India, Bangladesh, and Nepal. Reversal of the pattern of winds during the winter months causes these regions to have a dry season, but produces winter monsoon rains in the Philipine islands, Indonesia, New Guinea, and Australia.

Summer monsoon rains are responsible for tropical rain forests in Central Africa. The interface between this region and the Sahara Desert varies over time. When the boundary is relatively far north, rain falls on the Sahel desert region at the interface, crops grow, and the people do relatively well. When the boundary is more to the south, a condition which may last for several years, devastating droughts and even starvation may occur.

It is known that there are fluctuations, cycles, and cycles imposed on cycles in climate. The causes of these variations are not completely understood, but they are known to be substantial, and even devastating to civilization. The last **ice age**, which ended only about 10,000 years ago and which was preceded by several similar ice ages, produced conditions under which much of the present land mass of the Northern Hemisphere was buried under thick layers of ice and uninhabitable. A "mini-ice age" occurred during the 1300s, causing crop failures and severe hardship in northern Europe. In modern times the El-Niño-Southern Oscillation occurs with a period of several years when a large, semi-permanent tropical low pressure area shifts into the Central Pacific region from its more common location in the vicinity of Indonesia. This shift modifies prevailing winds, changes the pattern of ocean currents, and affects upwelling of ocean nutrients with profound effects on weather, rainfall, and fish and bird life over a vast area of the Pacific from Australia to the west coasts of South and North America.

Human Modifications of Climate

Although Earth's atmosphere is huge and has an enormous ability to resist and correct for detrimental change, it is possible that human activities are reaching a point at which they may be adversely affecting climate. One such way is by emission of large quantities of carbon dioxide and other greenhouse gases into the atmosphere, such that global warming may occur and cause substantial climatic change. Another way is through the release of gases, particularly chlorofluoro-carbons (Freons) that may cause destruction of vital stratospheric ozone. Human effects on climate are addressed in Chapter 14, "The Endangered Global Atmosphere."

Microclimate

The preceding section described climate on a large scale, ranging up to global dimensions. The climate that organisms and objects on the surface are exposed to close to the ground, under rocks, and surrounded by vegetation, is often quite different from the surrounding macroclimate. Such highly localized climatic conditions are termed the **microclimate**. Microclimate effects are largely determined by the uptake and loss of solar energy very close to earth's surface, and by the fact that air circulation due to wind is much lower at the surface. During the day, solar energy absorbed by relatively bare soil heats the surface, but is lost only slowly because of very limited air circulation at the surface. This provides a warm blanket of surface air several cm thick, and an even thinner layer of warm soil. At night, radiative loss of heat from the surface of soil and vegetation can result in surface temperatures several degrees colder than the air about 2 meters above ground level. These lower temperatures result in condensation of **dew** on vegetation and the soil surface, thus providing a relatively more moist microclimate near ground level. Heat absorbed during early morning evaporation of the dew tends to prolong the period of cold experienced right at the surface.

Vegetation substantially affects microclimate. In relatively dense growths, circulation may be virtually zero at the surface because vegetation severely limits convection and diffusion. The crown surface of the vegetation intercepts most of the solar energy, so that maximum solar heating may be a significant distance up from earth's surface. The region below the crown surface of vegetation thus becomes one of relatively stable temperature. In addition, in a dense growth of vegetation, most of the moisture loss is not from evaporation from the soil surface, but rather from transpiration from plant leaves. The net result is the creation of temperature and humidity conditions that provide a favorable living environment for a number of organisms, such as insects and rodents.

Another factor influencing microclimate is the degree to which the slope of land faces north or south. South-facing slopes of land in the northern hemisphere receive greater solar energy. Advantage has been taken of this phenomenon in restoring land strip-mined for brown coal in Germany by terracing the land such that the terraces have broad south slopes and very narrow north slopes. On the south-sloping portions of the terrace, the net effect has been to extend the short summer growing season by several days, thereby significantly increasing crop productivity. In areas where the growing season is longer, better growing conditions may exist on a north slope because it is less subject to temperature extremes and to loss of water by evaporation and transpiration.

Effects of Urbanization on Microclimate

A particularly marked effect on microclimate is that induced by urbanization. In a rural setting, vegetation and bodies of water have a moderating effect, absorbing modest amounts of solar energy and releasing it slowly. The stone, concrete, and asphalt pavement of cities have an opposite effect, strongly absorbing solar energy, and re-radiating heat back to the urban microclimate. Rainfall is not allowed to accumulate in ponds, but is drained away as rapidly and efficiently as possible. Human activities generate significant amounts of heat, and produce large quantities of CO_2 and other greenhouse gases that retain heat. The net result of these effects is that a city is capped by a **heat dome** in which the temperature is as much as 5°C warmer than in the surrounding rural areas, such that large cities have been described as "heat islands." The rising warmer air over a city brings in a breeze from the surrounding area and causes a local greenhouse effect that probably is largely counterbalanced by reflection of incoming solar energy by particulate matter above cities. Overall, compared to climatic conditions in nearby rural surroundings, the city microclimate is warmer, foggier and overlain with more cloud cover a greater percentage of the time, and is subject to more precipitation, though generally less humid.

9.8. CHEMICAL AND PHOTOCHEMICAL REACTIONS IN THE ATMOSPHERE

Figure 9.8. represents some of the major atmospheric chemical processes, which are discussed under the topic of **atmospheric chemistry**. The study of atmospheric chemical reactions is difficult. One of the primary obstacles encountered in studying atmospheric chemistry is that the chemist generally must deal with incredibly low concentrations, so that the detection and analysis of reaction products is quite difficult. Simulating high-altitude conditions in the laboratory can be extremely hard because of interferences, such as those from species given off from container walls

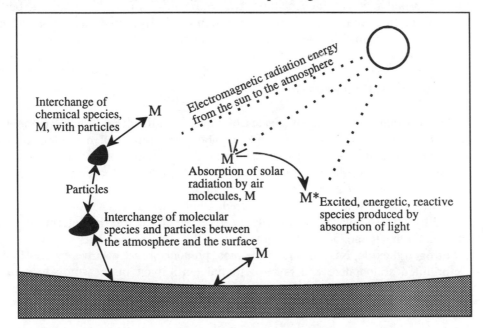

Figure 9.8. Representation of major atmospheric chemical processes.

under conditions of very low pressure. Many chemical reactions that require a third body to absorb excess energy occur very slowly in the upper atmosphere where there is a sparse concentration of third bodies, but occur readily in a container whose walls effectively absorb energy. Container walls may serve as catalysts for some important reactions, or they may absorb important species and react chemically with the more reactive ones.

Atmospheric chemistry involves the unpolluted atmosphere, highly polluted atmospheres, and a wide range of gradations in between. The same general phenomena govern all and produce one huge atmospheric cycle in which there are numerous subcycles. Gaseous atmospheric chemical species fall into the following somewhat arbitrary and overlapping classifications: Inorganic oxides (CO, CO_2, NO_2, SO_2), oxidants (O_3, H_2O_2, HO· radical, HO_2· radical, ROO· radicals, NO_3), reductants (CO, SO_2, H_2S), organics (also reductants; in the unpolluted atmosphere, CH_4 is the predominant organic species, whereas alkanes, alkenes, and aryl compounds are common around sources of organic pollution), oxidized organic species (carbonyls, organic nitrates), photochemically active species (NO_2, formaldehyde), acids (H_2SO_4), bases (NH_3), salts (NH_4HSO_4,), and unstable reactive species (electronically excited NO_2, HO· radical). In addition, both solid and liquid particles in atmospheric aerosols and clouds play a strong role in atmospheric chemistry as sources and sinks for gas-phase species, as sites for surface reactions (solid particles), and as bodies for aqueous-phase reactions (liquid droplets). Two constituents of utmost importance in atmospheric chemistry are radiant energy from the sun, predominantly in the ultraviolet region of the spectrum, and the hydroxyl radical, HO·. The former provides a way to pump a high level of energy into a single gas molecule to start a series of atmospheric chemical reactions, and the latter is the most important reactive intermediate and "currency" of daytime atmospheric chemical phenomena; NO_3 radicals are important intermediates in nightime atmospheric chemistry. These are addressed in more detail in this chapter and Chapters 10-14.

Photochemical Processes

The absorption by chemical species of light, broadly defined here to include ultraviolet radiation from the sun, can bring about reactions, called **photochemical reactions**, which do not otherwise occur under the conditions (particularly the temperature) of the medium in the absence of light. Thus, photochemical reactions, even in the absence of a chemical catalyst, occur at temperatures much lower than those which otherwise would be required. Photochemical reactions, which are induced by intense solar radiation, play a very important role in determining the nature and ultimate fate of a chemical species in the atmosphere.

Nitrogen dioxide, NO_2, is one of the most photochemically active species found in a polluted atmosphere and is an essential participant in the smog-formation process. A species such as NO_2 may absorb light of energy hv, producing an **electronically excited molecule**,

$$NO_2 + hv \rightarrow NO_2^*$$
(9.8.1)

designated in the reaction above by an asterisk, *. The photochemistry of nitrogen dioxide is discussed in greater detail in Chapters 11 and 13.

Electronically excited molecules are one of the three relatively reactive and unstable species that are encountered in the atmosphere and are strongly involved with atmospheric chemical processes. The other two species are atoms or molecular fragments with unshared electrons, called **free radicals**, and **ions** consisting of electrically-charged atoms or molecular fragments.

Electronically excited molecules are produced when stable molecules absorb energetic electromagnetic radiation in the ultraviolet or visible regions of the spectrum. A molecule may possess several possible excited states, but generally ultraviolet or visible radiation is energetic enough to excite molecules only to several of the lowest energy levels. The nature of the excited state may be understood by considering the disposition of electrons in a molecule. Most molecules have an even number of electrons. The electrons occupy orbitals, with a maximum of two electrons with opposite spin occupying the same orbital. The absorption of light may promote one of these electrons to a vacant orbital of higher energy. In some cases the electron thus promoted retains a spin opposite to that of its former partner, giving rise to an **excited singlet state**. In other cases the spin of the promoted electron is reversed, such that it has the same spin as its former partner; this gives rise to an **excited triplet state**.

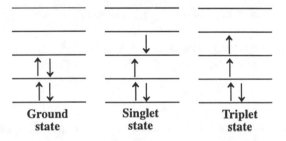

| Ground | Singlet | Triplet |
| state | state | state |

These excited states are relatively energized compared to the ground state and are chemically reactive species. Their participation in atmospheric chemical reactions, such as those involved in smog formation, will be discussed later in detail.

In order for a photochemical reaction to occur, light must be absorbed by the reacting species. If the absorbed light is in the visible region of the sun's spectrum, the absorbing species is colored. Colored NO_2 is a common example of such a species in the atmosphere. Normally, the first step in a photochemical process is the activation of the molecule by the absorption of a single unit of photochemical energy characteristic of the frequency of the light called a **quantum** of light. The energy of one quantum is equal to the product $h\nu$, where h is Planck's constant, 6.63×10^{-27} erg·s (6.63×10 J·s), and ν is the frequency of the absorbed light in s^{-1} (inversely proportional to its wavelength, λ).

The reactions that occur following absorption of a photon of light to produce an electronically excited species are largely determined by the way in which the excited species loses its excess energy. This may occur by one of the following processes:

- Loss of energy to another molecule or atom (M) by **physical quenching**, followed by dissipation of the energy as heat

$$O_2^* + M \rightarrow O_2 + M(\text{higher translational energy}) \qquad (9.8.2)$$

- **Dissociation** of the excited molecule (the process responsible for the predominance of atomic oxygen in the upper atmosphere)

$$O_2^* \rightarrow O + O \qquad (9.8.3)$$

- **Direct reaction** with another species

$$O_2^* + O_3 \rightarrow 2O_2 + O \qquad (9.8.4)$$

- **Luminescence** consisting of loss of energy by the emission of electromagnetic radiation

$$NO_2^* \rightarrow NO_2 + h\nu \qquad (9.8.5)$$

If the re-emission of light is almost instantaneous, luminescence is called **fluorescence**, and if it is significantly delayed, the phenomenon is **phosphorescence**. **Chemiluminescence** is said to occur when the excited species (such as NO_2^* below) is formed by a chemical process:

$$O_3 + NO \rightarrow NO_2^* + O_2(\text{higher energy}) \qquad (9.8.6)$$

- **Intermolecular energy transfer** in which an excited species transfers energy to another species which then becomes excited

$$O_2^* + Na \rightarrow O_2 + Na^* \qquad (9.8.7)$$

A subsequent reaction by the second species is called a **photosensitized** reaction.

- **Intramolecular transfer** in which energy is transferred within a molecule

$$XY^* \rightarrow XY^\dagger(\text{where } ^\dagger \text{ denotes another excited}$$
$$\text{state of the same molecule}) \qquad (9.8.8)$$

- **Spontaneous isomerization** as in the conversion of o-nitrobenzaldehyde to o-nitrosobenzoic acid, a reaction used in chemical actinometers to measure exposure to electromagnetic radiation:

$$(9.8.9)$$

- **Photoionization** through loss of an electron

$$N_2^* \rightarrow N_2^+ + e^- \qquad (9.8.10)$$

Electromagnetic radiation absorbed in the infrared region lacks the energy to break chemical bonds, but does cause the receptor molecules to gain vibrational and rotational energy. The energy absorbed as infrared radiation ultimately is dissipated as heat and raises the temperature of the whole atmosphere. As noted in Section 9.3, the absorption of infrared radiation is very important in the earth's acquiring heat from the sun and in the retention of energy radiated from the earth's surface.

Ions and Radicals in the Atmosphere

One of the characteristics of the upper atmosphere which is difficult to duplicate under laboratory conditions is the presence of significant levels of electrons and positive ions. Because of the rarefied conditions, these ions may exist in the upper atmosphere for long periods before recombining to form neutral species.

At altitudes of approximately 50 km and up, ions are so prevalent that the region is called the **ionosphere**. The presence of the ionosphere has been known since about 1901, when it was discovered that radio waves could be transmitted over long distances where the curvature of the Earth makes line-of-sight transmission impossible. These radio waves bounce off the ionosphere.

Ultraviolet light is the primary producer of ions in the ionosphere. In darkness, the positive ions slowly recombine with free electrons. The process is more rapid in the lower regions of the ionosphere where the concentration of species is relatively high. Thus, the lower limit of the ionosphere lifts at night and makes possible the transmission of radio waves over much greater distances.

The earth's magnetic field has a strong influence upon the ions in the upper atmosphere. Probably the best-known manifestation of this phenomenon is found in the Van Allen belts, discovered in 1958 consisting of two belts of ionized particles encircling the earth. They can be visualized as two doughnuts with the axis of earth's magnetic field extending through the holes in the doughnuts. The inner belt consists of protons and the outer belt consists of electrons. A schematic diagram of the Van Allen belts is shown in Figure 9.9.

Although ions are produced in the upper atmosphere primarily by the action of energetic electromagnetic radiation, they may also be produced in the troposphere by the shearing of water droplets during precipitation. The shearing may be caused by the compression of descending masses of cold air or by strong winds over hot, dry land masses. The last phenomenon is known as the foehn, sharav (in the Near East), or Santa Ana (in southern California). These hot, dry winds cause severe discomfort. The ions they produce consist of electrons and positively charged molecular species.

Free Radicals

In addition to forming ions by photoionization, energetic electromagnetic radiation in the atmosphere may produce atoms or groups of atoms with unpaired electrons called **free radicals**:

$$H_3C-\overset{\overset{\displaystyle O}{\|}}{C}-H \; + \; h\nu \; \longrightarrow \; H_3C\bullet \; + \; \bullet\overset{\overset{\displaystyle O}{\|}}{C}-H \tag{9.8.11}$$

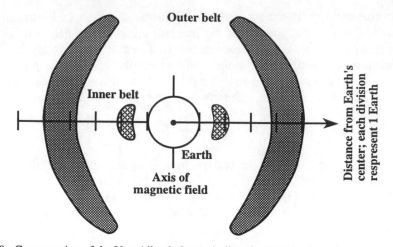

Figure 9.9. Cross section of the Van Allen belts encircling the Earth.

Free radicals are involved with most significant atmospheric chemical phenomena and are of the utmost importance in the atmosphere. Because of their unpaired electrons and the strong pairing tendencies of electrons under most circumstances, free radicals are highly reactive. The upper atmosphere is so rarefied, however, that at very high altitudes radicals may have half-lives of several minutes, or even longer. Radicals can take part in chain reactions in which one of the products of each reaction is a radical. Eventually, through processes such as reaction with another radical, one of the radicals in a chain is destroyed and the chain ends:

$$H_3C\bullet + H_3C\bullet \rightarrow C_2H_6 \qquad (9.8.12)$$

This process is a **chain-terminating reaction**. Reactions involving free radicals are responsible for smog formation, discussed in Chapter 13.

Free radicals are quite reactive; therefore, they generally have short lifetimes. It is important to distinguish between high reactivity and instability. A totally isolated free radical or atom would be quite stable. Therefore, free radicals and single atoms from diatomic gases tend to persist under the rarefied conditions of very high altitudes because they can travel long distances before colliding with another reactive species. However, electronically excited species have a finite, generally very short lifetime because they can lose energy through radiation without having to react with another species.

Hydroxyl and Hydroperoxyl Radicals in the Atmosphere

As illustrated in Figure 9.10, the hydroxyl radical, HO•, is the single most important reactive intermediate species in atmospheric chemical processes. It is formed by several mechanisms. At higher altitudes it is produced by photolysis of water:

$$H_2O + h\nu \rightarrow HO\bullet + H \qquad (9.8.13)$$

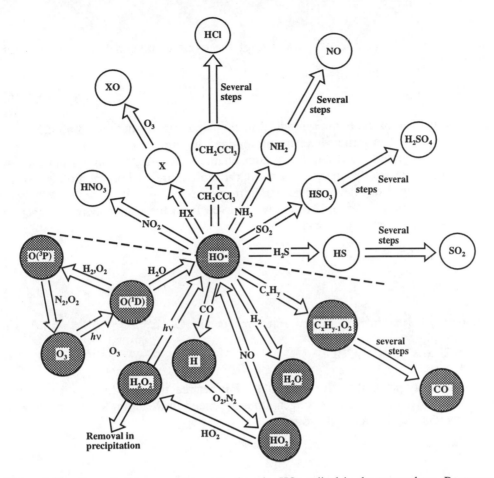

Figure 9.10. Control of trace gas concentrations by HO• radical in the troposphere. Processes below the dashed line are those largely involved in controlling the concentrations of HO• in the troposphere; those above the line control the concentrations of the associated reactants and products. Reservoirs of atmospheric species are shown in circles, reactions denoting conversion of one species to another are shown by arrows, and the reactants or photons needed to bring about a particular conversion are shown along the arrows. Hydrogen halides are denoted by HX and hydrocarbons by H_xY_y. (Source: D. D. Davis and W. L. Chameides, "Chemistry in the Troposphere," *Chemical and Engineering News*, October 4, 1982, pp. 39-52. Reprinted by permission of the American Chemical Society.)

In the presence of organic matter, hydroxyl radical is produced in abundant quantities as an intermediate in the formation of photochemical smog (see Chapter 13). To a certain extent in the atmosphere, and for laboratory experimentation, HO• is made by the photolysis of nitrous acid vapor:

$$HONO + h\nu \rightarrow HO\bullet + NO \tag{9.8.14}$$

In the relatively unpolluted troposphere, hydroxyl radical is produced as the result of the photolysis of ozone,

$$O_3 + h\nu(\lambda < 315 \text{ nm}) \rightarrow O^* + O_2 \tag{9.8.15}$$

followed by the reaction of a fraction of the excited oxygen atoms with water molecules:

$$O* + H_2O \rightarrow 2HO\bullet \tag{9.8.16}$$

Involvement of the hydroxyl radical in chemical transformations of a number of trace species in the atmosphere is summarized in Figure 9.10, and some of the pathways illustrated are discussed in later chapters. Among the important atmospheric trace species that react with hydroxyl radical are carbon monoxide, sulfur dioxide, hydrogen sulfide, methane, and nitric oxide.

Hydroxyl radical is most frequently removed from the troposphere by reaction with methane or carbon monoxide:

$$CH_4 + HO\bullet \rightarrow H_3C\bullet + H_2O \tag{9.8.17}$$

$$CO + HO\bullet \rightarrow CO_2 + H \tag{9.8.18}$$

The highly reactive methyl radical, $H_3C\bullet$, reacts with O_2,

$$H_3C\bullet + O_2 \rightarrow H_3COO\bullet \tag{9.8.19}$$

to form **methylperoxyl radical**, $H_3COO\bullet$. (Further reactions of this species are discussed in Chapter 13.) The hydrogen atom produced in Reaction 9.8.17 reacts with O_2 to produce **hydroperoxyl radical**:

$$H + O_2 \rightarrow HOO\bullet \tag{9.8.20}$$

The hydroperoxyl radical can undergo chain termination reactions, such as

$$HOO\bullet + HO\bullet \rightarrow H_2O + O_2 \tag{9.8.21}$$

$$HOO\bullet + HOO\bullet \rightarrow H_2O_2 + O_2 \tag{9.8.22}$$

or reactions that regenerate hydroxyl radical:

$$HOO\bullet + NO \rightarrow NO_2 + HO\bullet \tag{9.8.23}$$

$$HOO\bullet + O_3 \rightarrow 2O_2 + HO\bullet \tag{9.8.24}$$

The global concentration of hydroxyl radical, averaged diurnally and seasonally, is estimated to range from 2×10^5 to 1×10^6 radicals per cm^3 in the troposphere. Because of the higher humidity and higher incident sunlight which result in elevated $O*$ levels, the concentration of $HO\bullet$ is higher in tropical regions. The southern hemisphere probably has about a 20% higher level of $HO\bullet$ than does the northern hemisphere because of greater production of anthropogenic, $HO\bullet$-consuming CO in the northern hemisphere.

The hydroperoxyl radical, HOO•, is an intermediate in some important chemical reactions. In addition to its production by the reactions discussed above, in polluted atmospheres, hydroperoxyl radical is made by the following two reactions, starting with photolytic dissociation of formaldehyde to produce a reactive formyl radical:

$$HCHO + h\nu \rightarrow H + H\overset{\bullet}{C}O \tag{9.8.25}$$

$$H\overset{\bullet}{C}O + O_2 \rightarrow HOO^{\bullet} + CO \tag{9.8.25}$$

The hydroperoxyl radical reacts more slowly with other species than does the hydroxyl radical. The kinetics and mechanisms of hydroperoxyl radical reactions are difficult to study because it is hard to retain these radicals free of hydroxyl radicals.

Chemical and Biochemical Processes in Evolution of the Atmosphere

It is now widely believed that the earth's atmosphere originally was very different from its present state and that the changes were brought about by biological activity and accompanying chemical changes. Approximately 3.5 billion years ago, when the first primitive life molecules were formed, the atmosphere was probably free of oxygen and consisted of a variety of gases such as carbon dioxide, water vapor, and perhaps even methane, ammonia, and hydrogen. The atmosphere was bombarded by intense, bond-breaking ultraviolet light which, along with lightning and radiation from radionuclides, provided the energy to bring about chemical reactions that resulted in the production of relatively complicated molecules, including even amino acids and sugars. From the rich chemical mixture in the sea, life molecules evolved. Initially, these very primitive life forms derived their energy from fermentation of organic matter formed by chemical and photochemical processes, but eventually they gained the capability to produce organic matter, "{CH_2O}," by photosynthesis:

$$CO_2 + H_2O + h\nu \rightarrow \{CH_2O\} + O_2(g) \tag{9.8.27}$$

Photosynthesis released oxygen, thereby setting the stage for the massive biochemical transformation that resulted in the production of almost all the atmosphere's oxygen.

The oxygen initially produced by photosynthesis was probably quite toxic to primitive life forms. However, much of this oxygen was converted to iron oxides by reaction with soluble iron(II):

$$4Fe^{2+} + O_2 + 4H_2O \rightarrow 2Fe_2O_3 + 8H^+ \tag{9.8.28}$$

This resulted in the formation of enormous deposits of iron oxides, the existence of which provides major evidence for the liberation of free oxygen in the primitive atmosphere.

Eventually, enzyme systems developed that enabled organisms to mediate the reaction of waste-product oxygen with oxidizable organic matter in the sea. Later, this mode of waste-product disposal was utilized by organisms to produce energy by respiration, which is now the mechanism by which nonphotosynthetic organisms obtain energy.

In time, O_2 accumulated in the atmosphere, providing an abundant source of oxygen for respiration. It had an additional benefit in that it enabled the formation of an ozone shield (see Section 9.9). The ozone shield absorbs bond-rupturing ultraviolet light. With the ozone shield protecting tissue from destruction by high-energy ultraviolet radiation, the earth became a much more hospitable environment for life, and life forms were enabled to move from the sea to land.

9.9. ACID-BASE REACTIONS IN THE ATMOSPHERE

Acid-base reactions occur between acidic and basic species in the atmosphere. The atmosphere is normally at least slightly acidic because of the presence of a low level of carbon dioxide, which dissolves in atmospheric water droplets and dissociates slightly:

$$CO_2(g) \quad \xrightarrow{\text{water}} \quad CO_2(aq) \tag{9.9.1}$$

$$CO_2(aq) + H_2O \rightarrow H^+ + HCO_3^- \tag{9.9.2}$$

Atmospheric sulfur dioxide forms a somewhat stronger acid when it dissolves in water:

$$SO_2(g) + H_2O \rightarrow H^+ + HSO_3^- \tag{9.9.3}$$

In terms of pollution, however, strongly acidic HNO_3 and H_2SO_4 formed by the atmospheric oxidation of N oxides, SO_2, and H_2S are much more important because they lead to the formation of damaging acid rain (see Chapter 14).

As reflected by the generally acidic pH of rainwater, basic species are relatively less common in the atmosphere. Particulate calcium oxide, hydroxide, and carbonate can get into the atmosphere from ash and ground rock, and can react with acids such as in the following reaction:

$$Ca(OH)_2(s) + H_2SO_4(aq) \rightarrow CaSO_4(s) + 2H_2O \tag{9.9.4}$$

The most important basic species in the atmosphere is gas-phase ammonia, NH_3. The major source of atmospheric ammonia is from biodegradation of nitrogen-containing biological matter and from bacterial reduction of nitrate:

$$NO_3^-(aq) + 2\{CH_2O\}(biomass) + H^+ \rightarrow NH_3(g) + 2CO_2 + H_2O \tag{9.9.5}$$

Ammonia is particularly important as a base in the air because it is the only water-soluble base present at significant levels in the atmosphere. Dissolved in atmospheric water droplets, it plays a strong role in neutralizing atmospheric acids:

$$NH_3(aq) + HNO_3(aq) \rightarrow NH_4NO_3(aq) \tag{9.9.6}$$

$$NH_3(aq) + H_2SO_4(aq) \rightarrow NH_4HSO_4(aq) \qquad (9.9.7)$$

These reactions have three effects: (1) They result in the presence of NH_4^+ ion in the atmosphere as dissolved or solid salts, (2) they serve in part to neutralize acidic consituents of the atmosphere, and (3) they produce relatively corrosive ammonium salts.

9.10. REACTIONS OF ATMOSPHERIC OXYGEN

Some of the primary features of the exchange of oxygen among the atmosphere, geosphere, hydrosphere, and biosphere are summarized in Figure 9.13. The oxygen cycle is critically important in atmospheric chemistry, geochemical transformations, and life processes.

Oxygen in the troposphere plays a strong role in processes that occur on the earth's surface. Atmospheric oxygen takes part in energy-producing reactions, such as the burning of fossil fuels:

$$CH_4(\text{in natural gas}) + 2O_2 \rightarrow CO_2 + 2H_2O \qquad (9.10.1)$$

Atmospheric oxygen is utilized by aerobic organisms in the degradation of organic material. Some oxidative weathering processes consume oxygen, such as

$$4FeO + O_2 \rightarrow 2Fe_2O_3 \qquad (9.10.2)$$

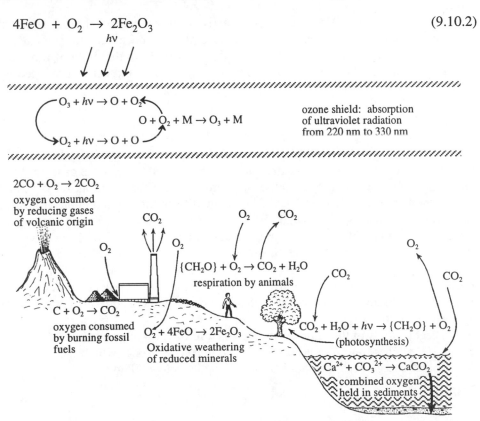

Figure 9.11. Oxygen exchange among the atmosphere, geosphere, hydrosphere, and biosphere.

Oxygen is returned to the atmosphere through plant photosynthesis:

$$CO_2 + H_2O + h\nu \rightarrow \{CH_2O\} + O_2 \tag{9.10.3}$$

All molecular oxygen now in the atmosphere is thought to have originated through the action of photosynthetic organisms, which shows the importance of photosynthesis in the oxygen balance of the atmosphere. It can be shown that most of the carbon fixed by these photosynthetic processes is dispersed in mineral formations as humic material (Section 3.17); only a very small fraction is deposited in fossil fuel beds. Therefore, although combustion of fossil fuels consumes large amounts of O_2, there is no danger of running out of atmospheric oxygen.

Molecular oxygen is somewhat unusual in that its ground state is a triplet state with two unpaired electrons, designated here as $^3O_2$, which can be excited to singlet molecular oxygen, designated here as $^1O_2$. The latter can be produced by several processes, including direct photochemical excitation, transfer of energy from other electronically excited molecules, ozone photolysis, and high-energy oxygen-producing reactions.

Because of the extremely rarefied atmosphere and the effects of ionizing radiation, elemental oxygen in the upper atmosphere exists to a large extent in forms other than diatomic O_2. In addition to O_2, the upper atmosphere contains oxygen atoms, O; excited oxygen molecules, O_2^*; and ozone, O_3.

Atomic oxygen, O, is stable primarily in the thermosphere, where the atmosphere is so rarefied that the three-body collisions necessary for the chemical reaction of atomic oxygen seldom occur (the third body in this kind of three-body reaction absorbs energy to stabilize the products). Atomic oxygen is produced by a photochemical reaction:

$$O_2 + h\nu \rightarrow O + O \tag{9.10.4}$$

The oxygen-oxygen bond is strong (120 kcal/mole) and ultraviolet radiation in the wavelength regions 135-176 nm and 240-260 nm is most effective in causing dissociation of molecular oxygen. Because of photochemical dissociation, O_2 is virtually nonexistent at very high altitudes and less than 10% of the oxygen in the atmosphere at altitudes exceeding approximately 400 km is present in the molecular form. At altitudes exceeding about 80 km, the average molecular weight of air is lower than the 28.97 g/mole observed at sea level because of the high concentration of atomic oxygen. The resulting division of the atmosphere into a lower section with a uniform molecular weight and a higher region with a nonuniform molecular weight is the basis for classifying these two atmospheric regions as the **homosphere** and **heterosphere**, respectively.

Oxygen atoms in the atmosphere can exist in the ground state (O) and in excited states (O*). These are produced by the photolysis of ozone, which has a relatively weak bond energy of 26 kcal/mole at wavelengths below 308 nm,

$$O_3 + h\nu(\lambda < 308 \text{ nm}) \rightarrow O^* + O_2 \tag{9.10.5}$$

or by highly energetic chemical reactions such as

$$O + O + O \rightarrow O_2 + O^* \tag{9.10.6}$$

Excited atomic oxygen emits visible light at wavelengths of 636 nm, 630 nm, and 558 nm. This emitted light is partially responsible for **airglow**, a very faint electromagnetic radiation continuously emitted by the earth's atmosphere. Although its visible component is extremely weak, airglow is quite intense in the infrared region of the spectrum.

Oxygen ion, O^+, which may be produced by ultraviolet radiation acting upon oxygen atoms,

$$O + h\nu \rightarrow O^+ + e^- \tag{9.10.7}$$

is the predominant positive ion in some regions of the ionosphere. It may react with molecular oxygen or nitrogen,

$$O^+ + O_2 \rightarrow O_2^+ + O \tag{9.10.8}$$

$$O^+ + N_2 \rightarrow NO^+ + N \tag{9.10.9}$$

to form other positive ions.

In intermediate regions of the ionosphere, O_2^+ is produced by absorption of ultraviolet radiation at wavelengths of 17-103 nm. This diatomic oxygen ion can also be produced by the photochemical reaction of low-energy X-rays,

$$O_2 + h\nu \rightarrow O_2^+ + e^- \tag{9.10.10}$$

and by the following reaction:

$$N_2^+ + O_2 \rightarrow N_2 + O_2^+ \tag{9.10.11}$$

Ozone, O_3, has an essential protective function because it absorbs harmful ultraviolet radiation in the stratosphere and serves as a radiation shield, protecting living beings on the earth from the effects of excessive amounts of such radiation. It is produced by a photochemical reaction,

$$O_2 + h\nu \rightarrow O + O \tag{9.10.12}$$

(where the wavelength of the exciting radiation must be less than 242.4 nm), followed by a three-body reaction,

$$O + O_2 + M \rightarrow O_3 + M(\text{increased energy}) \tag{9.10.13}$$

in which M is another species, such as a molecule of N_2 or O_2, which absorbs the excess energy given off by the reaction and enables the ozone molecule to stay together. The region of maximum ozone concentration is found within the range of 25-30 km high in the stratosphere where it may reach 10 ppm.

Ozone absorbs ultraviolet light very strongly in the region 220-330 nm. If this light were not absorbed by ozone, severe damage would result to exposed forms of life on the earth. Absorption of electromagnetic radiation by ozone converts the radiation's energy to heat and is responsible for the temperature maximum encountered at the boundary between the stratosphere and the mesosphere at an altitude of approximately 50 km. The reason that the temperature maximum occurs at a higher altitude than that of the maximum ozone concentration arises from the fact that ozone is such an effective absorber of ultraviolet light, that most of this radiation is absorbed in the upper stratosphere where it generates heat, and only a small fraction reaches the lower altitudes, which remain relatively cool.

The overall reaction,

$$2O_3 \rightarrow 3O_2 \tag{9.10.14}$$

is favored thermodynamically so that ozone is inherently unstable. Its decomposition in the stratosphere is catalyzed by a number of natural and pollutant trace constituents, including NO, NO_2, H, $HO^\bullet$, $HOO^\bullet$, ClO, Cl, Br, and BrO. Ozone decomposition also occurs on solid surfaces, such as metal oxides and salts produced by rocket exhausts.

Although the mechanisms and rates for the photochemical production of ozone in the stratosphere are reasonably well known, the natural pathways for ozone removal are less well understood. In addition to undergoing decomposition by the action of ultraviolet radiation, stratospheric ozone reacts with atomic oxygen, hydroxyl radical, and NO:

$$O_3 + h\nu \rightarrow O_2 + O \tag{9.10.15}$$

$$O_3 + O \rightarrow O_2 + O_2 \tag{9.10.16}$$

$$O_3 + HO^\bullet \rightarrow O_2 + HOO^\bullet \tag{9.10.17}$$

The $HO^\bullet$ radical is regenerated from $HOO^\bullet$ by the reaction,

$$HOO^\bullet + O \rightarrow HO^\bullet + O_2 \tag{9.10.18}$$

The NO consumed in this reaction is regenerated from NO_2,

$$O_3 + NO \rightarrow NO_2 + O_2 \tag{9.10.19}$$

$$NO_2 + O \rightarrow NO + O_2 \tag{9.10.20}$$

and some NO is produced from N_2O:

$$N_2O + O \rightarrow 2NO \tag{9.10.21}$$

Recall that N_2O is a natural component of the atmosphere and is a major product of the denitrification process by which fixed nitrogen is returned to the atmosphere in gaseous form. This is shown in the nitrogen cycle, Figures 1.6 and 6.10.

Ozone is an undesirable pollutant in the troposphere. It is toxic to animals and plants (see Chapter 23), and it also damages materials, particularly rubber.

9.11. REACTIONS OF ATMOSPHERIC NITROGEN

The 78% by volume of nitrogen contained in the atmosphere constitutes an inexhaustible reservoir of that essential element. The nitrogen cycle and nitrogen fixation by microorganisms were discussed in Chapter 6. A small amount of nitrogen is fixed in the atmosphere by lightning, and some is also fixed by combustion processes, particularly in internal combustion and turbine engines.

Before the use of synthetic fertilizers reached its current high levels, chemists were concerned that denitrification processes in the soil would lead to nitrogen depletion on the Earth. Now, with millions of tons of synthetically fixed nitrogen being added to the soil each year, major concern has shifted to possible excess accumulation of nitrogen in soil, fresh water, and the oceans.

Unlike oxygen, which is almost completely dissociated to the monatomic form in higher regions of the thermosphere, molecular nitrogen is not readily dissociated by ultraviolet radiation. However, at altitudes exceeding approximately 100 km, atomic nitrogen is produced by photochemical reactions:

$$N_2 + h\nu \rightarrow N + N \tag{9.11.1}$$

Other reactions which may produce monatomic nitrogen are:

$$N_2^+ + O \rightarrow NO^+ + N \tag{9.11.2}$$

$$NO^+ + e^- \rightarrow N + O \tag{9.11.3}$$

$$O^+ + N_2 \rightarrow NO^+ + N \tag{9.11.4}$$

As shown in Reactions 9.9.19 and 9.9.20, NO is involved in the removal of stratospheric ozone and is regenerated by the reaction of NO_2 with atomic O, itself a precursor to the formation of ozone. An ion formed from NO, the NO^+ ion, is one of the predominant ionic species in the so-called E region of the ionosphere. A plausible sequence of reactions by which NO^+ is formed is the following:

$$N_2 + h\nu \rightarrow N_2^+ + e^- \tag{9.11.5}$$

$$N_2^+ + O \rightarrow NO^+ + N \tag{9.11.6}$$

In the lowest (D) region of the ionosphere, which extends from approximately 50 km in altitude to approximately 85 km, NO^+ is produced directly by ionizing radiation:

$$NO + h\nu \rightarrow NO^+ + e^- \tag{9.11.7}$$

In the lower part of this region, the ionic species N_2^+ is formed through the action of galactic cosmic rays:

$$N_2 + h\nu \rightarrow N_2^+ + e^- \tag{9.11.8}$$

Pollutant oxides of nitrogen, particularly NO_2, are key species involved in air pollution and the formation of photochemical smog. For example, NO_2 is readily dissociated photochemically to NO and reactive atomic oxygen:

$$NO_2 + h\nu \rightarrow NO + O \tag{9.11.9}$$

This reaction is the most important primary photochemical process involved in smog formation. The roles played by nitrogen oxides in smog formation and other forms of air pollution are discussed in Chapters 11–14.

9.12. ATMOSPHERIC CARBON DIOXIDE

Although only about 0.035% (350 ppm) of air consists of carbon dioxide, it is the atmospheric "nonpollutant" species of most concern. As mentioned in Section 9.3, carbon dioxide, along with water vapor, is primarily responsible for the absorption of infrared energy re-emitted by the earth such that some of this energy is reradiated back to the earth's surface. Current evidence suggests that changes in the atmospheric carbon dioxide level will substantially alter the earth's climate through the greenhouse effect.

Valid measurements of overall atmospheric CO_2 can only be taken in areas remote from industrial activity. Such areas include Antarctica and the top of Mauna Loa Mountain in Hawaii. Measurements of carbon dioxide levels in these locations over the last 30 years suggest an annual increase in CO_2 of about 1 ppm per year.

The most obvious factor contributing to increased atmospheric carbon dioxide is consumption of carbon-containing fossil fuels. In addition, release of CO_2 from the biodegradation of biomass and uptake by photosynthesis are important factors determining overall CO_2 levels in the atmosphere. The role of photosynthesis is illustrated in Figure 9.12, which shows a seasonal cycle in carbon dioxide levels in the northern hemisphere. Maximum values occur in April and minimum values in late September or October. These oscillations are due to the "photosynthetic pulse," influenced most strongly by forests in middle latitudes. Forests have a much greater influence than other vegetation because trees carry out more photosynthesis. Furthermore, forests store enough fixed but readily oxidizable carbon in the form of wood and humus to have a marked influence on atmospheric CO_2 content. Thus, during the summer months, forest trees carry out enough photosynthesis to reduce the atmospheric carbon dioxide content markedly. During the winter, metabolism of biota, such as bacterial decay of humus, releases a significant amount of CO_2. Therefore, the current worldwide destruction of forests and conversion of forest lands to agricultural uses contributes substantially to a greater overall increase in atmospheric CO_2 levels.

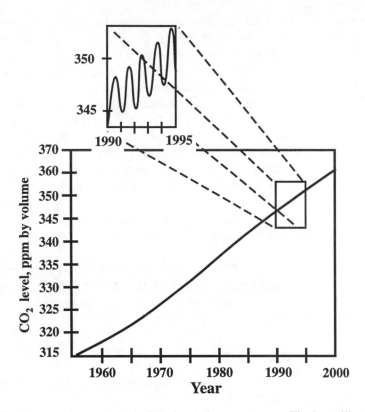

Figure 9.12. Increases in atmospheric CO_2 levels in recent years. The inset illustrates seasonal variations in the northern hemisphere.

With current trends, it is likely that global CO_2 levels will double by the middle of the next century, which may well raise the earth's mean surface temperature by 1.5–4.5°C. Such a change might have more potential to cause massive irreversible environmental changes than any other disaster short of global nuclear war or asteroid impact .

Chemically and photochemically, carbon dioxide is a comparatively insignificant species because of its relatively low concentrations and low photochemical reactivity. The one significant photochemical reaction that carbon dioxide undergoes, and a major source of CO at higher altitudes, is the photodissociation of CO_2 by energetic solar ultraviolet radiation in the stratosphere:

$$CO_2 + h\nu \rightarrow CO + O \tag{9.12.1}$$

9.13. ATMOSPHERIC WATER

The water vapor content of the troposphere is normally within a range of 1–3% by volume with a global average of about 1%. However, air can contain as little as 0.1% or as much as 5% water. The percentage of water in the atmosphere decreases rapidly with increasing altitude. Water circulates through the atmosphere in the hydrologic cycle as shown in Figure 3.1.

Water vapor absorbs infrared radiation even more strongly than does carbon dioxide, thus greatly influencing the earth's heat balance. Clouds formed from water vapor reflect light from the sun and have a temperature-lowering effect. On the other hand, water vapor in the atmosphere acts as a kind of "blanket" at night, retaining heat from the earth's surface by absorption of infrared radiation.

As discussed in Section 9.5, water vapor and the heat released and absorbed by transitions of water between the vapor state and liquid or solid are strongly involved in atmospheric energy transfer. Condensed water vapor in the form of very small droplets is of considerable concern in atmospheric chemistry. The harmful effects of some air pollutants—for instance, the corrosion of metals by acid-forming gases—requires the presence of water which may come from the atmosphere. Atmospheric water vapor has an important influence upon pollution-induced fog formation under some circumstances. Water vapor interacting with pollutant particulate matter in the atmosphere may reduce visibility to undesirable levels through the formation of very small atmospheric aerosol particles.

As noted in Section 9.2, the cold tropopause serves as a barrier to the movement of water into the stratosphere. Thus, little water is transferred from the troposphere to the stratosphere, and the main source of water in the stratosphere is the photochemical oxidation of methane:

$$CH_4 + 2O_2 + h\nu \xrightarrow{\text{(several steps)}} CO_2 + 2H_2O \tag{9.13.1}$$

The water thus produced serves as a source of stratospheric hydroxyl radical as shown by the following reaction:

$$H_2O + h\nu \rightarrow HO\bullet + H \tag{9.13.2}$$

SUPPLEMENTAL REFERENCES

Barker, John. R., "A Brief Introduction to Atmospheric Chemistry," *Advances Series in Physical Chemistry*, **3**, 1-33 (1995).

Barker, John R., Ed., *Progress and Problems in Atmospheric Chemistry*, World Scientific, 1995.

Berner, Elizabeth Kay, and Robert A. Berner, *Global Environment: Water, Air, and Geochemical Cycles*, Prentice Hall,Upper Saddle River, NJ, 1996.

Birks, John W., Jack G. Calvert, and Robert W. Sievers, Eds., *The Chemistry of the Atmosphere: Its Impact on Global Change*, American Chemical Society, Washington, D.C., 1993.

Board on Atmospheric Sciences and Climate, Commission on Geosciences, Environment, and Resources, National Research Council, *The Atmospheric Sciences: Entering the Twenty-First Century*, National Academy Press, Washington, D.C., 1998.

Boubel, Richard W., Donald L. Fox, and D. Bruce Turner, *Fundamentals of Air Pollution*, 3rd ed., Academic Press, Orlando, FL, 1994.

Bradley, Raymond S., and Philip D. Jones, Eds., *Climate Since A.D. 1500*, Routledge, Chapman and Hall, New York, 1992.

Brasseur, Guy P., John J. Orlando, and Geoffrey S. Tyndall, Eds., *Atmospheric Chemistry and Global Change*, Oxford University Press, New York, 1999.

Crutzen, Paul J. and Veerabhadran Ramanathan, *Clouds, Chemistry and Climate*, Springer, Berlin, 1996.

Finlayson-Pitts, Barbara J. and James N. Pitts, *Atmospheric Chemistry*, John Wiley & Sons, New York, 1986.

Goody, Richard, *Principles of Atmospheric Physics and Chemistry*, Oxford University Press, New York, 1995.

Graedel, Thomas E. and Paul J. Crutzen, *Atmosphere, Climate, and Change*, W.H. Freeman , New York, 1995.

Jacob, Daniel J., *Introduction to Atmospheric Chemistry*, Princeton University Press, Princeton, NJ, 1999.

Lamb, Hubert, and Knud Frydendahl, *Historic Storms of the North Sea, British Isles and Northwest Europe*, Cambridge University Press, New York, 1991.

Lutgens, Frederick K., Edward J. Tarbuck, and Dennis Tasa, *The Atmosphere: An Introduction to Meteorology*, 7th ed., Prentice Hall, Englewood Cliffs, NJ, (1997).

Morawska, L., N. D. Bofinger, and M. Maroni, *Indoor Air, an Integrated Approach*, Elsevier Science, Inc., Tarrytown, NY, 1995.

Neckers, Douglas C., David H. Volman, and Günther von Bünau, Eds., *Advances in Photochemistry*, Vol. 23, John Wiley and Sons, Inc., New York, 1997.

Singh, Hanwant B., Ed., Composition, *Chemistry, and Climate of the Atmosphere*, Van Nostrand Reinhold, New York, 1995.

Soroos, Marvin S., *The Endangered Atmosphere: Preserving a Global Commons*, University of South Carolina Press, Columbia, SC, 1997.

Spellman, Frank R., *The Science of Air: Concepts and Applications*, Technomic Publishing Company, Lancaster, PA, 1999.

Yung, Y. L. and William B. Demore, *Photochemistry of Planetary Atmospheres*, Oxford University Press, New York, 1998.

Zellner, R. and H. Hermann, "Free Radical Chemistry of the Aqueous Atmospheric Phase," *Advances in Spectroscopy*, **24**, 381-451 (1995).

QUESTIONS AND PROBLEMS

1. What phenomenon is responsible for the temperature maximum at the boundary of the stratosphere and the mesosphere?

2. What function does a third body serve in an atmospheric chemical reaction?

3. Why does the lower boundary of the ionosphere lift at night?

4. Considering the total number of electrons in NO_2, why might it be expected that the reaction of a free radical with NO_2 is a chain-terminating reaction?

5. The average atmospheric pressure at sea level is 1.012×10^6 dynes/m^2. The value of g (acceleration of gravity) at sea level is 980 cm/sec^2. What is the mass in kg of the column of air having a cross-sectional area of 1.00 cm^2 at the earth's surface and extending to the limits of the atmosphere? (Recall that the dyne is a unit of force, and force = mass x acceleration of gravity.)

6. Suppose that 22.4 liters of air at STP is used to burn 1.50 g of carbon to form CO_2, and that the gaseous product is adjusted to STP. What is the volume and the average molar mass of the resulting mixture?

7. If the pressure is 0.01 atm at an altitude of 38 km and 0.001 at 57 km, what is it at 19 km (ignoring temperature variations)?

8. Measured in μm, what are the lower wavelength limits of solar radiation reaching the earth; the wavelength at which maximum solar radiation reaches the earth; and the wavelength at which maximum energy is radiated back into space?

9. Of the species O, HO*•, NO_2*, H_3C•, and N^+, which could most readily revert to a nonreactive, "normal" species in total isolation?

10. Of the gases neon, sulfur dioxide, helium, oxygen, and nitrogen, which shows the most variation in its atmospheric concentration?

11. A 12.0-liter sample of air at 25°C and 1.00 atm pressure was collected and dried. After drying, the volume of the sample was exactly 11.50 L. What was the percentage *by weight* of water in the original air sample?

12. The sunlight incident upon a 1 square meter area perpendicular to the line of transmission of the solar flux just above the Earth's atmosphere provides energy at a rate most closely equivalent to: (a) that required to power a pocket calculator, (b) that required to provide a moderate level of lighting for a 40-person capacity classroom illuminated with fluorescent lights, (c) that required to propel a 2500 pound automobile at 55 mph, (d) that required to power a 100-watt incandescent light bulb, (e) that required to heat a 40-person classroom to 70°F when the outside temperature is -10°F.

13. At an altitude of 50 km, the average atmospheric temperature is essentially 0°C. What is the average number of air molecules per cubic centimeter of air at this altitude?

14. What is the distinction between chemiluminescence and luminescence caused when light is absorbed by a molecule or atom?

15. State two factors that make the stratosphere particularly important in terms of acting as a region where atmospheric trace contaminants are converted to other, chemically less reactive, forms.

16. What two chemical species are most generally responsible for the removal of hydroxyl radical from the unpolluted troposphere.

17. What is the distinction between the symbols * and • in discussing chemically active species in the atmosphere?

10 PARTICLES IN THE ATMOSPHERE

10.1. PARTICLES IN THE ATMOSPHERE

Particles in the atmosphere, which range in size from about one-half millimeter (the size of sand or drizzle) down to molecular dimensions, are made up of an amazing variety of materials and discrete objects that may consist of either solids or liquid droplets.[1] A number of terms are commonly used to describe atmospheric particles; the more important of these are summarized in Table 10.1. Particles abound in the atmosphere. Even the Arctic, remote from sources of industrial pollution, is afflicted with an "Arctic Haze" of airborne particles from October to May each year. **Particulates** is a term that has come to stand for particles in the atmosphere, although *particulate matter* or simply *particles*, is preferred usage.

Table 10.1. Important Terms Describing Atmospheric Particles

Term	Meaning
Aerosol	Colloidal-sized atmospheric particle
Condensation aerosol	Formed by condensation of vapors or reactions of gases
Dispersion aerosol	Formed by grinding of solids, atomization of liquids, or dispersion of dusts
Fog	Term denoting high level of water droplets
Haze	Denotes decreased visibility due to the presence of particles
Mists	Liquid particles
Smoke	Particles formed by incomplete combustion of fuel

Particulate matter makes up the most visible and obvious form of air pollution. Atmospheric **aerosols** are solid or liquid particles smaller than 100 µm in diameter. Pollutant particles in the 0.001 to 10 µm range are commonly suspended in the air near sources of pollution such as the urban atmosphere, industrial plants, highways, and power plants.

Very small, solid particles include carbon black, silver iodide, combustion nuclei, and sea-salt nuclei (see Figure 10.1). Larger particles include cement dust, wind-blown soil dust, foundry dust, and pulverized coal. Liquid particulate matter, **mist**, includes raindrops, fog, and sulfuric acid mist. Particulate matter may be organic or inorganic; both types are very important atmospheric contaminants.

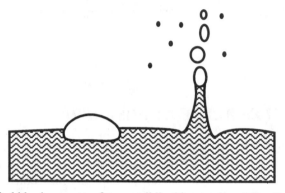

Figure 10.1. Bursting bubbles in seawater form small liquid aerosol particles. Evaporation of water from aerosol particles results in the formation of small solid particles of sea-salt nuclei.

Some particles are of biological origin, such as viruses, bacteria, bacterial spores, fungal spores, and pollen. In addition to organic materials, organisms may contribute to sulfate particulate matter in the atmosphere. Marine biological sources may contribute significantly to atmospheric aerosols.[2] Biogenic materials reacting in and on the surface of sea-salt aerosols produce some significant atmospheric chemical species, such as halogen radicals, and in so doing influence cycles involving atmospheric sulfur, nitrogen, and oxidants.[3]

As discussed later in this chapter, particulate matter originates from a wide variety of sources and processes, ranging from simple grinding of bulk matter to complicated chemical or biochemical syntheses. The effects of particulate matter are also widely varied. Possible effects on climate are discussed in Chapter 14. Either by itself, or in combination with gaseous pollutants, particulate matter may be detrimental to human health. Atmospheric particles may damage materials, reduce visibility, and cause undesirable esthetic effects. It is now recognized that very small particles have a particularly high potential for harm, including adverse health effects. In new ambient air quality standards for particulate matter issued in July 1997, the U. S. Environmental Protection Agency regulated particles 2.5 µm in diameter or less for the first time.

For the most part, aerosols consist of carbonaceous material, metal oxides and glasses, dissolved ionic species (electrolytes), and ionic solids. The predominant constituents are carbonaceous material, water, sulfate, nitrate, ammonium nitrogen, and silicon. The composition of aerosol particles varies significantly with size. The very small particles tend to be acidic and often originate from gases, such as from the conversion of SO_2 to H_2SO_4. Larger particles tend to consist of materials generated mechanically, such as by the grinding of limestone, and have a greater tendency to be basic.

10.2. PHYSICAL BEHAVIOR OF PARTICLES IN THE ATMOSPHERE

As shown in Figure 10.2, atmospheric particles undego a number of processes in the atmosphere. Small colloidal particles are subject to *diffusion processes*. Smaller particles *coagulate* together to form larger particles. *Sedimentation* or *dry deposition* of particles, which have often reached sufficient size to settle by coagulation, is one of two major mechanisms for particle removal from the atmosphere. The other is *scavenging* by raindrops and other forms of precipitation. Particles also react with atmospheric gases.

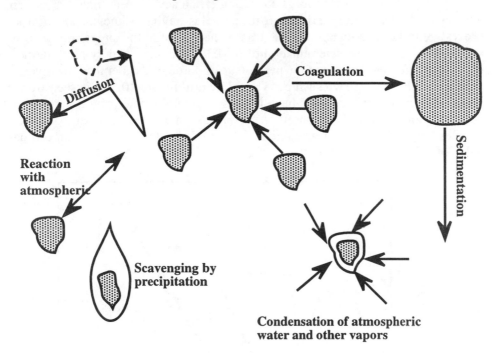

Figure 10.2. Processes that particles undergo in the atmosphere.

Particle size usually expresses the diameter of a particle, though sometimes it is used to denote the radius. The rate at which a particle settles is a function of particle diameter and density. The settling rate is important in determining the effect of the particle in the atmosphere. For spherical particles greater than approximately 1 μm in diameter, Stokes' law applies,

$$v = \frac{gd^2(\rho_1 - \rho_2)}{18\eta} \tag{10.2.1}$$

where ν is the settling velocity in cm/sec, g is the acceleration of gravity in cm/sec², ρ_1 is the density of the particle in g/cm³, ρ_2 is the density of air in g/cm³, and η is the viscosity of air in poise. Stokes' law can also be used to express the effective diameter of an irregular nonspherical particle. These are called **Stokes diameters** (aerodynamic diameters) and are normally the ones given when particle diameters are expressed. Furthermore, since the density of a particle is often not known, an arbitrary density of 1 g/cm³ is conventionally assigned to ρ_1; when this is done, the diameter calculated from Equation 10.2.1 is called the **reduced sedimentation diameter**.

Size and Settling of Atmospheric Particles

Most kinds of aerosol particles have unknown diameters and densities and occur over a range of sizes. For such particles, the term **mass median diameter (MMD)** may be used to describe aerodynamically equivalent spheres having an assigned density of 1 g/cm³ at a 50% mass collection efficiency, as determined in sampling devices calibrated with spherical aerosol particles having a known, uniform size. (Polystyrene latex is commonly used as a material for the preparation of such standard aerosols.) The determination of MMD is accomplished by plotting the log of particle size as a function of the percentage of particles smaller than the given size on a probability scale. Two such plots are shown in Figure 10.3. It is seen from the plot that particles of aerosol X have a mass median diameter of 2.0 μm (ordinate corresponding to 50% on the abscissa). In the case of aerosol Y, linear extrapolation to sizes below the lower measurable size limit of about 0.7 μm gives an estimated value of 0.5 μm for the MMD.

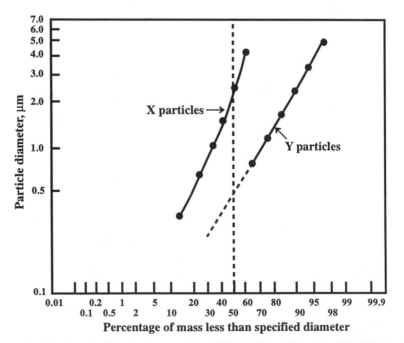

Figure 10.3. Particle size distribution for particles of X (MMD = 2.0 μm) and Y (MMD = 0.5 μm).

The settling characteristics of particles smaller than about 1 µm in diameter deviate from Stokes' Law because the settling particles "slip between" air molecules. Extremely small particles are subject to **Brownian motion** resulting from random movement due to collisions with air molecules and do not obey Stokes' Law. Deviations are also observed for particles above 10 µm in diameter because they settle rapidly and generate turbulence as they fall.

10.3. PHYSICAL PROCESSES FOR PARTICLE FORMATION

Dispersion aerosols, such as dusts, formed from the disintegration of larger particles are usually above 1 µm in size. Typical processes for forming dispersion aerosols include evolution of dust from coal grinding, formation of spray in cooling towers, and blowing of dirt from dry soil.

Many dispersion aerosols originate from natural sources such as sea spray, windblown dust, and volcanic dust. However, a vast variety of human activities break up material and disperse it to the atmosphere. "All terrain" vehicles churn across desert lands, coating fragile desert plants with layers of dispersed dust. Quarries and rock crushers spew out plumes of ground rock. Cultivation of land has made it much more susceptible to dust-producing wind erosion.

However, since much more energy is required to break material down into small particles than is required for or released by the synthesis of particles through chemical synthesis or the adhesion of smaller particles, most dispersion aerosols are relatively large. Larger particles tend to have fewer harmful effects than smaller ones. As examples, larger particles are less *respirable* in that they do not penetrate so far into the lungs as smaller ones, and larger particles are relatively easier to remove from air pollution effluent sources.

Huge volcanic eruptions can cause highly elevated levels of particles in the atmosphere. These can be from the physical process of simply blowing as much as several cubic kilometers of volcanic dust as high as the stratosphere and through chemical processes, usually involving the production of sulfuric acid and sulfates from volcanic SO_2 and H_2S. A study of the 1982 eruption of El Chichon volcano in Mexico showed that volcanic glass, sodium chloride, and sulfate from the volcano were deposited in snow in Greenland.[4] The June 15, 1991 eruption of Mount Pinatubo in the Philippines caused perceptible perturbations in Earth's atmospheric solar and infrared radiation transmission.[5]

10.4. CHEMICAL PROCESSES FOR PARTICLE FORMATION

Chemical processes in the atmosphere convert large quantities of atmospheric gases to particulate matter. Among the chemical species most responsible for this conversion are the organic pollutants and nitrogen oxides that cause formation of ozone and photochemical smog (see Chapter 13) in the troposphere.[6] To an extent, therefore, control of hydrocarbon and NO_x emissions to reduce smog will also curtail atmospheric particulate matter pollution.

A major fraction of ambient particulate matter arises from atmospheric gas-to-particle conversion. Attempts to reduce particulate matter levels require control of the same organic and nitrogen oxide (NO_x) emissions that are precursors to urban and regional ozone formation.

Most chemical processes that produce particles are combustion processes, including fossil-fuel-fired power plants; incinerators; home furnaces, fireplaces, and stoves; cement kilns; internal combustion engines; forest, brush, and grass fires; and active volcanoes. Particles from combustion sources tend to occur in a size range below 1 μm. Such very small particles are particularly important because they are most readily carried into the alveoli of lungs (see pulmonary route of exposure to toxicants in Chapter 22) and they are likely to be enriched in more hazardous constituents, such as toxic heavy metals and arsenic. The latter characteristic can enable use of small particle analysis for tracking sources of particulate pollutants.

Inorganic Particles

Metal oxides constitute a major class of inorganic particles in the atmosphere. These are formed whenever fuels containing metals are burned. For example, particulate iron oxide is formed during the combustion of pyrite-containing coal:

$$3FeS_2 + 8O_2 \rightarrow Fe_3O_4 + 6SO_2 \tag{10.4.1}$$

Organic vanadium in residual fuel oil is converted to particulate vanadium oxide. Part of the calcium carbonate in the ash fraction of coal is converted to calcium oxide and is emitted into the atmosphere through the stack:

$$CaCO_3 + heat \rightarrow CaO + CO_2 \tag{10.4.2}$$

A common process for the formation of aerosol mists involves the oxidation of atmospheric sulfur dioxide to sulfuric acid, a hygroscopic substance that accumulates atmospheric water to form small liquid droplets:

$$2SO_2 + O_2 + 2H_2O \rightarrow 2H_2SO_4 \tag{10.4.3}$$

In the presence of basic air pollutants, such as ammonia or calcium oxide, the sulfuric acid reacts to form salts:

$$H_2SO_4(droplet) + 2NH_3(g) \rightarrow (NH_4)_2SO_4 (droplet) \tag{10.4.4}$$

$$H_2SO_4(droplet) + CaO(s) \rightarrow CaSO_4(droplet) + H_2O \tag{10.4.5}$$

Under low-humidity conditions water is lost from these droplets and a solid aerosol is formed.

The preceding examples show several ways in which solid or liquid inorganic aerosols are formed by chemical reactions. Such reactions constitute an important general process for the formation of aerosols, particularly the smaller particles.

Organic Particles

A significant portion of organic particulate matter is produced by internal combustion engines in complicated processes that involve pyrosynthesis and nitrogenous compounds. These products may include nitrogen-containing compounds and oxidized hydrocarbon polymers. Lubricating oil and its additives may also contribute to organic particulate matter. A study of particulate matter emitted by gasoline auto engines (with and without catalysts) and diesel truck engines measured more than 100 compounds quantitatively.[7] Among the prominent classes of compounds found were n-alkanes, n-alkanoic acids, benzaldehydes, benzoic acids, azanaphthalenes, polycyclic aromatic hydrocarbons, oxygenated PAHs, pentacyclic triterpanes, and steranes (the last two classes of hydrocarbons are multiringed compounds characteristic of petroleum that enter exhaust gases from lubricating oil).

PAH Synthesis

The organic particles of greatest concern are PAH hydrocarbons, which consist of condensed ring aromatic (aryl) molecules. The most often cited example of a PAH compound is benzo(a)pyrene, a compound that the body can metabolize to a carcinogenic form:

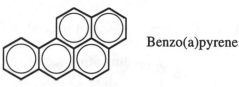 Benzo(a)pyrene

PAHs may be synthesized from saturated hydrocarbons under oxygen-deficient conditions. Hydrocarbons with very low molecular masses, including even methane, may act as precursors for the polycyclic aromatic compounds. Low-molar-mass hydrocarbons form PAHs by **pyrosynthesis**. This happens at temperatures exceeding approximately 500°C at which carbon-hydrogen and carbon-carbon bonds are broken to form free radicals. These radicals undergo dehydrogenation and combine chemically to form aryl ring structures which are resistant to thermal degradation. The basic process for the formation of such rings from pyrosynthesis starting with ethane is,

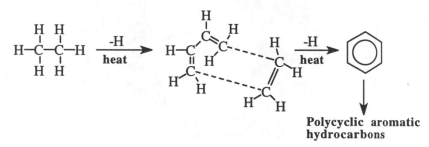

Polycyclic aromatic hydrocarbons

which results in the formation of stable PAH structures. The tendency of hydrocarbons to form PAHs by pyrosynthesis varies in the order aromatics > cycloolefins > olefins > paraffins. The existing ring structure of cyclic compounds is conducive to PAH formation. Unsaturated compounds are especially susceptible to the addition reactions involved in PAH formation.

Polycyclic aromatic compounds may be formed from higher alkanes present in fuels and plant materials by the process of **pyrolysis**, the "cracking" of organic compounds to form smaller and less stable molecules and radicals.

10.5. THE COMPOSITION OF INORGANIC PARTICLES

Figure 10.4 illustrates the basic factors responsible for the composition of inorganic particulate matter. In general, the proportions of elements in atmospheric particulate matter reflect relative abundances of elements in the parent material.

The source of particulate matter is reflected in its elemental compositon, taking into consideration chemical reactions that may change the composition. For example, particulate matter largely from an ocean spray origin in a coastal area receiving sulfur dioxide pollution may show anomalously high sulfate and corresponding low chloride content. The sulfate comes from atmospheric oxidation of sulfur dioxide to form nonvolatile ionic sulfate, whereas some chloride originally from the NaCl in the seawater may be lost from the solid aerosol as volatile HCl:

$$2SO_2 \ + \ O_2 \ + \ 2H_2O \ \rightarrow \ 2H_2SO_4 \tag{10.5.1}$$

$$H_2SO_4 \ + \ 2NaCl(particulate) \ \rightarrow \ Na_2SO_4(particulate) + 2HCl \tag{10.5.2}$$

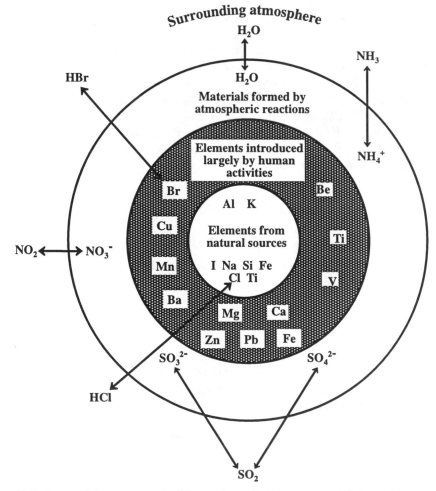

Figure 10.4. Some of the components of inorganic particulate matter and their origins.

Acids other than sulfuric acid can also be involved in the modification of sea-salt particles. An examination with an electron microscope equipped with an X-ray emission analyzer of sea-salt particles taken at Tsukuba, Japan, 50 km from the Pacific Coast, showed a marked deficiency of chloride.[8] This was attributed primarily to reaction with pollutant nitric acid as indicated by high levels of nitrate in the samples. The presence of significant levels of zinc in the particles indicated contamination from pollution sources. The chemical composition of atmospheric particulate matter is quite diverse. Among the constituents of inorganic particulate matter found in polluted atmospheres are salts, oxides, nitrogen compounds, sulfur compounds, various metals, and radionuclides. In coastal areas sodium and chlorine get into atmospheric particles as sodium chloride from sea spray. The major trace elements that typically occur at levels above 1 $\mu g/m^3$ in particulate matter are aluminum, calcium, carbon, iron, potassium, sodium, and silicon; note that most of these tend to originate from terrestrial sources. Lesser quantities of copper, lead, titanium, and zinc, and even lower levels of antimony, beryllium, bismuth, cadmium, cobalt, chromium, cesium, lithium, manganese, nickel, rubidium, selenium, strontium, and vanadium are commonly observed. The likely sources of some of these elements are given below:

- **Al, Fe, Ca, Si**: Soil erosion, rock dust, coal combustion

- **C**: Incomplete combustion of carbonaceous fuels

- **Na, Cl**: Marine aerosols, chloride from incineration of organohalide polymer wastes

- **Sb, Se**: Very volatile elements, possibly from the combustion of oil, coal, or refuse

- **V**: Combustion of residual petroleum (present at very high levels in residues from Venezuelan crude oil)

- **Zn**: Tends to occur in small particles, probably from combustion

- **Pb**: Combustion of leaded fuels and wastes containing lead

Particulate carbon as soot, carbon black, coke, and graphite originates from auto and truck exhausts, heating furnaces, incinerators, power plants, and steel and foundry operations, and composes one of the more visible and troublesome particulate air pollutants. Because of its good adsorbent properties, carbon can be a carrier of gaseous and other particulate pollutants. Particulate carbon surfaces may catalyze some heterogeneous atmospheric reactions, including the important conversion of SO_2 to sulfate.

Fly Ash

Much of the mineral particulate matter in a polluted atmosphere is in the form of oxides and other compounds produced during the combustion of high-ash fossil fuel. Much of the mineral matter in fossil fuels such as coal or lignite is converted during combustion to a fused, glassy bottom ash which presents no air pollution problems.

Smaller particles of **fly ash** enter furnace flues and are efficiently collected in a properly equipped stack system. However, some fly ash escapes through the stack and enters the atmosphere. Unfortunately, the fly ash thus released tends to consist of smaller particles that do the most damage to human health, plants, and visibility.

The composition of fly ash varies widely, depending upon the fuel. The predominant constituents are oxides of aluminum, calcium, iron, and silicon. Other elements that occur in fly ash are magnesium, sulfur, titanium, phosphorus, potassium, and sodium. Elemental carbon (soot, carbon black) is a significant fly ash constituent.

The size of fly ash particles is a very important factor in determining their removal from stack gas and their ability to enter the body through the respiratory tract. Fly ash from coal-fired utility boilers has shown a bimodal (two peak) distribution of size, with a peak at about 0.1 μm as illustrated in Figure 10.5.[9] Although only about 1-2 % of the total fly ash mass is in the smaller size fraction, it includes the vast majority of the total number of particles and particle surface area. Submicrometer particles probably result from a volatilization-condensation process during combustion, as reflected in a higher concentration of more volatile elements such as As, Sb, Hg, and Zn. Furthermore, the very small particles are the most difficult to remove by electrostatic precipitators and bag houses (see Section 10.11).

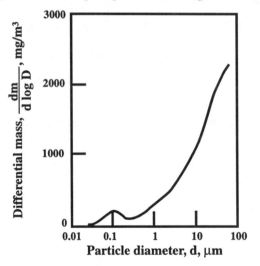

Figure 10.5. General appearance of particle-size distribution in coal-fired power plant ash. The data are given on differential mass coordinates, where M is the mass, so that the area under the curve in a given size range is the mass of the particles in that size range.

Asbestos

Asbestos is the name given to a group of fibrous silicate minerals, typically those of the serpentine group, approximate formula $Mg_3P(Si_2O_5)(OH)_4$. The tensile strength, flexibility, and nonflammability of asbestos have led to many uses including structural materials, brake linings, insulation, and pipe manufacture. In 1979, 560,000 metric tons of asbestos were used in the U.S. By 1988 annual consumption had dropped to 85,000 metric tons, most of it used for brake linings and pads, roofing products, cement/asbestos pipe, gaskets, heat-resistant packing, and specialty papers. In 1989, the U.S. Environmental Protection Agency announced regulations that phased out most uses of asbestos by 1996.

Asbestos is of concern as an air pollutant because when inhaled it may cause asbestosis (a pneumonia condition), mesothelioma (tumor of the mesothelial tissue lining the chest cavity adjacent to the lungs), and bronchogenic carcinoma (cancer originating with the air passages in the lungs). Therefore, uses of asbestos have been severely curtailed and widespread programs have been undertaken to remove the material from buildings.

10.6. TOXIC METALS

Some of the metals found predominantly as particulate matter in polluted atmospheres are known to be hazardous to human health. All of these except beryllium are so-called "heavy metals." Lead is the toxic metal of greatest concern in the urban atmosphere because it comes closest to being present at a toxic level; mercury ranks second. Others include beryllium, cadmium, chromium, vanadium, nickel, and arsenic (a metalloid).

Atmospheric Mercury

Atmospheric mercury is of concern because of its toxicity, volatility, and mobility. Some atmospheric mercury is associated with particulate matter. Much of the mercury entering the atmosphere does so as volatile elemental mercury from coal combustion and volcanoes. Volatile organomercury compounds such as dimethylmercury, $(CH_3)_2Hg$, and monomethylmercury salts, such as CH_3HgBr, are also encountered in the atmosphere.

Atmospheric Lead

With the reduction of leaded fuels, atmospheric lead is of less concern than it used to be. However, during the decades that leaded gasoline containing tetraethyllead was the predominant automotive fuel, particulate lead halides were emitted in large quantities. This occurs through the action of dichloroethane and dibromoethane added as halogenated scavengers to prevent the accumulation of lead oxides inside engines. The lead halides formed,

$$Pb(C_2H_5)_4 + O_2 + \text{halogenated scavengers} \rightarrow CO_2 + H_2O +$$
$$PbCl_2 + PbClBr + PbBr_2 \text{(unbalanced)} \quad (10.6.1)$$

are volatile enough to exit through the exhaust system but condense in the air to form particles. During the period of peak usage of leaded gasoline in the early 1970s, about 200,000 tons of lead were entering the atmosphere each year by this route in the U.S.

Atmospheric Beryllium

Only about 350 metric tons of beryllium are used each year in the U.S. for the formulation of specialty alloys used in electrical equipment, electronic instrumentation, space gear, and nuclear reactor components, so that distribution of beryllium is by no means comparable to that of other toxic metals such as lead or mercury. However, because of its "high tech" applications, consumption of beryllium may increase in the future.

During the 1940s and 1950s, the toxicity of beryllium and beryllium compounds became widely recognized; it has the lowest allowable limit in the atmosphere of all the elements. One of the main results of the recognition of beryllium toxicity hazards was the elimination of this element from phosphors (coatings which produce visible light from ultraviolet light) in fluorescent lamps.

10.7. RADIOACTIVE PARTICLES

Some of the radioactivity detected in atmospheric particles is of natural origin. This activity includes that produced when cosmic rays act on nuclei in the atmosphere to produce radionuclides, including ^7Be, ^{10}Be, ^{14}C, ^{39}Cl, ^3H, ^{22}Na, ^{32}P, and ^{33}P. A significant natural source of radionuclides in the atmosphere is **radon**, a noble gas product of radium decay. Radon may enter the atmosphere as either of two isotopes, ^{222}Rn (half-life 3.8 days) and ^{220}Rn (half-life 54.5 seconds). Both are alpha emitters in decay chains that terminate with stable isotopes of lead. The initial decay products, ^{218}Po and ^{216}Po, are nongaseous and adhere readily to atmospheric particulate matter.

The catastrophic 1986 meltdown and fire at the Chernobyl nuclear reactor in the former Soviet Union spread large quantities of radioactive materials over a wide area of Europe. Much of this radioactivity was in the form of particles.[10]

One of the more serious problems in connection with radon is that of radioactivity originating from uranium mine tailings that have been used in some areas as backfill, soil conditioner, and a base for building foundations. Radon produced by the decay of radium exudes from foundations and walls constructed on tailings. Higher than normal levels of radioactivity have been found in some structures in the city of Grand Junction, Colorado, where uranium mill tailings have been used extensively in construction. Some medical authorities have suggested that the rate of birth defects and infant cancer in areas where uranium mill tailings have been used in residential construction are significantly higher than normal. The combustion of fossil fuels introduces radioactivity into the atmosphere in the form of radionuclides contained in fly ash. Large coal-fired power plants lacking ash-control equipment may introduce up to several hundred millicuries of radionuclides into the atmosphere each year, far more than either an equivalent nuclear or oil-fired power plant.

The radioactive noble gas ^{85}Kr (half-life 10.3 years) is emitted into the atmosphere by the operation of nuclear reactors and the processing of spent reactor fuels. In general, other radionuclides produced by reactor operation are either chemically reactive and can be removed from the reactor effluent, or have such short half-lives that a short time delay prior to emission prevents their leaving the reactor. Widespread use of fission power will inevitably result in an increased level of ^{85}Kr in the atmosphere. Fortunately, biota cannot concentrate this chemically unreactive element.

The above-ground detonation of nuclear weapons can add large amounts of radioactive particulate matter to the atmosphere. Among the radioisotopes that have been detected in rainfall collected after atmospheric nuclear weapon detonation are ^{91}Y, ^{141}Ce, ^{144}Ce, ^{147}Nd, ^{147}Pm, ^{149}Pm, ^{151}Sm, ^{153}Sm, ^{155}Eu, ^{156}Eu, ^{89}Sr, ^{90}Sr,

[115m]Cd, [129m]Te, ^{131}I, [132]Te, and [140]Ba. (Note that "m" denotes a metastable state that decays by gamma-ray emission to an isotope of the same element.) The rate of travel of radioactive particles through the atmosphere is a function of particle size. Appreciable fractionation of nuclear debris is observed because of differences in the rates at which various components of nuclear variety move through the atmosphere.

10.8. THE COMPOSITION OF ORGANIC PARTICLES

Organic atmospheric particles occur in a wide variety of compounds. For analysis, such particles can be collected onto a filter; extracted with organic solvents; fractionated into neutral, acid, and basic groups; and analyzed for specific constituents by chromatography and mass spectrometry. The neutral group contains predominantly hydrocarbons, including aliphatic, aromatic, and oxygenated fractions. The aliphatic fraction of the neutral group contains a high percentage of long-chain hydrocarbons, predominantly those with 16-28 carbon atoms. These relatively unreactive compounds are not particularly toxic and do not participate strongly in atmospheric chemical reactions. The aromatic fraction, however, contains carcinogenic polycyclic aromatic hydrocarbons, which are discussed below. Aldehydes, ketones, epoxides, peroxides, esters, quinones, and lactones are found among the oxygenated neutral components, some of which may be mutagenic or carcinogenic. The acidic group contains long-chain fatty acids and nonvolatile phenols. Among the acids recovered from air-pollutant particulate matter are lauric, myristic, palmitic, stearic, behenic, oleic, and linoleic acids. The basic group consists largely of alkaline N-heterocyclic hydrocarbons such as acridine:

Acridine

Polycyclic Aromatic Hydrocarbons

Polycyclic aromatic hydrocarbons (PAH) in atmospheric particles have received a great deal of attention because of the known carcinogenic effects of some of these compounds, which are discussed in greater detail in Chapter 23. Prominent among these compounds are benzo(a)pyrene, benz(a)anthracene, chrysene, benzo-(e)pyrene, benz(e)acephenanthrylene, benzo(j)fluoranthene, and indenol. Some representative structures of PAH compounds are given below:

Benzo(a)pyrene   Chrysene   Benzo(j)fluoranthene

Elevated levels of PAH compounds of up to about 20 µg/m^3 are found in the atmosphere. Elevated levels of PAHs are most likely to be encountered in polluted urban atmospheres, and in the vicinity of natural fires such as forest and prairie fires. Coal furnace stack gas may contain over 1000 μg/m³ of PAH compounds, and cigarette smoke almost 100 μg/m³.

Atmospheric polycyclic aromatic hydrocarbons are found almost exclusively in the solid phase, largely sorbed to soot particles. Soot itself is a highly condensed product of PAHs. Soot contains 1–3% hydrogen and 5–10% oxygen, the latter due to partial surface oxidation. Benzo(a)pyrene adsorbed on soot disappears very rapidly in the presence of light yielding oxygenated products; the large surface area of the particle contributes to the high rate of reaction. Oxidation products of benzo(a)-pyrene include epoxides, quinones, phenols, aldehydes, and carboxylic acids as illustrated by the composite structures shown below:

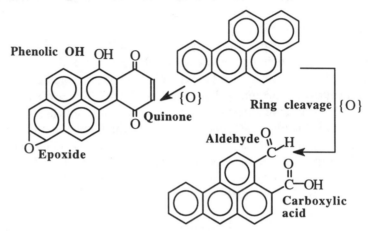

Carbonaceous Particles from Diesel Engines

Diesel engines emit significant levels of carbonaceous particles.[11] Although an appreciable fraction of these particles have aerodynamic diameters less than 1 μm, they may exist as aggregates of several thousand smaller particles in clusters up to 30 μm in diameter. This particulate matter consists largely of elemental carbon, although as much as 40% of the particle mass consists of organic extractable hydro-carbons and hydrocarbon derivatives including organosulfur and organonitrogen compounds.

10.9. EFFECTS OF PARTICLES

Atmospheric particles have numerous effects. The most obvious of these is reduction and distortion of visibility. They provide active surfaces upon which heterogeneous atmospheric chemical reactions can occur, and nucleation bodies for the condensation of atmospheric water vapor, thereby exerting a significant influ-ence upon weather and air pollution phenomena.

The most visible effects of aerosol particles upon air quality result from their optical effects. Particles smaller than about 0.1 μm in diameter scatter light much like molecules, that is, Rayleigh scattering. Generally, such particles have an insignificant effect upon visibility in the atmosphere. The light-scattering and intercepting properties of particles larger than 1 μm are approximately proportional to the particles' cross-sectional areas. Particles of 0.1 μm – 1 μm cause interference phenomena because they are about the same dimensions as the wavelengths of visible light, so their light-scattering properties are especially significant.

Atmospheric particles inhaled through the respiratory tract may damage health. Relatively large particles are likely to be retained in the nasal cavity and in the pharynx, whereas very small particles are likely to reach the lungs and be retained by them. The respiratory system possesses mechanisms for the expulson of inhaled particles. In the ciliated region of the respiratory system, particles are carried as far as the entrance to the gastrointestinal tract by a flow of mucus. Macrophages in the nonciliated pulmonary regions carry particles to the ciliated region.

The respiratory system may be damaged directly by particulate matter that enters the blood system or lymph system through the lungs. In addition, the particulate material or soluble components of it may be transported to organs some distance from the lungs and have a detrimental effect on these organs. Particles cleared from the respiratory tract are to a large extent swallowed into the gastrointestinal tract.

A strong correlation has been found between increases in the daily mortality rate and acute episodes of air pollution. In such cases, high levels of particulate matter are accompanied by elevated concentrations of SO_2 and other pollutants, so that any conclusions must be drawn with caution.

10.10. WATER AS PARTICULATE MATTER

Droplets of water are very widespread in the atmosphere. Although a natural phenomenon, such droplets can have significant and sometimes harmful effects. The most important such consequence is reduction of visibility, with accompanying detrimental effects on driving, flying, and boat navigation. Water droplets in fog act as carriers of pollutants. The most important of these are solutions of corrosive salts, particularly ammonium nitrates and sulfates, and solutions of strong acids. As discussed in Chapter 14, Section 14.3, the pH of water in acidic mist droplets collected during a Los Angeles acidic fog has been as low as 1.7, far below that of acidic precipitation. Such acidic mist can be especially damaging to the respiratory tract because it is very penetrating.

Arguably the most significant effect of water droplets in the atmosphere is as aquatic media in which important atmospheric chemical processes occur. The single most significant such process may well be the oxidation of S(IV) species to sulfuric acid and sulfate salts. The S(IV) species so oxidized include $SO_2(aq)$, HSO_3^-, and SO_3^{2-}. Another important oxidation that takes place in atmospheric water droplets is the oxidation of aldehydes to organic carboxylic acids.

The hydroxyl radical, $HO\cdot$, is very important in initiating atmospheric oxidation reactions such as those noted above. Hydroxyl radical as $HO\cdot$ can enter water droplets from the gas-phase atmosphere, it can be produced in water droplets photo-chemically, or it can be generated from H_2O_2 and $\cdot O_2^-$ radical-ion, which dissolve in water from the gas phase and then produce $HO\cdot$ by solution chemical reaction:

$$H_2O_2 + \cdot O_2^- \rightarrow HO\cdot + O_2 + OH^- \qquad\qquad (10.10.1)$$

Several solutes can react photochemically in aqueous solution (as opposed to the gas phase) to produce hydroxyl radical. One of these is hydrogen peroxide:

$$H_2O_2(aq) + h\nu \ \rightarrow \ 2HO\cdot(aq) \hspace{4cm} (10.10.2)$$

Nitrite as NO_2^- or HNO_2, nitrate (NO_3^-), and iron(III) as $Fe(OH)^{2+}(aq)$ can also react photochemically in aqueous solution to produce $HO\cdot$. It has been observed that ultraviolet radiation at 313 nm and simulated sunlight can react to produce $HO\cdot$ radical in authentic samples of water collected from cloud and fog sources.[12] Based on the results of this study and related investigations, it may be concluded that the aqueous-phase formation of hydroxyl radical is an important, and in some cases dominant means by which this key atmospheric oxidant is introduced into atmospheric water droplets.

Iron is an inorganic solute of particular importance in atmospheric water. This is because of the participation of iron(III) in the atmospheric oxidation of sulfur(IV) to sulfur(VI)—that is, the conversion of $SO_2(aq)$, HSO_3^-, and SO_3^{2-} to sulfates and H_2SO_4.

10.11. CONTROL OF PARTICULATE EMISSIONS

The removal of particulate matter from gas streams is the most widely practiced means of air pollution control. A number of devices have been developed for this purpose which differ widely in effectiveness, complexity, and cost. The selection of a particle removal system for a gaseous waste stream depends upon the particle loading, nature of particles (size distribution), and type of gas-scrubbing system used.

Particle Removal by Sedimentation and Inertia

The simplest means of particulate matter removal is **sedimentation**, a phenomenon that occurs continuously in nature. Gravitational settling chambers may be employed for the removal of particles from gas streams by simply settling under the influence of gravity. These chambers take up large amounts of space and have low collection efficiencies, particularly for small particles.

Gravitational settling of particles is enhanced by increased particle size, which occurs spontaneously by coagulation. Thus, over time, the sizes of particles increase and the number of particles decreases in a mass of air that contains particles. Brownian motion of particles less than about 0.1 µm in size is primarily responsible for their contact, enabling coagulation to occur. Particles greater than about 0.3 µm in radius do not diffuse appreciably and serve primarily as receptors of smaller particles.

Inertial mechanisms are effective for particle removal. These depend upon the fact that the radius of the path of a particle in a rapidly moving, curving air stream is larger than the path of the stream as a whole. Therefore, when a gas stream is spun by vanes, a fan, or a tangential gas inlet, the particulate matter may be collected on a separator wall because the particles are forced outward by centrifugal force. Devices utilizing this mode of operation are called **dry centrifugal collectors** (cyclones).

Particle filtration

Fabric filters, as their name implies, consist of fabrics that allow the passage of gas but retain particulate matter. These are used to collect dust in bags contained in structures called *baghouses*. Periodically, the fabric composing the filter is shaken to remove the particles and to reduce back-pressure to acceptable levels. Typically, the bag is in a tubular configuration as shown in Figure 10.6. Numerous other configurations are possible. Collected particulate matter is removed from bags by mechanical agitation, blowing air on the fabric, or rapid expansion and contraction of the bags.

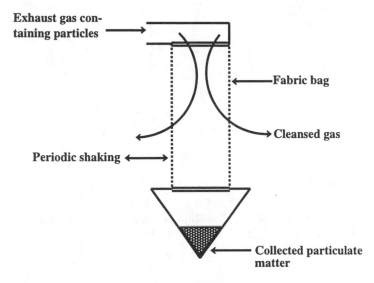

Exhaust gas containing particles

Fabric bag

Cleansed gas

Periodic shaking

Collected particulate matter

Figure 10.6. Baghouse collection of particulate emissions.

Although simple, baghouses are generally effective in removing particles from exhaust gas. Particles as small as 0.01 μm in diameter are removed, and removal efficiency is relatively high for particles down to 0.5 μm in diameter. Aided by the development of mechanically strong, heat-resistant fabrics from which the bags are fabricated, baghouse installations have increased significantly in the effort to control particulate emissions.

Scrubbers

A venturi scrubber passes gas through a device which leads the gas stream through a converging section, throat, and diverging section as shown in Figure 10.7. Injection of the scrubbing liquid at right angles to incoming gas breaks the liquid into very small droplets, which are ideal for scavenging particles from the gas stream. In the reduced-pressure (expanding and, therefore, cooling) region of the venturi, some condensation can occur of vapor from liquid initially evaporated in the generally hot waste gas, adding to the scrubbing efficiency. In addition to removing particles, venturis may serve as quenchers to cool exhaust gas, and as scrubbers for pollutant gases.

Ionizing wet scrubbers place an electrical charge on particles upstream from a wet scrubber. Larger particles and some gaseous contaminants are removed by scrubbing action. Smaller particles tend to induce opposite charges in water droplets in the scrubber and in its packing material and are removed by attraction of the opposite charges.

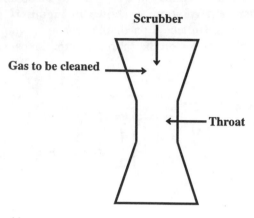

Figure 10.7. Venturi scrubber.

Electrostatic Removal

Aerosol particles may acquire electrical charges. In an electric field, such particles are subjected to a force, F (dynes), given by

$$F = Eq \tag{10.11.1}$$

where E is the voltage gradient (statvolt/cm) and q is the electrostatic charge on the particle (in esu). This phenomenon has been widely used in highly efficient **electrostatic precipitators**, as shown in Figure 10.8. The particles acquire a charge

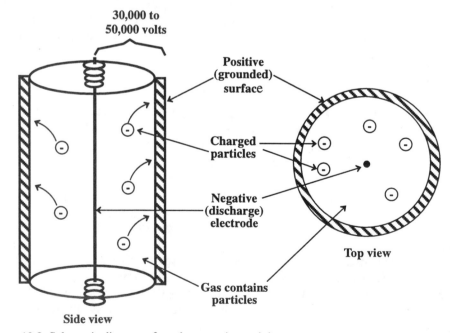

Figure 10.8. Schematic diagram of an electrostatic precipitator

when the gas stream is passed through a high-voltage, direct current corona. Because of the charge, the particles are attracted to a grounded surface from which they may be later removed. Ozone may be produced by the corona discharge. Similar devices used as household dust collectors may produce toxic ozone if not operated properly.

LITERATURE CITED

1. Harrison, Roy M. and Rene Van Grieken, Eds., *Atmospheric Particles*, Wiley, Chichester, U.K, 1998.

2. Matthias-Maser, Sabine, Martina Kramer, Jutta Brinkmann, and Wilhelm Schneider, "A Contribution of Primary Biological Aerosol Particles as Insoluble Component to the Atmospheric Aerosol over the South Atlantic Ocean, *Journal of Aerosol Science*, **28**, S3-S4 (1997).

3. Andreae, Meinrat O, and Paul J. Crutzen, "Atmospheric Aerosols: Biogeochemical Sources and Role in Atmospheric Chemistry," *Science*, 1052-1058 (1997).

4. Zielinski, Gregory A., Jack E. Dibb, Qinzhao Yang, Paul A. Mayewski, Sallie Whitlow, Mark S. Twickler, and Mark S. Germani, "Assessment of the Record of the 1982 El Chichon Eruption as Preserved in Greenland Snow," *Journal of Geophysical Research*, **102**, 30031-30045 (1997).

5. Stenchikov, Georgiy L., Ingo Kirchner, Alan Robock, Hans F. Graf, Juan Carlos Antuna, R. G. Grainger, Alyn Lambert, and Larry Thomason, "Radiative Forcing from the 1991 Mount Pinatubo Volcanic Eruption," *Journal of Geophysical Research*, **103**, 1383713857 (1998).

6. Meng, Z., D. Dabdub, and John H. Seinfeld, "Chemical Coupling between Atmospheric Ozone and Particulate Matter," *Science*, **277**, 116-119 (1997).

7. Rogge, Wolfgang, F., Lynn M. Hildemann, Monica A. Mazurek, Glen R. Cass, and Bernd R. T. Simonelt, "Sources of Fine Organic Aerosol. 2. Non-catalyst and Catalyst-Equipped Automobiles and Heavy Duty Diesel Trucks," *Environmental Science and Technology*, **27**, 636-651 (1993).

8. Roth, Beate and Kikuo Okdada, "On the Modification of Sea-Salt Particles in the Coastal Atmosphere," *Atmospheric Environment*, **32**, 1555-1569 (1998).

9. McElroy, M. W., R. C. Carr, D. S. Ensor, and G. R. Markowski, "Size Distribution of Fine Particles from Coal Combustion," *Science*, **215**, (1982) 13-19.

10. Pollanen, Roy, Ilka Valkama, and Harri Toivonen, "Transport of Radioactive Particles from the Chernobyl Accident," *Atmospheric Environment*, **31**, 3575-3590 (1997).

11. Morawska, Lidia, Neville D. Bofinger, Ladislaw Kocis, and Alwell Nwankwoala, "Submicrometer and Supermicrometer Particles from Diesel Vehicle Emissions," *Environmental Science and Technology*, **32**, 2033-2042 (1998).

12. Faust, Bruce C., and John M. Allen, "Aqueous-Phase Photochemical Formation of Hydroxyl Radical in Authentic Cloudwaters and Fogwaters," *Environmental Science and Technology* **27**, 1221-1224 (1993).

SUPPLEMENTARY REFERENCES

Air Quality Criteria for Particulate Matter, National Air Pollution Control Administration Publication No. AP-49, 1969.

Brimblecombe, Peter, *Air Composition and Chemistry*, 2nd ed., Cambridge University Press, Cambridge, U.K., 1996.

Botkin, Daniel B., Ed., *Changing the Global Environment: Perspectives on Human Involvement*, Academic Press, San Diego, CA, 1989.

Control Techniques for Particulate Air Pollutants, National Air Pollution Control Administration Publication No. AP-51, 1969.

d'Almeida, Guillaume A., Peter Koepke, and Eric P. Shettle, *Atmospheric Aerosols*, Deepak, Hampton, VA, 1991.

Götz, G., E. Méazarós and G. Vali, *Atmospheric Particles and Nuclei*, Akadémiai Kiadó, Budapest, 1991.

Hinds, W. C., *Aerosol Technology*, John Wiley and Sons, New York, 1982.

Kneip, T. J., and P. J. Lioy, Eds., *Aerosols: Anthropogenic and Natural Sources and Transport*, New York Academy of Sciences, New York, 1980.

Seinfeld, John H., *Atmospheric Chemistry and Physics*, John Wiley & Sons, Inc., New York, 1998.

Willeke, Klaus, and Paul A. Baron, Eds., *Aerosol Measurement: Principles, Techniques, and Applications*, Van Nostrand Reinhold, New York, 1993.

QUESTIONS AND PROBLEMS

1. The maximum electrical charge that an atmospheric particle may attain in dry air is about 8 esu/cm^2. Where the charge of an electron is 4.77×10^{-10} esu, equal to the *elementary quantum of electricity* (the charge on 1 electron or proton), how many electrons per square centimeter of surface area is this?

2. For small charged particles, those that are 0.1 μm or less in size, an average charge of 4.77×10^{-10} esu, is normally assumed for the whole particle. What is the surface charge in esu/cm^2 for a charged spherical particle with a radius of 0.1 μm?

3. What is the settling velocity of a particle having a Stokes diameter of 10 μm and a density of 1 g/cm^3 in air at 1.00 atm pressure and 0°C temperature? (The viscosity of air at 0°C is 170.8 micropoise. The density of air under these conditions is 1.29 g/L.)

4. A freight train that included a tank car containing anhydrous NH_3 and one containing concentrated HCl was wrecked, causing both of the tank cars to leak. In the region between the cars a white aerosol formed. What was it, and how was it produced?

5. Examination of aerosol fume particles produced by a welding process showed that 2% of the particles were greater than 7 μm in diameter and only 2% were less than 0.5 μm. What is the mass median diameter of the particles?

6. What two vapor forms of mercury might be found in the atmosphere?

7. Analysis of particulate matter collected in the atmosphere near a seashore shows considerably more Na than Cl on a molar basis. What does this indicate?

8. What type of process results in the formation of very small aerosol particles?

9. Which size range encompasses most of the particulate matter mass in the atmosphere?

10. Why are aerosols in the 0.1–1 μm size range especially effective in scattering light?

11. Per unit mass, why are smaller particles more effective catalysts for atmospheric chemical reactions?

12. In terms of origin, what are the three major categories of elements found in atmospheric particles?

13. What are the five major classes of material making up the composition of atmospheric aerosol particles?

14. The size distribution of particles emitted from coal-fired power plants is bimodal. What are some of the properties of the smaller fraction in terms of potential environmental implications?

11 GASEOUS INORGANIC AIR POLLUTANTS

11.1 INORGANIC POLLUTANT GASES

A number of gaseous inorganic pollutants enter the atmosphere as the result of human activities. Those added in the greatest quantities are CO, SO_2, NO, and NO_2. (These quantities are relatively small compared to the amount of CO_2 in the atmosphere. The possible environmental effects of increased atmospheric CO_2 levels are discussed in Chapter 14.) Other inorganic pollutant gases include NH_3, N_2O, N_2O_5, H_2S, Cl_2, HCl, and HF. Substantial quantities of some of these gases are added to the atmosphere each year by human activities. Globally, atmospheric emissions of carbon monoxide, sulfur oxides, and nitrogen oxides are of the order of one to several hundred million tons per year.

11.2. PRODUCTION AND CONTROL OF CARBON MONOXIDE

Carbon monoxide, CO, causes problems in cases of locally high concentrations because of its toxicity (see Chapter 23). The overall atmospheric concentration of carbon monoxide is about 0.1 ppm corresponding to a burden in the earth's atmosphere of approximately 500 million metric tons of CO with an average residence time ranging from 36 to 110 days. Much of this CO is present as an intermediate in the oxidation of methane by hydroxyl radical (see Section 9.8, Reaction 9.8.17). From Table 9.1 it may be seen that the methane content of the atmosphere is about 1.6 ppm, more than 10 times the concentration of CO. Therefore, any oxidation process for methane that produces carbon monoxide as an intermediate is certain to contribute substantially to the overall carbon monoxide burden, probably around two-thirds of the total CO.

Degradation of chlorophyll during the autumn months releases CO, amounting to perhaps as much as 20% of the total annual release. Anthropogenic sources account for about 6% of CO emissions. The remainder of atmospheric CO comes from largely unknown sources. These include some plants and marine organisms known as siphonophores, an order of Hydrozoa. Carbon monoxide is also produced by decay of plant matter other than chlorophyll.

Because of carbon monoxide emissions from internal combustion engines, the highest levels of this toxic gas tend to occur in congested urban areas at times when the maximum number of people are exposed, such as during rush hours. At such times, carbon monoxide levels in the atmosphere have become as high as 50-100 ppm.

Atmospheric levels of carbon monoxide in urban areas show a positive correlation with the density of vehicular traffic, and a negative correlation with wind speed. Urban atmospheres may show average carbon monoxide levels of the order of several ppm, much higher than those in remote areas.

Control of Carbon Monoxide Emissions

Since the internal combustion engine is the primary source of localized pollutant carbon monoxide emissions, control measures have been concentrated on the automobile. Carbon monoxide emissions may be lowered by employing a leaner air-fuel mixture, that is, one in which the weight ratio of air to fuel is relatively high. At air-fuel (weight:weight) ratios exceeding approximately 16:1, an internal combustion engine emits very little carbon monoxide.

Modern automobiles use catalytic exhaust reactors to cut down on carbon monoxide emissions. Excess air is pumped into the exhaust gas, and the mixture is passed through a catalytic converter in the exhaust system, resulting in oxidation of CO to CO_2.

11.3. FATE OF ATMOSPHERIC CO

It is generally agreed that carbon monoxide is removed from the atmosphere by reaction with hydroxyl radical, $HO^\bullet$:

$$CO + HO^\bullet \rightarrow CO_2 + H \tag{11.3.1}$$

The reaction produces hydroperoxyl radical as a product:

$$O_2 + H + M \rightarrow HOO^\bullet + M \tag{11.3.2}$$

$HO^\bullet$ is regenerated from $HOO^\bullet$ by the following reactions:

$$HOO^\bullet + NO \rightarrow HO^\bullet + NO_2 \tag{11.3.3}$$

$$HOO^\bullet + HOO^\bullet \rightarrow H_2O_2 + O_2 \tag{11.3.4}$$

The latter reaction is followed by photochemical dissociation of H_2O_2 to regenerate $HO^\bullet$:

$$H_2O_2 + h\nu \rightarrow 2HO^\bullet \tag{11.3.5}$$

Methane is also involved through the atmospheric $CO/HO^\bullet/CH_4$ cycle.

Soil microorganisms act to remove CO from the atmosphere. Therefore, soil is a sink for carbon monoxide.

11.4. SULFUR DIOXIDE SOURCES AND THE SULFUR CYCLE

Figure 11.1 shows the main aspects of the global sulfur cycle. This cycle involves primarily H_2S, $(CH_3)_2S$, SO_2, SO_3, and sulfates. There are many uncertainties regarding the sources, reactions, and fates of these atmospheric sulfur species. On a global basis, sulfur compounds enter the atmosphere to a very large extent through human activities. Approximately 100 million metric tons of sulfur per year enters the global atmosphere through anthropogenic activities, primarily as SO_2 from the com-

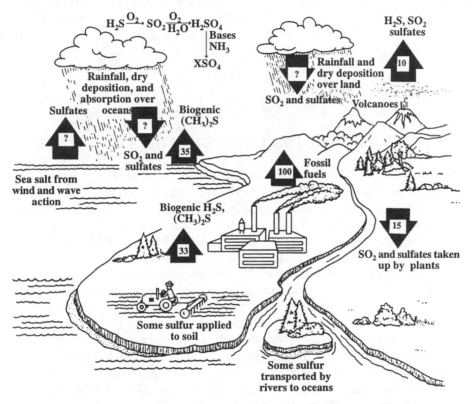

Figure 11.1. The global atmospheric sulfur cycle. Fluxes of sulfur represented by the arrows are in millions of metric tons per year. Those marked with a question mark are uncertain, but large, probably of the order of 100 million metric tons per year.

bustion of coal and residual fuel oil. The greatest uncertainties in the cycle have to do with nonanthropogenic sulfur, which enters the atmosphere largely as SO_2 and H_2S from volcanoes,[1] and as $(CH_3)_2S$ and H_2S from the biological decay of organic matter and reduction of sulfate. The single largest source of natural sulfur discharged to the atmosphere is now believed to be biogenic dimethyl sulfide, $(CH_3)_2S$, from marine sources.[2] Any H_2S that does get into the atmosphere is converted rapidly to SO_2 by the following overall process:

$$H_2S + {}^3/_2 O_2 \rightarrow SO_2 + H_2O \qquad\qquad (11.4.1)$$

The initial reaction is hydrogen ion abstraction by hydroxyl radical,

$$H_2S + HO\cdot \rightarrow HS\cdot + H_2O \tag{11.4.2}$$

followed by the following two reactions to give SO_2:

$$HS\cdot + O_2 \rightarrow HO\cdot + SO \tag{11.4.3}$$

$$SO + O_2 \rightarrow SO_2 + O \tag{11.4.4}$$

The primary source of anthropogenic sulfur dioxide is coal, from which sulfur must be removed at great expense to keep sulfur dioxide emissions at acceptable levels. Approximately half of the sulfur in coal is in some form of pyrite, FeS_2, and the other half is organic sulfur. The production of sulfur dioxide by the combustion of pyrite is given by the following reaction:

$$4FeS_2 + 11O_2 \rightarrow 2Fe_2O_3 + 8SO_2 \tag{11.4.5}$$

Essentially all of the sulfur is converted to SO_2 and only 1 or 2% to SO_3.

11.5. SULFUR DIOXIDE REACTIONS IN THE ATMOSPHERE

Many factors, including temperature, humidity, light intensity, atmospheric transport, and surface characteristics of particulate matter, may influence the atmospheric chemical reactions of sulfur dioxide. Like many other gaseous pollutants, sulfur dioxide reacts to form particulate matter, which then settles or is scavenged from the atmosphere by rainfall or other processes. It is known that high levels of air pollution normally are accompanied by a marked increase in aerosol particles and a consequent reduction in visibility. Reaction products of sulfur dioxide are thought to be responsible for some aerosol formation. Whatever the processes involved, much of the sulfur dioxide in the atmosphere is ultimately oxidized to sulfuric acid and sulfate salts, particularly ammonium sulfate and ammonium hydrogen sulfate. In fact, it is likely that these sulfates account for the turbid haze that covers much of the eastern part of the U.S. under all atmospheric conditions except those characterized by massive intrusions of Arctic air masses during the winter months. The potential of sulfates to induce climatic change is high and must be taken into account when considering control of sulfur dioxide.

Some of the possible ways in which sulfur dioxide may react in the atmosphere are (1) photochemical reactions; (2) photochemical and chemical reactions in the presence of nitrogen oxides and/or hydrocarbons, particularly alkenes; (3) chemical processes in water droplets, particularly those containing metal salts and ammonia; and (4) reactions on solid particles in the atmosphere. It should be kept in mind that the atmosphere is a highly dynamic system with great variations in temperature, composition, humidity, and intensity of sunlight; therefore, different processes may predominate under various atmospheric conditions.

Photochemical reactions are probably involved in some of the processes resulting in the atmospheric oxidation of SO_2. Light with wavelengths above 218 nm is not sufficiently energetic to bring about the photodissociation of SO_2, so direct photochemical reactions in the troposphere are of no significance. The oxidation of sulfur dioxide at the parts-per-million level in an otherwise unpolluted atmosphere is a slow process. Therefore, other pollutant species must be involved in the process in atmospheres polluted with SO_2.

The presence of hydrocarbons and nitrogen oxides greatly increases the oxidation rate of atmospheric SO_2. As discussed in Chapter 13, hydrocarbons, nitrogen oxides, and ultraviolet light are the ingredients necessary for the formation of photochemical smog. This disagreeable condition is characterized by high levels of various oxidizing species (photochemical oxidants) capable of oxidizing SO_2. In the smog-prone Los Angeles area, the oxidation of SO_2 ranges up to 5-10% per hour. Among the oxidizing species present which could bring about this fast reaction are $HO\bullet$, $HOO\bullet$, O, O_3, NO_3, N_2O_5, $ROO\bullet$, and $RO\bullet$. As discussed in Chapters 12 and 13, the latter two species are reactive organic free radicals containing oxygen. Although ozone, O_3, is an important product of photochemical smog, it is believed that the oxidation of SO_2 by ozone in the gas phase is too slow to be appreciable, but it is probably significant in water droplets.

The most important gas-phase reaction leading to the oxidation of SO_2 is the addition of $HO\bullet$ radical,[3]

$$HO\bullet + SO_2 \rightarrow HOSO_2\bullet \tag{11.5.1}$$

forming a reactive free radical which is eventually converted to a form of sulfate.

In all but relatively dry atmospheres, it is probable that sulfur dioxide is oxidized by reactions occurring inside water aerosol droplets. The overall process of sulfur dioxide oxidation in the aqueous phase is rather complicated. It involves the transport of gaseous SO_2 and oxidant to the aqueous phase, diffusion of species in the aqueous droplet, hydrolysis and ionization of SO_2, and oxidation of SO_2 by the following overall process, where {O} represents an oxidizing agent such as H_2O_2, $HO\bullet$, or O_3 and S(IV) is $SO_2(aq)$, $HSO_3^-(aq)$, and $SO_3^{2-}(aq)$.

$$\{O\}(aq) + S(IV)(aq) \rightarrow 2H^+ + SO_4^{2-} \text{ (unbalanced)} \tag{11.5.2}$$

In the absence of catalytic species, the reaction with dissolved molecular O_2,

$$\tfrac{1}{2}O_2(aq) + SO_2(aq) + H_2O \rightarrow H_2SO_4(aq) \tag{11.5.3}$$

is too slow to be significant. Hydrogen peroxide is an important oxidizing agent in the atmosphere. It reacts with dissolved sulfur dioxide through the overall reaction,

$$SO_2(aq) + H_2O_2(aq) \rightarrow H_2SO_4(aq) \tag{11.5.4}$$

to produce sulfuric acid. The major reaction is thought to be between hydrogen peroxide and HSO_3^- ion with peroxymonosulfurous acid, $HOOSO_2^-$, as an intermediate.

Ozone, O_3, oxidizes sulfur dioxide in water. The fastest reaction is with sulfite ion;

$$SO_3^{2-}(aq) + O_3(aq) \rightarrow SO_4^{2-}(aq) + O_2 \tag{11.5.5}$$

reactions are slower with $HSO_3^-(aq)$ and $SO_2(aq)$. The rate of oxidation of aqueous SO_2 species by ozone increases with increasing pH. The oxidation of sulfur dioxide in water droplets is faster in the presence of ammonia, which reacts with sulfur dioxide to produce bisulfite ion and sulfite ion in solution:

$$NH_3 + SO_2 + H_2O \rightarrow NH_4^+ + HSO_3^-$$ (11.5.6)

Some solutes dissolved in water catalyze the oxidation of aqueous SO_2. Both iron(III) and Mn(II) have this effect. The reactions catalyzed by these two ions are faster with increasing pH. Dissolved nitrogen species, NO_2 and HNO_2, oxidize aqueous sulfur dioxide in the laboratory. As noted in Section 10.10, nitrite dissolved in water droplets may react photochemically to produce $HO\cdot$ radical, and this species in turn could act to oxidize dissolved sulfite.

Heterogeneous reactions on solid particles may also play a role in the removal of sulfur dioxide from the atmosphere. In atmospheric photochemical reactions, such particles may function as nucleation centers. Thus, they act as catalysts and grow in size by accumulating reaction products. The final result would be production of an aerosol with a composition unlike that of the original particle. Soot particles, which consist of elemental carbon contaminated with polynuclear aromatic hydrocarbons (see Chapter 10, Section 10.4) produced in the incomplete combustion of carbonaceous fuels, can catalyze the oxidation of sulfur dioxide to sulfate as indicated by the presence of sulfate on the soot particles. Soot particles are very common in polluted atmospheres, so it is very likely that they are strongly involved in catalyzing the oxidation of sulfur dioxide.

Oxides of metals such as aluminum, calcium, chromium, iron, lead, or vanadium may also be catalysts for the heterogenous oxidation of sulfur dioxide. These oxides may also adsorb sulfur dioxide. However, the total surface area of oxide particulate matter in the atmosphere is very low so that the fraction of sulfur dioxide oxidized on metal oxide surfaces is relatively small.

Effects of Atmospheric Sulfur Dioxide

Though not terribly toxic to most people, low levels of sulfur dioxide in air do have some health effects. Its primary effect is upon the respiratory tract, producing irritation and increasing airway resistance, especially to people with respiratory weaknesses and sensitized asthmatics. Therefore, exposure to the gas may increase the effort required to breathe. Mucus secretion is also stimulated by exposure to air contaminated by sulfur dioxide. Although SO_2 causes death in humans at 500 ppm, it has not been found to harm laboratory animals at 5 ppm.

Sulfur dioxide has been at least partially implicated in several acute incidents of air pollution. In December 1930 a thermal inversion trapped waste products from a number of industrial sources in the narrow Meuse River Valley of Belgium. Sulfur dioxide levels reached 38 ppm. Approximately 60 people died in the episode, and some cattle were killed. In October 1948 a similar incident caused illness in over 40% of the population of Donora, Pennsylvania, and 20 people died. Sulfur dioxide concentrations of 2 ppm were recorded. During a five-day period marked by a temperature inversion and fog in London in December 1952, approximately 3500-4000 deaths in excess of normal occurred. Levels of SO_2 reached 1.3 ppm. Autopsies revealed irritation of the respiratory tract, and high levels of sulfur dioxide were suspected of contributing to excess mortality.

Atmospheric sulfur dioxide is harmful to plants, some species of which are affected more than others. Acute exposure to high levels of the gas kills leaf tissue, a condition called leaf necrosis. The edges of the leaves and the areas between the leaf veins show characteristic damage. Chronic exposure of plants to sulfur dioxide causes chlorosis, a bleaching or yellowing of the normally green portions of the leaf. Plant injury increases with increasing relative humidity. Plants incur most injury from sulfur dioxide when their stomata (small openings in plant surface tissue that allow interchange of gases with the atmosphere) are open. For most plants, the stomata are open during the daylight hours, and most damage from sulfur dioxide occurs then. Long-term, low-level exposure to sulfur dioxide can reduce the yields of grain crops such as wheat or barley. Sulfur dioxide in the atmosphere is converted to sulfuric acid, so that in areas with high levels of sulfur dioxide pollution, plants may be damaged by sulfuric acid aerosols. Such damage appears as small spots where sulfuric acid droplets have impinged on leaves.

One of the more costly effects of sulfur dioxide pollution is deterioration of building materials. Limestone, marble, and dolomite are calcium and/or magnesium carbonate minerals that are attacked by atmospheric sulfur dioxide to form products that are either water-soluble or composed of poorly adherent solid crusts on the rock's surface, adversely affecting the appearance, structural integrity, and life of the building. Although both SO_2 and NO_x attack such stone, chemical analysis of the crusts shows predominantly sulfate salts. Dolomite, a calcium/magnesium carbonate mineral, reacts with atmospheric sulfur dioxide as follows:

$$CaCO_3 \cdot MgCO_3 + 2SO_2 + O_2 + 9H_2O \rightarrow$$

$$CaSO_4 \cdot 2H_2O + MgSO_4 \cdot 7H_2O + 2CO_2 \qquad (11.5.7)$$

Sulfur Dioxide Removal

A number of processes are being used to remove sulfur and sulfur oxides from fuel before combustion and from stack gas after combustion. Most of these efforts concentrate on coal, since it is the major source of sulfur oxides pollution. Physical separation techniques may be used to remove discrete particles of pyritic sulfur from coal. Chemical methods may also be employed for removal of sulfur from coal. Fluidized bed combustion of coal promises to eliminate SO_2 emissions at the point of combustion. The process consists of burning granular coal in a bed of finely divided limestone or dolomite maintained in a fluid-like condition by air injection. Heat calcines the limestone,

$$CaCO_3 \rightarrow CaO + CO_2 \qquad (11.5.8)$$

and the lime produced absorbs SO_2:

$$CaO + SO_2 \rightarrow CaSO_3 \text{ (Which may be oxidized to } CaSO_4) \qquad (11.5.9)$$

Many processes have been proposed or studied for the removal of sulfur dioxide from stack gas. Table 11.1 summarizes major stack gas scrubbing systems including

Table 11.1. Major Stack Gas Scrubbing Systems[1]

Process	Reaction	Significant advantages or disadvantages
Lime slurry scrubbing[2]	$Ca(OH)_2 + SO_2 \rightarrow CaSO_3 + H_2O$	Up to 200 kg of lime are needed per metric ton of coal, producing huge quantities of wastes
Limestone slurry scrubbing[2]	$CaCO_3 + SO_2 \rightarrow CaSO_3 + CO_2(g)$	Lower pH than lime slurry, not so efficient
Magnesium oxide scrubbing	$Mg(OH)_2(slurry) + SO_2 \rightarrow$ $MgSO_3 + H_2O$	The sorbent can be regenerated, which can be done off site, if desired.
Sodium-base scrubbing	$Na_2SO_3 + H_2O + SO_2 \rightarrow 2NaHSO_3$ $2NaHSO_3 + heat \rightarrow Na_2SO_3 +$ $H_2O + SO_2$ (regeneration)	No major technological limitations. Relatively high annual costs.
Double alkali[2]	$2NaOH + SO_2 \rightarrow Na_2SO_3 + H_2O$ $Ca(OH)_2 + Na_2SO_3 \rightarrow CaSO_3(s) +$ $2NaOH$ (regeneration of NaOH)	Allows for regeneration of expensive sodium alkali solution with inexpensive lime.

[1] For details regarding these and more advanced processes see (1) Satriana, M., *New Developments in Flue Gas Desulfurization Technology*, Noyes Data Corp., Park Ridge, NJ, 1982, and (2) Takeshita, Mitsuru, and Herminé Soud, *FGD Performance and Experience on Coal-Fired Plants*, Gemini House, London, 1993.

[2] These processes have also been adapted to produce a gypsum product by oxidation of $CaSO_3$ in the spent scrubber medium:

$$CaSO_3 + \frac{1}{2}O_2 + 2H_2O \rightarrow CaSO_4 \cdot 2H_2O(s)$$

Gypsum has some commercial value, such as in the manufacture of plasterboard, and makes a relatively settleable waste product.

throwaway and recovery systems as well as wet and dry systems. A dry throwaway system used with only limited success involves injection of dry limestone or dolomite into the boiler followed by recovery of dry lime, sulfites, and sulfates. The overall reaction, shown here for dolomite, is the following:

$$CaCO_3 \cdot MgCO_3 + SO_2 + \frac{1}{2}O_2 \rightarrow CaSO_4 + MgO + 2CO_2 \qquad (11.5.10)$$

The solid sulfate and oxide products are removed by electrostatic precipitators or cyclone separators. The process has an efficiency of 50% or less for the removal of sulfur oxides.

As may be noted from the chemical reactions shown in Table 11.1, all sulfur dioxide removal processes, except for catalytic oxidation, depend upon absorption of SO_2 by an acid-base reaction. The first two processes listed are throwaway processes yielding large quantities of wastes; the others provide for some sort of sulfur product recovery.

Lime or limestone slurry scrubbing for SO_2 removal involves acid-base reactions with SO_2. Since they were introduced during the late 1960s, wet lime flue gas desulfurization processes have been the most widely used means for removing sulfur dioxide from flue gas and are likely to remain so for the foreseeable future.[4]

When sulfur dioxide dissolves in water as part of a wet scrubbing process, equilibrium is established between SO_2 gas and dissolved SO_2:

$$SO_2(g) \leftrightarrow SO_2(aq) \qquad\qquad (11.5.11)$$

This equilibrium is described by Henry's law (Section 5.3),

$$[SO_2(aq)] = K \times P_{SO_2} \qquad\qquad (11.5.12)$$

where $[SO_2(aq)]$ is the concentration of dissolved molecular sulfur dioxide; K is the Henry's law constant for SO_2; and P_{SO_2} is the partial pressure of sulfur dioxide gas. In the presence of base, Reaction 11.5.11 is shifted strongly to the right by the following reactions:

$$H_2O + SO_2(aq) \leftrightarrow H^+ + HSO_3^- \qquad\qquad (11.5.13)$$

$$HSO_3^- \leftrightarrow H^+ + SO_3^{2-} \qquad\qquad (11.5.14)$$

In the presence of calcium carbonate slurry (as in limestone slurry scrubbing), hydrogen ion is taken up by the reaction

$$CaCO_3 + H^+ \leftrightarrow Ca^{2+} + HCO_3^- \qquad\qquad (11.5.15)$$

The reaction of calcium carbonate with carbon dioxide from stack gas,

$$CaCO_3 + CO_2 + H_2O \leftrightarrow Ca^{2+} + 2HCO_3^- \qquad\qquad (11.5.16)$$

results in some sorption of CO_2. The reaction of sulfite and calcium ion to form highly insoluble calcium sulfite hemihydrate

$$Ca^{2+} + SO_3^{2-} + \tfrac{1}{2}H_2O \leftrightarrow CaSO_3 \cdot \tfrac{1}{2}H_2O(s) \qquad\qquad (11.5.17)$$

also shifts Reactions 11.5.13 and 11.5.14 to the right. Gypsum is formed in the scrubbing process by the oxidation of sulfite,

$$SO_3^{2-} + \tfrac{1}{2}O_2 \rightarrow SO_4^{2-} \qquad\qquad (11.5.18)$$

followed by reaction of sulfate ion with calcium ion:

$$Ca^{2+} + SO_4^{2-} + 2H_2O \leftrightarrow CaSO_4 \cdot 2H_2O(s) \qquad\qquad (11.5.19)$$

Formation of gypsum in the scrubber is undesirable because it creates scale in the scrubber equipment. However, gypsum is sometimes produced deliberately in the spent scrubber liquid downstream from the scrubber.

When lime, $Ca(OH)_2$, is used in place of limestone (lime slurry scrubbing), a source of hydroxide ions is provided for direct reaction with H^+:

$$H^+ + OH^- \rightarrow H_2O \qquad\qquad (11.5.20)$$

The reactions involving sulfur species in a lime slurry scrubber are essentially the same as those just discussed for limestone slurry scrubbing. The pH of a lime slurry is higher than that of a limestone slurry, so that the former has more of a tendency to react with CO_2, resulting in the absorption of that gas:

$$CO_2 + OH^- \rightarrow HCO_3^- \qquad\qquad (11.5.21)$$

Current practice with lime and limestone scrubber systems calls for injection of the slurry into the scrubber loop beyond the boilers. A number of power plants are now operating with this kind of system. Experience to date has shown that these scrubbers remove well over 90% of both SO_2 and fly ash when operating properly. (Fly ash is fuel combustion ash normally carried up the stack with flue gas, see Chapter 10.) In addition to corrosion and scaling problems, disposal of lime sludge poses formidable obstacles. The quantity of this sludge may be appreciated by considering that approximately one ton of limestone is required for each five tons of coal. The sludge is normally disposed of in large ponds, which can present some disposal problems. Water seeping through the sludge beds becomes laden with calcium sulfate and other salts. It is difficult to stabilize this sludge as a structurally stable, nonleachable solid.

Recovery systems in which sulfur dioxide or elemental sulfur are removed from the spent sorbing material, which is recycled, are much more desirable from an environmental viewpoint than are throwaway systems.[5] Many kinds of recovery processes have been investigated, including those that involve scrubbing with magnesium oxide slurry, sodium hydroxide solution, sodium sulfite solution, ammonia solution, or sodium citrate solution.

Sulfur dioxide trapped in a stack-gas-scrubbing process can be converted to hydrogen sulfide by reaction with synthesis gas (H_2, CO, CH_4),

$$SO_2 + (H_2, CO, CH_4) \rightarrow H_2S + CO_2 (H_2O) \qquad\qquad (11.5.22)$$

The Claus reaction is then employed to produce elemental sulfur:

$$2H_2S + SO_2 \rightarrow 2H_2O + 3S \qquad\qquad (11.5.23)$$

11.6. NITROGEN OXIDES IN THE ATMOSPHERE

The three oxides of nitrogen normally encountered in the atmosphere are nitrous oxide (N_2O), nitric oxide (NO), and nitrogen dioxide (NO_2). Nitrous oxide, a commonly used anesthetic known as "laughing gas," is produced by microbiological processes and is a component of the unpolluted atmosphere at a level of approximately 0.3 ppm (see Table 9.1). This gas is relatively unreactive and probably does not significantly influence important chemical reactions in the lower atmosphere. Its concentration decreases rapidly with altitude in the stratosphere due to the photochemical reaction

$$N_2O + h\nu \rightarrow N_2 + O \tag{11.6.1}$$

and some reaction with singlet atomic oxygen:

$$N_2O + O \rightarrow N_2 + O_2 \tag{11.6.2}$$

$$N_2O + O \rightarrow 2NO \tag{11.6.3}$$

These reactions are significant in terms of depletion of the ozone layer. Increased global fixation of nitrogen, accompanied by increased microbial production of N_2O, could contribute to ozone layer depletion.

Colorless, odorless nitric oxide (NO) and pungent red-brown nitrogen dioxide (NO_2) are very important in polluted air. Collectively designated NO_x, these gases enter the atmosphere from natural sources, such as lightning and biological processes, and from pollutant sources. The latter are much more significant because of regionally high NO_2 concentrations which can cause severe air quality deterioration. Practically all anthropogenic NO_x enters the atmosphere as a result of the combustion of fossil fuels in both stationary and mobile sources. Globally, somewhat less than 100 million metric tons of nitrogen oxides are emitted to the atmosphere from these sources each year, compared to several times that much from widely dispersed natural sources. United States production of nitrogen oxides is of the order of 20 million metric tons per year. The contribution of automobiles to nitric oxide production in the U.S. has become somewhat lower in the last decade as newer automobiles have replaced older models.

Most NO_x entering the atmosphere from pollution sources does so as NO generated from internal combustion engines. At very high temperatures, the following reaction occurs:

$$N_2 + O_2 \rightarrow 2NO \tag{11.6.4}$$

The speed with which this reaction takes place increases steeply with temperature. The equilibrium concentration of NO in a mixture of 3% O_2 and 75% N_2, typical of that which occurs in the combustion chamber of an internal combustion engine, is shown as a function of temperature in Figure 11.2. At room temperature (27°C) the equilibrium concentration of NO is only 1.1×10^{-10} ppm, whereas at high temperatures it is much higher. Therefore, high temperatures favor both a high equilibrium concentration and a rapid rate of formation of NO. Rapid cooling of the exhaust gas from combustion "freezes" NO at a relatively high concentration because equilibrium is not maintained. Thus, by its very nature, the combustion process both in the internal combustion engine and in furnaces produces high levels of NO in the combustion products.The mechanism for formation of nitrogen oxides from N_2 and O_2 during combustion is a complicated process. Both oxygen and nitrogen atoms are formed at the very high combustion temperatures by the reactions

$$O_2 + M \rightarrow O + O + M \qquad\qquad (11.6.5)$$

$$N_2 + M \rightarrow N^{\bullet} + N^{\bullet} + M \qquad\qquad (11.6.6)$$

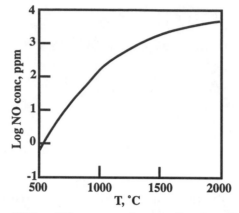

Figure 11.2. Log of equilibrium NO concentration as a function of temperature in a mixture containing 75% N_2 and 3% O_2.

where M is a highly energetic third body that imparts enough energy to the molecular N_2 and O_2 to break their chemical bonds. The energies required for these reactions are quite high because breakage of the oxygen bond requires 118 kcal/mole and breakage of the nitrogen bond requires 225 kcal/mole. Once formed, O and N atoms participate in the following chain reaction for the formation of nitric oxide from nitrogen and oxygen:

$$N_2 + O \rightarrow NO + N \qquad\qquad (11.6.7)$$

$$\underline{N + O_2 \rightarrow NO + O} \qquad\qquad (11.6.8)$$

$$N_2 + O_2 \rightarrow 2NO \qquad\qquad (11.6.9)$$

There are, of course, many other species present in the combustion mixture besides those shown. The oxygen atoms are especially reactive toward hydrocarbon fragments by reactions such as the following:

$$RH + O \rightarrow R^{\bullet} + HO^{\bullet} \qquad\qquad (11.6.10)$$

where RH represents a hydrocarbon fragment with an extractable hydrogen atom. These fragments compete with N_2 for oxygen atoms. It is partly for this reason that the formation of NO is appreciably higher at air/fuel ratios exceeding the stoichiometric ratio (lean mixture), as shown in Figure 13.3.

The hydroxyl radical itself can participate in the formation of NO. The reaction is

$$N + HO^{\bullet} \rightarrow NO + H^{\bullet} \qquad\qquad (11.6.11)$$

Nitric oxide, NO, is a product of the combustion of coal and petroleum containing chemically bound nitrogen. Production of NO by this route occurs at much lower temperatures than those required for "thermal" NO, discussed previously.

Atmospheric Reactions of NO$_x$

Atmospheric chemical reactions convert NO$_x$ to nitric acid, inorganic nitrate salts, organic nitrates, and peroxyacetyl nitrate (see Chapter 13). The principal reactive nitrogen oxide species in the troposphere are NO, NO$_2$, and HNO$_3$. These species cycle among each other, as shown in Figure 11.3. Although NO is the primary form in which NO$_x$ is released to the atmosphere, the conversion of NO to NO$_2$ is relatively rapid in the troposphere.

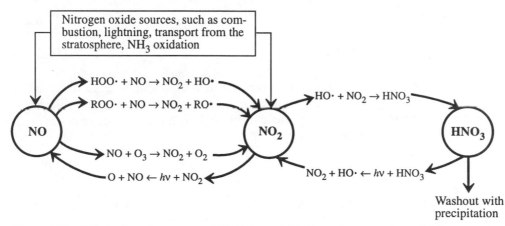

Figure 11.3. Principal reactions among NO, NO$_2$, and HNO$_3$ in the atmosphere. ROO• represents an organic peroxyl radical, such as the methylperoxyl radical, CH$_3$OO•.

Nitrogen dioxide is a very reactive and significant species in the atmosphere. It absorbs light throughout the ultraviolet and visible spectrum penetrating the troposphere. At wavelengths below 398 nm, photodissociation occurs,

$$NO_2 + h\nu \rightarrow NO + O \tag{11.6.12}$$

to produce ground state oxygen atoms. Above 430 nm, only excited molecules are formed,

$$NO_2 + h\nu \rightarrow NO_2^* \tag{11.6.13}$$

whereas at wavelengths between 398 nm and 430 nm, either process may occur. Photodissociation at these wavelengths requires input of rotational energy from rotation of the NO$_2$ molecule. The tendency of NO$_2$ to photodissociate is shown clearly by the fact that in direct sunlight the half-life of NO$_2$ is much shorter than that of any other common molecular atmospheric species. The photodissociation of nitrogen dioxide can give rise to the following significant inorganic reactions in addition to a host of atmospheric reactions involving organic species:

$$O + O_2 + M(\text{third body}) \rightarrow O_3 + M \tag{11.6.14}$$

$$NO + O_3 \rightarrow NO_2 + O_2 \tag{11.6.15}$$

$$NO_2 + O_3 \rightarrow NO_3 + O_2 \tag{11.6.16}$$

$$O + NO_2 \rightarrow NO + O_2 \tag{11.6.17}$$

$$O + NO_2 + M \rightarrow NO_3 + M \tag{11.6.18}$$

$$NO_2 + NO_3 \rightarrow N_2O_5 \tag{11.6.19}$$

$$NO + NO_3 \rightarrow 2NO_2 \tag{11.6.20}$$

$$O + NO + M \rightarrow NO_2 + M \tag{11.6.21}$$

Nitrogen dioxide ultimately is removed from the atmosphere as nitric acid, nitrates, or (in atmospheres where photochemical smog is formed) as organic nitrogen. Dinitrogen pentoxide formed in Reaction 11.6.19 is the anhydride of nitric acid, which it forms by reacting with water:

$$N_2O_5 + H_2O \rightarrow 2HNO_3 \tag{11.6.22}$$

In the stratosphere, nitrogen dioxide reacts with hydroxyl radicals to produce nitric acid:

$$HO\cdot + NO_2 \rightarrow HNO_3 \tag{11.6.23}$$

In this region, the nitric acid can also be destroyed by hydroxyl radicals,

$$HO\cdot + HNO_3 \rightarrow H_2O + NO_3 \tag{11.6.24}$$

or by a photochemical reaction,

$$HNO_3 + h\nu \rightarrow HO\cdot + NO_2 \tag{11.6.25}$$

so that HNO_3 serves as a temporary sink for NO_2 in the stratosphere. Nitric acid produced from NO_2 is removed as precipitation, or reacts with bases (ammonia, particulate lime) to produce particulate nitrates.

Harmful Effects of Nitrogen Oxides

Nitric oxide, NO, is less toxic than NO_2. Like carbon monoxide and nitrite, NO attaches to hemoglobin and reduces oxygen transport efficiency. However, in a polluted atmosphere, the concentration of nitric oxide normally is much lower than that of carbon monoxide so that the effect on hemoglobin is much less.

Acute exposure to NO_2 can be quite harmful to human health. For exposures ranging from several minutes to one hour, a level of 50-100 ppm of NO_2 causes inflammation of lung tissue for a period of 6-8 weeks, after which time the subject normally recovers. Exposure of the subject to 150-200 ppm of NO_2 causes *bronchiolitis fibrosa obliterans*, a condition fatal within 3-5 weeks after exposure. Death generally results within 2-10 days after exposure to 500 ppm or more of NO_2. "Silo-filler's disease," caused by NO_2 generated by the fermentation of ensilage containing nitrate, is a particularly striking example of nitrogen dioxide poisoning. Deaths have resulted from the inhalation of NO_2-containing gases from burning celluloid and nitrocellulose film, and from spillage of NO_2 oxidant (used with liquid hydrazine fuel) from missile rocket motors.

Although extensive damage to plants is observed in areas receiving heavy exposure to NO_2, most of this damage probably comes from secondary products of nitrogen oxides, such as PAN formed in smog (see Chapter 14). Exposure of plants to several parts per million of NO_2 in the laboratory causes leaf spotting and break-down of plant tissue. Exposure to 10 ppm of NO causes a reversible decrease in the rate of photosynthesis. The effect on plants of long-term exposure to a few tenths of a part per million of NO_2 is less certain.

Nitrogen oxides are known to cause fading of dyes and inks used in some textiles. This has been observed in gas clothes dryers and is due to NO_x formed in the dryer flame. Much of the damage to materials caused by NO_x comes from secondary nitrates and nitric acid. For example, stress-corrosion cracking of springs used in telephone relays occurs far below the yield strength of the nickel-brass spring metal because of the action of particulate nitrates and aerosol nitric acid formed from NO_x.

Concern has been expressed about the possibility that NO_x emitted to the atmosphere by supersonic transport planes could catalyze the partial destruction of the stratospheric ozone layer that absorbs damaging short-wavelength (240-300 nm) ultraviolet radiation. Detailed consideration of this effect is quite complicated, and only the main features are considered here.

In the upper stratosphere and in the mesosphere, molecular oxygen is photo-dissociated by ultraviolet light of less than 242-nm wavelength:

$$O_2 + h\nu \rightarrow O + O \tag{11.6.26}$$

In the presence of energy-absorbing third bodies, the atomic oxygen reacts with molecular oxygen to produce ozone:

$$O_2 + O + M \rightarrow O_3 + M \tag{11.6.27}$$

Ozone can be destroyed by reaction with atomic oxygen,

$$O_3 + O \rightarrow O_2 + O_2 \tag{11.6.28}$$

and its formation can be prevented by recombination of oxygen atoms:

$$O + O + M \rightarrow O_2 + M \tag{11.6.29}$$

Addition of the reaction of nitric oxide with ozone,

$$NO + O_3 \;\rightarrow\; NO_2 + O_2 \tag{11.6.30}$$

to the reaction of nitrogen dioxide with atomic oxygen,

$$NO_2 + O \;\rightarrow\; NO + O_2 \tag{11.6.31}$$

results in a net reaction for the destruction of ozone:

$$O + O_3 \;\rightarrow\; O_2 + O_2 \tag{11.6.32}$$

Along with NOx, water vapor is also emitted into the atmosphere by aircraft exhausts, which could accelerate ozone depletion by the following two reactions:

$$O + H_2O \;\rightarrow\; HO^\bullet + HO^\bullet \tag{11.6.33}$$

$$HO^\bullet + O_3 \;\rightarrow\; HOO^\bullet + O_2 \tag{11.6.34}$$

However, there are many natural stratospheric buffering reactions which tend to mitigate the potential ozone destruction from those reactions outlined above. Atomic oxygen capable of regenerating ozone is produced by the photochemical reaction,

$$NO_2 + h\nu \;\rightarrow\; NO + O \quad (\lambda < 420 \text{ nm}) \tag{11.6.35}$$

A competing reaction removing catalytic NO is

$$NO + HOO^\bullet \;\rightarrow\; NO_2 + HO^\bullet \tag{11.6.36}$$

Current belief is that supersonic aircraft emissions will not cause nearly as much damage to the ozone layer as chlorofluorocarbons.

Control of Nitrogen Oxides

The level of NO_x emitted from stationary sources such as power plant furnaces generally falls within the range of 50-1000 ppm. NO production is favored both kinetically and thermodynamically by high temperatures and by high excess oxygen concentrations. These factors must be considered in reducing NO emissions from stationary sources. Reduction of flame temperature to prevent NO formation is accomplished by adding recirculated exhaust gas, cool air, or inert gases. Unfortunately, this decreases the efficiency of energy conversion as calculated by the Carnot equation (see Chapter 18).

Low-excess-air firing is effective in reducing NO_x emissions during the combustion of fossil fuels. As the term implies, low-excess-air firing uses the minimum amount of excess air required for oxidation of the fuel, so that less oxygen is available for the reaction

$$N_2 + O_2 \;\rightarrow\; 2NO \tag{11.6.37}$$

in the high temperature region of the flame. Incomplete fuel burnout with the emission of hydrocarbons, soot, and CO is an obvious problem with low-excess-air firing. This may be overcome by a two-stage combustion process consisting of the following steps:

1. A first stage in which the fuel is fired at a relatively high temperature with a substoichiometric amount of air, for example, 90-95% of the stoichiometric requirement. NO formation is limited by the absence of excess oxygen.

2. A second stage in which fuel burnout is completed at a relatively low temperature in excess air. The low temperature prevents formation of NO.

In some power plants fired with gas, the emission of NO has been reduced by as much as 90% by a two-stage combustion process.

Removal of NO_x from stack gas presents some formidable problems. These problems arise largely from the low water solubility of NO, the predominant nitrogen oxide species in stack gas. Possible approaches to NO_x removal are catalytic decomposition of nitrogen oxides, catalytic reduction of nitrogen oxides, and sorption of NO_x by liquids or solids. Uptake of NO_x is facilitated by oxidation of NO to more water-soluble species including NO_2, N_2O_3, N_2O_4, HNO_2, and HNO_3.

A typical catalytic reduction of NO in stack gas involves methane:

$$CH_4 + 4NO \rightarrow 2N_2 + CO_2 + 2H_2O \qquad (11.6.38)$$

Production of undesirable by-products is a major concern in these processes. For example, sulfur dioxide reacts with carbon monoxide used to reduce NO to produce toxic carbonyl sulfide, COS:

$$SO_2 + 3CO \rightarrow 2CO_2 + COS \qquad (11.6.39)$$

Most sorption processes have been aimed at the simultaneous removal of both nitrogen oxides and sulfur oxides. Sulfuric acid solutions or alkaline scrubbing solutions containing $Ca(OH)_2$ or $Mg(OH)_2$ may be used. The species N_2O_3 produced by the reaction

$$NO_2 + NO \rightarrow N_2O_3 \qquad (11.6.40)$$

is most efficiently absorbed. Therefore, since NO is the primary combustion product, the introduction of NO_2 into the flue gas is required to produce the N_2O_3, which is absorbed efficiently.

11.7. ACID RAIN

As discussed in this chapter, much of the sulfur and nitrogen oxides entering the atmosphere are converted to sulfuric and nitric acids, respectively. When combined with hydrochloric acid arising from hydrogen chloride emissions, these acids cause acidic precipitation (acid rain) that is now a major pollution problem in some areas.

Headwater streams and high-altitude lakes are especially susceptible to the effects of acid rain and may sustain loss of fish and other aquatic life. Other effects include reductions in forest and crop productivity; leaching of nutrient cations and heavy metals from soils, rocks, and the sediments of lakes and streams; dissolution of metals such as lead and copper from water distribution pipes; corrosion of exposed metal; and dissolution of the surfaces of limestone buildings and monuments.

As a result of its widespread distribution and effects, acid rain is an air pollutant that may pose a threat to the global atmosphere. Therefore, it is discussed in greater detail in Chapter 14.

11.8 AMMONIA IN THE ATMOSPHERE

Ammonia is present even in unpolluted air as a result of natural biochemical and chemical processes. Among the various sources of atmospheric ammonia are microorganisms, decay of animal wastes, sewage treatment, coke manufacture, ammonia manufacture, and leakage from ammonia-based refrigeration systems. High concentrations of ammonia gas in the atmosphere are generally indicative of accidental release of the gas.

Ammonia is removed from the atmosphere by its affinity for water and by its action as a base. It is a key species in the formation and neutralization of nitrate and sulfate aerosols in polluted atmospheres. Ammonia reacts with these acidic aerosols to form ammonium salts:

$$NH_3 + HNO_3 \rightarrow NH_4NO_3 \tag{11.8.1}$$

$$NH_3 + H_2SO_4 \rightarrow NH_4HSO_4 \tag{11.8.2}$$

Ammonium salts are among the more corrosive salts in atmospheric aerosols.

11.9. FLUORINE, CHLORINE, AND THEIR GASEOUS COMPOUNDS

Fluorine, hydrogen fluoride, and other volatile fluorides are produced in the manufacture of aluminum, and hydrogen fluoride is a by-product in the conversion of fluorapatite (rock phosphate) to phosphoric acid, superphosphate fertilizers, and other phosphorus products. The wet process for the production of phosphoric acid involves the reaction of fluorapatite, $Ca_5F(PO_4)_3$, with sulfuric acid:

$$Ca_5F(PO_4)_3 + 5H_2SO_4 + 10H_2O \rightarrow 5CaSO_4 \cdot 2H_2O + HF + 3H_3PO_4 \tag{11.9.1}$$

It is necessary to recover most of the by-product fluorine from rock phosphate processing to avoid severe pollution problems. Recovery as fluorosilicic acid, H_2SiF_6, is normally practiced.

Hydrogen fluoride gas is a dangerous substance that is so corrosive it even reacts with glass. It is irritating to body tissues, and the respiratory tract is very sensitive to it. Brief exposure to HF vapors at the part-per-thousand level may be fatal. The acute toxicity of F_2 is even higher than that of HF. Chronic exposure to high levels of fluorides causes fluorosis, the symptoms of which include mottled teeth and pathological bone conditions.

Plants are particularly susceptible to the effects of gaseous fluorides. Fluorides from the atmosphere appear to enter the leaf tissue through the stomata. Fluoride is a cumulative poison in plants, and exposure of sensitive plants to even very low levels of fluorides for prolonged periods results in damage. Characteristic symptoms of fluoride poisoning are chlorosis (fading of green color due to conditions other than the absence of light), edge burn, and tip burn. Conifers (such as pine trees) afflicted with fluoride poisoning may have reddish-brown, necrotic needle tips. The sensitivity of some conifers to fluoride poisoning is illustrated by the fact that fluorine produced by aluminum plants in Norway has destroyed forests of *Pinus sylvestris* up to 8 miles distant; trees were damaged at distances as great as 20 miles from the plant.

Silicon tetrafluoride gas, SiF_4, is another gaseous fluoride pollutant produced during some steel and metal smelting operations that employ CaF_2, fluorspar. Fluorspar reacts with silicon dioxide (sand), releasing SiF_4 gas:

$$2CaF_2 + 3SiO_2 \rightarrow 2CaSiO_3 + SiF_4 \tag{11.9.2}$$

Another gaseous fluorine compound, sulfur hexafluoride, SF_6, occurs in the atmosphere at levels of about 0.3 parts per trillion. It is extremely unreactive with an atmospheric lifetime estimated at 3200 years,[6] and is used as an atmospheric tracer. It does not absorb ultraviolet light in either the troposphere or stratosphere, and is probably destroyed above 60 km by reactions beginning with its capture of free electrons. Current atmospheric levels of SF_6 are significantly higher than the estimated background level of 0.04 ppt in 1953 when commercial production of it began. The compound is very useful in specialized applications including gas-insulated electrical equipment and inert blanketing/degassing of molten aluminum and magnesium. Increasing uses of sulfur hexafluoride have caused concern because it is the most powerful greenhouse gas known, with a global warming potential (per molecule added to the atmosphere) about 23,900 times that of carbon dioxide.

Chlorine and Hydrogen Chloride

Chlorine gas, Cl_2, does not occur as an air pollutant on a large scale but can be quite damaging on a local scale. Chlorine was the first poisonous gas deployed in World War I. It is widely used as a manufacturing chemical in the plastics industry, for example, as well as for water treatment and as a bleach. Therefore, possibilities for its release exist in a number of locations. Chlorine is quite toxic and is a mucous-membrane irritant. It is very reactive and a powerful oxidizing agent. Chlorine dissolves in atmospheric water droplets, yielding hydrochloric acid and hypochlorous acid, an oxidizing agent:

$$H_2O + Cl_2 \rightarrow H^+ + Cl^- + HOCl \tag{11.9.3}$$

Spills of chlorine gas have caused fatalities among exposed persons. For example, the rupture of a derailed chlorine tank car at Youngstown, Florida, on February 25, 1978, resulted in the deaths of 8 people who inhaled the deadly gas, and a total of 89 people were injured.

Hydrogen chloride, HCl, is emitted from a number of sources. Incineration of chlorinated plastics, such as polyvinylchloride, releases HCl as a combustion product.

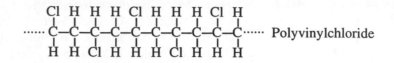

Some compounds released to the atmosphere as air pollutants hydrolyze to form HCl. One such incident occurred on April 26, 1974, when a storage tank containing 750,000 gallons of liquid silicon tetrachloride, $SiCl_4$, began to leak in South Chicago, Illinois. This compound reacted with water in the atmosphere to form a choking fog of hydrochloric acid droplets:

$$SiCl_4 + 2H_2O \rightarrow SiO_2 + 4HCl \tag{11.9.4}$$

Many people became ill from inhaling the vapor.

In February, 1981, in Stroudsburg, Pennsylvania, a wrecked truck dumped 12 tons of powdered aluminum chloride during a rainstorm. This compound produces HCl gas when wet,

$$AlCl_3 + 3H_2O \rightarrow Al(OH)_3 + 3HCl \tag{11.9.5}$$

and more than 1200 residents had to be evacuated from their homes because of the fumes generated.

11.10. HYDROGEN SULFIDE, CARBONYL SULFIDE, AND CARBON DISULFIDE

Hydrogen sulfide is produced by microbial processes including the decay of sulfur compounds and bacterial reduction of sulfate (see Chapter 6). Hydrogen sulfide is also released from geothermal steam, from wood pulping, and from a number of miscellaneous natural and anthropogenic sources. Because it is so readily oxidized, most atmospheric hydrogen sulfide is rapidly converted to SO_2. The organic homologs of hydrogen sulfide, the mercaptans, enter the atmosphere from decaying organic matter and have particularly objectionable odors.

Hydrogen sulfide pollution from artificial sources is not as much of an overall air pollution problem as sulfur dioxide pollution. However, there have been several acute incidents of hydrogen sulfide emissions resulting in damage to human health and even fatalities. The most notorious such incident occurred in Poza Rica, Mexico, in 1950. Accidental release of hydrogen sulfide from a plant used for the recovery of sulfur from natural gas caused the deaths of 22 people and the hospitalization of more than 300. The symptoms of poisoning included irritation of the respiratory tract and damage to the central nervous system. Unlike sulfur dioxide, which appears to affect older people and those with respiratory weaknesses, there was little evidence of correlation between the observed hydrogen sulfide poisoning and the ages or physical conditions of the victims.

In a tragic incident that occurred in February 1975, hydrogen sulfide gas leaking from an experimental secondary-recovery oil well near Denver City, Texas, killed nine people trying to flee the lethal fumes. A process was being tried in which carbon dioxide, rather than water, was injected under high pressure to recover petroleum. Leakage from the well released deadly hydrogen sulfide present in the oil-bearing formation. Efforts to tap very deep natural gas formations have increased the hazard from hydrogen sulfide. A pocket of H_2S was struck at 15,000 feet while drilling such a well near Athens, Texas, in 1978. Leakage of hydrogen sulfide on May 12, 1978, forced the evacuation of 50 families. As an emergency measure in such cases, the gas may be ignited, although that releases large quantities of pollutant SO_2 to the atmosphere.

Hydrogen sulfide at levels well above ambient concentrations destroys immature plant tissue. This type of plant injury is readily distinguished from that due to other phytotoxins. More sensitive species are killed by continuous exposure to around 3000 ppb H_2S, whereas other species exhibit reduced growth, leaf lesions, and defoliation.

Damage to certain kinds of materials is a very expensive effect of hydrogen sulfide pollution. Paints containing lead pigments, $2PbCO_3 \cdot Pb(OH)_2$ (no longer used), are particularly susceptible to darkening by H_2S. Darkening results from exposure over several hours to as little as 50 ppb H_2S. The lead sulfide originally produced by reaction of the lead pigment with hydrogen sulfide eventually may be converted to white lead sulfate by atmospheric oxygen after removal of the source of H_2S, thus partially reversing the damage.

A black layer of copper sulfide forms on copper metal exposed to H_2S. Eventually, this layer is replaced by a green coating of basic copper sulfate, $CuSO_4 \cdot 3Cu(OH)_2$. The green "patina," as it is called, is very resistant to further corrosion. Such layers of corrosion can seriously impair the function of copper contacts on electrical equipment. Hydrogen sulfide also forms a black sulfide coating on silver.

Carbonyl sulfide, COS, is now recognized as a component of the atmosphere at a tropospheric concentration of approximately 500 parts per trillion by volume, corresponding to a global burden of about 2.4 million tons. It is, therefore, a significant sulfur species in the atmosphere. It is possible that the $HO^\bullet$ radical-initiated oxidation of COS and carbon disulfide (CS_2) would yield 8-12 million tons as S in atmospheric sulfur dioxide per year. Though this is a small yield compared to pollution sources, the $HO^\bullet$-initiated process could account for much of the SO_2 burden in the remote troposphere.

Both COS and CS_2 are oxidized in the atmosphere by reactions initiated by the hydroxyl radical. The initial reactions are

$$HO^\bullet + COS \rightarrow CO_2 + HS^\bullet \tag{11.10.1}$$

$$HO^\bullet + CS_2 \rightarrow COS + HS^\bullet \tag{11.10.2}$$

These reactions with hydroxyl radical initiate oxidation processes that occur through a series of atmospheric chemical reactions. The sulfur-containing products that are initially formed as shown by Reactions 11.10.1 and 11.10.2 undergo further reactions to sulfur dioxide and, eventually, to sulfate species.

LITERATURE CITED

1. Brimblecombe, Peter, *Air Composition and Chemistry*, 2nd ed., Cambridge University Press, Cambridge, U.K., 1996.

2. Moeller, D., "Sulfate Aerosols and their Atmospheric Precursors," *Environmental Science Research Report* **17**, 73-90 (1995).

3. Seinfeld, John H. and Spyros N. Pandis "Chemistry of the Troposphere," Chapter 5 in *Atmospheric Chemistry and Physics*, John Wiley and Sons, New York, 1998, pp. 234–336.

4. Klingspor, J. S., and G. E. Bresowar, "Next Generation Low Cost Wet FGD System," *Proceedings of the 1995 International Joint Power Generation Conference*, **20**, 3-0 (1995).

5. Sell, Nancy J., Jack C. Norman, and John A. Ciriacks, "Flue Gas Desulfurization Scheme To Recover Elemental Sulfur," *Industrial Engineering Chemistry Research*, **34**, 1428-1433 (1995).

6. Maiss, Manfred, and Carl A. M. Brenninkmeijer, "Atmospheric SF_6: Trends, Sources, and Prospects," *Environmental Science and Technology*, **32**, 3077-3086 (1998).

SUPPLEMENTARY REFERENCES

Beim, Howard J., Jennifer Spero, and Louis Theodore, *Rapid Guide to Hazardous Air Pollutants*, John Wiley & Sons, New York, 1997.

Ghodish, Thad, *Air Quality*, 3rd ed., Lewis Publishers/CRC Press, Boca Raton, FL, 1997.

Heumann, William L., Ed., *Industrial Air Pollution Control Systems*, McGraw-Hill, New York, 1997.

Hocking, Martin B., *Handbook of Chemical Technology and Pollution Control*, Academic Press, San Diego, CA, 1998.

Mackenzie, James J., and Mohammed T. El-Ashry, Eds., *Air Pollution's Toll on Forests and Crops*, Yale University Press, New Haven, CT, 1992.

Matson, P., *Biogenic Trace Gases: Measuring Emissions from Soil and Water*, Blackwell, Carlton South, Australia, 1995.

Maynard, Douglas G., Ed., *Sulfur in the Environment*, Marcel Dekker, 270 Madison Ave., New York, 1998.

Mycock, John C., John D. McKenna, and Louis Theodore, Eds., *Handbook of Air Pollution Control Engineering and Technology*, Lewis Publishers/CRC Press, Boca Raton, FL, 1995.

Nriagu, Jerome. O., Ed., *Gaseous Pollutants: Characterization and Cycling*, John Wiley & Sons, New York, 1992.

Rogers, John E., and William B. Whitman, Eds., *Microbial Production and Consumption of Greenhouse Gases: Methane, Nitrogen Oxides, and Halomethanes*, American Society for Microbiology, Washington, D.C., 1991.

Schifftner, Kenneth C. and Howard E. Hesketh, *Wet Scrubbers*, Technomic Publishing Co., Lancaster, PA, 1996.

Stevens, Lem B., William L. Cleland, and E. Roberts Alley, *Air Quality Control Handbook*, McGraw-Hill, New York, 1998.

Takeshita, Mitsuru, and Herminé Soud, *FGD Performance and Experience on Coal-Fired Plants*, Gemini House, London, 1993.

Warner, Cecil F., Wayne T. Davis, and Kenneth Wark, *Air Pollution: Its Origin and Control*, 3rd ed., Addison-Wesley, Reading, MA, 1997.

QUESTIONS AND PROBLEMS

1. Why is it that "highest levels of carbon monoxide tend to occur in congested urban areas at times when the maximum number of people are exposed?"

2. Which unstable, reactive species is responsible for the removal of CO from the atmosphere?

3. Which of the following fluxes in the atmospheric sulfur cycle is smallest: (a) sulfur species washed out in rainfall over land, (b) sulfates entering the atmosphere as "sea salt," (c) sulfur species entering the atmosphere from volcanoes, (d) sulfur species entering the atmosphere from fossil fuels, (e) hydrogen sulfide entering the atmosphere from biological processes in coastal areas and on land?

4. Of the following agents, the one that would not favor conversion of sulfur dioxide to sulfate species in the atmosphere is: (a) ammonia, (b) water, (c) contaminant reducing agents, (d) ions of transition metals such as manganese, (e) sunlight.

5. Of the stack gas scrubber processes discussed in this chapter, which is the least effficient for the removal of SO_2?

6. The air inside a garage was found to contain 10 ppm CO by volume at standard temperature and pressure (STP). What is the concentration of CO in mg/L and in ppm by mass?

7. How many metric tons of 5%–S coal would be needed to yield the H_2SO_4 required to produce a 3.00–cm rainfall of pH 2.00 over a 100 km^2 area?

8. In what major respect is NO_2 a more significant species than SO_2 in terms of participation in atmospheric chemical reactions?

9. Assume that an incorrectly adjusted lawn mower is operated in a garage such that the combustion reaction in the engine is

$$C_8H_{18} + {}^{17}/_2 O_2 \rightarrow 8CO + 9H_2O$$

If the dimensions of the garage are $5 \times 3 \times 3$ meters, how many grams of gasoline must be burned to raise the level of CO in the air to 1000 ppm by volume at STP?

10. A 12.0-L sample of waste air from a smelter process was collected at 25°C and 1.00 atm pressure, and the sulfur dioxide was removed. After SO_2 removal, the volume of the air sample was 11.50 L. What was the percentage by weight of SO_2 in the original sample?

11. What is the oxidant in the Claus reaction?

12. Carbon monoxide is present at a level of 10 ppm by volume in an air sample taken at 15°C and 1.00 atm pressure. At what temperature (at 1.00 atm pressure) would the sample also contain 10 mg/m^3 of CO?

13. How many metric tons of coal containing an average of 2% S are required to produce the SO_2 emitted by fossil fuel combustion shown in Figure 11.1? (Note that the values given in the figure are in terms of elemental sulfur, S.) How many metric tons of SO_2 are emitted?

14. Assume that the wet limestone process requires 1 metric ton of $CaCO_3$ to remove 90% of the sulfur from 4 metric tons of coal containing 2% S. Assume that the sulfur product is $CaSO_4$. Calculate the percentage of the limestone converted to calcium sulfate.

15. Referring to the two preceding problems, calculate the number of metric tons of $CaCO_3$ required each year to remove 90% of the sulfur from 1 billion metric tons of coal (approximate annual U.S. consumption), assuming an average of 2% sulfur in the coal.

16. If a power plant burning 10,000 metric tons of coal per day with 10% excess air emits stack gas containing 100 ppm by volume of NO, what is the daily output of NO? Assume the coal is pure carbon.

17. How many cubic kilometers of air at 25°C and 1 atm pressure would be contaminated to a level of 0.5 ppm NO_x from the power plant discussed in the preceding question?

12 ORGANIC AIR POLLUTANTS

12.1. ORGANIC COMPOUNDS IN THE ATMOSPHERE

Organic pollutants may have a strong effect upon atmospheric quality. The effects of organic pollutants in the atmosphere may be divided into two major categories. The first consists of **direct effects**, such as cancer caused by exposure to vinyl chloride. The second is the formation of **secondary pollutants**, especially photochemical smog (discussed in detail in Chapter 13). In the case of pollutant hydrocarbons in the atmosphere, the latter is the more important effect. In some localized situations, particularly the workplace, direct effects of organic air pollutants may be equally important.

This chapter discusses the nature and distribution of organic compounds in the atmosphere. Chapter 13 deals with photochemical smog and addresses the mechanisms by which organic compounds undergo photochemical reactions in the atmosphere.

Loss of Organic Substances from the Atmosphere

Organic contaminants are lost from the atmosphere by a number of routes. These include dissolution in precipitation (rainwater), dry deposition, photochemical reactions, formation of and incorporation into particulate matter, and uptake by plants. Reactions of organic atmospheric contaminants are particularly important in determining their manner and rates of loss from the atmosphere. Such reactions are discussed in this chapter.

Forest trees present a large surface area to the atmosphere and are particularly important in filtering organic contaminants from air. Forest trees and plants contact the atmosphere through plant cuticle layers, the biopolymer "skin" on the leaves and needles of the plants. The cuticle layer is lipophilic, meaning that it has a particular affinity for organic substances, including those in the atmosphere. A study of the uptake of organic substances (lindane, triadimenol, bitertanol, 2,4-dichlorophenoxyacetic acid, and pentachlorophenol) by conifer needles[1] has shown that the process consists of two phases: (1) adsorption onto the needle surfaces and (2) transport through the cuticle layer into the needle and plant. Uptake increases with increasing lipophilicity of the compounds and with increasing surface area of the leaves. This phenomenon points to the importance of forests in atmospheric purification.

Global Distillation and Fractionation of Persistent Organic Pollutants

On a global scale, it is likely that persistent organic pollutants undergo a cycle of distillation and fractionation in which they are vaporized into the atmosphere in warmer regions of the Earth and condense and are deposited in colder regions.[2] The theory of this phenomenon holds that the distribution of such pollutants is governed by their physicochemical properties and the temperature conditions to which they are exposed. As a result, the least volatile persistent organic pollutants are deposited near their sources, those of relatively high volatility are distilled into polar regions, and those of intermediate volatility are deposited predominantly at mid latitudes. This phenomeonon has some important implications regarding the accumulation of persistent organic pollutants in environmentally fragile polar regions far from industrial sources.

12.2. ORGANIC COMPOUNDS FROM NATURAL SOURCES

Natural sources are the most important contributors of organics in the atmosphere, and hydrocarbons generated and released by human activities consitute only about 1/7 of the total hydrocarbons in the atmosphere. This ratio is primarily the result of the huge quantities of methane produced by anaerobic bacteria in the decomposition of organic matter in water, sediments, and soil:

$$2\{CH_2O\} \text{ (bacterial action)} \rightarrow CO_2(g) + CH_4(g) \tag{12.2.1}$$

Flatulent emissions from domesticated animals, arising from bacterial decomposition of food in their digestive tracts, add about 85 million metric tons of methane to the atmosphere each year. Anaerobic conditions in intensively cultivated rice fields produce large amounts of methane, perhaps as much as 100 million metric tons per year. Methane is a natural constituent of the atmosphere and is present at a level of about 1.4 parts per million (ppm) in the troposphere.

Methane in the troposphere contributes to the photochemical production of carbon monoxide and ozone. The photochemical oxidation of methane is a major source of water vapor in the stratosphere.

Atmospheric hydrocarbons produced by living sources are called **biogenic hydrocarbons**. Vegetation is the most important natural source of non-methane biogenic compounds. A compilation of organic compounds in the atmosphere[3] documented a total of 367 different compounds that are released to the atmosphere from vegetation sources. Other natural sources include microorganisms, forest fires, animal wastes, and volcanoes.

One of the simplest organic compounds given off by plants is ethylene, C_2H_4. This compound is produced by a variety of plants and released to the atmosphere in its role as a messenger species regulating plant growth. Because of its double bond, ethylene is highly reactive with hydroxyl radical, HO•, and with oxidizing species in the atmosphere. Ethylene from vegetation sources should be considered as an active participant in atmospheric chemical processes.

Most of the hydrocarbons emitted by plants are **terpenes**, which constitute a large class of organic compounds found in essential oils. Essential oils are obtained when parts of some types of plants are subjected to steam distillation. Most of the plants that produce terpenes are conifers (evergreen trees and shrubs such as pine and cypress), plants of the genus *Myrtus*, and trees and shrubs of the genus *Citrus*. One of the most common terpenes emitted by trees is α-pinene, a principal component of turpentine. The terpene limonene, found in citrus fruit and pine needles, is encountered in the atmosphere around these sources. Isoprene (2-methyl-1,3-butadiene), a hemiterpene, has been identified in the emissions from cottonwood, eucalyptus, oak, sweetgum, and white spruce trees. Linalool is a terpene with the chemical formula $(CH_3)_2C=CHCH_2CH_2C(CH_3)(OH)CH=CH_2$, that is given off by some plant species common to Italy and Austria, including the pine *Pinus pinea* and orange blossoms.[4] Other terpenes that are known to be given off by trees include β-pinene, myrcene, ocimene, and α-terpinene.

As exemplified by the structural formulas of α-pinene, β-pinene, Δ^3-carene, isoprene, and limonene, shown in Figure 12.1, terpenes contain alkenyl (olefinic) bonds, in some cases two or more per molecule. Because of these and other structural features, terpenes are among the most reactive compounds in the atmosphere. The reaction of terpenes with hydroxyl radical, HO•, is very rapid, and terpenes also react with other oxidizing agents in the atmosphere, particularly ozone, O_3. Turpentine, a mixture of terpenes, has been widely used in paint because it reacts with atmospheric oxygen to form a peroxide, then a hard resin. It is likely that compounds such as α-pinene and isoprene undergo similar reactions in the atmos-

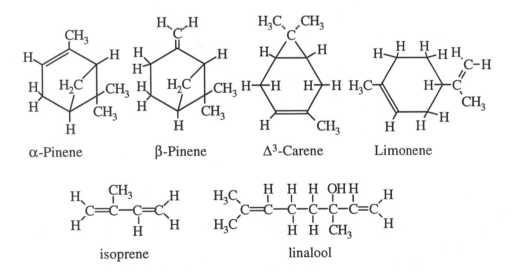

Figure 12.1. Some common terpenes emitted to the atmosphere by vegetation, primarily trees such as pine and citrus trees. These reactive compounds are involved in the formation of much of the small particulate matter encountered in the atmosphere.

phere to form particulate matter. The resulting Aitken nuclei aerosols (see Chapter 10) are a cause of the blue haze in the atmosphere above some heavy growths of vegetation.

Laboratory and smog-chamber experiments have been performed in an effort to determine the fates of atmospheric terpenes. Oxidation initiated by reaction with NO_3 of the four cyclic monoterpenes listed above, α-pinene, β-pinene, Δ^3-carene, and limonene, has given products containing carbonyl (C=O) functionality and organically bound nitrogen as organic nitrate.[5] When a mixture of α-pinene with NO and NO_2 in air is irradiated with ultraviolet light, pinonic acid is formed:

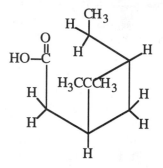

Found in forest aerosol particles, this compound is produced by photochemical processes acting upon α-pinene.

Although terpenes are highly reactive with hydroxyl radical, it is now believed that much of the atmospheric aerosol formed as the result of reactions of unsaturated biogenic hydrocarbons is the result of processes that start with reactions between the them and ozone. Pinonic acid (see above) is produced by the reaction of α-pinene with ozone.[6] Two of the products of the reaction of limonene with ozone are formaldehyde and 4-acetyl-1-methylcyclohexene:

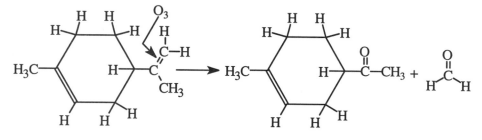

Limonene   4-acetyl-1-methylcyclohexene

Perhaps the greatest variety of compounds emitted by plants consists of **esters**. However, they are released in such small quantities that they have little influence upon atmospheric chemistry. Esters are primarily responsible for the fragrances associated with much vegetation. Some typical esters that are released by plants to the atmosphere are shown below:

$$H_3C-C=C-C-C-C-C-C-O-C-H \quad \textbf{Citronellyl formate}$$

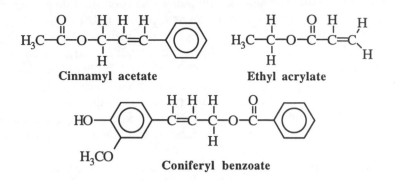

Cinnamyl acetate   Ethyl acrylate

Coniferyl benzoate

12.3. POLLUTANT HYDROCARBONS

Ethylene and terpenes, which were discussed in the preceding section, are **hydrocarbons**, organic compounds containing only hydrogen and carbon. The major classes of hydrocarbons are **alkanes** (formerly called paraffins), such as 2,2,3-trimethylbutane;

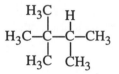

alkenes (olefins, compounds with double bonds between adjacent carbon atoms), such as ethylene; **alkynes** (compounds with triple bonds), such as acetylene;

$$H-C\equiv C-H$$

and **aryl** (aromatic) **compounds**, such as naphthalene:

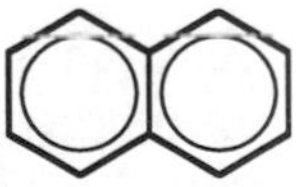

Because of their widespread use in fuels, hydrocarbons predominate among organic atmospheric pollutants. Petroleum products, primarily gasoline, are the source of most of the anthropogenic (originating through human activities) pollutant hydrocarbons found in the atmosphere. Hydrocarbons may enter the atmosphere either directly or as by-products of the partial combustion of other hydrocarbons. The latter are particularly important because they tend to be unsaturated and relatively reactive (see Chapter 13 for a discussion of hydrocarbon reactivity in photochemical smog formation). Most hydrocarbon pollutant sources produce about 15% reactive hydrocarbons, whereas those from incomplete combustion of gasoline are about 45% reactive. The hydrocarbons in uncontrolled automobile exhausts are only about 1/3 alkanes, with the remainder divided approximately equally between more reactive alkenes and aromatic hydrocarbons, thus accounting for the relatively high reactivity of automotive exhaust hydrocarbons.

Investigators who study smog formation in smog chambers have developed synthetic mixtures of hydrocarbons that mimic the smog-forming behavior of hydrocarbons in a polluted atmosphere. The compounds so used provide a simplified idea of the composition of pollutant hydrocarbons likely to lead to smog formation. A typical mixture consists of 0.556 mole fraction of alkanes, including 2-methylbutane, *n*-pentane, 2-methylpentane, 2,4-dimethylpentane, and 2,2,4-trimethylpentane, and 0.444 mole fraction alkenes, including 1-butene, *cis*-2-butene, 2-methyl-1-butene, 2-methyl-2-butene, ethene (ethylene), and propene (propylene).

Alkanes are among the more stable hydrocarbons in the atmosphere. Straight-chain alkanes with 1 to more than 30 carbon atoms, and branched-chain alkanes with 6 or fewer carbon atoms, are commonly present in polluted atmospheres. Because of their high vapor pressures, alkanes with 6 or fewer carbon atoms are normally present as gases, alkanes with 20 or more carbon atoms are present as aerosols or sorbed to atmospheric particles, and alkanes with 6 to 20 carbon atoms per molecule may be present either as vapor or particles, depending upon conditions.

In the atmosphere, alkanes (general formula C_xH_{2x+2}) are attacked primarily by hydroxyl radical, $HO^\bullet$, resulting in the loss of a hydrogen atom and formation of an **alkyl radical,**

$$C_xH_{2x+1}^\bullet$$

Subsequent reaction with O_2 causes formation of **alkylperoxyl radical,**

$$C_xH_{2x+1}O_2^\bullet$$

These radicals may act as oxidants, losing oxygen (usually to NO forming NO_2) to produce **alkoxyl radicals**:

$$C_xH_{2x+1}O^\bullet$$

As a result of these and subsequent reactions, lower-molecular-mass alkanes are eventually oxidized to species that can be precipitated from the atmosphere with particulate matter to ultimately undergo biodegradation in soil.

Alkenes enter the atmosphere from a variety of processes, including emissions from internal combustion engines and turbines, foundry operations, and petroleum refining. Several alkenes, including the ones shown below, are among the top 50 chemicals produced each year, with annual worldwide production of several billion kg:

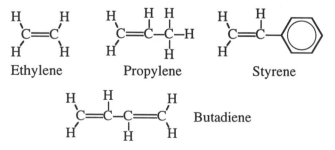

These compounds are used primarily as monomers, which are polymerized to create polymers for plastics (polyethylene, polypropylene, polystyrene), synthetic rubber (styrenebutadiene, polybutadiene), latex paints (styrenebutadiene), and other applications. All of these compounds, as well as others manufactured in lesser quantities, are released into the atmosphere. In addition to the direct release of alkenes, these hydrocarbons are commonly produced by the partial combustion and "cracking" at high temperatures of alkanes, particularly in the internal combustion engine.

Alkynes occur much less commonly in the atmosphere than do alkenes. Detectable levels are sometimes found of acetylene used as a fuel for welding torches, and 1-butyne used in synthetic rubber manufacture:

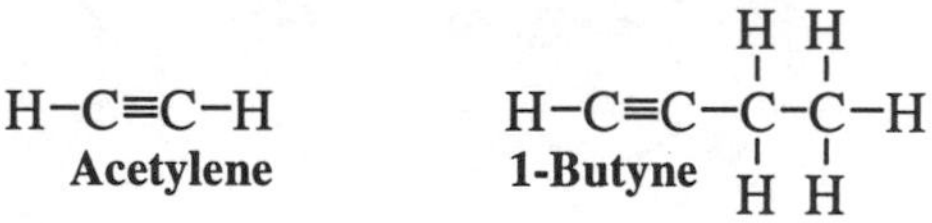

$$H-C\equiv C-H \qquad H-C\equiv C-\underset{\underset{H}{|}}{\overset{\overset{H}{|}}{C}}-\underset{\underset{H}{|}}{\overset{\overset{H}{|}}{C}}-H$$

Acetylene   **1-Butyne**

Unlike alkanes, alkenes are highly reactive in the atmosphere, especially in the presence of NO_x and sunlight. Hydroxyl radical reacts readily with alkenes, adding to the double bond and, to a much lesser extent, by abstracting a hydrogen atom. If hydroxyl radical adds to the double bond in propylene, for example, the product is:

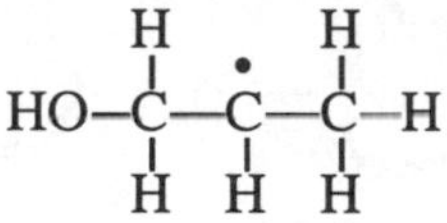

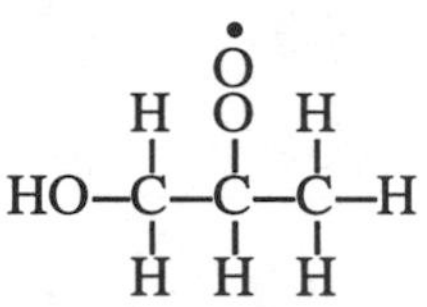

These radicals then participate in reaction chains, such as those discussed for the formation of photochemical smog in Chapter 13.

Ozone, O_3, adds across double bonds and is rather reactive with alkenes. As shown for the natural alkene limonene in Section 12.2, aldehydes are among the products of reactions between alkenes and ozone.

Although the reaction of alkenes with NO_3 is much slower than that with $HO\cdot$, the much higher levels of NO_3 relative to $HO\cdot$, especially at night, make it a significant reactant for atmospheric alkenes. The initial reaction with NO_3 is addition across the alkene double bond which, because NO_3 is a free radical species, forms a radical species. A typical reaction sequence is the following:

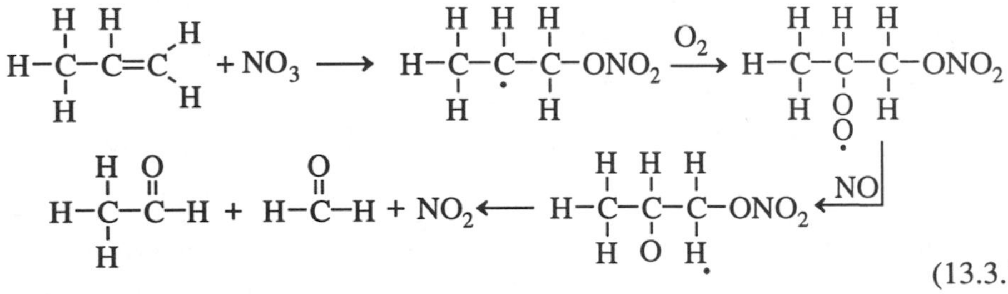

(13.3.1)

12.4. ARYL HYDROCARBONS

Aryl (aromatic) hydrocarbons may be divided into the two major classes of those that have only one benzene ring and those with multiple rings. As discussed in Chapter 10, the latter are *polycyclic aryl hydrocarbons, PAH*. Aryl hydrocarbons with two rings, such as naphthalene, are intermediate in their behavior. Some typical aryl hydrocarbons are:

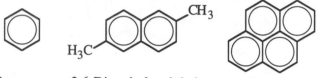

Benzene   2,6-Dimethylnaphthalene   Pyrene

The following aryl hydrocarbons are among the top 50 chemicals manufactured each year:

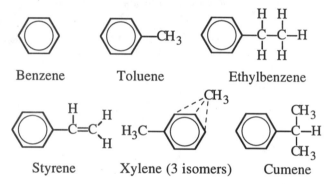

Benzene   Toluene   Ethylbenzene

Styrene   Xylene (3 isomers)   Cumene

Single-ring aryl compounds are important constituents of lead-free gasoline, which has largely replaced leaded gasoline. Aryl solvents are widely used in industry. Aryl hydrocarbons are raw materials for the manufacture of monomers and plasticizers in polymers. Styrene is a monomer used in the manufacture of plastics and synthetic rubber. Cumene is oxidized to produce phenol and acetone, which are valuable by-products. Because of these applications, plus production of these compounds as combustion by-products, aryl compounds are common atmospheric pollutants.

Many hydrocarbons containing a single benzene ring and a number of hydrocarbon derivatives of naphthalene have been found as atmospheric pollutants. In addition, several compounds containing two or more *unconjugated* rings (not sharing the same π electron cloud between rings) have been detected as atmospheric pollutants. One such compound is biphenyl,

 Biphenyl

which occurs in diesel smoke. It should be pointed out that many of these aryl hydrocarbons have been detected primarily as ingredients of tobacco smoke and are, therefore, of much greater significance in an indoor environment than in an outdoor one.

As discussed in Section 10.8, polycyclic aryl hydrocarbons are present as aerosols in the atmosphere because of their extremely low vapor pressures. These compounds are the most stable form of hydrocarbons having low hydrogen-to-carbon ratios and are formed by the combustion of hydrocarbons under oxygen-deficient conditions. The partial combustion of coal, which has a hydrogen-to-carbon ratio less than 1, is a major source of PAH compounds.

Reactions of Atmospheric Aryl Hydrocarbons

As with most atmospheric hydrocarbons, the most likely reaction of benzene and its derivatives is with hydroxyl radical. Addition of HO• to the benzene ring results in the formation of an unstable radical species,

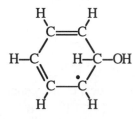

where the dot denotes an unpaired electron in the radical. The electron is not confined to one atom; therefore, it is **delocalized** and may be represented in the aryl radical structure by a half-circle with a dot in the middle. Using this notation for the radical above, its reaction with O_2 is,

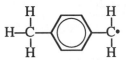

$$+ O_2 \longrightarrow \qquad + HO_2• \qquad\qquad (12.4.1)$$

to form stable phenol and reactive hydroperoxyl radical, HOO•. Alkyl-substituted aryls may undergo reactions involving the alkyl group. For example, abstraction of alkyl H by HO• from a compound such as *p*-xylene can result in the formation of a radical,

$$\text{H}-\overset{\overset{\text{H}}{|}}{\underset{\underset{\text{H}}{|}}{\text{C}}}-\bigcirc-\overset{\overset{\text{H}}{|}}{\underset{\underset{\text{H}}{|}}{\text{C}}}•$$

which can react further with O_2 to form a peroxyl radical, then enter chain reactions involved in the formation of photochemical smog (Chapter 13).

Although reaction with hydroxyl radical is the most common fate of aryl compounds during daylight, they react with NO_3 at night. This oxide of nitrogen is formed by the reaction of ozone with NO_2,

$$NO_2 + O_3 \rightarrow NO_3 + O_2 \qquad\qquad (12.4.2)$$

and can remain in the atmosphere for some time as its addition product with NO_2:

$$NO_2 + NO_3 + M \rightarrow N_2O_5 + M \qquad\qquad (12.4.2)$$

The importance of aryl compounds in the formation of secondary air pollutants was emphasized in a study of the aerosol formation potential of gasoline vapor.[7] This study showed that the potential of gasoline vapor to form atmospheric aerosols could be attributed entirely to the aromatic hydrocarbon fraction of the gasoline.

12.5. ALDEHYDES AND KETONES

Carbonyl compounds, consisting of aldehydes and ketones that have a carbonyl moiety, C=O, are often the first species formed, other than unstable reaction intermediates, in the photochemical oxidation of atmospheric hydrocarbons. The general formulas of aldehydes and ketones are represented by the following, where R and R' represent the hydrocarbon *moieties* (portions), such as the –CH$_3$ group.

$$
\begin{matrix} \underset{\textbf{Aldehyde}}{R\overset{\displaystyle \overset{O}{\|}}{-}C-H} & \underset{\textbf{Ketone}}{R\overset{\displaystyle \overset{O}{\|}}{-}C-R'} & \underset{\textbf{Carbonyl moiety}}{-\overset{\displaystyle \overset{O}{\|}}{C}-} \end{matrix}
$$

Carbonyl compounds are byproducts of the generation of hydroperoxyl radicals from organic alkoxyl radicals (see Section 12.3) by reactions such as the following:

$$\text{H–C–C–C–H} + O_2 \;\rightarrow\; \text{H–C–C–C–H} + HO_2\cdot \tag{12.5.1}$$

$$\text{H–C–C–C–H} + O_2 \;\rightarrow\; \text{H–C–C–C–H} + HO_2\cdot \tag{12.5.2}$$

The simplest and most widely produced of the carbonyl compounds is the lowest aldehyde, **formaldehyde**:

$$\underset{H}{\overset{\displaystyle \overset{O}{\|}}{C}}{\diagdown}H \qquad \text{Formaldehyde}$$

Formaldehyde is produced in the atmosphere as a product of the reaction of atmospheric hydrocarbons beginning with their reactions with hydroxyl radical. For example, formaldehyde is the product of the reaction of methoxyl radical with O$_2$:

$$H_3CO\cdot + O_2 \;\rightarrow\; H\overset{\displaystyle \overset{O}{\|}}{-}C-H + HOO\cdot \tag{12.5.3}$$

With annual global industrial production exceeding 1 billion kg, formaldehyde is used in the manufacture of plastics, resins, lacquers, dyes, and explosives. It is uniquely important because of its widespread distribution and toxicity. Humans may be exposed to formaldehyde in the manufacture and use of phenol, urea, and melamine resin plastics, and from formaldehyde-containing adhesives in pressed wood products such as particle board, used in especially large quantities in mobile home construction. However, significantly improved manufacturing processes have greatly reduced formaldehyde emissions from these synthetic building materials. Formaldehyde occurs in the atmosphere primarily in the gas phase.

The structures of some important aldehydes and ketones are shown below:

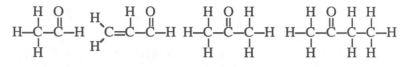

Acetaldehyde   Acrolein   Acetone   Methylethyl ketone

Acetaldehyde is a widely produced organic chemical used in the manufacture of acetic acid, plastics, and raw materials. Approximately a billion kg of acetone are produced each year as a solvent and for applications in the rubber, leather, and plastics industries. Methylethyl ketone is employed as a low-boiling solvent for coatings and adhesives, and for the synthesis of other chemicals.

In addition to their production from hydrocarbons by photochemical oxidation, carbonyl compounds enter the atmosphere from a large number of sources and processes. These include direct emissions from internal combustion engine exhausts, incinerator emissions, spray painting, polymer manufacture, printing, petrochemicals manufacture, and lacquer manufacture. Formaldehyde and acetaldehyde are produced by microorganisms, and acetaldehyde is emitted by some kinds of vegetation.

Aldehydes are second only to NO_2 as atmospheric sources of free radicals produced by the absorption of light. This is because the carbonyl group is a **chromophore**, a molecular group that readily absorbs light. It absorbs well in the near-ultraviolet region of the spectrum. The activated compound produced when a photon is absorbed by an aldehyde dissociates into a formyl radical,

$$\overset{\bullet}{H}CO$$

and an alkyl radical. The photodissociation of acetaldehyde illustrates this two-step process:

$$H-\underset{\underset{H}{|}}{\overset{\overset{H}{|}}{C}}-\overset{O}{\overset{\|}{C}}-H \; + \; h\nu \longrightarrow H-\underset{\underset{H}{|}}{\overset{\overset{H}{|}}{C}}-\overset{O}{\overset{\|}{C}}{}^{*}-H \longrightarrow H-\underset{\underset{H}{|}}{\overset{\overset{H}{|}}{C}}{}^{\bullet} \; + \; {}^{\bullet}\overset{O}{\overset{\|}{C}}-H \qquad (12.5.4)$$
(Photoexcited)

Photolytically excited formaldehyde, CH_2O^*, may dissociate in two ways. The first of these produces an H atom and HCO radical; the second produces chemically stable H_2 and CO.

As a result of their reaction with HO· followed by O_2 then NO_2, aldehydes are precursors to the production of strongly oxidizing peroxyacyl nitrates (PANs) such as peroxyacetyl nitrate. This process is discussed in Chapter 13, Section 13.5.

$$H-\underset{\underset{H}{|}}{\overset{\overset{H}{|}}{C}}-\overset{O}{\overset{\|}{C}}-OO-NO_2 \quad \text{Peroxyacetyl nitrate}$$

Because of the presence of both double bonds and carbonyl groups, olefinic aldehydes are especially reactive in the atmosphere. The most common of these found in the atmosphere is acrolein,

a powerful lachrymator (tear producer) used as an industrial chemical and produced as a combustion byproduct.

The most abundant atmospheric ketone is acetone, $CH_3C(O)CH_3$. About half of the acetone in the atmosphere is generated as a product of the atmospheric oxidation of propane, isobutane, isobutene, and other hydrocarbons. Most of the remainder comes about equally from direct biogenic emissions and biomass burning, with about 3% from direct anthropogenic emissions.

Acetone photolyzes in the atmosphere,

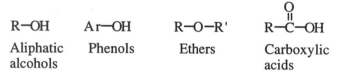

$$(12.5.5)$$

to produce the PAN precursor acetyl radical. It is believed that the mechanism for the removal of the higher ketones from the atmosphere involves an initial reaction with HO• radical.

12.6. MISCELLANEOUS OXYGEN-CONTAINING COMPOUNDS

Oxygen-containing aldehydes, ketones, and esters in the atmosphere were covered in preceding sections. This section discusses the oxygen-containing organic compounds consisting of **aliphatic alcohols, phenols, ethers**, and **carboxylic acids**. These compounds have the general formulas given below, where R and R' represent hydrocarbon moieties, and Ar stands specifically for an aryl moiety, such as the phenyl group (benzene less an H atom):

$$R\text{—}OH \qquad Ar\text{—}OH \qquad R\text{—}O\text{—}R' \qquad \overset{\displaystyle O}{\overset{\displaystyle \|}{R\text{—}C}}\text{—}OH$$

Aliphatic   Phenols   Ethers   Carboxylic
alcohols   acids

These classes of compounds include many important organic chemicals.

Alcohols

Of the alcohols, methanol, ethanol, isopropanol, and ethylene glycol rank among the top 50 chemicals with annual worldwide production of the order of a billion kg or more. The most common of the many uses of these chemicals is for the manufacture of other chemicals. Methanol is widely used in the manufacture of formaldehyde (see Section 12.5) as a solvent, and mixed with water as an antifreeze formulation. Ethanol is used as a solvent and as the starting material for the manufacture of acetaldehyde, acetic acid, ethyl ether, ethyl chloride, ethyl bromide, and several important esters. Both methanol and ethanol can be used as motor vehicle fuels, usually in mixtures with gasoline. Ethylene glycol is a common antifreeze compound.

Numerous aliphatic alcohols have been reported in the atmosphere. Because of their volatility, the lower alcohols, especially methanol and ethanol, predominate as atmospheric pollutants. Among the other alcohols released to the atmosphere are 1-propanol, 2-propanol, propylene glycol, 1-butanol, and even octadecanol, chemical formula $CH_3(CH_2)_{16}CH_2OH$, which is evolved by plants. Alcohols can undergo photochemical reactions, beginning with abstraction of hydrogen by hydroxyl radical. Mechanisms for scavenging alcohols from the atmosphere are relatively efficient because the lower alcohols are quite water soluble and the higher ones have low vapor pressures.

Some alkenyl alcohols have been found in the atmosphere, largely as by-products of combustion. Typical of these is 2-buten-1-ol,

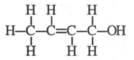

which has been detected in automobile exhausts. Some alkenyl alcohols are emitted by plants. One of these, *cis*-3-hexen-1-ol, $CH_3CH_2CH=CHCH_2CH_2OH$, is emitted from grass, trees and crop plants to the extent that it is known as "leaf alcohol." In addition to reacting with HO· radical, alkenyl radicals react strongly with atmospheric ozone, which adds across the double bond.

Phenols

Phenols are aryl alcohols that have an –OH group bonded to an aryl ring. They are more noted as water pollutants than as air pollutants. Some typical phenols that have been reported as atmospheric contaminants are the following:

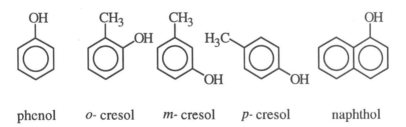

phenol   *o*- cresol   *m*- cresol   *p*- cresol   naphthol

The simplest of these compounds, phenol, is among the top 50 chemicals produced annually. It is most commonly used in the manufacture of resins and polymers, such as Bakelite, a phenol-formaldehyde copolymer. Phenols are produced by the pyrolysis of coal and are major by-products of coking. Thus, in local situations involving coal coking and similar operations, phenols can be troublesome air pollutants.

Ethers

Ethers are relatively uncommon atmospheric pollutants; however, the flammability hazard of diethyl ether vapor in an enclosed work space is well known. In addition to aliphatic ethers, such as dimethyl ether and diethyl ether, several alkenyl ethers, including vinylethyl ether, are produced by internal combustion engines. A cyclic ether and important industrial solvent, tetrahydrofuran, occurs as an air contaminant. Methyltertiarybutyl ether, MTBE, became the octane booster of choice to replace tetraethyllead in gasoline. Because of its widespread distribution, MTBE has the potential to be an air pollutant, although its hazard is limited by its low vapor pressure. Largely because of its potential to contaminate water, MTBE was proposed for phaseout by both the state of California and the U.S. Environmental Protection Agency in 1999. Another possible air contaminant because of its potential uses as an octane booster is diisopropyl ether (DIPE). The structural formulas of the ethers mentioned above are given below:

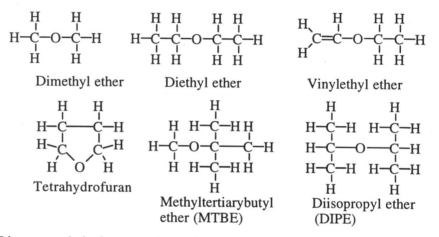

Dimethyl ether   Diethyl ether   Vinylethyl ether

Tetrahydrofuran   Methyltertiarybutyl   Diisopropyl ether
                    ether (MTBE)   (DIPE)

Ethers are relatively unreactive and not as water-soluble as the lower alcohols or carboxylic acids. The predominant process for their atmospheric removal begins with hydroxyl radical attack.

Oxides

Ethylene oxide and propylene oxide,

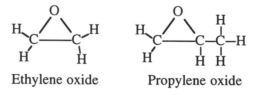

Ethylene oxide   Propylene oxide

rank among the 50 most widely produced industrial chemicals and have a limited potential to enter the atmosphere as pollutants. Ethylene oxide is a moderately to highly toxic, sweet-smelling, colorless, flammable, explosive gas used as a chemical intermediate, sterilant, and fumigant. It is a mutagen and a carcinogen to experimental animals. It is classified as hazardous for both its toxicity and ignitability.

Carboxylic Acids

Carboxylic acids have one or more of the functional groups,

$$\begin{matrix} & O \\ & \| \\ -C & -OH \end{matrix}$$

attached to an alkane, alkene, or aryl hydrocarbon moiety. A carboxylic acid, pinonic acid, produced by the photochemical oxidation of naturally-produced α-pinene, was discussed in Section 12.2. Many of the carboxylic acids found in the atmosphere probably result from the photochemical oxidation of other organic compounds through gas-phase reactions or by reactions of other organic compounds dissolved in aqueous aerosols. These acids are often the end products of photochemical oxidation because their low vapor pressures and relatively high water solubilities make them susceptible to scavenging from the atmosphere. The sources of relatively abundant atmospheric formic acid, HCOOH, and acetic acid, H_3CCOOH, are not known with certainty.

12.7. ORGANOHALIDE COMPOUNDS

Organohalides consist of halogen-substituted hydrocarbon molecules, each of which contains at least one atom of F, Cl, Br, or I. They may be saturated (**alkyl halides**), unsaturated (**alkenyl halides**), or aryl (**aryl halides**). The organohalides of environmental and toxicological concern exhibit a wide range of physical and chemical properties. Although most organohalide compounds of pollution concern are from anthropogenic sources, it is now known that a large variety of such compounds are generated by organisms, particularly those in marine environments.[8]

Structural formulas of several alkyl halides commonly encountered in the atmosphere are given below:

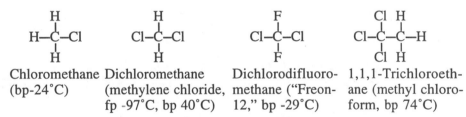

Chloromethane (bp-24°C)   Dichloromethane (methylene chloride, fp -97°C, bp 40°C)   Dichlorodifluoro- methane ("Freon- 12," bp -29°C)   1,1,1-Trichloroeth- ane (methyl chloro- form, bp 74°C)

Volatile **chloromethane** (methyl chloride) is consumed in the manufacture of silicones. **Dichloromethane** is a volatile liquid with excellent solvent properties for nonpolar organic solutes. It has been used as a solvent for the decaffeination of coffee, in paint strippers, as a blowing agent in urethane polymer manufacture, and to depress vapor pressure in aerosol formulations. **Dichlorodifluoromethane** is one of the chlorofluorocarbon compounds once widely manufactured as a refrigerant and involved in stratospheric ozone depletion. One of the more common industrial chlorinated solvents is **1,1,1-trichloroethane**.

Viewed as halogen-substituted derivatives of alkenes, the **alkenyl** or **olefinic organohalides** contain at least one halogen atom and at least one carbon-carbon double bond. The most significant of these are the lighter chlorinated compounds.

Vinyl chloride is consumed in large quantities as a raw material to manufacture pipe, hose, wrapping, and other products fabricated from polyvinyl chloride plastic. This highly flammable, volatile, sweet-smelling gas is known to cause angiosarcoma, a rare form of liver cancer. **Trichloroethylene** is a clear, colorless, nonflammable, volatile liquid. It is an excellent degreasing and dry-cleaning solvent, and has been used as a household solvent and for food extraction (for example, in decaffeination of coffee). **Allyl chloride** is an intermediate in the manufacture of allyl alcohol and other allyl compounds, including pharmaceuticals, insecticides, and thermosetting varnish and plastic resins.

Some commonly used aryl halide derivatives of benzene and toluene are shown below:

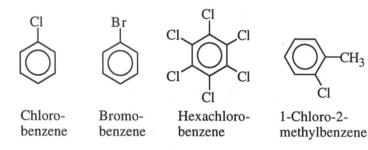

Chloro-   Bromo-   Hexachloro-   1-Chloro-2-
benzene   benzene   benzene   methylbenzene

Aryl halide compounds have many uses. The inevitable result of all these uses has been widespread occurrences of human exposure and environmental contamination. Polychlorinated biphenyls, PCBs, a group of compounds formed by the chlorination of biphenyl,

$$\text{(biphenyl)} + x\text{Cl}_2 \rightarrow \text{(chlorinated biphenyl)}_{\text{Cl}_x} + x\text{HCl} \tag{12.7.1}$$

have extremely high physical and chemical stabilities and other qualities that have led to their being used in many applications, including heat transfer fluids, hydraulic fluids, and dielectrics.

As expected from their high vapor pressures and volatilities, the lighter organohalide compounds are the most likely to be found at detectable levels in the atmosphere. On a global basis, the three most abundant organochlorine compounds in the atmosphere are methyl chloride, methyl chloroform, and carbon tetrachloride, which have tropospheric concentrations ranging from a tenth to several tenths of a part per billion. Methyl chloroform is relatively persistent in the atmosphere, with residence times of several years. Therefore, it may pose a threat to the stratospheric ozone layer in the same way as chlorofluorocarbons. Also found are methylene chloride; methyl bromide, CH_3Br; bromoform, $CHBr_3$; assorted chlorofluorocarbons; and halogen-substituted ethylene compounds such as trichloroethylene, vinyl chloride, perchloroethylene, ($CCl_2=CCl_2$), and solvent ethylene dibromide ($CHBr=CHBr$).

Chlorofluorocarbons

Chlorofluorocarbons (CFCs), such as dichlorodifluoromethane, commonly called Freons, are volatile 1- and 2-carbon compounds that contain Cl and F bonded to carbon. These compounds are notably stable and nontoxic. They have been widely used in recent decades in the fabrication of flexible and rigid foams and as fluids for refrigeration and air conditioning. The most widely manufactured of these compounds are CCl_3F (CFC-11, bp 24°C), CCl_2F_2 (CFC-12, bp - 28°C), $C_2Cl_3F_3$ (CFC-113), $C_2Cl_2F_4$ (CFC-114), and C_2ClF_5 (CFC-115).

Halons are related compounds that contain bromine and are used in fire extinguisher systems. The major commercial halons are $CBrClF_2$ (Halon-1211), $CBrF_3$ (Halon-1301), and $C_2Br_2F_4$ (Halon-2402), where the sequence of numbers denotes the number of carbon, fluorine, chlorine, and bromine atoms, respectively, per molecule. Halons are particularly effective fire extinguishing agents because of the way in which they stop combustion. Halons act by chain reactions that destroy hydrogen atoms which sustain combustion. The basic sequence of reactions involved is outlined below:

$$CBrClF_2 + H\cdot \longrightarrow CClF_2\cdot + HBr \qquad (12.7.2)$$

$$HBr + H\cdot \longrightarrow Br\cdot + H_2 \qquad (12.7.3)$$

$$Br\cdot + H\cdot \longrightarrow HBr \qquad (12.7.4)$$

Halons are used in automatic fire extinguishing systems, particularly those located in flammable solvent storage areas, and in specialty fire extinguishers, particularly those on aircraft. Because of their potential to destroy stratospheric ozone discussed below, the use of halons in fire extinguishers was severely curtailed in a ban imposed in developed nations on January 1, 1994. Since that time, however, the atmospheric burden of halons has increased.[9] The ban on halons has caused concern because of the favorable properties of halons in fire extinguishers, particularly on aircraft. It is possible that hydrogen-containing analogs of halons may be effective as fire extinguishers without posing a threat to ozone.

The nonreactivity of CFC compounds, combined with worldwide production of approximately one-half million metric tons per year and deliberate or accidental release to the atmosphere, has resulted in CFCs becoming homogeneous components of the global atmosphere. In 1974 it was convincingly suggested, in a classic work that earned the authors a Nobel Prize, that chlorofluoromethanes could catalyze the destruction of stratospheric ozone that filters out cancer-causing ultraviolet radiation from the sun.[10] More recent data on ozone levels in the stratosphere and on increased ultraviolet radiation at earth's surface have shown that the threat to stratospheric ozone posed by chlorofluorocarbons is real.[11] Although quite inert in the lower atmosphere, CFCs undergo photodecomposition by the action of high-energy ultraviolet radiation in the stratosphere, which is energetic enough to break their very strong C-Cl bonds through reactions such as,

$$Cl_2CF_2 + h\nu \rightarrow Cl\cdot + ClCF_2\cdot \tag{12.7.5}$$

thereby releasing Cl atoms. The Cl atoms are very reactive species. Under the rarefied conditions of the stratosphere, one of the most abundant reactive species available for them to react with is ozone, which they destroy through a process that generates ClO:

$$Cl + O_3 \rightarrow ClO + O_2 \tag{12.7.6}$$

In the stratosphere, there is an appreciable concentration of atomic oxygen by virtue of the reaction

$$O_3 + h\nu \rightarrow O_2 + O \tag{12.7.7}$$

Nitric oxide, NO, is also present. The ClO species may react with either O or NO, regenerating Cl atoms and resulting in chain reactions that cause the net destruction of ozone:

$$ClO + O \rightarrow Cl + O_2 \tag{12.7.8}$$

$$\underline{Cl + O_3 \rightarrow ClO + O_2} \tag{12.7.9}$$

$$O_3 + O \rightarrow 2O_2 \tag{12.7.10}$$

$$ClO + NO \rightarrow Cl + NO_2 \tag{12.7.11}$$

$$\underline{O_3 + Cl \rightarrow ClO + O_2} \tag{12.7.12}$$

$$O_3 + NO \rightarrow NO_2 + O_2 \tag{12.7.13}$$

Both ClO and Cl involved in the above chain reactions have been detected in the 25-45-km altitude region.

The effects of CFCs on the ozone layer may be the single greatest threat to the global atmosphere and is discussed as such in Chapter 14. U. S. Environmental Protection Agency regulations, imposed in accordance with the 1986 Montreal Protocol on Substances that Deplete the Ozone Layer, curtailed production of CFCs and halocarbons in the U. S. starting in 1989. The substitutes for these halocarbons are hydrogen-containing chlorofluorocarbons (HCFCs) and hydrogen-containing fluorocarbons (HFCs). These include CH_2FCF_3 (HFC-134a, 1,1,1,2-tetrafluoroethane, a substitute for CFC-12 in automobile air conditioners and refrigeration equipment), $CHCl_2CF_3$ (HCFC-123, substitute for CFC-11 in plastic foam-blowing), CH_3CCl_2F (HCFC-141b, substitute for CFC-11 in plastic foam-blowing) $CHClF_2$ (HCFC-22, air conditioners and manufacture of plastic foam food containers). Because of the more readily broken H-C bonds they contain, these compounds are more easily destroyed by atmospheric chemical reactions (particularly with hydroxyl radical) before they reach the stratosphere. As of 1999 the HFC 134a market was growing at about a 10 or 15% annual rate with increases of several tens of thousands of metric tons per year.[12] This compound and related compounds containing only fluorine and hydrogen bound to carbon are favored because they cannot generate any ozone-destroying chlorine atoms.

Atmospheric Reactions of Hydrofluorocarbons and Hydrochlorofluorocarbons

The atmospheric chemistry of hydrofluorocarbons and hydrochlorofluorocarbons is important, even though these compounds do not pose much danger to the ozone layer. Of particular importance is the photooxidation of these compounds and the fates and effects of their photooxidation products. Hydrofluorocarbon 134a, CF_3CH_2F, reacts as follows with hydroxyl radical in the troposphere:

$$CF_3CH_2F + HO^\bullet \rightarrow CF_3CHF^\bullet + H_2O \tag{12.7.14}$$

The alkyl radical produced by this reaction forms a peroxy radical with molecular oxygen,

$$CF_3CHF^\bullet + O_2 + M \rightarrow CF_3CHFO_2^\bullet + M \tag{12.7.15}$$

and the peroxy radical reacts with NO:

$$CF_3CHFO_2^\bullet + NO \rightarrow CF_3CHFO^\bullet + NO_2 \tag{12.7.16}$$

The product of that reaction can either decompose,

$$CF_3CHFO^\bullet \rightarrow CF_3^\bullet + HC(O)F \tag{12.7.17}$$

or react with molecular O_2:

$$CF_3CHFO^\bullet + O_2 \rightarrow CF_3C(O)F + HO_2^\bullet \tag{12.7.18}$$

These latter two processes are thought to occur to about equal extents.

Perfluorocarbons

Perfluorocarbons are completely fluorinated organic compounds, the simplest examples of which are carbon tetrafluoride (CF_4) and hexafluoroethane (C_2F_6). Several hundred metric tons of these compounds are produced annually as etching agents in the electronics industry. However, about 30,000 metric tons of CF_4 and about 10% that amount of C_2F_6 are emitted to the global atmosphere each year from aluminum production.[13]

Nontoxic perfluorocarbons do not react with hydroxyl radical, ozone, or other reactive substances in the atmosphere, and the only known significant mechanism by which they are destroyed in the atmosphere is photolysis by radiation less than 130 nm in wavelength. Because of their extreme lack of reactivity, they are involved in neither photochemical smog formation nor ozone layer depletion. As a result of this stability, perfluorocarbons are very long-lived in the atmosphere; the lifetime of CF_4 is estimated to be an astoundingly long 50,000 years! The major atmospheric concern with these compounds is their potential to cause greenhouse warming (see Chapter 14). Taking into account their nonreactivity and ability to absorb infrared radiation, perfluorocarbons have a potential to cause global warming over a very long time span with an aggregate effect per molecule several thousand times that of carbon dioxide.

Chlorinated Dibenzo-*p*-dioxins and Dibenzofurans

Polychlorinated dibenzo-*p*-dioxins (PCDDs) and polychlorinated dibenzofurans (PCDFs) are pollutant compounds with the general formulas shown below:

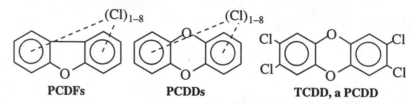

PCDFs   PCDDs   TCDD, a PCDD

As discussed in Chapters 7 and 22, these compounds are of considerable concern because of their toxicities. One of the more infamous environmental pollutant chemicals is 2,3,7,8-tetrachlorodibenzo-*p*-dioxin, TCDD, often known simply as "dioxin."

PCDDs and PCDFs enter the air from numerous sources, including automobile engines, waste incinerators, and steel and other metal production. A particularly important source may well be municipal solid waste incinerators. The formation of PCDDs and PCDFs in such incinerators results in part because of the presence of both chlorine (such as from polyvinylchloride plastic in municipal waste) and catalytic metals. Furthermore, PCDDs and PCDFs are produced by *de novo* synthesis on carbonaceous fly ash surfaces in the post-combustion region of an incinerator at relatively low temperatures of around 300°C in the presence of oxygen and sources of chlorine and hydrogen.

Atmospheric levels of PCDDs and PCDFs are quite low, in the range of 0.4–100 picograms per cubic meter of air. Because of their lower volatilities, the more highly chlorinated congeners of these compounds tend to occur in atmospheric particulate matter, in which they are relatively protected from photolysis and reaction with hydroxyl radical, which are the two main mechanisms by which PCDDs and PCDFs are eliminated from the atmosphere. Furthermore, the less highly chlorinated congeners are more reactive because of their C-H bonds, which are susceptible to attack from hydroxyl radical.

12.8. ORGANOSULFUR COMPOUNDS

Substitution of alkyl or aryl hydrocarbon groups such as phenyl and methyl for H on hydrogen sulfide, H_2S, leads to a number of different organosulfur thiols (mercaptans, R–SH) and sulfides, also called thioethers (R–S–R). Structural formulas of examples of these compounds are shown at the top of the following page.

The most significant atmospheric organosulfur compound is dimethylsulfide, produced in large quantities by marine organisms and introducing quantities of sulfur to the atmosphere comparable in magnitude to those introduced from pollution sources. Its oxidation produces most of the SO_2 in the marine atmosphere.

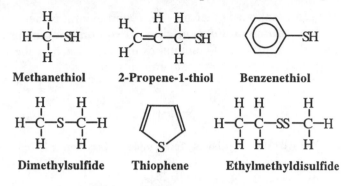

Methanethiol   2-Propene-1-thiol   Benzenethiol

Dimethylsulfide   Thiophene   Ethylmethyldisulfide

Methanethiol and other lighter alkyl thiols are fairly common air pollutants that have "ultragarlic" odors; both 1- and 2-butanethiol are associated with skunk odor. Gaseous methanethiol and volatile liquid ethanethiol are used as odorant leak-detecting additives for natural gas, propane, and butane, and are also employed as intermediates in pesticide synthesis. Allyl mercaptan (2-propene-1-thiol) is a toxic, irritating volatile liquid with a strong garlic odor. Benzenethiol (phenyl mercaptan) is the simplest of the aryl thiols. It is a toxic liquid with a severely "repulsive" odor.

Alkyl sulfides or thioethers contain the C-S-C functional group. The lightest of these compounds is dimethyl sulfide, a volatile liquid (bp 38°C) that is moderately toxic by ingestion. Cyclic sulfides contain the C-S-C group in a ring structure. The most common of these compounds is thiophene, a heat-stable liquid (bp 84°C) with a solvent action much like that of benzene that is used in the manufacture of pharmaceuticals, dyes, and resins.

Although not highly significant as atmospheric contaminants on a large scale, organic sulfur compounds can cause local air pollution problems because of their bad odors. Major sources of organosulfur compounds in the atmosphere include microbial degradation, wood pulping, volatile matter evolved from plants, animal wastes, packing house and rendering plant wastes, starch manufacture, sewage treatment, and petroleum refining.

Although the impact of organosulfur compounds on atmospheric chemistry is minimal in areas such as aerosol formation or production of acid precipitation components, these compounds are the worst of all in producing odor. Therefore, it is important to prevent their release into the atmosphere.

As with all hydrogen-containing organic species in the atmosphere, reaction of organosulfur compounds with hydroxyl radical is a first step in their atmospheric photochemical reactions. The sulfur from both mercaptans and sulfides ends up as SO_2. In both cases there is thought to be a readily oxidized SO intermediate, and HS• radical may also be an intermediate in the oxidation of mercaptans. Another possibility is the addition of O atoms to S, resulting in the formation of free radicals as shown below for methyl mercaptan:

$$CH_3SH + O \rightarrow H_3C\bullet + HSO\bullet \qquad (12.8.1)$$

The HSO• radical is readily oxidized by atmospheric O_2 to SO_2.

12.9. ORGANONITROGEN COMPOUNDS

Organic nitrogen compounds that may be found as atmospheric contaminants may be classified as **amines, amides, nitriles, nitro compounds,** or **heterocyclic nitrogen compounds.** Structures of common examples of each of these five classes of compounds reported as atmospheric contaminants are:

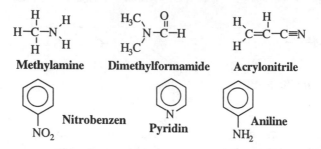

The organonitrogen compounds listed above can come from anthropogenic pollution sources. Significant amounts of anthropogenic atmospheric nitrogen may also come from reactions of inorganic nitrogen with reactive organic species. Examples include nitrates produced by the reaction of atmospheric NO_3.

Amines consist of compounds in which one or more of the hydrogen atoms in NH_3 has been replaced by a hydrocarbon moiety. Lower-molecular-mass amines are volatile. These are prominent among the compounds giving rotten fish their characteristic odor—an obvious reason why air contamination by amines is undesirable. The simplest and most important aryl amine is aniline, used in the manufacture of dyes, amides, photographic chemicals, and drugs. A number of amines are widely used industrial chemicals and solvents, so industrial sources have the potential to contaminate the atmosphere with these chemicals. Decaying organic matter, especially protein wastes, produce amines, so rendering plants, packing houses, and sewage treatment plants are important sources of these substances.

Aryl amines are of special concern as atmospheric pollutants, especially in the workplace, because some are known to cause urethral tract cancer (particularly of the bladder) in exposed individuals. Aryl amines are widely used as chemical intermediates, antioxidants, and curing agents in the manufacture of polymers (rubber and plastics), drugs, pesticides, dyes, pigments, and inks. In addition to aniline, some aryl amines of potential concern are the following:

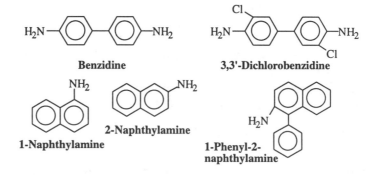

In the atmosphere, amines can be attacked by hydroxyl radicals and undergo further reactions. Amines are bases (electron-pair donors). Therefore, their acid-base chemistry in the atmosphere may be important, particularly in the presence of acids in acidic precipitation.

The amide most likely to be encountered as an atmospheric pollutant is dimethylformamide. It is widely used commercially as a solvent for the synthetic polymer, polyacrylonitrile (Orlon, Dacron). Most amides have relatively low vapor pressures, which limit their entry into the atmosphere.

Nitriles, which are characterized by the —C≡N group, have been reported as air contaminants, particularly from industrial sources. Both acrylonitrile and acetonitrile, CH_3CN, have been reported in the atmosphere as a result of synthetic rubber manufacture. As expected from their volatilities and levels of industrial production, most of the nitriles reported as atmospheric contaminants are low-molecular-mass aliphatic or olefinic nitriles, or aryl nitriles with only one benzene ring. Acrylonitrile, used to make polyacrylonitrile polymer, is the only nitrogen-containing organic chemical among the top 50 chemicals with annual worldwide production exceeding 1 billion kg.

Among the nitro compounds, RNO_2, reported as air contaminants are nitromethane, nitroethane, and nitrobenzene. These compounds are produced from industrial sources. Highly oxygenated compounds containing the NO_2 group, particularly peroxyacetyl nitrate (PAN, discussed in Chapter 13), are end products of the photochemical oxidation of hydrocarbons in urban atmospheres.

A large number of **heterocyclic nitrogen compounds** have been reported in tobacco smoke, and it is inferred that many of these compounds can enter the atmosphere from burning vegetation. Coke ovens are another major source of these compounds. In addition to the derivatives of pyridine, some of the heterocyclic nitrogen compounds are derivatives of pyrrole:

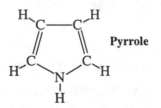

Pyrrole

Heterocyclic nitrogen compounds occur almost entirely in association with aerosols in the atmosphere.

Nitrosamines, which contain the N-N=O group, and having, therefore, the general formula,

$$\underset{R'}{\overset{R}{\diagdown}}N-N=O$$

deserve special mention as atmospheric contaminants because some are known carcinogens. As discussed in Chapter 22, nitrosamines include compounds that can attach alkyl groups to DNA resulting in modified DNA that can lead to cancer. Both N-nitrosodimethylamine and N-nitrosodiethylamine have been detected in the atmosphere.

LITERATURE CITED

1. Schreiber, Lukas, and Jörg Schönherr, "Uptake of Organic Chemicals in Conifer Needles: Surface Adsorption and Permeability of Cuticles," *Environmental Science and Technology* **26**, 153-159 (1992).

2. Ockenden, Wendy A., Andrew J. Sweetman, Harry F. Prest, Eiliv Steinnes, and Kevin C. Jones, "Toward an Understanding of the Global Atmospheric Distribution of Persistent Atmospheric Pollutants: The Use of Semipermeable Membrane Devices as Time-Integrated Passive Samplers," *Environmental Science and Technology*, **32**, 2795–2803 (1998).

3. Graedel, T. E., *Chemical Compounds in the Atmosphere*, Academic Press, San Diego, 1978.

4. Shu, Yonghui, Erik S. C. Kwok, Ernesto C. Tuazon, Roger Atkinson, and Janet Arey, "Products of the Gas-Phase Reactions of Linalool with OH Radicals, NO_3 Radicals, and O_3," *Environmental Science and Technology*, **31**, 896-904 (1997).

5. Hallquist, Mattias, Ingvar Wängberg, Evert Ljungström, Ian Barnes, and Karl-Heinz Becker, "Aerosol and Product Yields from NO_3 Radical Initiated Oxidation of Selected Monoterpenes," *Environmental Science and Technology* **33**, 553-559 (1999).

6. Yu, Jianzhen, Richard C. Flagan, and John H. Seinfeld, "Identification of Products Containing –COOH, –OH, and –C=O in Atmospheric Oxidation of Hydrocarbons," *Environmental Science and Technology* **32**, 2357–2370 (1998).

7. Odum, J. R., T. P. W. Jungkamp, Richard C. Flagan, and John H. Seinfeld, "The Atmospheric Aerosol-Forming Potential of Whole Gasoline Vapor," *Science*, **276**, 96-99 (1997).

8. Grimvall, Anders and Ed W. B. deLeer, *Naturally Produced Organohalogens*, Kluwer Academic Publishers, Dordrecht, The Netherlands, 1995.

9. Butler, James H., Stephen A. Montzka, Andrew D. Clarke, Jurgen M. Lobert, and James W. Elkins, "Growth and Distribution of Halons in the Atmosphere," *Journal of Geophysical Research*, **103**, 1503-1511 (1998).

10. Molina, Mario. J. and F. Sherwood Rowland, "Stratospheric Sink for Chlorofluoromethanes," *Nature*, **249**, 810-12 (1974).

11. Madronich, S., R. L. McKenzie, L. O. Bjorn, and M. M. Caldwell, "Changes in Biologically Active Ultraviolet Radiation Reaching the Earth's Surface," *Journal of Photochemistry and Photobiology*, **46**, 5-19 (1998).

12. "ICI to Boost Capacity of Hydrofluorocarbon," *Chemical and Engineering News*, June 21, 1999, p. 14.

13. Zurer, Pamela, "Perfluorocarbons Use, Emissions May Face Restriction," *Chemical and Engineering News*, August 9, 1993, p. 16.

SUPPLEMENTARY REFERENCES

Brimblecombe, Peter, *Air Composition and Chemistry*, 2nd ed., Cambridge University Press, Cambridge, U.K., 1996.

Hewitt, C. Nicholas, Ed., *Reactive Hydrocarbons in the Atmosphere*, Academic Press, San Diego, CA, 1998.

Kavouras, Ilias G., Nikolaos Mihalopoulos, and Euripides G. Stephanou, "Formation of Atmospheric Particles from Organic Acids Produced by Forests," *Nature*, **395**, 683-686 (1998).

Khalil, M. A. K., and M. J. Shearer, Guest Editors, "Atmospheric Methane: Sources, Sinks and Role in Global Change," special issue of *Chemosphere*, **26**(1-4) 1993.

Klessinger, Martin and Josef Michl, *Excited States and Photochemistry of Organic Molecules*, VCH Publishers, New York, 1995.

Neckers, Douglas C., David H. Volman, and Günther von Bünau, Eds., *Advances in Photochemistry*, Vol. 23, John Wiley and Sons, Inc., New York, NY, 1997.

Nriagu, Jerome O., Ed., *Gaseous Pollutants: Characterization and Cycling*, John Wiley and Sons, New York, 1992.

Seinfeld, John H. and Spyros N. Pandis, *Atmospheric Chemistry and Physics*, John Wiley and Sons, New York, 1998.

Smith, William H., *Air Pollution and Forests*, 2nd ed., Springer-Verlag, New York, 1990.

Wayne, Carol E., and Richard P. Wayne, *Photochemistry*, Oxford University Press, New York, 1996.

Yung, Y. L. and William B. Demore, *Photochemistry of Planetary Atmospheres*, Oxford University Press, New York, 1998.

QUESTIONS AND PROBLEMS

1. Match each organic pollutant in the left column with its expected effect in the right column, below:

    (a) CH_3SH 1. Most likely to have a secondary effect in the atmosphere

    (b) $CH_3CH_2CH_2CH_3$ 2. Most likely to have a direct effect

    (c)   3. Should have the least effect of these three

2. Why are hydrocarbon emissions from uncontrolled automobile exhaust particularly reactive?

3. Assume an accidental release of a mixture of gaseous alkanes and alkenes into an urban atmosphere early in the morning. If the atmosphere at the release site is monitored for these compounds, what can be said about their total and relative concentrations at the end of the day? Explain.

4. Match each radical in the left column with its type in the right column, below:

   (a) $H_3C^\bullet$ 1. Formyl radical

   (b) $CH_3CH_2O^\bullet$ 2. Alkylperoxyl radical

   (c) $H\overset{\bullet}{C}O$ 3. Alkyl radical

   (d) $CH_xCH_{2x+1}O_2{}^\bullet$ 4. Alkoxyl radical

5. When reacting with hydroxyl radical, alkenes have a reaction mechanism not available to alkanes, which makes the alkenes much more reactive. What is this mechanism?

6. What is the most stable type of hydrocarbon that has a very low hydrogen-to-carbon ratio?

7. In the sequence of reactions leading to the oxidation of hydrocarbons in the atmosphere, what is the first stable class of compounds generally produced?

8. Give a sequence of reactions leading to the formation of acetaldehyde from ethane starting with the reaction of hydroxyl radical.

9. What important photochemical property do carbonyl compounds share with NO_2?

13 PHOTOCHEMICAL SMOG

13.1. INTRODUCTION

This chapter discusses the **oxidizing smog** or **photochemical smog** that permeates atmospheres in Los Angeles, Mexico City, Zurich, and many other urban areas. Although **smog** is the term used in this book to denote a photochemically oxidizing atmosphere, the word originally was used to describe the unpleasant combination of smoke and fog laced with sulfur dioxide which was formerly prevalent in London when high-sulfur coal was the primary fuel used in that city. This mixture is characterized by the presence of sulfur dioxide, a reducing compound; therefore it is a **reducing smog** or **sulfurous smog**. In fact, sulfur dioxide is readily oxidized and has a short lifetime in an atmosphere where oxidizing photochemical smog is present.

Smog has a long history. Exploring what is now southern California in 1542, Juan Rodriguez Cabrillo named San Pedro Bay "The Bay of Smokes" because of the heavy haze that covered the area. Complaints of eye irritation from anthropogenically polluted air in Los Angeles were recorded as far back as 1868. Characterized by reduced visibility, eye irritation, cracking of rubber, and deterioration of materials, smog became a serious nuisance in the Los Angeles area during the 1940s. It is now recognized as a major air pollution problem in many areas of the world.

Smoggy conditions are manifested by moderate to severe eye irritation or visibility below 3 miles when the relative humidity is below 60%. The formation of oxidants in the air, particularly ozone, is indicative of smog formation. Serious levels of photochemical smog may be assumed to be present when the oxidant level exceeds 0.15 ppm for more than one hour. The three ingredients required to generate photochemical smog are ultraviolet light, hydrocarbons, and nitrogen oxides. Advanced techniques of analysis have shown a large variety of hydrocarbon precursors to smog formation in the atmosphere.[1]

The importance of ozone as an atmospheric pollutant in atmospheres contaminated with photochemical smog has been recognized by proposed regulations to lower allowable ozone concentrations in the U.S.[2] The revised ozone standard put forward by the U.S. Environmental Protection Agency was designed to change allowed levels of atmospheric ozone from the long-standing limit of 0.12 ppm measured over one hour to 0.08 ppm over 8 hours.

13.2. SMOG-FORMING AUTOMOTIVE EMISSIONS

Internal combustion engines used in automobiles and trucks produce reactive hydrocarbons and nitrogen oxides, two of the three key ingredients required for smog to form. Therefore, automotive air emissions are discussed next.

The production of nitrogen oxides was discussed in Section 11.6. At the high temperature and pressure conditions in an internal combustion engine, products of incompletely burned gasoline undergo chemical reactions which produce several hundred different hydrocarbons. Many of these are highly reactive in forming photochemical smog. As shown in Figure 13.1, the automobile has several potential sources of hydrocarbon emissions other than the exhaust. The first of these to be controlled was the crankcase mist of hydrocarbons composed of lubricating oil and "blowby." The latter consists of exhaust gas and unoxidized fuel/air mixture that enters the crankcase from the combustion chambers around the pistons. This mist is destroyed by recirculating it through the engine intake manifold by way of the positive crankcase ventilation (PCV) valve.

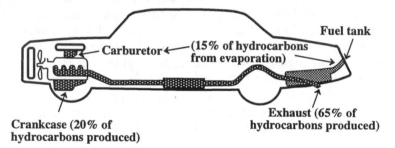

Figure 13.1. Potential sources of pollutant hydrocarbons from an automobile without pollution control devices.

A second major source of automotive hydrocarbon emissions is the fuel system, from which hydrocarbons are emitted through fuel tank and carburetor vents. When the engine is shut off and the engine heat warms up the fuel system, gasoline may be evaporated and emitted to the atmosphere. In addition, heating during the daytime and cooling at night causes the fuel tank to breathe and emit gasoline fumes. Such emissions are reduced by fuel formulated to reduce volatility. Automobiles are equipped with canisters of carbon which collect evaporated fuel from the fuel tank and fuel system, to be purged and burned when the engine is operating.

Control of Exhaust Hydrocarbons

In order to understand the production and control of automotive hydrocarbon exhaust products, it is helpful to understand the basic principles of the internal combustion engine. As shown in Figure 13.2, the four steps involved in one complete cycle of the four-cycle engine used in most vehicles are the following:

1. **Intake:** Air is drawn into the cylinder through the open intake valve. Gasoline is either injected with the intake air or injected separately into the cylinder.

2. **Compression:** The combustible mixture is compressed at a ratio of about 7:1. Higher compression ratios favor thermal efficiency and complete combustion of hydrocarbons. However, higher temperatures, premature combustion ("pinging"), and high production of nitrogen oxides also result from higher combustion ratios.

3. **Ignition and power stroke:** As the fuel-air mixture normally produced by injecting fuel into the cylinder is ignited by the spark plug near top-dead-center, a temperature of about 2,500°C is reached very rapidly at pressures up to 40 atm. As the gas volume increases with downward movement of the piston, the temperature decreases in a few milliseconds. This rapid cooling "freezes" nitric oxide in the form of NO without allowing it time to dissociate to N_2 and O_2, which are thermodynamically favored at the normal temperatures and pressures of the atmosphere.

4. **Exhaust:** Exhaust gases consisting largely of N_2 and CO_2, with traces of CO, NO, hydrocarbons, and O_2, are pushed out through the open exhaust valve, thus completing the cycle.

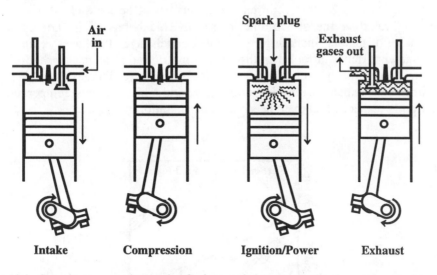

Figure 13.2. Steps in one complete cycle of a four-cycle internal combustion engine. Fuel is mixed with the intake air or injected separately into each cylinder.

The primary cause of unburned hydrocarbons in the engine cylinder is wall quench, wherein the relatively cool wall in the combustion chamber of the internal combustion engine causes the flame to be extinguished within several thousandths of a centimeter from the wall. Part of the remaining hydrocarbons may be retained as residual gas in the cylinder, and part may be oxidized in the exhaust system. The remainder is emitted to the atmosphere as pollutant hydrocarbons. Engine misfire due to improper adjustment and deceleration greatly increases the emission of hydrocarbons. Turbine engines are not subject to the wall quench phenomenon because their surfaces are always hot.

Several engine design characteristics favor lower exhaust hydrocarbon emissions. Wall quench, which is mentioned above, is diminished by design that decreases the combustion chamber surface/volume ratio through reduction of compression ratio, more nearly spherical combustion chamber shape, increased displacement per engine cylinder, and increased ratio of stroke relative to bore.

Spark retard also reduces exhaust hydrocarbon emissions. For optimum engine power and economy, the spark should be set to fire appreciably before the piston reaches the top of the compression stroke and begins the power stroke. Retarding the spark to a point closer to top-dead-center reduces the hydrocarbon emissions markedly. One reason for this reduction is that the effective surface/volume ratio of the combustion chamber is reduced, thus cutting down on wall quench. Second, when the spark is retarded, the combustion products are purged from the cylinders sooner after combustion. Therefore, the exhaust gas is hotter, and reactions consuming hydrocarbons are promoted in the exhaust system.

As shown in Figure 13.3, the air/fuel ratio in the internal combustion engine has a marked effect upon the emission of hydrocarbons. As the air/fuel ratio becomes richer in fuel than the stoichiometric ratio, the emission of hydrocarbons increases significantly. There is a moderate decrease in hydrocarbon emissions when the mixture becomes appreciably leaner in fuel than the stoichiometric ratio requires. The lowest level of hydrocarbon emissions occurs at an air/fuel ratio somewhat leaner in fuel than the stoichiometric ratio. This behavior is the result of a combination of factors, including minimum quench layer thickness at an air/fuel ratio somewhat richer in fuel than the stoichiometric ratio, decreasing hydrocarbon concentration in the quench layer with a leaner mixture, increasing oxygen concentration in the exhaust with a leaner mixture, and a peak exhaust temperature at a ratio slightly leaner in fuel than the stoichiometric ratio.

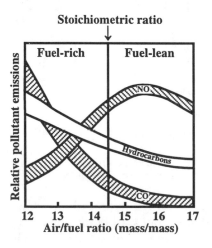

Figure 13.3. Effects of air/fuel ratio on pollutant emissions from an internal combustion piston engine.

Catalytic converters are now used to destroy pollutants in exhaust gases.[3] Currently, the most commonly used automotive catalytic converter is the three-way conversion catalyst, so called because a single catalytic unit destroys all three of the main class of automobile exhaust pollutants—hydrocarbons, carbon monoxide, and nitrogen oxides. This catalyst depends upon accurate sensing of oxygen levels in the exhaust combined with computerized engine control which cycles the air/fuel mixture several times per second back and forth between slightly lean and slightly rich relative to the stoichiometric ratio. Under these conditions carbon monoxide, hydrogen, and hydrocarbons ($C_c H_h$) are oxidized.

$$CO + \tfrac{1}{2}O_2 \rightarrow CO_2 \tag{13.2.1}$$

$$H_2 + \tfrac{1}{2}O_2 \rightarrow H_2O \tag{13.2.2}$$

$$C_c H_h + (c + {}^{h}/_4)O_2 \rightarrow cCO_2 + {}^{h}/_2 H_2O \tag{13.2.3}$$

Nitrogen oxides are reduced on the catalyst to N_2 by carbon monoxide, hydrocarbons, or hydrogen as shown by the following reduction with CO:

$$CO + NO \rightarrow \tfrac{1}{2}N_2 + CO_2 \tag{13.2.4}$$

Automotive exhaust catalysts are dispersed on a high surface area substrate, most commonly consisting of cordierite, a ceramic composed of alumina (Al_2O_3), silica, and magnesium oxide.[4] The substrate is formed as a honeycomb type structure providing maximum surface area to contact exhaust gases. The support needs to be mechanically strong to withstand vibrational stresses from the automobile, and it must resist severe thermal stresses in which the temperature may rise from ambient temperatures to approximately 900°C over an approximately two-minute period during "light-off" when the engine is started. The catalytic material, which composes only about 0.10-0.15% of the catalyst body, consists of a mixture of precious metals. Platinum and palladium catalyze the oxidation of hydrocarbons and carbon monoxide, and rhodium acts as a catalyst for the reduction of nitrogen oxides; presently, palladium is the most common precious metal in exhaust catalysts.

Since lead can poison auto exhaust catalysts, automobiles equipped with catalytic exhaust-control devices require lead-free gasoline, which has become the standard motor fuel. Sulfur in gasoline is also detrimental to catalyst performance, and a controversial topic in 1999 was a proposed change in regulations to greatly lower sulfur levels in fuel.

The internal combustion automobile engine has been developed to a remarkably high degree in terms of its emissions. The ultimate development of such an engine is one claimed to be so clean that when it is operated in a smoggy atmosphere, its exhaust is cleaner than the air that it is taking in![5]

The 1990 U. S. Clean Air Act calls for reformulating gasoline by adding more oxygenated compounds to reduce emissions of hydrocarbons and carbon monoxide. However, this measure has been rather controversial and questions have been raised regarding one of the major oxygenated additives, methyltertiarybutyl ether, MTBE, which has become a common water pollutant in some areas. Because of these concerns, California's governor announced in March 1999 that MTBE would be phased out of gasoline for sale in California, and later in the year the U.S. Environmental Protection Agency proposed the phaseout of MTBE in the entire U.S.

Automotive Emission Standards

Federal law and California state law mandate automotive emission standards. The allowable emissions have been on a downward trend since the first standards were imposed in the mid-1960s. Table 13.1. gives emissions prior to controls and those for various time intervals since 1970.

Table 13.1. Emission Standards for Light-Duty Motor Vehicles in the U.S.

Model year	Federal standards				California standards				
	HCs^1	CO^1	$NO_x{}^1$	Evap. HCs^1	HCs	CO	NO_x	Evap. HCs	Particulate matter[2]
Before controls	10.60	84.0	4.1	47	10.60	84.0	4.1	47	- - -
1970	4.1	34.0	- - -	- - -	4.1	34.0	- - -	- - -	- - -
1975	1.5	15.0	3.1	2	0.90	9	2.0	2	- - -
1980	0.41	7.0	2.0	6	0.39	9.0	1.0	2	- - -
1985	0.41	3.4	1.0	2	0.39	7.0	0.4	2	0.40
1990	0.41	3.4	1.0	2	0.39	7.0	0.4	2	0.08
1993	0.41	3.4	1.0	2	0.25	3.4	0.4	2	0.08
2000	0.41^3	3.4	0.4	- - -	Values not given				

[1] HCs, hydrocarbons from exhaust; CO, carbon monoxide; NO_x, sum of NO and NO_2; Evap. HCs, hydrocarbons evaporated from fuel system; all values in g/mile, except Evap. HCs, which are given in g/test.
[2] Particulate matter from diesel passenger car exhausts.
[3] Total hydrocarbons

13.3. SMOG-FORMING REACTIONS OF ORGANIC COMPOUNDS IN THE ATMOSPHERE

Hydrocarbons are eliminated from the atmosphere by a number of chemical and photochemical reactions. These reactions are responsible for the formation of many noxious secondary pollutant products and intermediates from relatively innocuous hydrocarbon precursors. These pollutant products and intermediates make up photochemical smog.

Hydrocarbons and most other organic compounds in the atmosphere are thermodynamically unstable toward oxidation and tend to be oxidized through a series of steps. The oxidation process terminates with formation of CO_2, solid organic particulate matter which settles from the atmosphere, or water-soluble products (for example, acids, aldehydes) which are removed by rain. Inorganic species such as ozone or nitric acid are by-products of these reactions.

Photochemical Reactions of Methane

Some of the major reactions involved in the oxidation of atmospheric hydrocarbons may be understood by considering the oxidation of methane, the most common and widely dispersed atmospheric hydrocarbon. Like other hydrocarbons, methane reacts with oxygen atoms (generally produced by the photochemical dissociation of NO_2 to O and NO) to generate the all-important hydroxyl radical and an alkyl (methyl) radical

$$CH_4 + O \rightarrow H_3C^{\bullet} + HO^{\bullet} \tag{13.3.1}$$

The methyl radical produced reacts rapidly with molecular oxygen to form very reactive peroxyl radicals,

$$H_3C^{\bullet} + O_2 + M \text{ (energy-absorbing third body,}$$
$$\text{usually a molecule of } N_2 \text{ or } O_2) \rightarrow H_3COO^{\bullet} + M \tag{13.3.2}$$

in this case, methyl peroxyl radical, $H_3COO^{\bullet}$. Such radicals participate in a variety of subsequent chain reactions, including those leading to smog formation. The hydroxyl radical reacts rapidly with hydrocarbons to form reactive hydrocarbon radicals,

$$CH_4 + HO^{\bullet} \rightarrow H_3C^{\bullet} + H_2O \tag{13.3.3}$$

in this case, the methyl radical, $H_3C^{\bullet}$. The following are more reactions involved in the overall oxidation of methane:

$$H_3COO^{\bullet} + NO \rightarrow H_3CO^{\bullet} + NO_2 \tag{13.3.4}$$

(This is a very important kind of reaction in smog formation because the oxidation of NO by peroxyl radicals is the predominant means of regenerating NO_2 in the atmosphere after it has been photochemically dissociated to NO.)

$$H_3CO^{\bullet} + O_3 \rightarrow \text{various products} \tag{13.3.5}$$

$$H_3CO^{\bullet} + O_2 \rightarrow CH_2O + HOO^{\bullet} \tag{13.3.6}$$

$$H_3COO^{\bullet} + NO_2 + M \rightarrow CH_3OONO_2 + M \tag{13.3.7}$$

$$H_2CO + h\nu \rightarrow \text{photodissociation products} \tag{13.3.8}$$

As will be seen throughout this chapter, hydroxyl radical, $HO^{\bullet}$, and hydroperoxyl radical, $HOO^{\bullet}$, are ubiquitous intermediates in photochemical chain-reaction processes. These two species are known collectively as odd hydrogen radicals.

Reactions such as (13.3.1) and (13.3.3) are **abstraction reactions** involving the removal of an atom, usually hydrogen, by reaction with an active species. **Addition reactions** of organic compounds are also common. Typically, hydroxyl radical reacts with an alkene such as propylene to form another reactive free radical:

$$HO\bullet \;+\; H-\underset{\underset{H}{|}}{\overset{\overset{H}{|}}{C}}-\underset{\underset{H}{|}}{\overset{\overset{H}{|}}{C}}=C\overset{H}{\underset{H}{\diagdown}} \;\;\rightarrow\;\; H-\underset{\underset{H}{|}}{\overset{\overset{H}{|}}{C}}-\overset{\bullet}{\underset{\underset{H}{|}}{C}}-\underset{\underset{H}{|}}{\overset{\overset{H}{|}}{C}}-OH \tag{13.3.9}$$

Ozone adds to unsaturated compounds to form reactive ozonides:

$$H-\underset{\underset{H}{|}}{\overset{\overset{H}{|}}{C}}-\underset{\underset{H}{|}}{\overset{\overset{H}{|}}{C}}=C\overset{H}{\underset{H}{\diagdown}} + O_3 \;\;\rightarrow\;\; H-\underset{\underset{H}{|}}{\overset{\overset{H}{|}}{C}}-\underset{\underset{H}{|}}{\overset{\overset{H}{|}}{C}}\overset{O-O}{\underset{O}{\diagdown\diagup}}C\overset{H}{\underset{H}{\diagup}} \tag{13.3.10}$$

Organic compounds (in the troposphere, almost exclusively carbonyls) can undergo primary photochemical reactions resulting in the direct formation of free radicals. By far the most important of these is the photochemical dissociation of aldehydes:

$$H_3C-\overset{\overset{O}{||}}{C}-H + h\nu \;\;\rightarrow\;\; H_3C\bullet + H\overset{\bullet}{C}O \tag{13.3.11}$$

Organic free radicals undergo a number of chemical reactions. Hydroxyl radicals may be generated from organic peroxyl reactions such as,

$$H_3C-\underset{\underset{H}{|}}{\overset{\overset{\overset{\bullet}{O}}{|}{O}}{|}}{C}-CH_3 \;\;\rightarrow\;\; H_3C-\overset{\overset{O}{||}}{C}-CH_3 + HO\bullet \tag{13.3.12}$$

leaving an aldehyde or ketone. The hydroxyl radical may react with other organic compounds, maintaining the chain reaction. Gas-phase reaction chains commonly have many steps. Furthermore, chain-branching reactions take place in which a free radical reacts with an excited molecule causing it to produce two new radicals. Chain termination may occur in several ways, including reaction of two free radicals,

$$2HO\bullet \rightarrow H_2O_2 \tag{13.3.13}$$

adduct formation with nitric oxide or nitrogen dioxide (which, because of their odd numbers of electrons, are themselves stable free radicals),

$$HO\bullet + NO_2 + M \rightarrow HNO_3 + M \tag{13.3.14}$$

or reaction of the radical with a solid particle surface.

Hydrocarbons may undergo heterogeneous reactions on particles in the atmosphere. Dusts composed of metal oxides or charcoal have a catalytic effect upon the oxidation of organic compounds. Metal oxides may enter into photochemical reactions. For example, zinc oxide photosensitized by exposure to light promotes oxidation of organic compounds.

The kinds of reactions just discussed are involved in the formation of photochemical smog in the atmosphere. Next, consideration is given to the smog-forming process.

13.4. OVERVIEW OF SMOG FORMATION

This section addresses the conditions that are characteristic of a smoggy atmosphere and the overall processes involved in smog formation. In atmospheres that receive hydrocarbon and NO pollution accompanied by intense sunlight and stagnant air masses, oxidants tend to form. In air-pollution parlance, **gross photochemical oxidant** is a substance in the atmosphere capable of oxidizing iodide ion to elemental iodine. Sometimes other reducing agents are used to measure oxidants. The primary oxidant in the atmosphere is ozone. Other atmospheric oxidants include H_2O_2, organic peroxides (ROOR'), organic hydroperoxides (ROOH), and peroxyacyl nitrates such as peroxyacetyl nitrate (PAN).

$$\underset{\displaystyle H}{\overset{\displaystyle H \quad O}{H-\underset{|}{\overset{|}{C}}-\overset{\|}{C}-O-O-NO_2}} \qquad \text{Peroxyacetyl nitrate (PAN)}$$

Nitrogen dioxide, NO_2, is not regarded as a gross photochemical oxidant. However, it is about 15% as efficient as O_3 in oxidizing iodide to iodine(0), and a correction is made in measurements for the positive interference of NO_2. Sulfur dioxide is oxidized by O_3 and produces a negative interference for which a measurement correction must also be made.

Peroxyacetyl nitrate and related compounds containing the -C(O)OONO$_2$ moiety, such as peroxybenzoyl nitrate (PBN),

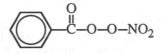

a powerful eye irritant and lachrymator, are produced photochemically in atmospheres containing alkenes and NO_x. PAN, especially, is a notorious organic oxidant. In addition to PAN and PBN, some other specific organic oxidants that may be important in polluted atmospheres are peroxypropionyl nitrate (PPN); peracetic acid, $CH_3(CO)OOH$; acetylperoxide, $CH_3(CO)OO(CO)CH_3$; butyl hydroperoxide, $CH_3CH_2CH_2CH_2OOH$; and *tert*-butylhydroperoxide, $(CH_3)_3COOH$.

As shown in Figure 13.4, smoggy atmospheres show characteristic variations with time of day in levels of NO, NO_2, hydrocarbons, aldehydes, and oxidants. Examination of the figure shows that shortly after sunrise the level of NO in the atmosphere decreases markedly, a decrease that is accompanied by a peak in the concentration of NO_2. During midday (significantly, after the concentration of NO has fallen to a very low level), the levels of aldehydes and oxidants become relatively high. The concentration of total hydrocarbons in the atmosphere peaks sharply in the morning, then decreases during the remaining daylight hours.

An overview of the processes responsible for the behavior just discussed is summarized in Figure 13.5. The chemical bases for the processes illustrated in this figure are explained in the following section.

388   **Environmental Chemistry**

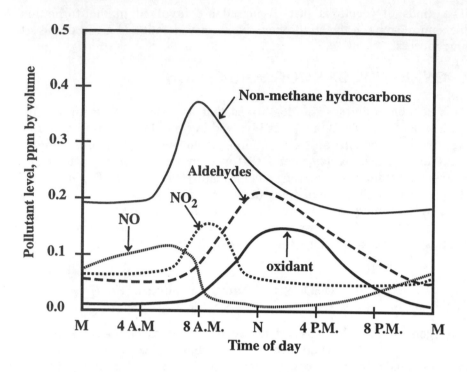

Figure 13.4. Generalized plot of atmospheric concentrations of species involved in smog formation as a function of time of day.

13.5. MECHANISMS OF SMOG FORMATION

Here are discussed some of the primary aspects of photochemical smog formation. For more details the reader is referred to books on atmospheric chemistry[6] and atmospheric chemistry and physics.[7] Since the exact chemistry of photochemical smog formation is very complex, many of the reactions are given as plausible illustrative examples rather than proven mechanisms.

The kind of behavior summarized in Figure 13.4 contains several apparent anomalies which puzzled scientists for many years. The first of these was the rapid increase in NO_2 concentration and decrease in NO concentration under conditions where it was known that photodissociation of NO_2 to O and NO was occurring. Furthermore, it could be shown that the disappearance of alkenes and other hydrocarbons was much more rapid than could be explained by their relatively slow reactions with O_3 and O. These anomalies are now explained by chain reactions involving the interconversion of NO and NO_2, the oxidation of hydrocarbons, and the generation of reactive intermediates, particularly hydroxyl radical (HO$^\bullet$).

Figure 13.5 shows the overall reaction scheme for smog formation, which is based upon the photochemically initiated reactions that occur in an atmosphere containing nitrogen oxides, reactive hydrocarbons, and oxygen. The time variations in levels of hydrocarbons, ozone, NO, and NO_2 are explained by the following overall reactions:

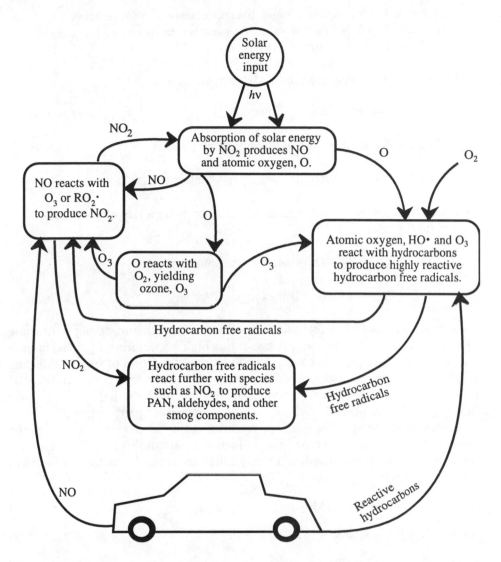

Figure 13.5. Generalized scheme for the formation of photochemical smog.

1. Primary photochemical reaction producing oxygen atoms:

$$NO_2 + h\nu \; (\lambda < 420 \, nm) \rightarrow NO + O \qquad (13.5.1)$$

2. Reactions involving oxygen species (M is an energy-absorbing third body):

$$O_2 + O + M \rightarrow O_3 + M \qquad (13.5.2)$$

$$O_3 + NO \rightarrow NO_2 + O_2 \qquad (13.5.3)$$

Because the latter reaction is rapid, the concentration of O_3 remains low until that of NO falls to a low value. Automotive emissions of NO tend to keep O_3 concentrations low along freeways.

3. Production of organic free radicals from hydrocarbons, RH:

$$O + RH \rightarrow R^\bullet + \text{other products} \qquad (13.5.4)$$

$$O_3 + RH \rightarrow R^\bullet + \text{and/or other products} \qquad (13.5.5)$$

($R^\bullet$ is a free radical which may or may not contain oxygen.)

4. Chain propagation, branching, and termination by a variety of reactions such as the following:

$$NO + ROO^\bullet \rightarrow NO_2 + \text{and/or other products} \qquad (13.5.6)$$

$$NO_2 + R^\bullet \rightarrow \text{products (for example, PAN)} \qquad (13.5.7)$$

The latter kind of reaction is the most common chain-terminating process in smog because NO_2 is a stable free radical present at high concentrations. Chains may terminate also by reaction of free radicals with NO or by reaction of two $R^\bullet$ radicals, although the latter is uncommon because of the relatively low concentrations of radicals compared to molecular species. Chain termination by radical sorption on a particle surface is also possible and may contribute to aerosol particle growth.

A large number of specific reactions are involved in the overall scheme for the formation of photochemical smog. The formation of atomic oxygen by a primary photochemical reaction (Reaction 13.5.1) leads to several reactions involving oxygen and nitrogen oxide species:

$$O + O_2 + M \rightarrow O_3 + M \qquad (13.5.8)$$

$$O + NO + M \rightarrow NO_2 + M \qquad (13.5.9)$$

$$O + NO_2 \rightarrow NO + O_2 \qquad (13.5.10)$$

$$O_3 + NO \rightarrow NO_2 + O_2 \qquad (13.5.11)$$

$$O + NO_2 + M \rightarrow NO_3 + M \qquad (13.5.12)$$

$$O_3 + NO_2 \rightarrow NO_3 + O_2 \qquad (13.5.13)$$

There are a number of significant atmospheric reactions involving nitrogen oxides, water, nitrous acid, and nitric acid:

$$NO_3 + NO_2 \rightarrow N_2O_5 \qquad (13.5.14)$$

$$N_2O_5 \rightarrow NO_3 + NO_2 \qquad (13.5.15)$$

$$NO_3 + NO \rightarrow 2NO_2 \qquad (13.5.16)$$

$$N_2O_5 + H_2O \rightarrow 2HNO_3 \qquad (13.5.17)$$

(This reaction is slow in the gas phase but may be fast on surfaces.)

Very reactive $HO^\bullet$ radicals can be formed by the reaction of excited atomic oxygen with water,

$$O^* + H_2O \rightarrow 2HO^\bullet \qquad (13.5.18)$$

by photodissociation of hydrogen peroxide,

$$H_2O_2 + h\nu \, (\lambda < 350 \, nm) \rightarrow 2HO^\bullet \qquad (13.5.19)$$

or by the photolysis of nitrous acid,

$$HNO_2 + h\nu \rightarrow HO^\bullet + NO \qquad (13.5.20)$$

Among the inorganic species with which the hydroxyl radical reacts are oxides of nitrogen,

$$HO^\bullet + NO_2 \rightarrow HNO_3 \qquad (13.5.21)$$

$$HO^\bullet + NO + M \rightarrow HNO_2 + M \qquad (13.5.22)$$

and carbon monoxide,

$$CO + HO^\bullet + O_2 \rightarrow CO_2 + HOO^\bullet \qquad (13.5.23)$$

The last reaction is significant in that it is responsible for the disappearance of much atmospheric CO (see Section 11.3) and because it produces the hydroperoxyl radical $HOO^\bullet$. One of the major inorganic reactions of the hydroperoxyl radical is the oxidation of NO:

$$HOO^\bullet + NO \rightarrow HO^\bullet + NO_2 \qquad (13.5.24)$$

For purely inorganic systems, kinetic calculations and experimental measurements cannot explain the rapid transformation of NO to NO_2 that occurs in an atmosphere undergoing photochemical smog formation and predict that the concentration of NO_2 should remain very low. However, in the presence of reactive hydrocarbons, NO_2 accumulates very rapidly by a reaction process beginning with its photodissociation! It may be concluded, therefore, that the organic compounds form species which react with NO directly rather than with NO_2.

A number of chain reactions have been shown to result in the general type of species behavior shown in Figure 13.4. When alkane hydrocarbons, RH, react with O, O_3, or $HO^\bullet$ radical,

$$RH + O + O_2 \rightarrow ROO^\bullet + HO^\bullet \tag{13.5.25}$$

$$RH + HO^\bullet + O_2 \rightarrow ROO^\bullet + H_2O \tag{13.5.26}$$

reactive oxygenated organic radicals, $ROO^\bullet$, are produced. Alkenes are much more reactive, undergoing reactions with hydroxyl radical,

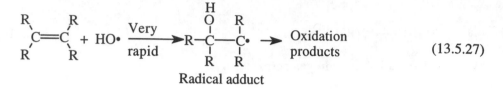

$$\tag{13.5.27}$$

Radical adduct

(where R may be one of a number of hydrocarbon moieties or an H atom) with oxygen atoms,

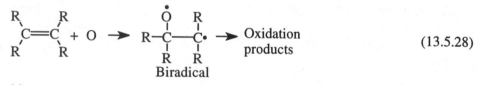

$$\tag{13.5.28}$$

Biradical

or with ozone:

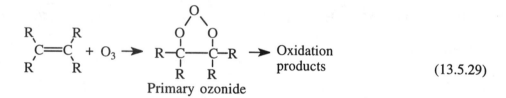

$$\tag{13.5.29}$$

Primary ozonide

Aromatic hydrocarbons, Ar-H, may also react with O and $HO^\bullet$. Addition reactions of aromatics with $HO^\bullet$ are favored. The product of the reaction of benzene with $HO^\bullet$ is phenol, as shown by the following reaction sequence:

$$\tag{13.5.30}$$

$$\tag{13.5.31}$$

In the case of alkyl benzenes, such as toluene, the hydroxyl radical attack may occur on the alkyl group, leading to reaction sequences such as those of alkanes.

Aldehydes react with $HO^\bullet$,

$$R-\overset{\overset{\displaystyle O}{\|}}{C}-H \; + \; HO\cdot + O_2 \; \rightarrow \; R-\overset{\overset{\displaystyle O}{\|}}{C}-OO\cdot \; + \; H_2O$$

$$(13.5.32)$$

$$\overset{H}{\underset{H}{\diagdown}}C{=}O \; + \; HO\cdot + \tfrac{3}{2}O_2 \; \rightarrow \; CO_2 + HOO\cdot + H_2O \qquad (13.5.33)$$

and undergo photochemical reactions:

$$R-\overset{\overset{\displaystyle O}{\|}}{C}-H + h\nu + 2O_2 \; \rightarrow \; ROO\cdot \; + \; CO \; + \; HOO\cdot \qquad (13.5.34)$$

$$\overset{H}{\underset{H}{\diagdown}}C{=}O + h\nu + 2O_2 \; \rightarrow \; CO + 2HOO\cdot \qquad (13.5.35)$$

Hydroxyl radical (HO$\cdot$), which reacts with some hydrocarbons at rates that are almost diffusion-controlled, is the predominant reactant in early stages of smog formation. Significant contributions are made by hydroperoxyl radical (HOO$\cdot$) and O_3 after smog formation is well underway.

One of the most important reaction sequences in the smog-formation process begins with the abstraction by HO$\cdot$ of a hydrogen atom from a hydrocarbon and leads to the oxidation of NO to NO_2 as follows:

$$RH + HO\cdot \; \rightarrow \; R\cdot \; + \; H_2O \qquad (13.5.36)$$

The alkyl radical, R$\cdot$, reacts with O_2 to produce a peroxyl radical, ROO$\cdot$:

$$R\cdot \; + \; O_2 \; \rightarrow \; ROO\cdot \qquad (13.5.37)$$

This strongly oxidizing species very effectively oxidizes NO to NO_2,

$$ROO\cdot \; + \; NO \; \rightarrow \; RO\cdot \; + \; NO_2 \qquad (13.5.38)$$

thus explaining the once-puzzling rapid conversion of NO to NO_2 in an atmosphere in which the latter is undergoing photodissociation. The alkoxyl radical product, RO$\cdot$, is not so stable as ROO$\cdot$. In cases where the oxygen atom is attached to a carbon atom that is also bonded to H, a carbonyl compound is likely to be formed by the following type of reaction:

$$H_3CO\cdot \; + \; O_2 \; \rightarrow \; H-\overset{\overset{\displaystyle O}{\|}}{C}-H \; + \; HOO\cdot \qquad (13.5.39)$$

The rapid production of photosensitive carbonyl compounds from alkoxyl radicals is an important stimulant for further atmospheric photochemical reactions. In the absence of extractable hydrogen, cleavage of a radical containing the carbonyl group occurs:

$$\underset{\underset{\displaystyle H_3C-\overset{\displaystyle \overset{O}{\|}}{C}-O\bullet}{}}{} \rightarrow H_3C\bullet + CO_2 \tag{13.5.40}$$

Another reaction that can lead to the oxidation of NO is of the following type:

$$R-\overset{\overset{\displaystyle O}{\|}}{C}-OO\bullet + NO + O_2 \rightarrow ROO\bullet + NO_2 + CO_2 \tag{13.5.41}$$

Peroxyacyl nitrates (PAN) are highly significant air pollutants formed by an addition reaction with NO_2:

$$R-\overset{\overset{\displaystyle O}{\|}}{C}-OO\bullet + NO_2 \rightarrow R-\overset{\overset{\displaystyle O}{\|}}{C}-OO-NO_2 \tag{13.5.42}$$

When R is the methyl group, the product is peroxyacetyl nitrate, mentioned in Section 13.4. Although it is thermally unstable, peroxyacetyl nitrate does not undergo photolysis rapidly, reacts only slowly with HO• radical, and has a low water solubility. Therefore, the major pathway by which it is lost from the atmosphere is thermal decomposition, the opposite of Reaction 13.5.42.

Alkyl nitrates and alkyl nitrites may be formed by the reaction of alkoxyl radicals (RO•) with nitrogen dioxide and nitric oxide, respectively:

$$RO\bullet + NO_2 \rightarrow RONO_2 \tag{13.5.43}$$

$$RO\bullet + NO \rightarrow RONO \tag{13.5.44}$$

Addition reactions with NO_2 such as these are important in terminating the reaction chains involved in smog formation. Since NO_2 is involved both in the chain initiation step (Reaction 13.5.1) and the chain termination step, moderate reductions in NO_x emissions alone may not curtail smog formation and in some circumstances may even increase it.

As shown in Reaction 13.5.39, the reaction of oxygen with alkoxyl radicals produces hydroperoxyl radical. Peroxyl radicals can react with one another to produce reactive hydrogen peroxide, alkoxyl radicals, and hydroxyl radicals:

$$HOO\bullet + HOO\bullet \rightarrow H_2O_2 + O_2 \tag{13.5.45}$$

$$HOO\bullet + ROO\bullet \rightarrow RO\bullet + HO\bullet + O_2 \tag{13.5.46}$$

$$ROO\bullet + ROO\bullet \rightarrow 2RO\bullet + O_2 \tag{13.5.47}$$

Nitrate Radical

First observed in the troposphere in 1980, nitrate radical, NO_3, is now recognized as being an important atmospheric chemical species, especially at night.[8] This species is formed by the reaction

$$NO_2 + O_3 \rightarrow NO_3 + O_2 \tag{13.5.48}$$

and exists in equilibrium with NO_2:

$$NO_2 + NO_3 + M \leftarrow\rightarrow N_2O_5 + M \text{ (energy-absorbing third body)} \tag{13.5.49}$$

Levels of NO_3 remain low during daylight, typically with a lifetime at noon of only about 5 seconds, because of the following two dissociation reactions:

$$NO_3 + h\nu \,(\lambda < 700\,\text{nm}) \rightarrow NO + O_2 \tag{13.5.50}$$

$$NO_3 + h\nu \,(\lambda < 580\,\text{nm}) \rightarrow NO_2 + O \tag{13.5.51}$$

However, at night the levels of NO_3 typically reach values of around 8×10^7 molecules $\times$ cm^{-3} compared to only about 1×10^6 molecules $\times$ cm^{-3} for hydroxyl radical. Although the hydroxyl radical reacts 10 to 1000 times faster than NO_3, the much higher concentration of the latter means that it is responsible for much of the atmospheric chemistry that occurs at night. The nitrate radical adds across the double bonds in alkenes leading to the formation of reactive radical species that participate in smog formation.

Photolyzable Compounds in the Atmosphere

It may be useful at this time to review the types of compounds capable of undergoing photolysis in the troposphere and thus initiating chain reaction. Under most tropospheric conditions, the most important of these is NO_2:

$$NO_2 + h\nu \,(\lambda < 420\,\text{nm}) \rightarrow NO + O \tag{13.5.1}$$

In relatively polluted atmospheres, the next most important photodissociation reaction is that of carbonyl compounds, particularly formaldehyde:

$$CH_2O + h\nu \,(\lambda < 335\,\text{nm}) \rightarrow H^\bullet + H\overset{\bullet}{C}O \tag{13.5.52}$$

Hydrogen peroxide photodissociates to produce two hydroxyl radicals:

$$HOOH + h\nu \,(\lambda < 350\,\text{nm}) \rightarrow 2HO^\bullet \tag{13.5.53}$$

Finally, organic peroxides may be formed and subsequently dissociate by the following reactions, starting with a peroxyl radical:

$$H_3COO^\bullet + HOO^\bullet \rightarrow H_3COOH + O_2 \tag{13.5.54}$$

$$H_3COOH + h\nu \,(\lambda < 350\,\text{nm}) \rightarrow H_3CO^\bullet + HO^\bullet \tag{13.5.55}$$

It should be noted that each of the last three photochemical reactions gives rise to two free radical species per photon absorbed. Ozone undergoes photochemical dissociation to produce excited oxygen atoms at wavelengths less than 315 nm. These atoms may react with H_2O to produce hydroxyl radicals.

13.6. REACTIVITY OF HYDROCARBONS

The reactivity of hydrocarbons in the smog formation process is an important consideration in understanding the process and in developing control strategies. It is useful to know which are the most reactive hydrocarbons so that their release can be minimized. Less reactive hydrocarbons, of which propane is a good example, may cause smog formation far downwind from the point of release.

Hydrocarbon reactivity is best based upon the interaction of hydrocarbons with hydroxyl radical. Methane, the least reactive common gas-phase hydrocarbon with an atmospheric half-life exceeding 10 days, is assigned a reactivity of 1.0. (Despite its low reactivity, methane is so abundant in the atmosphere that it accounts for a significant fraction of total hydroxyl radical reactions.) In contrast, β-pinene produced by conifer trees and other vegetation, is almost 9000 times as reactive as methane, and d-limonene, produced by orange rind, is almost 19000 times as reactive. Relative to their rates of reaction with hydroxyl radical, hydrocarbon reactivities may be classified from I through V as shown in Table 13.2.

13.7. INORGANIC PRODUCTS FROM SMOG

Two major classes of inorganic products from smog are sulfates and nitrates. Inorganic sulfates and nitrates, along with sulfur and nitrogen oxides, can contribute to acidic precipitation, corrosion, reduced visibility, and adverse health effects.

Although the oxidation of SO_2 to sulfate species is relatively slow in a clean atmosphere, it is much faster under smoggy conditions. During severe photochemical smog conditions, oxidation rates of 5-10% per hour may occur, as compared to only a fraction of a percent per hour under normal atmospheric conditions. Thus, sulfur dioxide exposed to smog can produce very high local concentrations of sulfate, which can aggravate already bad atmospheric conditions.

Several oxidant species in smog can oxidize SO_2. Among the oxidants are compounds, including O_3, NO_3, and N_2O_5, as well as reactive radical species, particularly $HO^\bullet$, $HOO^\bullet$, O, $RO^\bullet$, and $ROO^\bullet$. The two major primary reactions are oxygen transfer,

$$SO_2 + O \text{ (from O, RO}^\bullet, ROO^\bullet) \rightarrow SO_3 \rightarrow H_2SO_4, \text{ sulfates} \qquad (13.7.1)$$

or addition. As an example of the latter, $HO^\bullet$ adds to SO_2 to form a reactive species which can further react with oxygen, nitrogen oxides, or other species to yield sulfates, other sulfur compounds, or compounds of nitrogen:

$$HO^\bullet + SO_2 \rightarrow HOSOO^\bullet \qquad (13.7.2)$$

The presence of $HO^\bullet$ (typically at a level of 3×10^6 radicals/cm^3, but appreciably higher in a smoggy atmosphere) makes this a likely route. Addition of SO_2 to $RO^\bullet$ or $ROO^\bullet$ can yield organic sulfur compounds.

Table 13.2. Relative Reactivities of Hydrocarbons and CO with HO• Radical[1]

Reactivity class	Reactivity range[2]	Approximate half-life in the atmosphere	Compounds in increasing order of reactivity
I	<10	>10 days	Methane
II	10-100	24 h- 10 d	CO, acetylene, ethane
III	100-1000	2.4 - 24 h	Benzene, propane, *n*-butane, isopentane, methylethyl ketone, 2-methylpentane, toluene, *n*-propylbenzene, isopropylbenzene, ethene, *n*-hexane, 3-methylpentane, ethylbenzene
IV	1,000-10,000	15 min-2.4 h	*p*-xylene, *p*-ethyltoluene, *o*-ethyltoluene, *o*-xylene, methylisobutyl ketone, *m*-ethyltoluene, *m*-xylene, 1,2,3-trimethylbenzene, propene, 1,2,4-trimethylbenzene, 1,3,5-trimethylbenzene, *cis*-2-butene, *β*-pinene, 1,3-butadiene
V	>10,000	<15 min	2-methyl-2-butene, 2,4-dimethyl-2-butene, *d*-limonene

[1] Based on data from K. R. Darnall, A. C. Lloyd, A. M. Winer, and J. N. Pitts, Jr., "Reactivity Scale for Atmospheric Hydrocarbons Based on Reaction with Hydroxyl Radical," *Environmental Science and Technology*, **10**, 692-696, (1976).

[2] Based on an assigned reactivity of 1.0 for methane reacting with hydroxyl radical

It should be noted that the reaction of H_2S with HO• is quite rapid. As a result, the normal atmospheric half-life of H_2S of about one-half day becomes much shorter in the presence of photochemical smog.

Inorganic nitrates or nitric acid are formed by several reactions in smog. Among the important reactions forming nitric acid are the reaction of N_2O_5 with water (Reaction 13.5.17) and the addition of hydroxyl radical to NO_2 (Reaction 13.5.21). The oxidation of NO or NO_2 to nitrate species may occur after absorption of gas by an aerosol droplet. Nitric acid formed by these reactions reacts with ammonia in the atmosphere to form ammonium nitrate:

$$NH_3 + HNO_3 \rightarrow NH_4NO_3 \qquad (13.7.3)$$

Other nitrate salts may also be formed.

Nitric acid and nitrates are among the more damaging end products of smog. In addition to possible adverse effects on plants and animals, they cause severe corrosion problems. Electrical relay contacts and small springs associated with electrical switches are especially susceptible to damage from nitrate-induced corrosion.

13.8. EFFECTS OF SMOG

The harmful effects of smog occur mainly in the areas of (1) human health and comfort, (2) damage to materials, (3) effects on the atmosphere, and (4) toxicity to plants. The exact degree to which exposure to smog affects human health is not known, although substantial adverse effects are suspected. Pungent-smelling, smog-produced ozone is known to be toxic. Ozone at 0.15 ppm causes coughing, wheezing, bronchial constriction, and irritation to the respiratory mucous system in healthy, exercising individuals. Largely because of potential health effects, the U.S. Environmental Protection Agency proposed in July 1997 to phase in revised national ambient air quality standards for ground-level ozone that would phase out the existing one-hour ozone standard and set a new 8-hour standard of 0.08 ppm ozone.[9] In addition to ozone, oxidant peroxyacyl nitrates and aldehydes found in smog are eye irritants.

Materials are adversely affected by some smog components. Rubber has a high affinity for ozone and is cracked and aged by it. Indeed, the cracking of rubber used to be employed as a test for the presence of ozone. Ozone attacks natural rubber and similar materials by oxidizing and breaking double bonds in the polymer according to the following reaction:

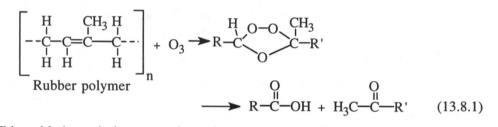

$$(13.8.1)$$

This oxidative scission type of reaction causes bonds in the polymer structure to break and results in deterioration of the polymer.

Aerosol particles that reduce visibility are formed by the polymerization of the smaller moleules produced in smog-forming reactions. Since these reactions largely involve the oxidation of hydrocarbons, it is not suprising that oxygen-containing organics make up the bulk of the particulate matter produced from smog. Ether-soluble aerosols collected from the Los Angeles atmosphere have shown an empirical formula of approximately CH_2O. Among the specific kinds of compounds identified in organic smog aerosols are alcohols, aldehydes, ketones, organic acids, esters, and organic nitrates. Hydrocarbons of plant origin are prominent among the precursors to particle formation in photochemical smog.[10]

Smog aerosols likely form by condensation on existing nuclei rather than by self-nucleation of smog reaction product molecules. In support of this view are electron micrographs of these aerosols showing that smog aerosol particles in the micrometer-size region consist of liquid droplets with an inorganic electron-opaque core (Figure 13.6). Thus, particulate matter from a source other than smog may have some influence on the formation and properties of smog aerosols.

In view of worldwide shortages of food, the known harmful effects of smog on plants is of particular concern. These effects are largely due to oxidants in the smoggy atmosphere. The three major oxidants involved are ozone, PAN, and nitro-

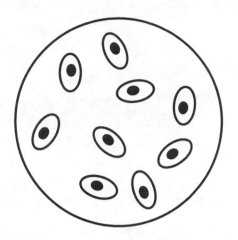

Figure 13.6. Representation of an electron micrograph of smog aerosol particles collected by a jet inertial impactor, showing electron-opaque nuclei in the centers of the impacted droplets.

gen oxides. Of these, PAN has the highest toxicity to plants, attacking younger leaves and causing "bronzing" and "glazing" of their surfaces. Exposure for several hours to an atmosphere containing PAN at a level of only 0.02-0.05 ppm will damage vegetation. The sulfhydryl group of proteins in organisms is susceptible to damage by PAN, which reacts with such groups as both an oxidizing agent and an acetylating agent. Fortunately, PAN is usually present at only low levels. Nitrogen oxides occur at relatively high concentrations during smoggy conditions, but their toxicity to plants is relatively low.

Short-chain alkyl hydroperoxides, which were mentioned in Section 13.4, occur at low levels under smoggy conditions, and even in remote atmospheres. It is possible that these species can oxidize DNA bases, causing adverse genetic effects. Alkyl hydroperoxides are formed under smoggy conditions by the reaction of alkyl peroxy radicals with hydroperoxy radical, $HO_2^\bullet$, as shown for the formation of methyl hydroperoxide below:

$$H_3CO_2^\bullet + HO_2^\bullet \rightarrow H_3COOH + O_2 \qquad\qquad (13.8.2)$$

Ames assays of methyl, ethyl, n-propyl, and n-butyl hydroperoxides (see Chapter 22) have shown some tendency toward mutagenicity on select strains of *Salmonella typhimurium*,[11] although any conclusions drawn from such studies on human health should be made with caution.

The low toxicity of nitrogen oxides and the usually low levels of PAN, hydroperoxides, and other oxidants present in smog leave ozone as the greatest smog-produced threat to plant life. Some plant species, including sword-leaf lettuce, black nightshade, quickweed, and double-fortune tomato, are so susceptible to the effects of ozone and other photochemical oxidants that they are used as bioindicators of the presence of smog.[12] Typical of the phytotoxicity of O_3, ozone damage to a lemon leaf is typified by chlorotic stippling (characteristic yellow spots on a green leaf), as represented in Figure 13.7. Reduction in plant growth may occur without visible lesions on the plant.

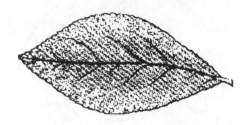

Figure 13.7. Representation of ozone damage to a lemon leaf. In color, the spots appear as yellow chlorotic stippling on the green upper surface caused by ozone exposure.

Brief exposure to approximately 0.06 ppm of ozone may temporarily cut photosynthesis rates in some plants in half. Crop damage from ozone and other photochemical air pollutants in California alone is estimated to cost millions of dollars each year. The geographic distribution of damage to plants in California is illustrated in Figure 13.8.

Figure 13.8. Geographic distribution of plant damage from smog in California.

LITERATURE CITED

1. Roscoe, Howard K. and Kevin C. Clemitshaw, "Measurement Techniques in Gas-phase Tropospheric Chemistry: A Selective View of the Past, Present, and Future," *Science*, **276**, 1065-1072 (1997).

2. Raber, Linda R., "Foes Take Aim at New Air Standards," *Chemical and Engineering News* August 11, 1997, pp. 26-27.

3. Mooney, John J., "Exhaust Control, Automotive," *Encyclopedia of Energy Technology and the Environment*, Vol. 2, John Wiley and Sons, New York, 1995, pp. 1326-1349.

4. Jacoby, Mitch, "Getting Auto Exhausts to Pristine," *Chemical and Engineering News*, January 25, 1999, pp. 36-44.

5. "Air Pollution: Honda Engine 'Sucks up Smog,'" *Chemistry and Industry* (London), **21**, 853 (1997).

6. Finlayson-Pitts, Barbara J., and James N. Pitts, *Atmospheric Chemistry*, John Wiley and Sons, Inc., New York, 1986, p. 478.

7. Seinfeld, John H. and Spyros N. Pandis, *Atmospheric Chemistry and Physics*, John Wiley & Sons, Inc., New York, 1998.

8. Bolzacchini, Ezio, Simone Meinardi, Marco Orlandi, Bruno Rindone, Jens Hjorth, and Gianbattista Restelli, "Nightime Tropospheric Chemistry: Kinetics and Product Studies in the Reaction of 4-Alkyl and 4-Alkoxytoluenes with NO_3 in Gas Phase," *Environmental Science and Technology* **33**, 461-468 (1999).

9. Raber, Linda Ross, "EPA Issues Guidance on Ozone, Particulates Rules," *Chemical and Engineering News*, September 7, 1998, p. 27.

10. Hoffmann, Thorsten, Jay R. Odum, Frank Bowman, Donald Collins, Dieter Klockow, Richard C. Flagan, and John H. Seinfeld, "Formation of Organic Aerosols from the Oxidation of Biogenic Hydrocarbons," *Journal of Atmospheric Chemistry*, **26**, 189-222 (1997).

11. Ball, James C., Willie C. Young, Timothy J. Wallington, and Steven M. Japar, "Mutagenic Properties of a Series of Alkyl Hydroperoxides," *Environmental Science and Technology* **27**, 397-399 (1993).

12. Sun, En-Jang and Mei-Luan Cheng, "Biomonitoring of Peroxyacetyl Nitrate (PAN) and Ozone with Indicator Plant Set," *Proceedings of the Annual Meeting of the Air Waste Management Association*, 90th, 1997.

SUPPLEMENTARY REFERENCES

Brimblecombe, Peter, *Air Composition and Chemistry*, 2nd ed., Cambridge University Press, Cambridge, U.K., 1996.

Bryce-Smith, D., and A. Gilbert, *Photochemistry*, Vol. 20, Royal Society of Chemistry, Letchworth, U.K., 1989.

Calvert, Jack G., *The Mechanisms of Atmospheric Oxidation of the Alkenes*, Oxford University Press, New York, 1999.

Chameides, W. L., and D. D. Davis, "Chemistry in the Troposphere," *Chemical and Engineering News*, October 4, 1982, pp. 38-52.

Finlayson-Pitts, Barbara J., and James N. Pitts, *Atmospheric Chemistry*, John Wiley and Sons, Inc., New York, 1986, p. 478.

Hewitt, C. Nicholas, Ed., *Reactive Hydrocarbons in the Atmosphere*, Academic Press, San Diego, CA, 1999.

Kavouras, Ilias G., Nikolaos Mihalopoulos, and Euripides G. Stephanou, "Formation of Atmospheric Particles from Organic Acids Produced by Forests," *Nature*, **395**, 683-686 (1998).

Lefohn, Allen S., *Surface-Level Ozone Exposures and Their Effects on Vegetation*, Lewis Publishers, Boca Raton, FL, 1991.

Neckers, Douglas C., David H. Volman, and Gunther Von Bunau, *Advances in Photochemistry* (Vol 24), Wiley-Interscience, New York, 1998.

Royal Society of Chemistry, *The Chemistry and Deposition of Nitrogen Species in the Troposphere*, Turpin Distribution Services, Ltd., Letchworth, Herts, U.K., 1993.

Schneider, T., Ed., *Atmospheric Ozone Research and its Policy Implications*, Elsevier Science Publishing Co., New York, 1989.

Seinfeld, John H. and Spyros N. Pandis, *Atmospheric Chemistry and Physics*, John Wiley & Sons, Inc., New York, 1998.

Warneck, Peter, *Chemistry of the Natural Atmosphere*, Academic Press, San Diego, 1988.

Wayne, Carol E. and Richard P. Wayne, *Photochemistry*, Oxford University Press, Oxford, U.K., 1996.

QUESTIONS AND PROBLEMS

1. Of the following species, the one which is the least likely product of the absorption of a photon of light by a molecule of NO_2 is: (a) O, (b) a free radical species, (c) NO, (d) NO_2*, (e) N atoms.

2. Which of the following statements is true: (a) RO• reacts with NO to form alkyl nitrates, (b) RO• is a free radical, (c) RO• is not a very reactive species, (d) RO• is readily formed by the action of stable hydrocarbons and ground state NO_2, (e) RO• is not thought to be an intermediate in the smog-forming process.

3. Of the following species, the one most likely to be found in reducing smog is: ozone, relatively high levels of atomic oxygen, SO_2, PAN, PBN.

4. Why are automotive exhaust pollutant hydrocarbons even more damaging to the environment than their quantities would indicate?

5. At what point in the smog-producing chain reaction is PAN formed?

6. What particularly irritating product is formed in the laboratory by the irradiation of a mixture of benzaldelhyde and NO_2 with ultraviolet light?

7. Which of the following species reaches its peak value last on a smog-forming day: NO, oxidants, hydrocarbons, NO_2?

8. What is the main species responsible for the oxidation of NO to NO_2 in a smoggy atmosphere?

9. Give two reasons why a turbine engine should have lower hydrocarbon emissions than an internal combustion engine.

10. What pollution problem does a lean mixture aggravate when employed to control hydrocarbon emissions from an internal combustion engine?

11. Why is a modern automotive catalytic converter called a "three-way conversion catalyst?"

12. What is the distinction between *reactivity* and *instability* as applied to some of the chemically active species in a smog-forming atmosphere?

13. Why might carbon monoxide be chosen as a standard against which to compare automotive hydrocarbon emissions in atmospheres where smog is formed? What are some pitfalls created by this choice?

14. What is the purpose of alumina in an automotive exhaust catalyst? What kind of material actually catalyzes the destruction of pollutants in the catalyst?

15. Some atmospheric chemical reactions are abstraction reactions and others are addition reactions. Which of these applies to the reaction of hydroxyl radical with propane? With propene (propylene)?

16. How might oxidants be detected in the atmosphere?

17. Although NO_x is necessary for smog formation, there are circumstances where it is plausible that moderate reductions of NO_x levels might actually increase the rate of smog formation. Suggest a reason why this might be so.

18. Why is ozone especially damaging to rubber?

19. Show how hydroxyl radical, HO•, might react differently with ethylene, $H_2C=CH_2$, and methane, CH_4.

20. Name the stable product that results from an initial addition reaction of hydroxyl radical, HO•, with benzene.

14 THE ENDANGERED GLOBAL ATMOSPHERE

14.1. ANTHROPOGENIC CHANGE IN THE ATMOSPHERE

There is a very strong connection between life forms on earth and the nature of earth's climate, which determines its suitability for life. As proposed by James Lovelock, a British chemist, this forms the basis of the **Gaia hypothesis**, which contends that the atmospheric O_2/CO_2 balance established and sustained by organisms determines and maintains earth's climate and other environmental conditions.[1]

Ever since life first appeared on earth, the atmosphere has been influenced by the metabolic processes of living organisms. When the first primitive life molecules were formed approximately 3.5 billion years ago, the atmosphere was very different from its present state. At that time it was chemically reducing and thought to contain nitrogen, methane, ammonia, water vapor, and hydrogen, but no elemental oxygen. These gases and water in the sea were bombarded by intense, bond-breaking ultraviolet radiation which, along with lightning and radiation from radionuclides, provided the energy to bring about chemical reactions that resulted in the production of relatively complicated molecules, including even amino acids and sugars. From this rich chemical mixture, life molecules evolved. Initially, these very primitive life forms derived their energy from fermentation of organic matter formed by chemical and photochemical processes, but eventually they gained the capability to produce organic matter, "$\{CH_2O\}$," by photosynthesis,

$$CO_2 + H_2O + h\nu \rightarrow \{CH_2O\} + O_2(g) \qquad (14.1.1)$$

and the stage was set for the massive biochemical transformation that resulted in the production of almost all the atmosphere's oxygen.

The oxygen initially produced by photosynthesis was probably quite toxic to primitive life forms. However, much of this oxygen was converted to iron oxides by reaction with soluble iron(II):

$$4Fe^{2+} + O_2 + 4H_2O \rightarrow 2Fe_2O_3 + 8H^+ \tag{14.1.2}$$

The enormous deposits of iron oxides thus formed provide convincing evidence for the liberation of free oxygen in the primitive atmosphere.

Eventually enzyme systems developed that enabled organisms to mediate the reaction of waste-product oxygen with oxidizable organic matter in the sea. Later this mode of waste-product disposal was utilized by organisms to produce energy for respiration, which is now the mechanism by which nonphotosynthetic organisms obtain energy.

In time, oxygen accumulated in the atmosphere, providing an abundant source of O_2 for respiration. It had an additional benefit in that it enabled the formation of an ozone shield against solar ultraviolet radiation (see Section 11.6). With this shield in place, the earth became a much more hospitable environment for life, and life forms were enabled to move from the protective surroundings of the sea to the more exposed environment of the land.

Other instances of climatic change and regulation induced by organisms can be cited. An example is the maintenance of atmospheric carbon dioxide at low levels through the action of photosynthetic organisms (note from Reaction 14.1.1 that photosynthesis removes CO_2 from the atmosphere). But, at an ever accelerating pace during the last 200 years, another organism, humankind, has engaged in a number of activities that are altering the atmosphere profoundly. As noted in Chapter 1, human influences are so strong that it is useful to invoke a fifth sphere of the environment, the anthrosphere. The effects of human activities and the anthrosphere on the atmosphere are summarized below:

- Industrial activities, which emit a variety of atmospheric pollutants including SO_2, particulate matter, photochemically reactive hydrocarbons, chlorofluorocarbons, and inorganic substances (such as toxic heavy metals)

- Burning of large quantities of fossil fuel, which can introduce CO_2, CO, SO_2, NO_X, hydrocarbons (including CH_4), and particulate soot, polycyclic aromatic hydrocarbons, and fly ash into the atmosphere

- Transportation practices, which emit CO_2, CO, NO_X, photochemically reactive (smog forming) hydrocarbons, and polycyclic aromatic hydrocarbons

- Alteration of land surfaces, including deforestation

- Burning of biomass and vegetation, including tropical and subtropical forests and savanna grasses, which produces atmospheric CO_2, CO, NO_X, and particulate soot and polycyclic aromatic hydrocarbons

- Agricultural practices, which produce methane (from the digestive tracts of domestic animals and from the cultivation of rice in waterlogged anaerobic soils) and N_2O from bacterial denitrification of nitrate-fertilized soils.

These kinds of human activities have significantly altered the atmosphere, particularly in regard to its composition of minor constituents and trace gases. Major effects have been the following:

- Increased acidity in the atmosphere

- Production of pollutant oxidants in localized areas of the lower troposphere (see Photochemical Smog, Chapter 13)

- Elevated levels of infrared-absorbing gases (greenhouse gases)

- Threats to the ultraviolet-filtering ozone layer in the stratosphere

- Increased corrosion of materials induced by atmospheric pollutants

In 1957 photochemical smog was only beginning to be recognized as a serious problem, acid rain and the greenhouse effect were scientific curiosities, and the ozone-destroying potential of chlorofluorocarbons had not even been imagined. In that year, Revelle and Suess[2] prophetically referred to human perturbations of the earth and its climate as a massive "geophysical experiment." The effects that this experiment may have on the global atmosphere are discussed in this chapter.

14.2. GREENHOUSE GASES AND GLOBAL WARMING

This section deals with infrared-absorbing trace gases (other than water vapor) in the atmosphere that contribute to global warming. These gases produce a "greenhouse effect" by allowing incoming solar radiant energy to penetrate to the earth's surface while reabsorbing infrared radiation emanating from it. Levels of these "greenhouse gases" have increased at a rapid rate during recent decades and are continuing to do so. Concern over this phenomenon has intensified since about 1980. This is because ever since accurate temperature records have been kept, the 1980s were the warmest 10-year period recorded and included several record warm years. In general, the 1990s continued the warming trend. All months in 1998 except for October (which missed by about 0.1°C) set record monthly temperature highs, and 1998 was the warmest year on record as of 1999.[3] Figure 4.1 shows global temperature trends since 1880. In addition to being a scientific issue, greenhouse warming of the atmosphere has also become a major policy, political, and economic issue.

The analysis of fossil ice provides evidence of past variations in temperature.[4] One characteristic of ice that indicates the temperature at which it was deposited is conductivity, which declines with declining temperature of ice formation. The other characteristic is the $^{18}O/^{16}O$ ratio, which is higher with increasing temperature of ice formation. Ice from the Vostok core taken in Antarctica dates back as far as 500,000 years, providing a valuable record of past climatic conditions.

There are many uncertainties surrounding the issue of greenhouse warming. However, several things about the phenomenon are certain. It is known that CO_2 and other greenhouse gases, such as CH_4, absorb infrared radiation by which earth loses heat. The levels of these gases have increased markedly since about 1850 as nations have become industrialized and as forest lands and grasslands have been converted to agriculture. As shown in Figure 14.2, per capita carbon dioxide emissions are high-

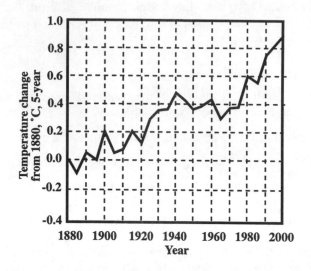

Figure 14.1. Global temperature trends. Earlier values are less certain because of the lack of sophisticated means of measuring temperature. More recent values are very accurate because of the use of satellite-based technologies for measuring temperature.

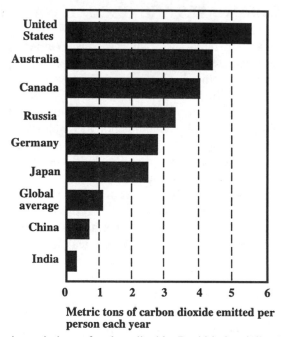

Figure 14.2. Per capita emissions of carbon dioxide. Rapid industrialization of high-population countries, particularly China and India, can be expected to add much larger quantities of carbon dioxide to the atmosphere in the future.

est for industrialized countries, and development of countries with high populations, such as China and India, can be expected to add large quantities of carbon dioxide to the atmosphere in the future. Chlorofluorocarbons, which also are greenhouse gases, were not even introduced into the atmosphere until the 1930s. Although trends in levels of these gases are well known, their effects on global temperature and climate are much less certain. The phenomenon has been the subject of much computer modeling. Most models predict global warming of at least 3.0° and up to 5.5°C occurring over a period of just a few decades. These estimates are sobering because they correspond to the approximate temperature increase since the last ice age 18,000 years past, which took place at a much slower pace of only about 1 or 2°C per 1,000 years. Such warming would have profound effects on rainfall, plant growth, and sea levels, which might rise as much as 0.5-1.5 meters.

Carbon dioxide is the gas most commonly thought of as a greenhouse gas; it is responsible for about half of the atmospheric heat retained by trace gases. It is produced primarily by the burning of fossil fuels, and deforestation accompanied by burning and biodegradation of biomass. On a molecule-for-molecule basis, methane, CH_4, is 20–30 times more effective in trapping heat than is CO_2. Other trace gases that contribute are chlorofluorocarbons and N_2O. The potential of such a gas to cause greenhouse warming may be expressed by a global warming potential (GWP), originally defined by the United Nations' Intergovernmental Panel on Climate Change, which is a function of both the infrared sorption characteristics and the lifetime of the gas.

Analyses of gases trapped in polar ice samples indicate that preindustrial levels of CO_2 and CH_4 in the atmosphere were approximately 260 parts per million and 0.70 ppm, respectively. Over the last 300 years these levels have increased to current values of around 360 ppm, and 1.8 ppm, respectively; most of the increase by far has taken place at an accelerating pace over the last 100 years. (A note of interest is the observation based upon analyses of gases trapped in ice cores that the atmospheric level of CO_2 at the peak of the last ice age about 18,000 years past was 25 percent below preindustrial levels.) About half of the increase in carbon dioxide in the last 300 years can be attributed to deforestation, which still accounts for approximately 20 percent of the annual increase in this gas. Carbon dioxide is increasing by about 1 ppm per year.

Methane is going up at a rate of almost 0.02 ppm/year. The comparatively very rapid increase in methane levels is attributed to a number of factors resulting from human activities. Among these are direct leakage of natural gas, byproduct emissions from coal mining and petroleum recovery, and release from the burning of savannas and tropical forests. Biogenic sources resulting from human activities produce large amounts of atmospheric methane. These include methane from bacteria degrading organic matter such as municipal refuse in landfills; methane evolved from anaerobic biodegradation of organic matter in rice paddies; and methane emitted as the result of bacterial action in the digestive tracts of ruminant animals.

In addition to acting as a greenhouse gas, methane has significant effects upon atmospheric chemistry. It produces atmospheric CO as an intermediate oxidation product and influences concentrations of atmospheric hydroxyl radicals and ozone. In the stratosphere, it produces hydrogen and H_2O, but acts to remove ozone-destroying chlorine.

A term called **radiative forcing** is used to describe the reduction in infrared radiation penetrating outward through the atmosphere per unit increase in the level of gas in the atmosphere. Using this measure it can be shown that the radiative forcing of CH_4 is about 25 times that of CO_2. Increases in the concentration of methane and several other greenhouse gases have such a disproportionate effect on retention of infrared radiation because their infrared absorption spectra fill gaps in the overall spectrum of outbound radiation not covered by much more abundant carbon dioxide. Therefore, whereas an increase in carbon dioxide concentration has a comparatively small incremental effect because the gas is already absorbing such a high fraction of infrared radiation in regions of the spectrum where it absorbs, an increase in the concentration of methane, chlorofluorocarbon, or other greenhouse gases has a comparatively much larger effect.

Both positive and negative feedback mechanisms may be involved in determining the rates at which carbon dioxide and methane build up in the atmosphere. Laboratory studies indicate that increased CO_2 levels in the atmosphere cause accelerated uptake of this gas by plants undergoing photosynthesis, which tends to slow the buildup of atmospheric CO_2. Given adequate rainfall, plants living in a warmer climate that would result from the greenhouse effect would grow faster and take up more CO_2. This could be an especially significant effect of forests, which have a high CO_2-fixing ability. However, the projected rate of increase in carbon dioxide levels is so rapid that forests would lag behind in their ability to fix additional CO_2. Similarly, higher atmospheric CO_2 concentrations will result in accelerated sorption of the gas by oceans. The amount of dissolved CO_2 in the oceans is about 60 times the amount of CO_2 gas in the atmosphere. However, the times for transfer of carbon dioxide from the atmosphere to the ocean are of the order of years. Because of low mixing rates, the times for transfer of oxygen from the upper approximately 100-meter layer of the oceans to ocean depths is much longer, of the order of decades. Therefore, like the uptake of CO_2 by forests, increased absorption by oceans will lag behind the emissions of CO_2. A concern with increased levels of CO_2 in the oceans is the lowering of ocean water pH that will result. Even though such an effect will be slight, of the order of one tenth to several tenths of a pH unit, it has the potential to strongly impact organisms that live in ocean water. Severe drought conditions resulting from climatic warming could cut down substantially on CO_2 uptake by plants. Warmer conditions would accelerate release of both CO_2 and CH_4 by microbial degradation of organic matter. (It is important to realize that about twice as much carbon is held in soil in dead organic matter—necrocarbon—potentially degradable to CO_2 and CH_4 as is present in the atmosphere.) Global warming might speed up the rates at which biodegradation adds these gases to the atmosphere.

It is certain that atmospheric CO_2 levels will continue to increase significantly. The degree to which this occurs depends upon future levels of CO_2 production and the fraction of that production that remains in the atmosphere. Given plausible projections of CO_2 production and a reasonable estimate that half of that amount will remain in the atmosphere, projections can be made that indicate that sometime during the middle part of the next century the concentration of this gas will reach 600 ppm in the atmosphere. This is well over twice the levels estimated for pre-industrial times. Much less certain are the effects that this change will have on climate. It is virtually impossible for the elaborate computer models used to estimate these effects to accurately take account of all variables, such as the degree and nature of cloud cover. Clouds both reflect incoming light radiation and absorb outgoing infrared radiation, with the former effect tending to predominate. The magnitudes of these effects depend upon the degree of cloud cover, brightness, altitude, and thickness. In the case of clouds, too, feedback phenomena occur; for example, warming induces formation of more clouds, which reflect more incoming energy.

Drought is one of the most serious problems that could arise from major climatic change resulting from greenhouse warming. Typically, a three-degree warming would be accompanied by a ten percent decrease in precipitation. Water shortages would be aggravated, not just from decreased rainfall, but from increased evaporation as well. Increased evaporation results in decreased runoff, thereby reducing water available for agricultural, municipal, and industrial use. Water shortages, in turn, lead to increased demand for irrigation and to the production of lower quality, higher salinity runoff water and wastewater. In the U. S. such a problem would be especially intense in the Colorado River basin, which supplies much of the water used in the rapidly growing U. S. Southwest.

A variety of other problems, some of them unforeseen as of now, could result from global warming. An example is the effect of warming on plant and animal pests —insects, weeds, diseases, and rodents. Many of these would certainly thrive much better under warmer conditions.

Interestingly, another air pollutant, acid-rain-forming sulfur dioxide (see Section 14.3), may have a counteracting effect on greenhouse gases.[5] This is because sulfur dioxide is oxidized in the atmosphere to sulfuric acid, forming a light-reflecting haze. Furthermore, the sulfuric acid and resulting sulfates act as condensation nuclei upon which atmospheric water vapor condenses, thereby increasing the extent, density, and brightness of light-reflecting cloud cover. Sulfate aerosols are particularly effective in counteracting greenhouse warming in central Europe and the eastern United States during the summer.

Some evidence of the effects of global warming may have been manifested by the powerful El Niño phenomenon that occurred during the late months of 1997 and early months of 1998. El Niño is the name given to the warming of surface water in the eastern Pacific Ocean that commonly takes place around Christmas time. The 1997/98 El Niño was particularly powerful and caused many marked weather phenomena. It also increased confidence in global climate models because of the generally accurate forecasts of its effect on climate. Forecasts of a warmer and wetter winter than normal in the continental United States with particularly heavy rains in California and the Gulf Coast regions were fulfilled. In fact, Los Angeles experienced more than 33 cm of rainfall during February 1998, setting a new record for the month, and the southeastern U. S. had the most rainfall in more than a century of record keeping. Eastern equatorial Africa experienced torrential rains, as did parts of Peru. Indonesia experienced a severe drought that resulted in extensive destruction of forests by fires. The central and northern continental U.S. benefitted from a warmer winter than usual with record warmth in some eastern and mid-western regions.

As it affected North America, the stronger El Niño enhanced the intensity of the eastward flowing high altitude jet stream, the result of a greater temperature differential between warmer southern tropical waters and the colder northern regions. The intense jet stream carried storms rapidly across the southern U. S., causing intense rainstorms in these regions and keeping cold Arctic air to the north. Some authorities contend that the effects of the 1997/1998 El Niño were increased by global warming.

Serious Concern Over Changes in Climate

As outlined in an article entitled "Storm Warnings Rattle Insurers,"[6] insurance companies have become quite concerned about the possibility of significant changes in global climate, especially because of potential effects on the frequency and severity of damaging storms. In 1996-97 there were at least six weather disasters that cost over a billion dollars each. These included (1) a catastrophic drought in the Southern Plains that began in the fall of 1995 and lasted through the summer of 1996; (2) a blizzard followed by flooding that occurred in the northeastern U.S., the mid-Atlantic states, and the Appalachian mountain areas in January of 1996; (3) flooding in the Pacific Northwest in February 1996; (4) Hurricane Fran, which caused 36 deaths and over $5 billion in damage during September 1996; (5) severe flooding in the northern west coast region of the U.S. in December 1996, and January, 1997; and (6) an unprecedented 500-year flood complicated by freezing weather and ice jams that hit the Dakotas and Minnesota in April 1997, virtually wiping out the city of Grand Forks, North Dakota. One of the most destructive storms of all time was Hurricane Mitch, which struck Central America in 1998. Estimates of loss of life in this storm ranged up to 13,000. Property damage in Honduras was approximately 3 billion dollars, and in Nicaragua the damage was about one billion dollars. On May 3, 1999, an F5 tornado, the largest class of this kind of treacherous storm, took a number of lives and caused about $1 billion damage in Central Oklahoma.

Although drought is the most frequently mentioned possible effect of greenhouse warming, the frequency and severity of storms, often accompanied by high levels of precipitation, have the insurance companies particularly concerned. During the 1990s "100-year" weather events, those that are expected to occur statistically only once each century, have become so common in the U.S. that the term has begun to lose its meaning. As shown in Figure 14.3, the last century has seen a significant increase in precipitation in the lower 48 continental United States. These observations are consistent with currently accepted models of the weather effects of greenhouse warming, which predict that more precipitation will come in the form of brief, heavy precipitation events such as thunderstorms (heavy convective storms) rather than through gentle rainfall that comes over a longer time period. The debate continues over whether the apparent weather anomalies observed during recent years denote a marked change in climate or are simply normal fluctuations in weather. However, the insurance companies, whose prosperity and even survival depend upon accurate statistical analysis of often subtle risk factors, seem to be concerned that human effects on climate are real and that the resulting losses may be substantial.

International concern over global warming led to a meeting of 160 nations in Kyoto, Japan, in December 1997. At that meeting, the U.S. proposed stabilizing emissions of greenhouse gases to 1990 levels during the period 2008-2012. If this were achieved, levels of greenhouse gases would be 23% below those projected from trends projected from the 1990s without remedial action. The proposed agreements resulting from this meeting have met with severe criticism in the U.S. because of exemptions for developing countries, which can be expected to produce increasing fractions of greenhouse gases in the future. Meanwhile, the increased popularity of large "sport utility vehicles" in the U.S., which emit disproportinate amounts of carbon dioxide per unit distance traveled, continue to add to levels of atmospheric carbon dioxide.

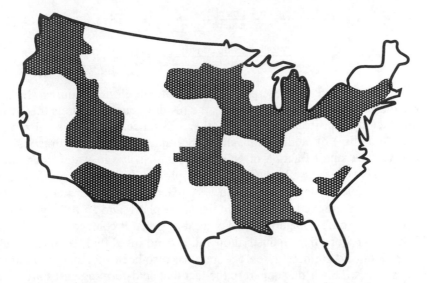

Figure 14.3. Sections of the lower 48 United States in which precipitation levels have increased by 10-20% since about 1900 (shown as shaded regions). Some areas, particularly North Dakota, eastern Montana, Wyoming, and California have experienced decreases in precipitation of a similar magnitude. This map is based on data gathered by the National Oceanic and Atmospheric Administration's National Climatic Data Center.

14.3. ACID RAIN

Precipitation made acidic by the presence of acids stronger than $CO_2(aq)$ is commonly called **acid rain**; the term applies to all kinds of acidic aqueous precipitation, including fog, dew, snow, and sleet. In a more general sense, **acid deposition** refers to the deposition on the earth's surface of aqueous acids, acid gases (such as SO_2), and acidic salts (such as NH_4HSO_4). According to this definition, deposition in solution form is *acid precipitation*, and deposition of dry gases and compounds is *dry deposition*. Although carbon dioxide is present at higher levels in the atmosphere, sulfur dioxide, SO_2, contributes more to the acidity of precipitation for two reasons. The first of these is that sulfur dioxide is significantly more soluble in water than is carbon dioxide, as indicated by its Henry's law constant (Section 5.3) of 1.2 $mol \times L^{-1} \times atm^{-1}$ compared to 3.38×10^{-2} $mol \times L^{-1} \times atm^{-1}$ for CO_2. Secondly, the value of K_{a1} for $SO_2(aq)$,

$$SO_2(aq) + H_2O \; \longleftarrow \!\! \rightarrow \; H^+ + HSO_3^- \tag{14.3.1}$$

$$K_{a1} = \frac{[H^+][HSO_3^-]}{[SO_2]} = 1.7 \times 10^{-2} \tag{14.3.2}$$

is more than four orders of magnitude higher than the value of 4.45×10^{-7} for CO_2.

Although acid rain can originate from the direct emission of strong acids, such as HCl gas or sulfuric acid mist, most of it is a secondary air pollutant produced by the atmospheric oxidation of acid-forming gases such as the following:

$$SO_2 + {}^1\!/_2 O_2 + H_2O \xrightarrow[\text{ing of several steps}]{\text{Overall reaction consist-}} \{2H^+ + SO_4{}^{2-}\}(aq) \qquad (14.3.3)$$

$$2NO_2 + {}^1\!/_2 O_2 + H_2O \xrightarrow[\text{ing of several steps}]{\text{Overall reaction consist-}} 2\{H^+ + NO_3{}^-\}(aq) \qquad (14.3.4)$$

Chemical reactions such as these play a dominant role in determining the nature, transport, and fate of acid precipitation. As the result of such reactions the chemical properties (acidity, ability to react with other substances) and physical properties (volatility, solubility) of acidic atmospheric pollutants are altered drastically. For example, even the small fraction of NO that does dissolve in water does not react significantly. However, its ultimate oxidation product, HNO_3, though volatile, is highly water-soluble, strongly acidic, and very reactive with other materials. Therefore, it tends to be removed readily from the atmosphere and to do a great deal of harm to plants, corrodible materials, and other things that it contacts.

Although emissions from industrial operations and fossil fuel combustion are the major sources of acid-forming gases, acid rain has also been encountered in areas far from such sources. This is due in part to the fact that acid-forming gases are oxidized to acidic constituents and deposited over several days, during which time the air mass containing the gas may have moved as much as several thousand km. It is likely that the burning of biomass, such as is employed in "slash-and-burn" agriculture evolves the gases that lead to acid formation in more remote areas. In arid regions, dry acid gases or acids sorbed to particles may be deposited with effects similar to those of acid rain deposition.

Acid rain spreads out over areas of several hundred to several thousand kilometers. This classifies it as a *regional* air pollution problem compared to a *local* air pollution problem for smog and a *global* one for ozone-destroying chlorofluorocarbons and greenhouse gases. Other examples of regional air pollution problems are those caused by soot, smoke, and fly ash from combustion sources and fires (forest fires). Nuclear fallout from weapons testing or from reactor fires (of which, fortunately, there has been only one major one to date—the one at Chernobyl in the Soviet Union) may also be regarded as a regional phenomenon.

Acid precipitation shows a strong geographic dependence, as illustrated in Figure 14.4, representing the pH of precipitation in the continental U.S. The preponderance of acidic rainfall in the northeastern U.S., which also affects southeastern Canada, is obvious. Analyses of the movements of air masses have shown a correlation between acid precipitation and prior movement of an air mass over major sources of anthropogenic sulfur and nitrogen oxides emissions. This is particularly obvious in southern Scandinavia, which receives a heavy burden of air pollution from densely populated, heavily industrialized areas in Europe.

Acid rain has been observed for well over a century, with many of the older observations from Great Britain. The first manifestations of this phenomenon were elevated levels of $SO_4{}^{2-}$ in precipitation collected in industrialized areas. More modern evidence was obtained from analyses of precipitation in Sweden in the 1950s, and of U.S. precipitation a decade or so later. A vast research effort on acid rain was conducted in North America by the National Acid Precipitation Assessment Program, which resulted from the U.S. Acid Precipitation Act of 1980.

Figure 14.4. Isopleths of pH illustrating a hypothetical precipitation-pH pattern in the lower 48 continental United States. Actual values found may vary with the time of year and climatic conditions.

The longest-term experimental study of acid precipitation in the U.S. has been conducted at the U.S. Forest Service Hubbard Brook Experimental Forest in New Hampshire's White Mountains. It is downwind from major U.S. urban and industrial centers and is, therefore, a prime candidate to receive acid precipitation. This is reflected by mean annual pH values ranging from 4.0 to 4.2 during the 1964-74 period. During this period, the annual hydrogen ion input ($[H^+]$ x volume) increased by 36%.

Table 14.1 shows typical major cations and anions in pH-4.25 precipitation. Although actual values encountered vary greatly with time and location of collection, this table does show some major features of ionic solutes in precipitation. From the predominance of sulfate anion, it is apparent that sulfuric acid is the major contributor to acid precipitation. Nitric acid makes a smaller but growing contribution to the acid present. Hydrochloric acid ranks third.

An important factor in the study of acid rain and sulfur pollution involves the comparison of primary sulfate species (those emitted directly by point sources) and secondary sulfate species (those formed from gaseous sulfur compounds, primarily by the atmospheric oxidation of SO_2). A low primary-sulfate content indicates transport of the pollutant from some distance, whereas a high primary-sulfate content indicates local emissions (the high water solubility of sulfate species means that they do not move far from their sources before being removed with precipitation). This information can be useful in determining the effectiveness of SO_2 control in reducing atmospheric sulfate, including sulfuric acid. Primary and secondary sulfates can be measured using the oxygen-18 content of the sulfates, which is higher in sulfate emitted directly from a power plant than it is in sulfate formed by the oxidation of SO_2. This technique can yield valuable information on the origins and control of acid rain.

Table 14.1. Typical Values of Ion Concentrations in Acidic Precipitation

Cations		Anions	
Ion	Concentration equivalents/L $\times 10^6$	Ion	Concentration equivalents/L $\times 10^6$
H^+	56	SO_4^{2-}	51
NH_4^+	10	NO_3^-	20
Ca^{2+}	7	Cl^-	12
Na^+	5	Total	83
Mg^{2+}	3		
K^+	2		
Total	83		

Ample evidence exists of the damaging effects of acid rain. The major effects are the following:

- Direct phytotoxicity to plants from excessive acid concentrations. (Evidence of direct or indirect phytoxicity of acid rain is provided by the declining health of Eastern U.S. and Scandinavian forests and especially by damage to Germany's Black Forest.)

- Phytotoxicity from acid-forming gases, particularly SO_2 and NO_2, that accompany acid rain

- Indirect phytotoxicity, such as from Al^{3+} liberated from soil

- Destruction of sensitive forests

- Respiratory effects on humans and other animals

- Acidification of lake water with toxic effects to lake flora and fauna, especially fish fingerlings

- Corrosion of exposed structures, electrical relays, equipment, and ornamental materials. Because of the effect of hydrogen ion,

$$2H^+ + CaCO_3(s) \rightarrow Ca^{2+} + CO_2(g) + H_2O$$

limestone, $CaCO_3$, is especially susceptible to damage from acid rain

- Associated effects, such as reduction of visibility by sulfate aerosols and the influence of sulfate aerosols on physical and optical properties of clouds. (As mentioned in Section 14.2, intensification of cloud cover and changes in the optical properties of cloud droplets—specifically, increased reflectance of light—resulting from acid sulfate in the atmosphere may even have a mitigating effect on greenhouse warming of the atmosphere.) A significant association exists between acidic sulfate in the atmosphere and haziness.

Soil sensitivity to acid precipitation can be estimated from cation exchange capacity (CEC, see Chapter 5). Soil is generally insensitive if free carbonates are present or if it is flooded frequently. Soils with a cation exchange capacity above 15.4 milliequivalents/100 g are also insensitive. Soils with cation exchange capacities between 6.2 meq/100 g and 15.4 meq/100 g are slightly sensitive. Soils with cation exchange capacities below 6.2 meq/100 g normally are sensitive if free carbonates are absent and the soil is not frequently flooded.

Forms of precipitation other than rainfall may contain excess acidity. Acidic fog can be especially damaging because it is very penetrating. In early December 1982, Los Angeles experienced a severe, two-day episode of acid fog. This fog consisted of a heavy concentration of acidic mist particles at ground level which reduced visibility and were very irritating to breathe. The pH of the water in these particles was 1.7, much lower than ever before recorded for acid precipitation. Another source of precipitation heavy in the ammonium, sulfate, and nitrate ions associated with atmospheric acid is **acid rime**. Rime is frozen cloudwater which may condense on snowflakes or exposed surfaces. Rime constitutes up to 60% of the snowpack in some mountainous areas, and the deposition of acidic constituents with rime may be a significant vector for the transfer of acidic atmospheric constituents to earth's surface in some cases.

14.4. OZONE LAYER DESTRUCTION

Recall from Section 9.9 that stratospheric ozone, O_3, serves as a shield to absorb harmful ultraviolet radiation in the stratosphere, protecting living beings on the earth from the effects of excessive amounts of such radiation. The two reactions by which stratospheric ozone is produced are,

$$O_2 + h\nu \rightarrow O + O \qquad (\lambda < 242.4 \text{ nm}) \tag{14.4.1}$$

$$O + O_2 + M \quad \rightarrow O_3 + M \text{ (energy-absorbing } N_2 \text{ or } O_2) \tag{14.4.2}$$

and it is destroyed by photodissociation,

$$O_3 + h\nu \rightarrow O_2 + O \qquad (\lambda < 325 \text{ nm}) \tag{14.4.3}$$

and a series of reactions from which the net result is the following:

$$O + O_3 \rightarrow 2O_2 \tag{14.4.4}$$

The concentration of ozone in the stratosphere is a steady-state concentration resulting from the balance of ozone production and destruction by the above processes. The quantities of ozone involved are interesting. A total of about 350,000 metric tons of ozone are formed and destroyed daily. Ozone never makes up more than a small fraction of the gases in the ozone layer. In fact, if all the atmosphere's ozone were in a layer at 273 K and 1 atm, it would be only 3 mm thick!

Ozone absorbs ultraviolet radiation very strongly in the region 220-330 nm. Therefore, it is effective in filtering out dangerous UV-B radiation, 290 nm < λ <

320 nm. (UV-A radiation, 320 nm-400 nm, is relatively less harmful and UV-C radiation, < 290 nm does not penetrate to the troposphere.) If UV-B were not absorbed by ozone, severe damage would result to exposed forms of life on the earth. Absorption of electromagnetic radiation by ozone converts the radiation's energy to heat and is responsible for the temperature maximum encountered at the boundary between the stratosphere and the mesosphere at an altitude of approximately 50 km. The reason that the temperature maximum occurs at a higher altitude than that of the maximum ozone concentration arises from the fact that ozone is such an effective absorber of ultraviolet light, so that most of this radiation is absorbed in the upper stratosphere where it generates heat and only a small fraction reaches the lower altitudes, which remain relatively cool.

Increased intensities of ground-level ultraviolet radiation caused by stratospheric ozone destruction would have some significant adverse consequences. One major effect would be on plants, including crops used for food. The destruction of microscopic plants that are the basis of the ocean's food chain (phytoplankton) could severely reduce the productivity of the world's seas. Human exposure would result in an increased incidence of cataracts. The effect of most concern to humans is the elevated occurrence of skin cancer in individuals exposed to ultraviolet radiation. This is because UV-B radiation is absorbed by cellular DNA (see Chapter 22) resulting in photochemical reactions that alter the function of DNA so that the genetic code is improperly translated during cell division. This can result in uncontrolled cell division leading to skin cancer. People with light complexions lack protective melanin, which absorbs UV-B radiation, and are especially susceptible to its effects. The most common type of skin cancer resulting from ultraviolet exposure is squamous cell carcinoma, which forms lesions that are readily removed and has little tendency to spread (metastasize). Readily metastasized malignant melanoma caused by absorption of UV-B radiation is often fatal. Fortunately, this form of skin cancer is relatively uncommon.

The major culprit in ozone depletion consists of chlorofluorocarbon (CFC) compounds, commonly known as "Freons." These volatile compounds have been used and released to a very large extent in recent decades. The major use associated with CFCs is as refrigerant fluids. Other applications have included solvents, aerosol propellants, and blowing agents in the fabrication of foam plastics. The same extreme chemical stability that makes CFCs nontoxic enables them to persist for years in the atmosphere and to enter the stratosphere. In the stratosphere, as discussed in Section 12.7, the photochemical dissociation of CFCs by intense ultraviolet radiation,

$$CF_2Cl_2 + h\nu \ \rightarrow \ Cl^\bullet + CClF_2^\bullet \tag{14.4.5}$$

yields chlorine atoms, each of which can go through chain reactions, particularly the following:

$$Cl^\bullet + O_3 \ \rightarrow \ ClO^\bullet + O_2 \tag{14.4.6}$$

$$ClO^\bullet + O \ \rightarrow \ Cl^\bullet + O_2 \tag{14.4.7}$$

The net effect of these reactions is catalysis of the destruction of several thousand molecules of O_3 for each Cl atom produced. Because of their widespread use and persistency, the two CFCs of most concern in ozone destruction are CFC-11 and CFC-12, $CFCl_3$, and CF_2Cl_2, respectively. Even in the intense ultraviolet radiation of the stratosphere the most persistent chlorofluorcarbons have lifetimes of the order of 100 years.

The most prominent instance of ozone layer destruction is the so-called "Antarctic ozone hole" that has shown up in recent years. This phenomenon is manifested by the appearance during the Antarctic's late winter and early spring of severely depleted stratospheric ozone (up to 50%) over the polar region. The reasons why this occurs are related to the normal effect of NO_2 in limiting Cl-atom-catalyzed destruction of ozone by combining with ClO,

$$ClO + NO_2 \rightarrow ClONO_2 \qquad (14.4.8)$$

In the polar regions, particularly Antarctica, NO_x gases are removed along with water by freezing in polar stratospheric clouds at temperatures below -70°C as compounds such as $ClONO_2$ and $HNO_3 \cdot 3H_2O$. During the Antarctic winter, HOCl and Cl_2 are generated and accumulate at the surfaces of the solid cloud particles by the reactions,

$$ClONO_2 + H_2O \rightarrow HOCl + HNO_3 \qquad (14.4.9)$$

$$ClONO_2 + HCl \rightarrow Cl_2 + HNO_3 \qquad (14.4.10)$$

where the HCl comes primarily from the reaction of stratospheric methane, CH_4, with Cl• atoms produced from chlorofluorocarbons. The preceding reactions are aided by the tendency of the HNO_3 product to become hydrogen-bonded with water in the cloud particles. The result of these processes is that over the winter months photoreactive Cl_2 and HOCl accumulate in the Antarctic stratospheric region in the absence of sunlight then undergo a burst of photochemical activity when spring arrives as shown by the following reactions:

$$HOCl + h\nu \rightarrow HO• + Cl• \qquad (14.4.11)$$

$$Cl_2 + h\nu \rightarrow Cl• + Cl• \qquad (14.4.12)$$

The Cl atoms react to destroy ozone according to Reaction 14.4.6. Under conditions of Antarctic spring, not enough O• atoms are available to regenerate Cl atoms from ClO• by Reaction 14.4.7. It is now known that ClO• forms the ClO-OCl dimer, which regenerates Cl• by the following reactions:

$$ClOOCl + h\nu \rightarrow ClOO• + Cl• \qquad (14.4.13)$$

$$ClOO• + h\nu \rightarrow O_2 + Cl• \qquad (14.4.14)$$

Chlorofluorocarbon Substitutes and Ozone Depletion

Currently, the substitutes for ozone-destroying chlorofluorocarbon compounds are **hydrohaloalkanes,** compounds that contain at least one H atom. Each molecule of this class of compounds has an H-C bond that is susceptible to attack by HO• radical in the troposphere, thereby eliminating the compound with its potential to produce ozone-depleting Cl atoms before it reaches the stratosphere. The tropospheric chemistry involved in the destruction of hydrohaloalkanes is discussed briefly in Section 12.7. The substitutes are either hydrochlorofluorocarbons (HCFCs) or hydrofluorocarbons (HFCs). The compounds used or proposed for use include HCFC-22 ($CHClF_2$), HCFC-123 ($CHCl_2CF_3$), HCFC-141b (CH_3CCl_2F), HCFC-124 ($CHClFCF_3$), HCFC-225ca, HCFC-225cb, HCFC-142b (CH_3CClF_2), HFC-134a (CH_2FCF_3), and HFC-152a (CH_3CHF_2).

The tropospheric lifetimes of hydrohalocarbons depend upon a number of factors. These are molar mass, number of hydrogen atoms (particularly important because of the H-C bond that is vulnerable to attack by HO•), number of carbon atoms, number of F atoms β to hydrogen, number of chlorine atoms α and β to hydrogen (where α and β designate positions on the same and on adjacent carbon atoms, respectively), rate of reaction with HO• radical, and photolytic cross section (tendency to undergo photolysis). Ozone depletion potentials of HCFCs and HFCs are compiled to express potential likelihood for the destruction of stratospheric ozone relative to a value of 1.0 for CFC-11, a non-hydrogen-containing chlorofluorocarbon with a formula of $CFCl_3$. The ozone-depletion potentials of some of the substitutes mentioned above are HCFC-22, 0.030, HCFC-123, 0.013, HCFC-141b, 0.10, HCFC-124, 0.035, and HCFC-142b, 0.038. Low ozone depletion potential correlates with short tropospheric lifetime, which means that the compound is destroyed in the troposphere before migrating to the stratosphere.

14.5. PHOTOCHEMICAL SMOG

Photochemical smog is a major air pollution phenomenon discussed in Chapter 13. It occurs in urban areas where the combination of pollution-forming emissions and appropriate atmospheric conditions are right for its formation. In order for high levels of smog to form, relatively stagnant air must be subjected to sunlight under low humidity conditions in the presence of pollutant nitrogen oxides and hydrocarbons. Although the automobile is the major source of these pollutants, hydrocarbons may come from biogenic sources, of which α-pinene and isoprene from trees are the most abundant. Stated succinctly, "The urban atmosphere is a giant chemical reactor in which pollutant gases such as hydrocarbons and oxides of nitrogen and sulfur react under the influence of sunlight to create a variety of products."[7] Although not as great a threat to the global atmosphere as some of the other air pollutants discussed in this chapter, smog does pose significant hazards to living things and materials in local urban areas in which millions of people are exposed.

Ironically, ozone, which serves an essential protective function in the stratosphere, is the major culprit in tropospheric smog. In fact, surface ozone levels are used as a measure of smog. Ozone's phytotoxicity raises particular concern with respect to trees and crops. Ozone is the smog constituent responsible for most of the respiratory system distress and eye irritation characteristic of human exposure to smog. Breathing is impaired at ozone levels approaching only about 0.1 ppm. Ozone is the "criterion" air pollutant that has been most resistant to control measures. Because of its strongly oxidizing nature, ozone attacks unsaturated bonds in fatty acid constituents of cell membranes. Other oxidants, such as PAN (Section 13.4), also contribute to the toxicity of smog, as do aldehydes produced as reactive intermediates in smog formation.

Smog is a secondary air pollutant that forms some time after and some distance from the injection into the atmosphere of the primary pollutant nitrogen oxides and reactive hydrocarbons required for its formation. The U.S. Environmental Protection Agency's Empirical Kinetic Modeling Approach uses the concept of an **air parcel** to model smog formation. This model utilizes the concept of a "parcel" of relatively unpolluted air moving across an urban area in which it becomes contaminated with smog-forming gases. When the upper boundary of this parcel is restricted to about 1000 meters by a temperature inversion and subjected to sunlight, the primary pollutants react to form smog in a system that involves photochemical reaction processes, transport, mixing, and dilution. As the hydrocarbons are consumed by photochemical oxidation processes in the air, and as nitrogen oxides are removed as nitrates and nitric acid (especially at nightime), ozone levels reach a peak concentration at a time and place some distance removed from the source of pollutants.

The most visible manifestation of smog is the **urban aerosol**, which greatly reduces visibility in smoggy urban atmospheres. Many of the particles composing this aerosol are condensation aerosols made from gases by chemical processes (see Chapter 10) and are therefore quite small, usually less than 2 μm. Particles of such a size are especially harmful because they scatter light most efficiently and are the most respirable. Aerosol particles formed from smog often contain toxic constituents, such as respiratory tract irritants and mutagens. The urban aerosol also contains particle constituents that originate from processes other than smog formation. Oxidation of pollutant sulfur dioxide by the strongly oxidizing conditions of photochemical smog,

$$SO_2 + 1/2O_2 + H_2O \rightarrow H_2SO_4 \quad \text{(overall process)} \qquad (14.5.1)$$

produces sulfuric acid and sulfate particles. Nitric acid and nitrates are produced at night when sunlight is absent, a process that involves intermediate NO_3 radical:

$$O_3 + NO_2 \rightarrow O_2 + NO_3 \qquad (14.5.2)$$

$$NO_3 + NO_2 + M \text{ (energy-absorbing third body)} \rightarrow N_2O_5 + M \qquad (14.5.3)$$

$$N_2O_5 + H_2O \rightarrow 2HNO_3 \qquad (14.5.4)$$

$$HNO_3 + NH_3 \rightarrow NH_4NO_3 \qquad (14.5.5)$$

As indicated by the last reaction above, ammonium salts are common constituents of urban aerosol particles; they tend to be particularly corrosive. Metals, which may contribute to the toxicity of urban aerosol particles and which may catalyze reactions on their surfaces, occur in the particles. Water is always present, even in low humidity atmospheres, and is usually present in urban aerosol particles. Carbon and polycyclic aromatic hydrocarbons from partial combustion and diesel engine emissions are usually abundant constituents; elemental carbon is usually the particulate constituent most responsible for absorbing light in the urban aerosol. If the air parcel originates over the ocean, it contains sea salt particles consisting largely of NaCl, from which some of the chloride may be lost as volatile HCl by the action of less volatile strong acids produced by smog. This phenomenon is responsible for Na_2SO_4 and $NaNO_3$ found in the urban aerosol.

Polycyclic aromatic hydrocarbons, PAH (see Section 10.8), are among the urban aerosol particle constituents of most concern, particularly because metabolites of some of these compounds (see the 7,8-diol-9,10-epoxide of benzo(a)pyrene in Figure 22.18) are carcinogenic. PAHs include unsubstituted compounds as well as those with alkyl, oxygen, or nitrogen substituents, or O or N hetero atoms:

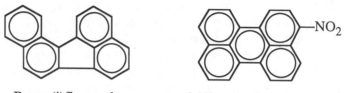

Benzo(j)fluoranthene   3-Nitroperylene (mutagenic)

These kinds of compounds are emitted by internal combustion engine exhausts and occur in both the gas and particulate phases. Numerous mechanisms exist for their destruction and chemical alteration, particularly reaction with oxidant species — $HO^\bullet$, O_3, NO_2, N_2O_5, and HNO_3. Direct photolysis is also possible. PAH compounds in the vapor phase are destroyed relatively rapidly by these means, whereas PAHs sorbed to particles are much more resistant to reaction.

Another kind of urban aerosol particulate matter of considerable concern is **acid fog**, which may have pH values below 2 due to the presence of H_2SO_4 or HNO_3. Acid fog formation covers a wide range of atmospheric chemical and physical phenomena. The gas-phase oxidation of SO_2 and NO_x produces strong acids, which form very small aerosol particles. These, in turn, act as condensation nuclei for water vapor. Acid-base phenomena occur in the droplets, and the droplets act as scavengers to remove ionic species from air. Because fog aerosol particles form in areas of intense acid gas pollution near the surface, the concentrations of acids and ionic species in fog aerosol droplets tend to be much higher than in cloud aerosol droplets at higher altitudes.

In addition to health effects and damage to materials, one of the greater problems caused by smog is destruction of crops and reduction of crop yields. The annual cost of these effects in California, alone is several billion dollars

Even lightly populated nonindustrial areas are subject to the effects of smog brought about by human activities. Particularly, the practice of burning savanna grasses for agricultural purposes causes smog. This burning produces NO_x and reactive hydrocarbons that are required for smog formation. Furthermore, these grasses grow in tropical regions which have the intense sunlight required for smog formation. The net result is rapid development of smoggy conditions as manifested by ozone levels several times normal background values.

Since about 1970, significant progress has been made in eliminating the emissions of organic compounds and NO_x that cause smog formation. Those efforts resulted in a less-than-anticipated reduction in smog; for example, urban ozone in the U.S. declined only about 8% during the 1980s. In an effort to reduce ozone and other manifestations of air pollution more rapidly, the U.S. Congress passed a new, more rigorous set of Clean Air Act Amendments in 1990. The Amendments were written to further reduce automotive exhaust emissions, require changes in automotive fuel formulations, reduce emissions from stationary sources, and mandate other changes designed to reduce photochemical smog and other forms of air pollution. As a result of the enforcement of regulations under the Clean Air Act and technological advances, particularly in the area of automobile emissions, substantial progress has been made in controlling smog in the U.S. during the 1990s.

14.6. NUCLEAR WINTER

Nuclear winter is a term used to describe a catastrophic atmospheric effect that might occur after a massive exchange of nuclear firepower between major powers.[8] The heat from the nuclear blasts and from resulting fires would cause powerful updrafts carrying sooty combustion products to stratospheric regions. This would result in several years of much lower temperatures and freezing temperatures even during summertime. There are several reasons for such an effect. First of all, the highly absorbent, largely black particulate matter would absorb solar radiation high in the atmosphere so that it would not reach earth's surface. Cooling would also occur from a phenomenon opposite to that of the greenhouse effect. That is because outgoing infrared radiation from particles high in the atmosphere would have to penetrate relatively much less of the atmosphere and, therefore, would be exposed to much less infrared-absorbing water vapor and carbon dioxide gas. This would deprive the lower atmosphere of the warming effect of outgoing infrared radiation and would mean that less infrared would be re-radiated from the atmosphere back to earth's surface. The cooling would also inhibit the evaporation of water, thereby reducing the amount of infrared-absorbing water vapor in the atmosphere and slowing the process by which particulate matter is scavenged from the atmosphere by rain.

Conditions similar to those of a nuclear winter occurred in 1816, "the year without a summer," following the astoundingly massive Tambora, Indonesia volcanic explosion of 1815. Brutally cold years around 210 B.C. that followed a similar volcanic incident in Iceland were recorded in ancient China. The June 1991 explosion of the Philippine volcano Pinatubo, which blasted millions of tons of material, including 15-30 million tons of sulfur dioxide, into the atmosphere resulted in an approximately 0.5°C cooling the following year. In addition to the direct suffering caused, massive starvation would result from crop failures accompanying years of nuclear winter. The incidents cited above clearly illustrate the climatic effects of huge quantities of particulate matter ejected high into the atmosphere.

Evidence exists to suggest that military explosives can result in the introduction of large quantities of particulate matter into the atmosphere. For example, carpet bombings of cities, such as the tragic bombing of Dresden, Germany, near the end of World War II, have caused huge firestorms that created their own wind causing a particle-laden updraft into the atmosphere. Of course, the effect of a full-scale nuclear exchange would be manyfold higher.

An idea of the potential climatic effect resulting from a full-scale nuclear exchange may be obtained by considering the magnitude of the blasts that might be involved. Only two nuclear bombs have been used in warfare, both dropped on cities in Japan in 1945. The Hiroshima fission bomb had the explosive force of 12 kilotons of TNT explosive. Its blast, fireball, and instantaneous emissions of neutrons and gamma radiation, followed by fires and exposure to radioactive fission products, killed about 100,000 people and destroyed the city on which it was dropped. By comparison with this 12-kiloton bomb, modern fusion bombs are typically rated at 500 kilotons, and 10-*mega*ton weapons are common. A full-scale nuclear exchange might involve a total of the order of 5,000 megatons of nuclear explosives. As a result, unimaginable quantities of soot from the partial combustion of wood, plastics, paving asphalt, petroleum, forests, and other combustibles would be carried to the stratosphere. At such high altitudes, tropospheric removal mechanisms for particles are not effective because there is not enough water in the stratosphere to produce rainfall to wash particles from the air, and convection processes are very limited. Much of the particulate matter would be in the μm size range in which light is reflected, scattered, and absorbed most effectively and settling is very slow. Therefore, vast areas of the earth would be overlain by a stable cloud of particles and the fraction of sunlight reaching the earth's surface would be drastically reduced, resulting in a dramatic cooling effect. There would be other effects as well. The extreme heat and pressure in the fireball would result in fixation of nitrogen by the following reaction:

$$O_2 + N_2 \rightarrow 2NO \tag{14.6.1}$$

The timing and location of nuclear blasts are very important in determining their climatic effects. Atmospheric testing of nuclear weapons, including a 58-megaton monster detonated by the Soviet Union, have had little atmospheric effect. Such tests were carried out at widely spaced intervals on deserts, small tropical islands, and other locations with minimal combustible matter. In contrast, military use of nuclear weapons would involve a high concentration of firepower, both in time and in space, on industrial and military targets consisting largely of combustibles. Furthermore, destruction of hardened military sites requires blasts that disrupt large quantities of soil, rock, and concrete, which are pulverized, vaporized, and blown into the atmosphere.

On a hopeful note, the East-West conflict that dominated world politics and threatened nuclear war from the mid-1900s until 1990 has now abated and the probability of nuclear warfare seems to have diminished. However, the outbreak of armed conflict in Kosovo and Yugoslavia in 1999, nuclear proliferation as manifested by tests of nuclear bombs by both India and Pakistan in 1998, disintegration of great powers with vast nuclear arsenals, racial hatred accompanied by a determination to perform "ethnic cleansing," and a "trigger-happy" state of mind among even educated people who should know better should still cause concern with respect to the prospect of "nuclear winter."

Visitors from Space

Of all the possible atmospheric catastrophes that can occur, arguably the most threatening would be one caused by collision of a large asteroid with earth. Convincing evidence now exists that mass extinctions of species in the past have resulted from earth being hit by asteroids several kilometers in diameter. Such an event would cause much the same effects as those from "nuclear winter," though with a large asteroid the effects would be much more pronounced.[9]

14.7. WHAT IS TO BE DONE?

Of all environmental hazards, there is little doubt that major disruptions in the atmosphere and climate have the greatest potential for catastrophic and irreversible environmental damage. If levels of greenhouse gases and reactive trace gases continue to increase at present rates, major environmental effects are virtually certain. On a hopeful note, the bulk of these emissions arise from industrialized nations which, in principle, can apply the resources needed to reduce them substantially. The best example to date has been the 1987 "Montreal Protocol on Substances that Deplete the Ozone Layer," an international treaty through which a large number of nations agreed to cut chlorofluorocarbon emissions by 50% by the year 2000. This agreement and subsequent ones, particularly the Copenhagen Amendment of 1992, may pave the way for more encompassing agreements covering carbon dioxide and other trace gases.

More ominous, however, is the combination of population pressure and desire for better living standards on a global basis. Consider, for example, the demand that these two factors place on energy resources, and the environmental disruption that may result. In many highly populated developing nations, high-sulfur coal is the most readily available, cheapest source of energy. It is understandably difficult to persuade populations faced with real hunger to forego short-term economic gain for the sake of long-term environmental quality. Destruction of rain forests by "slash-and-burn" agricultural methods does seem to make economic sense to those engaged in subsistence farming to obtain badly needed hard currency, which can be earned by converting forest to pasture land and exporting fast-food-hamburger beef to wealthier nations.

What is to be done? First of all, it is important to keep in mind that the atmosphere has a strong ability to cleanse itself of pollutant species. Water-soluble gases, including greenhouse-gas CO_2, acid-gas SO_2, and fine particulate matter are removed with precipitation. For most gaseous contaminants, oxidation precedes or accompanies removal processes. To a degree, oxidation is carried out by O_3. To a larger extent, the most active atmospheric oxidant is hydroxyl radical, $HO^\bullet$. As illustrated in Figure 9.10, this atmospheric scavenger species reacts with all important trace gas species except for CO_2 and chlorofluorocarbons. It is now generally recognized that $HO^\bullet$ is an almost universal atmospheric cleansing agent. Given this crucial role of $HO^\bullet$ radical, any pollutants that substantially reduce its concentration in the atmosphere are potentially troublesome. One concern over carbon monoxide emissions to the atmosphere is the reactivity of $HO^\bullet$ with CO,

$$CO + HO^\bullet \rightarrow CO_2 + H \qquad\qquad\qquad (14.7.1)$$

which could result in removal of HO$^\bullet$ from the atmosphere.

Of all the major threats to the global climate, it is virtually certain that humankind will have to try to cope with greenhouse warming and the climatic effects thereof. The measures to be taken in dealing with this problem fall into the three following categories:

- **Minimization** by reducing emissions of greenhouse gases, switching to alternate energy sources, increasing energy conservation, and reversing deforestation. It is especially sensible to use measures that have major benefits in addition to reduction of greenhouse warming. Such measures include, as examples, reforestation, restoration of grasslands, increased energy conservation, and a massive shift to solar energy sources.

- **Counteracting measures**, such as injecting light-reflecting particles into the upper atmosphere.

- **Adaptation**, particularly through increased efficiency and flexibility of the distribution and use of water, which might be in very short supply in many parts of the world as a consequence of greenhouse warming. Important examples are implementation of more efficient irrigation practices and changes in agriculture to grow crops that require less irrigation. Emphasis on adaptation is favored by those who contend that not enough is known about the types and severity of global warming to justify massive expenditures on minimization and counteractive measures. In any case, adaptation will certainly have to be employed as a means of coping with global warming.

Potentially, tax strategy can be very effective in reducing use of carbonaceous fuels and greenhouse CO_2 emissions. This is the rationale behind the **carbon tax**, which is tied with the carbon content of various fuels. Another option is to dispose of carbon dioxide to a sink other than the atmosphere. The most obvious such sink is the ocean; other possibilities are deep subterranean aquifers and exhausted oil and gas wells.

A common measure taken against the effects of another atmospheric hazard, ultraviolet radiation, provides an example of adaptation. This measure is the use of sunscreens placed on the skin as lotions to filter out UV-B radiation. The active ingredient of sunscreen must absorb ultraviolet light effectively, but this is not enough because it is the absorption of ultraviolet light by skin that makes it so dangerous in the first place. Therefore, active compounds in sunscreen formulations must also dissipate the absorbed energy in a harmless way.[10] The way in which this is done is illustrated below for *o*-hydroxybenzophenone contained in sunscreens, which reacts as shown by Reaction 14.7.2 at the top of the next page.

The first step in the above reaction sequence can be regarded as *intramolecular transfer* of energy and *internal (spontaneous) isomerization* (see Section 9.7) by which absorbed ultraviolet energy is accomodated within the molecule, which reacts to produce the more energized enol form. In Step 2 the molecule reverts back to the

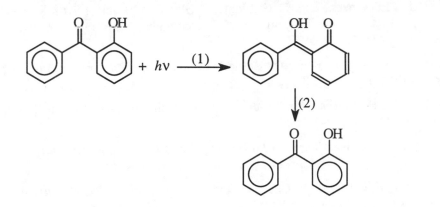

$$(14.7.?)$$

more stable keto form, losing energy thermally in the process. The net result is that energy is absorbed, the excited absorbing species reacts only with itself, then the energy is dissipated harmlessly as thermal energy, which does not cause additional reactions to occur.

The **"tie-in strategy"** has been proposed as a sensible approach to dealing with the kinds of global environmental problems discussed in this chapter. This approach was first enunciated in 1980.[11] It advocates taking measures consisting of "high-leverage actions" which are designed to prevent problems from occurring and which have substantial merit even if the major problems that they are designed to avoid do not materialize. An example is implementation of environmentally sound substitutes for fossil fuels to lower atmospheric CO_2 output and prevent greenhouse warming. Even if it turns out that the greenhouse effect is exaggerated, such substitutes would save the earth from other kinds of environmental damage, such as disruption of land by strip mining coal or preventing oil spills from petroleum transport. Definite economic and political benefits would also accrue from lessened dependence on uncertain, volatile petroleum supplies. Increased energy efficiency would diminish both greenhouse gas and acid rain production, while lowering costs of production and reducing the need for expensive and environmentally disruptive new power plants. The implementation of these kinds of tie-in strategies requires some degree of incentive beyond normal market forces and, therefore, is opposed by some on ideological grounds. A good example is opposition to mandatory fuel mileage standards for automobiles. However, to quote Schneider,[12] "a market that does not include the costs of environmental disruptions can hardly be called a free market."

LITERATURE CITED

1. Schneider, Stephen H., and Penelope J. Boston, *Scientists on Gaia*, The MIT Press, Cambridge, MA, 1993.

2. Revelle, R., and H. Suess, *Tellus*, **9**, 18 (1957).

3. Baum, Rudy, "Wintertime Reflections on Global Warming," *Chemical and Engineering News*, January 25, 1999, p. 45.

4. Hileman, Betty, "Data Rebut Interglacial Temperature Swing Idea," *Chemical and Engineering News*, December 20, 1993, p. 7.

5. Zellner, R., R. Sausen, and T. F. Mentel, "Global Aspects of Atmospheric Chemistry. Part 5. Global Change and Consequences," *Topics in Physical Chemistry*, **6**, 255-322 (1999).

6. Hileman, Bette, "Storm Warnings Rattle Insurers," *Chemical and Engineering News*, April 14, 1997, pp. 28-31.

7. Seinfeld, John H., "Urban Air Pollution: State of the Science," *Science*, **243**, February 10, 1989, pp. 745-752.

8. Sagan, Carl, and Richard Turco, *A Path Where No Man Thought: Nuclear Winter and the End of the Arms Race*, Random House, New York, 1990.

9. Cox, Donald W. and James H. Chestek, *Doomsday Asteroid: Can We Survive?*, Prometheus Books, Amherst, NY, 1996.

10. Gasparro, Francis P., Ed., *Sunscreen Photobiology: Molecular, Cellular, and Physiological Aspects*, Landes Bioscience, Austin, TX,1997.

11. E. Boulding, in *Carbon Dioxide Effects, Research and Assessment Program: Workshop on Environmental and Societal Consequence of a Possible CO_2-Induced Climatic Change*, Report 009, CONF-7904143, U. S. Department of Energy, U. S. Government Printing Office, Washington, DC, October 1980, pp. 79-10.

12. Schneider, S. H., "The Greenhouse Effect: Science and Policy," *Science*, **243**, February 10, 1989, pp. 751-781.

SUPPLEMENTARY REFERENCES

Adger, W. Neil and Katrina Brown, *Land Use and the Causes of Global Warming*, John Wiley & Sons, New York, 1995.

Barnes-Svarney, Patricia, *Asteroid: Earth Destroyer or New Frontier?*, Plenum Press, New York, 1996.

Beim, Howard J., Jennifer Spero, and Louis Theodore, *Rapid Guide to Hazardous Air Pollutants*, John Wiley & Sons, New York, 1997.

Brimblecombe, Peter, *Air Composition and Chemistry*, 2nd ed., Cambridge University Press, Cambridge, U.K., 1996.

Brown, Paul, *Global Warming: Can Civilization Survive?*, Blandford Press, London, 1997.

Cagin, Seth, and Philip Dray, *Between Earth and Sky: How CFCs Changed Our World and Endangered the Ozone Layer*, Pantheon, New York, 1993.

Dunnettee, David A., and Robert J. O'Brien, Eds., *The Science of Global Change: The Impact of Human Activities on the Environment*, American Chemical Society, Washington, D.C. 1992.

Graedel, T. E., and Paul J. Crutzen, *Atmospheric Change: An Earth System Perspective*, W. H. Freeman and Co., New York, 1993.

Johnson, Rebecca L., *The Greenhouse Effect: Life on a Warmer Planet*, Lerner Publications Company, Minneapolis, MN, 1995.

Lewis, John S., *Rain of Iron and Ice: The Very Real Threat of Comet and Asteroid Bombardment*, Perseus Press, Addison-Wesley Publishing Co, Reading, MA, 1996.

Mabey, Nick, Stephen Hall, Clare Smith, and Sujata Gupta, *Argument in the Greenhouse: The International Economics of Controlling Global Warming*, Routledge, London, 1997.

Moore, Thomas Gale, *Climate of Fear: Why We Shouldn't Worry About Global Warming*, Cato Insitute, 1998.

National Academy of Sciences, *Policy Implications of Greenhouse Warming*, National Academy Press, Washington, D.C., 1991.

Nilsson, Annika, *Ultraviolet Reflections: Life Under a Thinning Ozone Layer*, John Wiley & Sons, New York, 1996.

Philander, George, *Is the Temperature Rising?: The Uncertain Science of Global Warming*, Princeton Univ Press, Princeton, NJ, 1998.

Ruddiman, W. F., Ed., *Tectonic Uplift and Climate Change*, Plenum Pub Corp., New York, 1997.

Seinfeld, John H. and Spyros N. Pandis, *Atmospheric Chemistry and Physics*, John Wiley & Sons, Inc., New York, 1998.

Turco, Richard P., *Earth Under Siege: From Air Pollution to Global Change*, Oxford University Press, New York, 1996.

Turekian, Karl K., *Global Environmental Change: Past, Present, and Future*, Prentice Hall College Division, Upper Saddle River, NJ, 1996.

QUESTIONS AND PROBLEMS

1. How do modern transportation problems contribute to the kinds of atmospheric problems discussed in this chapter?

2. What is the rationale for classifying most acid rain as a secondary pollutant?

3. Distinguish among UV-A, UV-B, and UV-C radiation. Why does UV-B pose the greatest danger in the troposphere?

4. How does the extreme cold of stratospheric clouds in Antarctic regions contribute to the Antarctic ozone hole?

5. How does the oxidizing nature of ozone from smog contribute to the damage that it does to cell membranes?

6. What may be said about the time and place of the occurrence of maximum ozone levels from smog with respect to the origin of the primary pollutants that result in smog formation?

7. What is the basis for "nuclear winter"?

8. Discuss the analogies between the effects of a large asteroid hit on Earth with "nuclear winter."

9. What is meant by a "tie-in strategy"?

15 THE GEOSPHERE AND GEOCHEMISTRY

15.1. INTRODUCTION

The **geosphere**, or solid earth, is that part of the earth upon which humans live and from which they extract most of their food, minerals, and fuels. Once thought to have an almost unlimited buffering capacity against the perturbations of humankind, the geosphere is now known to be rather fragile and subject to harm by human activities. For example, some billions of tons of earth material are mined or otherwise disturbed each year in the extraction of minerals and coal. Two atmospheric pollutant phenomena—excess carbon dioxide and acid rain (see Chapter 14)—have the potential to cause major changes in the geosphere. Too much carbon dioxide in the atmosphere may cause global heating ("greenhouse effect"), which could significantly alter rainfall patterns and turn currently productive areas of the earth into desert regions. The low pH characteristic of acid rain can bring about drastic changes in the solubilities and oxidation-reduction rates of minerals. Erosion caused by intensive cultivation of land is washing away vast quantities of topsoil from fertile farmlands each year. In some areas of industrialized countries, the geosphere has been the dumping ground for toxic chemicals. Ultimately, the geosphere must provide disposal sites for the nuclear wastes of the more than 400 nuclear reactors that have operated worldwide. It may be readily seen that the preservation of the geosphere in a form suitable for human habitation is one of the greatest challenges facing humankind.

The interface between the geosphere and the atmosphere at earth's surface is very important to the environment. Human activities on the earth's surface may affect climate, most directly through the change of surface albedo, defined as the percentage of incident solar radiation reflected by a land or water surface. For example, if the sun radiates 100 units of energy per minute to the outer limits of the atmosphere, and the earth's surface receives 60 units per minute of the total, then reflects 30 units upward, the albedo is 50 percent. Some typical albedo values for different areas on the earth's surface are: evergreen forests, 7-15%; dry, plowed fields, 10-15%; deserts, 25-35%; fresh snow, 85-90%; asphalt, 8%. In some heavily developed areas, anthropogenic (human-produced) heat release is comparable to the solar input. The anthropogenic energy release over the 60 square kilometers of Manhattan Island averages about 4 times the solar energy falling on the area; over the 3500 km^2 of Los Angeles the anthropogenic energy release is about 13% of the solar flux.

One of the greater impacts of humans upon the geosphere is the creation of desert areas through abuse of land with marginal amounts of rainfall. This process, called **desertification**, is manifested by declining groundwater tables, salinization of topsoil and water, reduction of surface waters, unnaturally high soil erosion, and desolation of native vegetation. The problem is severe in some parts of the world, particularly Africa's Sahel (southern rim of the Sahara), where the Sahara advanced southward at a particularly rapid rate during the period 1968-73, contributing to widespread starvation in Africa during the 1980s. Large, arid areas of the western U.S. are experiencing at least some desertification as the result of human activities and a severe drought during the latter 1980s and early 1990s. As the populations of the western states increase, one of the greatest challenges facing the residents is to prevent additional conversion of land to desert.

The most important part of the geosphere for life on earth is soil. It is the medium upon which plants grow, and virtually all terrestrial organisms depend upon it for their existence. The productivity of soil is strongly affected by environmental conditions and pollutants. Because of the importance of soil, all of Chapter 16 is devoted to its environmental chemistry.

With increasing population and industrialization, one of the more important aspects of human use of the geosphere has to do with the protection of water sources. Mining, agricultural, chemical, and radioactive wastes all have the potential for contaminating both surface water and groundwater. Sewage sludge spread on land may contaminate water by release of nitrate and heavy metals. Landfills may likewise be sources of contamination. Leachates from unlined pits and lagoons containing hazardous liquids or sludges may pollute drinking water.

It should be noted, however, that many soils have the ability to assimilate and neutralize pollutants. Various chemical and biochemical phenomena in soils operate to reduce the harmful nature of pollutants. These phenomena include oxidation-reduction processes, hydrolysis, acid-base reactions, precipitation, sorption, and biochemical degradation. Some hazardous organic chemicals may be degraded to harmless products on soil, and heavy metals may be sorbed by it. In general, however, extreme care should be exercised in disposing of chemicals, sludges, and other potentially hazardous materials on soil, particularly where the possibility of water contamination exists.

15.2. THE NATURE OF SOLIDS IN THE GEOSPHERE

The earth is divided into layers, including the solid iron-rich inner core, molten outer core, mantle, and crust. Environmental chemistry is most concerned with the **lithosphere**, which consists of the outer mantle and the **crust**. The latter is the earth's outer skin that is accessible to humans. It is extremely thin compared to the diameter of the earth, ranging from 5 to 40 km thick.

Most of the solid earth crust consists of rocks. Rocks are composed of minerals, where a **mineral** is a naturally-occurring inorganic solid with a definite internal crystal structure and chemical composition.[1] A **rock** is a solid, cohesive mass of pure mineral or an aggregate of two or more minerals.

Structure and Properties of Minerals

The combination of two characteristics is unique to a particular mineral. These characteristics are a defined chemical composition, as expressed by the mineral's chemical formula, and a specific crystal structure. The **crystal structure** of a mineral refers to the way in which the atoms are arranged relative to each other. It cannot be determined from the appearance of visible crystals of the mineral, but requires structural methods such as X-ray structure determination. Different minerals may have the same chemical composition, or they may have the same crystal structure, but may not be identical for truly different minerals.

Physical properties of minerals can be used to classify them. The characteristic external appearance of a pure crystalline mineral is its **crystal form**. Because of space constrictions on the ways that minerals grow, the pure crystal form of a mineral is often not expressed. **Color** is an obvious characteristic of minerals, but can vary widely due to the presence of impurities. The appearance of a mineral surface in reflected light describes its **luster**. Minerals may have a metallic luster or appear partially metallic (or submetallic), vitreous (like glass), dull or earthy, resinous, or pearly. The color of a mineral in its powdered form as observed when the mineral is rubbed across an unglazed porcelain plate is known as **streak**. **Hardness** is expressed on Mohs scale, which ranges from 1 to 10 and is based upon 10 minerals that vary from talc, hardness 1, to diamond, hardness 10. **Cleavage** denotes the manner in which minerals break along planes and the angles in which these planes intersect. For example, mica cleaves to form thin sheets. Most minerals **fracture** irregularly, although some fracture along smooth curved surfaces or into fibers or splinters. **Specific gravity**, density relative to that of water, is another important physical characteristic of minerals.

Kinds of Minerals

Although over two thousand minerals are known, only about 25 **rock-forming minerals** make up most of the earth's crust. The nature of these minerals may be better understood with a knowledge of the elemental composition of the crust. Oxygen and silicon make up 49.5% and 25.7% by mass of the earth's crust, respectively. Therefore, most minerals are **silicates** such as quartz, SiO_2, or orthoclase, $KAlSi_3O_8$. In descending order of abundance; the other elements in the earth's crust are aluminum (7.4%), iron (4.7%), calcium (3.6%), sodium (2.8%), potassium (2.6%), magnesium (2.1%), and other (1.6%). Table 15.1 summarizes the major kinds of minerals in the earth's crust.

Secondary minerals are formed by alteration of parent mineral matter. **Clays** are silicate minerals, usually containing aluminum, that constitute one of the most significant classes of secondary minerals. Olivine, augite, hornblende, and feldspars all form clays. Clays are discussed in detail in Section 15.7.

Table 15.1. Major Mineral Groups in the Earth's Crust

Mineral group	Examples	Formula
Silicates	Quartz	SiO_2
	Olivine	$(Mg,Fe)_2SiO_4$
	Potassium feldspar	$KAlSi_3O_8$
Oxides	Corundum	Al_2O_3
	Magnetite	Fe_3O_4
Carbonates	Calcite	$CaCO_3$
	Dolomite	$CaCO_3 \cdot MgCO_3$
Sulfides	Pyrite	FeS_2
	Galena	PbS
Sulfates	Gypsum	$CaSO_4 \cdot 2H_2O$
Halides	Halite	$NaCl$
	Fluorite	CaF_2
Native elements	Copper	Cu
	Sulfur	S

Evaporites

Evaporites are soluble salts that precipitate from solution under special arid conditions, commonly as the result of the evaporation of seawater. The most common evaporite is **halite**, $NaCl$. Other simple evaporite minerals are sylvite (KCl), thenardite (Na_2SO_4), and anhydrite ($CaSO_4$). Many evaporites are hydrates, including bischofite ($MgCl_2 \cdot 6H_2O$), gypsum ($CaSO_4 \cdot 2H_2O$), kieserite ($MgSO_4 \cdot H_2O$), and epsomite ($MgSO_4 \cdot 7H_2O$). Double salts, such as carnallite ($KMgCl_3 \cdot 6H_2O$), kainite ($KMgClSO_4 \cdot 1^1/_4H_2O$), glaserite ($K_3Na(SO_4)_2$), polyhalite ($K_2MgCa_2(SO_4)_4 \cdot 2H_2O$), and loeweite ($Na_{12}Mg_7(SO_4)_{13} \cdot 15H_2O$), are very common in evaporites.

The precipitation of evaporites from marine and brine sources depends upon a number of factors. Prominent among these are the concentrations of the evaporite ions in the water and the solubility products of the evaporite salts. The presence of a common ion decreases solubility; for example, $CaSO_4$ precipitates more readily from a brine that contains Na_2SO_4 than it does from a solution that contains no other source of sulfate. The presence of other salts that do not have a common ion increases solubility because it decreases activity coefficients. Differences in temperature result in significant differences in solubility.

The nitrate deposits that occur in the hot and extraordinarily dry regions of northern Chile are chemically unique because of the stability of highly oxidized nitrate salts. The dominant salt, which has been mined for its nitrate content for use in explosives and fertilizers, is Chile saltpeter, $NaNO_3$. Traces of highly oxidized $CaCrO_4$ and $Ca(ClO_4)_2$ are also encountered in these deposits, and some regions contain enough $Ca(IO_3)_2$ to serve as a commercial source of iodine.

Volcanic Sublimates

A number of mineral substances are gaseous at the magmatic temperatures of volcanoes and are mobilized with volcanic gases. These kinds of substances condense near the mouths of volcanic fumaroles and are called **sublimates**. Elemental sulfur is a common sublimate. Some oxides, particularly of iron and silicon, are deposited as sublimates. Most other sublimates consist of chloride and sulfate salts. The cations most commonly involved are monovalent cations of ammonium ion, sodium, and potassium; magnesium; calcium; aluminum; and iron. Fluoride and chloride sublimates are sources of gaseous HF and HCl formed by their reactions at high temperatures with water, such as the following:

$$2H_2O + SiF_4 \rightarrow 4HF + SiO_2 \tag{15.2.1}$$

Igneous, Sedimentary, and Metamorphic Rock

At elevated temperatures deep beneath earth's surface, rocks and mineral matter melt to produce a molten substance called **magma**. Cooling and solidification of magma produces **igneous rock**. Common igneous rocks include granite, basalt, quartz (SiO_2), pyroxene ((Mg,Fe)SiO_3), feldspar ((Ca,Na,K)$AlSi_3O_8$), olivine ((Mg,Fe)$_2SiO_4$), and magnetite (Fe_3O_4). Igneous rocks are formed under water-deficient, chemically reducing conditions of high temperature and high pressure. Exposed igneous rocks are under wet, oxidizing, low-temperature, and low-pressure conditions. Since such conditions are opposite those conditions under which igneous rocks were formed, they are not in chemical equilibrium with their surroundings when they become exposed. As a result, such rocks disintegrate by a process called **weathering**.[2] Weathering tends to be slow because igneous rocks are often hard, nonporous, and of low reactivity. Erosion from wind, water, or glaciers picks up materials from weathering rocks and deposits it as **sediments** or **soil**. A process called **lithification** describes the conversion of sediments to **sedimentary rocks**. In contrast to the parent igneous rocks, sediments and sedimentary rocks are porous, soft, and chemically reactive. Heat and pressure convert sedimentary rock to **metamorphic rock**.

Sedimentary rocks may be **detrital rocks** consisting of solid particles eroded from igneous rocks as a consequence of weathering; quartz is the most likely to survive weathering and transport from its original location chemically intact. A second kind of sedimentary rocks consists of **chemical sedimentary rocks** produced by the precipitation or coagulation of dissolved or colloidal weathering products. **Organic sedimentary rocks** contain residues of plant and animal remains. Carbonate minerals of calcium and magnesium—**limestone** or **dolomite**—are especially abundant in sedimentary rocks. Important examples of sedimentary rocks are the following:

- Sandstone produced from sand-sized particles of minerals such as quartz

- Conglomerates made up of relatively larger particles of variable size

- Shale formed from very fine particles of silt or clay

- Limestone, $CaCO_3$, produced by the chemical or biochemical precipitation of calcium carbonate:

$$Ca^{2+} + CO_3^{2-} \rightarrow CaCO_3(s)$$

$$Ca^{2+} + 2HCO_3^- + h\nu(\text{algal photosynthesis}) \rightarrow \{CH_2O\}(\text{biomass}) \\ + CaCO_3(s) + O_2(g)$$

- Chert consisting of microcrystalline SiO_2

Rock Cycle

The interchanges and conversions among igneous, sedimentary, and metamorphic rocks, as well as the processes involved therein, are described by the **rock cycle**. A rock of any of these three types may be changed to a rock of any other type. Or a rock of any of these three kinds may be changed to a different rock of the same general type in the rock cycle. The rock cycle is illustrated in Figure 15.1.

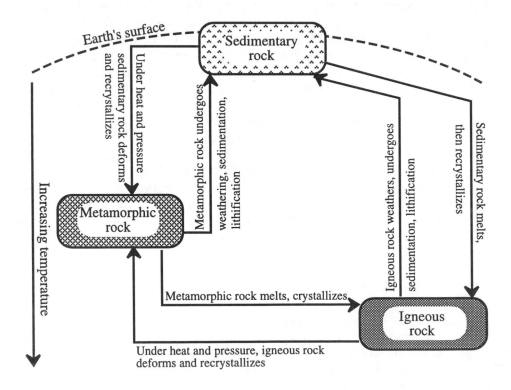

Figure 15.1. The rock cycle

Stages of Weathering

Weathering can be classified into **early**, **intermediate**, and **advanced stages**. The stage of weathering to which a mineral is exposed depends upon time; chemical conditions, including exposure to air, carbon dioxide, and water; and physical conditions such as temperature and mixing with water and air.

Reactive and soluble minerals such as carbonates, gypsum, olivine, feldspars, and iron(II)-rich substances can survive only early weathering. This stage is characterized by dry conditions, low leaching, absence of organic matter, reducing conditions, and limited time of exposure. Quartz, vermiculite, and smectites can survive the intermediate stage of weathering manifested by retention of silica, sodium, potassium, magnesium, calcium, and iron(II) not present in iron(II) oxides. These substances are mobilized in advanced-stage weathering, other characteristics of which are intense leaching by fresh water, low pH, oxidizing conditions (iron(II) → iron(III)), presence of hydroxy polymers of aluminum, and dispersion of silica.

15.3. PHYSICAL FORM OF THE GEOSPHERE

The most fundamental aspect of the physical form of the geosphere has to do with earth's shape and dimensions. The earth is shaped as a **geoid** defined by a surface corresponding to the average sea level of the oceans and continuing as hypothetical sea levels under the continents. This shape is not a perfect sphere because of variations in the attraction of gravity at various places on earth's surface. This slight irregularity in shape is important in surveying to precisely determine the locations of points on earth's surface according to longitude, latitude, and elevation above sea level. Of more direct concern to humans is the nature of landforms and the processes that occur on them. This area of study is classified as **geomorphology**.

Plate Tectonics and Continental Drift

The geosphere has a highly varied, constantly changing physical form. Most of the earth's land mass is contained in several massive continents separated by vast oceans. Towering mountain ranges spread across the continents, and in some places the ocean bottom is at extreme depths. Earthquakes, which often cause great destruction and loss of life, and volcanic eruptions, which sometimes throw enough material into the atmosphere to cause temporary changes in climate, serve as reminders that the earth is a dynamic, living body that continues to change. There is convincing evidence, such as the close fit between the western coast of Africa and the eastern coast of South America, that widely separated continents were once joined and have moved relative to each other. This ongoing phenomenon is known as **continental drift**. It is now believed that 200 million years ago much of earth's land mass was all part of a supercontinent, now called Gowandaland. This continent split apart to form the present-day continents of Antarctica, Australia, Africa, and South America, as well as Madagascar, the Seychelle Islands, and India.

The observations described above are explained by the theory of **plate tectonics**.[3] This theory views earth's solid surface as consisting of several rigid plates that move relative to each other. These plates drift at an average rate of several

438   **Environmental Chemistry**

centimeters per year atop a relatively weak, partially molten layer that is part of earth's upper mantle called the **asthenosphere**. The science of plate tectonics explains the large-scale phenomena that affect the geosphere, including the creation and enlargement of oceans as the ocean floors open up and spread, the collision and breaking apart of continents, the formation of mountain chains, volcanic activities, the creation of islands of volcanic origin, and earthquakes.

The boundaries between these plates are where most geological activity such as earthquakes and volcanic activity occur. These boundaries are of the three following types:

- **Divergent boundaries** where the plates are moving away from each other. Occurring on ocean floors, these are regions in which hot magma flows upward and cools to produce new solid lithosphere. This new solid material creates **ocean ridges**.

- **Convergent boundaries** where plates move toward each other. One plate may be pushed beneath the other in a **subduction zone** in which matter is buried in the asthenosphere and eventually remelted to form new magma. When this does not occur, the lithosphere is pushed up to form mountain ranges along a collision boundary.

- **Transform fault boundaries** in which two plates slide past each other. These boundaries create faults that result in earthquakes.

The phenomena described above are parts of the **tectonic cycle**, a geological cycle which describes how tectonic plates move relative to each other, magma rises to form new solid rocks, and solid lithospheric rocks sink to become melted thus forming new magma. The tectonic cycle is illustrated in Figure 15.2.

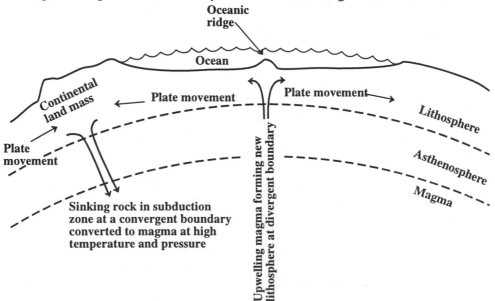

Figure 15.2. Illustration of the techtonic cycle in which upwelling magma along a boundary where two plates diverge creates new lithosphere on the ocean floor, and sinking rock in a subduction zone is melted to form magma.

Structural Geology

Earth's surface is constantly being reshaped by geological processes. The movement of rock masses during processes such as the formation of mountains results in substantial deformation of rock. At the opposite extreme of the size scale are defects in crystals at a microscopic level. **Structural geology** addresses the geometric forms of geologic structures over a wide range of size, the nature of structures formed by geological processes, and the formation of folds, faults, and other geological structures.

Primary structures are those that have resulted from the formation of a rock mass from its parent materials. Primary structures are modified and deformed to produce **secondary structures.** A basic premise of structural geology is that most layered rock formations were deposited in a horizontal configuration. Cracking of such a formation without displacement of the separate parts of the formation relative to each other produces a **joint**, whereas displacement produces a **fault** (see Figure 15.3).

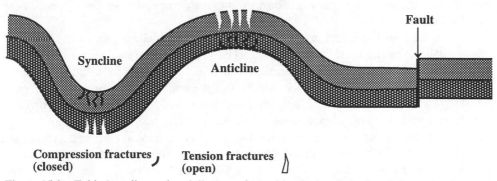

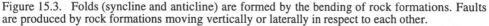

Figure 15.3. Folds (syncline and anticline) are formed by the bending of rock formations. Faults are produced by rock formations moving vertically or laterally in respect to each other.

An important relationship in structural geology is that between the force or **stress** placed upon a geological formation or object and the deformation resulting therefrom, called the **strain**. An important aspect of structural geology, therefore, is **rheology**, which deals with the deformation and flow of solids and semisolids. Whereas rocks tend to be strong, rigid, and brittle under the conditions at earth's surface, their rheology changes such that they may become weak and pliable under the extreme conditions of temperature and pressure at significant depths below earth's surface.

15.4. INTERNAL PROCESSES

The preceding section addressed the physical form of the geosphere. Related to the physical configuration of the geosphere are several major kinds of processes that occur that change this configuration and that have the potential to cause damage and even catastrophic effects. These can be divided into the two main categories of **internal processes** that arise from phenomena located significantly below the earth's surface, and **surface processes** that occur on the surface. Internal processes are addressed in this section, and surface processes in Section 15.5.

Earthquakes

Earthquakes usually arise from plate tectonic processes and originate along plate boundaries occurring as motion of ground resulting from the release of energy that accompanies an abrupt slippage of rock formations subjected to stress along a fault. Basically, two huge masses of rock tend to move relative to each other, but are locked together along a fault line. This causes deformation of the rock formations, which increases with increasing stress. Eventually, the friction between the two moving bodies is insufficient to keep them locked in place, and movement occurs along an existing fault, or a new fault is formed. Freed from constraints on their movement, the rocks undergo elastic rebound, causing the earth to shake. The serious damage that may ensue is discussed further in Section 15.11.

In addition to shaking of ground, which can be quite violent, earthquakes can cause the ground to rupture, subside, or rise.[4] **Liquefaction** is an important phenomenon that occurs during earthquakes with ground that is poorly consolidated and in which the water table may be high. Liquefaction results from separation of soil particles accompanied by water infiltration. when this occurs, the ground behaves like a fluid.

The location of the initial movement along a fault that causes an earthquake to occur is called the **focus** of the earthquake. The surface location directly above the focus is the **epicenter**. Energy is transmitted from the focus by **seismic waves**. Seismic waves that travel through the interior of the earth are called **body waves** and those that traverse the surface are **surface waves**. Body waves are further categorized as **P-waves**, compressional vibrations that result from the alternate compression and expansion of geospheric material, and **S-waves**, consisting of shear waves manifested by sideways oscillations of material. The motions of these waves are detected by a **seismograph**, often at great distances from the epicenter. The two types of waves move at different rates, with P-waves moving faster. From the arrival times of the two kinds of waves at different seismographic locations, it is possible to locate the epicenter of an earthquake.

Volcanoes

In addition to earthquakes, the other major subsurface process that has the potential to massively affect the environment consists of emissions of molten rock (lava), gases, steam, ash, and particles due to the presence of magma near the earth's surface. This phenomenon is called a **volcano** (Figure 13.5).[5] Volcanoes can be very destructive and damaging to the environment. Aspects of the potential harm from volcanoes are discussed in Section 15.12.

Volcanoes take on a variety of forms which are beyond the scope of this chapter to cover in detail. Basically, they are formed when magma rises to the surface. This frequently occurs in subduction zones created where one plate is pushed beneath another (see Figure 15.2). The downward movement of solid lithospheric material subjects it to high temperatures and pressures that causes the rock in it to melt and rise to the surface as magma. Molten magma issuing from a volcano at temperatures usually in excess of 500°C and often as high as 1,400°C, is called **lava**, and is one of the more common manifestations of volcanic activity.

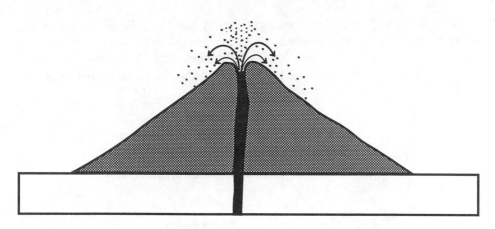

Figure 15.4. Volcanoes come in many shapes and forms. A classically shaped volcano may be a cinder cone formed by ejection of rock and lava, called pyroclastics, from the volcano to produce a relatively uniform cone.

15.5. SURFACE PROCESSES

Surface geological features are formed by upward movement of materials from earth's crust. With exposure to water, oxygen, freeze-thaw cycles, organisms, and other influences on the surface, surface features are subject to two processes that largely determine the landscape—weathering and erosion. As noted earlier in this chapter, weathering consists of the physical and chemical breakdown of rock and erosion is the removal and movement of weathered products by the action of wind, liquid water, and ice. Weathering and erosion work together in that one augments the other in breaking down rock and moving the products. Weathered products removed by erosion are eventually deposited as sediments and may undergo diagenesis and lithification to form sedimentary rocks.

One of the most common surface processes that can adversely affect humans consists of **landslides** that occur when soil or other unconsolidated materials slide down a slope.[6] Related phenomena include rockfalls, mudflows, and snow avalanches. As shown in Figure 15.5, a landslide typically consists of an upper slump that is prevented from sliding farther by a mass of material accumulated in a lower flow. Figure 15.5 illustrates what commonly happens in a landslide when a mass of earth moves along a slip plane under the influence of gravity. The stability of earthen material on a slope depends upon a balance between the mass of slope material and the resisting force of the shear strength of the slope material. There is a tendency for the earth to move along slip planes. In addition to the earthen material itself, water, vegetation, and structures constructed by humans may increase the driving force leading to a landslide. The shear strength is, of course, a function of the geological material along the slip plane and may be affected by other factors as well, such as the presence of various levels of water and the degree and kinds of vegetation growing on the surface.

The tendency of landslides to form is influenced by a number of outside factors. Climate is important because it influences the accumulation of water that often precedes a landslide as well as the presence of plants that can also influence soil stability. Although it would seem that plant roots should stabilize soil, the ability of some plants to add significant mass to the slope by accumulating water and to destabilize soil by aiding water infiltration may have an opposite effect. Disturbance of earth by road or other construction may cause landslides to occur. Earth may be shaken loose by earthquakes, causing landslides to occur.

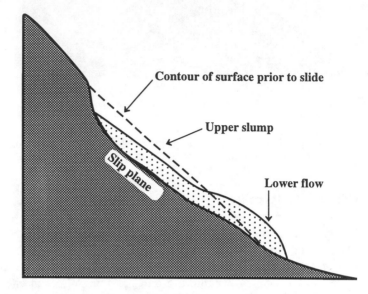

Figure 15.5. A landslide occurs when earth moves along a slip plane. Typically, a landslide consists of an upper slump and lower flow. The latter serves to stabilize the slide, and when it is disturbed, such as by cutting through it to construct a road, the earth may slide farther.

As discussed in Section 15.13, landslides can be very dangerous to human life and their costs in property damage can be enormous. In addition to destroying structures located on the surface of sliding land or covering structures or people with earth, landslides can have catastrophic indirect effects. For example, landslides that dump huge quantities of earth into reservoirs can raise water levels almost instantaneously and cause devastating waves and floods.

Subsidence occurs when the surface level of earth sinks over a significant area. The most spectacular evidence of subsidence is manifested as large sinkholes that may form rather suddenly, sometimes swallowing trees, automobiles, and even whole buildings in the process. Overall, much more damage is caused by gradual and less extreme subsidence, which may damage structures as it occurs or result in inundation of areas near water level. Such subsidence is frequently caused by the removal of fluids, such as petroleum, from below ground.

15.6. SEDIMENTS

Vast areas of land, as well as lake and stream sediments, are formed from sedimentary rocks. The properties of these masses of material depend strongly upon their origins and transport. Water is the main vehicle of sediment transport, although wind can also be significant. Hundreds of millions of tons of sediment are carried by major rivers each year.

The action of flowing water in streams cuts away stream banks and carries sedimentary materials for great distances. Sedimentary materials may be carried by flowing water in streams as the following:

- **Dissolved load** from sediment-forming minerals in solution
- **Suspended load** from solid sedimentary materials carried along in suspension
- **Bed load** dragged along the bottom of the stream channel.

The transport of calcium carbonate as dissolved calcium bicarbonate provides a straightforward example of dissolved load and is the most prevalent type of such load. Water with a high dissolved carbon dioxide content (usually present as the result of bacterial action) in contact with calcium carbonate formations contains Ca^{2+} and HCO_3^- ions. Flowing water containing calcium as such *temporary hardness* may become more basic by loss of CO_2 to the atmosphere, consumption of CO_2 by algal growth, or contact with dissolved base, resulting in the deposition of insoluble $CaCO_3$:

$$Ca^{2+} + 2HCO_3^- \rightarrow CaCO_3(s) + CO_2(g) + H_2O \qquad (15.6.1)$$

Most flowing water that contains dissolved load originates underground, where the water has had the opportunity to dissolve minerals from the rock strata that it has passed through.

Most sediments are transported by streams as suspended load, obvious in the observation of "mud" in the flowing water of rivers draining agricultural areas or finely divided rock in Alpine streams fed by melting glaciers. Under normal conditions, finely divided silt, clay, or sand make up most of the suspended load, although larger particles are transported in rapidly flowing water. The degree and rate of movement of suspended sedimentary material in streams are functions of the velocity of water flow and the settling velocity of the particles in suspension.

Bed load is moved along the bottom of a stream by the action of water "pushing" particles along. Particles carried as bed load do not move continuously. The grinding action of such particles is an important factor in stream erosion.

Typically, about $2/3$ of the sediment carried by a stream is transported in suspension, about $1/4$ in solution, and the remaining relatively small fraction as bed load. The ability of a stream to carry sediment increases with both the overall rate of flow of the water (mass per unit time) and the velocity of the water. Both of these are higher under flood conditions, so floods are particularly important in the transport of sediments.

Streams mobilize sedimentary materials through **erosion**, **transport** materials along with stream flow, and release them in a solid form during **deposition**. Deposits of stream-borne sediments are called **alluvium**. As conditions such as lowered stream velocity begin to favor deposition, larger, more settleable particles are released first. This results in **sorting** such that particles of a similar size and type tend to occur together in alluvial deposits. Much sediment is deposited in flood plains where streams overflow their banks.

15.7. CLAYS

Clays are extremely common and important in mineralogy. Furthermore, in general (see Chapter 16), clays predominate in the inorganic components of most soils and are very important in holding water and in plant nutrient cation exchange. All clays contain silicate and most contain aluminum and water. Physically, clays consist of very fine grains having sheet-like structures. For purposes of discussion here, **clay** is defined as a group of microcrystalline secondary minerals consisting of hydrous aluminum silicates that have sheet-like structures. Clay minerals are distinguished from each other by general chemical formula, structure, and chemical and physical properties. The three major groups of clay minerals are the following:

- **Montmorillonite**, $Al_2(OH)_2Si_4O_{10}$
- **Illite**, $K_{0-2}Al_4(Si_{8-6}Al_{0-2})O_{20}(OH)_4$
- **Kaolinite**, $Al_2Si_2O_5(OH)_4$

Many clays contain large amounts of sodium, potassium, magnesium, calcium, and iron, as well as trace quantities of other metals. Clays bind cations such as Ca^{2+}, Mg^{2+}, K^+, Na^+, and NH_4^+, which protects the cations from leaching by water but keeps them available in soil as plant nutrients. Since many clays are readily suspended in water as colloidal particles, they may be leached from soil or carried to lower soil layers.

Olivine, augite, hornblende, and feldspars are all parent minerals that form clays. An example is the formation of kaolinite ($Al_2Si_2O_5(OH)_4$) from potassium feldspar rock ($KAlSi_3O_8$):

$$2KAlSi_3O_8(s) + 2H^+ + 9H_2O \rightarrow Al_2Si_2O_5(OH)_4(s) + 2K^+(aq)$$
$$+ 4H_4SiO_4(aq) \quad (15.7.1)$$

The layered structures of clays consist of sheets of silicon oxide alternating with sheets of aluminum oxide. The silicon oxide sheets are made up of tetrahedra in which each silicon atom is surrounded by four oxygen atoms. Of the four oxygen atoms in each tetrahedron, three are shared with other silicon atoms that are components of other tetrahedra. This sheet is called the **tetrahedral sheet**. The aluminum oxide is contained in an **octahedral sheet**, so named because each aluminum atom is surrounded by six oxygen atoms in an octahedral configuration. The structure is such that some of the oxygen atoms are shared between aluminum atoms and some are shared with the tetrahedral sheet.

Structurally, clays may be classified as either **two-layer clays** in which oxygen atoms are shared between a tetrahedral sheet and an adjacent octahedral sheet, and **three-layer clays** in which an octahedral sheet shares oxygen atoms with tetrahedral sheets on either side. These layers composed of either two or three sheets are called **unit layers**. A unit layer of a two-layer clay typically is around 0.7 nanometers (nm) thick, whereas that of a three-layer clay exceeds 0.9 nm in thickness. The structure of the two-layer clay kaolinite is represented in Figure 15.6. Some clays, particularly the montmorillonites, may absorb large quantities of water between unit layers, a process accompanied by swelling of the clay.

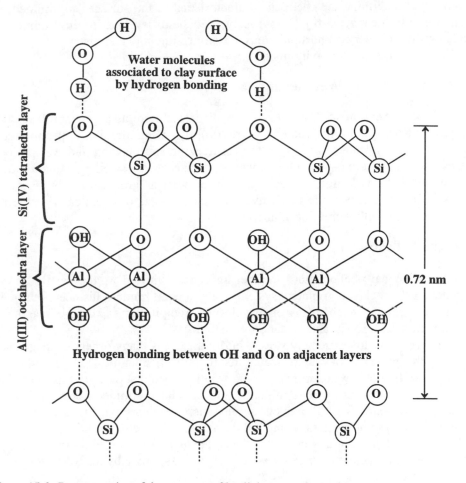

Figure 15.6. Representation of the structure of kaolinite, a two-layer clay.

As described in Section 5.4, clay minerals may attain a net negative charge by **ion replacement**, in which Si(IV) and Al(III) ions are replaced by metal ions of similar size but lesser charge. Compensation must be made for this negative charge by association of cations with the clay layer surfaces. Since these cations need not fit specific sites in the crystalline lattice of the clay, they may be relatively large ions, such as K^+, Na^+, or NH_4^+. These cations are called **exchangeable cations** and are exchangeable for other cations in water. The amount of exchangeable cations, expressed as milliequivalents (of monovalent cations) per 100 g of dry clay, is called the **cation-exchange capacity, CEC**, of the clay and is a very important characteristic of colloids and sediments that have cation-exchange capabilities.

15.8. GEOCHEMISTRY

Geochemistry deals with chemical species, reactions, and processes in the lithosphere and their interactions with the atmosphere and hydrosphere. The branch of geochemistry that explores the complex interactions among the rock/water/air/life systems that determine the chemical characteristics of the surface environment is **environmental geochemistry**.[7] Obviously, geochemistry and its environmental subdiscipline are very important areas of environmental chemistry with many applications related to the environment.

Physical Aspects of Weathering

Defined in Section 15.2, *weathering*, is discussed here as a geochemical phenomenon. Rocks tend to weather more rapidly when there are pronounced differences in physical conditions—alternate freezing and thawing and wet periods alternating with severe drying. Other mechanical aspects are swelling and shrinking of minerals with hydration and dehydration, as well as growth of roots through cracks in rocks. Temperature is involved in that the rates of chemical weathering (below) increase with increasing temperature.

Chemical Weathering

As a chemical phenomenon, weathering can be viewed as the result of the tendency of the rock/water/mineral system to attain equilibrium. This occurs through the usual chemical mechanisms of dissolution/precipitation, acid-base reactions, complexation, hydrolysis, and oxidation-reduction.

Weathering occurs extremely slowly in dry air but is many orders of magnitude faster in the presence of water. Water, itself, is a chemically active weathering substance and it holds weathering agents in solution such that they are transported to chemically active sites on rock minerals and contact the mineral surfaces at the molecular and ionic level. Prominent among such weathering agents are CO_2, O_2, organic acids (including humic and fulvic acids, see Section 3.17), sulfur acids ($SO_2(aq)$, H_2SO_4), and nitrogen acids (HNO_3, HNO_2). Water provides the source of H^+ ion needed for acid-forming gases to act as acids as shown by the following:

$$CO_2 + H_2O \rightarrow H^+ + HCO_3^- \tag{15.8.1}$$

$$SO_2 + H_2O \rightarrow H^+ + HSO_3^- \tag{15.8.2}$$

Rainwater is essentially free of mineral solutes. It is usually slightly acidic due to the presence of dissolved carbon dioxide or more highly acidic because of acid-rain forming constitutents. As a result of its slight acidity and lack of alkalinity and dissolved calcium salts, rainwater is *chemically aggressive* (see Section 8.7) toward some kinds of mineral matter, which it breaks down by chemical weathering processes. Because of this process, river water has a higher concentration of dissolved inorganic solids than does rainwater.

The processes involved in chemical weathering may be divided into the following major categories:

- **Hydration/dehydration**, for example:

$$CaSO_4(s) + 2H_2O \rightarrow CaSO_4 \cdot 2H_2O(s)$$

$$2Fe(OH)_3 \cdot xH_2O(s) \rightarrow Fe_2O_3(s) + (3 + 2x)H_2O$$

- **Dissolution**, for example:

$$CaSO_4 \cdot 2H_2O(s) \text{ (water)} \rightarrow Ca^{2+}(aq) + SO_4^{2-}(aq) + 2H_2O$$

- **Oxidation**, such as occurs in the dissolution of pyrite:

$$4FeS_2(s) + 15O_2(g) + (8 + 2x)H_2O \rightarrow$$
$$2Fe_2O_3 \cdot xH_2O + 8SO_4^{2-}(aq) + 16H^+(aq)$$

or in the following example in which dissolution of an iron(II) mineral is followed by oxidation of iron(II) to iron(III):

$$Fe_2SiO_4(s) + 4CO_2(aq) + 4H_2O \rightarrow 2Fe^{2+} + 4HCO_3^- + H_4SiO_4$$

$$4Fe^{2+} + 8HCO_3^- + O_2(g) \rightarrow 2Fe_2O_3(s) + 8CO_2 + 4H_2O$$

The second of these two reactions may occur at some distance from the first, resulting in net transport of iron from its original location. Iron, manganese, and sulfur are the major elements that undergo oxidation as part of the weathering process.

- **Dissolution with hydrolysis** as occurs with the hydrolysis of carbonate ion when mineral carbonates dissolve:

$$CaCO_3(s) + H_2O \rightarrow Ca^{2+}(aq) + HCO_3^-(aq) + OH^-(aq)$$

Hydrolysis is the major means by which silicates undergo weathering as shown by the following reaction of forsterite:

$$Mg_2SiO_4(s) + 4CO_2 + 4H_2O \rightarrow 2Mg^{2+} + 4HCO_3^- + H_4SiO_4$$

The weathering of silicates yields soluble silicon as species such as H_4SiO_4, and residual silicon-containing minerals (clay minerals).

- **Acid hydrolysis**, which accounts for the dissolution of significant amounts of $CaCO_3$ and $CaCO_3 \cdot MgCO_3$ in the presence CO_2-rich water:

$$CaCO_3(s) + H_2O + CO_2(aq) \rightarrow Ca^{2+}(aq) + 2HCO_3^-(aq)$$

- **Complexation**, as exemplified by the reaction of oxalate ion, $C_2O_4^{2-}$ with aluminum in muscovite, $K_2(Si_6Al_2)Al_4O_{20}(OH)_4$:

$$K_2(Si_6Al_2)Al_4O_{20}(OH)_4(s) + 6C_2O_4^{2-}(aq) + 20\, H^+ \rightarrow$$
$$6AlC_2O_4^+(aq) + 6Si(OH)_4 + 2K^+$$

Reactions such as these largely determine the kinds and concentrations of solutes in surface water and groundwater. Acid hydrolysis, especially, is the predominant process that releases elements such as Na^+, K^+, and Ca^{2+} from silicate minerals.

15.9. GROUNDWATER IN THE GEOSPHERE

Groundwater (Figure 15.7) is a vital resource in its own right that plays a crucial role in geochemical processes, such as the formation of secondary minerals. The nature, quality, and mobility of groundwater are all strongly dependent upon the rock formations in which the water is held. Physically, an important characteristic of such formations is their **porosity**, which determines the percentage of rock volume available to contain water. A second important physical characteristic is **permeability**, which describes the ease of flow of the water through the rock. High permeability is usually associated with high porosity. However, clays tend to have low permeability even when a large percentage of the volume is filled with water.

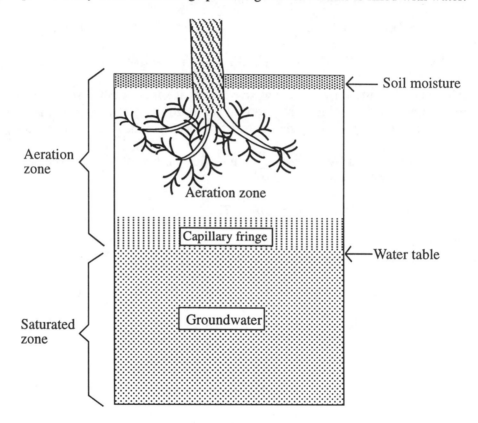

Figure 15.7. Some major features of the distribution of water underground.

Most groundwater originates as **meteoric** water from precipitation in the form of rain or snow. If water from this source is not lost by evaporation, transpiration, or to stream runoff, it may infiltrate into the ground. Initial amounts of water from precipitation onto dry soil are held very tightly as a film on the surfaces and in the micropores of soil particles in a **belt of soil moisture**. At intermediate levels, the soil particles are covered with films of water, but air is still present in larger voids in the soil. The region in which such water is held is called the **unsaturated zone** or **zone of aeration** and the water present in it is **vadose water**. At lower depths in the presence of adequate amounts of water, all voids are filled to produce a **zone of**

saturation, the upper level of which is the **water table**. Water present in a zone of saturation is called **groundwater**. Because of its surface tension, water is drawn somewhat above the water table by capillary-sized passages in soil in a region called the capillary fringe.

The water table (Figure 15.8) is crucial in explaining and predicting the flow of wells and springs and the levels of streams and lakes. It is also an important factor in determining the extent to which pollutant and hazardous chemicals underground are likely to be transported by water. The water table can be mapped by observing the equilibrium level of water in wells, which is essentially the same as the top of the saturated zone. The water table is usually not level, but tends to follow the general contours of the surface topography. It also varies with differences in permeability and water infiltration. The water table is at surface level in the vicinity of swamps and frequently above the surface where lakes and streams are encountered. The water level in such bodies may be maintained by the water table. **Influent** streams or reservoirs are located above the water table; they lose water to the underlying aquifer and cause an upward bulge in the water table beneath the surface water.

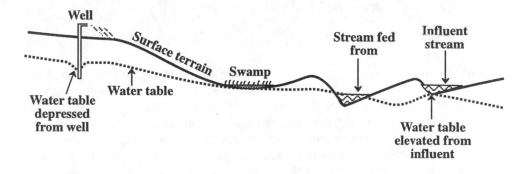

Figure 15.8. The water table and influences of surface features on it.

Groundwater **flow** is an important consideration in determining the accessibility of the water for use and transport of pollutants from underground waste sites. Various parts of a body of groundwater are in hydraulic contact so that a change in pressure at one point will tend to affect the pressure and level at another point. For example, infiltration from a heavy, localized rainfall may affect the water table at a point remote from the infiltration. Groundwater flow occurs as the result of the natural tendency of the water table to assume even levels by the action of gravity.

Groundwater flow is strongly influenced by rock permeability. Porous or extensively fractured rock is relatively highly **pervious**, meaning that water can migrate through the holes, fissures, and pores in such rock. Because water can be extracted from such a formation, it is called an **aquifer**. By contrast, an **aquiclude** is a rock formation that is too impermeable or unfractured to yield groundwater. Impervious rock in the unsaturated zone may retain water infiltrating from the surface to produce a **perched water table** that is above the main water table and from which water may be extracted. However, the amounts of water that can be extracted from such a formation are limited and the water is vulnerable to contamination.

Water Wells

Most groundwater is tapped for use by water wells drilled into the saturated zone. The use and misuse of water from this source has a number of environmental implications. In the U.S. about 2/3 of the groundwater pumped is consumed for irrigation; lesser amounts of ground water are used for industrial and municipal applications.

As water is withdrawn, the water table in the vicinity of the well is lowered. This **drawdown** of water creates a **zone of depression**. In extreme cases the groundwater is severely depleted and surface land levels can even subside (which is one reason that Venice, Italy is now very vulnerable to flooding). Heavy drawdown can result in infiltration of pollutants from sources such as septic tanks, municipal refuse sites, and hazardous waste dumps. When soluble iron(II) or manganese(II) are present in groundwater, exposure to air at the well wall can result in the formation of deposits of insoluble iron(III) and manganese(IV) oxides produced by bacterially catalyzed processes:

$$4Fe^{2+}(aq) + O_2(aq) + 10H_2O \rightarrow 4Fe(OH)_3(s) + 8H^+ \qquad (15.9.1)$$

$$2Mn^{2+}(aq) + O_2(aq) + (2x+2)H_2O \rightarrow 2MnO_2 \cdot xH_2O(s) + 4H^+ \qquad (15.9.2)$$

Deposits of iron(III) and manganese(IV) that result from the processes outlined above line the surfaces from which water flows into the well with a coating that is relatively impermeable to water. The deposits fill the spaces that water must traverse to enter the well. As a result, they can seriously impede the flow of water into the well from the water-bearing aquifer. This creates major water source problems for municipalities using groundwater for water supply. As a result of this problem, chemical or mechanical cleaning, drilling of new wells, or even acquisition of new water sources may be required.

15.10. ENVIRONMENTAL ASPECTS OF THE GEOSPHERE

Most of the remainder of this chapter deals specifically with the environmental aspects of geology and human interactions with the geosphere. It discusses how natural geological phenomena affect the environment through occurrences such as volcanic eruptions that may blast so much particulate matter and acid gas into the atmosphere that it may have a temporary effect on global climate, or massive earthquakes that disrupt surface topography and disturb the flow and distribution of groundwater and surface water. Also discussed are human influences on the geosphere and the strong connection between the geosphere and the anthrosphere.

Going back several billion years to its formation as a ball of dust particles collected from the universe and held together by gravitational forces, earth has witnessed constant environmental change and disruption. During its earlier eons earth was a most inhospitable place for humans and, indeed, for any form of life. Heat generated by gravitational compression of primitive earth and by radioactive elements in its interior caused much of the mass of the planet to liquify. Relatively high density iron sank into the core, and lighter minerals, primarily silicates, solidified and floated to the surface.

Although in the scale of a human lifetime earth changes almost imperceptibly, the planet is in fact in a state of constant change and turmoil. It is known that continents have formed, broken apart, and moved around. Rock formations produced in ancient oceans have been thrust up onto continental land and huge masses of volcanic rock exist where volcanic activity is now unknown. Today the angry bowels of earth unleash enormous forces that push molten rock to the surface and move continents continuously as evidenced from volcanic activity, and from earthquakes resulting from the movement of great land masses relative to each other. earth's surface is constantly changing as new mountain ranges are heaved up and old ones are worn down.

Humans have learned to work with, against, and around natural earth processes and phenomena to exploit earth's resources and to make these processes and phenomena work for the benefit of humankind. Human efforts have been moderately successful in mitigating some of the major hazards posed by natural geospheric phenomena, although such endeavors often have had unforeseen detrimental consequences, sometimes many years after they were first applied. The survival of modern civilization and, indeed, of humankind will depend upon how intelligently humans work with the earth. That is why it is so important for humans to have a fundamental understanding of the geospheric environment.

An important consideration in human interaction with the geosphere is the application of engineering to geology. Engineering geology takes account of the geological characteristics of soil and rock in designing buildings, dams, highways, and other structures in a manner compatible with the geological strata on which they rest. Engineering geology must consider a large number of geological factors including type, strength, and fracture characteristics of rock, tendency for landslides to occur, susceptibility to settling, and likelihood of erosion. Engineering geology is an important consideration in land-use planning.

Natural Hazards

Earth presents a variety of natural hazards to the creatures that dwell on it. Some of these are the result of internal processes that arise from the movement of land masses relative to each other and from heat and intrusions of molten rock from below the surface. The most common such hazards are earthquakes and volcanoes. Whereas internal processes tend to force matter upward, often with detrimental effects, surface processes are those that generally result from the tendency of matter to seek lower levels. Such processes include erosion, landslides, avalanches, mudflows, and subsidence.

A number of natural hazards result from the interaction and conflict between solid Earth and liquid and solid water. Perhaps the most obvious such hazard consists of floods when too much water falls as precipitation and seeks lower levels through streamflow. Wind can team with water to increase destructive effects, such as beach erosion and destruction of beachfront property resulting from wind-driven seawater. Ice, too, can have some major effects on solid earth. Evidence of such effects of the Ice Age times include massive glacial moraines left over from deposition of till from melting glaciers, and landscape features carved by advancing ice sheets.

Anthropogenic Hazards

All too often, attempts to control and reshape the geosphere to human demands have been detrimental to the geosphere and dangerous to human life and well-being. Such attempts may exacerbate damaging natural phenomena. A prime example of this interaction occurs when efforts are made to control the flow of rivers by straightening them and building levees. The initial results can be deceptively favorable in that a modified stream may exist for decades, flowing smoothly and staying within the confines imposed by humans. But eventually, the forces of nature are likely to overwhelm the efforts of humans to control them, such as when a record flood breaks levees and destroys structures constructed in flood-prone areas. Landslides of mounds of earthen material piled up from mining can be very destructive. Destruction of wetlands in an effort to provide additional farmland can have some detrimental effects upon wildlife and upon the overall health of ecosystems.

15.11. EARTHQUAKES

The loss of life and destruction of property by earthquakes makes them some of nature's more damaging natural phenomena. The destructive effects of an earthquake are due to the release of energy. The released energy moves from the quake's focus as seismic waves, discussed in Section 15.4. Literally millions of lives have been lost in past earthquakes, and damage from an earthquake in a developed urban area can easily run into billions of dollars. A tragic August 1999 earthquake in Turkey killed several thousand people and caused massive damage. Earthquakes may cause catastrophic secondary effects, especially large, destructive ocean waves called tsunamis (discussed below).

Adding to the terror of earthquakes is their lack of predictability. An earthquake can strike at any time—during the calm of late night hours or in the middle of busy rush hour traffic. Although the exact prediction of earthquakes has so far eluded investigators, the locations where earthquakes are most likely to occur are much more well-known. These are located in lines corresponding to boundaries along which tectonic plates collide and move relative to each other, building up stresses that are suddenly released when earthquakes occur. Such interplate boundaries are locations of preexisting faults and breaks. Occasionally, however, an earthquake will occur within a plate, made more massive and destructive because for it to occur the thick lithosphere has to be ruptured.

The scale of earthquakes can be estimated by the degree of motion that they cause and by their destructiveness. The former is termed the **magnitude** of an earthquake and is commonly expressed by the **Richter scale**. The Richter scale is open-ended, and each unit increase in the scale reflects a ten-fold increase in magnitude. Several hundred thousand earthquakes with magnitudes from two to three occur each year; they are detected by seismographs, but are not felt by humans. Minor earthquakes range from four to five on the Richter scale, and earthquakes cause damage at a magnitude greater than about five. Great earthquakes, which occur about once or twice a year, register over eight on the Richter scale.

The **intensity** of an earthquake is a subjective estimate of its potential destructive effect. On the Mercalli intensity scale, an intensity III earthquake feels like the passage of heavy vehicles; one with an intensity of VII causes difficulty in standing, damage to plaster, and dislodging of loose brick, whereas a quake with an intensity of XII causes virtually total destruction, throws objects upward, and shifts huge masses of earthen material. Intensity does not correlate exactly with magnitude.

Distance from the epicenter, the nature of underlying strata, and the types of structures affected may all result in variations in intensity from the same earthquake. In general, structures built on bedrock will survive with much less damage than those constructed on poorly consolidated material. Displacement of ground along a fault can be substantial, for example, up to six meters along the San Andreas fault during the 1906 San Francisco earthquake. Such shifts can break pipelines and destroy roadways. Highly destructive surface waves can shake vulnerable structures apart.

The shaking and movement of ground are the most obvious means by which earthquakes cause damage. In addition to shaking it, earthquakes can cause the ground to rupture, subside, or rise. **Liquefaction** is an important phenomenon that occurs during earthquakes with ground that is poorly consolidated and in which the water table may be high. Liquefaction results from separation of soil particles accompanied by water infiltration such that the ground behaves like a fluid.

Another devastating phenomenon consists of **tsunamis**, large ocean waves resulting from earthquake-induced movement of ocean floor.[8] Tsunamis sweeping onshore at speeds up to 1000 km/hr have destroyed many homes and taken many lives, often large distances from the epicenter of the earthquake, itself.[9] This effect occurs when a tsunami approaches land and forms huge breakers, some as high as 10-15 meters, or even higher. On April 1, 1946, an earthquake off the coast of Alaska generated a Tsunami estimated to be more than 30 meters high that killed 5 people on a nearby lighthouse. About 5 hours later a Tsunami generated by the same earthquake reached Hilo, Hawaii, and killed 159 people with a wave exceeding 15 meters high. The March 27, 1964, Alaska earthquake generated a tsunami over 10 meters high that hit a freighter docked at Valdez, tossing it around like matchwood. Miraculously, nobody on the freighter was killed, but 28 people on the dock died.

Literally millions of lives have been lost in past earthquakes, and damage from an earthquake in a developed urban area can easily run into billions of dollars. As examples, a massive earthquake in Egypt and Syria in 1201 A.D. took over 1 million lives, one in Tangshan, China, in 1976 killed about 650,000, and the 1989 Loma Prieta earthquake in California cost about 7 billion dollars.

Significant progress has been made in designing structures that are earthquake-resistant. As evidence of that, during a 1964 earthquake in Niigata, Japan, some buildings tipped over on their sides due to liquefaction of the underlying soil, but remained structurally intact! Other areas of endeavor that can lessen the impact of earthquakes is the identification of areas susceptible to earthquakes, discouraging development in such areas, and educating the public about earthquake hazards. Accurate prediction would be a tremendous help in lessening the effects of earthquakes, but so far has been generally unsuccessful. Most challenging of all is the possibility of preventing major earthquakes. One unlikely possibility would be to detonate nuclear explosives deep underground along a fault line to release stress before it builds up to an excessive level. Fluid injection to facilitate slippage along a fault has also been considered.

15.12. VOLCANOES

On May 18, 1980, Mount St. Helens, a volcano in Washington State erupted, blowing out about 1 cubic kilometer of material. This massive blast spread ash over half the United States, causing about $1 billion in damages and killing an estimated 62 people, many of whom were never found. Many volcanic disasters have been recorded throughout history. Perhaps the best known of these is the 79 A.D. eruption of Mount Vesuvius, which buried the Roman city of Pompei with volcanic ash.

Temperatures of **lava**, molten rock flowing from a volcano, typically exceed 500°C and may get as high as 1400°C or more. Lava flows destroy everything in their paths, causing buildings and forests to burn and burying them under rock that cools and becomes solid. Often more dangerous than a lava flow are the **pyroclastics** produced by volcanoes and consisting of fragments of rock and lava. Some of these particles are large and potentially very damaging, but they tend to fall quite close to the vent. Ash and dust may be carried for large distances and, in extreme cases, as was the case in ancient Pompei, may bury large areas to some depth with devastating effects. The explosion of Tambora volcano in Indonesia in 1815 blew out about 30 cubic kilometers of solid material. The ejection of so much solid into the atmosphere had such a devastating effect on global climate that the following year was known as "the year without a summer," causing widespread hardship and hunger because of global crop failures.

A special kind of particularly dangerous pyroclastic consists of **nuée ardente**. This term, French for "glowing cloud," refers to a dense mixture of hot toxic gases and fine ash particles reaching temperatures of 1000°C that can flow down the slopes of a volcano at speeds of up to 100 km/hr. In 1902 a nuée ardente was produced by the eruption of Mont Pelée on Martinique in the Caribbean. Of as many as 40,000 people in the town of St. Pierre, the only survivor was a terrified prisoner shielded from the intense heat by the dungeon in which he was imprisoned.

One of the more spectacular and potentially damaging volcanic phenomena is a **phreatic eruption** that occurs when infiltrating water is superheated by hot magma and causes a volcano to literally explode. This happened in 1883 when uninhabited Krakatoa in Indonesia blew up with an energy release of the order of 100 megatons of TNT. Dust was blown 80 kilometers into the stratosphere, and a perceptible climatic cooling was noted for the next 10 years. As is the case with earthquakes, volcanic eruptions may cause the devastating tsunamis. Krakatoa produced a tsunami 40 meters high that killed 30 to 40 thousand people on surrounding islands.

Some of the most damaging health and environmental effects of volcanic eruptions are caused by gases released to the atmosphere. Huge quantities of water vapor are often evolved. Dense carbon dioxide gas can suffocate people near the point of release. Highly toxic H_2S and CO gases may be released by volcanoes. Volcanoes tend to give off acid gases such as hydrogen chloride produced by the subduction and heating of sodium chloride entrained in ocean sediment. Sulfur oxides released by volcanoes may affect the atmosphere. In 1982 El Chichón erupted in Mexico, producing comparatively little dust but huge quantities of sulfur oxides. These gases were converted to sulfuric acid droplets in the atmosphere, which reflected enough sunlight to cause a perceptible cooling in climate. Eventually the sulfuric acid released fell as acidic precipitation, "acid rain."

Volcanic activity could change the global environment dramatically. Massive volcanic eruptions many millions of years ago were probably responsible for widespread extinctions of organisms on earth's surface. These effects occur primarily by the ejection of particles and sulfuric acid precursors into the atmosphere causing global cooling and potential harm to the protective stratospheric ozone layer. Although such an extinction event is unlikely in modern times, a volcanic eruption such as that of the Tambora volcano described above could certainly happen. With humankind "living on the edge" as far as grain supplies are concerned, widespread starvation resulting from a year or two of crop failures would almost certainly occur.

15.13. SURFACE EARTH MOVEMENT

Mass movements are the result of gravity acting upon rock and soil on earth's surface. This produces a shearing stress on earthen materials located on slopes that can exceed the shear strength of the material and produce landslides and related phenomena involving the downward movement of geological materials. Such phenomena are affected by several factors, including the kinds and, therefore, strengths of materials, slope steepness, and degree of saturation with water. Usually, a specific event initiates mass movement. This can occur when excavation by humans steepens the slopes, by the action of torrential rains, or by earthquakes.

Described in Section 15.5 and illustrated in Figure 15.5, landslides refer to events in which large masses of rock and dirt move downslope rapidly. Such events occur when material resting on a slope at an **angle of repose** is acted upon by gravity to produce a **shearing stress**. This stress may exceed the forces of friction or **shear strength**. Weathering, fracturing, water, and other factors may induce the formation of **slide planes** or **failure planes** such that a landslide results.

Loss of life and property from landslides can be substantial. In 1970 a devastating avalanche of soil, mud, and rocks initiated by an earthquake slid down Mt. Huascaran in Peru killing an estimated 20,000 people. Sometimes the effects are indirect. In 1963 a total of 2600 people were killed near the Vaiont Dam in Italy. A sudden landslide filled the reservoir behind the dam with earthen material and, although the dam held, the displaced water spilled over its abutments as a wave 90 meters high, wiping out structures and lives in its path.

Although often ignored by developers, the tendency toward landslides is predictable and can be used to determine areas in which homes and other structures should not be built. Slope stability maps based upon the degree of slope, the nature of underlying geological strata, climatic conditions, and other factors can be used to assess the risk of landslides. Evidence of a tendency for land to slide can be observed from effects on existing structures, such as walls that have lost their alignment, cracks in foundations, and poles that tilt. The likelihood of landslides can be minimized by moving material from the upper to the lower part of a slope, avoiding the loading of slopes, and avoiding measures that might change the degree and pathways of water infiltration into slope materials. In cases where the risk is not too severe, retaining walls may be constructed that reduce the effects of landslides.

Several measures can be used to warn of landslides. Simple visual observations of changes in the surface can be indicative of an impending landslide. More sophisticated measures include tilt meters and devices that sense vibrations accompanying the movement of earthen materials.

In addition to landslides, there are several other kinds of mass movements that have the potential to be damaging. **Rockfalls** occur when rocks fall down slopes so steep that at least part of the time the falling material is not in contact with the ground. The fallen material accumulates at the bottom of the fall as a pile of **talus**. A much less spectacular event is **creep**, in which movement is slow and gradual. The action of frost—frost heaving—is a common form of creep. Though usually not life-threatening, over a period of time creep may ruin foundations and cause misalignment of roads and railroads with significant property damage often the result.

Special problems are presented by permanently frozen ground in arctic climates such as Alaska or Siberia. In such areas the ground may remain permanently frozen, thawing to only a shallow depth during the summer. This condition is called **permafrost**. Permafrost poses particular problems for construction, particularly where the presence of a structure may result in thawing such that the structure rests in a pool of water-saturated muck resting on a slick surface of frozen water and soil. The construction and maintenance of highways, railroads, and pipelines, such as the Trans-Alaska pipeline in Alaska, can become quite difficult in the presence of permafrost.

Some types of soils, particularly so-called expansive clays, expand and shrink markedly as they become saturated with water and dry out. Although essentially never life-threatening, the movement of structures and the damage caused to them by expansive clays can be very high. Aside from years when catastrophic floods and earthquakes occur, the monetary damage done by the action of expansive soil exceeds that of earthquakes, landslides, floods, and coastal erosion combined!

Sinkholes are a kind of earth movement resulting when surface earth falls into an underground cavity. They rarely injure people but may cause spectacular property damage. Cavities that produce sinkholes may form by the action of water containing dissolved carbon dioxide on limestone (See Chapter 3, Reaction 3.7. 6); loss of underground water during drought or from heavy pumping, thus removing support that previously kept soil and rock from collapsing; heavy underground water flow; and other factors that remove solid material from underground strata.

15.14. STREAM AND RIVER PHENOMENA

A **stream** consists of water flowing through a channel. The area of land from which water is drawn that flows into a stream is the stream's **drainage basin**. The sizes of streams are described by **discharge** defined as the volume of water flowing past a given point on the stream per unit time. Discharge and **gradient**, the steepness of the downward slope of a stream determine the stream **velocity**.

Internal processes raise masses of land and whole mountain ranges, which in turn are shaped by the action of streams. Streams cut down mountain ranges, create valleys, form plains, and produce great deposits of sediment, thus playing a key role in shaping the geospheric environment. Streams spontaneously develop bends and curves by cutting away the outer parts of stream banks and depositing materials on the inner parts. These curved features of streams are known as **meanders**. Left undisturbed, a stream forms meanders across a valley in a constantly changing pattern. The cutting away of material by the stream and the deposition of sediment eventually forms a generally flat area. During times of high stream flow, the stream leaves its banks, inundating parts or all of the valley, thus creating a **floodplain**.

A **flood** occurs when a stream develops a high flow such that it leaves its banks and spills out onto the floodplain. Floods are arguably the most common and damaging of surface phenomena in the geosphere. Though natural and in many respects beneficial occurrences, floods cause damage to structures located in their paths, and the severity of their effects is greatly increased by human activities.

A number of factors determine the occurrence and severity of floods. One of these is the tendency of particular geographic areas to receive large amounts of rain within short periods of time. One such area is located in the middle of the continental United States where warm, moisture-laden air from the Gulf of Mexico is carried northward during the spring months to collide with cold air from the north; the resultant cooling of the moist air can cause torrential rains to occur, resulting in severe flooding. In addition to season and geography, geological conditions have a strong effect on flooding potential. Rain falling on a steep surface tends to run off rapidly, creating flooding. A watershed can contain relatively massive quantities of rain if it consists of porous, permeable materials that allow a substantial rate of infiltration, assuming that it is not already saturated. Plants in a watershed tend to slow runoff and loosen soil, enabling additional infiltration. Through transpiration (see Chapter 16, Section 16.2), plants release moisture to the atmosphere quickly, enabling soil to absorb more moisture.

Several terms are used to describe flooding. When the **stage** of a stream, that is, the elevation of the water surface, exceeds the stream bank level, the stream is said to be at **flood stage**. The highest stage attained defines the flood **crest**. **Upstream** floods occur close to the inflow from the drainage basin, usually the result of intense rainfall. Whereas upstream floods usually affect smaller streams and watersheds, **downstream floods** occur on larger rivers that drain large areas. Widespread spring snowmelt and heavy, prolonged spring rains, often occurring together, cause downstream floods.

Floods are made more intense by higher fractions and higher rates of runoff, both of which may be aggravated by human activities. This can be understood by comparing a vegetated drainage basin to one that has been largely denuded of vegetation and paved over. In the former case, rainfall is retained by vegetation, such as grass cover. Thus the potential flood water is delayed, the time span over which it enters a stream is extended, and a higher proportion of the water infiltrates into the ground. In the latter case, less rainfall infiltrates, and the runoff tends to reach the stream quickly and to be discharged over a shorter time period, thus leading to more severe flooding. These factors are illustrated in Figure 15.9.

The conventional response to the threat of flooding is to control a river, particularly by the construction of raised banks called **levees**. In addition to raising the banks to contain a stream, the stream channel may be straightened and deepened to increase the volume and velocity of water flow, a process called **channelization**. Although effective for common floods, these measures may exacerbate extreme floods by confining and increasing the flow of water upstream such that the capacity to handle water downstream is overwhelmed. Another solution is to construct dams to create reservoirs for flood control upstream. Usually such reservoirs are multipurpose facilities designed for water supply, recreation, and to control river flow for navigation in addition to flood control. The many reservoirs constructed for flood control in recent decades have been reasonably successful. There are, however, conflicts in the goals for their uses. Ideally, a reservoir for flood control should remain largely empty until needed to contain a large volume of floodwater, an approach that is obviously inconsistent with other uses. Another concern is that of exceeding the capacity of the reservoir, or dam failure, the latter of which can lead to catastrophic flooding.

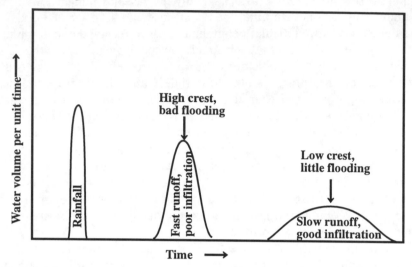

Figure 15.9. Influence of runoff on flooding.

15.15. PHENOMENA AT THE LAND/OCEAN INTERFACE

The coastal interface between land masses and the ocean is an important area of environmental activity. The land along this boundary is under constant attack from the waves and currents from the ocean, so that most coastal areas are always changing. The most common structure of the coast is shown in cross section in Figure 15.10. The beach, consisting of sediment, such as sand formed by wave action on coastal rock, is a sloping area that is periodically inundated by ocean waves. Extend-

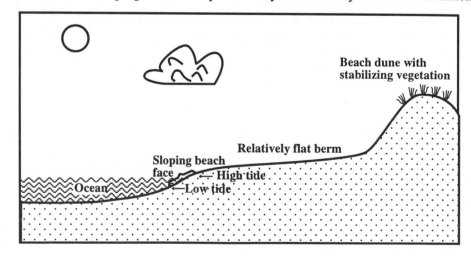

Figure 15.10. Cross section of the ocean/land interface along a beach.

ing from approximately the high tide mark to the dunes lining the landward edge of the beach is a relatively level area called the **berm**, which is usually not washed over by ocean water. The level of water to which the beach is subjected varies with the tides. Through wind action the surface of the water is in constant motion as undulations called **ocean waves**. As these waves reach the shallow water along the beach, they "touch bottom" and are transformed to **breakers**, characterized by crested tops. These breakers crashing onto a beach give it much of its charm, but can also be extremely destructive.

Coastlines exhibit a variety of features. Steep valleys carved by glacial activity, then filled with rising seawater, constitute the fjords along the coast of Norway. Valleys, formerly on land, now filled with seawater, constitute **drowned valley Estuaries** occur where tidal salt water mixes with inflowing fresh water.

Erosion is a constant feature of a beachfront. Unconsolidated beach sand can be shifted readily, sometimes spectacularly through great distances over short periods of time, by wave action. Sand, pebbles, and rock in the form of rounded cobbles constantly wear against the coast by wave action, exerting a constant abrasive action called **milling**. This action is augmented by the chemical weathering effects of seawater, in which the salt content may play a role.

Some of the more striking alterations to coastlines occur during storms, such as hurricanes and typhoons. The low pressure that accompanies severe storms tends to suck ocean water upward. This effect, usually combined with strong winds blowing onshore and coinciding with high tide, can cause ocean water to wash over the berm on a beach to attack dunes or cliffs inland. Such a **storm surge** can remove large quantities of beach, damage dune areas, and wash away structures unwisely constructed too close to the shore. A storm surge associated with a hurricane washed away most of the structures in Galveston, Texas, in 1900, claiming 6000 lives.

An especially vulnerable part of the coast consists of barrier islands, long, low strips of land roughly paralleling the coast some distance offshore. High storm surges may wash completely over barrier islands, partially destroying them and shifting them around. Many dwellings unwisely constructed on barrier islands, such as the outer banks of North Carolina, have been destroyed by storm surges during hurricanes.

The Threat of Rising Sea Levels

Large numbers of people live at a level near, or in some cases below, sea level. As a result, any significant temporary or permanent rise in sea level poses significant risks to lives and property. Such an event occurred on February 1, 1953, when high tides and strong winds combined to breach the system of dikes protecting much of the Netherlands from seawater. About one sixth of the country was flooded as far inland as 64 kilometers from the coast, killing about 2000 people and leaving approximately 100,000 without homes.

Although isolated instances of flooding by seawater caused by combinations of tidal and weather phenomena will continue to occur, a much more long-lasting threat is posed by long-term increases in sea level. These could result from global warming due to the greenhouse gas emissions discussed in Chapter 14. Several factors could raise ocean levels to destructive highs, also a result of greenhouse warming. Simple expansion of warmed oceanic water could raise sea levels by about 1/3 meter over the next century. The melting of glaciers, such as those in the Alps, has probably raised ocean levels about 5 cm during the last century, and the process is continuing. The greatest concern, however, is that global warming could cause the great West Antarctic ice sheet to melt, which would raise sea levels by as much as 6 meters.

A great deal of uncertainty exists regarding the possibility of the West Antarctic ice sheet melting with consequent rises in sea level. Current computer models show a compensating effect in that hotter air produced by greenhouse warming could carry much more atmospheric moisture to the Antarctic regions where the moisture would be deposited as snow. The net result could well be an *increase* in solid snow and ice in the Antarctic, and an accompanying *decrease* in sea levels.[10] Some of the uncertainty regarding the status of the West Antarctic ice sheet may be alleviated in the future by highly accurate space satellite measurements. The measurement of sea levels has proven to be a difficult task because the levels of the surface of land keep changing. Land most recently covered with Ice Age glaciers in areas such as Scandinavia is still "springing back" from the immense mass of the glaciers, so that sea levels measured by gauges fixed on land actually appear to be dropping by several millimeters per year in such locations. An opposite situation exists on the east coast of North America where land was pushed outward and raised around the edge of the enormous sheet of ice that covered Canada and the northern U.S. about 20,000 years ago and is now settling back. Factors such as these illustrate the advantages of remarkably accurate satellite technology now used in the determination of sea levels.

15.16. PHENOMENA AT THE LAND/ATMOSPHERE INTERFACE

The interface between the atmosphere and land is a boundary of intense environmental activity. The combined effects of air and water tend to cause significant changes to the land materials at this interface. The top layer of exposed land is especially susceptible to physical and chemical weathering. Here, air laden with oxidant oxygen contacts rock, originally formed under reducing conditions, causing oxidation reactions to occur. Acid naturally present in rainwater as dissolved CO_2 or present as pollutant sulfuric, sulfurous, nitric, or hydrochloric acid, can dissolve portions of some kinds of rocks. Organisms such as lichens, which consist of fungi and algae growing symbiotically on rock surfaces, drawing carbon dioxide, oxygen, or nitrogen from air can grow on rock surfaces at the boundary of the atmosphere and geosphere, causing additional weathering to take place.

One of the most significant agents affecting exposed geospheric solids at the atmosphere/geosphere boundary is wind. Consisting of air moving largely in a horizontal fashion, wind both erodes solids and acts as an agent to deposit solids on geospheric surfaces. The influence of wind is especially pronounced in dry areas. A major factor in wind erosion is wind **abrasions** in which solid particles of sand and rock carried by wind tend to wear away exposed rock and soil. Loose, unconsolidated sand and soil may be removed in large volumes by wind, a process called **deflation**.

The potential for wind to move matter is illustrated by the formation of large deposits of **loess**, consisting of finely divided soil carried by wind. Loess particles are typically several tens of micrometers in size, small enough to be carried great distances by wind. Especially common are loess deposits that originated with matter composed of rock ground to a fine flour by ice age glaciers. This material was first deposited in river valleys by flood waters issuing from melting glaciers, then blown some distance from the rivers by strong winds after drying out.

One of the more common geospheric features created by wind is a **dune**, consisting of a mound of debris, usually sand, dropped when wind slows down. When a dune begins to form, it forms an obstruction that slows wind even more, so that more sediment is dropped. The result is that in the presence of sediment-laden wind, dunes several meters or more high may form rapidly. In forming a dune, heavier, coarser particles settle first so that the matter in dunes is sorted according to size, just like sediments deposited by flowing streams. In areas in which winds are prevalently from one direction, as is usually the case, dunes show a typical shape as illustrated in Figure 15.11. It is seen that the steeply sloping side, called the **slip face**, is downwind.

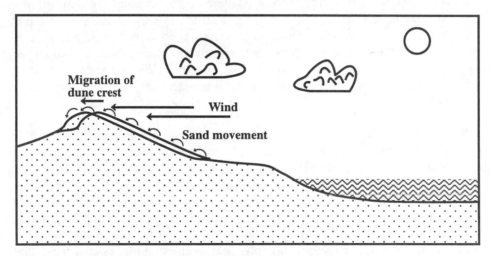

Figure 15.11. Shape and migration of a dune as determined by prevailing wind direction.

Some of the environmental effects of dunes result from their tendency to migrate with the prevailing winds. Migration occurs as matter is blown by the wind up the gently sloping face of the dune and falls down the slip face. Migrating sand dunes have inundated forest trees, and dust dunes in drought-stricken agricultural areas have filled road ditches, causing severely increased maintenance costs.

15.17. EFFECTS OF ICE

The power of ice to alter the geosphere is amply demonstrated by the remains of past glacial activity from the Ice Age. Those large areas of the earth's surface that were once covered with layers of glacial ice 1 or 2 kilometers in thickness show evidence of how the ice carved the surface, left massive piles of rock and gravel, and left rich deposits of fresh water. The enormous weight of glaciers on earth's surface compressed it, and in places it is still springing back 10,000 or so years after the glaciers retreated. Today the influence of ice on earth's surface is minimal, and there is substantial concern that melting of glaciers by greenhouse warming will raise sea levels so high that coastal areas will be inundated.

Glaciers form at sufficiently high latitudes and altitudes such that snow does not melt completely each summer. This occurs when snow becomes compacted over several to several thousand years such that the frozen water turns to crystals of true ice. Huge masses of ice with areas of several thousand square kilometers or more, and often around 1 kilometer thick, occur in polar regions and are called **continental glaciers**. Both Greenland and the Antarctic are covered by continental glaciers. **Alpine glaciers** occupy mountain valleys.

Glaciers on a slope flow as a consequence of their mass. This rate of flow is usually only a few meters per year, but may reach several kilometers per year. If a glacier flows into the sea, it may lose masses of ice as icebergs, a process called **calving**. Ice may also be lost by melting along the edges. The processes by which ice is lost is termed **ablation**.

Glacial ice affects the surface of the geosphere by both erosion and deposition. It is easy to imagine that a flowing mass of glacial ice is very efficient in scraping away the surface over which it flows, a process called **abrasion**. Adding to the erosive effect is the presence of rocks frozen into the glaciers which can act like tools to carve the surface of the underlying rock and soil. Whereas abrasion tends to wear rock surfaces away producing a fine rock powder, larger bits of rock can be dislodged from the surface over which the glacier flows and be carried along with the glacial ice.

When glacial ice melts, the rock that has been incorporated into it is left behind. This material is called **till**, or if it has been carried for some distance by water running off the melting glacier it is called **outwash**. Piles of rock left by melting glaciers produce unique structures called **moraines**.

Although the effects of glaciers described above are the most spectacular manifestations of the action of ice on the geosphere, at a much smaller level ice can have some very substantial effects. Freezing and expansion of water in pores and small crevices in rock is a major contributor to physical weathering processes. Freeze/thaw cycles are also very destructive to some kinds of structures, such as stone buildings.

15.18. EFFECTS OF HUMAN ACTIVITIES

Human activities have profound effects on the geosphere. Such effects may be obvious and direct, such as strip mining, or rearranging vast areas for construction projects, such as roads and dams. Or the effects may be indirect, such as pumping so much water from underground aquifers that the ground subsides, or abusing soil such that it no longer supports plant life well and erodes. As the source of minerals and other resources used by humans, the geosphere is dug up, tunnelled, stripped bare, rearranged, and subjected to many other kinds of indignities. The land is often severely disturbed, air can be polluted with dust particles during mining, and water may be polluted. Many of these effects, such as soil erosion caused by human activities, are addressed elsewhere in this book.

Extraction of Geospheric Resources: Surface Mining

Many human effects on the geosphere are associated with the extraction of resources from earth's crust. This is done in a number of ways, the most damaging of which can be surface mining. Surface mining is employed in the United States to extract virtually all of the rock and gravel that is mined, well over half of the coal, and numerous other resources. Properly done, with appropriate restoration practices, surface mining does minimal damage and may even be used to improve surface quality, such as by the construction of surface reservoirs where rock or gravel have been extracted. In earlier times before strict reclamation laws were in effect, surface mining, particularly of coal, left large areas of land scarred, devoid of vegetation, and subject to erosion.

Several approaches are employed in surface mining. Sand and gravel located under water are extracted by **dredging** with draglines or chain buckets attached to large conveyers. In most cases resources are covered with an **overburden** of earthen material that does not contain any of the resource that is being sought. This material must be removed as **spoil**. **Open-pit mining** is, as the name implies, a procedure in which gravel, building stone, iron ore, and other materials are simply dug from a big hole in the ground. Some of these pits, such as several from which copper ore has been taken in the U. S., are truly enormous in size.

The most well known (sometimes infamous) method of surface mining is **strip mining**, in which strips of overburden are removed by draglines and other heavy earth-moving equipment to expose seams of coal, phosphate rock, or other materials. Heavy equipment is used to remove a strip of overburden, and the exposed mineral resource is removed and hauled away. Overburden from a parallel strip is then removed and placed over the previously mined strip, and the procedure is repeated numerous times. Older practices left the replaced overburden as relatively steep erosion-prone banks. On highly sloping terrain, overburden is removed on progressively higher terraces and placed on the terrace immediately below.

Environmental Effects of Mining and Mineral Extraction

Some of the environmental effects of surface mining have been mentioned above. Although surface mining is most often considered for its environmental effects, subsurface mining may also have a number of effects, some of which are not immediately apparent and may be delayed for decades. Underground mines have a tendency to collapse leading to severe subsidence. Mining disturbs groundwater aquifers. Water seeping through mines and mine tailings may become polluted. One of the more common and damaging effects of mining on water occurs when pyrite, FeS_2, commonly associated with coal, is exposed to air and becomes oxidized to sulfuric acid by bacterial action to produce acid mine water (see Section 6.14). Some of the more damaging environmental effects of mining are the result of the processing of mined materials. Usually, ore is only part, often a small part, of the material that must be excavated. Various **beneficiation** processes are employed to separate the useful fraction of ore, leaving a residue of **tailings**. A number of adverse effects can result from environmental exposure of tailings. For example, residues left from the beneficiation of coal are often enriched in pyrite, FeS_2, which is oxidized microbiologically and chemically to produce damaging acidic drainage (acid mine water). Uranium ore tailings unwisely used as fill material have contaminated buildings with radioactive radon gas.

15.19. AIR POLLUTION AND THE GEOSPHERE

The geosphere can be a significant source of air pollutants. Of these geospheric sources, volcanic activity is one of the most common. Volcanic eruptions, fumaroles, hot springs, and geysers can emit toxic and acidic gases, including carbon monoxide, hydrogen chloride, and hydrogen sulfide. Greenhouse gases that tend to increase global climatic warming — carbon dioxide and methane — can come from volcanic sources. Massive volcanic eruptions may inject huge amounts of particulate matter into the atmosphere. The incredibly enormous 1883 eruption of the East Indies volcano Krakatoa blew about 2.5 cubic kilometers of solid matter into the atmosphere, some of which penetrated well into the stratosphere. This material stayed aloft long enough to circle the earth several times, causing red sunsets and a measurable lowering of temperature worldwide.

The 1982 eruption of the southern Mexico volcano El Chicón showed the importance of the type of particulate matter in determining effects on climate. The matter given off by this eruption was unusually rich in sulfur, so that an aerosol of sulfuric acid formed and persisted in the atmosphere for about three years, during which time the mean global temperature was lowered by several tenths of a degree due to the presence of atmospheric sulfuric acid. By way of contrast, the eruption of Mt. St. Helens in Washington State in the U.S. two years earlier had little perceptible effect on climate, although the amount of material blasted into the atmosphere was about the same as that from El Chicón. The material from the Mt. St. Helens eruption had comparatively little sulfur in it, so the climatic effects were minimal.

Thermal smelting processes used to convert metal fractions in ore to usable forms have caused a number of severe air pollution problems that have affected the geosphere. Many metals are present in ores as sulfides, and smelting can release large quantities of sulfur dioxide, as well as particles that contain heavy metals such as arsenic, cadmium, or lead. The resulting acid and heavy metal pollution of surrounding land can cause severe damage to vegetation so that devastating erosion occurs. One such area is around a large nickel smelter in Sudbury, Ontario, Canada, where a large area of land has become denuded of vegetation. Similar dead zones have been produced by copper smelters in Tennessee and in eastern Europe, including the former Soviet Union.

Soil and its cultivation produces significant quantities of atmospheric emissions. Waterlogged soil, particularly that cultivated for rice, generates significant quantities of methane, a greenhouse gas. The microbial reduction of nitrate in soil releases nitrous oxide, N_2O, to the atmosphere. However, soil and rock can also remove atmospheric pollutants. It is believed that microorganisms in soil account for the loss from the atmosphere of some carbon monoxide, which some fungi and bacteria can metabolize. Carbonate rocks, such as calcium carbonate, $CaCO_3$, can neutralize acid from atmospheric sulfuric acid and acid gases.

As discussed in Section 9.6, masses of atmospheric air can become trapped and stagnant under conditions of a temperature inversion in which the vertical circulation of air is limited by the presence of a relatively warm layer of air overlaying a colder layer at ground level. The effects of inversions can be aggravated by topographical conditions that tend to limit circulation of air. Figure 15.12 shows such a condition in which surrounding mountain ridges limit horizontal air movement. Air pollutants may be forced up a mountain ridge from a polluted area to significantly higher altitudes than they would otherwise reach. Because of this "chimney effect," air pollutants may reach mountain pine forests that are particularly susceptible to damage from air pollutants such as ozone formed along with photochemical smog.

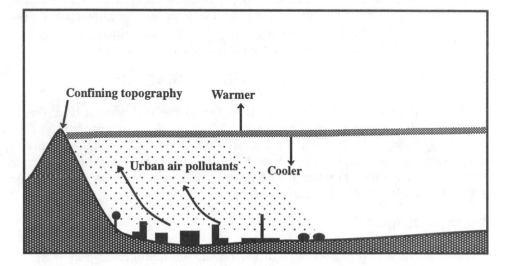

Figure 15.12. Topographical features, such as confining mountain ridges, may work with temperature inversions to increase the effects of air pollution.

15.20. WATER POLLUTION AND THE GEOSPHERE

Water pollution is addressed in detail elsewhere in this book. Much water pollution arises from interactions of groundwater and surface water with the geosphere. These aspects are addressed briefly here.

The relationship between water and the geosphere is twofold. The geosphere may be severely damaged by water pollution. This occurs, for example, when water pollutants produce contaminated sediments, such as those contaminated by heavy metals or PCBs. In some cases the geosphere serves as a source of water pollutants. Examples include acid produced by exposed metal sulfides in the geosphere or synthetic chemicals improperly discarded in landfills.

The sources of water pollution are divided into two main categories. The first of these consists of **point sources**, which enter the environment at a single, readily identified entry point. An example of a point source would be a sewage water outflow. Point sources tend to be those directly identified as to their origins from human activities. **Nonpoint sources** of pollution are those from broader areas. Such a source is water contaminated by fertilizer from fertilized agricultural land, or water contaminated with excess alkali leached from alkaline soils. Nonpoint sources are relatively harder to identify and monitor. Pollutants associated with the geosphere are usually nonpoint sources.

An especially common and damaging geospheric source of water pollutants consists of sediments carried by water from land into the bottoms of bodies of water. Most such sediments originate with agricultural land that has been disturbed such that soil particles are eroded from land into water. The most common manifestation of sedimentary material in water is opacity, which seriously detracts from the esthetics of the water. Sedimentary material deposited in reservoirs or canals can clog them and eventually make them unsuitable for water supply, flood control, navigation, and recreation. Suspended sediment in water used as a water supply can clog filters and add significantly to the cost of treating the water. Sedimentary material can devastate wildlife habitats by reducing food supplies and ruining nesting sites. Turbidity in water can severely curtail photosynthesis, thus reducing primary productivity necessary to sustain the food chains of aquatic ecosystems.

15.21. WASTE DISPOSAL AND THE GEOSPHERE

The geosphere receives many kinds and large amounts of wastes. Its ability to cope with such wastes with minimal damage is one of its most important characteristics and is dependent upon the kinds of wastes disposed on it. A variety of wastes, ranging from large quantities of relatively innocuous municipal refuse to much smaller quantities of potentially lethal radioactive wastes, are deposited on land or in landfills. These are addressed briefly in this section.

Municipal Refuse

The currently favored method for disposing of municipal solid wastes—household garbage—is in **sanitary landfills** (Figure 15.13) consisting of refuse piled on top of the ground or into a depression such as a valley, compacted, and covered at frequent intervals by soil. This approach permits frequent covering of the refuse so that loss of blowing trash, water contamination, and other undesirable effects are minimized. A completed landfill can be put to beneficial uses, such as a recreational area; because of settling, gas production, and other factors, landfill surfaces are generally not suitable for building construction. Modern sanitary landfills are much preferable to the open dump sites that were once the most common means of municipal refuse disposal.

Although municipal refuse is much less dangerous than hazardous chemical waste, it still poses some hazards. Despite prohibitions against the disposal of cleaners, solvents, lead storage batteries, and other potentially hazardous materials in landfills, materials that pose some environmental hazards do find their way into landfills and can contaminate their surroundings.

Landfills produce both gaseous and aqueous emissions. Biomass in landfills quickly depletes oxygen by aerobic biodegradation of microorganisms in the landfill,

$$\{CH_2O\}(biomass) + O_2 \rightarrow CO_2 + H_2O \qquad (15.21.1)$$

emitting carbon dioxide. Over a period of many decades the buried biodegradable materials undergo anaerobic biodegradation,

$$2\{CH_2O\} \rightarrow CO_2 + CH_4 \qquad (15.21.2)$$

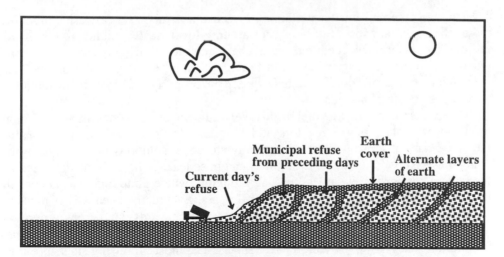

Figure 15.13. Structure of a sanitary landfill.

releasing methane as well as carbon dioxide. Although often impractical and too expensive, it is desirable to reclaim the methane as fuel, and some large sanitary landfills are major sources of methane. Released methane is a greenhouse gas and can pose significant explosion hazards to structures built on landfills. Although produced in much smaller quantities than methane, hydrogen sulfide, H_2S, is also generated by anaerobic biodegradation. This gas is toxic and has a bad odor. In a properly designed sanitary landfill, hydrogen sulfide releases are small and the gas tends to oxidize before it reaches the atmosphere in significant quantities.

Water infiltrating into sanitary landfills dissolves materials from the disposed refuse and runs off as **leachate**. Contaminated leachate is the single greatest potential pollution problem with refuse disposal sites, so it is important to minimize its production by designing landfills in a way that keeps water infiltration as low as possible. The anaerobic degradation of biomass produces organic acids that give the leachate a tendency to dissolve acid-soluble solutes, such as heavy metals. Leachate can infiltrate into groundwater posing severe contamination problems. This is minimized by siting sanitary landfills over formations of poorly permeable clay or depositing layers of clay in the landfill before refuse is put in it. In addition, impermeable synthetic polymer liners may be placed in the bottom of the landfill. In areas of substantial rainfall, infiltration into the landfill exceeds its capacity to hold water so that leachate flows out. In order to prevent water pollution downstream, this leachate should be controlled and treated.

Hazardous chemical wastes are disposed in so-called **secure landfills**, which are designed to prevent leakage and geospheric contamination of toxic chemicals disposed in them. Such a landfill is equipped with a variety of measures to prevent contamination of groundwater and the surrounding geosphere. The base of the landfill is made of compacted clay that is largely impermeable to leachate. An impermeable polymer liner is placed over the clay liner. The surface of the landfill is covered with material designed to reduce water infiltration, and the surface is designed with slopes that also minimize the amount of water running in. Elaborate drainage systems are installed to collect and treat leachate.

The most pressing matter pertaining to geospheric disposal of wastes involves radioactive wastes. Most of these wastes are **low-level** wastes, including discarded radioactive laboratory chemicals and pharmaceuticals, filters used in nuclear reactors, and ion-exchange resins used to remove small quantities of radionuclides from nuclear reactor cooler water. Disposed in properly designed landfills, such wastes pose minimal hazards.

Of greater concern are the **high-level** radioactive wastes, primarily fission products of nuclear power reactors and by-products of nuclear weapons manufacture. Many of these wastes are currently stored as solutions in tanks, many of which have outlived their useful lifetimes and pose leakage hazards, at sites such as the federal nuclear facility at Hanford, Washington, where plutonium was generated in large quantities during post-World War II years. Eventually, such wastes must be placed in the geosphere such that they will pose no hazards. Numerous proposals have been advanced for their disposal, including disposal in salt formations, subduction zones in the seafloor, and ice sheets. The most promising sites appear to be those in poorly permeable formations of igneous rock. Among these are basalts, which are strong, glassy igneous types of rock found in the Columbia River plateau. Granite and pyroclastic welded tuffs fused by past high temperature volcanic eruptions are also likely possibilities as sites for disposing of nuclear wastes and keeping them isolated for tens of thousands of years.

LITERATURE CITED

1. Montgomery, Carla W., Brian J. Skinner, and Stephen J. Porter, *Environmental Geology*, 5th ed., McGraw-Hill, Boston, 1999.

2. Rolls, David and Will J. Bland, *Weathering: An Introduction to the Basic Principles*, Edward Arnold Publishing Co., New York, 1998.

3. Condie, Kent C., *Plate Tectonics and Crustal Evolution*, 4th ed., Butterworth-Heinemann, Newton, MA, 1997.

4. Keller, Edward A., *Active Tectonics: Earthquakes, Uplift, and Landscape*, Prentice Hall, Upper Saddle River, NJ, 1996.

5. Llamas-Ruiz, Andres and Ali Garousi, *Volcanos and Earthquakes*, Sterling Publishing Co., New York, 1997.

6. Goodwin, Peter, *Landslides, Slumps, and Creep*, Franklin Watts Publishing Co., New York, 1997.

7. Plant, Jane A., and Robert Raiswell, "Principles of Environmental Geochemistry," Chapter 1 in *Applied Environmental Geochemistry*, Iain Thornton, Ed., Academic Press, New York, 1983, pp. 1-40.

8. Dudley, Walter and Min Lee, *Tsunami!*, 2nd ed., University of Hawaii Press, Honolulu, 1998.

9. Satake, Kenji, Fumihiko Imamura, and Fumihi Imamura, *Tsunamis: Their Generation, Dynamics, and Hazard*, Birkhauser, Basel, Switzerland, 1995.

10. Schneider, David, "The Rising Seas," *Scientific American*, March, 1997, pp. 112-117.

SUPPLEMENTARY REFERENCES

Bell, Frederick G.,*Environmental Geology: Principles and Practice*, Blackwell Science, Malden, MA, 1998.

Bennett, Matthew R. and Peter Doyle, *Environmental Geology: Geology and the Human Environment*, John Wiley & Sons, New York, 1997.

Berthelin, J., Ed., *Diversity of Environmental Biogeochemistry*, Elsevier Science Publishing, New York, 1991.

Brownlow, Arthur H., *Geochemistry*, 2nd ed., Prentice Hall, Inc., Upper Saddle River, NJ, 1996.

Coch, Nicholas K., *Geohazards: Natural and Human*, Prentice Hall, Upper Saddle River, NJ, 1995.

Colley, H., *Introduction to Environmental Geology* Stanley Thornes Publishing Ltd., Cheltenham, Glucestershire, U.K., 1999.

Craig, P. J., *The Natural Environment and The Biogeochemical Cycles*, Springer-Verlag, Inc., New York, 1980.

Faure, Gunter, *Principles and Applications of Geochemistry: A Comprehensive Textbook for Geology Students*, Prentice Hall, Upper Saddle River, NJ, 1998.

Foley, Duncan, Garry D. McKenzie, Russell O. Utgard, *Investigations in Environmental Geology*, Prentice Hall, Upper Saddle River, NJ, 1999.

Keller, Edward A., *Environmental Geology*, 7th ed., Prentice Hall, Upper Saddle River, NJ, 1996.

Langmuir, Donald, *Aqueous Environmental Geochemistry*, Prentice-Hall, Inc., Upper Saddle River, N.J., 1997.

Lundgren, Lawrence W., *Environmental Geology,* 2nd ed., Prentice Hall, Upper Saddle River, N.J., 1999.

Marshall, Clare P. and Rhodes Whitmore Fairbridge, Eds., *Encyclopedia of Geochemistry (Encyclopedia of Earth Sciences)*, Kluwer Academic Publishing Co., Hingham, MA, 1998.

Merritts, Dorothy J., Andrew De Wet, and Kirsten Menking, *Environmental Geology: An Earth System Science Approach*, W. H. Freeman & Co., New York, 1998.

Montgomery, Carla W., Brian J. Skinner, and Stephen J. Porter, *Environmental Geology*, 5th ed, McGraw-Hill, Boston, MA 1999.

Murck, Barbara W., Brian Skinner, and Stephen Porter, *Environmental Geology*, John Wiley & Sons, New York, 1995.

Ottonello. Giulio, *Principles of Geochemistry*, Columbia University Press, New York, 1997.

Pipkin, Bernard W. and D. D. Trent, *Geology and the Environment* West/Wadsworth, Belmont, CA, 1997.

Soliman, Mostafa M., Philip E. Lamoreaux, and Bashir A. Memon, Eds., *Environmental Hydrogeology*, CRC Press/Lewis Publishers, Boca Raton, FL, 1997.

Tobin, Graham A. and Burrell E. Montz, *Natural Hazards: Explanation and Integration*, Guilford Publications, New York, 1997.

QUESTIONS AND PROBLEMS

1. Of the following, the one that is **not** a manifestation of desertification is (a) declining groundwater tables, (b) salinization of topsoil and water, (c) production of deposits of MnO_2 and $Fe_2O_3 \cdot H_2O$ from anaerobic processes, (d) reduction of surface waters, (e) unnaturally high soil erosion.

2. Give an example of how each of the following chemical or biochemical phenomena in soils operates to reduce the harmful nature of pollutants: (a) Oxidation-reduction processes, (b) hydrolysis, (c) acid-base reactions, (d) precipitation, (e) sorption, (f) biochemical degradation.

3. Why do silicates and oxides predominate among earth's minerals?

4. Give the common characteristic of the minerals with the following formulas: $NaCl$, Na_2SO_4, $CaSO_4 \cdot 2H_2O$, $MgCl_2 \cdot 6H_2O$, $MgSO_4 \cdot 7H_2O$, $KMgClSO_4 \cdot ^{11}/_4 H_2O$, $K_2MgCa_2(SO_4)_4 \cdot 2H_2O$.

5. Explain how the following are related: weathering, igneous rock, sedimentary rock, soil.

6. Match the following:

   1. Metamorphic rock
   2. Chemical sedimentary rocks
   3. Detrital rock
   4. Organic sedimentary rocks

   (a) Produced by the precipitation or coagulation of dissolved or colloidal weathering products
   (b) Contain residues of plant and animal remains
   (c) Formed from action of heat and pressure on sedimentary rock
   (d) Formed from solid particles eroded from igneous rocks as a consequence of weathering

7. Where does most flowing water that contains dissolved load originate? Why does it tend to come from this source?

8. What role might be played by water pollutants in the production of dissolved load and in the precipitation of secondary minerals from it?

9. As defined in this chapter, are the ions involved in ion replacement the same as exchangeable cations? If not, why not?

10. Speculate regarding how water present in poorly consolidated soil might add to the harm caused by earthquakes.

11. In what sense may volcanoes contribute to air pollution? What possible effects may this have on climate?

12. Explain how excessive pumping of groundwater might adversely affect streams, particularly in regard to the flow of small streams.

13. Which three elements are most likely to undergo oxidation as part of chemical weathering process? Give example reactions of each.

14. Match the following:

1. Groundwater	(a) Water from precipitation in the form of rain or snow
2. Vadose water	(b) Water present in a zone of saturation
3. Meteoric water	(c) Water held in the unsaturated zone or zone of aeration
4. Water in capillary fringe	(d) Water drawn somewhat above the water table by surface tension

16 SOIL ENVIRONMENTAL CHEMISTRY

16.1. SOIL AND AGRICULTURE

Soil and agricultural practices are strongly tied with the environment. Some of these considerations are addressed later in this chapter along with a discussion of soil erosion and conservation. Cultivation of land and agricultural practices can influence both the atmosphere and the hydrosphere. Although this chapter deals primarily with soil, the topic of agriculture in general is introduced for perspective.

Agriculture

Agriculture, the production of food by growing crops and livestock, provides for the most basic of human needs. No other industry impacts as much as agriculture does on the environment. Agriculture is absolutely essential to the maintenance of the huge human populations now on earth. The displacement of native plants, destruction of wildlife habitat, erosion, pesticide pollution, and other environmental aspects of agriculture have enormous potential for environmental damage. Survival of humankind on earth demands that agricultural practice become as environmentally friendly as possible. On the other hand, growth of domestic crops removes (at least temporarily) greenhouse gas carbon dioxide from the atmosphere and provides potential sources of renewable resources of energy and fiber that can substitute for petroleum-derived fuels and materials.

Agriculture can be divided into the two main categories of **crop farming**, in which plant photosynthesis is used to produce grain, fruit, and fiber, and **livestock farming**, in which domesticated animals are grown for meat, milk, and other animal products. The major divisions of crop farming include production of cereals, such as wheat, corn or rice; animal fodder, such as hay; fruit; vegetables; and specialty crops, such as sugarcane, pineapple, sugar beets, tea, coffee, tobacco, cotton, and cacao. Livestock farming involves the raising of cattle, sheep, goats, swine, asses, mules, camels, buffalo, and various kinds of poultry. In addition to meat, livestock produce dairy products, eggs, wool, and hides. Freshwater fish and even crayfish are raised on "fish farms." Beekeeping provides honey.

Agriculture is based on domestic plants engineered by early farmers from their wild plant ancestors. Without perhaps much of an awareness of what they were doing, early farmers selected plants with desired characteristics for the production of food. This selection of plants for domestic use brought about a very rapid evolutionary change, so profound that the products often barely resemble their wild ancestors. Plant breeding based on scientific principles of heredity is a very recent development dating from early in the present century. One of the major objectives of plant breeding has been to increase yield. An example of success in this area is the selection of dwarf varieties of rice, which yield much better and mature faster than the varieties that they replaced. Such rice was largely responsible for the "green revolution" dating from about the 1950s. Yields of crops can also be increased by selecting for resistance to insects, drought, and cold. In some cases the goal is to increase nutritional value, such as in the development of corn high in lysine, an amino acid essential for human nutrition, such that corn becomes a more complete food.

The development of hybrids has vastly increased yields and other desired characteristics of a number of important crops. Basically, **hybrids** are the offspring of crosses between two different **true-breeding** strains. Often quite different from either parent strain, hybrids tend to exhibit "hybrid vigor" and to have significantly higher yields. The most success with hybrid crops has been obtained with corn (maize). Corn is one of the easiest plants to hybridize because of the physical separation of the male flowers, which grow as tassels on top of the corn plant, from female flowers, which are attached to incipient ears on the side of the plant. Despite past successes by more conventional means and some early disappointments with "genetic engineering," application of recombinant DNA technology (Section 16.10) will probably eventually overshadow all the advances ever made in plant breeding.

In addition to plant strains and varieties, numerous other factors are involved in crop production. Weather is an obvious factor, and shortages of water, chronic in many areas of the world, are mitigated by irrigation. Here, automated techniques and computer control are beginning to play an important, more environmentally friendly role by minimizing the quantities of water required. The application of chemical fertilizer has vastly increased crop yields. The judicious application of pesticides, especially herbicides, but including insecticides and fungicides as well, has increased crop yields and reduced losses greatly. Use of herbicides has had an environmental benefit in reducing the degree of mechanical cultivation of soil required. Indeed, "no-till" and "low-till" agriculture are now widely practiced on a large scale.

The crops that provide for most of human caloric food intake, as well as much food for animals, are **cereals**, which are harvested for their starch-rich seeds. In addition to corn, mentioned above, wheat used for making bread and related foods, and rice consumed directly, other major cereal crops include barley, oats, rye, sorghum, and millet. As applied to agriculture and food, **vegetables** are plants or their products that can be eaten directly by humans. A large variety of different parts of plants are consumed as vegetables. These include leaves (lettuce), stems (asparagus), roots (carrots), tubers (potato), bulb (onion), immature flower (broccoli), immature fruit (cucumber), mature fruit (tomato), and seeds (pea). Fruits, which are bodies of plant tissue containing the seed, may be viewed as a subclassification of vegetables. Common fruits include apple, peach, apricot, citrus (orange, lemon, lime, grapefruit), banana, cherry, and various kinds of berries.

The rearing of domestic animals may have significant environmental effects. The Netherlands' pork industry has been so productive that hog manure and its by-products have caused serious problems. Overflow from waste lagoons associated with hog production caused very damaging water pollution in Eastern North Carolina following the record rainfall from Hurricane Floyd in September 1999. Goats and sheep have destroyed pastureland in the Near East, Northern Africa, Portugal, and Spain. Of particular concern are the environmental effects of raising cattle. Significant amounts of forest land have been converted to marginal pasture land to raise beef. Production of one pound of beef requires about four times as much water and four times as much feed as does production of one pound of chicken. An interesting aspect of the problem is emission of greenhouse-gas methane by anaerobic bacteria in the digestive systems of cattle and other ruminant animals; cattle rank right behind wetlands and rice paddies as producers of atmospheric methane. However, because of the action of specialized bacteria in their stomachs, cattle and other ruminant animals are capable of converting otherwise unusable cellulose to food.

Pesticides and Agriculture

Pesticides, particularly insecticides and herbicides, are an integral part of modern agricultural production. In the United States, agricultural pesticides are regulated under the Federal Insecticide, Fungicide, and Rodenticide (FIFRA) act, first passed in 1947, revised in a major way in 1972, and subjected to several amendments since then.[1] Starting in 1989 under FIFRA, pre-1972 pesticides were required to undergo reregistration. Since this process began, manufacturers have withdrawn from the market several thousand products because of the expense of the safety review process. The problem has been especially severe for **minor-use pesticides** for which the market is not very large. In contrast to pesticides used on approximately 220 million acres of major crops in the U.S.—corn, soybeans, wheat, and cotton—minor-use pesticides are applied on only about 8 million acres of orchards, trees, ornamental plants, turf grass, fruits, nuts, and vegetables. Despite their limited use, about 40% of the monetary value of agricultural pesticides resides with minor-use pesticides. Complicating the reregistration process for many years was the well-intentioned but unrealistic 1958 Delaney Amendment to the Food, Drug, and Cosmetic Act which prohibited in food any chemical that had been shown to cause cancer in animals and humans. Many chemicals that are almost certainly safe as trace-level contaminants in food will, indeed, cause cancer when fed in massive doses to experimental animals. This effect has been attributed in part to the fact that huge doses of some chemicals destroy tissue, which the organism attempts to replace by growing new cells. Cells that are reproducing rapidly in an attempt to make up for tissue loss are more likely to become cancerous.

An interesting development regarding the use of herbicides in the late 1990s was the production of transgenic crops resistant to specific herbicides. The Monsanto company pioneered this approach with the development of "Roundup Ready" crops that resist the herbicidal effects of Monsanto's flagship Roundup® herbicide, a glyphosate compound with an annual market exceeding $2 billion. The seedlings of crops resistant to the herbicide are not harmed by exposure to it, whereas competing weeds are killed. In an interesting marketing strategy, Monsanto has licensed manufacturing and distribution rights for Roundup to other companies for application onto "Roundup Ready" crops sold by Monsanto.[2]

Soil

Soil, consisting of a finely divided layer of weathered minerals and organic matter upon which plants grow, is the most fundamental requirement for agriculture. To humans and most terrestrial organsms, soil is the most important part of the geosphere. Though only a tissue-thin layer compared to the earth's total diameter, soil is the medium that produces most of the food required by most living things. Good soil—and a climate conducive to its productivity—is the most valuable asset a nation can have.

In addition to being the site of most food production, soil is the receptor of large quantities of pollutants, such as particulate matter from power plant smokestacks. Fertilizers, pesticides, and some other materials applied to soil often contribute to water and air pollution. Therefore, soil is a key component of environmental chemical cycles.

Soils are formed by the weathering of parent rocks as the result of interactive geological, hydrological, and biological processes (see Chapter 15). Soils are porous and are vertically stratified into horizons as the result of downward-percolating water and biological processes, including the production and decay of biomass. Soils are open systems that undergo continual exchange of matter and energy with the atmosphere, hydrosphere, and biosphere.

16.2. NATURE AND COMPOSITION OF SOIL

Soil is a variable mixture of minerals, organic matter, and water capable of supporting plant life on the earth's surface.[3] It is the final product of the weathering action of physical, chemical, and biological processes on rocks, which largely produces clay minerals. The organic portion of soil consists of plant biomass in various stages of decay. High populations of bacteria, fungi, and animals such as earthworms may be found in soil. Soil contains air spaces and generally has a loose texture (Figure 16.1).

The solid fraction of typical productive soil is approximately 5% organic matter and 95% inorganic matter. Some soils, such as peat soils, may contain as much as 95% organic material. Other soils contain as little as 1% organic matter.

Typical soils exhibit distinctive layers with increasing depth (Figure 16.2). These layers are called **horizons**. Horizons form as the result of complex interactions among processes that occur during weathering. Rainwater percolating through soil carries dissolved and colloidal solids to lower horizons where they are deposited. Biological processes, such as bacterial decay of residual plant biomass, produces slightly acidic CO_2, organic acids, and complexing compounds that are carried by rainwater to lower horizons where they interact with clays and other minerals, altering the properties of the minerals. The top layer of soil, typically several inches in thickness, is known as the A horizon, or **topsoil**. This is the layer of maximum biological activity in the soil and contains most of the soil organic matter. Metal ions and clay particles in the A horizon are subject to considerable leaching. The next layer is the B horizon, or **subsoil**. It receives material such as organic matter, salts, and clay particles leached from the topsoil. The C horizon is composed of weathered parent rocks from which the soil originated.

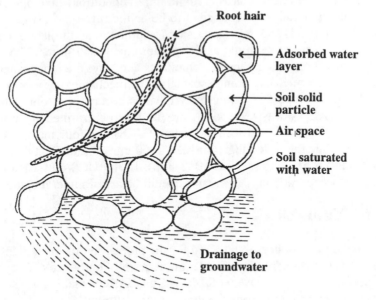

Figure 16.1. Fine structure of soil, showing solid, water, and air phases.

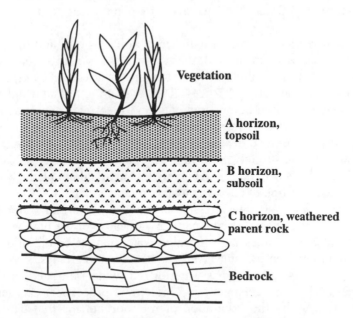

Figure 16.2. Soil profile showing soil horizons.

Soils exhibit a large variety of characteristics that are used for their classification for various purposes, including crop production, road construction, and waste disposal. Soil profiles are discussed above. The parent rocks from which soils are formed obviously play a strong role in determining the composition of soils. Other soil characteristics include strength, workability, soil particle size, permeability, and degree of maturity. One of the more important classes of productive soils is the **podzol** type of soil formed under relatively high rainfall conditions in temperate zones of the world. These generally rich soils tend to be acidic (pH 3.5–4.5) such that alkali and alkaline earth metals and, to a lesser extent aluminum and iron, are leached from their A horizons, leaving kaolinite as the predominant clay mineral. At somewhat higher pH in the B horizons, hydrated iron oxides and clays are redeposited.

From the engineering standpoint, especially, the mechanical properties of soil are emphasized. These properties, which may have important environmental implications in areas such as waste disposal, are largely determined by particle size. According to the United Classification System (UCS), the four major categories of soil particle sizes are the following: Gravels (2–60 mm) > sands (0.06–2 mm) > silts (0.06–0.006 mm) > clays (less than 0.002 mm). In the UCS classification scheme clays represent a size fraction rather than a specific class of mineral matter.

Water and Air in Soil

Large quantities of water are required for the production of most plant materials. For example, several hundred kg of water are required to produce one kg of dry hay. Water is part of the three-phase, solid-liquid-gas system making up soil. It is the basic transport medium for carrying essential plant nutrients from solid soil particles into plant roots and to the farthest reaches of the plant's leaf structure (Figure 16.3). The water enters the atmosphere from the plant's leaves, a process called **transpiration**.

Normally, because of the small size of soil particles and the presence of small capillaries and pores in the soil, the water phase is not totally independent of soil solid matter. The availability of water to plants is governed by gradients arising from capillary and gravitational forces. The availability of nutrient solutes in water depends upon concentration gradients and electrical potential gradients. Water present in larger spaces in soil is relatively more available to plants and readily drains away. Water held in smaller pores or between the unit layers of clay particles is held much more strongly. Soils high in organic matter may hold appreciably more water than other soils, but it is relatively less available to plants because of physical and chemical sorption of the water by the organic matter.

There is a very strong interaction between clays and water in soil. Water is absorbed on the surfaces of clay particles. Because of the high surface/volume ratio of colloidal clay particles, a great deal of water may be bound in this manner. Water is also held between the unit layers of the expanding clays, such as the montmorillonite clays. As soil becomes waterlogged (water-saturated) it undergoes drastic changes in physical, chemical, and biological properties. Oxygen in such soil is rapidly used up by the respiration of microorganisms that degrade soil organic matter. In such soils, the bonds holding soil colloidal particles together are broken, which causes disruption of soil structure. Thus, the excess water in such soils is detrimental to plant growth, and the soil does not contain the air required by most plant roots. Most useful crops, with the notable exception of rice, cannot grow on waterlogged soils.

One of the most marked chemical effects of waterlogging is a reduction of pE by the action of organic reducing agents acting through bacterial catalysts. Thus, the redox condition of the soil becomes much more reducing, and the soil pE may drop

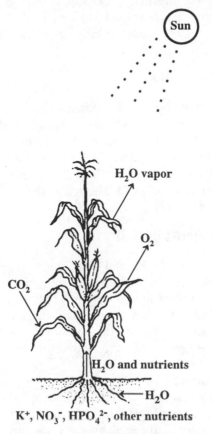

Figure 16.3. Plants transport water from the soil to the atmosphere by transpiration. Nutrients are also carried from the soil to the plant extremities by this process. Plants remove CO_2 from the atmosphere and add O_2 by photosynthesis. The reverse occurs during plant respiration.

from that of water in equilibrium with air (+13.6 at pH 7) to 1 or less. One of the more significant results of this change is the mobilization of iron and manganese as soluble iron(II) and manganese(II) through reduction of their insoluble higher oxides:

$$MnO_2 + 4H^+ + 2e^- \rightarrow Mn^{2+} + 2H_2O \qquad (16.2.1)$$

$$Fe_2O_3 + 6H^+ + 2e^- \rightarrow 2Fe^{2+} + 3H_2O \qquad (16.2.2)$$

Although soluble manganese generally is found in soil as Mn^{2+} ion, soluble iron(II) frequently occurs as negatively charged iron-organic chelates. Strong chelation of iron(II) by soil fulvic acids (Chapter 3) apparently enables reduction of iron(III) oxides at more positive pE values than would otherwise be possible. This causes an upward shift in the Fe(II)-Fe(OH)$_3$ boundary shown in Figure 4.4.

Some soluble metal ions such as Fe^{2+} and Mn^{2+} are toxic to plants at high levels. Their oxidation to insoluble oxides may cause formation of deposits of Fe_2O_3 and MnO_2, which clog tile drains in fields.

Roughly 35% of the volume of typical soil is composed of air-filled pores. Whereas the normal dry atmosphere at sea level contains 21% O_2 and 0.03% CO_2 by volume, these percentages may be quite different in soil air because of the decay of organic matter:

$$\{CH_2O\} + O_2 \rightarrow CO_2 + H_2O \qquad (16.2.3)$$

This process consumes oxygen and produces CO_2. As a result, the oxygen content of air in soil may be as low as 15%, and the carbon dioxide content may be several percent. Thus, the decay of organic matter in soil increases the equilibrium level of dissolved CO_2 in groundwater. This lowers the pH and contributes to weathering of carbonate minerals, particularly calcium carbonate (see Reaction 3.7.6). As discussed in Section 16.3, CO_2 also shifts the equilibrium of the process by which roots absorb metal ions from soil.

The Inorganic Components of Soil

The weathering of parent rocks and minerals to form the inorganic soil components results ultimately in the formation of inorganic colloids. These colloids are repositories of water and plant nutrients, which may be made available to plants as needed. Inorganic soil colloids often absorb toxic substances in soil, thus playing a role in detoxification of substances that otherwise would harm plants. The abundance and nature of inorganic colloidal material in soil are obviously important factors in determining soil productivity.

The uptake of plant nutrients by roots often involves complex interactions with the water and inorganic phases. For example, a nutrient held by inorganic colloidal material has to traverse the mineral/water interface and then the water/root interface. This process is often strongly influenced by the ionic structure of soil inorganic matter.

As noted in Section 15.2, the most common elements in the earth's crust are oxygen, silicon, aluminum, iron, calcium, sodium, potassium, and magnesium. Therefore, minerals composed of these elements — particularly silicon and oxygen — constitute most of the mineral fraction of the soil. Common soil mineral constituents are finely divided quartz (SiO_2), orthoclase ($KAlSi_3O_8$), albite ($NaAlSi_3O_8$), epidote ($4CaO\cdot3(AlFe)_2O_3\cdot6SiO_2\cdot H_2O$), geothite ($FeO(OH)$), magnetite ($Fe_3O_4$), calcium and magnesium carbonates ($CaCO_3$, $CaCO_3\cdot MgCO_3$), and oxides of manganese and titanium.

Organic Matter in Soil

Though typically comprising less than 5% of a productive soil, organic matter largely determines soil productivity. It serves as a source of food for microorganisms, undergoes chemical reactions such as ion exchange, and influences the physical properties of soil. Some organic compounds even contribute to the weathering of mineral matter, the process by which soil is formed. For example, $C_2O_4^{2-}$, oxalate ion, produced as a soil fungi metabolite, occurs in soil as the calcium salts whewellite and weddelite. Oxalate in soil water dissolves minerals, thus speeding the weathering process and increasing the availability of nutrient ion species. This weathering process involves oxalate complexation of iron or aluminum in minerals, represented by the reaction

$$3H^+ + M(OH)_3(s) + 2CaC_2O_4(s) \rightarrow M(C_2O_4)_2^-(aq) +$$
$$2Ca^{2+}(aq) + 3H_2O \quad (16.2.4)$$

in which M is Al or Fe. Some soil fungi produce citric acid and other chelating organic acids which react with silicate minerals and release potassium and other nutrient metal ions held by these minerals.

The strong chelating agent 2-ketogluconic acid is produced by some soil bacteria. By solubilizing metal ions, it may contribute to the weathering of minerals. It may also be involved in the release of phosphate from insoluble phosphate compounds.

Biologically active components of the organic soil fraction include polysaccharides, amino sugars, nucleotides, and organic sulfur and phosphorus compounds. Humus, a water-insoluble material that biodegrades very slowly, makes up the bulk of soil organic matter. The organic compounds in soil are summarized in Table 16.1.

Table 16.1. Major Classes of Organic Compounds in Soil

Compound type	Composition	Significance
Humus	Degradation-resistant residue from plant decay, largely C, H, and O	Most abundant organic component, improves soil physical properties, exchanges nutrients, reservoir of fixed N
Fats, resins, and waxes	Lipids extractable by organic solvents	Generally, only several percent of soil organic matter, may adversely affect soil physical properties by repelling water, perhaps phytotoxic
Saccharides	Cellulose, starches, hemicellulose, gums	Major food source for soil microorganisms, help stabilize soil aggregates
N-containing organics	Nitrogen bound to humus, amino acids, amino sugars, other compounds	Provide nitrogen for soil fertility
Phosphorus compounds	Phosphate esters, inositol phosphates (phytic acid), phospholipids	Sources of plant phosphate

The accumulation of organic matter in soil is strongly influenced by temperature and by the availability of oxygen. Since the rate of biodegradation decreases with decreasing temperature, organic matter does not degrade rapidly in colder climates and tends to build up in soil. In water and in waterlogged soils, decaying vegetation does not have easy access to oxygen, and organic matter accumulates. The organic content may reach 90% in areas where plants grow and decay in soil saturated with water.

The presence of naturally occurring polynuclear aromatic (PAH) compounds is an interesting feature of soil organic matter. These compounds, some of which are carcinogenic, are discussed as air pollutants in Sections 10.8 and 12.4. PAH compounds found in soil include fluoranthene, pyrene, and chrysene. PAH compounds in soil result in part from combustion from both natural sources (grass fires) or pollutant sources. Terpenes also occur in soil organic matter. Extraction of soil with ether and alcohol yields the pigments β-carotene, chlorophyll, and xanthophyll.

Soil Humus

Of the organic components listed in Table 16.1, **soil humus** is by far the most significant.[4] Humus, composed of a base-soluble fraction called humic and fulvic acids (described in Section 3.17), and an insoluble fraction called humin, is the residue left when bacteria and fungi biodegrade plant material. The bulk of plant biomass consists of relatively degradable cellulose and degradation-resistant lignin, which is a polymeric substance with a higher carbon content than cellulose. Among lignin's prominent chemical components are aromatic rings connected by alkyl chains, methoxyl groups, and hydroxyl groups. These structural artifacts occur in soil humus and give it many of its characteristic properties.

The process by which humus is formed is called **humification**. Soil humus is similar to its lignin precursors, but has more carboxylic acid groups. Part of each molecule of humic substance is nonpolar and hydrophobic, and part is polar and hydrophilic. Such molecules are called **amphiphiles**, and they form micelles (see Section 5.4 and Figure 5.4) in which the nonpolar parts compose the inside of small colloidal particles and the polar functional groups are on the outside. Amphiphilic humic substances probably also form bilayer surface coatings on mineral grains in soil.

An increase in nitrogen/carbon ratio is a significant feature of the transformation of plant biomass to humus through the humification process. This ratio starts at approximately 1/100 in fresh plant biomass. During humification, microorganisms convert organic carbon to CO_2 to obtain energy. Simultaneously, the bacterial action incorporates bound nitrogen with the compounds produced by the decay processes. The result is a nitrogen/carbon ratio of about 1/10 upon completion of humification. As a general rule, therefore, humus is relatively rich in organically bound nitrogen.

Humic substances influence soil properties to a degree out of proportion to their small percentage in soil. They strongly bind metals, and serve to hold micronutrient metal ions in soil. Because of their acid-base character, humic substances serve as buffers in soil. The water-holding capacity of soil is significantly increased by humic substances. These materials also stabilize aggregates of soil particles, and increase the sorption of organic compounds by soil.

Humic materials in soil strongly sorb many solutes in soil water and have a particular affinity for heavy polyvalent cations. Soil humic substances may contain levels of uranium more than 10^4 times that of the water with which they are in equilibrium. Thus, water becomes depleted of its cations (or purified) in passing through humic-rich soils. Humic substances in soils also have a strong affinity for organic compounds with low water-solubility such as DDT or Atrazine, a herbicide widely used to kill weeds in corn fields.

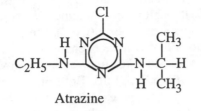

Atrazine

In some cases, there is a strong interaction between the organic and inorganic portions of soil. This is especially true of the strong complexes formed between clays and humic (fulvic) acid compounds. In many soils, 50-100% of soil carbon is complexed with clay. These complexes play a role in determining the physical properties of soil, soil fertility, and stabilization of soil organic matter. One of the mechanisms for the chemical binding between clay colloidal particles and humic organic particles is probably of the flocculation type (see Chapter 5) in which anionic organic molecules with carboxylic acid functional groups serve as bridges in combination with cations to bind clay colloidal particles together as a floc. Support is given to this hypothesis by the known ability of NH_4^+, Al^{3+}, Ca^{2+}, and Fe^{3+} cations to stimulate clay-organic complex formation. The synthesis, chemical reactions, and biodegradation of humic materials are affected by interaction with clays. The lower-molecular-weight fulvic acids may be bound to clay, occupying spaces in layers in the clay.

The Soil Solution

The **soil solution** is the aqueous portion of soil that contains dissolved matter from soil chemical and biochemical processes in soil and from exchange with the hydrosphere and biosphere.[5] This medium transports chemical species to and from soil particles and provides intimate contact between the solutes and the soil particles. In addition to providing water for plant growth, it is an essential pathway for the exchange of plant nutrients between roots and solid soil.

Obtaining a sample of soil solution is often very difficult because the most significant part of it is bound in capillaries and as surface films. The most straightforward means is collection of drainage water. Soil solution can be isolated from moist solid soil by displacement with a water-immiscible fluid, mechanical separation by centrifugation, or pressure or vacuum treatment.

Dissolved mineral matter in soil is largely present as ions. Prominent among the cations are H^+, Ca^{2+}, Mg^{2+}, K^+, Na^+, and usually very low levels of Fe^{2+}, Mn^{2+}, and Al^{3+}. The last three cations may be present in partially hydrolized form, such as $FeOH^+$, or complexed by organic humic substance ligands. Anions that may be present are HCO_3^-, CO_3^{2-}, HSO_4^-, SO_4^{2-}, Cl^-, and F^-. In addition to being bound to H^+ in species such as bicarbonate, anions may be complexed with metal ions, such as in AlF^{2+}. Multivalent cations and anions form ion pairs with each other in soil solutions. Examples of these are $CaSO_4$ and $FeSO_4$.

16.3. ACID-BASE AND ION EXCHANGE REACTIONS IN SOILS

One of the more important chemical functions of soils is the exchange of cations. As discussed in Chapter 5, the ability of a sediment or soil to exchange cations is expressed as the cation-exchange capacity (CEC), the number of milliequivalents (meq) of monovalent cations that can be exchanged per 100 g of soil (on a dry-weight basis). The CEC should be looked upon as a conditional constant since it may vary with soil conditions such as pE and pH. Both the mineral and organic portions of soils exchange cations. Clay minerals exchange cations because of the presence of negatively charged sites on the mineral, resulting from the substitution of an atom of lower oxidation number for one of higher number, for example, magnesium for aluminum. Organic materials exchange cations because of the presence of the carboxylate group and other basic functional groups. Humus typically has a very high cation-exchange capacity. The cation-exchange capacity of peat may range from 300-400 meq/100 g. Values of cation-exchange capacity for soils with more typical levels of organic matter are around 10-30 meq/100 g.

Cation exchange in soil is the mechanism by which potassium, calcium, magnesium, and essential trace-level metals are made available to plants. When nutrient metal ions are taken up by plant roots, hydrogen ion is exchanged for the metal ions. This process, plus the leaching of calcium, magnesium, and other metal ions from the soil by water containing carbonic acid, tends to make the soil acidic:

$$\text{Soil}\}Ca^{2+} + 2CO_2 + 2H_2O \rightarrow \text{Soil}\}(H^+)_2 + Ca^{2+}(root) + 2HCO_3^- \quad (16.3.1)$$

Soil acts as a buffer and resists changes in pH. The buffering capacity depends upon the type of soil.

Production of Mineral Acid in Soil

The oxidation of pyrite in soil causes formation of acid-sulfate soils sometimes called "cat clays":

$$FeS_2 + {}^7/_2O_2 + H_2O \rightarrow Fe^{2+} + 2H^+ + 2SO_4^{2-} \quad\quad\quad (16.3.2)$$

Cat clay soils may have pH values as low as 3.0. These soils, which are commonly found in Delaware, Florida, New Jersey, and North Carolina, are formed when neutral or basic marine sediments containing FeS_2 become acidic upon oxidation of pyrite when exposed to air. For example, soil reclaimed from marshlands and used for citrus groves has developed high acidity detrimental to plant growth. In addition, H_2S released by reaction of FeS_2 with acid is very toxic to citrus roots.

Soils are tested for potential acid-sulfate formation using a peroxide test. This test consists of oxidizing FeS_2 in the soil with 30% H_2O_2,

$$FeS_2 + {}^{15}/_2H_2O_2 \rightarrow Fe^{3+} + H^+ + 2SO_4^{2-} + 7H_2O \quad\quad\quad (16.3.3)$$

then testing for acidity and sulfate. Appreciable levels of sulfate and a pH below 3.0 indicate potential to form acid-sulfate soils. If the pH is above 3.0, either little FeS_2 is present or sufficient $CaCO_3$ is in the soil to neutralize the H_2SO_4 and acidic Fe^{3+}.

Pyrite-containing mine spoils (residue left over from mining) also form soils similar to acid-sulfate soils of marine origin. In addition to high acidity and toxic H_2S, a major chemical species limiting plant growth on such soils is Al(III). Aluminum ion liberated in acidic soils is very toxic to plants.

Adjustment of Soil Acidity

Most common plants grow best in soil with a pH near neutrality. If the soil becomes too acidic for optimum plant growth, it may be restored to productivity by liming, ordinarily through the addition of calcium carbonate:

$$Soil\}(H^+)_2 + CaCO_3 \rightarrow Soil\}Ca^{2+} + CO_2 + H_2O \tag{16.3.4}$$

In areas of low rainfall, soils may become too basic (alkaline) due to the presence of basic salts such as Na_2CO_3. Alkaline soils may be treated with aluminum or iron sulfate, which release acid on hydrolysis:

$$2Fe^{3+} + 3SO_4^{2-} + 6H_2O \rightarrow 2Fe(OH)_3(s) + 6H^+ + 3SO_4^{2-} \tag{16.3.5}$$

Sulfur added to soils is oxidized by bacterially mediated reactions to sulfuric acid:

$$S + 3/2O_2 + H_2O \rightarrow 2H^+ + SO_4^{2-} \tag{16.3.6}$$

and sulfur is used, therefore, to acidify alkaline soils. The huge quantities of sulfur now being removed from fossil fuels to prevent air pollution by sulfur dioxide may make the treatment of alkaline soils by sulfur much more attractive economically.

Ion Exchange Equilibria in Soil

Competition of different cations for cation exchange sites on soil cation exchangers may be described semiquantitatively by exchange constants. For example, soil reclaimed from an area flooded with seawater will have most of its cation exchange sites occupied by Na^+, and restoration of fertility requires binding of nutrient cations such as K^+:

$$Soil\}Na^+ + K^+ \leftrightarrow Soil\}K^+ + Na^+ \tag{16.3.7}$$

The exchange constant is K_C,

$$K_C = \frac{N_K[Na^+]}{N_{Na}[K^+]} \tag{16.3.8}$$

which expresses the relative tendency of soil to retain K^+ and Na^+. In this equation, N_K and N_{Na} are the equivalent ionic fractions of potassium and sodium, respectively, bound to soil, and $[Na^+]$ and $[K^+]$ are the concentrations of these ions in the surrounding soil water. For example, a soil with all cation exchange sites occupied by Na^+ would have a value of 1.00 for N_{Na}; with one-half of the cation exchange sites occupied by Na^+, N_{Na} is 0.5; etc. The exchange of anions by soil is not nearly so clearly defined as is the exchange of cations. In many cases, the exchange of anions does not involve a simple ion-exchange process. This is true of the strong retention of orthophosphate species by soil. At the other end of the scale, nitrate ion is very weakly retained by the soil.

Anion exchange may be visualized as occurring at the surfaces of oxides in the mineral portion of soil. A mechanism for the acquisition of surface charge by metal oxides is shown in Chapter 5, Figure 5.5, using MnO_2 as an example. At low pH, a metal oxide surface may have a net positive charge enabling it to hold anions, such as chloride, by electrostatic attraction as shown below where M represents a metal:

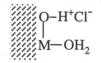

At higher pH values, the metal oxide surface has a net negative charge due to the formation of OH^- ion on the surface caused by loss of H^+ from the water molecules bound to the surface:

In such cases, it is possible for anions such as HPO_4^{2-} to displace hydroxide ion and bond directly to the oxide surface:

$$\text{M}-\text{OH}^- + \text{HPO}_4^{2-} \rightarrow \text{M}-\text{OPO}_3\text{H}^{2-} + \text{OH}^- \qquad (16.3.9)$$

16.4. MACRONUTRIENTS IN SOIL

One of the most important functions of soil in supporting plant growth is to provide essential plant nutrients—macronutrients and micronutrients. Macronutrients are those elements that occur in substantial levels in plant materials or in fluids in the plant. Micronutrients (Section 16.6) are elements that are essential only at very low levels and generally are required for the functioning of essential enzymes.

The elements generally recognized as essential macronutrients for plants are carbon, hydrogen, oxygen, nitrogen, phosphorus, potassium, calcium, magnesium, and sulfur. Carbon, hydrogen, and oxygen are obtained from the atmosphere. The other essential macronutrients must be obtained from soil. Of these, nitrogen, phosphorus, and potassium are the most likely to be lacking and are commonly added to soil as fertilizers. Because of their importance, these elements are discussed separately in Section 16.5.

Calcium-deficient soils are relatively uncommon. Application of lime, a process used to treat acid soils (see Section 16.3), provides a more than adequate calcium supply for plants. However, calcium uptake by plants and leaching by carbonic acid

(Reaction 16.3.1) may produce a calcium deficiency in soil. Acid soils may still contain an appreciable level of calcium which, because of competition by hydrogen ion, is not available to plants. Treatment of acid soil to restore the pH to near-neutrality generally remedies the calcium deficiency. In alkaline soils, the presence of high levels of sodium, magnesium, and potassium sometimes produces calcium deficiency because these ions compete with calcium for availability to plants.

Most of the 2.1% of magnesium in earth's crust is rather strongly bound in minerals. Exchangeable magnesium held by ion-exchanging organic matter or clays is considered available to plants. The availability of magnesium to plants depends upon the calcium/magnesium ratio. If this ratio is too high, magnesium may not be available to plants and magnesium deficiency results. Similarly, excessive levels of potassium or sodium may cause magnesium deficiency.

Sulfur is assimilated by plants as the sulfate ion, SO_4^{2-}. In addition, in areas where the atmosphere is contaminated with SO_2, sulfur may be absorbed as sulfur dioxide by plant leaves. Atmospheric sulfur dioxide levels have been high enough to kill vegetation in some areas (see Chapter 11). However, some experiments designed to show SO_2 toxicity to plants have resulted in increased plant growth where there was an unexpected sulfur deficiency in the soil used for the experiment.

Soils deficient in sulfur do not support plant growth well, largely because sulfur is a component of some essential amino acids and of thiamin and biotin. Sulfate ion is generally present in the soil as immobilized insoluble sulfate minerals, or as soluble salts which are readily leached from the soil and lost as soil water runoff. Unlike the case of nutrient cations such as K^+, little sulfate is adsorbed to the soil (that is, bound by ion exchange binding) where it is resistant to leaching while still available for assimilation by plant roots.

Soil sulfur deficiencies have been found in a number of regions of the world. Whereas most fertilizers used to contain sulfur, its use in commercial fertilizers has declined. With continued use of sulfur-deficient fertilizers, it is possible that sulfur will become a limiting nutrient in more cases.

As noted in Section 16.3, the reaction of FeS_2 with acid in acid-sulfate soils may release H_2S, which is very toxic to plants and which also kills many beneficial microorganisms. Toxic hydrogen sulfide can also be produced by reduction of sulfate ion through microorganism-mediated reactions with organic matter. Production of hydrogen sulfide in flooded soils may be inhibited by treatment with oxidizing compounds, one of the most effective of which is KNO_3.

16.5. NITROGEN, PHOSPHORUS, AND POTASSIUM IN SOIL

Nitrogen, phosphorus, and potassium are plant nutrients that are obtained from soil. They are so important for crop productivity that they are commonly added to soil as fertilizers. The environmental chemistry of these elements is discussed here and their production as fertilizers in Section 16.7.

Nitrogen

Figure 16.4 summarizes the primary sinks and pathways of nitrogen in soil. In most soils, over 90% of the nitrogen content is organic. This organic nitrogen is primarily the product of the biodegradation of dead plants and animals. It is eventually hydrolyzed to NH_4^+, which can be oxidized to NO_3^- by the action of bacteria in the soil.

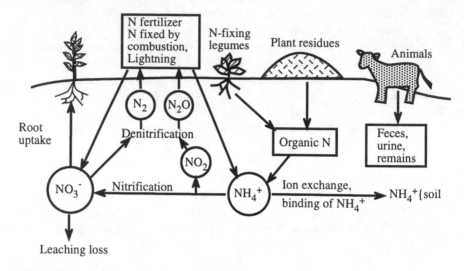

Figure 16.4. Nitrogen sinks and pathways in soil.

Nitrogen bound to soil humus is especially important in maintaining soil fertility. Unlike potassium or phosphate, nitrogen is not a significant product of mineral weathering. Nitrogen-fixing organisms ordinarily cannot supply sufficient nitrogen to meet peak demand. Inorganic nitrogen from fertilizers and rainwater is often largely lost by leaching. Soil humus, however, serves as a reservoir of nitrogen required by plants. It has the additional advantage that its rate of decay, hence its rate of nitrogen release to plants, roughly parallels plant growth—rapid during the warm growing season, slow during the winter months.

Nitrogen is an essential component of proteins and other constituents of living matter. Plants and cereals grown on nitrogen-rich soils not only provide higher yields, but are often substantially richer in protein and, therefore, more nutritious. Nitrogen is most generally available to plants as nitrate ion, NO_3^-. Some plants such as rice may utilize ammonium nitrogen; however, other plants are poisoned by this form of nitrogen. When nitrogen is applied to soils in the ammonium form, nitrifying bacteria perform an essential function in converting it to available nitrate ion.

Plants may absorb excessive amounts of nitrate nitrogen from soil. This phenomenon occurs particularly in heavily fertilized soils under drought conditions. Forage crops containing excessive amounts of nitrate can poison ruminant animals such as cattle or sheep. Plants having excessive levels of nitrate can endanger people when used for ensilage, an animal food consisting of finely chopped plant material such as partially matured whole corn plants, fermented in a structure called a silo. Under the reducing conditions of fermentation, nitrate in ensilage may be reduced to toxic NO_2 gas, which can accumulate to high levels in enclosed silos. There have been many cases reported of persons being killed by accumulated NO_2 in silos.

Nitrogen fixation is the process by which atmospheric N_2 is converted to nitrogen compounds available to plants. Human activities are resulting in the fixation of a great deal more nitrogen than would otherwise be the case. Artificial sources now account for 30-40% of all nitrogen fixed. These include chemical fertilizer manufacture, nitrogen fixed during fuel combustion, combustion of nitrogen-containing fuels, and the increased cultivation of nitrogen-fixing legumes (see the following paragraph). A major concern with this increased fixation of nitrogen is the possible effect upon the atmospheric ozone layer by N_2O released during denitrification of fixed nitrogen.

Before the widespread introduction of nitrogen fertilizers, soil nitrogen was provided primarily by legumes. These are plants such as soybeans, alfalfa, and clover, which contain on their root structures bacteria capable of fixing atmospheric nitrogen. Leguminous plants have a symbiotic (mutually advantageous) relationship with the bacteria that provide their nitrogen. Legumes may add significant quantities of nitrogen to soil, up to 10 pounds per acre per year, which is comparable to amounts commonly added as synthetic fertilizers. Soil fertility with respect to nitrogen may be maintained by rotating plantings of nitrogen-consuming plants with plantings of legumes, a fact recognized by agriculturists as far back as the Roman era.

The nitrogen-fixing bacteria in legumes exist in special structures on the roots called root nodules (see Fig. 16.5). The rod-shaped bacteria that fix nitrogen are members of a special genus, *Rhizobium*. These bacteria may exist independently, but cannot fix nitrogen except in symbiotic combination with plants. Although all species of *Rhizobium* appear to be very similar, they exhibit a great deal of specificity in their choice of host plants. Curiously, legume root nodules also contain a form of hemoglobin, which must somehow be involved in the nitrogen-fixation process.

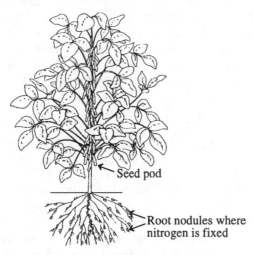

Figure 16.5. A soybean plant, showing root nodules where nitrogen is fixed.

Nitrate pollution of some surface waters and groundwater has become a major problem in some agricultural areas (see Chapter 7). Although fertilizers have been implicated in such pollution, there is evidence that feedlots are a major source of nitrate pollution. The growth of livestock populations and the concentration of livestock in feedlots have aggravated the problem. Such concentrations of cattle, coupled with the fact that a steer produces approximately 18 times as much waste material as a human, have resulted in high levels of water pollution in rural areas with small human populations. Streams and reservoirs in such areas frequently are just as polluted as those in densely populated and highly industrialized areas.

Nitrate in farm wells is a common and especially damaging manifestation of nitrogen pollution from feedlots because of the susceptibility of ruminant animals to nitrate poisoning. The stomach contents of ruminant animals such as cattle and sheep constitute a reducing medium (low pE) and contain bacteria capable of reducing nitrate ion to toxic nitrite ion:

$$NO_3^- + 2H^+ + 2e^- \rightarrow NO_2^- + H_2O \tag{16.5.1}$$

The origin of most nitrate produced from feedlot wastes is amino nitrogen present in nitrogen-containing waste products. Approximately one-half of the nitrogen excreted by cattle is contained in the urine. Part of this nitrogen is proteinaceous and the other part is in the form of urea, NH_2CONH_2. As a first step in the degradation process, the amino nitrogen is probably hydrolyzed to ammonia, or ammonium ion:

$$RNH_2 + H_2O \rightarrow R\text{-}OH + NH_3 (NH_4^+) \tag{16.5.2}$$

This product is then oxidized through microorganism-catalyzed reactions to nitrate ion:

$$NH_3 + 2O_2 \rightarrow H^+ + NO_3^- + H_2O \tag{16.5.3}$$

Under some conditions, an appreciable amount of the nitrogen originating from the degradation of feedlot wastes is present as ammonium ion. Ammonium ion is rather strongly bound to soil (recall that soil is a generally good cation exchanger), and a small fraction is fixed as nonexchangeable ammonium ion in the crystal lattice of clay minerals. Because nitrate ion is not strongly bound to soil, it is readily carried through soil formations by water. Many factors, including soil type, moisture, and level of organic matter, affect the production of ammonia and nitrate ion originating from feedlot wastes, and a marked variation is found in the levels and distributions of these materials in feedlot areas.

Phosphorus

Although the percentage of phosphorus in plant material is relatively low, it is an essential component of plants. Phosphorus, like nitrogen, must be present in a simple inorganic form before it can be taken up by plants. In the case of phosphorus, the utilizable species is some form of orthophosphate ion. In the pH range that is present in most soils, $H_2PO_4^-$ and HPO_4^{2-} are the predominant orthophosphate species.

Orthophosphate is most available to plants at pH values near neutrality. It is believed that in relatively acidic soils, orthophosphate ions are precipitated or sorbed by species of Al(III) and Fe(III). In alkaline soils, orthophosphate may react with calcium carbonate to form relatively insoluble hydroxyapatite:

$$3HPO_4^{2-} + 5CaCO_3(s) + 2H_2O \rightarrow Ca_5(PO_4)_3(OH)(s)$$
$$+ 5HCO_3^- + OH^- \tag{16.5.4}$$

In general, because of these reactions, little phosphorus applied as fertilizer leaches from the soil. This is important from the standpoint of both water pollution and utilization of phosphate fertilizers.

Potassium

Relatively high levels of potassium are utilized by growing plants. Potassium activates some enzymes and plays a role in the water balance in plants. It is also essential for some carbohydrate transformations. Crop yields are generally greatly reduced in potassium-deficient soils. The higher the productivity of the crop, the more potassium is removed from soil. When nitrogen fertilizers are added to soils to increase productivity, removal of potassium is enhanced. Therefore, potassium may become a limiting nutrient in soils heavily fertilized with other nutrients.

Potassium is one of the most abundant elements in the earth's crust, of which it makes up 2.6%; however, much of this potassium is not easily available to plants. For example, some silicate minerals such as leucite, $K_2O \cdot Al_2O_3 \cdot 4SiO_2$, contain strongly bound potassium. Exchangeable potassium held by clay minerals is relatively more available to plants.

16.6. MICRONUTRIENTS IN SOIL

Boron, chlorine, copper, iron, manganese, molybdenum (for N-fixation), and zinc are considered essential plant **micronutrients**. These elements are needed by plants only at very low levels and frequently are toxic at higher levels. There is some chance that other elements will be added to this list as techniques for growing plants in environments free of specific elements improve. Most of these elements function as components of essential enzymes. Manganese, iron, chlorine, and zinc may be involved in photosynthesis. Though not established for all plants, it is possible that sodium, silicon, and cobalt may also be essential plant nutrients.

Iron and manganese occur in a number of soil minerals. Sodium and chlorine (as chloride) occur naturally in soil and are transported as atmospheric particulate matter from marine sprays (see Chapter 10). Some of the other micronutrients and trace elements are found in primary (unweathered) minerals that occur in soil. Boron is substituted isomorphically for Si in some micas and is present in tourmaline, a mineral with the formula $NaMg_3Al_6B_3Si_6O_{27}(OH,F)_4$. Copper is isomorphically substituted for other elements in feldspars, amphiboles, olivines, pyroxenes, and micas; it also occurs as trace levels of copper sulfides in silicate minerals. Molybdenum occurs as molybdenite (MoS_2). Vanadium is isomorphically substituted for Fe or Al in oxides, pyroxenes, amphiboles, and micas. Zinc is present as the result of isomorphic substitution for Mg, Fe, and Mn in oxides, amphiboles, olivines, and pyroxenes and as trace zinc sulfide in silicates. Other trace elements that occur as specific minerals, sulfide inclusions, or by isomorphic substitution for other elements in minerals are chromium, cobalt, arsenic, selenium, nickel, lead, and cadmium.

The trace elements listed above may be coprecipitated with secondary minerals (see Section 15.2) that are involved in soil formation. Such secondary minerals include oxides of aluminum, iron, and manganese (precipitation of hydrated oxides of iron and manganese very efficiently removes many trace metal ions from solution); calcium and magnesium carbonates; smectites; vermiculites; and illites.

Some plants accumulate extremely high levels of specific trace metals. Those accumulating more than 1.00 mg/g of dry weight are called **hyperaccumulators**. Nickel and copper both undergo hyperaccumulation in some plant species. As an example of a metal hyperaccumulater, *Aeolanthus biformifolius DeWild* growing in copper-rich regions of Shaba Province, Zaire, contains up to 1.3% copper (dry weight) and is known as a "copper flower".

The hyperaccumulation of metals by some plants has led to the idea of **phytoremediation** in which plants growing on contaminated ground accumulate metals, which are then removed with the plant biomass. *Brassica juncea* and *Brassica chinensis* (Chinese cabbage) have been shown to hyperaccumulate as much as 5 grams of uranium per kg plant dry weight when grown on uranium-contaminated soil.[6] Uranium accumulation in the plants was enhanced by the addition of citrate, which complexes uranium and makes it more soluble.

16.7. FERTILIZERS

Crop fertilizers contain nitrogen, phosphorus, and potassium as major components. Magnesium, sulfate, and micronutrients may also be added. Fertilizers are designated by numbers, such as 6-12-8, showing the respective percentages of nitrogen expressed as N (in this case 6%), phosphorus as P_2O_5 (12%), and potassium as K_2O (8%). Farm manure corresponds to an approximately 0.5-0.24-0.5 fertilizer. The organic fertilizers such as manure must undergo biodegradation to release the simple inorganic species (NO_3^-, $H_XPO_4^{X-3}$, K^+) assimilable by plants.

Most modern nitrogen fertilizers are made by the Haber process, in which N_2 and H_2 are combined over a catalyst at temperatures of approximately 500°C and pressures up to 1000 atm:

$$N_2 + 3H_2 \rightarrow 2NH_3 \tag{16.7.1}$$

The anhydrous ammonia product has a very high nitrogen content of 82%. It may be added directly to the soil, for which it has a strong affinity because of its water solubility and formation of ammonium ion:

$$NH_3(g) \text{ (water)} \rightarrow NH_3(aq) \tag{16.7.2}$$

$$NH_3(aq) + H_2O \rightarrow NH_4^+ + OH^- \tag{16.7.3}$$

Special equipment is required, however, because of the toxicity of ammonia gas. Aqua ammonia, a 30% solution of NH_3 in water, may be used with much greater safety. It is sometimes added directly to irrigation water. It should be pointed out that ammonia vapor is toxic and NH_3 is reactive with some substances. Improperly discarded or stored ammonia can be a hazardous waste.

Ammonium nitrate, NH_4NO_3, is a common solid nitrogen fertilizer. It is made by oxidizing ammonia over a platinum catalyst, converting the nitric oxide product to nitric acid, and reacting the nitric acid with ammonia. The molten ammonium nitrate product is forced through nozzles at the top of a *prilling tower* and solidifies to form small pellets while falling through the tower. The particles are coated with a water repellent. Ammonium nitrate contains 33.5% nitrogen. Although convenient to apply to soil, it requires considerable care during manufacture and storage because it is explosive. Ammonium nitrate also poses some hazards. It is mixed with fuel oil to form an explosive that serves as a substitute for dynamite in quarry blasting and construction. This mixture was used to devastating effect in the dastardly bombing of the Oklahoma City Federal Building in 1995.

Urea,

$$H_2N-\overset{\overset{\textstyle O}{\|}}{C}-NH_2$$

is easier to manufacture and handle than ammonium nitrate. It is now the favored solid nitrogen-containing fertilizer. The overall reaction for urea synthesis is

$$CO_2 + 2NH_3 \rightarrow CO(NH_2)_2 + H_2O \qquad (16.7.4)$$

involving a rather complicated process in which ammonium carbamate, chemical formula $NH_2CO_2NH_4$, is an intermediate.

Other compounds used as nitrogen fertilizers include sodium nitrate (obtained largely from Chilean deposits, see Section 15.2), calcium nitrate, potassium nitrate, and ammonium phosphates. Ammonium sulfate, a by-product of coke ovens, used to be widely applied as fertilizer. The alkali metal nitrates tend to make soil alkaline, whereas ammonium sulfate leaves an acidic residue.

Phosphate minerals are found in several states, including Idaho, Montana, Utah, Wyoming, North Carolina, South Carolina, Tennessee, and Florida. The principal mineral is fluorapatite, $Ca_5(PO_4)_3F$. The phosphate from fluorapatite is relatively unavailable to plants, and fluorapatite is frequently treated with phosphoric or sulfuric acids to produce superphosphates:

$$2Ca_5(PO_4)_3F(s) + 14H_3PO_4 + 10H_2O \rightarrow 2HF(g)$$
$$+ 10Ca(H_2PO_4)_2 \cdot H_2O \qquad (16.7.5)$$

$$2Ca_5(PO_4)_3F(s) + 7H_2SO_4 + 3H_2O \rightarrow 2HF(g)$$
$$+ 3Ca(H_2PO_4)_2 \cdot H_2O + 7CaSO_4 \qquad (16.7.6)$$

The superphosphate products are much more soluble than the parent phosphate minerals. The HF produced as a byproduct of superphosphate production can create air pollution problems.

Phosphate minerals are rich in trace elements required for plant growth, such as boron, copper, manganese, molybdenum, and zinc. Ironically, these elements are lost to a large extent when the phosphate minerals are processed to make fertilizer. Ammonium phosphates are excellent, highly soluble phosphate fertilizers. Liquid ammonium polyphosphate fertilizers consisting of ammonium salts of pyrophosphate, triphosphate, and small quantities of higher polymeric phosphate anions in aqueous solution work very well as phosphate fertilizers. The polyphosphates are believed to have the additional advantage of chelating iron and other micronutrient metal ions, thus making the metals more available to plants.

Potassium fertilizer components consist of potassium salts, generally KCl. Such salts are found as deposits in the ground or may be obtained from some brines. Very large deposits are found in Saskatchewan, Canada. These salts are all quite soluble in water. One problem encountered with potassium fertilizers is the luxury uptake of potassium by some crops, which absorb more potassium than is really needed for their maximum growth. In a crop where only the grain is harvested, leaving the rest of the plant in the field, luxury uptake does not create much of a problem because most of the potassium is returned to the soil with the dead plant. However, when hay or forage is harvested, potassium contained in the plant as a consequence of luxury uptake is lost from the soil.

16.8. WASTES AND POLLUTANTS IN SOIL

Soil receives large quantities of waste products. Much of the sulfur dioxide emitted in the burning of sulfur-containing fuels ends up as soil sulfate. Atmospheric nitrogen oxides are converted to nitrates in the atmosphere, and the nitrates eventually are deposited on soil. Soil sorbs NO and NO_2, and these gases are oxidized to nitrate in the soil. Carbon monoxide is converted to CO_2 and possibly to biomass by soil bacteria and fungi. Particulate lead from automobile exhausts is found at elevated levels in soil along heavily traveled highways. Elevated levels of lead from lead mines and smelters are found on soil near such facilities.

Soil is the receptor of many hazardous wastes from landfill leachate, lagoons, and other sources (see Section 19.13). In some cases, land farming of degradable hazardous organic wastes is practiced as a means of disposal and degradation. The degradable material is worked into the soil, and soil microbial processes bring about its degradation. As discussed in Chapter 8, sewage and fertilizer-rich sewage sludge may be applied to soil.

Volatile organic compounds (VOC) such as benzene, toluene, xylenes, dichloromethane, trichloroethane, and trichloroethylene, may contaminate soil in industrialized and commercialized areas, particularly in countries in which enforcement of regulations is not very stringent. One of the more common sources of these contaminants is leaking underground storage tanks. Landfills built before current stringent regulations were enforced and improperly discarded solvents are also significant sources of soil VOCs.

Measurements of levels of polychlorinated biphenyls (PCBs) in soils that have been archived for several decades provide interesting insight into the contamination of soil by pollutant chemicals and subsequent loss of these substances from soil.[7] Analyses of soils from the United Kingdom dating from the early 1940s to 1992 showed that the PCB levels increased sharply from the 1940s, reaching peak levels around 1970. Subsequently, levels fell sharply and now are back to early 1940s concentrations. This fall was accompanied by a shift in distribution to the more highly chlorinated PCBs, which was attributed by those doing the study to volatilization and long range transport of the lighter PCBs away from the soil. These trends parallel levels of PCB manufacture and use in the United Kingdom from the early 1940s to the present. This is consistent with the observation that relatively high concentrations of PCBs have been observed in remote Arctic and sub-Arctic regions, attributed to condensation in colder climates of PCBs volatilized in warmer regions.

Some pollutant organic compounds are believed to become bound with humus during the humification process that occurs in soil. This largely immobilizes and detoxifies the compounds. Binding of pollutant compounds by humus is particularly likely to occur with compounds that have structural similarities to humic substances, such as phenolic and anilinic compounds illustrated by the following two examples:

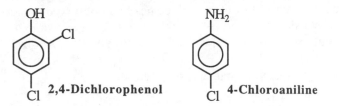

2,4-Dichlorophenol   4-Chloroaniline

Such compounds can become covalently bonded to humic substance molecules, largely through the action of microbial enzymes. After binding they are known as **bound residues** and are highly resistant to extraction with solvents by procedures that would remove unbound parent compounds.[8] Compounds in the bound residues are resistant to biological and chemical attack.

Soil receives enormous quantities of pesticides as an inevitable result of their application to crops. The degradation and eventual fate of these pesticides on soil largely determines their ultimate environmental effects. Detailed knowledge of these effects are now required for licensing of a new pesticide (in the U.S. under the Federal Insecticide, Fungicide, and Rodenticide act, FIFRA). Among the factors to be considered are the sorption of the pesticide by soil; leaching of the pesticide into water, as related to its potential for water pollution; effects of the pesticide on microorganisms and animal life in the soil; and possible production of relatively more toxic degradation products.

Adsorption by soil is a key aspect of pesticide degradation and plays a strong role in the speed and degree of degradation. The degree of adsorption and the speed and extent of ultimate degradation are influenced by a number of other factors. Some of these, including solubility, volatility, charge, polarity, and molecular structure and size, are properties of the medium. Adsorption of a pesticide by soil components may have several effects. Under some circumstances, it retards degradation by separating the pesticide from the microbial enzymes that degrade it, whereas under other circumstances the reverse is true. Purely chemical degradation reactions may be catalyzed by adsorption. Loss of the pesticide by volatilization or leaching is diminished. The toxicity of a herbicide to plants may be reduced by sorption on soil. The forces holding a pesticide to soil particles may be of several types. Physical adsorption involves van der Waals forces arising from dipole-dipole interactions between the pesticide molecule and charged soil particles. Ion exchange is especially effective in holding cationic organic compounds, such as the herbicide paraquat,

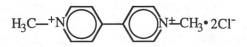

to anionic soil particles. Some neutral pesticides become cationic by protonation and are bound as the protonated positive form. Hydrogen bonding is another mechanism by which some pesticides are held to soil. In some cases a pesticide may act as a ligand coordinating to metals in soil mineral matter.

The three primary ways in which pesticides are degraded in or on soil are *chemical degradation, photochemical reactions*, and, most important, *biodegradation*. Various combinations of these processes may operate in the degradation of a pesticide.

Chemical degradation of pesticides has been observed experimentally in soils and clays sterilized to remove all microbial activity. For example, clays have been shown to catalyze the hydrolysis of *o, o*-dimethyl-*o*-2,4,5-trichlorophenyl thiophosphate (also called Trolene, Ronnel, Etrolene, or trichlorometafos), an effect attributed to -OH groups on the mineral surface:

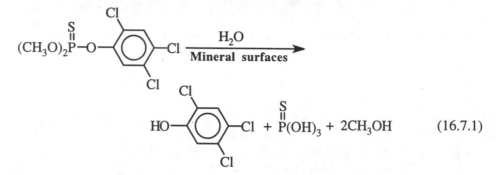

$$(16.7.1)$$

Many other purely chemical hydrolytic reactions of pesticides occur in soil.

A number of pesticides have been shown to undergo **photochemical reactions**, that is, chemical reactions brought about by the absorption of light (see Chapter 9). Frequently, isomers of the pesticides are formed as products. Many of the studies reported apply to pesticides in water or on thin films, and the photochemical reactions of pesticides on soil and plant surfaces remain largely a matter of speculation.

Biodegradation and the Rhizosphere

Although insects, earthworms, and plants may play roles in the **biodegradation** of pesticides and other pollutant organic chemicals, microorganisms have the most important role. Several examples of microorganism-mediated degradation of organic chemical species are given in Chapter 6.

The **rhizosphere**, the layer of soil in which plant roots are especially active, is a particularly important part of soil with respect to biodegradation of wastes. It is a zone of increased biomass and is strongly influenced by the plant root system and the microorganisms associated with plant roots. The rhizosphere may have more than ten times the microbial biomass per unit volume compared to nonrhizospheric zones of soil. This population varies with soil characteristics, plant and root characteristics, moisture content, and exposure to oxygen. If this zone is exposed to pollutant compounds, microorganisms adapted to their biodegradation may also be present.

Plants and microorganisms exhibit a strong synergistic relationship in the rhizosphere, which benefits the plant and enables highly elevated populations of rhizospheric microorganisms to exist. Epidermal cells sloughed from the root as it grows and carbohydrates, amino acids, and root-growth-lubricant mucigel secreted from the roots all provide nutrients for microorganism growth. Root hairs provide a hospitable biological surface for colonization by microorganisms.

The biodegradation of a number of synthetic organic compounds has been demonstrated in the rhizosphere. Understandably, studies in this area have focused on herbicides and insecticides that are widely used on crops. Among the organic species for which enhanced biodegradation in the rhizosphere has been demonstrated are the following (associated plant or crop shown in parentheses): 2,4-D herbicide (wheat, African clover, sugarcane, flax), parathion (rice, bush bean), carbofuran (rice), atrazine (corn), diazinon (wheat, corn, peas), volatile aromatic alkyl and aryl hydrocarbons and chlorocarbons (reeds), and surfactants (corn, soybean, cattails). It is interesting to note that enhanced biodegradation of polycyclic aromatic hydrocarbons (PAH) was observed in the rhizospheric zones of prairie grasses. This observation is consistent with the fact that in nature such grasses burn regularly and significant quantities of PAH compounds are deposited on soil as a result.

16.9. SOIL LOSS AND DEGRADATION

Soil is a fragile resource that can be lost by erosion or become so degraded that it is no longer useful to support crops. The physical properties of soil and, hence, its susceptibility to erosion, are strongly affected by the cultivation practices to which the soil is subjected.[9] **Desertification** refers to the process associated with drought and loss of fertility by which soil becomes unable to grow significant amounts of plant life. Desertification caused by human activities is a common problem globally, occurring in diverse locations such as Argentina, the Sahara, Uzbekistan, the U.S. Southwest, Syria, and Mali. It is a very old problem dating back many centuries to the introduction of domesticated grazing animals to areas where rainfall and groundcover were marginal. The most notable example is desertification aggravated by domesticated goats in the Sahara region. Desertification involves a number of interrelated factors, including erosion, climate variations, water availability, loss of fertility, loss of soil humus, and deterioration of soil chemical properties.

A related problem is **deforestation**, loss of forests. The problem is particularly acute in tropical regions, where the forests contain most of the existing plant and animal species. In addition to extinction of these species, deforestation can cause devastating deterioration of soil through erosion and loss of nutrients.

Soil erosion can occur by the action of both water and wind, although water is the primary source of erosion. Millions of tons of topsoil are carried by the Mississippi River and swept from its mouth each year. About one-third of U.S. topsoil has been lost since cultivation began on the continent. At the present time, approximately one-third of U.S. cultivated land is eroding at a rate sufficient to reduce soil productivity. It is estimated that 48 million acres of land, somewhat more than 10 percent of that under cultivation, is eroding at unacceptable levels, taken to mean a loss of more than 14 tons of topsoil per acre each year. Specific areas in which the greatest erosion is occurring include northern Missouri, southern Iowa, west Texas, western Tennessee, and the Mississippi Basin. Figure 16.6 shows the pattern of soil erosion in the continental U.S. in 1977.

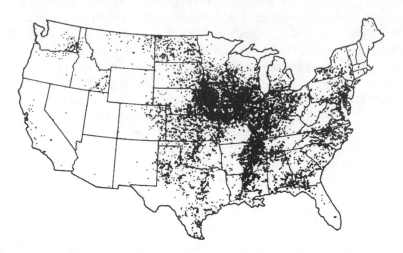

Figure 16.6. Pattern of soil erosion in the continental U.S. as of 1977. The dark areas indicate locations where the greatest erosion is occurring.

Problems involving soil erosion were aggravated in the 1970s and early 1980s when high prices for farmland resulted in the intensive cultivation of high-income crops, particularly corn and soybeans. These crops grow in rows with bare soil in between, which tends to wash away with each rainfall. Furthermore, the practice of planting corn and soybeans year after year without intervening plantings of soil-restoring clover or grass became widespread. The problem of decreased productivity due to soil erosion has been masked somewhat by increased use of chemical fertilizers.

Wind erosion, such as occurs on the generally dry, high plains soils of eastern Colorado, poses another threat. After the Dust Bowl days of the 1930s, much of this land was allowed to revert to grassland, and the topsoil was held in place by the strong root systems of the grass cover. However, in an effort to grow more wheat and improve the sale value of the land, much of it was later returned to cultivation. For example, from 1979 through 1982, more than 450,000 acres of Colorado grasslands were plowed. Much of this was done by speculators who purchased grassland at a low price of $100-$200 per acre, broke it up, and sold it as cultivated land at more than double the original purchase price. Although freshly cultivated grassland may yield well for one or two years, the nutrients and soil moisture are rapidly exhausted and the land becomes very susceptible to wind erosion.

The preservation of soil from erosion is commonly termed **soil conservation**. There are a number of solutions to the soil erosion problem. Some are old, well-known agricultural practices such as terracing, contour plowing, and periodically planting fields with cover crops such as clover. For some crops **no-till agriculture**, now commonly called **conservation tillage**,[10,11] greatly reduces erosion. This practice consists of planting a crop among the residue of the previous year's crop without plowing.[12] Weeds are killed in the newly planted crop row by application of a herbicide prior to planting. The surface residue of plant material left on top of the soil prevents erosion.

Another, more experimental, solution to the soil erosion problem is the cultivation of perennial plants which develop large root systems and come up each spring after being harvested the previous fall. For example, a perennial corn plant has been developed by crossing corn with a distant, wild relative, teosinte, which grows in Central America. Unfortunately, the resulting plant does not give outstanding grain yields. It should be noted that an annual plant's ability to propagate depends upon producing large quantities of seeds, whereas a perennial plant must develop a strong root system with bulbous growths called rhizomes which store food for the coming year. However, it is possible that the application of genetic engineering (see Section 16.9) may result in the development of perennial crops with good seed yields. The cultivation of such a crop would cut down on a great deal of soil erosion.

The best known perennial plants are trees, which are very effective in stopping soil erosion. Wood from trees can be used as biomass fuel, as a source of raw materials, and as food (see below). There is a tremendous unrealized potential for an increase in the production of biomass from trees. For example, the production of biomass from natural forests of loblolly pine trees in South Carolina has been about three dry tons per hectare per year. This has now been increased at least four-fold through selection of superior trees, and 30 tons may eventually be possible. In Brazil, experiments were conducted with a species of Eucalyptus, which has a seven-year growth cycle. With improved selection of trees, the annual yields for three successive cycles of these trees in dry tons per hectare per year were 23, 33, and 40.

The most important use for wood is, of course, as lumber for construction. This use will remain important as higher energy prices increase the costs of other construction materials such as steel, aluminum, and cement. Wood is about 50 percent cellulose, which can be hydrolyzed by rapidly improving enzyme processes to yield glucose sugar. The glucose can be used directly as food, fermented to ethyl alcohol for fuel (gasohol), or employed as a carbon and energy source for protein-producing yeasts. Given these and other potential uses, the future of trees as an environmentally desirable and profitable crop is very bright.

Soil and Water Resources

The conservation of soil and the protection of water resources are strongly inter-related. Most fresh water falls initially on soil, and the condition of the soil largely determines the fate of the water and how much is retained in a usable condition. The land area upon which rainwater falls is called a **watershed**. In addition to collecting the water, the watershed determines the direction and rate of flow and the degree of water infiltration into groundwater aquifers (see the hydrologic cycle in Figure 2.1). Excessive rates of water flow prevent infiltration, lead to flash floods, and cause soil erosion. Measures taken to enhance the utility of land as a watershed also fortunately help prevent erosion. Some of these measures involve modification of the contour of the soil, particularly terracing, construction of waterways, and construction of water-retaining ponds. Waterways are planted with grass to prevent erosion, and water-retaining crops and bands of trees can be planted on the contour to achieve much the same goal. Reforestation and control of damaging grazing practices conserve both soil and water.

16.10. GENETIC ENGINEERING AND AGRICULTURE

The nuclei of living cells contain the genetic instruction for cell reproduction. These instructions are in the form of a special material called deoxyribonucleic acid, DNA. In combination with proteins, DNA makes up the cell chromosomes. During the 1970s the ability to manipulate DNA through genetic engineering became a reality, and during the 1980s it became the basis of a major industry. Such manipulation falls into the category of recombinant DNA technology. Recombinant DNA gets its name from the fact that it contains DNA from two different organisms, recombined together. This technology promises some exciting developments in agriculture and, indeed, is expected to lead to a "second green revolution."[13]

The first "green revolution" of the mid-1960s used conventional plant-breeding techniques of selective breeding, hybridization, cross-pollination, and back-crossing to develop new strains of rice, wheat, and corn which, when combined with chemical fertilizers, yielded spectacularly increased crop yields. For example, India's output of grain increased 50 percent. By working at the cellular level, however, it is now possible to greatly accelerate the process of plant breeding. Thus, plants may be developed that resist particular diseases, grow in seawater, or have much higher productivity. The possibility exists for developing entirely new kinds of plants.

One exciting possibility with genetic engineering is the development of plants other than legumes that fix their own nitrogen. For example, if nitrogen-fixing corn could be developed, the savings in fertilizer would be enormous. Furthermore, since the nitrogen is fixed in an organic form in plant root structures, there would be no pollutant runoff of chemical fertilizers.

Another promising possibility with genetic engineering is increased efficiency of photosynthesis. Plants utilize only about one percent of the sunlight striking their leaves, so there is appreciable room for improvement in that area.

Cell-culture techniques can be applied in which billions of cells are allowed to grow in a medium and develop mutants which, for example, might be resistant to particular viruses or herbicides or have other desirable qualities. If the cells with the desired qualities can be regenerated into whole plants, results can be obtained that might have taken decades using conventional plant-breeding techniques.

Despite the enormous potential of the "green revolution," genetic engineering, and more intensive cultivation of land to produce food and fiber, these technologies cannot be relied upon to support an uncontrolled increase in world population, and may even simply postpone an inevitable day of reckoning with the consequences of population growth. Changes in climate resulting from global warming (greenhouse effect, Section 14.2), ozone depletion (by chlorofluorocarbons, Section 14.4), or natural disasters such as massive volcanic eruptions or collisions with large meteorites can, and almost certainly will, result in worldwide famine conditions in the future that no agricultural technology will be able to alleviate.

There has been strong resistance to transgenic crops and livestock in some quarters. This is especially true in Europe. Reports that pollen from transgenic corn producing an insecticide of bacterial origin has killed monarch butterfly caterpillars led the European Commission, the executive body of the the European Union, to suspend action on seven transgenic crops that it was considering for approval.[14] The Commission had already approved 16 varieties of transgenic corn, potatoes, rape, sugar beets, turnips, and other crops. As of 1999 it appeared that major agricultural product processors in the U.S. would have to certify products as "GM free" (free of genetically modified products) for export to Europe.

16.11. AGRICULTURE AND HEALTH

Some authorities hold that soil has an appreciable effect upon health. An obvious way in which such an effect might be manifested is the incorporation into food of micronutrient elements essential for human health. One such nutrient (which is toxic at overdose levels) is selenium. It is definitely known that the health of animals is adversely affected in selenium-deficient areas as it is in areas of selenium excess. Human health might be similarly affected.

There are some striking geographic correlations with the occurrence of cancer, some of which may be due to soil type. A high incidence of stomach cancer occurs in areas with certain types of soil in the Netherlands, the United States, France, Wales, and Scandinavia. These soils are high in organic matter content, are acidic, and are frequently waterlogged. A "stomach cancer-prone lifestyle" has been described,[15] which includes consumption of home-grown food, consumption of water from one's own well, and reliance on native and uncommon foodstuffs.

One possible reason for the existence of "stomach cancer-producing soils" is the production of cancer-causing secondary metabolites by plants and microorganisms. Secondary metabolites are biochemical compounds that are of no apparent use to the organism producing them. It is believed that they are formed from the precursors of primary metabolites when the primary metabolites accumulate to excessive levels.

The role of soil in environmental health is not well known, nor has it been extensively studied. The amount of research on the influence of soil in producing foods that are more nutritious and lower in content of naturally occurring toxic substances is quite small compared to research on higher soil productivity. It is to be hoped that the environmental health aspects of soil and its products will receive much greater emphasis in the future.

Chemical Contamination

Human activities, most commonly pesticide application, may contaminate food grown on soil. An interesting example of such contamination occurred in Hawaii in early 1982.[16] It was found that milk from several sources on Oahu contained very high levels of heptachlor (see Table 7.5). This pesticide causes cancer and liver disorders in mice; therefore, it is a suspected human carcinogen. Remarkably, in this case it was not until 57 days after the initial discovery that the public was informed of the contamination by the Department of Health. The source of heptachlor was traced to contaminated "green chop," chopped-up pineapple leaves fed to cattle. Although heptachlor was banned for most applications, Hawaiian pineapple growers had obtained special federal permission to use it to control mealybug wilt. Although it was specified that green chop could not be collected within one year of the last application of the pesticide, this regulation was apparently violated, and the result was distribution of contaminated milk to consumers.

In the late 1980s, Alar residues on food caused considerable controversy in the marketplace. **Alar**, daminozide, is a growth regulator that was widely used on apples to bring about uniform ripening of the fruit, and to improve the firmness and color of the apples. It was discontinued for this purpose after 1988 because of concerns that it might cause cancer, particularly in those children who consume relatively large amounts of apples, apple juice, and other apple products. Dire predictions were made in the industry of crop losses and financial devastation. However, the 1989 apple crop, which was the first without Alar in the U.S., had a value of $1.0 billion, only $0.1 billion less than that of the 1988 crop. Production of apples has continued since then without serious problems from the unavailability of Alar.

A possible source of soil contamination results from recycling industrial wastes for fertilizer.[17] According to data compiled by the Environmental Working Group (EWG) during the 1990s, approximately 25 million kg per year of potentially toxic wastes were used to prepare fertilizers that contained elevated levels of arsenic, cadmium, lead, radioactive materials, and dioxins. A potential source of heavy metal pollution in fertilizers is ash from furnaces used to recycle steel, commonly processed to provide zinc in zinc-deficient soils.

LITERATURE CITED

1. Miller, Marshall L., "Pesticides," Chapter 13 in *Environmental Law Handbook*, 13th ed., Thomas F. P. Sullivan, Ed., Government Institutes, Inc., Rockville, MD, 1995.

2. Thayer, Ann, "Monsanto Expands Licensing Strategy for Flagship Herbicide," *Chemical and Engineering News*, January 25, 1999, p. 8.

3. Brady, Nyle C. and Ray R. Weil, *The Nature and Properties of Soils*, Prentice Hall, Upper Saddle River, N.J., 1996.

4. Stevenson, F. J., *Humus Chemistry*, 2nd ed., John Wiley and Sons, Somerset, NJ, 1994.

5. Wolt, Jeffrey D., *Soil Solution Chemistry: Applications to Environmental Science and Agriculture*, John Wiley & Sons, New York, 1994.

6. Huang, Jianwei W., Michael J. Blaylock, Yoram Kapulnik, and Burt W. Ensley, "Phytoremediation of Uranium-Contaminated Soils: Role of Organic Acids in Triggering Uranium Hyperaccumulation in Plants," *Environmental Science and Technology*, **32**, 2004-2008 (1998).

7. Alcock, R. E., A. E. Johnston, S. P. McGrath, M. L. Berrow, and Kevin C. Jones, "Long-Term Changes in the Polychlorinated Biphenyl Content of United Kingdom Soils," *Environmental Science and Technology*, **27**, 1918-1923 (1993).

8. Nieman, Karl C., Ronald C. Sims, Judith L. Sims, Darwin L. Sorensen, Joan E. McLean, and James A. Rice, "[^{14}C]Pyrene Bound Residue Evaluation Using MIBK Fractionation Method for Creosote-Contaminated Soil," *Environmental Science and Technology*, **33**, 776-781 (1999).

9. Carter, M. R., D. A. Angers, and G. C. Topp, "Characterizing Equilibrium Physical Condition Near the Surface of a Fine Sandy Loam under Conservation Tillage in a Humid Climate," *Canadian Journal of Soil Science*, **164**, 101-110 (1999).

10. Uri, Noel D., *Conservation Tillage in U.S. Agriculture: Environmental, Economic, and Policy Issues,* Food Products Press, New York, 1999.

11. Uri, Noel D., "The Environmental Consequences of the Conservation Tillage Adoption Decision in Agriculture in the United States," *Water, Air, and Soil Pollution,* **103**, 9-33 (1998).

12. Uri, N. D., J. D. Atwood, and J. Sanabria, "The Environmental Benefits and Costs of Conservation Tillage," *Science of the Total Environment,* **216**, 13-32 (1998).

13. Conway, Gordon, *The Doubly Green Revolution: Food for All in the 21st Century,* Cornell University Press, Ithaca, NY, 1999.

14. Hileman, Betty, "Butterfly Concerns Halt Crop Approvals," *Chemical and Engineering News,* May 31, 1999, p. 5.

15. Adams, R. S., Jr., "Soil Variability and Cancer," *Chemical and Engineering News,* June 12, 1978, p. 84.

16. Smith, R. J., "Hawaiian Milk Contamination Creates Alarm," *Science,* **217**, 137-140 (1982).

17. "Waste Recycled into Fertilizer Contains Toxics," *Chemical and Engineering News,* March 30, 1998, p. 29.

SUPPLEMENTARY REFERENCES

Agassi, Menachem, Ed., *Soil Erosion, Conservation, and Rehabilitation,* Marcel Dekker, New York, 1996.

Alef, Kassem, and Paolo Nannipieri, *Methods in Applied Soil Microbiology and Biochemistry,* Academic Press, San Diego, CA, 1995.

Arntzen, Charles J., and Ellen M. Ritter, Eds., *Encyclopedia of Agricultural Science,* Academic Press, San Diego, CA, 1994.

Biondo, Ronald J. and Jasper S. Lee, *Introduction to Plant and Soil Science,* Interstate Publishers, Danville, IL, 1997.

Birkeland, Peter, W., Soils and Geomorphology, Oxford University Press, New York,1999.

Boardman, John and Favis-Mortlock, David, Eds., *Modelling Soil Erosion by Water,* Springer-Verlag, Berlin, 1998.

Boulding, J. Russell, *Practical Handbook of Soil, Vadose Zone, and Ground-Water Contamination: Assessment, Prevention, and Remediation,* CRC Press/Lewis Publishers, Boca Raton, FL, 1995.

Brady, Nyle C. and Ray R. Weil, *The Nature and Properties of Soils,* Prentice Hall, Upper Saddle River, NJ, 1999.

Buol, S. W., F. D. Hole, and R. J. McCracken, *Soil Genesis and Classification*, 4th ed., Iowa State University Press, Ames, IA, 1997.

Carter, Martin R., Ed., *Conservation Tillage in Temperate Agroecosystems*, CRC Press/Lewis Publishers, Boca Raton, FL,1994.

Conway, Gordon R., and Edward B. Barbier, *After the Green Revolution. Sustainable Agriculture for Development*, Earthscan Publications, London, 1990.

Domenech, Xavier, *Quimica del Suelo: El Impacto de los Contaminantes*, Miraguano Ediciones, Madrid, 1995.

Evangelou, V. P. and Bill Evangelou, *Environmental Soil and Water Chemistry: Principles and Applications*, John Wiley & Sons, New York, 1998.

Gates, Jane Potter, *Conservation Tillage, January 1991-December 1993*, National Agricultural Library, Beltsville, MD, 1994.

Grifo, Francesca, and Joshua Rosenthal, Eds., *Biodiversity and Human Health*, Island Press, Washington, D.C., 1997.

Harpstead, Milo I., Thomas J. Sauer, and William F. Bennett, *Soil Science Simplified*, Iowa State University Press, Ames, IA, 1997.

Hoddinott, Keith B., Ed., *Superfund Risk Assessment in Soil Contamination Studies*, 2nd volume, ASTM, West Conshohocken, PA, 1996.

Hudson, Norman, *Soil Conservation*, Iowa State University Press, Ames, IA, 1995.

Kohnke, Helmut and D. P. Franzmeier, *Soil Science Simplified*, Waveland Press, Prospect Heights, IL, 1995.

Lal, Rattan, *Soil Quality and Soil Erosion*, CRC Press, Boca Raton, FL, 1999

Lal, Rattan., *Methods for Assessment of Soil Degradation*, CRC Press, Boca Raton, FL,1998.

Lal, Rattan and B. A. Stewart, Eds., *Soil Degradation*, Springer-Verlag, New York, 1990.

Lefohn, Allen S., *Surface-Level Ozone Exposures and Their Effects on Vegetation*, CRC Press/Lewis Publishers, Boca Raton, FL, 1991.

Marschner, Horst, *Mineral Nutrition of Higher Plants*, 2nd ed., Academic Press, Orlando, FL, 1995.

Mansour, Mohammed, Ed., *Fate and Prediction of Environmental Chemicals in Soils, Plants, and Aquatic Systems;* CRC Press/Lewis Publishers, Boca Raton, FL, 1993.

MidWest Plan Service, Agricultural and Biosystems Engineering Dept., *Conservation Tillage Tystems and Management: Crop Residue Management With No-Till, Ridge-Till, Mulch-Till*, Iowa State University, Ames, IA, 1992.

Miller, Raymond W., Duane T. Gardiner and Joyce U. Miller, *Soils in Our Environment*, Prentice Hall, Upper Saddle River, NJ, 1998.

Montgomery, John H., Ed., *Agrochemicals Desk Reference*, 2nd ed., CRC Press/Lewis Publishers, Boca Raton, FL, 1997.

Paul, Eldor Alvin, and Francis E. Clark, Eds., *Soil Microbiology and Biochemistry*, 2nd ed., Academic Press, San Diego, CA, 1996.

Plaster, Edward J., *Soil Science and Management*, Delmar Publishers, Albany, NY, 1997.

Singer, Michael J. and Donald N. Munns, *Soils: An Introduction*, 3rd ed., Prentice Hall, Upper Saddle River, NJ, 1996.

Soil and Water Conservation Society, *Soil Quality and Soil Erosion*, CRC Press, Boca Raton, FL, 1999.

Sparks, Donald L., *Environmental Soil Chemistry*, Academic Press, Orlando, FL, 1995.

Tan, Kim H., *Soil Sampling, Preparation, and Analysis*, Marcel Dekker, New York, 1996.

Tan, Kim H., *Environmental Soil Science*, Marcel Dekker, Inc., New York, 1994.

Tan, Kim H., *Principles of Soil Chemistry*, 3rd ed., Marcel Dekker, Inc., New York, 1998.

United States. Soil Conservation Service, *No-Till in the United States: 1990*, U.S. Department of Agriculture, Soil Conservation Service, USDA-SCS-National Cartographic Center, Fort Worth, TX, 1991.

Uri, Noel D., "The Role of Public Policy in the Use of Conservation Tillage in the USA," *Science of the Total Environment*, **216**, 89-102 (1998).

White, Robert Edwin, *Principles and Practice of Soil Science: The Soil as a Natural Resource*, Blackwell Science, Oxford, U. K., 1997.

Winegardner, Duane L., *An Introduction to Soils for Environmental Professionals*, CRC Press/Lewis Publishers, Boca Raton, FL, 1996.

QUESTIONS AND PROBLEMS

1. Give two examples of reactions involving manganese and iron compounds that may occur in waterlogged soil.

2. What temperature and moisture conditions favor the buildup of organic matter in soil?

3. "Cat clays" are soils containing a high level of iron pyrite, FeS_2. Hydrogen peroxide, H_2O_2, is added to such a soil, producing sulfate as a test for cat clays. Suggest the chemical reaction involved in this test.

4. What effect upon soil acidity would result from heavy fertilization with ammonium nitrate accompanied by exposure of the soil to air and the action of aerobic bacteria?

5. How many moles of H^+ ion are consumed when 200 kilograms of $NaNO_3$ undergo denitrification in soil?

6. What is the primary mechanism by which organic material in soil exchanges cations?

7. Prolonged waterlogging of soil does **not** (a) increase NO_3^- production, (b) increase Mn^{2+} concentration, (c) increase Fe^{2+} concentration, (d) have harmful effects upon most plants, (e) increase production of NH_4^+ from NO_3^-.

8. Of the following phenomena, the one that eventually makes soil more basic is (a) removal of metal cations by roots, (b) leaching of soil with CO_2-saturated water, (c) oxidation of soil pyrite, (d) fertilization with $(NH_4)_2SO_4$, (e) fertilization with KNO_3.

9. How many metric tons of farm manure are equivalent to 100 kg of 10-5-10 fertilizer?

10. How are the chelating agents that are produced from soil microorganisms involved in soil formation?

11. What specific compound is both a particular animal waste product and a major fertilizer?

12. What happens to the nitrogen/carbon ratio as organic matter degrades in soil?

13. To prepare a rich potting soil, a greenhouse operator mixed 75% "normal" soil with 25% peat. Estimate the cation-exchange capacity in milliequivalents/100 g of the product.

14. Explain why plants grown on either excessively acidic or excessively basic soils may suffer from calcium deficiency.

15. What are two mechanisms by which anions may be held by soil mineral matter?

16. What are the three major ways in which pesticides are degraded in or on soil?

17. Lime from lead mine tailings containing 0.5% lead was applied at a rate of 10 metric tons per acre of soil and worked in to a depth of 20 cm. The soil density was 2.0 g/cm. To what extent did this add to the burden of lead in the soil? There are 640 acres per square mile and 1,609 meters per mile.

18. Match the soil or soil-solution constituent in the left column with the soil condition described on the right, below:

    (1) High Mn^{2+} content in   (a) "Cat clays" containing initially high levels of
        soil solution   pyrite, FeS_2
    (2) Excess H^+ (b) Soil in which biodegradation has not occurred
    (3) High H^+ and SO_4^{2-} to a great extent
        content   (c) Waterlogged soil
    (4) High organic content   (d) Soil, the fertility of which can be improved by
                                          adding limestone.

19. What are the processes occurring in soil that operate to reduce the harmful effects of pollutants?

20. Under what conditions do the reactions,

$$MnO_2 + 4H^+ + 2e^- \rightarrow Mn^{2+} + 2H_2O$$

and

$$Fe_2O_3 + 6H^+ + 2e^- \rightarrow 2Fe^{2+} + 3H_2O$$

occur in soil? Name two detrimental effects that can result from these reactions.

21. What are four important effects of organic matter in soil?

22. How might irrigation water treated with fertilizer containing potassium and ammonia become depleted of these nutrients in passing through humus-rich soil?

10. Match the following:

1. Subsoil	(a) Weathered parent rocks from which the soil originated
2. Gravels	(b) Largest particle size fraction (2–60 mm) according to the United Classification System
3. Topsoil	(c) B horizon of soil
4. C horizon	(d) Layer of maximum biological activity in soil that contains most of the soil organic matter

17 PRINCIPLES OF INDUSTRIAL ECOLOGY

17.1. INTRODUCTION AND HISTORY

Industrial ecology *is an approach based upon systems engineering and ecological principles that integrates the production and consumption aspects of the design, production, use, and termination (decommissioning) of products and services in a manner that minimizes environmental impact while optimizing utilization of resources, energy, and capital.*[1] The practice of industrial ecology represents an environmentally acceptable, sustainable means of providing goods and services. The meaning of industrial ecology, a relatively new concept, has been outlined in a book dealing with the topic and its implementation.[2]

Industrial ecology mimics natural ecosystems which, usually driven by solar energy and photosynthesis, consist of an assembly of mutually interacting organisms and their environment in which materials are interchanged in a largely cyclical manner. An ideal system of industrial ecology follows the flow of energy and materials through several levels, uses wastes from one part of the system as raw material for another part, and maximizes the efficiency of energy utilization. Whereas wastes, effluents, and products used to be regarded as leaving an industrial system at the point where a product or service was sold to a consumer, industrial ecology regards such materials as part of a larger system that must be considered until a complete cycle of manufacture, use, and disposal is completed.

From the discussion above and in the remainder of this book, it may be concluded that industrial ecology is all about *cyclization of materials*. This approach is summarized in a statement attributed to Kumar Patel of the University of California at Los Angles, "The goal is *cradle to reincarnation*, since if one is practicing industrial ecology correctly there is no grave." For the practice of industrial ecology to be as efficient as possible, cyclization of materials should occur at the highest possible level of material purity and stage of product development. For example, it is much more efficient in terms of materials, energy, and monetary costs to bond a new rubber tread to a large, expensive tire used on heavy earth moving equipment than it is to try to separate the rubber from the tire and remold it into a new one.

The basis of industrial ecology is provided by the phenomenon of **industrial metabolism**, which refers to the ways in which an industrial system handles materials and energy, extracting needed materials from sources such as ores, using energy to assemble materials in desired ways, and disassembling materials and components. In this respect, an industrial ecosystem operates in a manner analogous to biological organisms, which act on biomolecules to perform anabolism (synthesis) and catabolism (degradation).

Just as occurs with biological systems, industrial enterprises can be assembled into **industrial ecosystems**. Such systems consist of a number (preferably large and diverse) of industrial enterprises acting synergistically and, for the most part, with each utilizing products and potential wastes from other members of the system. Such systems are best assembled through natural selection and, to a greater or lesser extent, such selection has occurred throughout the world. However, recognition of the existence and desirability of smoothly functioning industrial ecosystems can provide the basis for laws and regulations (or the repeal thereof) that give impetus to the establishment and efficient operation of such systems.

The term **sustainable development** has been used to describe industrial development that can be sustained without environmental damage and to the benefit of all people. Although the term has become widely used, it has been pointed out[3] that some consider the term to be "an oxymoron without substance." Clearly, if humankind is to survive with a reasonable standard of living, something like "sustainable development" must evolve in which use of nonrenewable resources is minimized insofar as possible, and the capability to produce renewable resources (for example, by promoting soil conservation to maintain the capacity to grow biomass) is enhanced.[4] This will require significant behavioral changes, particularly in limiting population growth and curbing humankind's appetite for increasing consumption of goods and energy.

17.2. INDUSTRIAL ECOSYSTEMS

A group of firms that practice industrial ecology through a system of industrial metabolism that is efficient in the use of both materials and resources constitute a functional **industrial ecosystem**. Such a system may be defined as a regional cluster of industrial firms and other entities linked together in a manner that enables them to utilize by-products, materials, and energy between various concerns in a mutually advantageous manner.

Figure 17.1 shows the main attributes of a functional industrial ecosystem, which, in the simplest sense, processes materials powered by a relatively abundant source of energy. Materials enter the system from a raw materials source and are put in a usable form by a primary materials producer. From there the materials go into manufacturing goods for consumers. Associated with various sectors of the operation are waste processors that can take by-product materials, upgrade them, and feed them back into the system. An efficient, functional transportation system is required for the system to work efficiently, and good communications links must exist among the various sectors. A key material in the system is water.

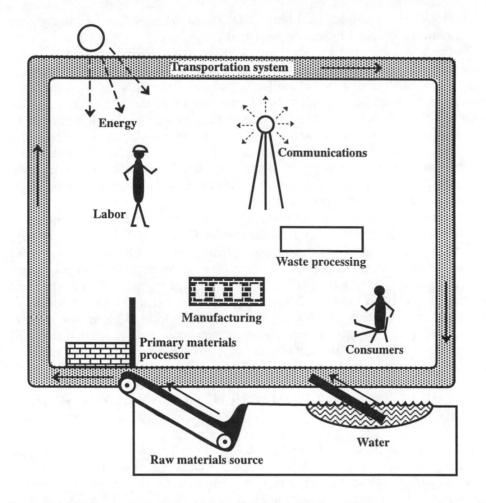

Figure 17.1. Major components required for an industrial system. When these components exist symbiotically utilizing waste materials from one concern as feedstock for another, they comprise a functioning industrial ecosystem.

A successfully operating industrial ecosystem provides several benefits. Such a system *reduces pollution*. It results in *high energy efficiency* compared to systems of firms that are not linked and it *reduces consumption of virgin materials* because it *maximizes materials recycle. Reduction of amounts of wastes* is another advantage of a functional system of industrial ecology. Finally, a key measure of the success of a system of industrial ecology is *increased market value of products* relative to material and energy consumption.

In setting up an industrial ecosystem, there are two basic, complementary approaches that may be pursued. Within an industry, emphasis may be placed upon product and service characteristics that are compatible with the practice of industrial ecology. Products may be designed for increased durability and amenability to repair and recycle. Instead of selling products, a concern may emphasize leasing so that it can facilitate recycling. The second approach emphasizes interactions between concerns so that they operate in keeping with good practice of industrial ecology. This approach facilitates materials and energy flow, exchange, and recycle between various firms in the industrial ecosystem.

An important aspect of an industrial ecosystem is the practice of a high degree of **industrial symbiosis**. Symbiotic relationships in natural biological systems occur when two often very dissimilar organisms live together in a mutually advantageous manner, each contributing to the welfare of the other. Examples are nitrogen-fixing *Rhizobium* bacteria living in root nodules on leguminous plants, or lichens consisting of fungi and bacteria living on a rock surface. Analogous symbiotic relationships in which firms utilize each others' residual materials form the basis of relationships between firms in a functional industrial ecosystem. Examples of industrial symbiosis are cited in Section 17.14 in the discussion of the Kalundborg, Denmark, industrial ecosystem.

An important consideration in the establishment and function of an industrial ecosystem is the geographical scope of the system. Often a useful way to view such a system is on the basis of a transportation network, such as a length of a navigable river or an interconnected highway system. The Houston Ship Channel, which stretches for many kilometers, is bordered by a large number of petrochemical concerns that exist to mutual advantage through the exchange of materials and energy. The purification of natural gas by concerns located along the channel yields lower molecular mass hydrocarbons such as ethane and propane that can be used by other concerns, for example, in polymers manufacture. Sulfur removed from natural gas and petroleum can be used to manufacture sulfuric acid.

17.3. THE FIVE MAJOR COMPONENTS OF AN INDUSTRIAL ECOSYSTEM

Industrial ecosystems can be broadly defined to include all types of production, processing, and consumption. These include, for example, agricultural production as well as purely industrial operations. It is useful to define five major components of an industrial ecosystem as shown in Figure 17.2. These are (1) a primary materials producer, (2) a source or sources of energy, (3) a materials processing and manufacturing sector, (4) a waste processing sector, and (5) a consumer sector. In such an idealized system, the flow of materials among the four major hubs is very high. Each constituent of the system evolves in a manner that maximizes the efficiency with which the system utilizes materials and energy.

Primary Materials and Energy Producers

It is convenient to consider the primary materials producers and the energy generators together because both materials and energy are required in order for the industrial ecosystem to operate. The primary materials producer or producers may consist of one or several enterprises devoted to providing the basic materials that sustain the industrial ecosystem. Most generally, in any realistic industrial ecosystem a significant fraction of the material processed by the system consists of virgin materials. In a number of cases, and increasingly so as pressures build to recycle materials, significant amounts of the materials come from recycling sources.

The processes that virgin materials entering the system are subjected to vary with the kind of material, but can generally be divided into several major steps. Typically, the first step is extraction, designed to remove the desired substance as completely as possible from the other substances with which it occurs. This stage of materials processing can produce large quantities of waste material requiring disposal, as is the case with some metal ores in which the metal makes up a small percentage of the ore that is mined. In other cases, such as corn grain providing the basis of a corn products industry, the "waste,"—in this specific example the cornstalks associated with the grain—can be left in place (cornstalks returned to soil serve to add humus and improve soil quality). A concentration step may follow extraction to put the desired material into a purer form. After concentration, the material may be put through additional refining steps that may involve separations. Following these steps, the material is usually subjected to additional processing and preparation leading to the finished materials. Throughout the various steps of extraction, concentration, separation, refining, processing, preparation, and finishing, various physical and chemical operations are used, and wastes requiring disposal may be produced. Recycled materials may be introduced at various parts of the process, although they are usually introduced into the system following the concentration step.

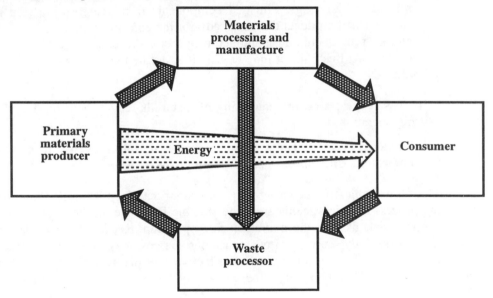

Figure 17.2. The major constituents or "hubs" of an industrial ecosystem.

The extraction and preparation of energy sources can follow many of the steps outlined above for the extraction and preparation of materials. For example, the processes involved in extracting uranium from ore, enriching it in the fissionable uranium-235 isotope, and casting it into fuel rods for nuclear fission power production include all of those outlined above for materials. On the other hand, some rich sources of coal are essentially scooped from a coal seam and sent to a power plant for power generation with only minimal processing, such as sorting and grinding.

Recycled materials added to the system at the primary materials and energy production phase may be from both pre- and postconsumer sources. As examples, recycled paper may be macerated and added at the pulping stage of paper manufacture. Recycled aluminum may be added at the molten metal stage of aluminum metal production.

Materials Processing and Manufacturing Sector

Finished materials from primary materials producers are fabricated to make products in the materials processing and manufacturing sector, which is often a very complex system. For example, the manufacture of an automobile requires steel for the frame, plastic for various components, rubber in tires, lead in the battery, and copper in the wiring, along with a large number of other materials. Typically, the first step in materials manufacturing and processing is a forming operation. For example, sheet steel suitable for making automobile frames may be cut, pressed, and welded into the configuration needed to make a frame. At this step some wastes may be produced that require disposal. An example of such wastes consists of carbon fiber/epoxy composites left over from forming parts such as jet aircraft engine housings. Finished components from the forming step are fabricated into finished products that are ready for the consumer market.

The materials processing and manufacturing sector presents several opportunities for recycling. At this point it may be useful to define two different streams of recycled materials:

- **Process recycle streams** consisting of materials recycled in the manufacturing operation itself

- **External recycle streams** consisting of materials recycled from other manufacturers or from postconsumer products

Materials suitable for recycling can vary significantly. Generally, materials from the process recycle streams are quite suitable for recycling because they are the same materials used in the manufacturing operation. Recycled materials from the outside, especially those from postconsumer sources, may be quite variable in their characteristics because of the lack of effective controls over recycled postconsumer materials. Therefore, manufacturers may be reluctant to use such substances.

The Consumer Sector

In the consumer sector, products are sold or leased to the consumers who use them. The duration and intensity of use vary widely with the product; paper towels are used only once, whereas an automobile may be used thousands of times over many years. In all cases, however, the end of the useful lifetime of the product is reached and it is either (1) discarded or (2) recycled. The success of a total industrial ecology system may be measured largely by the degree to which recycling predominates over disposal.

Waste Processing Sector

Recycling has become so widely practiced that an entirely separate waste processing sector of an economic system may now be defined. This sector consists of enterprises that deal specifically with the collection, separation, and processing of recyclable materials and their distribution to end users. Such operations may be entirely private or they may involve cooperative efforts with governmental sectors. They are often driven by laws and regulations that provide penalties against simply discarding used items and materials, as well as positive economic and regulatory incentives for their recycle.

17.4. INDUSTRIAL METABOLISM

Industrial metabolism in its entirety follows the flows of materials and energy from their initial sources through an industrial system, to the consumer, and to their ultimate disposal.[5,6] In biological systems, metabolism may be studied at any level ranging from the molecular processes that occur in individual cells through the multiple processes and metabolic cycles that occur in individual organs, and to the overall process of metabolism that takes place in the whole organism. Similarly, industrial metabolism can be examined as individual unit operations within an industrial operation, at the factory level, at the industry level, and globally. For an industrial ecology approach it is often most useful to view industrial metabolic processes at the regional level, large enough to have a number of industries with a variety of potential waste products that might be used by other industries, but small enough to permit transport and exchange of materials among various industries. To minimize pollution it can be useful to consider units consisting of environmental domains, such as atmospheric basins or watersheds.

Unlike the living metabolic processes that occur in natural systems where true waste products are very rare, industrial metabolism as it is now practiced has a vexing tendency to dilute, degrade, and disperse materials to an extent that they are no longer useful but are still harmful to the environment. Indeed, waste has been defined as *dissipative use of natural resources*.[7] In addition to simple loss from dilution and dispersion in the environment, materials may be lost by being tied up in low energy forms or by being put into a chemical form from which they are very difficult to retrieve.

An example of dissipation of material resulting in environmental pollution, now a very much diminished problem, was the widespread use of lead in tetraethyl lead antiknock additive in gasoline. The net result of this use was to disperse lead throughout the environment with auto exhaust gas, with no hope of recovery.

Industrial Metabolism and Biological Analogies

The strong analogy between natural ecosystems and efficient industrial systems was first clearly stated in 1989 by Frosch and Gallopoulos.[8] A natural ecosystem, which is usually driven by solar energy and photosynthesis, consists of an assembly of mutually interacting organisms and their environment in which materials are interchanged in a largely cyclical manner.[9] It is possible to visualize an analogous industrial ecosystem in which materials are cycled, driven by an energy source.

Biological metabolism is defined as biochemical processes that involve the alteration of biomolecules. Metabolic processes can be divided into the two major categories of anabolism (synthesis) and catabolism (degradation). It is useful to view the metabolic processes of an ecosystem as a whole, rather than just observing each individual organism. An industrial ecosystem likewise synthesizes substances, thus performing anabolism, and it degrades substances, thereby performing in a manner analogous to biological catabolism. Typically, a large amount of a material, such as an ore or petroleum source, is metabolized to yield a relatively small quantity of a finished product. The objective of a properly designed and operated industrial ecosystem is to perform industrial metabolism in the most efficient manner possible so that the least possible raw material is used, the maximum amounts of materials are recycled through the system, and the most efficient possible use is made of the energy that sustains the industrial ecosystem.

An ideal biological ecosystem involves many organisms living in harmony with their environment without any net consumption of resources or production of waste products. The only input required for such an ecosystem is solar energy. This energy is captured and used by photosynthetic primary producers to convert carbon dioxide, water, and other inorganic materials into biomass. Herbivores ingest this biomass and use it for their energy and to synthesize their own biomass. Carnivores, of which there may be several levels, consume herbivores, and a food chain exists that may consist of several levels of organisms. Parasites exist on or in other organisms. Saprophytes and bacteria and fungi responsible for decay utilize and degrade biomass, eventually converting it back to simple inorganic constituents through the process of mineralization. Symbiotic and synergistic relationships abound in a natural ecosystem. Thus, an ideal ecosystem exists indefinitely in a steady-state condition without causing any net degradation of its environment.

A natural ecosystem can be visualized as having compartments in which various *stocks* of materials are kept, connected by *flows* of materials. Examples of such compartments include soil, which is a repository of plant nutrients; a body of water, such as a lake; and the atmosphere, which is a repository of carbon dioxide required for photosynthesis. In an undisturbed natural ecosystem, the quantities of the materials in each of these compartments remains relatively stable because such systems are inherently recycling. In contrast, the quantities of materials in the compartments of an industrial system are not constant. Reservoirs of raw materials, such as essential minerals, are constantly diminishing, although with new discoveries of mineral resources they may *appear* to increase. Furthermore, reservoirs of wastes in an industrial system continually increase as materials traverse an essentially one-way path through the system. Sustainable industrial systems maximize recycling so that quantities of materials in the reservoirs remain constant as much as possible.

Systems of biological metabolism are self-regulating. At the level of the individual organism, regulation is accomplished internally by biological regulatory mechanisms, such as those that employ hormones. At the ecosystem level, regulation occurs through competition among organisms for available resources. In a manner analogous to natural ecosystems, industrial systems can be designed in principle to operate in a similar steady-state manner, ideally neither consuming nonrenewable resources nor producing useless waste products. Such systems can be self-regulating with the economic system operating under the laws of supply and demand as the regulatory mechanism.

A comparison of the metabolisms of natural ecosystems with that of industrial systems as they are commonly encountered shows a marked contrast. These contrasts are highlighted in Table 17.1.

Table 17.1. Metabolic Characteristics of Natural Ecosystems and Industrial Systems

Characteristic	Natural ecosystems	Current industrial systems
Basic unit	Organism	Firm
Material pathways	Closed loops	Largely one-way
Recycling	Essentially complete	Often very low
Material fate	Tend to concentrate, such as atmospheric CO_2 converted to biomass by photosynthesis	Dissipative to produce materials too dilute to use, but concentrated enough to pollute
Reproduction	A major function of organisms is reproduction	Production of goods and services is the prime objective, not reproduction *per se*

Attractive as the idea may sound; in a modern society and especially in a "global economy," a complete industrial ecosystem carrying out industrial metabolism in an idealized fashion is not practical or even desirable. Essentially all modern communities produce at least one product that is exported to the outside, and must bring in materials and energy sources from elsewhere. A more realistic model of an industrial ecosystem is one in which raw materials are imported from the outside and at least one major product is exported from the system, but in which a number of enterprises coexist synergistically, utilizing each others' products and services to mutual advantage. In such a system, typically, raw materials flow into a primary raw material processor, which converts them to a processed material. The processed material then goes to one or more fabricators that make a product for distribution and sale. Associated with the enterprise as a whole are suppliers of a variety of materials, items, or services, and processors of secondary materials or wastes. To meet the criteria of an industrial ecosystem, as many of the by-products and wastes as possible must be utilized and processed within the system.

Although industrial systems are self-regulating, they are not necessarily so in a manner conducive to sustainability; indeed, the opposite is frequently the case. Left to their own devices and operating under the principles of traditional economics, industrial systems tend toward a state of equilibrium or maximum entropy in which essential materials have been exploited, run through the system, and dissipated to the environment in dilute, useless forms. A central question is, therefore, the time scale on which this irreversible dissipation will occur. If it is a few decades, modern civilization is in real trouble; if it is on a scale of thousands of years, there is ample time to take corrective action to maintain sustainability. A challenge to modern industrialized societies is to modify industrial systems to maximize the time spans under which sustainability may be achieved.

17.5. LEVELS OF MATERIALS UTILIZATION

There are two extremes in levels of materials utilization in industrial systems. At the most inefficient level, as shown in Figure 17.3, raw materials are viewed as being unlimited and no consideration is given to limiting wastes. Such an approach was typical of industrial development in the U.S. in the 1800s and early 1900s when the prevailing view was that there were no limits to ores, fossil energy resources, and other kinds of raw materials; furthermore, it was generally held that the continent had an unlimited capacity to absorb industrial wastes.

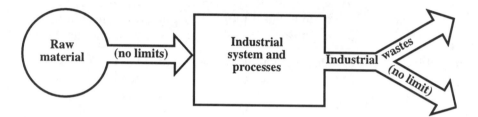

Figure 17.3. An industrial system without limits on either raw materials consumed or wastes produced.

A second kind of industrial system in which both raw materials and wastes are limited to greater or lesser extents is illustrated in Figure 17.4. Such a system has a relatively large circulation of materials within the industrial system as a whole, compared with reduced quantities of material going into the system and relatively lower production of wastes. Such systems are typical of those in industrialized nations and modern economic systems in which shortages of raw materials and limits to the places to put wastes are beginning to be felt. Even with such constraints,

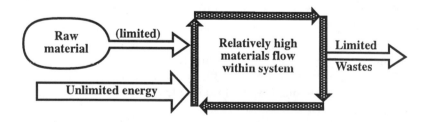

Figure 17.4. Illustration of an industrial system in which both the utilization of raw materials and the production of wastes are limited to a certain degree.

large quantities of materials are extracted, processed, and used, then either disposed in the environment in concentrated form (hazardous wastes) or dispersed. In recent years regulations and other constraints have markedly decreased point source pollution from industrial activity. However, because of the sheer volume of materials processed through industrial societies, dissipative pollution continues to be a problem.

An industrial ecosystem with no materials input and no wastes is illustrated in Figure 17.5. The energy requirements of such a system are rather high and the material flows within the system itself are quite high. Such a system is an idealized one that can never be realized in practice, but it serves as a useful goal around which more practical and achievable systems can be based.

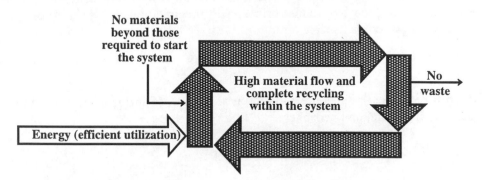

Figure 17.5. Idealized industrial ecosystem in which no materials are required for input beyond those needed to start the system. Energy requirements are relatively high, and the material flow within the system is high and continuous.

17.6. LINKS TO OTHER ENVIRONMENTAL SPHERES

Having addressed industrial ecology largely from the anthrospheric viewpoint, it is now appropriate to consider how the anthrosphere and the practice of indusrial ecology influence the other four spheres of the environment—the atmosphere, the hydrosphere, the geosphere, and the biosphere. Influences of industrial activities, broadly defined to include energy and agricultural production as well as manufacture of goods and provision of essential services, can range from minor to major effects that may pose significant threats to earth's ability to support life. Such effects can range from highly localized ones occurring for only a brief period of time to global effects that have the potential to last for centuries. An example of the former would be an isolated incidence of water pollution by oxygen-consuming organic matter in a reservoir; only the reservoir is affected and only for the relatively short period of time required to degrade the wastes and replenish the oxygen supply. The prime example of a long term global effect is the emission of greenhouse gases, which has the potential to change earth's entire climate for thousands of years.

The major goal of industrial ecology, therefore, must be to minimize or eliminate detrimental effects of anthrospheric activities on other spheres of the environment. Beyond environmental preservation, the practice of industrial ecology should also improve and enhance environmental conditions. Listed below are the major anthrospheric activities along with their potential effects on other environmental spheres.

Fossil fuel combustion

Atmosphere: The greatest potential effect is greenhouse warming. Emission of partially combusted hydrocarbons and nitrogen oxides can cause formation of photochemical oxidants (photochemical smog). Acid precipitation may be caused by emissions of sulfur oxides from fuel combustion. General deterioration of atmospheric quality may occur through reduced visibility.

Hydrosphere: The potential exists for water pollution from acid mine water, petroleum production by-product brines, acid precipitation, and heating of water used to cool power plants.

Geosphere: The greatest potential effects are disturbance of land from coal mining.

Biosphere: Most effects are indirect as the result of influences on the atmosphere, hydrosphere, and geosphere.

Industrial manufacturing and processing

Atmosphere: Greatest potential effects are due to emissions of gases, vapors, and particles. These include greenhouse gases, acid gases, particles, precursors to photochemical smog formation, and species with the potential to deplete stratospheric ozone.

Hydrosphere: Industrial activities may contaminate water with a variety of pollutants. Consumptive uses of water may put pressure on limited water supplies, especially in arid regions. Water used for cooling may be thermally polluted.

Geosphere: The greatest effect results from the extractive industries through which minerals are recovered. The geosphere may be contaminated by solid and hazardous wastes, and available landfill space may become depleted.

Biosphere: The greatest direct effect is from the distribution of toxic substances as the result of industrial activities. There may also be significant indirect effects resulting from deterioration of the atmosphere, hydrosphere, and geosphere.

Crop production

Atmosphere: A major potential effect is emission of greenhouse gases as the result of deforestation and "slash and burn" agriculture to grow more crops. Significant amounts of greenhouse gas methane are emitted into the atmosphere as the result of methane-generating bacteria growing in rice paddies.

Hydrosphere: Large quantities of water are used for irrigation. Some of the water is lost by transpiration from plants, and some by infiltration to groundwater. Water returned to the hydrosphere from irrigation may have an excessively high salinity. Surface water and groundwater may become contaminated by solids, fertilizers, and herbicides from crop production.

Geosphere: Large areas of the geosphere may be disturbed by cultivation to produce crops. Topsoil can be lost from water and wind erosion. Proper agricultural practices, such as contour farming and low-tillage agriculture, minimize these effects and may even enhance soil quality.

Biosphere: Organisms are profoundly affected by agricultural practices designed to produce crops. Entire ecosystems are destroyed and replaced by other "anthrospheric" ecosystems. The greatest effect on the biosphere is loss of species diversity from the destruction of natural ecosystems, and from the cultivation of only limited strains of crops.

Livestock production (domestic animals)

Atmosphere: Ruminant animals are significant producers of greenhouse gas methane as the result of methane-producing bacteria in their digestive systems.

Hydrosphere: Livestock production requires large quantities of water. Large amounts of oxygen-consuming wastes that may contaminate surface water are produced by livestock. Nitrogen wastes from the manure and urine of animals in feedlots may cause nitrate contamination of groundwater.

Geosphere: The production of a unit mass of food from livestock sources requires much more crop production than is required for grains consumed directly by humans. A major impetus behind destruction of rain forests has been to grow forage and other foods for livestock. Rangeland has deteriorated because of over-grazing.

Biosphere: A major effect is loss of species diversity. This occurs even within domestic strains of livestock where modern breeding practices have resulted in the loss of entire breeds of livestock. The ultimate loss of domestic diversity occurs when animals are cloned.

The most environmentally damaging effects of human activities are those that are cumulative. As noted previously, the most significant of these at present is likely the accumulation of greenhouse gases that have the potential to cause global warming. Some environmental problems, such as those resulting from the emission of photochemical smog-forming pollutants into the atmosphere are potentially reversible. However, by the time that global warming has been demonstrated to be a genuine problem, if such turns out to be the case, the damage will have been done, and little if anything will be able to reverse it.

17.7. CONSIDERATION OF ENVIRONMENTAL IMPACTS IN INDUSTRIAL ECOLOGY

By its nature, industrial production has an impact upon the environment. Whenever raw materials are extracted, processed, used, and eventually discarded, some environmental impacts will occur. In designing an industrial ecological system, several major kinds of environmental impacts must be considered in order to minimize them and keep them within acceptable limits. These impacts and the measures taken to alleviate them are discussed below.

For most industrial processes, the first environmental impact is that of extracting raw materials. This can be a straightforward case of mineral extraction, or it can be less direct, such as utilization of biomass grown on forest or crop land. A basic decision, therefore, is the choice of the kind of material to be used. Wherever possible, materials should be chosen that are not likely to be in short supply in the foreseeable future. As an example, the silica used to make the lines employed for fiber-optics communication is in unlimited supply and a much better choice for communication lines than copper wire made from limited supplies of copper ore.

Industrial ecology systems should be designed to reduce or even totally eliminate air pollutant emissions. Among the most notable recent progress in that area has been the marked reduction and even total elimination of solvent vapor emissions (volatile organic carbon, VOC), particularly those from organochlorine solvents. Some progress in this area has been made with more effective trapping of solvent vapors. In other cases, the use of the solvents has been totally eliminated. This is the case for chlorofluorocarbons (CFCs), which are no longer used in plastic foam blowing and parts cleaning because of their potential to affect stratospheric ozone. Other air pollutant emissions that should be eliminated are hydrocarbon vapors, including those of methane, CH_4, and oxides of nitrogen or sulfur.

Discharges of water pollutants should be entirely eliminated wherever possible. For many decades, efficient and effective water treatment systems have been employed that minimize water pollution. However, these are "end of pipe" measures, and it is much more desirable to design industrial systems such that potential water pollutants are not even generated.

Industrial ecology systems should be designed to prevent production of liquid wastes that may have to be sent to a waste processor. Such wastes fall into the two broad categories of water-based wastes and those contained in organic liquids. Under current conditions the largest single constituent of so-called "hazardous wastes" is water. Elimination of water from the waste stream automatically prevents pollution and reduces amounts of wastes requiring disposal. The solvents in organic wastes largely represent potentially recyclable or combustible constituents. A properly designed industrial ecosystem does not allow such wastes to be generated or to leave the factory site.

In addition to liquid wastes, many solid wastes must be considered in an industrial ecosystem. The most troublesome are toxic solids that must be placed in a secure hazardous waste landfill. The problem has become especially acute in some industrialized nations in which the availability of landfill space is severely limited. In a general sense, solid wastes are simply resources that have not been properly utilized. Closer cooperation among suppliers, manufacturers, consumers, regulators, and recyclers can minimize quantities and hazards of solid wastes.

Whenever energy is expended, there is a degree of environmental damage. Therefore, energy efficiency must have a high priority in a properly designed industrial ecosystem. Significant progress has been made in this area in recent decades, as much because of the high costs of energy as for environmental improvement. More efficient devices, such as electric motors, and approaches, such as congeneration of electricity and heat, that make the best possible use of energy resources are highly favored. An important side benefit of more efficient energy utilization is the lowered emissions of air pollutants, including greenhouse gases.

17.8. THREE KEY ATTRIBUTES: ENERGY, MATERIALS, DIVERSITY

By analogy with biological ecosystems, a successful industrial ecosystem should have (1) renewable energy, (2) complete recyling of materials, and (3) species diversity for resistance to external shocks. These three key characteristics of industrial ecosystems are addressed here.

Unlimited Energy

Energy is obviously a key ingredient of an industrial ecosystem. Unlike materials, the flow of energy in even a well-balanced closed industrial ecosystem is essentially one-way in that energy enters in a concentrated, highly usable form, such as chemical energy in natural gas, and leaves in a dilute, disperse form as waste heat. An exception is the energy that is stored in materials. This can be in the form of energy that can be obtained from materials, such as by burning rubber tires, or it can be in the form of what might be called "energy credit," which means that by using a material in its refined form, energy is not consumed in making the material from its raw material precursors. A prime example of this is the "energy credit" in metals, such as that in aluminum metal, which can be refined into new aluminum objects requiring only a fraction of the energy consumed to refine the metal from aluminum ore. On the other hand, recycling and reclaiming some materials can require a lot of energy, and the energy consumption of a good closed industrial ecosystem can be rather high.

Given the needed elements, any material can be made if a sufficient amount of energy is available. The key energy requirement is a source that is abundant and of high quality, that can be used efficiently, and that does not produce unacceptable by-products.

Although energy is ultimately dissipated from an industrial ecosystem, it may go through two or more levels of use before it is wasted. An example of this would be energy from natural gas burned in a turbine linked to a generator, the exhaust gases used to raise steam in a power plant to run a steam turbine, and the relatively cool steam from the turbine used to heat buildings.

Natural ecosystems run on unlimited, renewable energy from the sun or, in some specialized cases, from geochemical sources. Successful industrial ecosystems must also have unlimited sources of energy in order to be sustained for an indefinite period of time. The obvious choice for such an energy source would seem to be solar energy. However, solar sources present formidable problems, not the least of which is that they work poorly during those times of the day and seasons of the year when the sun does not shine. Even under optimum conditions, solar energy has a low power density necessitating collection and distribution systems of an unprecedented scale if they are going to displace present fossil energy sources. Other renewable sources, such as wind, tidal, geothermal, biomass, and hydropower present similar challenges. It is likely, therefore, that fossil energy sources will provide a large share of the energy for industrial ecosystems in the foreseeable future. This assumes that a way can be found to manage greenhouse gases. At the present time it appears that injection of carbon dioxide from combustion into deep ocean regions is the only viable alternative for sequestering carbon dioxide, and this approach remains an unproven technology on a large scale. (One potential problem is that the slight increase in ocean water pH that would result, though only of the order of a tenth of a pH unit, could be detrimental to many of the organisms that live in the ocean.)

Nuclear fusion power remains a tantalizing possibility for unlimited energy, but so far practical nuclear fusion reactors for power generation have proven an elusive target. Unattractive as it is to many, the only certain, environmentally acceptable energy source that can without question fill the energy needs of modern industrial ecology systems is nuclear fission energy. With breeder reactors that can generate additional fissionable material from essentially unlimited supplies of uranium-238, nuclear fission can meet humankind's energy needs for the foreseeable future. Of course, there are problems with nuclear fission—more political and regulatory than technical. The solution to these problems remains a central challenge for humans in the modern era.

Industrial Ecology and Material Resources

A system of industrial ecology is successful if it reduces demand for materials from virgin sources. Strategies for reduced material use may be driven by technology, by economics, or by regulation. The four major ways in which material consumption may be reduced are (1) using less of a material for a specific application, an approach called **dematerialization**; (2) **substitution** of a relatively more abundant and safe material for one that is scarce and/or toxic; (3) **recycling**, broadly defined; and (4) extraction of useful materials from wastes, sometimes called **waste mining**. These four facets of efficient materials utilization are outlined in this section.

Dematerialization

There are numerous recent examples of reduced uses of materials for specific applications. One example of dematerialization is the transmission of greater electrical power loads with less copper wire by using higher voltages on long distance transmission lines. Copper is also used much more efficiently for communications transmission than it was in the early days of telegraphy and telephone communication. Amounts of silver used per roll of photographic film have decreased significantly in recent years. The layer of tin plated onto the surface of a "tin can" used for food preservation and storage is much lower now. In response to the need for greater fuel economy, the quantities of materials used in automobiles have decreased significantly over the last two decades, a trend reversed, unfortunately, by the more recent increased demand for large "sport utility vehicles." Automobile storage batteries now use much less lead for the same amount of capacity than they did in former years. The switch from 6-volt to 12-volt auto batteries in the 1950s enabled use of lighter wires, such as those from the battery to the electrical starter. Somewhat later, the change to steel-belted radial tires enabled use of lighter tires and resulted in greatly increased tire lifetimes so that much less rubber was used for tires.

One of the most commonly cited examples of dematerialization is that resulting from the change from vacuum tubes to solid state circuit devices. Actually, this conversion should be regarded as material substitution as transistors replaced vacuum tubes, followed by spectacular mass reductions as solid state circuit technology advanced.

Dematerialization can be expected to continue as technical advances, some rapid and spectacular, others slow and incremental, continue to be made. Some industries lead the way out of necessity. Aircraft weight has always played a crucial role in determining performance, so the aircraft manufacturing sector is one of the leaders in dematerialization.

Substitution of Materials

Substitution and dematerialization are complementary approaches to reducing materials use. The substitution of solid state components for electronic vacuum tubes and the accompanying reduction in material quantities has already been cited. The substitution of polyvinylchloride (PVC) siding in place of wood on houses has resulted in dematerialization over the long term because the plastic siding does not require paint.

Technology and economics combined have been leading factors in materials substitution. For example, the technology to make PVC pipe for water and drain lines has enabled its use in place of more expensive cast iron, copper, and even lead pipe (in the last case, toxicity from lead contamination of water is also a factor to be considered).

A very significant substitution that has taken place over recent decades is that of aluminum for copper and other substances. Copper, although not a strategically short metal resource, nevertheless is not one of the more abundant metals in relation to the demand for it. Considering its abundance in the geosphere and in sources such as coal ash, aluminum is a very abundant metal. Now aluminum is used in place of copper in many high voltage electrical transmission applications. Aluminum is also used in place of brass, a copper-containing alloy, in a number of applications. Aluminum roofing substitutes for copper in building construction. Aluminum cans are used for beverages in place of tin-plated steel cans.

There have been a number of subsitutions of chemicals in recent years, many of them driven by environmental concerns and regulations resulting from those concerns. One of the greater of these has been the substitution of hydrochlorofluorocarbons (HCFCs) and hydrofluorocarbons (HFCs) for chlorofluorocarbons (Freons or CFCs) driven by concerns over stratospheric ozone depletion. Substitutions of nonhalogenated solvents, supercritical fluid carbon dioxide, and even water with appropriate additives for chlorinated hydrocarbon solvents will continue as environmental concerns over these solvents increase.

Substitutions for metal-containing chemicals promise to reduce costs and toxicities. One such substitution that has greatly reduced the possibilities for lead poisoning is the use of titanium-based pigments in place of lead for white paints. In addition to lead, cadmium, chromium, and zinc are also used in pigments, and substitution of organic pigments for these metals in paints has reduced toxicity risks. Copper, chromium, and arsenic are used in treated wood (CCA lumber).

Because of the toxicity of arsenic, particularly, it would be advisable to develop substitutes for these metals in wood. It should be pointed out, however, that the production of practically indestructable CCA lumber has resulted in much less use of wood, and has saved the materials and energy required to replace wood that has rotted or been damaged by termites.

Recycling

For a true and complete industrial ecosystem, close to 100% recycling of materials must be realized. In principle, given a finite supply of all the required elements and abundant energy, essentially complete recycling can be achieved. A central goal of industrial ecology is to develop efficient technologies for recycling that reduce the need for virgin materials to the lowest possible levels. Another goal must be to implement process changes that eliminate dissipative uses of toxic substances, such as heavy metals, that are not biodegradable and that pose a threat to the environment when they are discarded.

For consideration of recycling, matter can be put into four separate categories. The first of these consists of elements that occur abundantly and naturally in essentially unlimited quantities in consumable products. Food is the ultimate consumable product. Soap is consumed for cleaning purposes, discarded down the drain, precipitated as its insoluble calcium salt, then finally biodegraded. Materials in this category of recyclables are discharged into the environment and recycled through natural processes or for very low value applications, such as sewage sludge used as fertilizer on soil.

A second category of recyclable materials consists of elements that are not in short supply, but are in a form that is especially amenable to recycling. Wood is one such commodity. At least a portion of wood taken from buildings that are being razed could and should be recycled. The best example of a kind of commodity in this class is paper. Paper fibers can be recycled up to five times, and the nature of paper is such that it is readily recycled. More than 1/3 of world paper production is currently from recycled sources, and that fraction should exceed 50% within the next several decades. The major impetus for paper recycling is not a shortage of wood to make virgin paper, but rather a shortage of landfill space for waste paper.

A third category of recyclables consists of those elements, mostly metals, for which world resources are low. Chromium and the platinum group of precious metals are examples of such elements. Given maximum incentives to recycle, especially through the mechanism of higher prices, it is likely that virgin sources of these metals can make up any shortfall not met by recycling in the foreseeable future.

A fourth category of materials to consider for recycling consists of parts and apparatus, such as auto parts discussed previously. In many cases such parts can be refurbished and reused. Even when this is not the case, substantial deposits at the time of purchase can provide incentives for recycling. In order for components to be recycled efficiently, they must be designed with reuse in mind in aspects such as facile disassembly. Such an approach has been called "design for environment," DFE, and is discussed in more detail in Section 17.10.

Combustion to produce energy can be a form of recycling. For some kinds of materials, combustion in a power plant is the most cost-effective and environmentally safe way of dealing with materials. This is true, for example, of municipal refuse that contains a significant energy value because of combustible materials in it as well as a variety of items that potentially could be recycled for the materials in them. However, once such items become mixed in municipal refuse and contaminated with impurities, the best means of dealing with them is simply combustion.

It should be noted that recycling comes with its own set of environmental concerns. One of the greatest of these is contamination of recycled materials with toxic substances. In some cases, motor oil, especially that collected from the individual consumer sector, can be contaminated with organohalide solvents and other troublesome impurities. Food containers pick up an array of contaminants and, as a consequence, recycled plastic is not generally regarded as a good material for food applications. Substances may become so mixed with use that recycling is not practical. This occurs particularly with synthetic fibers, but it may be a problem with plastics, glass, and other kinds of recyclable materials.

Extraction of Useful Materials from Wastes

Sometimes called waste mining, the extraction of useful materials from wastes has some significant, largely unrealized potential for the reduction in use of virgin materials. Waste mining can often take advantage of the costs that must necessarily be incurred in treating wastes, such as flue gases. Sulfur is one of the best examples of a material that is now commonly recovered from wastes. Sulfur is a constituent of all coal and can be recovered from flue gas produced by coal combustion. It would not be cost-effective to use flue gas simply as a source of sulfur. However, since removal of sulfur dioxide from flue gas is now required by regulation, the incremental cost of recovering sulfur as a commodity, rather than simply discarding it, can make sulfur recovery economically feasible.

There are several advantages to recovering a useful resource from wastes. One of these is the reduced need to extract the resource from a primary source. Therefore, every kilogram of sulfur recovered from flue gas means one less kg of sulfur that must be extracted from sulfur ore sources. By using waste sources, the primary source is preserved for future use. Another advantage is that extraction of a resource from a waste stream can reduce the toxicity or potential environmental harm from the waste stream. As noted previously, arsenic is a by-product of the refining of some other metals. The removal of arsenic from the residues of refining such metals significantly reduces the toxicities and potential environmental harm by the wastes. Coal ash, the residue remaining after the combustion of coal for power generation, has a significant potential as a source of iron (ferrosilicon), silicon, and aluminum, and perhaps several other elements as well. An advantage of using coal ash in such applications is its physical form. For most power applications, the feed coal is finely ground, so that the ash is in the form of a powder. This means that coal ash is already in the physical form most amenable to processing for by-products recovery. For a particular coal feedstock, coal ash is homogeneous, which offers some definite advantages in processing and resource recovery. A third advantage of coal ash is that it is anhydrous so, no additional energy needs to be expended in removing water from an ore.

Diversity and Robust Character of Industrial Ecosystems

Successful natural ecosystems are highly diverse, as a consequence of which they are also very robust. Robustness means that if one part of the system is perturbed, there are others that can take its place. Consider what happens if the numbers of a top predator at the top of a food chain in a natural ecosystem are severely reduced because of disease. If the system is well balanced, another top predator is available to take its place.

The energy sector of industrial ecosystems often suffers from a lack of robustness. Examples of energy vulnerability have become obvious with several "energy crises" during recent history. Another requirement of a healthy industrial ecology system that is vulnerable in some societies is water. In some regions of the world, both the quantity and quality of water are severely limited. A lack of self-sufficiency in food is a third example of vulnerability. Vulnerabililty in food and water are both strongly dependent upon climate, which in turn is tied to environmental concerns as a whole.

17.9. LIFE CYCLES: EXPANDING AND CLOSING THE MATERIALS LOOP

In a general sense, the traditional view of product utilization is the one-way process of extraction → production → consumption → disposal shown in the upper portion of Figure 17.6. Materials that are extracted and refined are incorporated into the production of useful items, usually by processes that produce large quantities of waste by-products. After the products are worn out, they are discarded. This essentially one-way path results in a relatively large exploitation of resources, such as metal ores, and a constant accumulation of wastes. As shown at the bottom of Figure 17.6, however, the one-way path outlined above can become a cycle in which manufactured goods are used, then recycled at the end of their life spans. As

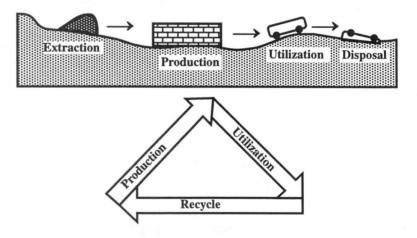

Figure 17.6. The one-way path of conventional utilization of resources to make manufactured goods followed by disposal of the materials and goods at the end consumes large quantities of materials and makes large quantities of wastes (top). In an ideal industrial ecosystem (bottom), the loop is closed and spent products are recycled to the production phase.

one aspect of such a cyclic system, it is often useful for manufacturers to assume responsibility for their products, to maintain "stewardship." Ideally, in such a system a product and/or the material in it would have a never-ending life cycle; when its useful lifetime is exhausted, it is either refurbished or converted into another product.

In considering life cycles, it is important to note that commerce can be divided into the two broad categories of **products** and **services**. Whereas most commercial activity used to be concentrated on providing large quantities of goods and products, demand has been largely satisfied for some segments of the population, and the wealthier economies are moving more to a service-based system. Much of the commerce required for a modern society consists of a mixture of services and goods. The trend toward a service economy offers two major advantages with respect to waste minimization. Obviously, a pure service involves little material. Secondly, a service provider is in a much better position to control materials to ensure that they are recycled and to control wastes, ensuring their proper disposal. A commonly cited example is that of photocopy machines. They provide a service, and a heavily used copy machine requires frequent maintenance and cleaning. The parts of such a machine and the consumables, such as toner cartridges, consist of materials that eventually will have to be discarded or recycled. In this case it is often reasonable for the provider to lease the machine to users, taking responsibility for its maintenance and ultimate fate. The idea could even be expanded to include recycling of the paper processed by the copier, with the provider taking responsibility for recyclable paper processed by the machine.

It is usually difficult to recycle products or materials within a single, relatively narrow industry. In most cases, to be practical, recycling must be practiced on a larger scale than simply that of a single industry or product. For example, recycling plastics used in soft drink bottles to make new soft drink bottles is not allowed because of the possibilities for contamination. However, the plastics can be used as raw material for auto parts. Usually, different companies are involved in making auto parts and soft drink bottles.

Product Stewardship

The degree to which products are recycled is strongly affected by the custody of the products. For example, batteries containing cadmium or mercury pose significant pollution problems when they are purchased by the public; used in a variety of devices, such as calculators and cameras; then discarded through a number of channels, including municipal refuse. However, when such batteries are used within a single organization, it is possible to ensure that almost all of them are returned for recycling. In cases such as this, systems of stewardship can be devised in which marketers and manufacturers exercise a high degree of control of the product. This can be done through several means. One is for the manufacturer to retain ownership of the product, as is commonly practiced with photocopy machines. Another mechanism is one in which a significant part of the purchase price is refunded for trade-in of a spent item. This approach could work very well with batteries containing cadmium or mercury. The normal purchase price could be doubled, then discounted to half with the trade-in of a spent battery.

Embedded Utility

Figure 17.7 can be regarded as an "energy/materials pyramid" showing that the amounts of energy and materials involved decrease going from the raw material to the finished product. The implication of this diagram is that significantly less energy, and certainly no more materials, are involved when recycling is performed near the top of the materials flow chain rather than near the bottom.

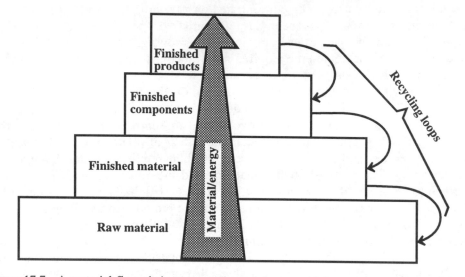

Figure 17.7. A material flow chain or energy/materials pyramid. Less energy and materials are involved when recycling is done near the end of the flow chain, thus retaining embedded utility.

To give a simple example, relatively little energy is required to return a glass beverage bottle from the consumer to the bottler, whereas returning the bottle to the glass manufacturer where it must be melted down and refabricated as a glass container obviously takes a greater amount of energy.

From a thermodynamic standpoint, a final product is relatively more ordered and it is certainly more usable for its intended purpose. The greater usability and lower energy requirements for recycling products higher in the order of material flow are called **embedded utility**. One of the major objectives of a system of industrial ecology and, therefore, one of the main reasons for performing life-cycle assessments is to retain the embedded utility in products by measures such as recycling as near to the end of the material flow as possible, and replacing only those components of systems that are worn out or obsolete. An example of the latter occurred during the 1960s when efficient and safe turboprop engines were retrofitted to still serviceable commercial aircraft airframes to replace relatively complex piston engines, thus extending the lifetime of the aircraft by a decade or more.

17.10. LIFE-CYCLE ASSESSMENT

From the beginning, industrial ecology must consider process/product design in the management of materials, including the ultimate fates of materials when they are discarded. The product and materials in it should be subjected to an entire **life-cycle assessment** or analysis. A life-cycle assessment applies to products, processes, and services through their entire life cycles from extraction of raw materials—through manufacturing, distribution, and use—to their final fates from the viewpoint of determining, quantifying, and ultimately minimizing their environmental impacts. It takes account of manufacturing, distribution, use, recycling, and disposal. Life-cycle assessment is particularly useful in determining the relative environmental merits of alternative products and services. At the consumer level, this could consist of an evaluation of paper vs. styrofoam drinking cups. On an industrial scale, life-cycle assessment could involve evaluation of nuclear vs. fossil energy-based electrical power plants.

A basic step in life-cycle analysis is **inventory analysis** which provides qualitative and quantitative information regarding consumption of material and energy resources (at the beginning of the cycle) and releases to the anthrosphere, hydrosphere, geosphere, and atmosphere (during or at the end of the cycle). It is based upon various materials cycles and budgets, and it quantifies materials and energy required as input and the benefits and liabilities posed by products. The related area of **impact analysis** provides information about the kind and degree of environmental impacts resulting from a complete life cycle of a product or activity. Once the environmental and resource impacts have been evaluated, it is possible to do an **improvement analysis** to determine measures that can be taken to reduce impacts on the environment or resources.

In making a life-cycle analysis the following must be considered:

- If there is a choice, selection of the kinds of materials that will minimize waste

- Kinds of materials that can be reused or recycled

- Components that can be recycled

- Alternate pathways for the manufacturing process or for various parts of it

Although a complete life-cycle analysis is expensive and time-consuming, it can yield significant returns in lowering environmental impacts, conserving resources, and reducing costs. This is especially true if the analysis is performed at an early stage in the development of a product or service. Improved computerized techniques are making significant advances in the ease and efficacy of life-cycle analyses. Until now, life-cycle assessments have been largely confined to simple materials and products such as reusable cloth vs. disposable paper diapers. A major challenge now is to expand these efforts to more complex products and systems such as aircraft or electronics products.

Scoping in Life-Cycle Assessment

A crucial early step in life-cycle assessment is **scoping** the process by determining the boundaries of time, space, materials, processes, and products to be considered. Consider as an example the manufacture of parts that are rinsed with an organochloride solvent in which some solvent is lost by evaporation to the atmosphere, by staying on the parts, during the distillation and purification process by which the solvent is made suitable for recycling, and by disposal of waste solvent that cannot be repurified. The scope of the life-cycle assessment could be made very narrow by confining it to the process as it exists. An assessment could be made of the solvent losses, the impacts of these losses, and means for reducing the losses, such as reducing solvent emissions to the atmosphere by installation of activated carbon air filters or reducing losses during purification by employing more efficient distillation processes. A more broadly scoped life-cycle assessment would be to consider alternatives to the organochloride solvent. An even broader scope would consider whether the parts even need to be manufactured; are there alternatives to their use?

17.11. CONSUMABLE, RECYCLABLE, AND SERVICE (DURABLE) PRODUCTS

In industrial ecology, most treatments of life-cycle analysis make the distinction between **consumable products**, which are essentially used up and dispersed to the environment during their life cycle and **service or durable products**, which essentially remain in their original form after use. Gasoline is clearly a consumable product, whereas the automobile in which it is burned is a service product. It is useful, however, to define a third category of products that clearly become "worn out" when employed for their intended purpose, but which remain largely undispersed to the environment. The motor oil used in an automobile is such a substance in that most of the original material remains after use. Such a category of material may be called a **recyclable commodity**.

Desirable Characteristics of Consumables

Consumable products include laundry detergents, hand soaps, cosmetics, windshield washer fluids, fertilizers, pesticides, laser printer toners, and all other materials that are impossible to reclaim after they are used. The environmental implications of the use of consumables are many and profound. In the late 1960s and early 1970s, for example, nondegradable surfactants in detergents caused severe foaming and esthetic problems at water treatment plants and sewage outflows, and the phosphate builders in the detergents promoted excessive algal growth in receiving waters, resulting in a condition known as eutrophication. Lead in consumable leaded gasoline was widely dispersed to the environment when the gasoline was burned. These problems have now been remedied with the adoption of phosphate-free detergents employing biodegradable surfactants and the mandatory use of unleaded gasoline.

Since they are destined to be dispersed into the environment, consumables should meet several "environmentally friendly" criteria, including the following:

- **Degradability**. This usually means biodegradability, such as that of household detergent constituents that occurs in waste treatment plants and in the environment. Chemical degradation may also occur.

- **Nonbioaccumulative**. Lipid-soluble, poorly biodegradable substances, such as DDT and PCBs tend to accumulate in organisms and to be magnified through the food chain. This characteristic should be avoided in consumable substances.

- **Nontoxic**. To the extent possible, consumables should not be toxic in the concentrations that organisms are likely to be exposed to them. In addition to their not being acutely toxic, consumables should not be mutagenic, carcinogenic, or teratogenic (cause birth defects).

Desirable Characteristics of Recyclables

Recyclables is used here to describe materials that are not used up in the sense that laundry detergents or photocopier toners are consumed, but are not durable items. In this context, recyclables can be understood to be chemical substances and formulations. The hydrochlorofluorocarbons (HCFCs) used as refrigerant fluids fall into this category, as does ethylene glycol mixed with water in automobile engine antifreeze/antiboil formulations (although rarely recycled in practice).

Insofar as possible, recyclables should be minimally hazardous with respect to toxicity, flammability, and other hazards. For example, both volatile hydrocarbon solvents and organochloride (chlorinated hydrocarbon) solvents are recyclable after use for parts degreasing and other applications requiring a good solvent for organic materials. The hydrocarbon solvents have relatively low toxicities, but may present flammability hazards during use and reclamation for recycling. The organochloride solvents are less flammable, but may present a greater toxicity hazard. An example of such a solvent is carbon tetrachloride, which is so nonflammable that it was once used in fire extinguishers, but the current applications of which are highly constrained because of its high toxicity.

An obviously important characteristic of recyclables is that they should be designed and formulated to be amenable to recycling. In some cases there is little leeway in formulating potentially recyclable materials; motor oil, for example, must meet certain criteria, including the ability to lubricate, stand up to high temperatures, and other attributes, regardless of its ultimate fate. In other cases formulations can be modified to enhance recyclability. For example, the use of bleachable or removable ink in newspapers enhances the recyclability of the newsprint, enabling it to be restored to an acceptable level of brightness.

For some commodities, the potential for recycling is enormous. This can be exemplified by lubricating oils. The volume of motor oil sold in the U.S. each year for gasoline engines is about 2.5 billion liters, a figure that is doubled if all lubricating oils are considered. A particularly important aspect of utilizing recyclables is their collection. In the case of motor oil, collection rates are low from consumers who change their own oil, and they are responsible for the dispersion of large amounts of waste oil to the environment.

Desirable Characteristics of Service Products

Since, in principle at least, service products are destined for recycling, they have comparatively lower constraints on materials and higher constraints on their ultimate disposal. A major impediment to the recycling of service products is the lack of convenient channels through which they can be put into the recycling loop. Television sets and major appliances such as washing machines or ovens have many recyclable components, but often end up in landfills and waste dumps simply because there is no handy means for getting them from the user and into the recycling loop. In such cases, government intervention may be necessary to provide appropriate channels. One partial remedy to the disposal/recycling problem consists of leasing arrangements or payment of deposits on items such as batteries to ensure their return to a recycler. The terms "de-shopping" or "reverse shopping" describe a process by which service commodities would be returned to a location such as a parking lot where they could be collected for recycling. According to this scenario, the analogy to a supermarket would be a facility in which service products are disassembled for recycling.

Much can be done in the design of service products to facilitate their recycle. One of the main characteristics of recyclable service products must be ease of disassembly so that remanufacturable components and recyclable materials, such as copper wire, can be readily removed and separated for recycling.

17.12. DESIGN FOR ENVIRONMENT

Design for environment is the term given to the approach of designing and engineering products, processes, and facilities in a manner that minimizes their adverse environmental impacts and, where possible, maximizes their beneficial environmental effects. In modern industrial operations, design for environment is part of a larger scheme termed "design for X," where "X" can be any one of a number of characteristics such as assembly, manufacturability, reliability, and serviceability. In making such a design, numerous desired characteristics of the product must be considered, including ultimate use, properties, costs, and appearance. Design for environment requires that the designs of the product, the process by which it is made, and the facilities involved in making it conform to appropriate environmental goals and limitations imposed by the need to maintain environmental quality. It must also consider the ultimate fate of the product, particularly whether or not it can be recycled at the end of its normal life span.

Products, Processes, and Facilities

In discussing design for environment, the distinctions among products, processes, and facilities must be kept in clear perspective. **Products**—automobile tires, laundry detergents, and refrigerators—are items sold to consumers. **Processes** are the means of producing products and services. For example, tires are made by a process in which hydrocarbon monomers are polymerized to produce rubber molded in the shape of a tire with a carcass reinforced by synthetic fibers and steel wires. A **facility** is where processes are carried out to produce or deliver products or services. In cases where services are regarded as products, the distinction between products and processes becomes blurred. For example, a lawn care service delivers products in the forms of fertilizers, pesticides, and grass seeds, but also delivers pure services including mowing, edging, and sod aeration.

Although *products* tend to get the most public attention in consideration of environmental matters, *processes* often have more environmental impact. Successful process designs tend to stay in service for many years and to be used to make a wide range of products. While the product of a process may have minimal environmental impact, the process by which the product is made may have marked environmental effects. An example is the manufacture of paper. The environmental impact of paper as a product, even when improperly discarded, is not terribly great, whereas the process by which it is made involves harvesting wood from forests, high use of water, potential emission of a wide range of air pollutants, and other factors with profound environmental implications.

Processes develop symbiotic relationships when one provides a product of service utilized in another. An example of such a relationship is that between steel making and the process for the production of oxygen required in the basic oxygen process by which carbon and silicon impurities are oxidized from molten iron to produce steel. The long lifetimes and widespread applicability of popular processes make their design for environment of utmost importance.

The nature of a properly functioning system of industrial ecology is such that processes are even more interwoven than would otherwise be the case, because byproducts from some processes are used by other processes. Therefore, the processes employed in such a system and the interrelationships and interpendencies among them are particularly important. A major change in one process may have a "domino effect" on the others.

Key Factors in Design for Environment

Two key choices that must be made in design for environment are those involving materials and energy. The choices of materials in an automobile illustrate some of the possible tradeoffs. Steel as a component of automobile bodies requires relatively large amounts of energy and involves significant environmental disruption in the mining and processing of iron ore. Steel is a relatively heavy material, so more energy is involved in moving automobiles made of steel. However, steel is durable, has a high rate of recycling, and is produced initially from abundant sources of iron ore. Aluminum is much lighter than steel and quite durable. It has an excellent percentage of recycling. Good primary sources of aluminum, bauxite ores, are not as abundant as iron ores, and large amounts of energy are required in the primary production of aluminum. Plastics are another source of automotive components. The light weight of plastic reduces automotive fuel consumption, plastics with desired properties are readily made, and molding and shaping plastic parts is a straightforward process. However, plastic automobile components have a low rate of recycling.

Three related characteristics of a product that should be considered in design for environment are durability, repairability, and recyclability. **Durability** simply refers to how well the product lasts and resists breakdown in normal use. Some products are notable for their durability; ancient two-cylinder John Deere farm tractors from the 1930s and 1940s are legendary in farming circles for their durability, enhanced by the affection engendered in their owners who tend to preserve them. **Repairability** is a measure of how easy and inexpensive it is to repair a product. A product that can be repaired is less likely to be discarded when it ceases to function for some reason. **Recyclability** refers to the degree and ease with which a product or components of it may be recycled. An important aspect of recyclability is the ease with which a product can be disassembled into constituents consisting of a single material that can be recycled. It also considers whether the components are made of materials that can be recycled.

Hazardous Materials in Design for Environment

A key consideration in the practice of design for environment is the reduction of the dispersal of hazardous materials and pollutants. This can entail the reduction or elimination of hazardous materials in manufacture, an example of which was the replacement of stratospheric ozone-depleting chlorofluorocarbons (CFCs) in foam blowing of plastics. If appropriate substitutes can be found, somewhat toxic and persistent chlorinated solvents should not be used in manufacturing applications such as parts washing. The use of hazardous materials in the product—such as batteries containing toxic cadmium, mercury, and lead—should be eliminated or minimized. Pigments containing heavy metal cadmium or lead should not be used if there are any possible substitutes. The substitution of hydrochlorofluorocarbons and hydrofluorocarbons for ozone-depleting CFCs in products (refrigerators and air conditioners) is an example of a major reduction in environmentally damaging materials in products. The elimination of extremely persistent polychlorinated biphenyls (PCBs) from electrical transformers removed a major hazardous waste problem due to the use of a common product (although PCB spills and contamination from the misuse and disassembly of old transformers has remained a persistent problem even up to the present).

17.13. OVERVIEW OF AN INTEGRATED INDUSTRIAL ECOSYSTEM

Figure 17.8 provides an overview of an integrated industrial ecosystem including all the components defined and discussed earlier in this chapter. Such a system may be divided into three separate, somewhat overlapping sectors controlled by the following: (1) the raw materials supply and processing sector, (2) the manufacturing sector, and (3) the consumer sector.

There are several important aspects of a complete industrial ecosystem. One of these is that, as discussed in the preceding section, there are several points at which materials may be recycled in the system. A second aspect is that there are several points at which wastes are produced. The potential for the greatest production of waste lies in the earlier stages of the cycle in which large quantities of materials with essentially no use associated with the raw material, such as ore tailings, may require disposal. In many cases, little if anything of value can be obtained from such wastes and the best thing to do with them is to return them to their source (usually a mine), if possible. Another big source of potential wastes, and often the one that causes the most problems, consists of postconsumer wastes generated when a product's life cycle is finished. With a properly designed industrial ecology cycle, such wastes can be be minimized and, ideally, totally eliminated.

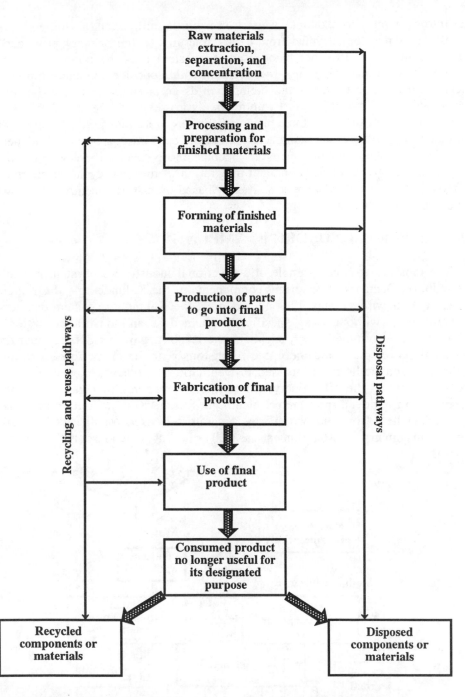

Figure 17.8. Outline of materials flow through a complete industrial ecosystem.

In general, the amount of waste per unit output decreases in going through the industrial ecology cycle from virgin raw material to final consumer product. Also, the amount of energy expended in dealing with waste or in recycling decreases farther into the cycle. For example, waste iron from the milling and forming of automobile parts may be recycled from a manufacturer to the primary producer of iron as scrap steel. In order to be used, such steel must be remelted and run through the steel manufacturing process again, with a considerable consumption of energy. However, a postconsumer item, such as an engine block, may be refurbished and recycled to the market with relatively less expenditure of energy.

At the present time, the three major enterprises in an industrial ecology cycle, the materials producer, the manufacturer, and the consumer, act largely independently of each other. As raw materials become more scarce, there will be more economic incentives for recycling and integration of the total cycle. Furthermore, there is a need for better, more scientifically based regulatory incentives leading to the practice of industrial ecology.

17.14. THE KALUNDBORG EXAMPLE

The most often cited example of a functional industrial ecosystem is that of Kalundborg, Denmark. The various components of the Kalundborg industrial ecosystem are shown in Figure 17.9. To a degree, the Kalundborg system developed spontaneously, without being specifically planned as an industrial ecosystem. It is based upon two major energy suppliers, the 1,500-megawatt ASNAES coal-fired electrical power plant and the 4–5 million tons/year Statoil petroleum refining complex, each the largest of its kind in Denmark. The electric power plant sells process steam to the oil refinery, from which it receives fuel gas and cooling water. Sulfur removed from the petroleum goes to the Kemira sulfuric acid plant. By-product heat from the two energy generators is used for district heating of homes and commercial establishments, as well as to heat greenhouses and a fish farm-

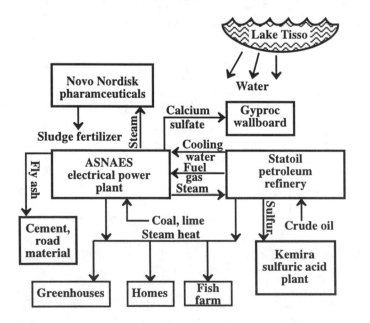

Figure 17.9. Schematic of the industrial ecosystem in Kalundborg, Denmark.

ing operation. Steam from the electrical power plant is used by the $2 billion/year Novo Nordisk pharmaceutical plant, a firm that produces industrial enzymes and 40% of the world's supply of insulin. This plant generates a biological sludge that is used by area farms for fertilizer. Calcium sulfate produced as a by-product of sulfur removal by lime scrubbing from the electrical plant is used by the Gyproc company to make wallboard. The wallboard manufacturer also uses clean-burning gas from the petroleum refinery as fuel. Fly ash generated from coal combustion goes into cement and roadbed fill. Lake Tisso serves as a fresh water source. Other examples of efficient materials utilization associated with Kalundborg include use of sludge from the plant that treats water and wastes from the fish farm's processing plant for fertilizer, and blending of excess yeast from Novo Nordisk's insulin production as a supplement to swine feed.

The development of the Kalundborg complex occurred over a long period of time, beginning in the 1960s, and provides some guidelines for the way in which an industrial ecosystem can grow naturally. The first of many synergistic (mutually advantageous) arrangements was cogeneration of usable steam along with electricity by the ASNAES electrical power plant. The steam was first sold to the Statoil petroleum refinery, then, as the advantages of large-scale, centralized production of steam became apparent, steam was also provided to homes, greenhouses, the pharmaceutical plant, and the fish farm. The need to produce electricity more cleanly than was possible simply by burning high sulfur coal resulted in two more synergies. Installation of a lime-scrubbing unit for sulfur removal on the power plant stack resulted in the production of large quantities of calcium sulfate, which found a market in the manufacture of gypsum wallboard. It was also found that a clean-burning gas by-product of the petroleum refining operation could be substituted in part for the coal burned in the power plant, further reducing pollution.

The implementation of the Kalundborg ecosystem occurred largely because of the close personal contact among the managers of the various facilities in a relatively close social and professional network over a long period of time. All the contracts have been based upon sound business fundamentals and have been bilateral. Each company has acted upon its perceived self-interest, and there has been no master plan for the system as a whole. The regulatory agencies have been cooperative, but not coercive in promoting the system. The industries involved in the agreements have fit well, with the needs of one matching the capabilities of the other in each of the bilateral agreements. The physical distances involved have been small and manageable; it is not economically feasible to ship commodities such as steam or fertilizer sludges for long distances.

17.15. SOCIETAL FACTORS AND THE ENVIRONMENTAL ETHIC

The "consumer society" in which people demand more and more goods, energy-consuming services, and other amenities that are in conflict with resource conservation and environmental improvement runs counter to a good workable system of industrial ecology. Much of the modern lifestyle and corporate ethic is based upon persuading usually willing consumers that they need and deserve more things, and that they should adopt lifestyles that are very damaging to the environment. The conventional wisdom is that consumers are unwilling to significantly change their lifestyles and lessen their demands on world resources for the sake of environmental preservation. However, in the few examples in which consumers have been given a chance to exercise good environmental citizenship, there are encouraging examples that they will do so willingly. A prime example of this is the success of paper, glass, and can recycling programs in connection with municipal refuse collection, implemented to extend landfill lifetimes.

Two major requirements for the kind of public ethic that must accompany any universal adoption of systems of industrial ecology are **education** and **opportunity**. Starting at an early age, people need to be educated about the environment and its crucial importance in maintaining the quality of their lives. They need to know about realistic ways, including the principles of industrial ecology, by which their environment may be maintained and improved. The electronic and print media have a very important role to play in educating the public regarding the environment and resources. Given the required knowledge, the majority of people will do the right thing for the environment.

People also need good opportunities for recycling and for general environmental improvement. It is often said that people will not commute by public transit, but of course they will not do so if public transit is not available, or if it is shabby, unreliable, and even dangerous. They will not recycle cans, paper, glass, and other consumer commodities if convenient, well-maintained collection locations are not accessible to them. There are encouraging examples, including some from the United States, that opportunities to contribute to environmental protection and resource conservation will be met with a positive response from the public.

LITERATURE CITED

1. Manahan, Stanley E., *Industrial Ecology: Environmental Chemistry and Hazardous Waste*, CRC Press/Lewis Publishers, Boca Raton, FL, 1999.

2. Allenby, Braden R., *Industrial Ecology: Policy Framework and Implementation*, Prentice Hall, Upper Saddle River, NJ, 1998.

3. Mazur, Allan, "Reconsidering Sustainable Development," *Chemical and Engineering News*, January 3, 1994, pp 26-27.

4. Meadows, Donella, Dennis Meadows, and Jørgen Randers, *Beyond the Limits: Confronting Global Collapse, Envisioning a Sustainable Future*, Chelsea Green Publishing, Post Mills, VT, 1992.

5. Ayres, Robert U., "Industrial Metabolism," in *Technology and Environment*, J. H. Ausubel and H. E. Sladovich, Eds., National Academy Press, Washington, D.C., 1989, pp. 23–49.

6. Ayres, Robert U., "Industrial Metabolism: Theory and Policy," in *The Greening of Industrial Ecosystems*, National Academy Press, Washington, D.C., 1994, pp. 23–37.

7. "Industrial Ecology Methods and Tools," Chapter 3 in *Discovering Industrial Ecology*, John L. Warren and Stephen R. Moran, Battelle Press, Columbus, OH, 1997, pp. 37–73.

8. Frosch, Robert A. and Nicholas E. Gallopoulos, "Strategies for Manufacturing," *Scientific American*, **261**, 94–102, 1989.

9. Manahan, Stanley E., *Environmental Science and Technology*, CRC Press, Boca Raton, FL, 1997.

SUPPLEMENTARY REFERENCES

Ausubel, Jesse, "The Virtual Ecology of Industry," *Journal of Industrial Ecology*, **1**(1), 10–11 (1997).

Ayres, Robert U. and Udo E. Simonis, Eds., *Industrial Metabolism: Restructuring for Sustainable Development*, United Nations University Press, New York, 1994.

Ayres, Robert U. and Leslie W. Ayres, *Industrial Ecology: Towards Closing the Materials Cycle*, Edward Elgar Publishers, Cheltenham, U.K., 1996.

Bisio, Attilio, and Sharon R. Boots, *Energy Technology and the Environment* , John Wiley and Sons, New York, 1995.

Cote, Ray, "Industrial Ecosystems: Evolving and Maturing," *Journal of Industrial Ecology*, **1**(3), 9–11 (1998).

Curran, Mary Ann, Ed., *Environmental Life-Cycle Assessment*, McGraw-Hill, New York, 1997.

Davis John B., *Product Stewardship and the Coming Age of Takeback: What Your Company Can Learn from The Electronic Industry's Experience*, Cutter Information Corporation, Arlington, MA, 1996.

DeSimone, Livio D., and Frank Popoff, *Eco-efficiency: The Business Link to Sustainable Development*, The MIT Press, Cambridge, MA, 1997.

Fiksel, Joseph, Ed., *Design for Environment: Creating Eco-Efficient Products and Processes*, McGraw-Hill, New York, 1996.

Frosch, Robert A. and Nicholas E. Gallopoulos, "Strategies for Manufacturing," *Scientific American*, **261**, 94–102 (1989).

Graedel, Thomas E. and B. R. Allenby, *Industrial Ecology*, Prentice Hall, Englewood Cliffs, NJ, 1995.

Graedel, Thomas E. and Braden R. Allenby, *Industrial Ecology and the Automobile*, Prentice Hall, Upper Saddle River, NJ, 1998.

Graham, John D. and Jennifer K. Hartwell, *The Greening of Industry*, Harvard University Press, Cambridge, MA, 1997.

"Industrial Ecology," *Environmental Science and Technology*, **31**, 1997, p. 26A.

Klostermann, Judith E. M. and Arnold Tukker, Eds., *Product Innovation and Eco-Efficiency: Twenty-Three Industry Efforts to Reach the Factor 4*, Kluwer Academic Publishing Co., Hingham, MA, 1998.

Leff, Enrique, *Green Production: Toward an Environmental Rationality*, Guilford Press, New York, 1995.

Lifset, Reid, "Relating Industry to Ecology," *Journal of Industrial Ecology*, **1**(2), 1–2 (1997).

Lowe, Ernest and John L. Warren, *The Source of Value: An Executive Briefing and Sourcebook on Industrial Ecology*, Battelle, Pacific Northwest National Laboratory, Richland, WA, 1997.

Lowe, Ernest A. John L. Warren, and Stephen R. Moran, *Discovering Industrial Ecology: An Executive Briefing and Sourcebook*, Battelle Press, Columbus, OH, 1997.

Nemerow, Nelson L., *Zero Pollution for Industry: Waste Minimization Through Industrial Complexes*, John Wiley & Sons, New York, 1995.

Peck, Steven and Elain Hardy, *The Eco-Efficiency Resource Manual*, Fergus, Ontario, Canada, 1997.

Smith, Ronald S., *Profit Centers in Industrial Ecology*, Quorum Books, Westport, CT, 1998.

Socolow, Robert, Clinton Andrews, Frans Berkhout, and Valerie Thomas, Eds., *Industrial Ecology and Global Change*, Cambridge University Press, New York, 1994.

Townsend, Mardie, *Making Things Greener: Motivations and Influences in the Greening of Manufacturing*, Ashgate, Publishing, Aldershot, U.K., 1998.

von Weizsäcker, Ernst U., Amory B. Lovins, and L. Hunter Lovins, *Factor Four: Doubling Wealth, Halving Resource Use*, Earthscan, London, 1997.

QUESTIONS AND PROBLEMS

1. In biological ecosystems a process called mineralization occurs as defined in this book. Name and describe a process analogous to mineralization that occurs in an industrial ecosystem.

2. How are the terms industrial metabolism, industrial ecosystem, and sustainable development related to industrial ecology?

3. How is industrial symbiosis related to industrial ecology?

4. Justify or refute the statement that in an operational industrial ecosystem only energy is consumed.

5. In what sense is the consumer sector the most difficult part of an industrial eco-system?

6. In what sense might a "moon station" or a colony on Mars advance the practice of industrial ecology?

7. In what sense do modern solid state electronic devices illustrate both dematerialization and material substitution?

8. As applied to material resources, what is the distinction between dematerialization and material substitution? Use the automobile as an example.

9. How does "design for recycling" (DFR) relate to embedded utility?

10. Distinguish among consumable, durable (service), and recyclable products.

11. List some of the "environmentally friendly" criteria met by soap as a consumable commodity.

12. What are the enterprises that serve to underpin the Kalundborg industrial ecosystem? How might they compare to the basic enterprises of an industrial ecosystem consisting of rural counties in the state of Iowa?

18 INDUSTRIAL ECOLOGY, RESOURCES, AND ENERGY

18.1. INTRODUCTION

Modern civilization depends upon a wide variety of resources consisting largely of minerals that are processed to recover the materials needed for industrial activities. The most common type of mineral material so used, and one that all people depend upon for their existence, is soil, used to grow plants for food. Also of crucial importance are metal ores. Some of these metal sources are common and abundant, such as iron ore; others, such as sources of chromium, are rare and will not last long at current rates of consumption. There are also some crucial sources of nonmetals. Sulfur, for example, is abundant and extracted in large quantities as a by-product of sulfur-rich fuels. Phosphorus, a key fertilizer element, will last only for several generations at current rates of consumption.

The materials needed for modern societies can be provided from either **extractive** (nonrenewable) or **renewable** sources. Extractive industries remove irreplaceable mineral resources from the earth's crust. The utilization of mineral resources is strongly tied with technology, energy, and the environment. Perturbations in one usually cause perturbations in the others. For example, reductions in automotive exhaust pollutant levels to reduce air pollution have made use of catalytic devices that require platinum-group metals, a valuable and irreplaceable natural resource. Furthermore, automotive pollution control devices result in greater gasoline consumption than would be the case if exhaust emissions were not a consideration (a particularly pronounced effect in the earlier years of emissions control). The availability of many metals depends upon the quantity of energy used and the amount of environmental damage tolerated in the extraction of low-grade ores. Many other such examples could be cited. Because of these intimate interrelationships, technology, resources, and energy must all be considered together. The practice of industrial ecology has a significant potential to improve environmental quality with reduced consumption of nonrenewable resources and energy.

In discussing nonrenewable sources of minerals and energy, it is useful to define two terms related to available quantities. The first of these is **resources**, defined as quantities that are estimated to be *ultimately* available. The second term is **reserves**, which refers to well-identified resources that can be profitably utilized with existing technology.

18.2. MINERALS IN THE GEOSPHERE

There are numerous kinds of mineral deposits that are used in various ways. These are, for the most part, sources of metals which occur in **batholiths** composed of masses of igneous rock that have been extruded in a solid or molten state into the surrounding rock strata. In addition to deposits formed directly from solidifying magma, associated deposits are produced by water interacting with magma. Hot aqueous solutions associated with magma can form rich **hydrothermal** deposits of minerals. Several important metals, including lead, zinc, and copper, are often associated with hydrothermal deposits.

Some useful mineral deposits are formed as **sedimentary deposits** along with the formation of sedimentary rocks. **Evaporites** are produced when seawater is evaporated. Common mineral evaporites are halite ($NaCl$), sodium carbonates, potassium chloride, gypsum ($CaSO_4 \cdot 2H_2O$), and magnesium salts. Many significant iron deposits consisting of hematite (Fe_2O_3) and magnetite (Fe_3O_4) were formed as sedimentary bands when earth's atmosphere was changed from reducing to oxidizing as photosynthetic organisms produced oxygen, precipitating the oxides from the oxidation of soluble Fe^{2+} ion.

Deposition of suspended rock solids by flowing water can cause segregation of the rocks according to differences in size and density. This can result in the formation of useful **placer** deposits that are enriched in desired minerals. Gravel, sand, and some other minerals, such as gold, often occur in placer deposits.

Some mineral deposits are formed by the enrichment of desired constituents when other fractions are weathered or leached away. The most common example of such a deposit is bauxite, Al_2O_3, remaining after silicates and other more soluble constituents have been dissolved by the weathering action of water under the severe conditions of hot tropical climates with very high levels of rainfall. This kind of material is called a **laterite.**

Evaluation of Mineral Resources

In order to make its extraction worthwhile, a mineral must be enriched at a particular location in earth's crust relative to the average crustal abundance. Normally applied to metals, such an enriched deposit is called an **ore**. The value of an ore is expressed in terms of a **concentration factor**:

$$\text{Concentration factor} = \frac{\text{Concentration of material in ore}}{\text{Average crustal concentration}} \qquad (18.2.1)$$

Obviously, higher concentration factors are always desirable. Required concentration factors decrease with average crustal concentrations and with the value of the commodity extracted. A concentration factor of 4 might be adequate for iron, which makes up a relatively high percentage of earth's crust. Concentration factors must be several hundred or even several thousand for less expensive metals that are not present at very high percentages in earth's crust. However, for an extremely valuable metal, such as platinum, a relatively low concentration factor is acceptable because of the high financial return obtained from extracting the metal.

Acceptable concentration factors are a sensitive function of the price of a metal. Shifts in price can cause significant changes in which deposits are mined. If the price of a metal increases by, for example, 50%, and the increase appears to be long term, it becomes profitable to mine deposits that had not been mined previously. The opposite can happen, as is often the case when substitute materials are found or newly discovered, richer sources go into production.

In addition to large variations in the concentration factors of various ores, there are extremes in the geographic distribution of mineral resources. The United States is perhaps about average for all nations in terms of its mineral resources, possessing significant resources of copper, lead, iron, gold, and molybdenum, but virtually without resources of some important strategic metals, including chromium, tin, and platinum-group metals. For its size and population, South Africa is particularly blessed with some important metal mineral resources.

18.3. EXTRACTION AND MINING

Minerals are usually extracted from Eearth's crust by various kinds of mining procedures, but other techniques may be employed as well. The raw materials so obtained include inorganic compounds such as phosphate rock, sources of metal such as lead sulfide ore, clay used for firebrick, and structural materials, such as sand and gravel.

Surface mining, which can consist of digging large holes in the ground, or strip mining, is used to extract minerals that occur near the surface. A common example of surface mining is quarrying of rock. Vast areas have been dug up to extract coal. Because of past mining practices, surface mining got a well-deserved bad name. With modern reclamation practices, however, topsoil is first removed and stored. After the mining is complete, the topsoil is spread on top of overburden that has been replaced such that the soil surface has gentle slopes and proper drainage. Topsoil spread over the top of the replaced spoil, often carefully terraced to prevent erosion, is seeded with indigenous grass and other plants, fertilized, and watered, if necessary, to provide vegetation. The end result of carefully done **mine reclamation** projects is a well-vegetated area suitable for wildlife habitat, recreation, forestry, grazing, and other beneficial purposes.

Extraction of minerals from placer deposits formed by deposition from water has obvious environmental implications. Mining of placer deposits can be accomplished by dredging from a boom-equipped barge. Another means that can be used is hydraulic mining with large streams of water. One interesting approach for more coherent deposits is to cut the ore with intense water jets, then suck up the resulting small particles with a pumping system. These techniques have a high potential to pollute water and disrupt waterways.

For many minerals, underground mining is the only practical means of extraction. An underground mine can be very complex and sophisticated. The structure of the mine depends upon the nature of the deposit. It is of course necessary to have a shaft that reaches to the ore deposit. Horizontal tunnels extend out into the deposit, and provision must be made for sumps to remove water and for ventilation. Factors that must be considered in designing an underground mine include the depth, shape, and orientation of the ore body, as well as the nature and strength of the rock in and around it; thickness of overburden; and depth below the surface.

Usually, significant amounts of processing are required before a mined product is used or even moved from the mine site. Such processing, and the by-products of it, can have significant environmental effects. Even rock to be used for aggregate and for road construction must be crushed and sized, a process that has the potential to emit air-polluting dust particles into the atmosphere. Crushing is also a necessary first step for further processing of ore. Some minerals occur to an extent of a few percent or even less in the rock taken from the mine and must be concentrated on site so that the residue does not have to be hauled far. For metals mining, these processes—as well as roasting, extraction—and similar operations, are covered under the category of **extractive metallurgy**.

One of the more environmentally troublesome by-products of mineral refining consists of waste **tailings**. By the nature of the mineral processing operations employed, tailings are usually finely divided and, as a result, subject to chemical weathering processes. Heavy metals associated with metal ores can be leached from tailings, producing water runoff contaminated with cadmium, lead, and other pollutants. Adding to the problem are some of the processes used to refine ore. Large quantities of cyanide solution are used in some processes to extract low levels of gold from ore, posing obvious toxicological hazards.

Environmental problems resulting from exploitation of extractive resources—including disturbance of land, air pollution from dust and smelter emissions, and water pollution from disrupted aquifers are aggravated by the fact that the general trend in mining involves utilization of less rich ore. This is illustrated in Figure 18.1, showing the average percentage of copper in copper ore mined since 1900. The average percentage of copper in ore mined in 1900 was about 4%, but by 1982 it was about 0.6% in domestic ore, and 1.4% in richer foreign ore. Ore as low as

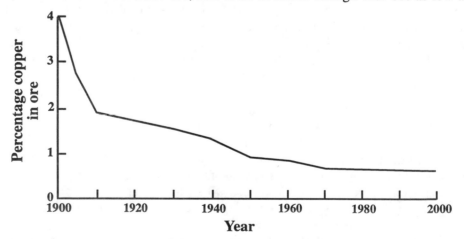

Figure 18.1. Average percentage of copper in ore that has been mined.

0.1% copper may eventually be processed. Increased demand for a particular metal, coupled with the necessity to utilize lower grade ore, has a vicious multiplying effect upon the amount of ore that must be mined and processed and accompanying environmental consequences.

The proper practice of industrial ecology can be used to significantly reduce the effects of mining and mining by-products. One way in which this can be done is to entirely eliminate the need for mining, utilizing alternate sources of materials. An example of such utilization, widely hypothesized but not yet put into practice to a large extent, is the extraction of aluminum from coal ash. This would have the double advantage of reducing amounts of waste ash and reducing the need to mine scarce aluminum ore.

18.4. METALS

The majority of elements are metals, most of which are of crucial importance as resources. The availability and annual usage of metals vary widely with the kind of metal. Some metals are abundant and widely used in structural applications; iron and aluminum are prime examples. Other metals, especially those of the platinum group (platinum, palladium, iridium, rhodium) are very precious and their use is confined to applications such as catalysts, filaments, or electrodes for which only small quantities are required. Some metals are considered to be "crucial" because of their applications for which no substitutes are available and shortages or uneven distribution in supply that occur. Such a metal is chromium, used to manufacture stainless steel (especially for parts exposed to high temperatures and corrosive gases), jet aircraft, automobiles, hospital equipment, and mining equipment. The platinum-group metals are used as catalysts in the chemical industry, in petroleum refining, and in automobile exhaust antipollution devices.

Metals exhibit a wide variety of properties and uses. They come from a number of different compounds; in some cases two or more compounds are significant mineral sources of the same metal. Usually these compounds are oxides or sulfides. However, other kinds of compounds and, in the cases of gold and platinum-group metals, the elemental (native) metals themselves serve as metal ore. Table 18.1 lists the important metals, their properties, major uses, and sources.

18.5. METAL RESOURCES AND INDUSTRIAL ECOLOGY

Considerations of industrial ecology are very important in extending and efficiently utilizing metal resources.[1] More than any other kind of resource, metals lend themselves to recycling and to the practice of industrial ecology. This section briefly addresses the industrial ecology of metals.

Aluminum

Aluminum metal has a remarkably wide range of uses resulting from its properties of low density, high strength, ready workability, corrosion resistance, and high electrical conductivity. Unlike some metals, such as toxic cadmium or lead, the use and disposal of aluminum presents no environmental problems. Further-ore, it is one of the most readily recycled of all metals.

Table 18.1. Worldwide and Domestic Metal Resources

Metals	Properties[a]	Major uses	Ore, aspects of resources[b]
Aluminum	mp 660°C, bp 2467°C, sg 2.70, malleable, ductile	Metal products, including autos, aircraft, electrical equipment. Conducts electricity better than copper per unit mass and is used in electrical transmission lines.	From bauxite ore containing 35-55% Al_2O_3. About 60 million metric tons of bauxite produced worldwide per year. U. S. resources of bauxite are 40 million metric tons, world resources about 15 billion metric tons.
Cadmium	Soft, ductile, silvery-white	Corrosion-resistant plating on steel and iron, alloys, bearings, pigments, rechargeable batteries.	Byproduct of zinc production, so annual production of cadmium parallels that of zinc. Abundant supply of toxic cadmium from this source has resulted in excessive dissipative uses.
Chromium	mp 1903°C, bp 2642°C, sg 7.14, hard, silvery color	Metal plating, stainless steel, wear-resistant and cutting tool alloys, chromium chemicals, including chromates.	From chromite, an oxide mineral containing Cr, Mg, Fe, Al. Resources of 1 billion metric tons in South Africa and Zimbabwe, large deposits in Russia, virtually none in the U.S.
Cobalt	mp 1495°C, bp 2880°C, sg 8.71, bright, silvery	Manufacture of hard, heat-resistant alloys, permanent magnet alloys, driers, pigments, animal feed additive.	From a variety of minerals, such as linnaeite, Co_3S_4, and as a by-product of other metals. Abundant global and U.S. resources.

Table 18.1. (Cont.)

Copper	mp 1083°C, bp 2582°C, sg 8.96, ductile, malleable	Electrical conductors, alloys, chemicals. Many uses.	Occurs in low percentages as sulfides, oxides, and carbonates. U.S. consumption 1.5 million metric tons per year. World resources of 344 million metric tons, including 78 million in U.S.
Gold	mp 1063°C, bp 2660°C, sg 19.3	Jewelry, basis of currency, electronics, increasing industrial uses.	In various minerals at only around 10 ppm for ore currently processed in the U.S.; byproduct of copper refining. World resources of 1 billion oz., 80 million in U.S.
Iron	mp 1535°C, bp 2885°C, sg 7.86, silvery metal, in (rare) pure form	Most widely produced metal, usually as steel, a high-tensile-strength material containing 0.3-1.7% C. Made into many specialized alloys.	Occurs as hematite (Fe_2O_3), goethite ($Fe_2O_3 \cdot H_2O$), and magnetite (Fe_3O_4), abundant global and U.S. resources.
Lead	mp 327°C, bp 1750°C, sg 11.35, silvery color	Fifth most widely used metal, storage batteries, chemicals; uses in gasoline, pigments, and ammunition decreasing for environmental reasons.	Major source is galena, PbS. World-wide consumption about 3.5 million metric tons, 1/3 in U.S. Global reserves about 140 million metric tons, 39 million metric tons U.S.
Manganese	mp 1244°C, bp 2040°C, sg 7.3, hard, brittle, gray-white	Sulfur and oxygen scavenger in steel, manufacture of alloys, dry cells, gasoline additive, chemicals	Found in several oxide minerals. About 20 million metric tons per year produced globally, 2 million consumed in U.S., no U.S. production, world reserves 6.5 billion metric tons.

Table 18.1. (Cont.)

Mercury	mp -38°C, bp 357°C, sg 13.6, shiny, liquid metal	Instruments, electronic apparatus electrodes, chemicals.	From cinnabar, HgS. Annual world production 11,500 metric tons, 1/3 used in U.S. World resources 275,000 metric tons, 6,600 U.S.
Molyb-denum	mp 2620°C, bp 4825°C, sg 9.01, ductile, silvery-gray	Alloys, pigments, catalysts, chemicals, lubricants.	Molybdenite (MoS_2) and wulfenite ($PbMoO_4$) are major kinds or ore. About 2/3 global Mo production in U.S., large global resources.
Nickel	mp 1455°C, bp 2835°C, sg 8.90, silvery color	Alloys, coins, storage batteries, catalysts (such as for hydrogenation of vegetable oil).	Found in ore associated with iron. U.S. consumes 150,000 metric tons per year, 10% from domestic production, large domestic reserves of low-grade ore.
Platinum-group[c]	Resist chemical attack, perform well at high temperatures, good electrical properties, catalytic properties	Jewelry, alloys, catalysts, electrodes, filaments	In alluvial deposits produced by weathering and gravity separation. Most resources in Russia, South Africa, and Canada.
Silver	mp 961°C, bp 2193°C, sg 10.5, shiny metal	Photographic film, electronics, sterling ware, jewelry, bearings, dentistry.	Found with sulfide minerals, a by-product of Cu, Pb, Zn. Annual U.S. consumption of 150 million troy ounces, short supply.

Table 18.1. (Cont.)

Tin	mp 232°C, bp 2687°C, sg 7.31	Coatings, solders, bearing alloys, bronze, chemicals, organo-metallic biocides.	Many forms associated with granitic rocks and chrysolites. Global consumption 190,000 metric tons/year, U.S. 60,000 metric tons/year, world resources 10 million metric tons.
Titanium	mp 1677°C, bp 3277°C, sg 4.5, silvery color	Strong, corrosion-resistant, used in aircraft, valves, pumps, paint pigments.	Commonly as TiO_2, ninth in elemental abundance, no shortages.
Tungsten	mp 3380°C, bp 5530°C, sg 19.3, gray	Very strong, high boiling point, used in alloys, tungsten carbide, drill bits, turbines, nuclear reactors.	Found as tungstates, such as scheelite ($CaWO_4$); U.S. has 7% world reserves, China 60%.
Vanadium	mp 1917°C, bp 3375°C, sg 5.87, gray	Used to make strong steel alloys.	In igneous rocks, primarily a by-product of other metals. U.S. consumption of 5,000 metric tons per year equals production.
Zinc	mp 420°C, bp 907°C, sg 7.14, bluish-white	Widely used in alloys (brass), galvanized steel, paint pigments, chemicals. Fourth in world metal production.	Found in many ore minerals. World production is 5 million metric tons per year, U.S. consumes 1.5 million metric tons per year. World resources 235 million metric tons, 20% in U.S.

a Abbreviations: mp, melting point; bp, boiling point; sg, specific gravity.

b All figures are approximate; quantities of minerals considered available depend upon price, technology, recent discoveries, and other factors, so that quantities quoted are subject to fluctuation.

c The platinum-group metals consist of platinum, palladium, iridium, ruthenium, and osmium, all of which are very valuable.

The environmental problems associated with aluminum result from the mining and processing of aluminum ore. It occurs as a mineral called **bauxite**, which contains 40–60% alumina, Al_2O_3, associated with water molecules. Hydrated alumina is concentrated in bauxite, particularly in high-rainfall regions of the tropics, by the weathering away of more water-soluble constituents of soil (see laterites in Section 18.2). Bauxite ore is commonly strip mined from thin seams, so its mining causes significant disturbance to the geosphere. The commonly used Bayer process for aluminum refining dissolves alumina, shown below as the hydroxide $Al(OH)_3$, from bauxite at high temperatures with sodium hydroxide as sodium aluminate,

$$Al(OH)_3 + NaOH \rightarrow NaAlO_2 + 2H_2O \qquad (18.5.1)$$

leaving behind large quantities of caustic "red mud." This residue, which is rich in oxides of iron, silicon, and titanium, has virtually no uses and a high potential to produce pollution. Aluminum hydroxide is then precipitated in the pure form at lower temperatures and calcined at about 1200°C to produce pure anhydrous Al_2O_3. The anhydrous alumina is then electrolyzed in molten cyrolite, Na_3AlF_6, at carbon electrodes to produce aluminum metal.

All aspects of aluminum production from bauxite are energy intensive. Large amounts of heat energy are required to heat the bauxite treated with caustic to extract sodium aluminate, and heat is required to calcine the hydrated alumina before it can be electrolyzed. Very large amounts of electrical energy are required to reduce aluminum to the metal in the electrolytic process for aluminum production.

An interesting possibility that could avoid many of the environmental problems associated with aluminum production is the use of coal fly ash as a source of the metal. Fly ash is produced in large quantities as a by-product of electricity generation, so it is essentially a free resource. As a raw material, coal fly ash is very attractive because it is anhydrous, thus avoiding the expense of removing water; it is finely divided, and it is homogeneous. Aluminum, along with iron, manganese, and titanium, can be extracted from coal fly ash with acid. If aluminum is extracted as the chloride salt, $AlCl_3$, it can be electrolyzed as the chloride by the ALCOA process. Although this process has not yet been proven to be competitive with the Bayer process, it may become so in the future.

Gallium is a metal that commonly occurs with aluminum ore and may be produced as a byproduct of aluminum manufacture. Gallium combined with arsenic or with indium and arsenic is useful in semiconducter applications, including integrated circuits, photoelectric devices, and lasers. Although important, these applications require only miniscule amounts of gallium compared to major metals.

Chromium

Chromium is of crucial importance because of its use in stainless steel and superalloys. These materials are vitally important to industrialized societies because of their applications in jet engines, nuclear power plants, chemical-resistant valves, and other applications in which a material that resists heat and chemical attack is required.

As noted in Table 18.1, supplies of chromium are poorly distributed around the earth. It is important that chromium be handled according to good practices of industrial ecology. Several measures may be taken in this respect. Chromium is almost impossible to recover from chrome-plated objects, and this use should be eliminated insofar as possible, as has been done with much of the decorative chrome-plated adornments formerly put on automobiles. Chromium(VI) (chromate) is a toxic form of the metal and its uses should be eliminated wherever possible. The use of chromium in leather tanning and miscellaneous chemical applications should be curtailed. One important use of chromium is in the preparation of treated CCA lumber, which resists fungal decay and termites. The widespread use of this lumber has greatly extended the lives of wood products, which is in keeping with the practice of industrial ecology. However, its use of toxic arsenic, scarce copper, and even more scarce chromium are negatives, and alternative means of preserving lumber still need to be found.

Copper

Copper is a low-toxicity, corrosion-resistant metal widely used because of its workability (ductility and malleability), electrical conductivity, and ability to conduct heat. In addition to its use in electrical wire, where in some applications it is now challenged by aluminum, copper is also used in tubing, copper pipe, shims, gaskets, and other applications.

There are at least two major environmental problems associated with the extraction and refining of copper. The first of these is the dilute form in which copper ore now occurs (see Figure 6.1), such that in the U.S. 150–175 tons of inert material (not counting overburden removed in strip mining) must be processed and discarded to produce a ton of copper metal. The second problem is the occurrence of copper as the sulfide so that in the production of copper, large amounts of sulfur must be recovered as a by-product or, unfortunately in some less developed countries, released into the atmosphere as pollutant SO_2.

An advantage to copper for recycling is that it is used primarily as the metal, which represents "stored energy" in that it does not require energy for reduction to the metal. Recycling rates of scrap copper appear low in part because so much of the inventory of copper metal is tied up in long-lasting electrical wire, in structures, and other places where the lifetime of the metal is long. (This is in contrast to lead, where the main source of recycled metal is storage batteries, which last only 2–4 years.) An impediment to copper recycling is the difficulty of recovering copper components from circuits, plumbing, and other applications.

Cobalt

Cobalt is a "strategic" metal with very important applications in alloys, particularly in heat-resistant applications, such as jet engines. The major source of cobalt is as a byproduct of copper refining, although it can also be obtained as a byproduct of nickel and lead. As much as 50% of the cobalt in these sources is lost to tailings, slag, or other wastes, so there is a significant potential to improve the recovery of cobalt. Relatively low percentages of cobalt are recycled as scrap.

Lead

The industrial ecology of lead is very important because of the widespread use of this metal and its toxicity. Global fluxes of lead from the anthrosphere are shown in Figure 18.2.

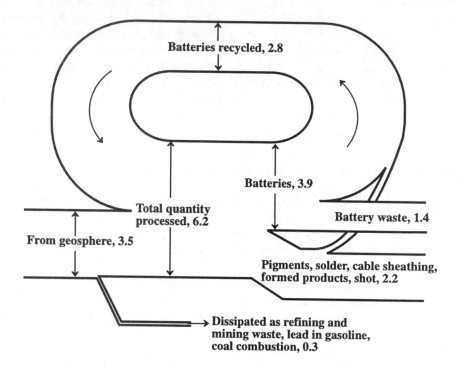

Figure 18.2. Flux of lead in the anthrosphere, globally, on an annual basis in millions of tons per year. Lead from the geosphere includes metal mined and a small quantity dissipated by coal combustion.

Somewhat more than half of the lead processed by humans comes from the geosphere, mostly as lead mined for the metal, and with a very small fraction contained in coal that is burned. By far the greatest use of lead is in batteries, and the amount of battery lead recycled each year approaches that taken from the geosphere. A small fraction of lead is dissipated as wastes associated with the mining and refining of the metal and as lead in gasoline, an amount that is decreasing as use of unleaded gasoline becomes prevalent around the world. A significant quantity of lead goes into various uses other than batteries, including pigments, solder, cable sheathing, formed products, and shot in ammunition. Only a small fraction of lead from these uses is recycled, and this represents a potential improvement in the conservation of lead. Another area in which improvements can be made is to eliminate or greatly reduce nonbattery uses of lead, as has been done in the case of lead shot and pigments. Although a large fraction of lead in batteries is recycled, about 1/3 of the lead used in batteries is lost; this represents another area of potential improvement in the utilization of lead.

Zinc

Zinc is relatively abundant and not particularly toxic, so its industrial ecology is of less concern than that of toxic lead or scarce chromium. As with other metals, the mining and processing of zinc can pose some environmental concerns. Zinc occurs as ZnS (a mineral called sphalerite), and the sulfur must be reclaimed in the smelting of zinc. Zinc minerals often contain significant fractions of lead and copper, as well as significant amounts of toxic arsenic and cadmium.

Zinc is widely used as the metal, and lesser amounts are used to make zinc chemicals. One of the larger uses for zinc is as a corrosion-resistant coating on steel. This application, refined to a high degree in the automotive industry in recent years, has significantly lengthened the life span of automotive bodies and frames. It is difficult to reclaim zinc from zinc plating. However, zinc is a volatile element and it can be recovered in baghouse dust from electric arc furnaces used to reprocess scrap steel.

Zinc is used along with copper to make the alloy called brass. Brass is very well adapted to the production of various parts and objectives. It is recyclable, and significant quantities of brass are recycled as wastes from casting, machining, and as postconsumer waste.

Although a number of zinc compounds are synthesized and used, by far the most important of them is zinc oxide, ZnO. Formerly widely used as a paint pigment, this white substance is now employed as an accelerating and activating agent for hardening rubber products, particularly tires. Tire wear is a major vector for the transfer of zinc to the environment and, since it occurs with zinc, toxic cadmium is also dissipated to the environment by tire wear. The other two major compounds of zinc employed commercially are zinc chloride used in dry cells, as a disinfectant, and to vulcanize rubber, and zinc sulfide, used in zinc electroplating baths and to manufacture zinc-containing insecticides, particularly Zineb.®

Two aspects of zinc may be addressed in respect to its industrial ecology. The first of these is that, although it is not very toxic to animals, zinc is phytotoxic (toxic to plants) and soil can be "poisoned" by exposure to zinc from zinc smelting or from application of zinc-rich sewage sludge. The second of these is that the recycling of zinc is complicated by its dispersal as a plating on other metals. However, means do exist to reclaim significant fractions of such zinc, such as from electric arc furnaces as mentioned above.

Potassium

Potassium deserves special mention as a metal because the potassium ion, K^+, is an essential element required for plant growth. It is mined as potassium minerals and applied to soil as plant fertilizer. Potassium minerals consist of potassium salts, generally KCl. Such salts are found as deposits in the ground or may be obtained from some brines. Very large deposits are found in Saskatchewan, Canada. These salts are all quite soluble in water.

18.6.   NONMETAL MINERAL RESOURCES

A number of minerals other than those used to produce metals are important resources. There are so many of these that it is impossible to discuss them all in this chapter; however, mention will be made of the major ones. As with metals, the environmental aspects of mining many of these minerals are quite important. Typically, even the extraction of ordinary rock and gravel can have important environmental effects.

Clays are secondary minerals formed by weathering processes on parent minerals (see Chapter 15, Section 15.7). Clays have a variety of uses. About 70% of the clays used are miscellaneous clays of variable composition that have uses for a number of applications including filler (such as in paper), brick manufacture, tile manufacture, and portland cement production. Somewhat more than 10% of the clay used is fireclay, which has the characteristic of being able to withstand firing at high temperatures without warping. This clay is used to make a variety of refractories, pottery, sewer pipe, tile, and brick. Somewhat less than 10% of the clay that is used is kaolin, which has the general formula $Al_2(OH)_4Si_2O_5$. Kaolin is a white mineral that can be fired without losing shape or color. It is employed to make paper filler, refractories, pottery, dinnerware, and as a petroleum-cracking catalyst. About 7% of clay mined consists of bentonite and fuller's earth, a clay of variable composition used to make drilling muds, petroleum catalyst, carriers for pesticides, sealers, and clarifying oils. Very small quantities of a highly plastic clay called ball clay are used to make refractories, tile, and whiteware. U.S. production of clay is about 60 million metric tons per year, and global and domestic resources are abundant.

Fluorine compounds are widely used in industry. Large quantities of fluorspar, CaF_2, are required as a flux in steel manufacture. Synthetic and natural cryolite, Na_3AlF_6, is used as a solvent for aluminum oxide in the electrolytic preparation of aluminum metal. Sodium fluoride is added to water to help prevent tooth decay, a measure commonly called water fluoridation. World reserves of high-grade fluorspar are around 190 million metric tons, about 13% of which is in the United States. This is sufficient for several decades at projected rates of use. A great deal of by-product fluorine is recovered from the processing of fluorapatite, $Ca_5(PO_4)_3F$, used as a source of phosphorus (see below).

Micas are complex aluminum silicate minerals that are transparent, tough, flexible, and elastic. Muscovite, $K_2O\cdot3Al_2O_3\cdot6\ SiO_2\cdot2H_2O$, is a major type of mica. Better grades of mica are cut into sheets and used in electronic apparatus, capacitors, generators, transformers, and motors. Finely divided mica is widely used in roofing, paint, welding rods, and many other applications. Sheet mica is imported into the United States, and finely divided "scrap" mica is recycled domestically. Shortages of this mineral are unlikely.

Pigments and **fillers** of various kinds are used in large quantities. The only naturally occurring pigments still in wide use are those containing iron. These minerals are colored by limonite, an amorphous brown-yellow compound with the formula $2Fe_2O_3\cdot3H_2O$, and hematite, composed of gray-black Fe_2O_3. Along with varying quantities of clay and manganese oxides, these compounds are found in ocher, sienna, and umber. Manufactured pigments include carbon black, titanium dioxide, and zinc pigments. About 1.5 million metric tons of carbon black, manufactured by the partial combustion of natural gas, are used in the U.S. each year, primarily as a reinforcing agent in tire rubber.

Over 7 million metric tons of minerals are used in the U.S. each year as fillers for paper, rubber, roofing, battery boxes, and many other products. Among the minerals used as fillers are carbon black, diatomite, barite, fuller's earth, kaolin (see clays, above), mica, limestone, pyrophyllite, and wollastonite ($CaSiO_3$).

Although **sand** and **gravel** are the cheapest of mineral commodities per ton, the average annual dollar value of these materials is greater than all but a few mineral products because of the huge quantities involved. In tonnage, sand and gravel production is by far the greatest of nonfuel minerals. Almost 1 billion tons of sand and gravel are employed in construction in the U.S. each year, largely to make concrete structures, road paving, and dams. Slightly more than that amount is used to manufacture portland cement and as construction fill. Although ordinary sand is predominantly silica, SiO_2, about 30 million tons of a more pure grade of silica are consumed in the U.S. each year to make glass, high-purity silica, silicon semiconductors, and abrasives.

At present, old river channels and glacial deposits are used as sources of sand and gravel. Many valuable deposits of sand and gravel are covered by construction and lost to development. Transportation and distance from source to use are especially crucial for this resource. Environmental problems involved with defacing land can be severe, although bodies of water used for fishing and other recreational activities frequently are formed by removal of sand and gravel.

18.7. PHOSPHATES

Phosphate minerals are of particular importance because of their essential use in the manufacture of fertilizers applied to land to increase crop productivity. In addition, phosphorus is used for supplementation of animal feeds, synthesis of detergent builders, and preparation of chemicals such as pesticides and medicines. The most common phosphate minerals are hydroxyapatite, $Ca_5(PO_4)_3(OH)$, and fluorapatite, $Ca_5(PO_4)_3F$. Ions of Na, Sr, Th, and U are found substituted for calcium in apatite minerals. Small amounts of PO_4^{3-} may be replaced by AsO_4^{3-} and the arsenic must be removed for food applications. Approximately 17% of world phosphate production is from igneous minerals, primarily fluorapatites. About three-fourths of world phosphate production is from sedimentary deposits, generally of marine origin. Vast deposits of phosphate, accounting for approximately 5% of world phosphate production, are derived from guano droppings of seabirds and bats. Current U.S. production of phosphate rock is around 40 million metric tons per year, most of it from Florida. Idaho, Montana, Utah, Wyoming, North Carolina, South Carolina, and Tennessee also have sources of phosphate. Reserves of phosphate minerals in the United States amount to 10.5 billion metric tons, containing approximately 1.4 billion metric tons of phosphorus.

Phosphate in the naturally occurring minerals is not sufficiently available to be used as fertilizer. For commercial phosphate fertilizer production, these minerals are treated with phosphoric or sulfuric acids to produce more soluble superphosphates.

$$2Ca_5(PO_4)_3F(s) + 14H_3PO_4 + 10H_2O \rightarrow$$
$$2HF(g) + 10Ca(H_2PO_4)_2 \cdot H_2O \qquad (18.7.1)$$

$$2Ca_5(PO_4)_3F(s) + 7H_2SO_4 + 3H_2O \rightarrow$$
$$2HF(g) + 3Ca(H_2PO_4)_2 \cdot H_2O + 7CaSO_4 \qquad (18.7.2)$$

The HF produced as a byproduct of superphosphate production can create air pollution problems, and the recovery of fluorides is an important aspect of the industrial ecology of phosphate production.

Phosphate minerals are rich in trace elements required for plant growth, such as boron, copper, manganese, molybdenum, and zinc. Ironically, these elements are lost in processing phosphate for fertilizers and are sometimes added later.

Ammonium phosphates are excellent, highly soluble phosphate fertilizers. Liquid ammonium polyphosphate fertilizers consisting of ammonium salts of pyro-phosphate, triphosphate, and small quantities of higher polymeric phosphate anions in aqueous solution can be used as phosphate fertilizers. The polyphosphates are believed to have the additional advantage of chelating iron and other micronutrient metal ions, thus making the metals more available to plants.

There are at least two major reasons why the industrial ecology of phosphorus is particularly important. The first of these is that current rates of phosphate use would exhaust known reserves of phosphate within two or three generations. Although additional sources of phosphorus will be found and exploited, it is clear that this essential mineral is in distressingly short supply relative to human consumption; phosphate shortages, along with sharply higher prices, will eventually cause a crisis in food production. The second significant aspect of the industrial ecology of phosphorus is the pollution of waterways by waste phosphate, a plant and algal nutrient. This results in excessive growth of algae in the water, followed by decay of the plant biomass, consumption of dissolved oxygen, and an undesirable condition of eutrophication.

Excessive use of phosphate coupled with phosphate pollution suggests that phosphate wastes, such as from sewage treatment, should be substituted as sources of plant fertilizer. Several other partial solutions to the problem of phosphate shortages are the following:

- Development and implementation of methods of fertilizer application that maximize efficient utilization of phosphate

- Genetic engineering of plants that have minimal phosphate requirements and that utilize phosphorus with maximum efficiency

- Development of systems to maximize the utilization of phosphorus-rich animal wastes

18.8. SULFUR

Sulfur is an important nonmetal; its greatest single use is in the manufacture of sulfuric acid. However, the element is employed in a wide variety of other industrial and agricultural products. Current consumption of sulfur amounts to approximately 10 million metric tons per year in the United States. The four most important sources of sulfur are (in decreasing order) deposits of elemental sulfur, H_2S recovered from sour natural gas, organic sulfur recovered from petroleum, and pyrite (FeS_2). Recovery of sulfur from coal used as a fuel is a huge potential, largely untapped source of this important nonmetal.

The resource situation for sulfur differs from that of phosphorus in several significant respects. Although sulfur is an essential nutrient like phosphorus, most soils contain sufficient amounts of nutrient sulfur, and the major uses of sulfur are in the industrial sector. The sources of sulfur are varied and abundant and supply is no problem either in the United States or worldwide; sulfur recovery from fossil fuels as a pollution control measure could even result in surpluses of this element.

About 90% of the use of sulfur in the world is for the manufacture of sulfuric acid. Almost 2/3 of the sulfuric acid consumed is used to make phosphate fertilizers as discussed in Section 18.7, in which case the phosphorus ends up as waste "phospho-gypsum," $CaSO_4 \cdot xH_2O$. Other uses of sulfur include lead storage batteries, steel pickling, petroleum refining, extraction of copper from copper ore, and the chemical industry.

The industrial ecology of sulfur needs to emphasize reduction of wastes and sulfur pollution, rather than supply of this element. Unlike many resources, such as most common metals, the uses of sulfur are for the most part dissipative, and the sulfur is "lost" to agricultural land, paper products, petroleum products, or other environmental sinks. There are two major environmental concerns with sulfur. One of these is the emission of sulfur into the atmosphere, which occurs mostly as pollutant sulfur dioxide and is largely manifested by production of acidic precipitation and dry deposition. The second major environmental concern with sulfur is that it is used mostly as sulfuric acid and is not incorporated into products, thus posing the potential to pollute water and create acidic wastes. Acid purification units are available to remove significant amounts of sulfuric acid from waste acid solutions for recycling.

Gypsum

Calcium sulfate in the form of the dihydrate $CaSO_4 \cdot 2H_2O$ is the mineral **gypsum**, one of the most common forms in which waste sulfur is produced. As noted, large quantities of this material are produced as a by-product of phosphate fertilizer manufacture. Another major source of gypsum is its production when lime is used to remove sulfur dioxide from power plant stack gas,

$$Ca(OH)_2 + SO_2 \rightarrow CaSO_3 + H_2O \tag{18.8.1}$$

to produce a calcium sulfite product that can be oxidized to calcium sulfate. About 100 million metric tons of gypsum are mined each year for a variety of uses, including production of portland cement, to produce wallboard, as a soil conditioner to loosen tight clay soils, and numerous other applications.

Calcium sulfate from industrial or natural (gypsum) sources can be calcined at a very low temperature of only 159°C to produce $CaSO_4 \cdot 1/2H_2O$, a material known as plaster of Paris, which was once commonly used for the manual application of plaster to walls. Plaster of Paris mixed with water forms a plastic material which sets up as the solid dihydrate,

$$CaSO_4 \cdot {}^1/_2H_2O + {}^3/_2H_2O \rightarrow CaSO_4 \cdot 2H_2O \tag{18.8.2}$$

Cast into sheets coated with paper, this material produces plasterboard commonly used for the interior walls of homes and other buildings. Historically, plaster of Paris was used for mortar and other structural applications, and it has the potential for similar applications today.

The very large quantities of gypsum that are mined suggest that by-product calcium sulfate, especially that produced with phosphate fertilizers and from flue gas desulfurization, should be a good candidate for reclamation through the practice of industrial ecology. The low temperature (see above) required to convert hydrated calcium sulfate to $CaSO_4 \cdot {}^1/_2H_2O$, which can be set up as a solid by mixing with water, suggests that the energy requirements for a gypsum-based by-products industry should be modest. Low-density gypsum blown as a foam and used as a filler in composites along with sturdy reinforcing materials should have good insulating, fire-resistant, and structural properties for building construction.

18.9. WOOD—A MAJOR RENEWABLE RESOURCE

Fortunately, one of the major natural resources in the world, wood, is a renewable resource. Production of wood and wood products is the fifth largest industry in the United States, and forests cover one-third of the United States surface area. Wood ranks first worldwide as a raw material for the manufacture of other products, including lumber, plywood, particle board, cellophane, rayon, paper, methanol, plastics, and turpentine.

Chemically, wood is a complicated substance consisting of long cells having thick walls composed of polysaccharides such as cellulose,

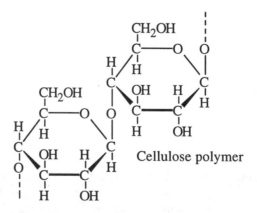

Cellulose polymer

The polysaccharides in cell walls account for approximately three-fourths of *solid wood*, wood from which extractable materials have been removed by an alcohol-benzene mixture. Wood typically contains a few tenths of a percent of ash (mineral residue left from the combustion of wood).

A wide variety of organic compounds can be extracted from wood by water, alcohol-benzene, ether, and steam distillation. These compounds include tannins, pigments, sugars, starch, cyclitols, gums, mucilages, pectins, galactans, terpenes, hydrocarbons, acids, esters, fats, fatty acids, aldehydes, resins, sterols, and waxes. Substantial amounts of methanol (sometimes called *wood alcohol*) are obtained from wood, particularly when it is pyrolyzed. Methanol, once a major source of liquid fuel, is now being used to a limited extent as an ingredient of some gasoline blends (see gasohol in Section 18.19).

A major use of wood is in paper manufacture. The widespread use of paper is a mark of an industrialized society. The manufacture of paper is a highly advanced technology. Paper consists essentially of cellulosic fibers tightly pressed together. The lignin fraction must first be removed from the wood, leaving the cellulosic fraction. Both the sulfite and alkaline processes for accomplishing this separation have resulted in severe water and air pollution problems, now significantly alleviated through the application of advanced treatment technologies.

Wood fibers and particles can be used for making fiberboard, paper-base laminates (layers of paper held together by a resin and formed into the desired structures at high temperatures and pressures), particle board (consisting of wood particles bonded together by a phenol-formaldehyde or urea-formaldehyde resin), and nonwoven textile substitutes consisting of wood fibers held together by adhesives. Chemical processing of wood enables the manufacture of many useful products, including methanol and sugar. Both of these substances are potential major products from the 60 million metric tons of wood wastes produced in the U.S. each year.

18.10. THE ENERGY PROBLEM

Since the 1973-74 "energy crisis," much has been said and written, many learned predictions have gone awry, and some concrete action has even taken place. Catastrophic economic disruption, people "freezing in the dark," and freeways given over to bicycles (perhaps a good idea) have not occurred. Nevertheless, uncertainties over petroleum availability and price and disruptions such as the 1990 Gulf War have caused energy to be one of the major problems of modern times.

In the U.S. concern over energy supplies and measures taken to ensure alternate supplies reached a peak in the late 1970s. Significant programs on applied energy research were undertaken in the areas of renewable energy sources, efficiency, and fossil fuels. The financing of these efforts reached a peak around 1980, then dwindled significantly after that date. By the year 2000, an abundance of fossil energy had resulted in a false sense of security regarding energy sources.

The solutions to energy problems are strongly tied to environmental considerations. For example, a massive shift of the energy base to coal in nations that now rely largely on petroleum for energy would involve much more strip mining, potential production of acid mine water, use of scrubbers, and release of greenhouse gases (carbon dioxide from coal combustion and methane from coal mining). Similar examples could be cited for most other energy alternatives.

Dealing with the energy problem requires a heavy reliance on technology, which is discussed in numerous places in this book. Computerized control of transportation and manufacturing processes enables much more efficient utilization of energy. New and improved materials enable higher peak temperatures and therefore greater extraction of usable energy in thermal energy conversion processes. Innovative manufacturing processes have greatly lowered the costs of photovoltaic cells used to convert sunlight directly to energy.

18.11. WORLD ENERGY RESOURCES

At present, most of the energy consumed by humans is produced from fossil fuels. Estimates of the amounts of fossil fuels available differ; those of the quantities of recoverable fossil fuels in the world before 1800 are given in Figure 18.3. By far the greatest recoverable fossil fuel is in the form of coal and lignite. Furthermore, only a small percentage of this energy source has been utilized to date, whereas much of the recoverable petroleum and natural gas has already been consumed. Projected use of these latter resources indicates rapid depletion.

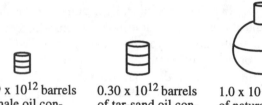

0.19 x 10^{12} barrels of shale oil containing 0.32 x 10^{15} kw-hr energy

0.30 x 10^{12} barrels of tar-sand oil containing 0.51 x 10^{15} kw-hr energy

1.0 x 10^{16} cubic feet of natural gas containing 2.94 x 10^{15} kw-hr energy

2.0 x 10^{12} barrels of liquid petroleum containing 3.25 x 10^{15} kw-hr energy

7.6 x 10^{12} metric tons of coal and lignite, containing 55.9 x 10^{15} kw-hr of energy

Figure 18.3. Original amounts of the world's recoverable fossil fuels (quantities in thermal kilowatt hours of energy based upon data taken from M. K. Hubbert, "The Energy Resources of the Earth," in *Energy and Power*, W. H. Freeman and Co., San Francisco, 1971).

Although world coal resources are enormous and potentially can fill energy needs for a century or two, their utilization is limited by environmental disruption from mining and emissions of carbon dioxide and sulfur dioxide. These would become intolerable long before coal resources were exhausted. Assuming only uranium-235 as a fission fuel source, total recoverable reserves of nuclear fuel are roughly about the same as fossil fuel reserves. These are many orders of magnitude higher if the use of breeder reactors is assumed. Extraction of only 2% of the deuterium present in the earth's oceans would yield about a billion times as much energy by controlled nuclear fusion as was originally present in fossil fuels! This prospect is tempered by the lack of success in developing a controlled nuclear fusion reactor. Geothermal power, currently utilized in northern California, Italy, and New Zealand, has the potential for providing a significant percentage of energy worldwide. The same limited potential is characteristic of several renewable energy resources, including hydroelectric energy, tidal energy, and especially wind power. All of these will continue to contribute significant, but relatively small, amounts of energy. Renewable, nonpolluting solar energy comes as close to being an ideal energy source as any available. It almost certainly has a bright future.

18.12. ENERGY CONSERVATION

Any consideration of energy needs and production must take energy conservation into consideration. This does not have to mean cold classrooms with thermostats set at 60°F in mid-winter, nor swelteringly hot homes with no air-conditioning, nor total reliance on the bicycle for transportation, although these and even more severe conditions are routine in many countries. The fact remains that the United States has wasted energy at a deplorable rate. For example, U.S. energy consumption is higher per capita than that of some other countries that have equal, or significantly better, living standards. Obviously, a great deal of potential exists for energy conservation that will ease the energy problem.

Transportation is the economic sector with the greatest potential for increased efficiencies. The private auto and airplane are only about one-third as efficient as buses or trains for transportation. Transportation of freight by truck requires about 3800 Btu/ton-mile, compared to only 670 Btu/ton-mile for a train. It is terribly inefficient compared to rail transport (as well as dangerous, labor-intensive, and environmentally disruptive). Major shifts in current modes of transportation in the U.S. will not come without anguish, but energy conservation dictates that they be made.

Household and commercial uses of energy are relatively efficient. Here again, appreciable savings can be made. The all-electric home requires much more energy (considering the percentage wasted in generating electricity) than a home heated with fossil fuels. The sprawling ranch-house style home uses much more energy per person than does an apartment unit or row house. Improved insulation, sealing around the windows, and other measures can conserve a great deal of energy. Electric generating plants centrally located in cities can provide waste heat for commercial and residential heating and cooling and, with proper pollution control, can use municipal refuse for part of their fuel, thus reducing quantities of solid wastes requiring disposal. As scientists and engineers undertake the crucial task of developing alternative energy sources to replace dwindling petroleum and natural gas supplies, energy conservation must receive proper emphasis. In fact, zero energy-use growth, at least on a per capita basis, is a worthwhile and achievable goal. Such a policy would go a long way toward solving many environmental problems. With ingenuity, planning, and proper management, it could be achieved while increasing the standard of living and quality of life.

18.13. ENERGY CONVERSION PROCESSES

As shown in Figure 18.4, energy occurs in several forms and must be converted to other forms. The efficiencies of conversion vary over a wide range. Conversion of electrical energy to radiant energy by incandescent light bulbs is very inefficient—less than 5% of the energy is converted to visible light and the remainder is wasted as heat. At the other end of the scale, a large electrical generator is around 80% efficient in producing electrical energy from mechanical energy. The once much-publicized Wankel rotary engine converts chemical to mechanical energy with an efficiency of about 18%, compared to 25% for a gasoline-powered piston engine and about 37% for a diesel engine. A modern coal-fired steam-generating power plant converts chemical energy to electrical energy with an overall efficiency of about 40%.

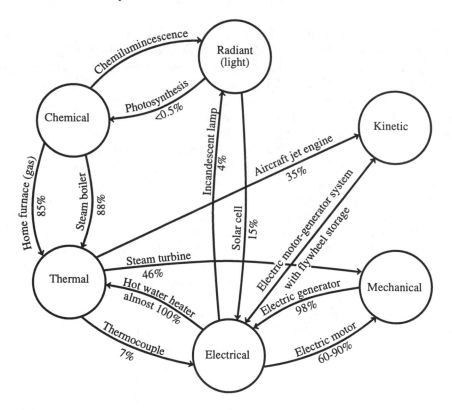

Figure 18.4.   Kinds of energy and examples of conversion between them, with conversion efficiency percentages.

One of the most significant energy conversion processes is that of thermal energy to mechanical energy in a heat engine such as a steam turbine. The Carnot equation,

$$\text{Percent efficiency} = \frac{T_1 - T_2}{T_1} \times 100 \qquad (18.13.1)$$

states that the percent efficiency is given by a fraction involving the inlet temperature (for example, of steam), T_1, and the outlet temperature, T_2. These temperatures are expressed in Kelvin (°C + 273). Typically, a steam turbine engine operates with approximately 810 K inlet temperature and 330 K outlet temperature. These temperatures substituted into the Carnot equation give a maximum theoretical efficiency of 59%. However, because it is not possible to maintain the incoming steam at the maximum temperature and because mechanical energy losses occur, overall efficiency of conversion of thermal energy to mechanical energy in a modern steam power plant is approximately 47%. Taking into account losses from conversion of chemical to thermal energy in the boiler, the total efficiency is about 40%.

Some of the greatest efficiency advances in the conversion of chemical to mechanical or electrical energy have been made by increasing the peak inlet temperature in heat engines. The use of superheated steam has raised T_1 in a steam power plant from around 550 K in 1900 to about 850 K at present. Improved materials and engineering design, therefore, have resulted in large energy savings.

The efficiency of nuclear power plants is limited by the maximum temperatures attainable. Reactor cores would be damaged by the high temperatures used in fossil-fuel-fired boilers and have a maximum temperature of approximately 620 K. Because of this limitation, the overall efficiency of conversion of nuclear energy to electricity is about 30%.

Most of the 60% of energy from fossil-fuel-fired power plants and 70% of energy from nuclear power plants that is not converted to electricity is dissipated as heat, either into the atmosphere or into bodies of water and streams. The latter is thermal pollution, which may either harm aquatic life or, in some cases, actually increase bioactivity in the water to the benefit of some species. This waste heat is potentially very useful in applications like home heating, water desalination, and aquaculture (growth of plants in water).

Some devices for the conversion of energy are shown in Figure 18.5. Substantial advances have been made in energy conversion technology over many decades and more can be projected for the future. The use of higher temperatures and larger generating units have increased the overall efficiency of fossil-fueled electrical power generation from less than 4% in 1900 to more than 40%. An approximately four-fold increase in the energy-use efficiency of rail transport occurred during the 1940s and 1950s with the replacement of steam locomotives with diesel locomotives. During the coming decades, increased efficiency can be anticipated from such techniques as combined power cycles in connection with generation of electricity. Magnetohydrodynamics (Figure 18.7) may be developed as a very efficient energy source used in combination with conventional steam generation. Entirely new devices such as thermonuclear reactors for the direct conversion of nuclear fusion energy to electricity will possibly be developed.

18.13. PETROLEUM AND NATURAL GAS

Since its first commercial oil well in 1859, the United States has produced somewhat more than 100 billion barrels of oil, most of it in recent years. In 1994 world petroleum consumption was at a rate of about 65 million barrels per day.

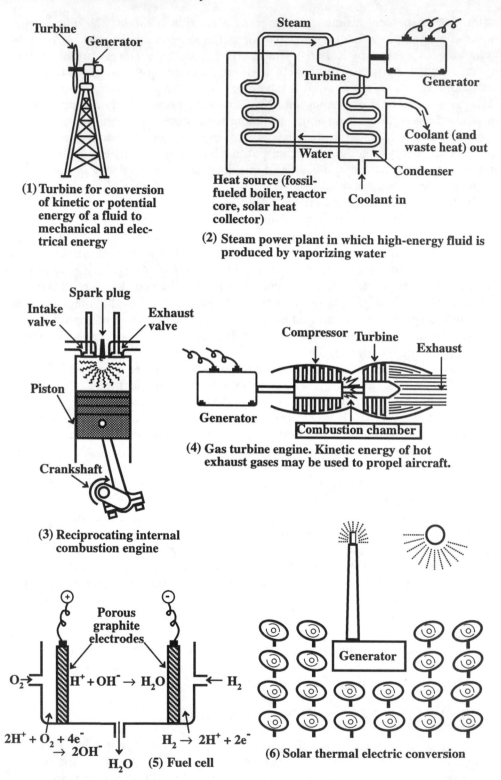

(1) Turbine for conversion of kinetic or potential energy of a fluid to mechanical and electrical energy

(2) Steam power plant in which high-energy fluid is produced by vaporizing water

(3) Reciprocating internal combustion engine

(4) Gas turbine engine. Kinetic energy of hot exhaust gases may be used to propel aircraft.

(5) Fuel cell

$2H^+ + O_2 + 4e^- \rightarrow 2OH^-$

$H^+ + OH^- \rightarrow H_2O$

$H_2 \rightarrow 2H^+ + 2e^-$

(6) Solar thermal electric conversion

Figure 18.5. Some energy conversion devices.

Liquid petroleum is found in rock formations ranging in porosity from 10 to 30%. Up to half of the pore space is occupied by water. The oil in these formations must flow over long distances to an approximately 15-cm diameter well from which it is pumped. The rate of flow depends on the permeability of the rock formation, the viscosity of the oil, the driving pressure behind the oil, and other factors. Because of limitations in these factors, **primary recovery** of oil yields an average of about 30% of the oil in the formation, although it is sometimes as little as 15%. More oil can be obtained using **secondary recovery** techniques, which involve forcing water under pressure into the oil-bearing formation to drive the oil out. Primary and secondary recovery together typically extract somewhat less than 50% of the oil from a formation. Finally, **tertiary recovery** can be used to extract even more oil, normally through the injection of pressurized carbon dioxide, which forms a mobile solution with the oil and allows it to flow more easily to the well. Other chemicals, such as detergents, may be used to aid in tertiary recovery. Currently, about 300 billion barrels of U.S. oil are not available through primary recovery alone. A recovery efficiency of 60% through secondary or tertiary techniques could double the amount of available petroleum. Much of this would come from fields which have already been abandoned or essentially exhausted using primary recovery techniques.

Shale oil is a possible substitute for liquid petroleum. Shale oil is a pyrolysis product of oil shale, a rock containing organic carbon in a complex structure of biological origin from eons past called kerogen. Oil shale is believed to contain approximately 1.8. trillion barrels of shale oil that could be recovered from deposits in Colorado, Wyoming, and Utah. In the Colorado Piceance Creek basin alone, more than 100 billion barrels of oil could be recovered from prime shale deposits.

Shale oil may be recovered from the parent mineral by retorting the mined shale in a surface retort. A major environmental disadvantage is that this process requires the mining of enormous quantities of mineral and disposal of the spent shale, which has a volume greater than the original mineral. *In situ* retorting limits the control available over infiltration of underground water and resulting water pollution. Water passing through spent shale becomes quite saline, so there is major potential for saltwater pollution.

During the late 1970s and early 1980s, several corporations began building facilities for shale oil extraction in northwestern Colorado. Large investments were made in these operations, and huge expenditures were projected for commercialization. Falling crude oil prices caused all these operations to be canceled. A large project for the recovery of oil from oil sands in Alberta, Canada, was also canceled in the 1980s.

Natural gas, consisting almost entirely of methane, has become more attractive as an energy source with recent discoveries and development of substantial new sources of this premium fuel. In addition to its use as a fuel, natural gas can be converted to many other hydrocarbon materials. It can be used as a raw material for the Fischer-Tropsch synthesis of gasoline. New unconventional sources of natural gas, such as may exist in geopressurized zones, could provide abundant energy reserves for the U.S., though at substantially increased prices.

18.14. COAL

From Civil War times until World War II, coal was the dominant energy source behind industrial expansion in most nations. However, after World War II, the greater convenience of lower-cost petroleum resulted in a decrease in the use of coal for energy in the U.S. and in a number of other countries. Annual coal production in the U.S. fell by about one third, reaching a low of approximately 400 million tons in 1958. Since that time U.S. production has increased. Several statistics illustrate the importance of coal as a source of energy by earth's population. Overall, about one-third of the energy used by humankind is provided from coal. The percentage of electricity generated by coal is even higher, around 45%. Almost three-fourths of the energy and coke used to make steel, the commodity commonly taken as a measure of industrial development, is provided by coal.

The general term *coal* describes a large range of solid fossil fuels derived from partial degradation of plants. Table 18.2 shows the characteristics of the major classes of coal found in the U.S., differentiated largely by percentage of fixed carbon, percentage of volatile matter, and heating value (*coal rank*). Chemically, coal is a very complex material and is by no means pure carbon. For example, a chemical formula expressing the composition of Illinois No. 6 bituminous coal is $C_{100}H_{85}S_{2.1}N_{1.5}O_{9.5}$.

Table 18.2. Major Types of Coal Found in the United States

| Type of Coal | Proximate analysis, percent[1] | | | | Range of heating value (Btu/pound) |
	Fixed carbon	Volatile matter	Moisture	Ash	
Anthracite	82	5	4	9	13,000 - 16,000
Bituminous					
Low-volatile	66	20	2	12	11,000 - 15,000
Medium-volatile	64	23	3	10	11,000 - 15,000
High-volatile	46	44	6	4	11,000 - 15,000
Subbituminous	40	32	19	9	8,000 - 12,000
Lignite	30	28	37	5	5,500 - 8,000

[1] These values may vary considerably with the source of coal.

Figure 18.6 shows areas in the U.S. with major coal reserves. Anthracite, a hard, clean-burning, low-sulfur coal, is the most desirable of all coals. Approximately half of the anthracite originally present in the United States has been mined. Bituminous coal found in the Appalachian and north central coal fields has been widely used. It is an excellent fuel with a high heating value. Unfortunately, most bituminous coals have a high percentage of sulfur (an average of 2-3%), so the use of this fuel presents environmental problems. Huge reserves of virtually untouched subbituminous and lignite coals are found in the Rocky Mountain states and in the northern plains of the Dakotas, Montana, and Wyoming. Despite some disadvantages, the low sulfur content and ease of mining these low-polluting fuels are resulting in a rapid increase in their use, and the sight of long unit trains carrying these fuels from western states to power plants in the eastern U. S. have become very common.

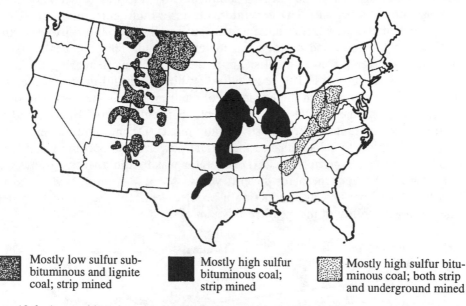

Mostly low sulfur subbituminous and lignite coal; strip mined

Mostly high sulfur bituminous coal; strip mined

Mostly high sulfur bituminous coal; both strip and underground mined

Figure 18.6. Areas with major coal reserves in the coterminous United States.

The extent to which coal can be used as a fuel depends upon solutions to several problems, including (1) minimizing the environmental impact of coal mining; (2) removing ash and sulfur from coal prior to combustion; (3) removing ash and sulfur dioxide from stack gas after combustion; (4) conversion of coal to liquid and gaseous fuels free of ash and sulfur; and, most important, (5) whether or not the impact of increased carbon dioxide emissions upon global climate can be tolerated. Progress is being made on minimizing the environmental impact of mining. As more is learned about the processes by which acid mine water is formed, measures can be taken to minimize the production of this water pollutant. Particularly on flatter lands, strip-mined areas can be reclaimed with relative success. Inevitably, some environmental damage will result from increased coal mining, but the environmental impact can be reduced by various control measures. Washing, flotation, and chemical processes can be used to remove some of the ash and sulfur prior to burning. Approximately half of the sulfur in the average coal occurs as pyrite, FeS_2, and half as organic sulfur. Although little can be done to remove the latter, much of the pyrite can be separated from most coals by physical and chemical processes.

The maintenance of air pollution emission standards requires the removal of sulfur dioxide from stack gas in coal-fired power plants. Stack gas desulfurization presents some economic and technological problems; the major processes available for it are summarized in Chapter 11, Section 11.5.

Magnetohydrodynamic power combined with conventional steam generating units has the potential for a major breakthrough in the efficiency of coal utilization. A schematic diagram of magnetohydrodynamic (MHD) generator is shown in Figure 18.7. This device uses a plasma of ionized gas at around 2400°C blasting through a very strong magnetic field of at least 50,000 gauss to generate direct current. The ionization of the gas is accomplished by injecting a "seed" of cesium or potassium salts. In an MHD generator, the ultra-high-temperature gas issuing through a supersonic nozzle contains ash, sulfur dioxide, and nitrogen oxides, which severely erode and corrode the materials used. This hot gas is used to generate steam for a conventional steam power plant, thus increasing the overall efficiency of the process. The seed salts combine with sulfur dioxide and are recovered along with ash in the exhaust. Pollutant emissions are low. The overall efficiency of combined MHD-steam power plants should reach 60%, one and one-half times the maximum of present steam-only plants. Despite some severe technological difficulties, there is a chance that MHD power could become feasible on a large scale, and an experimental MHD generator was tied to a working power grid in the former Soviet Union for several years. As of the early 1990s, the U.S. Department of Energy was conducting a proof-of-concept project to help determine the practicability of magnetohydrodynamics.

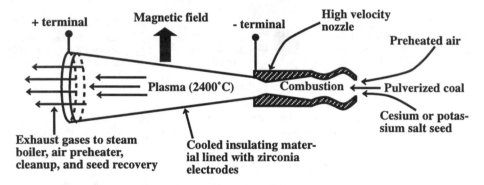

Figure 18.7. A magnetohydrodynamic power generator.

Coal Conversion

As shown in Figure 18.8, coal can be converted to gaseous, liquid, or low-sulfur, low-ash solid fuels such as coal char (coke) or solvent-refined coal (SRC). Coal conversion is an old idea; a house belonging to William Murdock at Redruth, Cornwall, England, was illuminated with coal gas in 1792. The first municipal coal-gas system was employed to light Pall Mall in London in 1807. The coal-gas industry began in the U.S. in 1816. The early coal-gas plants used coal pyrolysis (heating in the absence of air) to produce a hydrocarbon-rich product particularly useful for illumination. Later in the 1800s the water-gas process was developed, in which steam was added to hot coal to produce a mixture consisting primarily of H_2 and CO. It was necessary to add volatile hydrocarbons to this "carbureted" water-gas to bring its illuminating power up to that of gas prepared by coal pyrolysis. The U.S. had 11,000 coal gasifiers operating in the 1920s. At the peak of its use in

1947, the water-gas method accounted for 57% of U.S.-manufactured gas. The gas was made in low-pressure, low-capacity gasifiers that by today's standards would be inefficient and environmentally unacceptable (many sites of these old plants have been designated as hazardous waste sites because of residues of coal tar and other wastes). During World War II, Germany developed a major synthetic petroleum industry based on coal, which reached a peak capacity of 100,000 barrels per day in 1944. A synthetic petroleum plant operating in Sasol, South Africa, reached a capacity of several tens of thousands of tons of coal per day in the 1970s.

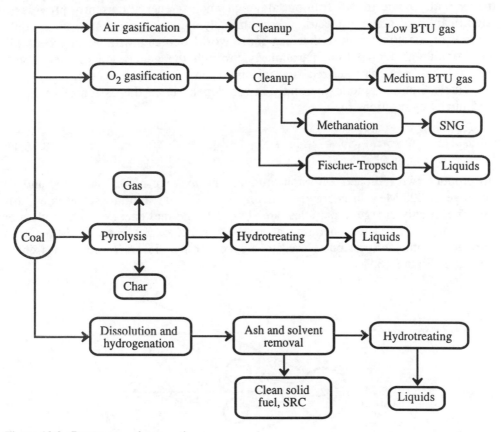

Figure 18.8. Routes to coal conversion.

The two broadest categories of coal conversion are gasification and lique-faction. Arguably the most developed route for coal gasification is the **Texaco process**, which gasifies a water slurry of coal at temperatures of 1250°C to 1500°C and pressures of 350 to 1200 pounds per square inch. Chemical addition of hydrogen to coal can liquefy it and produce a synthetic petroleum product. This can be done with a hydrogen donor solvent, which is recycled and itself hydrogenated with H_2 during part of the cycle. Such a process forms the basis of the successful **Exxon Donor Solvent process**, which has been used in a 250 ton/day pilot plant.

A number of environmental implications are involved in the widespread use of coal conversion. These include strip mining, water consumption in arid regions, lower overall energy conversion compared to direct coal combustion, and increased output of atmospheric carbon dioxide. These plus economic factors have prevented coal conversion from being practiced on a very large scale.

18.15. NUCLEAR FISSION POWER

The awesome power of the atom revealed at the end of World War II held out enormous promise for the production of abundant, cheap energy. This promise has never really come to full fruition, although nuclear energy currently provides a significant percentage of electric energy in many countries, and it may be the only source of electrical power that can meet world demand without unacceptable environmental degradation, particularly through the generation of greenhouse gases. It has been characterized as a "misunderstood" source of electricity.[2]

Nuclear power reactors currently in use depend upon the fission of uranium-235 nuclei by reactions such as

$$^{235}_{92}U + {}^{1}_{0}n \longrightarrow {}^{133}_{51}Sb + {}^{99}_{41}Nb + 4{}^{1}_{0}n \qquad (18.15.1)$$

to produce two radioactive fission products, an average of 2.5 neutrons, and an average of 200 MeV of energy per fission. The neutrons, initially released as fast-moving, highly energetic particles, are slowed to thermal energies in a moderator medium. For a reactor operating at a steady state, exactly one of the neutron products from each fission is used to induce another fission reaction in a chain reaction (Figure 18.9):

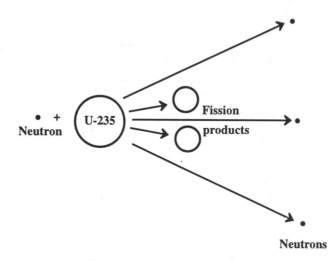

Figure 18.9. Fission of a uranium-235 nucleus.

The energy from these nuclear reactions is used to heat water in the reactor core and produce steam to drive a steam turbine, as shown in Figure 18.10.[3] As noted in Section 18.13 , temperature limitations make nuclear power less efficient in converting heat to mechanical energy and, therefore, to electricity, than fossil energy conversion processes.

A limitation of fission reactors is the fact that only 0.71% of natural uranium is fissionable uranium-235. This situation could be improved by the development of **breeder reactors**, which convert uranium-238 (natural abundance 99.28%) to fissionable plutonium-239.

A major consideration in the widespread use of nuclear fission power is the production of large quantities of highly radioactive waste products. These remain lethal for thousands of years. They must either be stored in a safe place or disposed of permanently in a safe manner. At the present time, spent fuel elements are being stored under water at the reactor sites. Eventually, the wastes from this fuel will have to be buried.

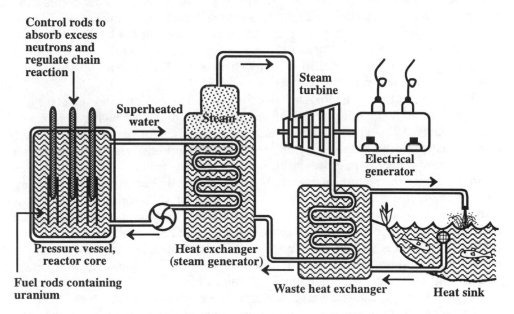

Figure 18.10. A typical nuclear fission power plant.

Another problem to be faced with nuclear fission reactors is their eventual decommissioning. There are three possible solutions. One is dismantling soon after shutdown, in which the fuel elements are removed, various components are flushed with cleaning fluids, and the reactor is cut up by remote control and buried. "Safe storage" involves letting the reactor stand 30-100 years to allow for radioactive decay, followed by dismantling. The third alternative is entombment, encasing the reactor in a concrete structure.

The course of nuclear power development was altered drastically by two accidents. The first of these occurred on March 28, 1979, with a partial loss of coolant water from the Metropolitan Edison Company's nuclear reactor located on Three Mile Island in the Susquehanna River, 28 miles outside of Harrisburg, Pennsylvania. The result was a loss of control, overheating, and partial disintegration of the reactor core. Some radioactive xenon and krypton gases were released and some radioactive water was dumped into the Susquehanna River. In August of 1993, cleanup workers finished evaporating the water from about 8 million liters of water solution contaminated by the reactor accident, enabling the reactor building to be sealed. A much worse accident occurred at Chernobyl in the Soviet

Union in April of 1986 when a reactor blew up spreading radioactive debris over a wide area and killing a number of people (officially 31, but probably many more).[4] Thousands of people were evacuated and the entire reactor structure had to be entombed in concrete. Food was seriously contaminated as far away as northern Scandinavia.

A much less serious, but still troublesome, nuclear accident occurred at the Tokaimura uranium processing plant in Japan on September 30, 1999, when a critical mass of enriched uranium was produced resulting in a chain reaction that exposed three workers to potentially lethal levels of radiation and contaminated 55 other workers and medical personnel. In violation of procedures, the workers used buckets to mix uranyl nitrate in a mixing tank, leading to an accumulation of about 16 kg of solution, greatly exceeding the tank's safety limit of 2.4 kg. The uranium, prepared for use in Japan's Nuclear Cycle Development Institutute's experimental fast-breeder reactor, was enriched to 18.8% fissionable ^{235}U, much higher than the 3–4% normally used in power reactors.

As of 1999, 21 years had passed since a new nuclear electric power plant had been ordered in the U.S., in large part because of the projected high costs of new nuclear plants. Although this tends to indicate hard times for the nuclear industry, pronouncements of its demise may be premature. Properly designed nuclear fission reactors can generate large quantities of electricity reliably and safely. For example, during the record summer 1993 Mississippi/Missouri River floods, many large fossil-fueled power plants were on the verge of shutting down because of disruptions of fuel supply normally delivered by river barge and train. During that time Union Electric's large Callaway nuclear plant in central Missouri ran continuously at full capacity, immune to the effects of the flood, probably saving a large area from a devastating, long-term power outage. The single most important factor that may lead to renaissance of nuclear energy is the threat to the atmosphere from greenhouse gases produced in large quantities by fossil fuels. It can be argued that nuclear energy is the only proven alternative that can provide the amounts of energy required within acceptable limits of cost, reliability, and environmental effects.

New designs for nuclear power plants should enable construction of power reactors that are much safer and environmentally acceptable than those built with older technologies. The proposed new designs incorporate built-in passive safety features that work automatically in the event of problems that could lead to incidents such as TMI or Chernobyl with older reactors. These devices—which depend upon phenomena such as gravity feeding of coolant, evaporation of water, or convection flow of fluids—give the reactor the desirable characteristics of **passive stability**. They have also enabled significant simplification of hardware, with only about half as many pumps, pipes, and heat exchangers as are contained in older power reactors.

18.16. NUCLEAR FUSION POWER

The two main reactions by which energy can be produced from the fusion of two light nuclei into a heavier nucleus are the deuterium-deuterium reaction,

$$_1^2H + {}_1^2H \rightarrow {}_2^3He + {}_0^1n + 1 \text{ Mev (energy released per fusion)} \qquad (18.16.1)$$

and the deuterium-tritium reaction:

$$_1^2H + {}_1^3H \rightarrow {}_2^4He + {}_0^1n + 17.6 \text{ Mev (energy released per fusion)} \qquad (18.16.2)$$

The second reaction is more feasible because less energy is required to fuse the two nuclei than to fuse two deuterium nuclei. However, the total energy from deuterium-tritium fusion is limited by the availability of tritium, which is made from nuclear reactions of lithium-6 (natural abundance, 7.4%). The supply of deuterium, however, is essentially unlimited; one out of every 6700 atoms of hydrogen is the deuterium isotope. The ^3He byproduct of the fusion of two deuterium nuclei, Reaction 18.16.1, reacts with neutrons, which are abundant in a nuclear fusion reactor, to produce tritium required for Reaction 18.16.2.

The power of nuclear fusion has not yet been harnessed in a sustained, controlled reaction of appreciable duration that produces more power than it consumes. Most approaches have emphasized **magnetic confinement**, the "squeezing" of a plasma (ionized gas) of fusionable nuclei in a strong magnetic field. In 1994 a record power level pulse of 10.7 megawatts (MW) was achieved from the fission of deuterium with tritium by the Tokamak Fusion Test Reactor operated by Princeton University for the U.S. Department of Energy.[5] This level exceeds 20% of the power put into the reactor to achieve fusion, which of course must be boosted to well over 100% for a self-sustained fusion reactor. Within three years after this record power pulse was achieved the Princeton Tokamak Fusion Test Reactor was shut down for lack of funding, although experiments on controlled nuclear fusion have continued at the facility and others around the world. The United States also withdrew from a huge international Tokamak project after $1 billion had been spent on the undertaking, although a consortium of European countries, Russia, and Japan continue to support the project.

An alternative to magnetic confinement is **inertial confinement** in which a pellet composed of deuterium and tritium frozen on the inside of a plastic-coated pellet smaller than a pinhead is bombarded by laser beams or X-rays, heating the fuel pellet to a temperature of about 100 million °C and causing fission of the deuterium and tritium nuclei. Each such event, literally a miniature thermonuclear explosion, can release energy equivalent to the explosion of about 45 kg of TNT.

With both magnetic confinement and inertial confinement reactors, a central challenge is in harnessing the energy once it is released. Much of the energy is in the form of neutrons, which react with nuclei, such as those of iron and copper composing the reactor structure, making it radioactive and causing metal embrittlement, which would rapidly destroy the reactor. Most proposed power reactors now call for a replaceable lining, probably composed of lithium, which would absorb the neutrons and produce heat energy.

A great flurry of excitement over the possibility of a cheap, safe, simple fusion power source was generated by an announcement from the University of Utah in 1989 of the attainment of "cold fusion" in the electrolysis of deuterium oxide (heavy water). Funding was appropriated and laboratories around the world were thrown into frenetic activity in an effort to duplicate the reported results. Some investigators reported evidence, particularly the generation of anomalously large amounts of heat, to support the idea of cold fusion, whereas others scoffed at the idea. Since that time cold fusion has been disproven, and the whole saga of it, described in a detailed book about the topic,[6] stands as a classic case of science gone astray.

Controlled nuclear fusion processes could be designed to produce almost no radioactive waste products. However, tritium used in the deuterium-tritium reaction is very difficult to contain, and some release of the isotope would occur. The deuterium-deuterium reaction promises an unlimited source of energy. Either of these reactions would be preferable to fission in terms of environmental considerations. Therefore, despite the possibility of insurmountable technical problems involved in harnessing fusion energy, the promise of this abundant, relatively nonpolluting energy source makes its pursuit well worth a massive effort.

18.17. GEOTHERMAL ENERGY

Underground heat in the form of steam, hot water, or hot rock used to produce steam has been used as an energy resource for about a century. This energy was first harnessed for the generation of electricity at Larderello, Italy, in 1904, and has since been developed in Japan, Russia, New Zealand, the Phillipines, and at the Geysers in northern California.

Underground dry steam is relatively rare, but is the most desirable from the standpoint of power generation. More commonly, energy reaches the surface as superheated water and steam. In some cases, the water is so pure that it can be used for irrigation and livestock; in other cases, it is loaded with corrosive, scale-forming salts. Utilization of the heat from contaminated geothermal water generally requires that the water be reinjected into the hot formation after heat removal to prevent contamination of surface water.

The utilization of hot rocks for energy requires fracturing of the hot formation, followed by injection of water and withdrawal of steam. This technology is still in the experimental state, but promises approximately ten times as much energy production as steam and hot-water sources.

Land subsidence and seismic effects are environmental factors that may hinder the development of geothermal power. However, this energy source holds considerable promise, and its development continues.

18.18. THE SUN: AN IDEAL ENERGY SOURCE

Solar power is an ideal source of energy that is unlimited in supply, widely available, and inexpensive. It does not add to the earth's total heat burden or produce chemical air and water pollutants. On a global basis, utilization of only a small fraction of solar energy reaching the earth could provide for all energy needs. In the United States, for example, with conversion efficiencies ranging from 10-30%, it would only require collectors ranging in area from one tenth down to one thirtieth that of the state of Arizona to satisfy present U.S. energy needs. (This is still an enormous amount of land, and there are economic and environmental problems related to the use of even a fraction of this amount of land for solar energy collection. Certainly, many residents of Arizona would not be pleased at having so much of the state devoted to solar collectors, and some environmental groups would protest the resultant shading of rattlesnake habitat.)

Solar power cells (photovoltaic cells) for the direct conversion of sunlight to electricity have been developed and are widely used for energy in space vehicles. With present technology, however, they remain too expensive for large-scale generation of electricity, although the economic gap is narrowing. Most schemes for the utilization of solar power depend upon the collection of thermal energy followed by conversion to electrical energy. The simplest such approach involves focusing sunlight on a steam-generating boiler (see Illustration 6 in Figure 18.5). Parabolic reflectors can be used to focus sunlight on pipes containing heat-transporting fluids. Selective coatings on these pipes can be used so that most of the incident energy is absorbed.

The direct conversion of energy in sunlight to electricity is accomplished by special solar voltaic cells. Such devices based on crystalline silicon have operated with a 15% efficiency for experimental cells and 11-12% for commercial units, at a cost of 25-50 cents per kilowatt-hour (kWh), about 5 times the cost of conventionally generated electricity. Part of the high cost results from the fact that the silicon used in the cells must be cut as small wafers from silicon crystals for mounting on the cell surfaces. Significant advances in costs and technology are being made with thin-film photovoltaics, which use an amorphous silicon alloy. A new approach to the design and construction of amorphous silicon film photovoltaic devices uses three layers of amorphous silicon to absorb, successively, short wavelength ("blue"), intermediate wavelength ("green"), and long wavelength ("red") light, as shown in Figure 18.11. Thin-film solar panels constructed with this approach have achieved solar-to-electricity energy conversion efficiencies just over 10%, lower than those using crystalline silicon, but higher than other amorphous

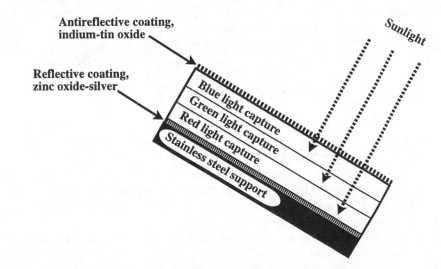

Figure 18.11. High-efficiency thin-film solar photovoltaic cell using amorphous silicon.

film devices. The low cost and relatively high conversion efficiencies of these solar panels should enable production of electricity at only about twice the cost of conventional electrical power, which would be competitive in some situations

A major disadvantage of solar energy is its intermittent nature. However, flexibility inherent in an electric power grid would enable it to accept up to 15% of its total power input from solar energy units without special provision for energy storage. Existing hydroelectric facilities may be used for pumped-water energy storage in conjunction with solar electricity generation. Heat or cold can be stored in water, in a latent form in water (ice) or eutectic salts, or in beds of rock. Enormous amounts of heat can be stored in water as a supercritical fluid contained at high temperatures and very high pressures deep underground. Mechanical energy can be stored with compressed air or flywheels.

Hydrogen gas, H_2, is an ideal chemical fuel that may serve as a storage medium for solar energy. Solar-generated electricity can be used to electrolyze water:

$$2H_2O + \text{electrical energy} \rightarrow 2H_2(g) + O_2(g) \qquad (18.18.1)$$

The hydrogen fuel product, and even oxygen, can be piped some distance and the hydrogen burned without pollution, or it may be used in a fuel cell (Illustration 5 in Figure 18.5). This may, in fact, make possible a "hydrogen economy." Disadvantages of using hydrogen as a fuel include its low heating value per unit volume and the wide range of explosive mixtures it forms with air. Although not yet economical, photochemical processes can be used to split water to H_2 and O_2 that can be used to power fuel cells.

No really insurmountable barriers exist to block the development of solar energy, such as might be the case with fusion power. In fact, the installation of solar space and water heaters became widespread in the late 1970s, and research on solar energy was well supported in the U.S. until after 1980, when it became fashionable to believe that free-market forces had solved the "energy crisis." With the installation of more heating devices and the probable development of some cheap, direct solar electrical generating capacity, it is likely that during the coming century solar energy will be providing an appreciable percentage of energy needs in areas receiving abundant sunlight.

The Surprising Success of Wind Power

Wind power is mentioned here because it is an indirect form of solar energy. During the 1990s, wind power emerged as a cost-competitive source of renewable energy with a remarkably high growth rate. Denmark has led other countries in establishing wind power as a significant fraction of its electrical generating capacity. Even in the United States wind power is gaining popularity,[7] and in 1999 the U. S. set a goal of providing a significant fraction of its electricity from wind within the next two decades.

In October 1996, the largest wind farm established up to that time in Europe was opened in Carno, Wales, by National Wind Power, Ltd. This was the 32nd wind farm in Britain, which was already generating enough electricity from wind to power 150,000 homes. Producing 33.6 megawatts of power, the 3-bladed turbines used to generate power at the Welsh facility are 56 meters in diameter and are mounted on towers 64 m high.

Northern regions, including parts of Alaska, Canada, the Scandanavian countries, and Russia often have consistently strong wind conditions conducive to the generation of wind power. Isolation from other sources of energy makes wind power attractive for many of these regions.[8] Severe climate conditions in these regions pose special challenges for wind generators. One problem can be the buildup of rime consisting of ice condensed directly on structures from super-cooled fog in air.[9]

18.19. ENERGY FROM BIOMASS

All fossil fuels originally came from photosynthetic processes. Photosynthesis does hold some promise of producing combustible chemicals to be used for energy production and could certainly produce all needed organic raw materials. It suffers from the disadvantage of being a very inefficient means of solar energy collection (a collection efficiency of only several hundredths of a percent by photosynthesis is typical of most common plants). However, the overall energy conversion efficiency of several plants, such as sugarcane, is around 0.6%. Furthermore, some plants, such as *Euphorbia lathyrus* (gopher plant), a small bush growing wild in California, produce hydrocarbon emulsions directly. The fruit of the Philippine plant, *Pittsosporum reiniferum*, can be burned for illumination due to its high content of hydrocarbon terpenes (see Section 12.2), primarily α-pinene and myrcene. Conversion of agricultural plant residues to energy could be employed to provide some of the energy required for agricultural production. Indeed, until about 80 years ago, virtually all of the energy required in agriculture—hay and oats for horses, home-grown food for laborers, and wood for home heating—originated from plant materials produced on the land. (An interesting exercise is to calculate the number of horses required to provide the energy used for transportation at the present time in the Los Angeles basin. It can be shown that such a large number of horses would fill the entire basin with manure at a rate of several feet per day.)

Annual world production of biomass is estimated at 146 billion metric tons, mostly from uncontrolled plant growth. Many farm crops and trees can produce 10-20 metric tons per acre per year of dry biomass, and some algae and grasses can produce as much as 50 metric tons per acre per year. The heating value of this biomass is 5000-8000 Btu/lb for a fuel having virtually no ash or sulfur (compare heating values of various coals in Table 18.2). Current world demand for oil and gas could be met with about 6% of the global production of biomass. Meeting U.S. demands for oil and gas would require that about 6-8% of the land area of the contiguous 48 states be cultivated intensively for biomass production. Another advantage of this source of energy that is becoming increasingly important as more is learned about potential greenhouse warming is that use of biomass for fuel would not add any net carbon dioxide to the atmosphere.

As it has been throughout history, biomass is significant as heating fuel, and in some parts of the world is the fuel most widely used for cooking. For example, as of the early 1990s, about 15% of Finland's energy needs were provided by wood and wood products (including black liquor by-product from pulp and paper manufacture), about 1/3 of which was from solid wood. Despite the charm of a wood fire and the sometimes pleasant odor of wood smoke, air pollution from wood-burning stoves and furnaces is a significant problem in some areas. Currently, wood provides about 8% of world energy needs. This percentage could increase through the development of "energy plantations" consisting of trees grown solely for their energy content.

Seed oils show promise as fuels, particularly for use in diesel engines. The most common plants producing seed oils are sunflowers and peanuts. More exotic species include the buffalo gourd, cucurbits, and Chinese tallow tree.

Biomass could be used to replace much of the 100 million metric tons of petroleum and natural gas currently consumed in the manufacture of primary chemicals in the world each year. Among the sources of biomass that could be used for chemical production are grains and sugar crops (for ethanol manufacture), oilseeds, animal by-products, manure, and sewage (the last two for methane generation). The biggest potential source of chemicals is the lignocellulose making up the bulk of most plant material. For example, both phenol and benzene might be produced directly from lignin. Brazil has had a program for the production of chemicals from fermentation-produced ethanol.

Gasohol

A major option for converting photosynthetically produced biochemical energy to forms suitable for internal combustion engines is the production of either methanol or ethanol. Either can be used by itself as fuel in a suitably designed internal combustion engine. More commonly, these alcohols are blended in proportions of up to 20% with gasoline to give **gasohol**, a fuel that can be used in existing internal combustion engines with little or no adjustment.

Gasohol boosts octane rating and reduces emissions of carbon monoxide. From a resource viewpoint, because of its photosynthetic origin, alcohol may be considered a renewable resource rather than a depletable fossil fuel. The manufacture of alcohol can be accomplished by the fermentation of sugar obtained from the hydrolysis of cellulose in wood wastes and crop wastes. Fermentation of these waste products offers an excellent opportunity for recycling. Cellulose has significant potential for the production of renewable fuels.

Ethanol is most commonly manufactured by fermentation of carbohydrates. Brazil, a country rich in potential to produce biomass such as sugarcane, has been a leader in the manufacture of ethanol for fuel uses, with 4 billion liters produced in 1982. At one time Brazil had over 450,000 automobiles that could run on pure alcohol, although many of these were converted back to gasoline during the era of relatively low petroleum prices since about 1980. Significant amounts of gasoline in the United States are supplemented with ethanol, more as an octane-ratings booster than as a fuel supplement.

Methanol, which can be blended with gasoline, can also be produced from biomass by the destructive distillation of wood (Section 18.9) or by converting biomass, such as wood, to CO and H_2, and synthesizing methanol from these gases.

18.20. FUTURE ENERGY SOURCES

As discussed in this chapter, a number of options are available for the supply of energy in the future. The major possibilities are summarized in Table 18.3.

18.21. EXTENDING RESOURCES THROUGH THE PRACTICE OF INDUSTRIAL ECOLOGY

A tremendous potential exists for applying the practice of industrial ecology to lower the burden on virgin raw materials and sources of energy. As discussed in Chapter 17, Section 17.8, these approaches include using less material (dematerialization), substitution of a relatively more abundant and safe material for one that is scarce and/or toxic, extracting useful materials from wastes (waste mining), and recycling materials and items. Properly applied, these measures can not only conserve increasingly scarce raw materials, but can increase wealth as it is conventionally defined.[10] Corresponding measures can also be applied to energy resources. In recent decades energy conservation ("de-energization"); substitution of energy sources, such as inexhaustible wind power for coal in the generation of electricity; and burning of municipal refuse to raise steam for electricity generation have reduced the need to utilize diminishing fossil energy resources and to build new power plants.

The greatest potential for extending material resources is by recycling through the practice of industrial ecology. In a sense, too, energy resources can be recycled by using otherwise waste materials to generate energy and by using heat that might otherwise go to waste for beneficial purposes, such as heating buildings.

Materials vary in their amenability to recycling. Arguably the most recyclable materials are metals in a relatively pure form. Such metals are readily melted and recast into other useful components. Among the least recyclable materials are mixed polymers or composites, the individual constituents of which cannot be readily separated. The chemistry of some polymers is such that, once they are prepared from monomers, they are not readily broken down again and reformed to a useful form. This section briefly addresses the kinds of materials that are recycled or that are candidates for recycling in a functional system of industrial ecology.

An important aspect of industrial ecology applied to recycling materials consists of the separation processes that are employed to "unmix" materials for recycling at the end of a product cycle. An example of this is the separation of graphite carbon fibers from the epoxy resins used to bind them together in carbon fiber composites. The chemical industry provides many examples where separations are required. For example, the separation of toxic heavy metals from solutions or sludges can yield a valuable metal product, leaving nontoxic water and other materials for safe disposal or reuse.

Metals

Pure metals are easily recycled, and the greatest challenge is to separate the metals into a pure state. The recycling process commonly involves reduction of metal oxides to the metal. One of the more difficult problems with metals recycling is the mixing of metals, such as occurs with metal alloys when a metal is plated

Table 18.3. Possible Future Sources of Energy

Source	Principles
Coal conversion	Manufacture of gas, hydrocarbon liquids, alcohol, or solvent-refined coal (SRC) from coal
Oil shale	Retorting petroleum-like fuel from oil shale
Geothermal	Utilization of underground heat
Gas-turbine electric	Utilization of hot combustion gases in a turbine, followed by a topping cycle involving steam generation
MHD	Electricity generated by passing a hot gas plasma through a magnetic field
Thermionics	Electricity generated across a thermal gradient
Fuel cells	Conversion of chemical to electrical energy
Solar heating and cooling	Direct use of solar energy for heating and cooling through the application of solar collectors
Solar cells	Use of silicon semiconductor sheets for the direct generation of electricity from sunlight
Solar thermal electric	Conversion of solar energy to heat followed by conversion to electricity
Wind	Conversion of wind energy to electricity
Ocean thermal electric	Use of ocean thermal gradients to convert heat energy to electricity
Nuclear fission	Conversion of energy released from fission of heavy nuclei to electricity
Breeder reactors	Nuclear fission combined with conversion of nonfissionable nuclei to fissionable nuclei
Nuclear fusion	Conversion of energy released by the fusion of light nuclei to electricity
Bottoming cycles	Utilization of waste heat from power generation for various purposes
Solid waste	Combustion of trash to produce heat and electricity
Photosynthesis	Use of plants for the conversion of solar energy to other forms by a biomass intermediate
Hydrogen	Generation of H_2 by thermochemical or photochemical means for use as an energy-transporting medium

onto another metal, or with components made of two or more metals in which it is hard to separate the metals. A common example of the complications from mixing metals is the contamination of iron with copper from copper wiring or other components made from copper. As an impurity, copper produces steel with inferior mechanical characteristics. Another problem is the presence of toxic cadmium used as plating on steel parts.

Recycling metals can take advantage of the technology developed over many years of technology for the separation of metals that occur together in ores. Examples of byproduct metals recovered during the refining of other metals are gallium from aluminum; arsenic from lead or copper; precious metal iridium, osmium, palladium, rhodium, and ruthenium from platinum; and cadmium, germanium, indium, and thorium from zinc.

Plastics

Much attention has been given to the recycling of plastics in recent years. Compared to metals, plastics are much less recyclable because recycling is technically difficult and plastics are less valuable than metals. There are two general classes of plastics, a fact that has a strong influence upon their recyclability. Thermoplastics are those that become fluid when heated and solid when cooled. Since they can be heated and reformed multiple times, thermoplastics are generally amenable to recycling. Recyclable thermoplastics include polyalkenes (low-density and high-density polyethylene and polypropylene); polyvinylchloride (PVC), used in large quantities to produce pipe, house siding, and other durable materials; polyethylene terephthalate; and polystyrene. Plastic packaging materials are commonly made from thermoplastics and are potentially recyclable. Fortunately, from the viewpoint of recycling, thermoplastics make up most of the quantities of plastics used.

Thermosetting plastics are those that form molecular cross linkages between their polymeric units when they are heated. These bonds set the shape of the plastic, which does not melt when it is heated. Therefore, thermosetting plastics cannot be simply remolded; they are not very amenable to recycling, and often burning them for their heat content is about the only use to which they may be put. An important class of thermosetting plastics consists of the epoxy resins, characterized by an oxygen atom bonded between adjacent carbons (1,2-epoxide or oxirane). Epoxies are widely used in composite materials combined with fibers of glass or graphite. Other thermosetting plastics include cross-linked phenolic polymers, some kinds of polyesters, and silicones. When recycling is contemplated, the best use for thermosetting plastics is for the fabrication of entire components that can be recycled.

Contaminants are an important consideration in recycling plastics. A typical kind of contaminant is paint used to color the plastic object. Adhesives and coatings of various kinds may also act as contaminants. Such materials may weaken the recycled material or decompose to produce gases when the plastic is heated for recycling. Toxic cadmium used to enable polymerization of plastics, a "tramp element" in recycling parlance, can hinder recycling of plastics and restrict the use of the recycled products.

Lubricating Oil

Lubricating oils are used in vast quantities and are prime candidates for recycling. The most simplistic means of recycling lubricating oil is to burn it, and large volumes of oil are burned for fuel. This is a very low level of recycling and will not be addressed further here.

For many years the main process for reclaiming waste lubricating oil used treatment with sulfuric acid followed by clay. This process generated large quantities of acid sludge and spent clay contaminated with oil. These undesirable byproducts contributed substantial amounts of wastes to hazardous waste disposal sites. Current state-of-the-art practices of lubricating oil reclamation do not utilize large quantities of clay for cleanup, but instead use solvents, vacuum distillation, and catalytic hydrofinishing to produce a usable material from spent lubricating oil.[11] The first step is dehydration to remove water and stripping to remove contaminant fuel (gasoline) fractions. If solvent treatment is used, the oil is then extracted with a solvent, such as isopropyl or butyl alcohols or methylethyl ketone. After treatment with a solvent, the waste oil is commonly centrifuged to remove impurities that are not soluble in the solvent. The solvent is then stripped from the oil. The next step is a vacuum distillation that removes a light fraction useful for fuel and a heavy residue. The lubricating oil can then be subjected to hydrofinishing over a catalyst to produce a suitable lubricating oil product.

LITERATURE CITED

1. Chapters 3–6 in *Industrial Ecology: Towards Closing the Materials Cycle*, Robert U. Ayres and Leslie W. Ayres, Edward Elgar, Cheltenham, U.K., 1996, pp. 32–96.

2. Carbon, Max W., *Nuclear Power: Villain or Victim? (Our Most Misunderstood Source of Electricity*, Pebble Beach Publishers, Madison, WI, 1997

3. Collier, John G. and Geoffrey F. Hewitt, *Introduction to Nuclear Power*, 2nd ed., Taylor & Francis, Washington, D.C., 1997.

4. Ebel, Robert E., *Chernobyl and its Aftermath: A Chronology of Events*, Center for Strategic & International Studies, Washington, DC, 1994.

5. "Reviving Quest to Tame Energy of the Stars," *New York Times*, June 8, 1999, pp. D1-D2.

6. Taubes, Gary, *Bad Science: The Short Life and Weird Times of Cold Fusion*, Random House, New York, 1993.

7. Giovando, CarolAnn, "Wind Energy Catches its 'Second Wind' in the US," *Power*, **142**, 92-95 (1998).

8. Gaudiosi, Gaetano, "Wind Farms in Northern Climates," *Environmental Engineering and Renewable Energy, Proceedings of the First International Conference (1998)*, Renato Gavasci and Sarantuyaa Zandaryaa, Eds., Elsevier Science, Oxford, U.K., 1999, pp. 161-170.

9. W. J., Jasinski, S. C. Noe, M. S. Selig, and M. B. Bragg, "Wind Turbine Performance under Icing Conditions," *Journal of Solar Energy Engineering*, **120**, 60-65 (1998).

10. von Weizsäcker, Ernst U., Amory B. Lovins, and L. Hunter Lovins, *Factor Four: Doubling Wealth, Halving Resource Use*, Earthscan, London, 1997.

11. McCabe, Mark M. and William Newton, "Waste Oil," Section 4.1 in *Standard Handbook of Waste Treatment and Disposal*, 2nd ed., Harry M. Freeman, Ed., McGraw-Hill, New York, 1998, pp. 4.3–4.13.

SUPPLEMENTARY REFERENCES

Anderson, Ewan W. and Liam D. Anderson, *Strategic Minerals: Resource Geopolitics and Global Geo-economics*, John Wiley & Sons, New York, 1998.

Aubrecht, Gordon J., *Energy*, 2nd ed., Prentice Hall, Upper Saddle River, NJ, 1995.

Auty, Richard M. and Raymond F. Mikesell, *Sustainable Development in Mineral Economies*, Clarendon, Oxford, U. K., 1998.

Azcue, Jose M., Ed., *Environmental Impacts of Mining Activities: Emphasis on Mitigation and Remedial Measures*, Springer-Verlag, Berlin, 1999.

Bisio, Attilio, and Sharon R. Boots, *Energy Technology and the Environment*, John Wiley and Sons, New York, 1995.

Cohn, Steven Mark, *Too Cheap to Meter: An Economic and Philosophical Analysis of the Nuclear Dream*, State University of New York Press, Albany, NY, 1997.

Daley, Michael J., *Nuclear Power: Promise or Peril?*, Lerner Publications Co., Minneapolis, MN, 1997.

Featherstone, Jane, *Energy*, Raintree Steck-Vaughn, Austin, TX, 1999.

Gallios, G. P. and K.A. Matis, Eds., *Mineral Processing and the Environment*, Kluwer Academic Publishers, Boston, 1998.

Giovando, CarolAnn, "Wind Energy Catches its 'Second Wind' in the U. S.," *Power*, 92-95, **142** (1998) .

Gipe, Paul, *Wind Energy Basics: A Guide to Small and Micro Wind Systems*, Chelsea Green Pub. Co., White River Junction, VT, 1999.

Graham, Ian, *Nuclear Power*, Raintree Steck-Vaughn Publishers, Austin, TX, 1999.

Hinrichs, Roger, *Energy: Its Use and the Environment*, 2nd ed., Saunders College Publishing, Ft. Worth, 1996.

Hudson, Travis, *Environmental Research Needs of Metal Mining, Society for Mining, Metallurgy, and Exploration*, American Geological Institute, Littleton, CO, 1998.

Kursunoglu, Behram N., Stephan L. Mintz, and Arnold Perlmutter, Eds., *Environment and Nuclear Energy*, Plenum Press, New York, 1998.

Lee, Kai N., *The Compass and Gyroscope: Integrating Science and Politics for the Environment*, Island Press, Washington, D.C., 1993.

Liu, Paul I., *Introduction to Energy and the Environment*, Van Nostrand Reinhold, New York, 1993.

Mineral Resource Surveys Program (U.S.), *The National Mineral Resource Surveys Program: A Plan for Mineral-Resource and Mineral-Environmental Research for National Land-Use, Environmental, and Mineral-Supply Decision Making*, U.S. Department of the Interior, U. S. Geological Survey, Reston, VA, 1995.

Oxlade, Chris, *Energy*, Heinemann Library, Des Plaines, IL, 1999.

Papp, John F., *Chromium Metal*, U.S. Department of the Interior, U.S. Bureau of Mines, Washington, D.C., 1995.

Paul E. Queneau International Symposium (1993: Denver, CO), *Extractive Metallurgy of Copper, Nickel, and Cobalt: Proceedings of the Paul E. Queneau International Symposium*, TMS, Warrendale, PA., 1993.

Power Surge: Guide to the Coming Energy Revolution, Worldwatch Publications, Washington, D.C., 1998.

Snedden, Robert, *Energy*, Heinemann Library, Des Plaines, IL, 1999.

Socolow, Robert H., Dennis Anderson, and John Harte, Eds., *Annual Review of Energy and Environment*, Annual Reviews, Inc., Palo Alto, CA, 1997.

Steen, Athena Swentzell, *The Straw Bale House*, Chelsea Green Pulishing Co., White River Junction, VT, 1994

United Nations Conference on Trade and Development, *Handbook of World Mineral Trade Statistics, 1991-1996*, United Nations, New York, 1997.

Walker, Graham, *The Stirling Alternative: Power Systems, Refrigerants, and Heat Pumps*, Gordon and Breach Publishers, Langhorne, PA, 1994.

Walker, J. F. and N. Jenkins, *Wind Energy*, John Wiley & Sons, New York, 1997.

QUESTIONS AND PROBLEMS

1. What pollution control measures may produce a shortage of platinum metals?

2. List and discuss some of the major environmental concerns related to the mining and utilization of metal ores.

3. What are the major phosphate minerals?

4. Arrange the following energy conversion processes in order from the least to the most efficient: (a) electric hot water heater, (b) photosynthesis, (c) solar cell, (d) electric generator, (e) aircraft jet engine.

5. Considering the Carnot equation and common means for energy conversion, what might be the role of improved materials (metal alloys, ceramics) in increasing energy conversion efficiency?

6. Why is shale oil, a possible substitute for petroleum in some parts of the world, considered to be a pyrolysis product?

7. List some coal ranks and describe what is meant by coal rank.

8. Why was it necessary to add hydrocarbons to gas produced by reacting steam with hot carbon from coal in order to make a useful gas product?

9. What is the principle of the Exxon Donor Solvent process for producing liquid hydrocarbons from coal?

10. As it is now used, what is the principle or basis for the production of energy from uranium by nuclear fission? Is this process actually used for energy production? What are some of its environmental disadvantages? What is one major advantage?

11. What would be at least two highly desirable features of nuclear fusion power if it could ever be achieved in a controllable fashion on a large scale?

12. Justify describing the sun as "an ideal energy source." What are two big disadvantages of solar energy?

13. What are some of the greater implications of the use of biomass for energy? How might such widespread use affect greenhouse warming? How might it affect agricultural production of food?

14. Describe how gasohol is related to energy from biomass.

15. How does the trend toward utilization of less rich ores affect the environment? What does it have to do with energy utilization?

16. Of the resources listed in this chapter, list and discuss those that are largely from by-product sources.

17. Why is the total dollar value of "cheap" sand and gravel so high? What does this fact imply for environmental protection?

19 NATURE, SOURCES, AND ENVIRONMENTAL CHEMISTRY OF HAZARDOUS WASTES

19.1. INTRODUCTION

A **hazardous substance** is a material that may pose a danger to living organisms, materials, structures, or the environment by explosion or fire hazards, corrosion, toxicity to organisms, or other detrimental effects. What, then, is a hazardous waste? Although it has has been stated that,[1] "The discussion on this question is as long as it is fruitless," a simple definition of a **hazardous waste** is that it is a hazardous substance that has been discarded, abandoned, neglected, released or designated as a waste material, or one that may interact with other substances to be hazardous. The definition of hazardous waste is addressed in greater detail in Section 19.2, but in a simple sense it is a material that has been left where it should not be and that may cause harm to one if one encounters it.

History of Hazardous Substances

Humans have always been exposed to hazardous substances, going back to prehistoric times when they inhaled noxious volcanic gases or succumbed to carbon monoxide from inadequately vented fires in cave dwellings sealed too well against Ice-Age cold. Slaves in Ancient Greece developed lung disease from weaving mineral asbestos fibers into cloth to make it more degradation-resistant. Some archaeological and historical studies have concluded that lead wine containers were a leading cause of lead poisoning in the more affluent ruling class of the Roman Empire, leading to erratic behavior such as fixation on spectacular sporting events, chronic unmanageable budget deficits, speculative purchases of overvalued stock, illicit trysts in governmental offices, and ill-conceived, overly ambitious military ventures in remote foreign lands. Alchemists who worked during the Middle Ages often suffered debilitating injuries and illnesses resulting from the hazards of their explosive and toxic chemicals. During the 1700s, runoff from mine spoils piles began to create serious contamination problems in Europe. As the production of dyes and other organic chemicals developed from the coal tar industry in Germany during the 1800s, pollution and poisoning from coal tar by-products was observed. By around 1900 the quantity and variety of chemical wastes produced each year was increasing sharply with the addition of wastes such as spent steel and iron pickling liquor, lead battery wastes, chromic wastes, petroleum refinery wastes, radium wastes, and fluoride wastes from aluminum ore refining. As the century progressed into the World War II era, the wastes and hazardous by-products of manufacturing increased markedly from sources such as chlorinated solvents manufacture, pesticides synthesis, polymers manufacture, plastics, paints, and wood preservatives.

The Love Canal affair of the 1970s and 1980s brought hazardous wastes to public attention as a major political issue in the U.S. Starting around 1940, this site in Niagara Falls, New York, had received about 20,000 metric tons of chemical wastes containing at least 80 different chemicals. By 1994 state and federal governments had spent well over $100 million to clean up the site and relocate residents.

Other areas containing hazardous wastes that received attention included an industrial site in Woburn, Massachusetts, that had been contaminated by wastes from tanneries, glue-making factories, and chemical companies dating back to about 1850; the Stringfellow Acid Pits near Riverside, California; the Valley of the Drums in Kentucky; and Times Beach, Missouri, an entire town that was abandoned because of contamination by TCDD (dioxin).

The problem of hazardous wastes is truly international in scope.[2] As the result of the problem of dumping such wastes in developing countries, the 1989 Basel Convention on the Control of Transboundary Movement of Hazardous Wastes and their Disposal was held in Basel, Switzerland in 1989, and by 1998 had been signed by more than 100 countries. This treaty defines a long List A of hazardous wastes, a List B of nonhazardous wastes, and a List C of as yet unclassified materials. An example of a material on List C is polyvinyl chloride (PVC) coated wire, which is harmless, itself, but may release dioxins or heavy metals when thermally treated.

Legislation

Governments in a number of nations have passed legislation to deal with hazardous substances and wastes. In the U.S. such legislation has included the following:

- Toxic Substances Control Act of 1976

- Resource Conservation and Recovery Act (RCRA) of 1976 (amended and and reauthorized by the Hazardous and Solid Wastes Amendments Act (HSWA) of 1984)

- Comprehensive Environmental Response, Compensation, and Liability Act (CERCLA) of 1980

RCRA legislation charged the U.S. Environmental Protection Agency (EPA) with protecting human health and the environment from improper management and disposal of hazardous wastes by issuing and enforcing regulations pertaining to such wastes. RCRA requires that hazardous wastes and their characteristics be listed and controlled from the time of their origin until their proper disposal or destruction. Regulations pertaining to firms generating and transporting hazardous wastes require that they keep detailed records, including reports on their activities and manifests to ensure proper tracking of hazardous wastes through transportation systems. Approved containers and labels must be used, and wastes can only be delivered to facilities approved for treatment, storage, and disposal. There are about 290 million tons of wastes regulated by RCRA. In the U.S. about 3,000 facilities are involved in the treatment, storage, or disposal of RCRA wastes.

CERCLA (Superfund) legislation deals with actual or potential releases of hazardous materials that have the potential to endanger people or the surrounding environment at uncontrolled or abandoned hazardous waste sites in the U.S. The act requires responsible parties or the government to clean up waste sites. Among CERCLA's major purposes are the following:

- Site identification

- Evaluation of danger from waste sites

- Evaluation of damages to natural resources

- Monitoring of release of hazardous substances from sites

- Removal or cleanup of wastes by responsible parties or government

CERCLA was extended for five years by the passage of the Superfund Amendments and Reauthorization Act (SARA) of 1986, legislation with greatly increased scope and additonal in funding. Actually longer than CERCLA, SARA encouraged the development of alternatives to land disposal that favor permanent solutions reducing volume, mobility, and toxicity of wastes; increased emphasis upon public health, research, training, and state and citizen involvement; and establishment of a new program for leaking underground (petroleum) storage tanks. After 1986 few new legislative initiatives dealing with hazardous wastes were forthcoming in the U.S. As of 1999, the U.S. Congress had gone for six years without reauthorizing Superfund. During 1999 both the House and Senate were considering passage of legislation to reauthorize the Superfund act in an amended form.[3]

As of 1998, there were 1,359 designated Superfund sites, of which 509 sites, 37 percent of the total, had been cleaned up at a total cost to companies of $15 billion.[4] Assuming the same cost for each of the remaining sites, an additional $25 billion in private sector funds would be required to clean them up. The DuPont company, for example, listed an accrued liability of $561 million for waste site cleanup on its 1997 annual report, down from $602 million in 1995.

19.2 CLASSIFICATION OF HAZARDOUS SUBSTANCES AND WASTES

Many specific chemicals in widespread use are hazardous because of their chemical reactivities, fire hazards, toxicities, and other properties. There are numerous kinds of hazardous substances, usually consisting of mixtures of specific chemicals. These include such things as explosives; flammable liquids; flammable solids, such as magnesium metal and sodium hydride; oxidizing materials; such as peroxides; corrosive materials, such as strong acids; etiologic agents that cause disease; and radioactive materials.

Characteristics and Listed Wastes

For regulatory and legal purposes in the U.S., hazardous substances are listed specifically and are defined according to general characteristics. Under the authority of the Resource Conservation and Recovery Act (RCRA), the United States Environmental Protection Agency (EPA) defines hazardous substances in terms of the following **characteristics**:

- **Ignitability**, characteristic of substances that are liquids, the vapors of which are likely to ignite in the presence of ignition sources; nonliquids that may catch fire from friction or contact with water and which burn vigorously or persistently; ignitable compressed gases; and oxidizers

- **Corrosivity**, characteristic of substances that exhibit extremes of acidity or basicity or a tendency to corrode steel

- **Reactivity**, characteristic of substances that have a tendency to undergo violent chemical change (examples are explosives, pyrophoric materials, water-reactive substances, or cyanide- or sulfide-bearing wastes)

- **Toxicity**, defined in terms of a standard extraction procedure followed by chemical analysis for specific substances.

In addition to classification by characteristics, EPA designates more than 450 **listed wastes** which are specific substances or classes of substances known to be hazardous. Each such substance is assigned an EPA **hazardous waste number** in the format of a letter followed by 3 numerals, where a different letter is assigned to substances from each of the four following lists:

- **F-type wastes from nonspecific sources**: For example, quenching wastewater treatment sludges from metal heat treating operations where cyanides are used in the process (F012).

- **K-type wastes from specific sources**: For example, heavy ends from the distillation of ethylene dichloride in ethylene dichloride production (K019).

- **P-type acute hazardous wastes**: Wastes that have been found to be fatal to humans in low doses, or capable of causing or significantly contributing to an increase in serious irreversible or incapacitating reversible illness. These are mostly specific chemical species such as fluorine (P056) or 3-chloropropane nitrile (P027).

- **U-Type miscellaneous hazardous wastes**: These are predominantly specific compounds such as calcium chromate (U032) or phthalic anhydride (U190).

Compared to RCRA, CERCLA gives a rather broad definition of hazardous substances that includes the following:

- Any element, compound, mixture, solution, or substance, the release of which may substantially endanger public health, public welfare, or the environment

- Any element, compound, mixture, solution, or substance in reportable quantities designated by CERCLA Section 102

- Certain substances or toxic pollutants designated by the Federal Water Pollution Control Act

- Any hazardous air pollutant listed under Section 112 of the Clean Air Act

- Any imminently hazardous chemical substance or mixture that has been the subject of government action under Section 7 of the Toxic Substances Control Act (TSCA)

- With the exception of those suspended by Congress under the Solid Waste Disposal Act, any hazardous waste listed or having characteristics identified by RCRA § 3001

Hazardous Wastes

Three basic approaches to defining hazardous wastes are (1) a qualitative description by origin, type, and constituents; (2) classification by characteristics largely based upon testing procedures; and (3) by means of concentrations of specific hazardous substances. Wastes may be classified by general type such as "spent halogenated solvents," or by industrial sources such as "pickling liquor from steel manufacturing."

Hazardous Wastes and Air and Water Pollution Control

Somewhat paradoxically, measures taken to reduce air and water pollution (Figure 19.1) have had a tendency to increase production of hazardous wastes. Most water treatment processes yield sludges or concentrated liquors that require stabilization and disposal. Air scrubbing processes likewise produce sludges. Baghouses and precipitators used to control air pollution all yield significant quantities of solids, some of which are hazardous.

19.3. SOURCES OF WASTES

Quantities of hazardous wastes produced each year are not known with certainty and depend upon the definitions used for such materials. In 1988 the figure for RCRA-regulated wastes in the U.S. was placed at 290 million tons. However, most of this material was water, with only a few million tons consisting of solids. Some high-water-content wastes are generated directly by processes that require large quantities of water in waste treatment, and other aqueous wastes are produced by mixing hazardous wastes with wastewater.

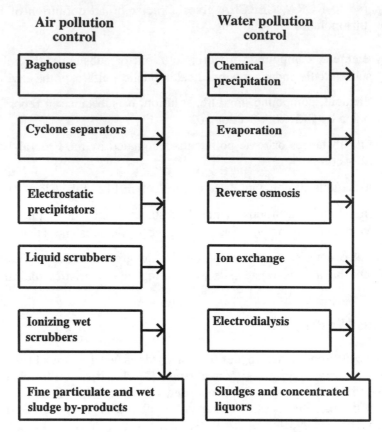

Potentially hazardous waste substances

Figure 19.1. Potential contributions of air and water pollution control measures to hazardous wastes production.

Some wastes that might exhibit a degree of hazard are exempt from RCRA regulation by legislation. These exempt wastes include the following:

- Fuel ash and scrubber sludge from power generation by utilities
- Oil and gas drilling muds
- By-product brine from petroleum production
- Cement kiln dust
- Waste and sludge from phosphate mining and beneficiation
- Mining wastes from uranium and other minerals
- Household wastes

Types of Hazardous Wastes

In terms of quantity by weight, the greatest quantities are those from categories designated by hazardous waste numbers preceded by F and K, respectively. The former are those from nonspecific sources and include the following examples:

- F001   The spent halogenated solvents used in degreasing: tetrachloroethylene, trichloroethylene, methylene chloride, 1,1,1-trichloroethane, carbon tetrachloride, and the chlorinated fluorocarbons; and sludges from the recovery of these solvents in degreasing operations

- F004   The spent nonhalogenated solvents: cresols, cresylic acid, and nitro-benzene; and still bottoms from the recovery of these solvents

- F007 Spent plating-bath solutions from electroplating operations

- F010   Quenching-bath sludge from oil baths from metal heat treating operations

The "K-type" hazardous wastes are those from specific sources produced by industries such as the manufacture of inorganic pigments, organic chemicals, pesticides, explosives, iron and steel, and nonferrous metals, and from processes such as petroleum refining or wood preservation; some examples are given below:

- K001   Bottoms sediment sludge from the treatment of wastewaters from wood-preserving processes that use creosote and/or pentachlorophenol

- K002   Wastewater treatment sludge from the production of chrome yellow and orange pigments

- K020   Heavy ends (residue) from the distillation of vinyl chloride in vinyl chloride monomer production

- K043   2,6-Dichlorophenol waste from the production of 2,4-D

- K047   Pink/red water from TNT operations

- K049   Slop oil emulsion solids from the petroleum refining industry

- K060   Ammonia lime still sludge from coking operations

- K067   Electrolytic anode slimes/sludges from primary zinc production

The remainder of wastes consist of reactive wastes, corrosive wastes, toxic wastes, ignitable wastes, and "P" wastes (discarded commercial chemical products, off-specification species, containers, and spill residues), "U" wastes, and unspecified types.

Hazardous Waste Generators

Several hundred thousand companies generate hazardous wastes in the U. S., but most of these generators produce only small quantities. Hazardous waste generators are unevenly distributed geographically across the continental U. S., with a relatively large number located in the industrialized upper Midwest, including the states of Illinois, Indiana, Ohio, Michigan, and Wisconsin.

Industry types of hazardous waste generators can be divided among the seven following major categories, each containing of the order of 10-20 percent of hazardous waste generators: chemicals and allied products manufacture, petroleum-related industries, fabricated metals, metal-related products, electrical equipment manufacture, "all other manufacturing," and nonmanufacturing and nonspecified generators. About 10% of the generators produce more than 95% of all hazardous wastes. Whereas, as noted above, the number of hazardous waste generators is distributed relatively evenly among several major types of industries, 70-85% of the *quantities* of hazardous wastes are generated by the chemical and petroleum industries. Of the remainder, about 3/4 comes from metal-related industries, and about 1/4 from all other industries.

19.4. FLAMMABLE AND COMBUSTIBLE SUBSTANCES

Most chemicals that are likely to burn accidentally are liquids. Liquids form **vapors** which are usually more dense than air, and thus tend to settle. The tendency of a liquid to ignite is measured by a test in which the liquid is heated and periodically exposed to a flame until the mixture of vapor and air ignites at the liquid's surface. The temperature at which ignition occurs under these conditions is called the **flash point**.

With these definitions in mind it is possible to divide ignitable materials into four major classes. A **flammable solid** is one that can ignite from friction or from heat remaining from its manufacture, or which may cause a serious hazard if ignited. Explosive materials are not included in this classification. A **flammable liquid** is one having a flash point below 60.5°C (141°F). A **combustible liquid** has a flash point in excess of 60.5°C, but below 93.3°C (200°F). Where gases are substances that exist entirely in the gaseous phase at 0°C and 1 atm pressure, a **flammable compressed gas** meets specified criteria for lower flammability limit, flammability range (see below), and flame projection.

In considering the ignition of vapors, two important concepts are those of flammability limit and flammability range. Values of the vapor/air ratio below which ignition cannot occur because of insufficient fuel define the **lower flammability limit (LFL)**. Similarly, values of the vapor/air ratio above which ignition cannot occur because of insufficient air define the **upper flammability limit (UFL)**. The difference between upper and lower flammability limits at a specified temperature is the **flammability range**. Table 19.1 gives some examples of these values for common liquid chemicals. The percentage of flammable substance for best combustion (most explosive mixture) is labeled "optimal." In the case of acetone, for example, the optimal flammable mixture is 5.0% acetone.

One of the more disastrous problems that can occur with flammable liquids is a boiling liquid expanding vapor explosion, BLEVE. These are caused by rapid pressure buildup in closed containers of flammable liquids heated by an external source. The explosion occurs when the pressure buildup is sufficient to break the container walls.

Combustion of Finely Divided Particles

Finely divided particles of combustible materials are somewhat analogous to vapors with respect to flammability. One such example is a spray or mist of hydrocarbon liquid in which oxygen has the opportunity for intimate contact with the liquid particles causing the liquid to ignite at a temperature below its flash point.

Table 19.1. Flammabilities of Some Common Organic Liquids

Liquid	Flash point (°C)[1]	Volume percent in air LFL[2]	UFL[2]
Diethyl ether	-43	1.9	36
Pentane	-40	1.5	7.8
Acetone	-20	2.6	13
Toluene	4	1.27	7.1
Methanol	12	6.0	37
Gasoline (2,2,4-tri- methylpentane)	---	1.4	7.6
Naphthalene	157	0.9	5.9

[1] Closed-cup flash point test
[2] LFL, lower flammability limit; UFL, upper flammability limit at 25°C.

Dust explosions can occur with a large variety of solids that have been ground to a finely divided state. Many metal dusts, particularly those of magnesium and its alloys, zirconium, titanium, and aluminum, can burn explosively in air. In the case of aluminum, for example, the reaction is the following:

$$4Al(powder) + 3O_2(from\ air) \rightarrow 2Al_2O_3 \tag{19.4.1}$$

Coal dust and grain dusts have caused many fatal fires and explosions in coal mines and grain elevators, respectively. Dusts of polymers such as cellulose acetate, polyethylene, and polystyrene can also be explosive.

Oxidizers

Combustible substances are reducing agents that react with **oxidizers** (oxidizing agents or oxidants) to produce heat. Diatomic oxygen, O_2, from air is the most common oxidizer. Many oxidizers are chemical compounds that contain oxygen in their formulas. The halogens (periodic table group 7A) and many of their compounds are oxidizers. Some examples of oxidizers are given in Table 19.2.

An example of a reaction of an oxidizer is that of concentrated HNO_3 with copper metal, which gives toxic NO_2 gas as a product:

$$4HNO_3 + Cu \rightarrow Cu(NO_3)_2 + 2H_2O + 2NO_2 \tag{19.4.2}$$

Spontaneous Ignition

Substances that catch fire spontaneously in air without an ignition source are called **pyrophoric**. These include several elements—white phosphorus, the alkali metals (group 1A), and powdered forms of magnesium, calcium, cobalt, manganese,

Table 19.2. Examples of Some Oxidizers

Name	Formula	State of matter
Ammonium nitrate	NH_4NO_3	Solid
Ammonium perchlorate	NH_4ClO_4	Solid
Bromine	Br_2	Liquid
Chlorine	Cl_2	Gas (stored as liquid)
Fluorine	F_2	Gas
Hydrogen peroxide	H_2O_2	Solution in water
Nitric acid	HNO_3	Concentrated solution
Nitrous oxide	N_2O	Gas (stored as liquid)
Ozone	O_3	Gas
Perchloric acid	$HClO_4$	Concentrated solution
Potassium permanganate	$KMnO_4$	Solid
Sodium dichromate	$Na_2Cr_2O_7$	Solid

iron, zirconium, and aluminum. Also included are some organometallic compounds, such as ethyllithium (LiC_2H_5) and phenyllithium (LiC_6H_5), and some metal carbonyl compounds such as iron pentacarbonyl, $Fe(CO)_5$. Another major class of pyrophoric compounds consists of metal and metalloid hydrides, including lithium hydride, LiH; pentaborane, B_5H_9; and arsine, AsH_3. Moisture in air is often a factor in spontaneous ignition. For example, lithium hydride undergoes the following reaction with water from moist air:

$$LiH \; + \; H_2O \; \rightarrow \; LiOH \; + \; H_2 \; + \; heat \qquad (19.4.4)$$

The heat generated from this reaction can be sufficient to ignite the hydride so that it burns in air:

$$2LiH \; + \; O_2 \; \rightarrow \; Li_2O \; + \; H_2O \qquad (19.4.5)$$

Some compounds with organometallic character are also pyrophoric. An example of such a compound is diethylethoxyaluminum:

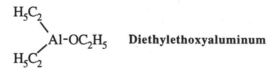

Diethylethoxyaluminum

Many mixtures of oxidizers and oxidizable chemicals catch fire spontaneously and are called **hypergolic mixtures**. Nitric acid and phenol form such a mixture.

Toxic Products of Combustion

Some of the greater dangers of fires are from toxic products and by-products of combustion. The most obvious of these is carbon monoxide, CO, which can cause serious illness or death because it forms carboxyhemoglobin with hemoglobin in the blood so that the blood no longer carries oxygen to body tissues. Toxic SO_2, P_4O_{10}, and HCl are formed by the combustion of sulfur, phosphorus, and organochlorine compounds, respectively. A large number of noxious organic compounds such as aldehydes are generated as by-products of combustion. In addition to forming carbon monoxide, combustion under oxygen-deficient conditions produces polycyclic aromatic hydrocarbons consisting of fused ring structures. Some of these compounds, such as benzo[a]pyrene, below, are precarcinogens that are acted upon by enzymes in the body to yield cancer-producing metabolites.

Benzo(a)pyrene

19.5. REACTIVE SUBSTANCES

Reactive substances are those that tend to undergo rapid or violent reactions under certain conditions. Such substances include those that react violently or form potentially explosive mixtures with water. An example is sodium metal, which reacts strongly with water as follows:

$$2Na\ +\ 2H_2O\ \rightarrow\ 2NaOH\ +\ H_2\ +\ heat \tag{19.5.1}$$

This reaction usually generates enough heat to ignite the sodium and hydrogen. Explosives constitute another class of reactive substances. For regulatory purposes substances are also classified as reactive that produce toxic gases or vapors when they react with water, acid, or base. Hydrogen sulfide or hydrogen cyanide are the most common toxic substances released in this manner.

Heat and temperature are usually very important factors in reactivity. Many reactions require energy of activation to get them started. The rates of most reactions tend to increase sharply with increasing temperature and most chemical reactions give off heat. Therefore, once a reaction is started in a reactive mixture lacking an effective means of heat dissipation, the rate may increase exponentially with time, leading to an uncontrollable event. Other factors that may affect reaction rate include physical form of reactants (for example, a finely divided metal powder that reacts explosively with oxygen, whereas a single mass of metal barely reacts), rate and degree of mixing of reactants, degree of dilution with nonreactive media (solvent), presence of a catalyst, and pressure.

Some chemical compounds are self-reactive in that they contain oxidant and reductant in the same compound. Nitroglycerin, a strong explosive with the formula $C_3H_5(ONO_2)_3$ decomposes spontaneously to CO_2, H_2O, O_2, and N_2 with a rapid release of a very high amount of energy. Pure nitroglycerin has such a high inherent instability that only a slight blow may be sufficient to detonate it. Trinitrotoluene (TNT) is also an explosive with a high degree of reactivity. However, it is inherently relatively stable in that some sort of detonating device is required to cause it to explode.

Chemical Structure and Reactivity

As shown in Table 19.3, some chemical structures are associated with high reactivity. High reactivity in some organic compounds results from unsaturated bonds in the carbon skeleton, particularly where multiple bonds are adjacent (allenes, C=C=C) or separated by only one carbon-carbon single bond (dienes, C=C-C=C). Some organic structures involving oxygen are very reactive. Examples are oxiranes, such as ethylene oxide,

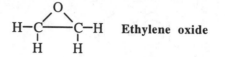   Ethylene oxide

hydroperoxides (ROOH), and peroxides (ROOR'), where R and R' stand for hydrocarbon moieties such as the methyl group, -CH$_3$. Many organic compounds containing nitrogen along with carbon and hydrogen are very reactive. Included are triazenes containing a functionality with three nitrogen atoms (R-N=N-N), some azo

Table 19.3. Examples of Reactive Compounds and Structures

Name	Structure or formula
Organic	
Allenes	C=C=C
Dienes	C=C-C=C
Azo compounds	C-N=N-C
Triazenes	C-N=N-N
Hydroperoxides	R-OOH
Peroxides	R-OO-R'
Alkyl nitrates	R-O-NO$_2$
Nitro compounds	R-NO$_2$
Inorganic	
Nitrous oxide	N$_2$O
Nitrogen halides	NCl$_3$, NI$_3$
Interhalogen compounds	BrCl
Halogen oxides	ClO$_2$
Halogen azides	ClN$_3$
Hypohalites	NaClO

compounds (R-N=N-R'), and some nitriles in which a nitrogen atom is triply bonded to a carbon atom:

$$R-C\equiv N \quad \text{Nitrile}$$

Functional groups containing both oxygen and nitrogen tend to impart reactivity to an organic compound. Examples of such functional groups are alkyl nitrates ($R–NO_2$), alkyl nitrites ($R–O–N=O$), nitroso compounds ($R–N=O$), and nitro compounds ($R–NO_2$).

Many different classes of inorganic compounds are reactive. These include some of the halogen compounds of nitrogen (shock-sensitive nitrogen triiodide, NI_3, is an outstanding example), compounds with metal-nitrogen bonds (NaN_3), halogen oxides (ClO_2), and compounds with oxyanions of the halogens. An example of the last group of compounds is ammonium perchlorate, NH_4ClO_4, which was involved in a series of massive explosions that destroyed 8 million lb of the compound and demolished a 40 million lb/year U.S. rocket fuel plant near Henderson, Nevada, in 1988. (By late 1989 a new $92 million plant for the manufacture of ammonium perchlorate had been constructed near Cedar City in a remote region of southwest Utah. Prudently, the buildings at the new plant have been placed at large distances from each other!)

Explosives such as nitroglycerin or TNT are single compounds containing both oxidizing and reducing functions in the same molecule. Such substances are commonly called **redox compounds**. Some redox compounds have even more oxygen than is needed for a complete reaction and are said to have a positive balance of oxygen, some have exactly the stoichiometric quantity of oxygen required (zero balance, maximum energy release), and others have a negative balance and require oxygen from outside sources to completely oxidize all components. Trinitrotoluene has a substantial negative balance of oxygen; ammonium dichromate (($NH_4)_2Cr_2O_7$) has a zero balance, reacting with exact stoichiometry to H_2O, N_2, and Cr_2O_3; and treacherously explosive nitroglycerin has a positive balance as shown by the following reaction:

$$4C_3H_5N_3O_9 \rightarrow 12CO_2 + 10H_2O + 6N_2 + O_2 \qquad (19.5.2)$$

19.6. CORROSIVE SUBSTANCES

Corrosive substances are regarded as those that dissolve metals or cause oxidized material to form on the surface of metals—rusted iron is a prime example and, more broadly, cause deterioration of materials, including living tissue, that they contact.[5] Most corrosives belong to at least one of the four following chemical classes: (1) strong acids, (2) strong bases, (3) oxidants, (4) dehydrating agents. Table 19.4 lists some of the major corrosive substances and their effects.

Sulfuric Acid

Sulfuric acid is a prime example of a corrosive substance. As well as being a strong acid, concentrated sulfuric acid is also a dehydrating agent and oxidant. The tremendous affinity of H_2SO_4 for water is illustrated by the heat generated when water

Table 19.4. Examples of Some Corrosive Substances

Name and formula	Properties and effects
Nitric acid, HNO_3 reacts	Strong acid and strong oxidizer, corrodes metal, with protein in tissue to form yellow xanthoproteic acid, lesions are slow to heal
Hydrochloric acid, HCl vapor,	Strong acid, corrodes metals, gives off HCl gas which can damage respiratory tract tissue
Hydrofluoric acid, HF	Corrodes metals, dissolves glass, causes particularly bad burns to flesh
Alkali metal hydroxides, NaOH and KOH	Strong bases, corrode zinc, lead, and aluminum, substances that dissolve tissue and cause severe burns
Hydrogen peroxide, H_2O_2	Oxidizer, all but very dilute solutions cause severe burns
Interhalogen compounds such as ClF, BrF_3	Powerful corrosive irritants that acidify, oxidize, and dehydrate tissue
Halogen oxides such as OF_2, Cl_2O, Cl_2O_7	Powerful corrosive irritants that acidify, oxidize, and dehydrate tissue
Elemental fluorine, chlorine, bromine (F_2, Cl_2, Br_2,)	Very corrosive to mucous membranes and moist tissue, strong irritants

and concentrated sulfuric acid are mixed. If this is done incorrectly by adding water to the acid, localized boiling and spattering can occur that result in personal injury. The major destructive effect of sulfuric acid on skin tissue is removal of water with accompanying release of heat. Sulfuric acid decomposes carbohydrates by removal of water. In contact with sugar, for example, concentrated sulfuric acid reacts to leave a charred mass. The reaction is

$$C_{12}H_{22}O_{11} \xrightarrow{H_2SO_4} 11H_2O\ (H_2SO_4) + 12C + heat \tag{19.6.1}$$

Some dehydration reactions of sulfuric acid can be very vigorous. For example, the reaction with perchloric acid produces unstable Cl_2O_7, and a violent explosion can result. Concentrated sulfuric acid produces dangerous or toxic products with a number of other substances, such as toxic carbon monoxide (CO) from reaction with oxalic acid, $H_2C_2O_4$; toxic bromine and sulfur dioxide (Br_2, SO_2) from reaction with sodium bromide, NaBr; and toxic, unstable chlorine dioxide (ClO_2) from reaction with sodium chlorate, $NaClO_3$.

Contact with sulfuric acid causes severe tissue destruction resulting in a severe burn that may be difficult to heal. Inhalation of sulfuric acid fumes or mists damages tissues in the upper respiratory tract and eyes. Long-term exposure to sulfuric acid fumes or mists has caused erosion of teeth!

19.7. TOXIC SUBSTANCES

Toxicity is of the utmost concern in dealing with hazardous substances. This includes both long term chronic effects from continual or periodic exposures to low levels of toxicants and acute effects from a single large exposure. Toxic substances are covered in greater detail in Chapters 22 and 23.

Toxicity Characteristic Leaching Procedure

For regulatory and remediation purposes a standard test is needed to measure the likelihood of toxic substances getting into the environment and causing harm to organisms. The U.S. Environmental Protection Agency specifies a test called the **Toxicity Characteristic Leaching Procedure** (**TCLP**) designed to determine the toxicity hazard of wastes.[6] The test was designed to estimate the availability to organisms of both inorganic and organic species in hazardous materials present as liquids, solids, or multiple phase mixtures and does not test for the direct toxic effects of wastes. Basically, the procedure consists of leaching a material with a solvent designed to mimic leachate generated in a municipal waste disposal site, followed by chemical analysis of the leachate. The procedure is discussed in more detail in Chapter 25.

19.8. PHYSICAL FORMS AND SEGREGATION OF WASTES

Three major categories of wastes based upon their physical forms are **organic materials**, **aqueous wastes**, and **sludges**. These forms largely determine the course of action taken in treating and disposing of the wastes. The **level of segregation**, a concept illustrated in Figure 19.2, is very important in treating, storing, and disposing of different kinds of wastes. It is relatively easy to deal with wastes that are not mixed with other kinds of wastes, that is, those that are highly segregated. For example, spent hydrocarbon solvents can be used as fuel in boilers. However, if these solvents are mixed with spent organochlorine solvents, the production of contaminant hydrogen chloride during combustion may prevent fuel use and require disposal in special hazardous waste incinerators. Further mixing with inorganic sludges adds mineral matter and water. These impurities complicate the treatment processes required by producing mineral ash in incineration or lowering the heating value of the material incinerated because of the presence of water. Among the most difficult types of wastes to handle and treat are those with the least segregation, of which a "worst case scenario" would be "dilute sludge consisting of mixed organic and inorganic wastes," as shown in Figure 19.2.

Concentration of wastes is an important factor in their management. A waste that has been concentrated or, preferably, never diluted is generally much easier and more economical to handle than one that is dispersed in a large quantity of water or soil. Dealing with hazardous wastes is greatly facilitated when the original quantities of wastes are minimized and the wastes remain separated and concentrated insofar as possible.

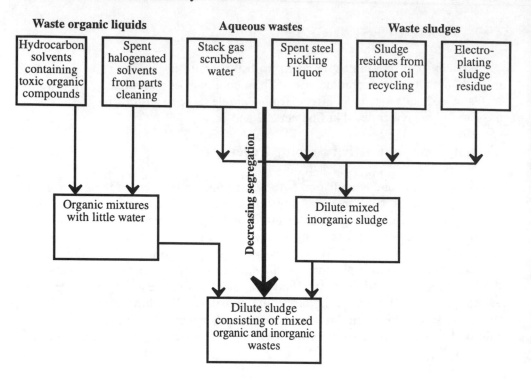

Figure 19.2. Illustration of waste segregation.

19.9. ENVIRONMENTAL CHEMISTRY OF HAZARDOUS WASTES

The properties of hazardous materials, their production, and what makes a hazardous substance a hazardous waste were discussed in preceding parts of this chapter. Hazardous materials normally cause problems when they enter the environment and have detrimental effects on organisms or other parts of the environment. Therefore, the present chapter deals with the environmental chemistry of hazardous materials. In discussing the environmental chemistry of hazardous materials, it is convenient to consider the following five aspects based upon the definition of environmental chemistry:

• Origins   • Transport   • Reactions   • Effects   • Fates

It is also useful to consider the five environmental spheres as defined and outlined in Chapter 1:

• Anthrosphere   • Geosphere   • Hydrosphere • Atmosphere • Biosphere

Hazardous materials almost always originate in the anthrosphere, are often discarded into the geosphere, and are frequently transported through the hydrosphere or the atmosphere. The greatest concern for their effects is usually on the biosphere, particularly human beings. Figure 19.3 summarizes these relationships.

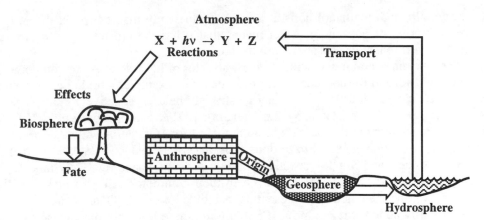

Figure 19.3. Scheme of interactions of hazardous wastes in the environment.

There are a variety of ways in which hazardous wastes get into the environment. Although now much more controlled by pollution prevention laws, hazardous substances have been deliberately added to the environment by humans. Wastewater containing a variety of toxic substances has been discharged in large quantities into waterways. Hazardous gases and particulate matter have been discharged into the atmosphere through stacks from power plants, incinerators, and a variety of industrial operations. Hazardous wastes have been deliberately spread on soil or placed in landfills in the geosphere. Evaporation and wind erosion may move hazardous materials from wastes dumps into the atmosphere, or they may be leached from waste dumps into groundwater or surface waters. Underground storage tanks or pipelines have leaked a variety of materials into soil. Accidents, fires, and explosions may distribute dangerous materials into the environment. Another source of such materials consists of improperly operated waste treatment or storage facilities.

19.10. PHYSICAL AND CHEMICAL PROPERTIES OF HAZARDOUS WASTES

Having considered the generation of hazardous wastes from the anthrosphere, the next thing to consider is their properties, which determine movement and other kinds of behaviors. These properties can be generally divided into physical and chemical properties.

The behavior of waste substances in the atmosphere is largely determined by their volatilities. In addition, their solubilities in water determine the degree to which they are likely to be removed with precipitation. Water solubility is the most important physical property in the hydrosphere. The movement of substances through the action of water in the geosphere is largely determined by the degree of sorption to soil, mineral strata, and sediments.

Volatility is a function of the vapor pressure of a compound. Vapor pressures at a particular temperature can vary over many orders of magnitude. Of common organic liquids, diethyl ether has one of the highest vapor pressures, whereas those of polychlorinated biphenyls (PCBs) are very low. When a volatile liquid is present in soil or in water, its water solubility also determines how well it evaporates. For example, although methanol boils at a lower temperature than benzene, the much lower solubility of benzene in water means that it has the greater tendency to go from the hydrosphere or geosphere into the atmosphere.

The environmental movement, effects, and fates of hazardous waste compounds are strongly related to their chemical properties. For example, a toxic heavy metal cationic species, such as Pb^{2+} ion, may be strongly held by negatively charged soil solids. If the lead is chelated by the chelating EDTA anion, represented Y^{4-}, it becomes much more mobile as PbY^{2-}, an anionic form. Oxidation state can be very important in the movement of hazardous substances. The reduced states of iron and manganese, Fe^{2+} and Mn^{2+}, respectively, are water soluble and relatively mobile in the hydrosphere and geosphere. However, in their common oxidized states, Fe(III) and Mn(IV), these elements are present as insoluble $Fe_2O_3 \cdot xH_2O$ and MnO_2, which have virtually no tendency to move. Furthermore, these iron and manganese oxides will sequester heavy metal ions, such as Pb^{2+} and Cd^{2+}, preventing their movement in the soluble form.

The major properties of hazardous substances and their surroundings that determine the environmental transport of such substances are the following:

- Physical properties of the substances, including vapor pressure and solubility.

- Physical properties of the surrounding matrix.

- Physical conditions to which wastes are subjected. Higher temperatures and erosive wind conditions enable volatile substances to move more readily.

- Chemical and biochemical properties of wastes. Substances that are less chemically reactive and less biodegradable will tend to move farther before breaking down.

19.11. TRANSPORT, EFFECTS, AND FATES OF HAZARDOUS WASTES

The transport of hazardous wastes is largely a function of their physical properties, the physical properties of their surrounding matrix, the physical conditions to which they are subjected, and chemical factors. Highly volatile wastes are obviously more likely to be transported through the atmosphere, and more soluble ones to be carried by water. Wastes will move farther faster in porous, sandy formations than in denser soils. Volatile wastes are more mobile under hot, windy conditions, and soluble ones during periods of heavy rainfall. Wastes that are more chemically and biochemically reactive will not move so far as less reactive wastes before breaking down.

Physical Properties of Wastes

The major physical properties of wastes that determine their amenability to transport are volatility, solubility, and the degree to which they are sorbed to solids, including soil and sediments.

The distribution of hazardous waste compounds between the atmosphere and the geosphere or hydrosphere is largely a function of compound volatility. Usually, in the hydrosphere, and often in soil, hazardous waste compounds are dissolved in water; therefore, the tendency of water to hold the compound is a factor in its mobility. For example, although ethyl alcohol has a higher vapor pressure and lower boiling temperature (77.8°C) than toluene (110.6 °C), vapor of the latter compound is more readily evolved from soil because of its limited solubility in water compared to ethanol, which is totally miscible with water.

Chemical Factors

As an illustration of chemical factors involved in transport of wastes, consider cationic inorganic species consisting of common metal ions. These inorganic species can be divided into three groups based upon their retention by clay minerals. Elements that tend to be highly retained by clay include cadmium, mercury, lead, and zinc. Potassium, magnesium, iron, silicon, and NH_4^+ ions are moderately retained by clay, whereas sodium, chloride, calcium, manganese, and boron ions are poorly retained. The retention of the last three elements is probably biased in that they are leached from clay, so that negative retention (elution) is often observed. It should be noted, however, that the retention of iron and manganese is a strong function of oxidation state in that the reduced forms of Mn and Fe are relatively poorly retained, whereas the oxidized forms of $Fe_2O_3 \cdot xH_2O$ and MnO_2 are very insoluble and stay on soil as solids.

Effects of Hazardous Wastes

The effects of hazardous wastes in the environment may be divided among effects on organisms, effects on materials, and effects on the environment. These are addressed briefly here and in greater detail in later sections.

The ultimate concern with wastes has to do with their toxic effects on animals, plants, and microbes. Virtually all hazardous waste substances are poisonous to a degree, some extremely so. The toxicity of a waste is a function of many factors, including the chemical nature of the waste, the matrix in which it is contained, circumstances of exposure, the species exposed, manner of exposure, degree of exposure, and time of exposure. The toxicities of hazardous wastes are discussed in more detail in Chapters 22 and 23.

As defined in Section 19.6, many hazardous wastes are *corrosive* to materials, usually because of extremes of pH or because of dissolved salt content. Oxidant wastes can cause combustible substances to burn uncontrollably. Highly reactive wastes can explode, causing damage to materials and structures. Contamination by wastes, such as by toxic pesticides in grain, can result in substances becoming unfit for use.

In addition to their toxic effects in the biosphere, hazardous wastes can damage air, water, and soil. Wastes that get into air can cause deterioration of air quality, either directly or by the formation of secondary pollutants. Hazardous waste compounds dissolved in, suspended in, or floating as surface films on the surface of water can render it unfit for use and for sustenance of aquatic organisms.

Soil exposed to hazardous wastes can be severely damaged by alteration of its physical and chemical properties and ability to support plants. For example, soil exposed to concentrated brines from petroleum production may become unable to support plant growth so that the soil becomes extremely susceptible to erosion.

Fates of Hazardous Wastes

The fates of hazardous waste substances are addressed in more detail in subsequent sections. As with all environmental pollutants, such substances eventually reach a state of physical and chemical stability, although that may take many centuries to occur. In some cases the fate of a hazardous waste material is a simple function of its physical properties and surroundings.

The fate of a hazardous waste substance in water is a function of the substance's solubility, density, biodegradability, and chemical reactivity. Dense, water-immiscible liquids may simply sink to the bottoms of bodies of water or aquifers and accumulate there as "blobs" of liquid. This has happened, for example, with hundreds of tons of PCB wastes that have accumulated in sediments in the Hudson River in New York State. Biodegradable substances are broken down by bacteria, a process for which the availability of oxygen is an important variable. Substances that readily undergo bioaccumulation are taken up by organisms, exchangeable cationic materials become bound to sediments, and organophilic materials may be sorbed by organic matter in sediments.

The fates of hazardous waste substances in the atmosphere are often determined by photochemical reactions. Ultimately, such substances may be converted to nonvolatile, insoluble matter and precipitate from the atmosphere onto soil or plants.

19.12. HAZARDOUS WASTES AND THE ANTHROSPHERE

As the part of the environment where humans process substances, the anthrosphere is the source of most hazardous wastes. These materials may come from manufacturing, transportation activities, agriculture, and any one of a number of activities in the anthrosphere. Hazardous wastes may be in any physical form and may include liquids, such as spent halogenated solvents used in degreasing parts; semisolid sludges, such as those generated from the gravitation separation of oil/water/solids mixtures in petroleum refining; and solids, such as baghouse dusts from the production of pesticides.

Releases of hazardous wastes from the anthrosphere commonly occur through incidents such as spills of liquids, accidental discharge of gases or vapors, fires, and explosions.[7] Resource Conservation and Recovery Act (RCRA) regulations designed to minimize such accidental releases from the anthrosphere and to deal with them when they occur are contained in 40 CFR 265.31 (Title 40 of the Code of Federal Regulations, Part 265.31). Under these regulations, hazardous waste generators are required to have specified equipment, trained personnel, and procedures that protect human health in the event of a release, and that facilitate remediation if a release occurs. An effective means of communication for summoning help and giving emergency instruction must be available. Also required are firefighting capabilities including fire extinguishers and adequate water. To deal with spills, a facility is required to have on hand absorbents, such as granular vermiculite clay, or absorbents in the form of pillows or pads. Neutralizing agents for corrosive substances that may be used should be available as well.

As noted above, hazardous wastes originate in the anthrosphere. However, to a large extent, they move, have effects, and end up in the anthrosphere as well. Large quantities of hazardous substances are moved by truck, rail, ship, and pipeline. Spills and releases from such movement, ranging from minor leaks from small containers to catastrophic releases of petroleum from wrecked tanker ships, are a common occurrence. Much effort in the area of environmental protection can be profitably devoted to minimizing and increasing the safety of the transport of hazardous substances through the anthrosphere.

In the United States the transportation of hazardous substances is regulated through the U.S. Department of Transportation (DOT). One of the ways in which this is done is through the **manifest** system of documentation designed to accomplish the following goals:

- Acts as a tracking device to establish responsibility for the generation, movement, treatment, and disposal of the waste

- By requiring the manifest to accompany the waste, such as during truck transport, it provides information regarding appropriate actions to take during emergencies such as collisions, spills, fires, or explosions

- Acts as the basic documentation for recordkeeping and reporting

Many of the adverse effects of hazardous substances occur in the anthrosphere. One of the main examples of such effects occurs as corrosion of materials that are strongly acidic or basic or that otherwise attack materials. Fire and explosion of hazardous materials can cause severe damage to anthrospheric infrastructure.

The fate of hazardous materials is often in the anthrosphere. One of the main examples of a material dispersed in the anthrosphere consists of lead-based anti-corrosive paints that are spread on steel structural members.

19.13. HAZARDOUS WASTES IN THE GEOSPHERE

The sources, transport, interactions, and fates of contaminant hazardous wastes in the geosphere involve a complex scheme, some aspects of which are illustrated in Figure 19.4. As illustrated in the figure, there are numerous vectors by which hazardous wastes can get into groundwater. Leachate from a landfill can move as a waste plume carried along by groundwater, in severe cases draining into a stream or into an aquifer where it may contaminate well water. Sewers and pipelines may leak hazardous substances into the geosphere. Such substances seep from waste lagoons into geological strata, eventually contaminating groundwater. Wastes leaching from sites where they have been spread on land for disposal or as a means of treatment can contaminate the geosphere and groundwater. In some cases wastes are pumped into deep wells as a means of disposal.

The movement of hazardous waste constituents in the geosphere is largely by the action of flowing water in a waste plume as shown in Figure 19.4. The speed and degree of waste flow depend upon numerous factors. Hydrologic factors such as water gradient and permeability of the solid formations through which the waste plume moves are important. The rate of flow is usually rather slow, typically several centimeters per day. An important aspect of the movement of wastes through the geosphere is **attenuation** by the mineral strata. This occurs because waste compounds are sorbed to solids by various mechanisms. A measure of the attenuation can be expressed by a **distribution coefficient**, K_d,

$$K_d = \frac{C_S}{C_W} \tag{9.13.1}$$

where C_S and C_W are the equilibrium concentrations of the constituent on solids and in solution, respectively. This relationship assumes relatively ideal behavior of the hazardous substance that is partitioned between water and solids (the sorbate). A more empirical expression is based on the Freundlich equation,

$$C_S = K_F C_{eq}^{1/n} \tag{9.13.2}$$

where and K_F and $1/n$ are empirical constants.

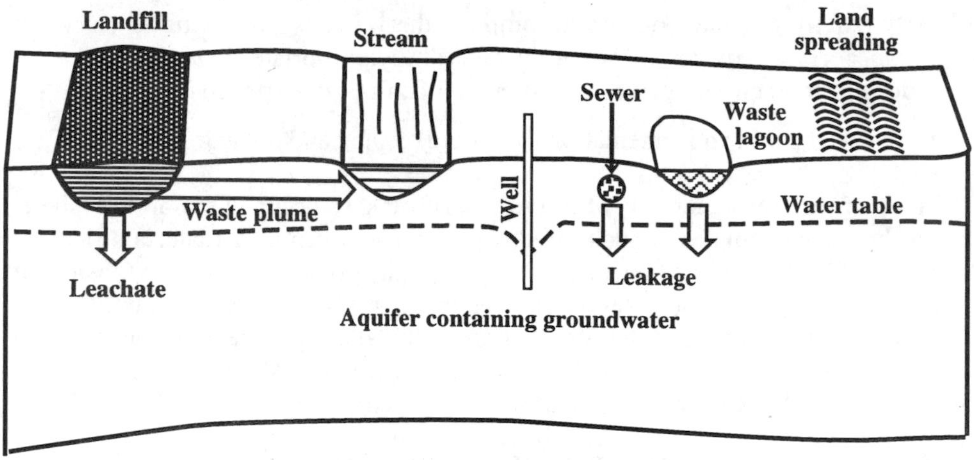

Figure 19.4. Sources and movement of hazardous wastes in the geosphere.

Several important properties of the solid determine the degree of sorption. One obvious factor is surface area. The chemical nature of the surface is also important. Among the important chemical factors are presence of sorptive clays, hydrous metal oxides, and humus (particularly important for the sorption of organic substances).

In general, sorption of hazardous waste solutes is higher above the water table in the unsaturated zone of soil. This region tends to have a higher surface area and to favor aerobic biodegradation processes.

The chemical nature of the leachate is important in sorptive processes of hazardous substances in the geosphere. Organic solvents or detergents in leachates will solubilize organic materials, preventing their retention by solids. Acidic leachates tend to dissolve metal oxides,

$$M(OH)_2(s) + 2H^+ \rightarrow M^{2+} + 2H_2O \tag{9.13.3}$$

thus preventing sorption of metals in insoluble forms. This is a reason that leachates from municipal landfills, which contain weak organic acids, are particularly prone to transport metals. Solubilization by acids is particularly important in the movement of heavy-metal ions.

Heavy metals are among the most dangerous hazardous waste constituents that are transported through the geosphere. Many factors affect their movement and attenuation. The temperature, pH, and reducing nature (as expressed by the negative log of the electron activity, pE) of the solvent medium are important. The nature of the solids, especially the inorganic and organic chemical functional groups on the surface, the cation-exchange capacity, and the surface area of the solids largely determine the attenuation of heavy metal ions. In addition to being sorbed and undergoing ion exchange with geospheric solids, heavy metals may undergo oxidation-reduction processes, precipitate as slightly soluble solids (especially sulfides), and in some cases, such as occurs with mercury, undergo microbial methylation reactions that produce mobile organometallic species.

The importance of chelating agents interacting with metals and increasing their mobilities has been illustrated by the effects of chelating ethylenediaminetetraacetic acid (EDTA) on the mobility of radioactive heavy metals, especially ^{60}Co.[8] The EDTA and other chelating agents, such as diethylenetriaminepentaacetic acid (DTPA) and nitrilotriacetic acid (NTA), were used to dissolve metals in the decontamination of radioactive facilities and were codisposed with radioactive materials at Oak Ridge National Laboratory (Tennessee) during the period 1951–1965. Unexpectedly high rates of radioactive metal mobility were observed, which was attributed to the formation of anionic species such as ^{60}CoT$^-$ (where T^{3-} is the chelating NTA anion). Whereas unchelated cationic metal species are strongly retained on soil by precipitation reactions and cation exchange processes,

$$Co^{2+} + 2OH^- \rightarrow Co(OH)_2(s) \tag{9.13.4}$$

$$2Soil\}^-H^+ + Co^{2+} \rightarrow (Soil\}^-)_2Co^{2+} + 2H^+ \tag{9.13.5}$$

anion bonding processes are very weak, so that the chelated anionic metal species are not strongly bound. Naturally occurring humic acid chelating agents may also be involved in the subsurface movement of radioactive metals. It is now generally accepted that poorly biodegradable, strong chelating agents, of which EDTA is the prime example, will facilitate transport of metal radionuclides, whereas oxalate and citrate will not do so because they form relatively weak complexes and are readily biodegraded.[9] Biodegradation of chelating agents tends to be a slow process under subsurface conditions.

Soil can be severely damaged by hazardous waste substances. Such materials may alter the physical and chemical properties of soil and thus its ability to support plants. Some of the more catastrophic incidents in which soil has been damaged by exposure to hazardous materials have arisen from soil contamination from SO_2 emitted from copper or lead smelters, or from brines from petroleum production. Both of these contaminants stop the growth of plants and, without the binding effects of viable plant root systems, topsoil is rapidly lost by erosion.

Unfortunate cases of the improper disposal of hazardous wastes into the geosphere have occurred throughout the world. For example, in December 1998 it was alleged that Formosa Plastics had disposed of 3000 tons of mercury-containing wastes in Cambodia, the result of by-product sludge generated by the chloralkali electrolytic prcess for generating chlorine and sodium hydroxide. Subsequently, illegal dump sites containing mercury were found in many places in Taiwan, causing major environmental concerns.[10]

19.14. HAZARDOUS WASTES IN THE HYDROSPHERE

Hazardous waste substances may enter the hydrosphere as leachate from waste landfills, drainage from waste ponds, seepage from sewer lines, or runoff from soil. Deliberate release into waterways also occurs, and is a particular problem in countries with lax environmental enforcement. There are, therefore, numerous ways by which hazardous materials may enter the hydrosphere.

For the most part, the hydrosphere is a dynamic, moving system, so that it provides perhaps the most important variety of pathways for moving hazardous waste species in the environment. Once in the hydrosphere, hazardous waste species can undergo a number of processes by which they are degraded, retained, and transformed. These include the common chemical processes of precipitation-dissolution, acid-base reactions, hydrolysis, and oxidation-reduction reactions. Also included are a wide variety of biochemical processes which, in most cases, reduce hazards, but in some cases, such as the biomethylation of mercury, greatly increase the risks posed by hazardous wastes.

The unique properties of water have a strong influence on the environmental chemistry of hazardous wastes in the hydrosphere. Aquatic systems are subject to constant change. Water moves with groundwater flow, stream flow, and convection currents. Bodies of water become stratified so that low-oxygen reducing conditions may prevail in the bottom regions of a body of water, and there is a constant interaction of the hydrosphere with the other environmental spheres. There is a continuing exchange of materials between water and the other environmental spheres. Organisms in water may have a strong influence on even poorly biode-gradable hazardous waste species through bioaccumulation mechanisms.

Figure 19.5 shows some of the pertinent aspects of hazardous waste materials in bodies of water, with emphasis upon the strong role played by sediments. An interesting kind of hazardous waste material that may accumulate in sediments consists of dense, water-immiscible liquids that may sink to the bottom of bodies of water or aquifers and remain there as "blobs" of liquid. Hundreds of tons of PCB wastes have accumulated in sediments in the Hudson River in New York State and are the subject of a heated debate regarding how to remediate the problem.

Hazardous waste species undergo a number of physical, chemical, and biochem-ical processes in the hydrosphere which strongly influence their effects and fates. The major ones of these are listed below:

- **Hydrolysis reactions** are those in which a molecule is cleaved with addition of a molecule of H_2O. An example of a hydrolysis reaction is the hydrolysis of dibutyl phthalate, Hazardous Waste Number U069:

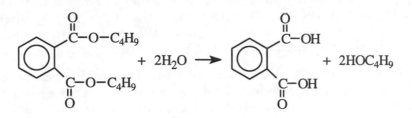

Another example is the hydrolysis of bis(chloromethyl)ether to produce HCl and formaldehyde:

$$
\underset{\substack{| \\ H}}{\overset{\substack{H \\ |}}{Cl-C}}-O-\underset{\substack{| \\ H}}{\overset{\substack{H \\ |}}{C}}-Cl + H_2O \longrightarrow 2H-\overset{\substack{O \\ ||}}{C}-H + 2HCl
$$

Compounds that hydrolyze are normally those, such as esters and acid anhydrides, originally formed by joining two other molecules with the loss of H_2O.

- **Precipitation reactions**, such as the formation of insoluble lead sulfide from soluble lead(II) ion in the anaerobic regions of a body of water:

$$Pb^{2+} + HS^- \rightarrow PbS(s) + H^+$$

An important part of the precipitation process is normally **aggregation** of the colloidal particles first formed to produce a cohesive mass. Precipitates are often relatively complicated species, such as the basic salt of lead carbonate, $2PbCO_3 \cdot Pb(OH)_2$. Heavy metals, a common ingredient of hazardous waste species precipitated in the hydrosphere, tend to form hydroxides, carbonates, and sulfates with the OH^-, HCO_3^-, and SO_4^{2-} ions

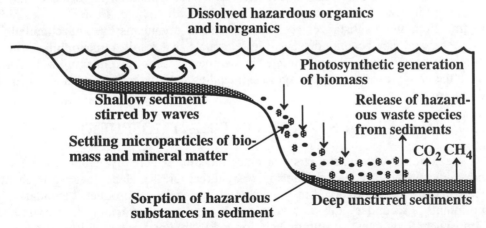

Figure 19.5. Aspects of hazardous wastes in surface water in the hydrosphere. The deep unstirred sediments are anaerobic and the site of hydrolysis reactions and reductive processes that may act on hazardous waste constituents sorbed to the sediment.

that commonly are present in water, and sulfides are likely to be formed in bottom regions of bodies of water where sulfide is generated by anaerobic bacteria. Heavy metals are often coprecipitated as a minor constituent of some other compound, or are sorbed by the surface of another solid.

- **Oxidation-reduction reactions** commonly occur with hazardous waste materials in the hydrosphere, generally mediated by microorganisms. An example of such a process is the oxidation of ammonia to toxic nitrite ion mediated by *Nitrosomonas* bacteria:

$$NH_3 + {}^3/_2O_2 \rightarrow H^+ + NO_2^-(s) + H_2O$$

- **Biochemical processes**, which often involve hydrolysis and oxidation-reduction reactions. Organic acids and chelating agents, such as citrate, produced by bacterial action may solubilize heavy metal ions. Bacteria also produce methylated forms of metals, particularly mercury and arsenic.

- **Photolysis reactions** and miscellaneous chemical phenomena. Photolysis of hazardous waste compounds in the hydrosphere commonly occurs on surface films exposed to sunlight on the top of water.

Hazardous waste compounds have a number of effects on the hydrosphere. Perhaps the most serious of these is the contamination of groundwater, which in some cases can be almost irreversible. Waste compounds accumulate in sediments, such as river or estuary sediments. Hazardous waste compounds dissolved in, suspended in, or floating as surface films on the surface of water can render it unfit for use and for sustenance of aquatic organisms.

Many factors determine the fate of a hazardous waste substance in water. Among these are the substance's solubility, density, biodegradability, and chemical reactivity. As discussed above and in Section 19.16, biodegradation largely determines the fates of hazardous waste substances in the hydrosphere. In addition to biodegradation, some substances are concentrated in organisms by bioaccumulation processes and may become deposited in sediments as a result. Organophilic materials may be sorbed by organic matter in sediments. Cation-exchanging sediments have the ability to bind cationic species, including cationic metal ions and organics that form cations.

19.15. HAZARDOUS WASTES IN THE ATMOSPHERE

Hazardous waste chemicals can enter the atmosphere by evaporation from hazardous waste sites, by wind erosion, or by direct release. Hazardous waste chemicals usually are not evolved in large enough quantities to produce secondary air pollutants. (Secondary air pollutants are formed by chemical processes in the atmosphere. Examples are sulfuric acid formed from emissions of sulfur oxides and oxidizing photochemical smog formed under sunny conditions from nitrogen oxides and hydrocarbons.) Therefore, species from hazardous waste sources are usually of most concern in the atmosphere as primary pollutants emitted in localized areas at a hazardous waste site. Plausible examples of primary air pollutant hazardous waste chemicals include corrosive acid gases, particularly HCl; toxic organic vapors, such as vinyl chloride (U043); and toxic inorganic gases, such as HCN potentially released by the accidental mixing of waste cyanides:

$$H_2SO_4 + 2NaCN \rightarrow Na_2SO_4 + 2HCN(g) \tag{19.15.1}$$

Primary air pollutants such as these are almost always of concern only adjacent to the site or to workers involved in site remediation. One such substance that has been responsible for fatal poisonings at hazardous waste sites, usually tanks that are undergoing cleanup or demolition, is highly toxic hydrogen sulfide gas, H_2S.

An important characteristic of a hazardous waste material that enters the atmosphere is its **pollution potential**. This refers to the degree of environmental threat posed by the substance acting as a primary pollutant, or to its potential to cause harm from secondary pollutants.

Another characteristic of a hazardous waste material that determines its threat to the atmosphere is its **residence time**, which can be expressed by an estimated atmospheric half-life, $\tau_{1/2}$. Among the factors that go into estimating atmospheric half-lives are water solubilities, rainfall levels, and atmospheric mixing rates.

Hazardous waste compounds in the atmosphere that have significant water solubilities are commonly removed from the atmosphere by **dissolution** in water. The water may be in the form of very small cloud or fog particles or it may be present as rain droplets.

Some hazardous waste species in the atmosphere are removed by **adsorption onto aerosol particles**. Typically, the adsorption process is rapid so that the lifetime of the species is that of the aerosol particles (typically a few days). Adsorption onto solid particles is the most common removal mechanism for highly nonvolatile constituents such as benzo[a]pyrene.

Dry deposition is the name given to the process by which hazardous waste species are removed from the atmosphere by impingement onto soil, water, or plants on the earth's surface. These rates are dependent upon the type of substance, the nature of the surface that they contact, and weather conditions.

A significant number of hazardous waste substances leave the atmosphere much more rapidly than predicted by dissolution, adsorption onto particles, and dry deposition, meaning that chemical processes must be involved. The most important of these are photochemical reactions, commonly involving hydroxyl radical, HO·. Other reactive atmospheric species that may act to remove hazardous waste compounds are ozone (O_3), atomic oxygen (O), peroxyl radicals (HOO·), alkyl-peroxyl radicals (ROO·), and NO_3. Although its concentration in the troposphere is relatively low, HO· is so reactive that it tends to predominate in the chemical processes that remove hazardous waste species from air. Hydroxyl radical undergoes *abstraction reactions* that remove H atoms from organic compounds,

$$R\text{-}H + HO^\bullet \rightarrow R^\bullet + H_2O \tag{9.15.2}$$

and may react with those containing unsaturated bonds by addition as illustrated by the following reaction:

$$R\!\!\diagdown\!\!C\!\!=\!\!C\!\!\diagup\!\!H_{} \quad + \quad HO\cdot \quad \longrightarrow \quad H\!-\!\overset{R}{\underset{\cdot}{C}}\!-\!\overset{H}{\underset{H}{C}}\!-\!OH \tag{9.15.3}$$

The free radical products are very reactive. They react further to form oxygenated species, such as aldehydes, ketones, and dehalogenated organics, eventually leading to the formation of particles or water-soluble materials that are readily scavenged from the atmosphere.

Direct photodissociation of hazardous waste compounds in the atmosphere may occur by the action of the shorter wavelength light that reaches to the troposphere and is absorbed by a molecule with a light-absorbing group called a **chromophore**:

$$R\!-\!X + h\nu \rightarrow R\!\cdot\! + X\!\cdot \tag{9.15.4}$$

Among the factors involved in assessing the effectiveness of direct absorption of light to remove species from the atmosphere are light intensity, quantum yields (chemical reactions per quantum absorbed), and atmospheric mixing. The requirement of a suitable chromophore limits direct photolysis as a removal mechanism for most compounds other than conjugated alkenes, carbonyl compounds, some halides, and some nitrogen compounds, particularly nitro compounds, all of which commonly occur in hazardous wastes.

19.16. HAZARDOUS WASTES IN THE BIOSPHERE

Microorganisms, bacteria, fungi, and, to a certain extent, protozoa may act metabolically on hazardous wastes substances in the environment. Most of these subtances are *anthropogenic* (made by human activities), and most are classified as *xenobiotic* molecules that are foreign to living systems. Although by their nature xenobiotic compounds are degradation resistant, almost all classes of them—nonhalogenated alkanes, halogenated alkanes (trichloroethane, dichloromethane), nonhalogenated aryl compounds (benzene, naphthalene, benzo[a]pyrene), halogenated aryl compounds (hexachlorobenzene, pentachlorophenol), phenols (phenol, cresols), polychlorinated biphenyls, phthalate esters, and pesticides (chlordane, parathion)—can be at least partially degraded by various microorganisms.

Bioaccumulation occurs in which wastes are concentrated in the tissue of organisms. It is an important mechanism by which wastes enter food chains. **Biodegradation** occurs when wastes are converted by biological processes to generally simpler molecules; the complete conversion to simple inorganic species, such as CO_2, NH_3, SO_4^{2-}, and $H_2PO_4^-/HPO_4^-$, is called **mineralization.** The production of a less toxic product by biochemical processes is called **detoxification**. An example is the bioconversion of highly toxic organophosphate paraoxon to *p*-nitrophenol, which is only about 1/200 as toxic:

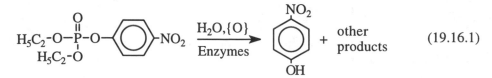

$$\tag{19.16.1}$$

Microbial Metabolism in Waste Degradation

The following terms and concepts apply to the metabolic processes by which microorganisms biodegrade hazardous wastes substances:

- **Biotransformation** is the enzymatic alteration of a substance by microorganisms.

- **Metabolism** is the biochemical process by which biotransformation is carried out.

- **Catabolism** is an enzymatic process by which more complex molecules are broken down into less complex ones.

- **Anabolism** is an enzymatic process by which simple molecules are assembled into more complex biomolecules.

Two major divisions of biochemical metabolism that operate on hazardous waste species are **aerobic processes** that use molecular O_2 as an oxygen source and **anaerobic processes**, which make use of another oxidant. For example, when sulfate ion acts as an oxidant (electron receptor) the transformation $SO_4^{2-} \rightarrow H_2S$ occurs. (This has the benefit of providing sulfide, which precipitates insoluble metal sulfides in the presence of hazardous waste heavy metals.) Because molecular oxygen does not penetrate to such depths, anaerobic processes predominate in the deep sediments as shown in Figure 19.5.

For the most part, anthropogenic compounds resist biodegradation much more strongly than do naturally occurring compounds. Given the nature of xenobiotic substances, there are very few enzyme systems in microorganisms that act directly on these substances, especially in making an intial attack on the molecule. Therefore, most xenobiotic compounds are acted upon by a process called **cometabolism**, which occurs concurrently with normal metabolic processes. An interesting example of cometabolism is provided by the white rot fungus, *Phanerochaete chrysosporium*, which has been promoted for the treatment of hazardous organochlorides such as PCBs, DDT, and chlorodioxins. This fungus uses dead wood as a carbon source and has an enzyme system that breaks down wood lignin, a degradation-resistant biopolymer that binds the cellulose in wood. Under appropriate conditions, this enzyme system attacks organochloride compounds and enables their mineralization.

The susceptibility of a xenobiotic hazardous waste compound to biodegradation depends upon its physical and chemical characteristics. Important physical characteristics include water solubility, hydrophobicity (aversion to water), volatility, and lipophilicity (affinity for lipids). In organic compounds, certain structural groups—branched carbon chains, ether linkages, meta-substituted benzene rings, chlorine, amines, methoxy groups, sulfonates, and nitro groups—impart particular resistance to biodegradation.

Microorganisms vary in their ability to degrade hazardous waste compounds; virtually never does a single microorganism have the ability to completely mineralize a waste compound. Abundant aerobic bacteria of the ***Pseudomonas*** family are particularly adept at degrading synthetic compounds such as biphenyl, naphthalene, DDT, and many other compounds. *Actinomycetes*, microorganisms that are morphologically similar to both bacteria and fungi, degrade a variety of organic compounds including degradation-resistant alkanes and lignocellulose, as well as pyridines, phenols, nonchlorinated aryls, and chlorinated aryls.

Because of their requirement for oxygen-free (anoxic) conditions, anaerobic bacteria are fastidious and difficult to study. However, they can play an important role in degrading biomass, particularly through hydrolytic processes in which molecules are cleaved with addition of H_2O. Anaerobic bacteria reduce oxygenated organic functional groups. As examples, they convert nitro compounds to amines, degrade nitrosamines, promote reductive dechlorination, reduce epoxide groups to alkenes, and break down aryl structures. Partial dechlorination of PCBs has been reported by bacteria growing anaerobically in PCB-contaminated river sediments, such as those in New York's Hudson River.[11] PCB waste remediation schemes have been proposed that make use of anaerobic dechlorination of the more highly chlorinated PCBs and aerobic degradation of the less highly chlorinated products.[12]

Fungi are particularly noted for their ability to attack long-chain and complex hydrocarbons, and are more successful than bacteria in the initial attack on PCB compounds. The potential of the white rot fungus, *Phanerochaete chrysosporium*, to degrade biodegradation-resistant compounds, especially organochloride species, was previously noted.

Phototrophic microorganisms, algae, photosynthetic bacteria, and cyanobacteria that perform photosynthesis have lipid bodies that accumulate lipophilic compounds. There is some evidence to suggest that these organisms can induce photochemical degradation of the stored compounds.

Biologically, the greatest concern with wastes has to do with their toxic effects on animals, plants, and microbes. Virtually all hazardous waste substances are poisonous to a degree, some extremely so. Toxicities vary markedly with the physical and chemical nature of the waste, the matrix in which it is contained, the type and condition of the species exposed, and the manner, degree, and time of exposure.

LITERATURE CITED

1. Wolbeck, Bernd, "Political Dimensions and Implications of Hazardous Waste Disposal," in *Hazardous Waste Disposal*, Lehman, John P., Ed., Plenum Press, New York, 1982, pp. 7-18

2. Hileman, Bette, "Treaty Grows Less Contentious," *Chemical and Engineering News*, April 6, 1998, pp. 29–30.

3. Long, Janice, "Congress is Giving Superfund Another Look," *Chemical and Engineering News* May 31, 1999, pp. 26-7.

4. Petersen, Melody, "Cleaning Up in the Dark," *New York Times*, May 14, 1998, p. C1.

5. Manahan, Stanley E., *Toxicological Chemistry*, 2nd ed., CRC Press/Lewis Publishers, Boca Raton, FL, 1992.

6. "Toxicity Characteristic Leaching Procedure," Test Method 1311 in *Test Methods for Evaluating Solid Waste, Physical/Chemical Methods*, EPA Publication SW-846, 3rd ed., (November, 1986), as amended by Updates I, II, IIA, U.S. Government Printing Office, Washington, D.C.

7. "Spills, Fires, Explosions, and Other Releases," Chapter 9 in *Hazardous Waste Management Compliance Handbook*, 2nd ed., Van Nostrand Reinhold Publishing Company, New York, 1997, pp. 107–116.

8. Means, J. L., D. A. Crear, and J. O. Duguid, "Migration of Radioactive Wastes: Radionuclide Mobilization by Complexing Agents," *Science*, **200**, 1477–81 (1978).

9. Serne, R. J., A. R. Felmy, K. J. Cantrell, H. Bolton, J. K. Fredrickson, K. M. Krupka, and J. A. Campbell, "Characterization of Radionuclide-Chelating Agent Complexes Found in Low-Level Radioactive Decontamination Waste: Literature Review," PNL-8856, Battelle Pacific Northwest Laboratories, Richland, WA (1996).

10. Tremblay, Jean-Francois, "Environmental Mess in Taiwan," *Chemical and Engineering News*, May 31, 1999, pp. 19-24.

11. Rhee, G.-Yull, Roger C. Sokol, Brian Bush, and Charlotte M. Bethoney, "Long-Term Study of the Anaerobic Dechlorination of Arochlor 1254 with and without Biphenyl Enrichment," *Environmental Science and Technology*, **27**, 714–719 (1993).

12. Liou, Raycharn, James H. Johnson, and John P. Tharakan, "Anaerobic Dechlorination and Aerobic Degradation of PCBs in Soil Columns and Slurries," *Hazardous Industrial Wastes*, **29**, 414-423 (1997).

SUPPLEMENTAL REFERENCES

Blackman, William C., Jr., *Basic Hazardous Waste Management*, CRC Press/Lewis Publishers, Boca Raton, FL, 1996.

Clark, Nancy and Anna Crull, *Bioremediation of Hazardous Wastes, Wastewater, and Municipal Waste*, Business Communications Co., Norwalk, CT, 1997.

Cheremisinoff, Nicholas P., *Handbook of Industrial Toxicology and Hazardous Materials*, Marcel Dekker, New York, 1999.

Davletshina, Tatyana A., *Industrial Fire Safety Guidebook*, Noyes Publications, Westwood, NJ, 1998.

Davletshina, Tatyana A. and Nicholas P. Cheremisinoff, *Fire and Explosion Hazards Handbook of Industrial Chemicals*, Noyes Publications, Westwood, NJ, 1998.

Flick, Ernest W., Ed., *Industrial Solvents Handbook*, 5th ed., Noyes Publications, Westwood, NJ, 1998.

Freeman, Harry M. and Eugene F. Harris, Eds., *Hazardous Waste Remediation : Innovative Treatment Technologies*, Technomic Publishing Co., Lancaster, PA, 1995.

General Accounting Office, *Hazardous Waste: Information on Potential Superfund Sites: Report to the Ranking Minority Member, Committee on Commerce, House of Representatives*, United States General Accounting Office, Washington, D.C., 1998.

General Accounting Office, *Hazardous Waste: Unaddressed Risks at Many Potential Superfund Sites: Report to the Ranking Minority Member, Committee on Commerce, House of Representatives*, United States General Accounting Office, Washington, D.C., 1998.

Haas, Charles N. and Richard J. Vamos, *Hazardous and Industrial Waste Treatment*, Prentice Hall, Upper Saddle River, NJ, 1995.

Hasan, Syed, *Geology and Hazardous Waste Management*, Prentice Hall College Div., Upper Saddle River, NJ, 1996.

Hazardous and Industrial Wastes: Proceedings of the Mid-Atlantic Industrial and Hazardous Waste Conference, Technomic Publishing Co., Lancaster, PA, 1996.

Hinchee, Robert E., Rodney S. Skeel, and Gregory D. Sayles, Eds., *Biological Unit Processes for Hazardous Waste Treatment*, Battelle Press,Columbus, OH, 1995.

Hocking, Martin B., *Handbook of Chemical Technology and Pollution Control*, Academic Press, San Diego, CA, 1998.

Howard, Philip H., *Handbook of Environmental Fate and Exposure Data for Organic Chemicals*, Vol. V. Solvents 3, CRC Press/Lewis Publishers, Boca Raton, FL, 1997.

Kindschy, Jon W., Marilyn Kraft, and Molly Carpenter, *Guide to Hazardous Materials and Waste Management*, Solano Press Books, Point Arena, CA, 1997.

Meyer, *Chemistry of Hazardous Materials*, Prentice-Hall, Upper Saddle River, NJ, 1998.

Nemerow, Nelson Leonard and Franklin J. Agardy, *Strategies of Industrial and Hazardous Waste Management*, Van Nostrand Reinhold, New York, 1998.

Task Force on Hazardous Waste Site Remediation Management, *Hazardous Waste Site Remediation Management*, Water Environment Federation, Alexandria, VA, 1999.

Tedder, D. William and Frederick G. Pohland, Eds., *Emerging Technologies in Hazardous Waste Management 7*, Plenum Press, New York, 1997.

Wagner, Travis P., *The Complete Guide to the Hazardous Waste Regulations : RCRA, TSCA, HMTA, OSHA, and Superfund*, John Wiley & Sons, New York, 1999.

Woodside, Gayle, *Hazardous Materials and Hazardous Waste Management*, John Wiley & Sons, New York, 1999.

QUESTIONS AND PROBLEMS

1. Match the following kinds of hazardous substances on the left with a specific example of each from the right, below:

   1. Explosives   (a)   Oleum, sulfuric acid, caustic soda

   2. Compressed gases   (b)   White phosphorus

   3. Radioactive materials   (c) NH_4ClO_4

   4. Flammable solids   (d)   Hydrogen, sulfur dioxide

   5. Oxidizing materials   (e)   Nitroglycerin

   6. Corrosive materials   (f)   Plutonium, cobalt-60

2. Of the following, the property that is **not** a member of the same group as the other properties listed is (a) substances that are liquids whose vapors are likely to ignite in the presence of ignition sources, (b) nonliquids that may catch fire from friction or contact with water and which burn vigorously or persistently, (c) ignitable compressed gases, (d) oxidizers, (e) substances that exhibit extremes of acidity or basicity.

3. In what respects may it be said that measures taken to alleviate air and water pollution tend to aggravate hazardous waste problems?

4. Why is attenuation of metals likely to be very poor in acidic leachate? Why is attenuation of anionic species in soil less than that of cationic species?

5. Discuss the significance of LFL, UFL, and flammability range in determining the flammability hazards of organic liquids.

6. Concentrated HNO_3 and its reaction products pose several kinds of hazards. What are these?

7. What are substances called that catch fire spontaneously in air without an ignition source?

8. Name four or five hazardous products of combustion and specify the hazards posed by these materials.

9. What kind of property tends to be imparted to a functional group of an organic compound containing both oxygen and nitrogen?

10. Match the corrosive substance from the column on the left, below, with one of its major properties from the right column:

    1. Alkali metal hydroxides   (a) Reacts with protein in tissue to form yellow

    2. Hydrogen peroxide   xanthoproteic acid

    3. Hydrofluoric acid, HF   (b) Dissolves glass

    4. Nitric acid, HNO_3 (c) Strong bases

                                         (d) Oxidizer

11. Rank the following wastes in increasing order of segregation (a) mixed halogenated and hydrocarbon solvents containing little water, (b) spent steel pickling liquor, (c) dilute sludge consisting of mixed organic and inorganic wastes, (d) spent hydrocarbon solvents free of halogenated materials, (e) dilute mixed inorganic sludge.

12. Inorganic species may be divided into three major groups based upon their retention by clays. What are the elements commonly listed in these groups? What is the chemical basis for this division? How might anions (Cl^-, NO_3^-) be classified?

13. In what form would a large quantity of hazardous waste PCB likely be found in the hydrosphere?

14. The Toxicity Characteristic Leaching Procedure was originally devised to mimic a "mismanagement scenario" in which hazardous wastes were disposed along with biodegradable organic municipal refuse. Discuss how this procedure reflects the conditions that might arise from circumstances in which hazardous wastes and actively decaying municipal refuse were disposed together.

15. What are three major properties of wastes that determine their amenability to transport?

16. List and discuss the significance of major sources for the origin of hazardous wastes, that is, their main modes of entry into the environment. What are the relative dangers posed by each of these? Which part of the environment would each be most likely to contaminate?

17. What is the influence of organic solvents in leachates upon attenuation of organic hazardous waste constituents?

18. What features or characteristics should a compound possess in order for direct photolysis to be a significant factor in its removal from the atmosphere?

19. Describe the particular danger posed by codisposal of strong chelating agents with radionuclide wastes. What may be said about the chemical nature of the latter with regard to this danger?

20. Describe a beneficial effect that might result from the precipitation of either $Fe_2O_3 \cdot xH_2O$ or $MnO_2 \cdot xH_2O$ from hazardous wastes in water.

21. Why are secondary air pollutants from hazardous waste sites usually of only limited concern as compared to primary air pollutants? What is the distinction between the two?

22. Match the following physical, chemical, and biochemical processes dealing with the transformations and ultimate fates of hazardous chemical species in the hydrosphere on the left with the description of the process on the right, below:

1. Precipitation reactions (a) Molecule is cleaved with the addition of H_2O
2. Biochemical processes (b) Generally accompanied by aggregation of colloidal particles suspended in water
3. Oxidation-reduction
4. Hydrolysis reactions (c) Generally mediated by microorganisms

5. Sorption   (d) By sediments and by suspended matter

                                    (e) Often involve hydrolysis and oxidation-reduction

23. As applied to hazardous wastes in the biosphere, distinguish among biodegradation, biotransformation, detoxification, and mineralization.

24. What is the potential role of *Phanerochaete chrysosporium* in treatment of hazardous waste compounds? For which kinds of compounds might it be most useful?

25. Which part of the hydrosphere is most subject to long-term, largely irreversible contamination from the improper disposal of hazardous wastes in the environment?

26. Several physical and chemical characteristics are involved in determining the amenability of a hazardous waste compound to biodegradation. These include hydrophobicity, solubility, volatility, and affinity for lipids. Suggest and discuss ways in which each one of these factors might affect biodegradability.

27. List and discuss some of the important processes determining the transformations and ultimate fates of hazardous chemical species in the hydrosphere.

20 INDUSTRIAL ECOLOGY FOR WASTE MINIMIZATION, UTILIZATION, AND TREATMENT

20.1. INTRODUCTION

Chapter 19 has addressed the nature and sources of hazardous wastes and their environmental chemistry and has pointed out some of the major problems associated with such wastes. Chapter 20 deals with means for minimizing wastes, utilizing materials that might go into wastes, and treating and disposing of wastes, the generation of which cannot be avoided. The practice of industrial ecology is all about not producing wastes and, instead, utilizing wastes for useful purposes. Therefore, in dealing with wastes, it is essential in the modern age to consider the potential contribution of industrial ecology.

Since the 1970s, efforts to reduce and clean up hazardous wastes have been characterized by:

- Legislation
- Procrastination
- Regulation
- Modeling
- Litigation
- Analysis
- Cleanup of a few select sites

It is perhaps fair to say that in proportion to the magnitude of the problems and the amount of money devoted to them so far, insufficient progress has been made in coping with hazardous wastes. In the U.S., huge amounts of time have been devoted to promulgating hazardous waste regulations, instrument manufacturers have prospered as more and more chemical analyses have been required, and computationists, some of whom would be offended at the sight of any chemical, much less a hazardous one, have consumed thousands of hours of computer time to model hazardous waste systems. This is to say nothing of the vast expense of litigation that has gone into lawsuits dealing with hazardous waste sites. In the future, a higher percentage of the effort and resources devoted to hazardous wastes needs to be placed on remediation of existing problems and preventive action to avoid future problems.

The U. S. Superfund act, a particular target of criticism by industrial and other groups, has been under consideration for renewal each year since 1994. A total of nine bills dealing with Superfund were introduced in the U. S. Congress in 1999, none with much of a chance of passing![1] Critics contend that Superfund's efforts have been directed toward high-cost solutions for minimal risks. Many Superfund critics contend that a much greater emphasis must be placed on institutional controls and waste isolation to deal effectively with improperly disposed hazardous wastes.

This chapter discusses how environmental chemistry can be applied to hazardous waste management to develop measures by which chemical wastes can be minimized, recycled, treated, and disposed. In descending order of desirability, hazardous waste management attempts to accomplish the following:

- Do not produce it

- If making it cannot be avoided, produce only minimum quantities

- Recycle it

- If it is produced and cannot be recycled, treat it, preferably in a way that makes it nonhazardous

- If it cannot be rendered nonhazardous, dispose of it in a safe manner

- Once it is disposed, monitor it for leaching and other adverse effects

The **effectiveness** of a hazardous waste management system is a measure of how well it reduces the quantities and hazards of wastes. As shown in Figure 20.1, the best management option consists of measures that prevent generation of wastes. Next in order of desirability is recovery and recycling of waste constituents. Next is destruction and treatment with conversion to nonhazardous waste forms. The least desirable option is disposal of hazardous materials in storage or landfill.

20.2. WASTE REDUCTION AND MINIMIZATION

The initial sections of this chapter address waste reduction and minimization. During recent years, substantial efforts have been made in reducing the quantities of wastes and, therefore, the burden of dealing with wastes. Much of this effort has been the result of legislation and regulations restricting wastes, along with the resulting concerns over possible legal actions and lawsuits. In many cases—and ideally in all—minimizing the quantities of wastes produced is simply good business. Wastes are materials, materials have value and, therefore, all materials should be used for some beneficial purpose and not discarded as wastes, usually at a high cost for waste disposal.

Industrial ecology is all about the efficient use of materials. Therefore, by its nature, a system of industrial ecology is also a system of waste reduction and minimization. In reducing quantities of wastes, it is important to take the broadest possible view. This is because dealing with one waste problem in isolation may simply create another. Early efforts to control air and water pollution resulted in problems from hazardous wastes isolated from industrial operations. A key aspect of industrial ecology is its approach based upon industrial systems as a whole, making a system of industrial ecology by far the best means of dealing with wastes by avoiding their production.

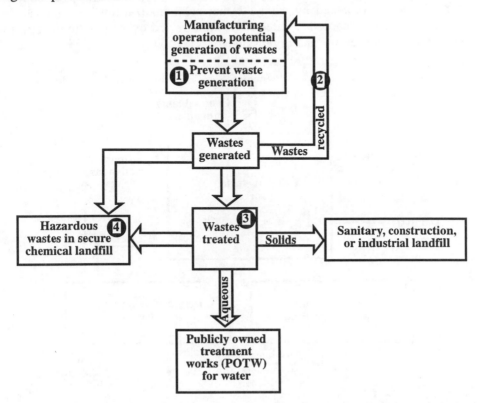

Figure 20.1. Order of effectiveness of waste treatment management options. The darkened circles indicate the degree of effectiveness from the most desirable (1) to the least (4).

Many hazardous waste problems can be avoided at early stages by **waste reduction** (cutting down quantities of wastes from their sources) and **waste minimization** (utilization of treatment processes which reduce the quantities of wastes requiring ultimate disposal). This section outlines basic approaches to waste minimization and reduction.

There are several ways in which quantities of wastes can be reduced, including source reduction, waste separation and concentration, resource recovery, and waste recycling. The most effective approaches to minimizing wastes center around careful control of manufacturing processes, taking into consideration discharges and the potential for waste minimization at every step of manufacturing. Viewing the process as a whole (as outlined for a generalized chemical manufacturing process in Figure 20.2) often enables crucial identification of the source of a waste, such as a raw material impurity, catalyst, or process solvent. Once a source is identified, it is much easier to take measures to eliminate or reduce the waste. The most effective approach to minimizing wastes is to emphasize waste minimization as an integral part of plant design.[2]

Modifications of the manufacturing process can yield substantial waste reduction. Some such modifications are of a chemical nature. Changes in chemical reaction conditions can minimize production of by-product hazardous substances. In some cases potentially hazardous catalysts, such as those formulated from toxic substances, can be replaced by catalysts that are nonhazardous or that can be recycled rather than discarded. Wastes can be minimized by volume reduction, for example, through dewatering and drying sludge.

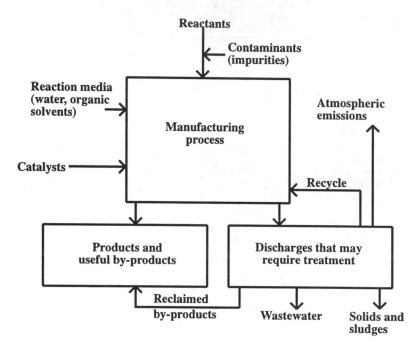

Figure 20.2. Chemical manufacturing process from the viewpoint of discharges and waste minimization.

Many kinds of waste streams are candidates for minimization. As examples, such waste streams identified at U. S. Government federal facilities have included solvents used for cleaning and degreasing, spent motor oil from gasoline and diesel engines, leftover and waste paint thinners, antifreeze/antiboil engine cooling formulations, batteries, inks, exposed photographic film, and pathology wastes.[3] The sources of the wastes are as varied as the waste streams themselves. Motor pool maintenance garages generate used motor oil and spent coolants. Hospitals, clinics, and medical laboratories generate pathology wastes. Aircraft maintenance depots where aircraft are cleaned, chemically stripped of paint and coatings, repainted, and electroplated generate large quantities of effluents, including organic materials.[4] Other facilities generating wastes include equipment and arms maintenance facilities, photo developing and printing laboratories, paint shops, and arts and crafts shops.

A crucial part of the process for reducing and minimizing wastes is the development of a material balance, which is an integral part of the practice of industrial ecology.[5] Such a balance addresses various aspects of waste streams, including sources, identification, and quantities of wastes and methods and costs of handling, treatment, recycling, and disposal. Priority waste streams can then be subjected to detailed process investigations to obtain the information needed to reduce wastes.

There are encouraging signs of progress in the area of waste minimization. All major companies have initiated programs to minimize quantities of wastes produced. A typical success story is a 97% reduction of landfilled wastes from Mobil's Torrance petroleum refinery from 1989 to 1993. During the same period the refinery went from less than 1% waste recycle to more than 70%. One of the technologies used for waste reduction was the Mobil Oil Sludge Coking Process.[6] Similar success stories in reducing wastes can be cited by a number of concerns in the U. S. and throughout the world.

20.3. RECYCLING

Wherever possible, recycling and reuse should be accomplished on-site because it avoids having to move wastes, and because a process that produces recyclable materials is often the most likely to have use for them. The four broad areas in which something of value may be obtained from wastes are the following:

- Direct recycle as raw material to the generator, as with the return to feed-stock of raw materials not completely consumed in a synthesis process

- Transfer as a raw material to another process; a substance that is a waste product from one process may serve as a raw material for another, sometimes in an entirely different industry

- Utilization for pollution control or waste treatment, such as use of waste alkali to neutralize waste acid

- Recovery of energy, for example, from the incineration of combustible hazardous wastes.

Examples of Recycling

Recycling of scrap industrial impurities and products occurs on a large scale with a number of different materials. Most of these materials are not hazardous but, as with most large-scale industrial operations, their recycling may involve the use or production of hazardous substances. Some of the more important examples are the following:

- **Ferrous metals** composed primarily of iron and used largely as feedstock for electric-arc furnaces

- **Nonferrous metals**, including aluminum (which ranks next to iron in terms of quantities recycled), copper and copper alloys, zinc, lead, cadmium, tin, silver, and mercury

- **Metal compounds**, such as metal salts

- **Inorganic substances** including alkaline compounds (such as sodium hydroxide used to remove sulfur compounds from petroleum products), acids (steel pickling liquor where impurities permit reuse), and salts (for example, ammonium sulfate from coal coking used as fertilizer)

- **Glass**, which makes up about 10 percent of municipal refuse

- **Paper**, commonly recycled from municipal refuse

- **Plastic**, consisting of a variety of moldable polymeric materials and composing a major constituent of municipal wastes

- **Rubber**

- **Organic substances**, especially solvents and oils, such as hydraulic and lubricating oils

- **Catalysts** from chemical synthesis or petroleum processing

- Materials with **agricultural uses**, such as waste lime or phosphate-containing sludges used to treat and fertilize acidic soils

Waste Oil Utilization and Recovery

Waste oil generated from lubricants and hydraulic fluids is one of the more commonly recycled materials. Annual production of waste oil in the U.S. is of the order of 4 billion liters. Around half of this amount is burned as fuel and lesser quantities are recycled or disposed as waste. The collection, recycling, treatment, and disposal of waste oil are all complicated by the fact that it comes from diverse, widely dispersed sources and contains several classes of potentially hazardous contaminants. These are divided between organic constituents (polycyclic aromatic hydrocarbons, chlorinated hydrocarbons) and inorganic constituents (aluminum, chromium, and iron from wear of metal parts; barium and zinc from oil additives; lead from leaded gasoline).

Recycling Waste Oil

The processes used to convert waste oil to a feedstock hydrocarbon liquid for lubricant formulation are illustrated in Figure 20.3. The first of these uses distillation to remove water and light ends that have come from condensation and contaminant fuel. The second, or processing, step may be a vacuum distillation in which the three

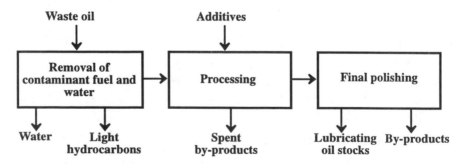

Figure 20.3. Major steps in reprocessing waste oil.

products are oil for further processing, a fuel oil cut, and a heavy residue. The processing step may also employ treatment with a mixture of solvents including isopropyl and butyl alcohols and methylethyl ketone to dissolve the oil and leave contaminants as a sludge; or contact with sulfuric acid to remove inorganic contaminants followed by treatment with clay to take out acid and contaminants that cause odor and color. The third step shown in Figure 20.3 employs vacuum distillation to separate lubricating oil stocks from a fuel fraction and heavy residue. This phase of treatment may also involve hydrofinishing, treatment with clay, and filtration.

Waste Oil Fuel

For economic reasons, waste oil that is to be used for fuel is given minimal treatment of a physical nature, including settling, removal of water, and filtration. Metals in waste fuel oil become highly concentrated in its fly ash, which may be hazardous.

Waste Solvent Recovery and Recycle

The recovery and recycling of waste solvents has some similarities to the recycling of waste oil and is also an important enterprise. Among the many solvents listed as hazardous wastes and recoverable from wastes are dichloromethane, tetrachloroethylene, trichloroethylene, 1,1,1-trichloroethane, benzene, liquid alkanes, 2-nitropropane, methylisobutyl ketone, and cyclohexanone. For reasons of both economics and pollution control, many industrial processes that use solvents are equipped for solvent recycle. The basic scheme for solvent reclamation and reuse is shown in Figure 20.4.

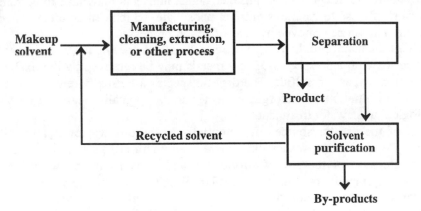

Figure 20.4. Overall process for recycling solvents.

A number of operations are used in solvent purification. Entrained solids are removed by settling, filtration, or centrifugation. Drying agents may be used to remove water from solvents, and various adsorption techniques and chemical treatment may be required to free the solvent from specific impurities. Fractional distillation, often requiring several distillation steps, is the most important operation in solvent purification and recycling. It is used to separate solvents from impurities, water, and other solvents.

Recovery of Water from Wastewater

It is often desirable to reclaim water from wastewater.[7] This is especially true in regions where water is in short supply. Even where water is abundant, water recycling is desirable to minimize the amount of water that is discharged.

A little more than half of the water used in the U.S. is consumed by agriculture, primarily for irrigation. Steam-generating power plants consume about one-fourth of the water, and other uses, including manufacturing and domestic uses, account for the remainder.

The three major manufacturing consumers of water are chemicals and allied products, paper and allied products, and primary metals. These industries use water for cooling, processing, and boilers. Their potential for water reuse is high and their total consumption of water is projected to drop in future years as recycling becomes more common.

The degree of treatment required for reuse of wastewater depends upon its application. Water used for industrial quenching and washing usually requires the least treatment, and wastewater from some other processes may be suitable for these purposes without additional treatment. At the other end of the scale, boiler makeup water, potable (drinking) water, water used to directly recharge aquifers, and water that people will directly contact (in boating, water skiing, and similar activities) must be of very high quality.

The treatment processes applied to wastewater for reuse and recycle depend upon both the characteristics of the wastewater and its intended uses. Solids can be removed by sedimentation and filtration. Biochemical oxygen demand is reduced by biological treatment, including trickling filters and activated sludge treatment. For uses conducive to the growth of nuisance algae, nutrients may have to be removed. The easiest of these to handle is nutrient phosphate, which can be precipitated with lime. Nitrogen can be removed by denitrification processes.

Two of the major problems with industrial water recycling are heavy metals and dissolved toxic organic species. Heavy metals may be removed by ion exchange or precipitation by base or sulfide. The organic species are usually removed with activated carbon filtration. Some organic species are biologically degraded by bacteria in biological wastewater treatment.

One of the greater sources of potentially hazardous wastewater arises from oil/water separators at wash racks where manufactured parts and materials are washed. Because of the use of surfactants and solvents in the washwater, the separated water tends to contain emulsified oil that is incompletely separated in an oil/water separator. In addition, the sludge that accumulates at the bottom of the separator may contain hazardous constituents, including heavy metals and some hazardous organic constituents. Several measures that include the incorporation of good industrial ecology practice may be taken to eliminate these problems.[8] One such measure is to eliminate the use of surfactants and solvents that tend to contaminate the water, and to use surfactants and solvents that are more amenable to separation and treatment. Another useful measure is to treat the water to remove harmful constituents and recycle it. This not only conserves water and reduces disposal costs, but also enables recycling of surfactants and other additives.

The ultimate water quality is achieved by processes that remove solutes from water, leaving pure H_2O. A combination of activated carbon treatment to remove organics, cation exchange to remove dissolved cations, and anion exchange for dissolved anions can provide very high quality water from wastewater. Reverse osmosis (see Chapter 8) can accomplish the same objective. However, these processes generate spent activated carbon, ion exchange resins that require regeneration, and concentrated brines (from reverse osmosis) that require disposal, all of which have the potential to end up as hazardous wastes.

20.4. PHYSICAL METHODS OF WASTE TREATMENT

This section addresses predominantly physical methods for waste treatment and the following section addresses methods that utilize chemical processes. It should be kept in mind that most waste treatment measures have both physical and chemical aspects. The appropriate treatment technology for hazardous wastes obviously depends upon the nature of the wastes. These may consist of volatile wastes (gases, volatile solutes in water, gases or volatile liquids held by solids, such as catalysts), liquid wastes (wastewater, organic solvents), dissolved or soluble wastes (water-soluble inorganic species, water-soluble organic species, compounds soluble in organic solvents), semisolids (sludges, greases), and solids (dry solids, including granular solids with a significant water content, such as dewatered sludges, as well as solids suspended in liquids). The type of physical treatment to be applied to wastes depends strongly upon the physical properties of the material treated, including state of matter, solubility in water and organic solvents, density, volatility, boiling point, and melting point.

As shown in Figure 20.5, waste treatment may occur at three major levels—**primary**, **secondary**, and **polishing**—somewhat analogous to the treatment of wastewater (see Chapter 8). Primary treatment is generally regarded as preparation for further treatment, although it can result in the removal of by-products and reduction of the quantity and hazard of the waste. Secondary treatment detoxifies, destroys, and removes hazardous constituents. Polishing usually refers to treatment of water that is removed from wastes so that it may be safely discharged. However, the term can be broadened to apply to the treatment of other products as well so that they may be safely discharged or recycled.

Methods of Physical Treatment

Knowledge of the physical behavior of wastes has been used to develop various unit operations for waste treatment that are based upon physical properties. These operations include the following:

- Phase separation
  Filtration
- Phase transition
  Distillation
  Evaporation
  Physical precipitation
- Phase transfer
  Extraction
  Sorption
- Membrane separations
  Reverse osmosis
  Hyper- and ultrafiltration

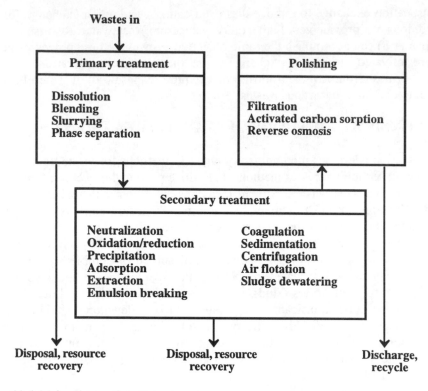

Figure 20.5. Major phases of waste treatment.

Phase Separations

The most straightforward means of physical treatment involves separation of components of a mixture that are already in two different phases. **Sedimentation** and **decanting** are easily accomplished with simple equipment. In many cases the separation must be aided by mechanical means, particularly **filtration** or **centrifugation**. **Flotation** is used to bring suspended organic matter or finely divided particles to the surface of a suspension. In the process of **dissolved air flotation** (DAF), air is dissolved in the suspending medium under pressure and comes out of solution when the pressure is released as minute air bubbles attached to suspended particles, which causes the particles to float to the surface.

An important and often difficult waste treatment step is **emulsion breaking** in which colloidal-sized **emulsions** are caused to aggregate and settle from suspension. Agitation, heat, acid, and the addition of **coagulants** consisting of organic polyelectrolytes, or inorganic substances such as an aluminum salt, may be used for this purpose. The chemical additive acts as a flocculating agent to cause the particles to stick together and settle out.

Phase Transition

A second major class of physical separation is that of **phase transition** in which a material changes from one physical phase to another. It is best exemplified by

distillation, which is used in treating and recycling solvents, waste oil, aqueous phenolic wastes, xylene contaminated with paraffin from histological laboratories, and mixtures of ethylbenzene and styrene. Distillation produces **distillation bottoms** (still bottoms), which are often hazardous and polluting. These consist of unevaporated solids, semisolid tars, and sludges from distillation. Specific examples with their hazardous waste numbers are distillation bottoms from the production of acetaldehyde from ethylene (hazardous waste number K009), and still bottoms from toluene reclamation distillation in the production of disulfoton (K036). The landfill disposal of these and other hazardous distillation bottoms used to be widely practiced but is now severely limited.

Evaporation is usually employed to remove water from an aqueous waste to concentrate it. A special case of this technique is **thin-film evaporation** in which volatile constituents are removed by heating a thin layer of liquid or sludge waste spread on a heated surface.

Drying—removal of solvent or water from a solid or semisolid (sludge) or the removal of solvent from a liquid or suspension—is a very important operation because water is often the major constituent of waste products, such as sludges obtained from emulsion breaking. In **freeze drying**, the solvent, usually water, is sublimed from a frozen material. Hazardous waste solids and sludges are dried to reduce the quantity of waste, to remove solvent or water that might interfere with subsequent treatment processes, and to remove hazardous volatile constituents. Dewatering can often be improved with addition of a filter aid, such as diatomaceous earth, during the filtration step.

Stripping is a means of separating volatile components from less volatile ones in a liquid mixture by the partitioning of the more volatile materials to a gas phase of air or steam (steam stripping). The gas phase is introduced into the aqueous solution or suspension containing the waste in a stripping tower that is equipped with trays or packed to provide maximum turbulence and contact between the liquid and gas phases. The two major products are condensed vapor and a stripped bottoms residue. Examples of two volatile components that can be removed from water by air stripping are benzene and dichloromethane. Air stripping can also be used to remove ammonia from water that has been treated with a base to convert ammonium ion to volatile ammonia.

Physical precipitation is used here as a term to describe processes in which a solid forms from a solute in solution as a result of a physical change in the solution, as compared to chemical precipitation (see Section 20.5) in which a chemical reaction in solution produces an insoluble material. The major changes that can cause physical precipitation are cooling the solution, evaporation of solvent, or alteration of solvent composition. The most common type of physical precipitation by alteration of solvent composition occurs when a water-miscible organic solvent is added to an aqueous solution so that the solubility of a salt is lowered below its concentration in the solution.

Phase Transfer

Phase transfer consists of the transfer of a solute in a mixture from one phase to another. An important type of phase transfer process is **solvent extraction,** a process in which a substance is transferred from solution in one solvent (usually water) to another (usually an organic solvent) without any chemical change taking place. When solvents are used to leach substances from solids or sludges, the process is called **leaching**. Solvent extraction and the major terms applicable to it are summarized in Figure 20.6. The same terms and general principles apply to leaching. The major application of solvent extraction to waste treatment has been in the removal of phenol from by-product water produced in coal coking, petroleum refining, and chemical syntheses that involve phenol.

One of the more promising approaches to solvent extraction and leaching of hazardous wastes is the use of **supercritical fluids**, most commonly CO_2, as extraction solvents. A supercritical fluid is one that has characteristics of both liquid and gas and consists of a substance above its critical temperature and pressure (31.1°C and 73.8 atm, respectively, for CO_2). After a substance has been extracted from a waste into a supercritical fluid at high pressure, the pressure can be released, resulting in separation of the substance extracted. The fluid can then be compressed again and recirculated through the extraction system. Some possibilities for treatment of hazardous wastes by extraction with supercritical CO_2 include removal of organic contaminants from wastewater, extraction of organohalide pesticides from soil, extraction of oil from emulsions used in aluminum and steel processing, and regeneration of spent activated carbon. Waste oils contaminated with PCBs, metals, and water can be purified using supercritical ethane.

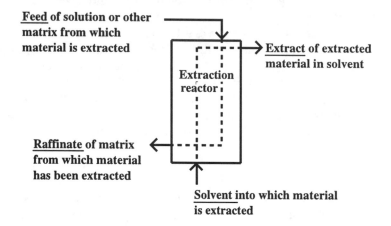

Figure 20.6. Outline of solvent extraction/leaching process with important terms underlined.

Transfer of a substance from a solution to a solid phase is called **sorption**. The most important sorbent is **activated carbon** used for several purposes in waste treatment; in some cases it is adequate for complete treatment. It can also be applied to pretreatment of waste streams going into processes such as reverse osmosis to improve treatment efficiency and reduce fouling. Effluents from other treatment processes, such as biological treatment of degradable organic solutes in water, can be polished with activated carbon. Activated carbon sorption is most effective for removing from water those hazardous waste materials that are poorly water-soluble and that have high molar masses, such as xylene, naphthalene (U165), cyclohexane (U056); chlorinated hydrocarbons, phenol (U188), aniline (U012), dyes, and surfactants. Activated carbon does not work well for organic compounds that are highly water-soluble or polar.

Solids other than activated carbon can be used for sorption of contaminants from liquid wastes. These include synthetic resins composed of organic polymers and mineral substances. Of the latter, clay is employed to remove impurities from waste lubricating oils in some oil recycling processes.

Molecular Separation

A third major class of physical separation is **molecular separation**, often based upon **membrane processes** in which dissolved contaminants or solvent pass through a size-selective membrane under pressure. The products are a relatively pure solvent phase (usually water) and a concentrate enriched in the solute impurities. **Hyperfiltration** allows passage of species with molecular masses of about 100 to 500, whereas **ultrafiltration** is used for the separation of organic solutes with molar masses of 500 to 1,000,000. With both of these techniques, water and lower molar mass solutes under pressure pass through the membrane as a stream of purified **permeate**, leaving behind a stream of **concentrate** containing impurities in solution or suspension. Ultrafiltration and hyperfiltration are especially useful for concentrating suspended oil, grease, and fine solids in water. They also serve to concentrate solutions of large organic molecules and heavy metal ion complexes.

Reverse osmosis is the most widely used of the membrane techniques. Although superficially similar to ultrafiltration and hyperfiltration, it operates on a different principle in that the membrane is selectively permeable to water and excludes ionic solutes. Reverse osmosis uses high pressures to force permeate through the membrane, producing a concentrate containing high levels of dissolved salts.

Electrodialysis, sometimes used to concentrate plating wastes, employs membranes alternately permeable to cations and to anions. The driving force for the separation is provided by electrolysis with a direct current between two electrodes. Alternate layers between the membranes contain concentrate (brine) and purified water.

20.5. CHEMICAL TREATMENT: AN OVERVIEW

The applicability of chemical treatment to wastes depends upon the chemical properties of the waste constituents, particularly acid-base, oxidation-reduction, precipitation, and complexation behavior; reactivity; flammability/combustibility; corrosivity; and compatibility with other wastes. The chemical behavior of wastes translates to various unit operations for waste treatment that are based upon chemical properties and reactions. These include the following:

- Acid/base neutralization
- Chemical extraction and leaching
- Ion exchange
- Chemical precipitation
- Oxidation
- Reduction

Some of the more sophisticated means available for treatment of wastes have been developed for pesticide disposal.

Acid/Base Neutralization

Waste acids and bases are treated by **neutralization**:

$$H^+ + OH^- \rightarrow H_2O \tag{20.5.1}$$

Although simple in principle, neutralization can present some problems in practice. These include evolution of volatile contaminants, mobilization of soluble substances, excessive heat generated by the neutralization reaction, and corrosion to apparatus. By adding too much or too little of the neutralizing agent, it is possible to get a product that is too acidic or basic. Lime, $Ca(OH)_2$, is widely used as a base for treating acidic wastes. Because of lime's limited solubility, solutions of excess lime do not reach extremely high pH values. Sulfuric acid, H_2SO_4, is a relatively inexpensive acid for treating alkaline wastes. However, addition of too much sulfuric acid can produce highly acidic products; for some applications, acetic acid, CH_3COOH, is preferable. As noted above, acetic acid is a weak acid and an excess of it does little harm. It is also a natural product and biodegradable.

Neutralization, or pH adjustment, is often required prior to the application of other waste treatment processes. Processes that may require neutralization include oxidation/reduction, activated carbon sorption, wet air oxidation, stripping, and ion exchange. Microorganisms usually require a pH in the range of 6-9, so neutralization may be required prior to biochemical treatment.

Chemical Precipitation

Chemical precipitation is used in hazardous waste treatment primarily for the removal of heavy metal ions from water as shown below for the chemical precipitation of cadmium:

$$Cd^{2+}(aq) + HS^-(aq) \rightarrow CdS(s) + H^+(aq) \tag{20.5.2}$$

Precipitation of Metals

The most widely used means of precipitating metal ions is by the formation of hydroxides such as chromium(III) hydroxide:

$$Cr^{3+} + 3OH^- \rightarrow Cr(OH)_3 \tag{20.5.3}$$

The source of hydroxide ion is a base (alkali), such as lime $(Ca(OH)_2)$, sodium hydroxide (NaOH), or sodium carbonate (Na_2CO_3). Most metal ions tend to produce basic salt precipitates, such as basic copper(II) sulfate, $CuSO_4 \cdot 3Cu(OH)_2$, which is formed as a solid when hydroxide is added to a solution containing Cu^{2+} and SO_4^{2-} ions. The solubilities of many heavy metal hydroxides reach a minimum value, often at a pH in the range of 9-11, then increase with increasing pH values due to the formation of soluble hydroxo complexes, as illustrated by the following reaction:

$$Zn(OH)_2(s) + 2OH^-(aq) \rightarrow Zn(OH)_4^{2-}(aq) \tag{20.5.4}$$

The chemical precipitation method that is used most is precipitation of metals as hydroxides and basic salts with lime. Sodium carbonate can be used to precipitate hydroxides ($Fe(OH)_3 \cdot xH_2O$) carbonates ($CdCO_3$), or basic carbonate salts ($2PbCO_3 \cdot Pb(OH)_2$). The carbonate anion produces hydroxide by virtue of its hydrolysis reaction with water:

$$CO_3^{2-} + H_2O \rightarrow HCO_3^- + OH^- \tag{20.5.5}$$

Carbonate, alone, does not give as high a pH as do alkali metal hydroxides, which may have to be used to precipitate metals that form hydroxides only at relatively high pH values.

The solubilities of some heavy metal sulfides are extremely low, so precipitation by H_2S or other sulfides (see Reaction 20.5.2) can be a very effective means of treatment. Hydrogen sulfide is a toxic gas that is itself considered to be a hazardous waste (U135). Iron(II) sulfide (ferrous sulfide) can be used as a safe source of sulfide ion to produce sulfide precipitates with other metals that are less soluble than FeS. However, toxic H_2S can be produced when metal sulfide wastes contact acid:

$$MS + 2H^+ \rightarrow M^{2+} + H_2S \tag{20.5.6}$$

Some metals can be precipitated from solution in the elemental metal form by the action of a reducing agent such as sodium borohydride,

$$4Cu^{2+} + NaBH_4 + 2H_2O \rightarrow 4Cu + NaBO_2 + 8H^+ \tag{20.5.7}$$

or with more active metals in a process called **cementation**:

$$Cd^{2+} + Zn \rightarrow Cd + Zn^{2+} \tag{20.5.8}$$

Regardless of the method used to precipitate a metal, the form of the metal in the waste solution can be an important consideration. Chelated metals can be especially difficult to remove. For example, difficulties encountered in precipitating copper from waste printed circuit board formulations revealed that chelation of the copper by EDTA chelant was responsible for the problem.[9]

Coprecipitation of Metals

In some cases advantage may be taken of the phenomenon of coprecipitation to remove metals from wastes. A good example of this application is the coprecipitation of lead from battery industry wastewater with iron(III) hydroxide. Raising the pH of such a wastewater consisting of dilute sulfuric acid and contaminated with Pb^{2+} ion precipitates lead as several species, including $PbSO_4$, $Pb(OH)_2$, and $Pb(OH)_2 \cdot 2PbCO_3$. In the presence of iron(III), gelatinous $Fe(OH)_3$ forms, which coprecipitates the lead, leading to much lower values of lead concentration than would otherwise be possible. Effective removal of lead from battery industry wastewater to below 0.2 ppm has been achieved by first adding an optimum quantity of iron(III), adjustment of the pH to a range of 9-9.5, addition of a polyelectrolyte to aid coagulation, and filtration.[10]

Oxidation/Reduction

As shown by the reactions in Table 20.1, **oxidation** and **reduction** can be used for the treatment and removal of a variety of inorganic and organic wastes. Some waste oxidants can be used to treat oxidizable wastes in water and cyanides. Ozone, O_3, is a strong oxidant that can be generated on-site by an electrical discharge through dry air or oxygen. Ozone employed as an oxidant gas at levels of 1-2 wt% in air and 2-5 wt% in oxygen has been used to treat a large variety of oxidizable contaminants, effluents, and wastes including wastewater and sludges containing oxidizable constituents.

Table 20.1. Oxidation/Reduction Reactions Used to Treat Wastes

Waste Substance	Reaction with Oxidant or Reductant
Oxidation of Organics	
Organic matter {CH$_2$O}	$\{CH_2O\} + 2\{O\} \rightarrow CO_2 + H_2O$
Aldehyde	$CH_3CHO + \{O\} \rightarrow CH_3COOH$ (acid)
Oxidation of Inorganics	
Cyanide	$2CN^- + 5OCl^- + H_2O \rightarrow N_2 + 2HCO_3^- + 5Cl^-$
Iron(II)	$4Fe^{2+} + O_2 + 10H_2O \rightarrow 4Fe(OH)_3 + 8H^+$
Sulfur dioxide	$2SO_2 + O_2 + 2H_2O \rightarrow 2H_2SO_4$
Reduction of Inorganics	
Chromate	$2CrO_4^{2-} + 3SO_2 + 4H^+ \rightarrow Cr_2(SO_4)_3 + 2H_2O$
Permanganate	$MnO_4^- + 3Fe^{2+} + 7H_2O \rightarrow MnO_2(s) +$ $3Fe(OH)_3(s) + 5H^+$

Electrolysis

As shown in Figure 20.7, **electrolysis** is a process in which one species in solution (usually a metal ion) is reduced by electrons at the **cathode** and another gives up electrons to the **anode** and is oxidized there. In hazardous waste applications electrolysis is most widely used in the recovery of cadmium, copper, gold, lead, silver, and zinc. Metal recovery by electrolysis is made more difficult by the presence of cyanide ion, which stabilizes metals in solution as the cyanide complexes, such as $Ni(CN)_4^{2-}$.

Electrolytic removal of contaminants from solution can be by direct electrodeposition, particularly of reduced metals, and as the result of secondary reactions of electrolytically generated precipitating agents. A specific example of both is the electrolytic removal of both cadmium and nickel from wastewater contaminated by nickel/cadmium battery manufacture using fibrous carbon electrodes.[11] At the cathode, cadmium is removed directly by reduction to the metal:

$$Cd^{2+} + 2e^- \rightarrow Cd \tag{20.5.9}$$

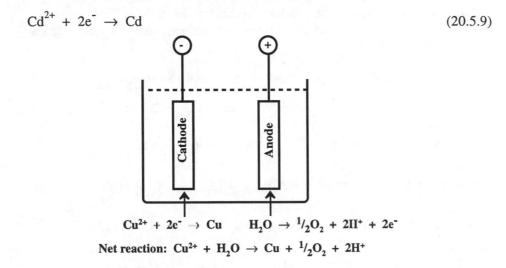

$$Cu^{2+} + 2e^- \rightarrow Cu \qquad H_2O \rightarrow {}^1\!/_2O_2 + 2H^+ + 2e^-$$

Net reaction: $Cu^{2+} + H_2O \rightarrow Cu + {}^1\!/_2O_2 + 2H^+$

Figure 20.6. Electrolysis of copper solution.

At relatively high cathodic potentials, hydroxide is formed by the electrolytic reduction of water,

$$2H_2O + 2e^- \rightarrow 2OH^- + H_2 \tag{20.5.10}$$

or by the reduction of molecular oxygen, if it is present:

$$2H_2O + O_2 + 4e^- \rightarrow 4OH^- \tag{20.5.11}$$

If the localized pH at the cathode surface becomes sufficiently high, cadmium can be precipitated and removed as colloidal $Cd(OH)_2$. The direct electrodeposition of nickel is too slow to be significant, but it is precipitated as solid $Ni(OH)_2$ at pH values above 7.5.

Cyanide, which is often present as an ingredient of electroplating baths with metals such as cadmium and nickel, can be removed by oxidation with electrolytically generated elemental chlorine at the anode. Chlorine is generated by the anodic oxidation of added chloride ion:

$$2Cl^- \rightarrow Cl_2 + 2e^- \tag{20.5.12}$$

The electrolytically generated chlorine then breaks down cyanide by a series of reactions for which the overall reaction is the following:

$$2CN^- + 5Cl_2 + 8OH^- \rightarrow 10Cl^- + N_2 + 2CO_2 + 4H_2O \tag{20.5.13}$$

Hydrolysis

One of the ways to dispose of chemicals that are reactive with water is to allow them to react with water under controlled conditions, a process called **hydrolysis**. Inorganic chemicals that can be treated by hydrolysis include metals that react with water; metal carbides, hydrides, amides, alkoxides, and halides; and nonmetal oxy-halides and sulfides. Examples of the treatment of these classes of inorganic species are given in Table 20.2.

Table 20.2. Inorganic Chemicals That May Be Treated by Hydrolysis

Class of chemical	Reaction with oxidant or reductant
Active metals (calcium)	$Ca + 2H_2O \rightarrow H_2 + Ca(OH)_2$
Hydrides (sodium aluminum hydride)	$NaAlH_4 + 4H_2O \rightarrow 4H_2 + NaOH + Al(OH)_3$
Carbides (calcium carbide)	$CaC_2 + 2H_2O \rightarrow Ca(OH)_2 + C_2H_2$
Amides (sodium amide)	$NaNH_2 + H_2O \rightarrow NaOH + NH_3$
Halides (silicon tetrachloride)	$SiCl_4 + 2H_2O \rightarrow SiO_2 + 4HCl$
Alkoxides (sodium ethoxide)	$NaOC_2H_5 + H_2O \rightarrow NaOH + C_2H_5OH$

Organic chemicals may also be treated by hydrolysis. For example, toxic acetic anhydride is hydrolyzed to relatively safe acetic acid:

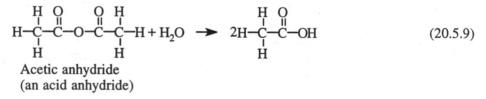

$$(20.5.9)$$

Acetic anhydride
(an acid anhydride)

Chemical Extraction and Leaching

Chemical extraction or **leaching** in hazardous waste treatment is the removal of a hazardous constituent by chemical reaction with an extractant in solution. Poorly soluble heavy metal salts can be extracted by reaction of the salt anions with H^+ as illustrated by the following:

$$PbCO_3 + H^+ \rightarrow Pb^{2+} + HCO_3^- \qquad (20.5.10)$$

Acids also dissolve basic organic compounds such as amines and aniline. Extraction with acids should be avoided if cyanides or sulfides are present to prevent formation of toxic hydrogen cyanide or hydrogen sulfide. Nontoxic weak acids are usually the safest to use. These include acetic acid, CH_3COOH, and the acid salt, NaH_2PO_4.

Chelating agents, such as dissolved ethylenediaminetetraacetate (EDTA, HY^{3-}), dissolve insoluble metal salts by forming soluble species with metal ions:

$$FeS + HY^{3-} \rightarrow FeY^{2-} + HS^- \qquad (20.5.11)$$

Heavy metal ions in soil contaminated by hazardous wastes may be present in a coprecipitated form with insoluble iron(III) and manganese(IV) oxides, Fe_2O_3 and MnO_2, respectively. These oxides can be dissolved by reducing agents, such as solutions of sodium dithionate/citrate or hydroxylamine. This results in the production of soluble Fe^{2+} and Mn^{2+} and the release of heavy metal ions, such as Cd^{2+} or Ni^{2+}, which are removed with the water.

Ion Exchange

Ion exchange is a means of removing cations or anions from solution onto a solid resin, which can be regenerated by treatment with acids, bases, or salts. The greatest use of ion exchange in hazardous waste treatment is for the removal of low levels of heavy metal ions from wastewater:

$$2H^{+-}\{CatExchr\} + Cd^{2+} \rightarrow Cd^{2+-}\{CatExchr\}_2 + 2H^+ \qquad (20.5.12)$$

Ion exchange is employed in the metal plating industry to purify rinsewater and spent plating bath solutions. Cation exchangers are used to remove cationic metal species, such as Cu^{2+}, from such solutions. Anion exchangers remove anionic cyanide metal complexes (for example, $Ni(CN)_4^{2-}$) and chromium(VI) species, such as CrO_4^{2-}. Radionuclides may be removed from radioactive wastes and mixed waste by ion exchange resins.

20.6. PHOTOLYTIC REACTIONS

Photolytic reactions were discussed in Chapter 9. *Photolysis* can be used to destroy a number of kinds of hazardous wastes. In such applications it is most useful in breaking chemical bonds in refractory organic compounds. TCDD (see Section 7.10), one of the most troublesome and refractory of wastes, can be treated by ultraviolet light in the presence of hydrogen atom donors {H} resulting in reactions such as the following:

$$(20.6.1)$$

As photolysis proceeds, the H-C bonds are broken, the C-O bonds are broken, and the final product is a harmless organic polymer.

An initial photolysis reaction can result in the generation of reactive intermediates that participate in **chain reactions** that lead to the destruction of a compound. One of the most important reactive intermediates is free radical HO•. In some cases **sensitizers** are added to the reaction mixture to absorb radiation and generate reactive species that destroy wastes.

Hazardous waste substances other than TCDD that have been destroyed by photolysis are herbicides (atrazine), 2,4,6-trinitrotoluene (TNT), and polychlorinated biphenyls (PCBs). The addition of a chemical oxidant, such as potassium peroxydisulfate, $K_2S_2O_8$, enhances destruction by oxidizing active photolytic products.

20.7. THERMAL TREATMENT METHODS

Thermal treatment of hazardous wastes can be used to accomplish most of the common objectives of waste treatment—volume reduction; removal of volatile, combustible, mobile organic matter; and destruction of toxic and pathogenic materials. The most widely applied means of thermal treatment of hazardous wastes is **incineration**. Incineration utilizes high temperatures, an oxidizing atmosphere, and often turbulent combustion conditions to destroy wastes. Methods other than incineration that make use of high temperatures to destroy or neutralize hazardous wastes are discussed briefly at the end of this section.

Incineration

Hazardous waste incineration will be defined here as a process that involves exposure of the waste materials to oxidizing conditions at a high temperature, usually in excess of 900°C. Normally, the heat required for incineration comes from the oxidation of organically bound carbon and hydrogen contained in the waste material or in supplemental fuel:

$$C(organic) + O_2 \rightarrow CO_2 + heat \tag{20.7.1}$$

$$4H(organic) + O_2 \rightarrow 2H_2O + heat \tag{20.7.2}$$

These reactions destroy organic matter and generate heat required for endothermic reactions, such as the breaking of C-Cl bonds in organochlorine compounds.

Incinerable Wastes

Ideally, incinerable wastes are predominantly organic materials that will burn with a heating value of at least 5,000 Btu/lb and preferably over 8,000 Btu/lb. Such heating values are readily attained with wastes having high contents of the most commonly incinerated waste organic substances, including methanol, acetonitrile, toluene, ethanol, amyl acetate, acetone, xylene, methyl ethyl ketone, adipic acid, and ethyl acetate. In some cases, however, it is desirable to incinerate wastes that will not burn alone and which require **supplemental fuel,** such as methane and petroleum liquids. Examples of such wastes are nonflammable organochlorine wastes, some aqueous wastes, or soil in which the elimination of a particularly troublesome contaminant is worth the expense and trouble of incinerating it. Inorganic matter, water, and organic hetero element contents of liquid wastes are important in determining their incinerability.

Hazardous Waste Fuel

Many industrial wastes, including hazardous wastes, are burned as **hazardous waste fuel** for energy recovery in industrial furnaces and boilers and in incinerators for nonhazardous wastes, such as sewage sludge incinerators. This process is called **coincineration**, and more combustible wastes are utilized by it than are burned solely for the purpose of waste destruction. In addition to heat recovery from combustible wastes, it is a major advantage to use an existing on-site facility for waste disposal rather than a separate hazardous waste incinerator.

Incineration Systems

The four major components of hazardous waste incineration systems are shown in Figure 20.8.

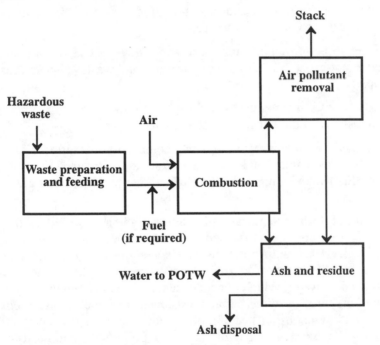

Figure 20.8. Major components of a hazardous waste incinerator system.

Waste preparation for liquid wastes may require filtration, settling to remove solid material and water, blending to obtain the optimum incinerable mixture, or heating to decrease viscosity. Solids may require shredding and screening. Atomization is commonly used to feed liquid wastes. Several mechanical devices, such as rams and augers, are used to introduce solids into the incinerator.

The most common kinds of **combustion chambers** are liquid injection, fixed hearth, rotary kiln, and fluidized bed. These types are discussed in more detail later in this section.

Often the most complex part of a hazardous waste incineration system is the **air pollution control system**, which involves several operations. The most common operations in air pollution control from hazardous waste incinerators are combustion gas cooling, heat recovery, quenching, particulate matter removal, acid gas removal, and treatment and handling of by-product solids, sludges, and liquids.

Hot ash is often quenched in water. Prior to disposal it may require dewatering and chemical stabilization. A major consideration with hazardous waste incinerators and the types of wastes that are incinerated is the disposal problem posed by the ash, especially with respect to potential leaching of heavy metals.

Types of Incinerators

Hazardous waste incinerators may be divided among the following, based upon type of combustion chamber:

- **Rotary kiln** (about 40% of U.S. hazardous waste incinerator capacity) in which the primary combustion chamber is a rotating cylinder lined with refractory materials, and an afterburner downstream from the kiln to complete destruction of the wastes

- **Liquid injection** incinerators (also about 40% of U.S. hazardous waste incinerator capacity) that burn pumpable liquid wastes dispersed as small droplets

- **Fixed-hearth incinerators** with single or multiple hearths upon which combustion of liquid or solid wastes occurs

- **Fluidized-bed incinerators** in which combustion of wastes is carried out on a bed of granular solid (such as limestone) maintained in a suspended (fluid-like) state by injection of air to remove pollutant acid gas and ash products

- **Advanced design incinerators** including **plasma incinerators** that make use of an extremely hot plasma of ionized air injected through an electrical arc;[12] **electric reactors** that use resistance-heated incinerator walls at around 2200°C to heat and pyrolyze wastes by radiative heat transfer; **infrared systems**, which generate intense infrared radiation by passing electricity through silicon carbide resistance heating elements; **molten salt combustion** that uses a bed of molten sodium carbonate at about 900°C to destroy the wastes and retain gaseous pollutants; and **molten glass processes** that use a pool of molten glass to transfer heat to the waste and to retain products in a poorly leachable glass form

Combustion Conditions

The key to effective incineration of hazardous wastes lies in the combustion conditions. These require (1) sufficient free oxygen in the combustion zone to ensure combustion of the wastes; (2) turbulence for thorough mixing of waste, oxidant, and (in cases where the waste does not have sufficient fuel content to be burned alone) supplemental fuel; (3) high combustion temperatures above about 900°C to ensure that thermally resistant compounds do react; and (4) sufficient residence time (at least 2 seconds) to allow reactions to occur.

Effectiveness of Incineration

EPA standards for hazardous waste incineration are based upon the effectiveness of destruction of the **principal organic hazardous constituents** (POHC). Measurement of these compounds before and after incineration gives the **destruction removal efficiency** (DRE) according to the formula,

$$DRE = \frac{W_{in} - W_{out}}{W_{in}} \times 100 \qquad (20.7.3)$$

where W_{in} and W_{out} are the mass flow rates of the principal organic hazardous constituent (POHC) input and output (at the stack downstream from emission controls), respectively. United States EPA regulations call for destruction of 99.99% of POHCs and 99.9999% ("six nines") destruction of 2,3,7,8-tetrachlorodibenzo-p-dioxin, commonly called TCDD or "dioxin."

Wet Air Oxidation

Organic compounds and oxidizable inorganic species can be oxidized by oxygen in aqueous solution. The source of oxygen usually is air. Rather extreme conditions of temperature and pressure are required, with a temperature range of 175-327°C and a pressure range of 300-3000 psig (2070-20,700 kPa). The high pressures allow a high concentration of oxygen to be dissolved in the water, and the high temperatures enable the reaction to occur.

Wet air oxidation has been applied to the destruction of cyanides in electroplating wastewaters. The oxidation reaction for sodium cyanide is the following:

$$2Na^+ + 2CN^- + O_2 + 4H_2O \rightarrow 2Na^+ + 2HCO_3^- + 2NH_3 \qquad (20.7.4)$$

Organic wastes can be oxidized in supercritical water, taking advantage of the ability of supercritical fluids to dissolve organic compounds (see Section 20.4). Wastes are contacted with water and the mixture raised to a temperature and pressure required for supercritical conditions for water. Oxygen is then pumped in sufficient to oxidize the wastes. The process produces only small quantities of CO, and no SO_2 or NO_x. It can be used to degrade PCBs, dioxins, organochlorine insecticides, benzene, urea, and numerous other materials.

UV-Enhanced Wet Oxidation

Hydrogen peroxide (H_2O_2) can be used as an oxidant in solution assisted by ultraviolet radiation ($h\nu$). For the oxidation of organic species represented in general as {CH_2O}, the overall reaction is

$$2H_2O_2 + \{CH_2O\} + h\nu \rightarrow CO_2 + 3H_2O \qquad (20.7.5)$$

The ultraviolet radiation breaks chemical bonds and serves to form reactive oxidant species, such as HO•.

20.8. BIODEGRADATION OF WASTES

Biodegradation of wastes is a term that describes the conversion of wastes by enzymatic biological processes to simple inorganic molecules (mineralization) and, to a certain extent, to biological materials. Usually the products of biodegradation are molecular forms that tend to occur in nature and that are in greater thermodynamic equilibrium with their surroundings than are the starting materials. **Detoxification** refers to the biological conversion of a toxic substance to a less toxic species. Microbial bacteria and fungi possessing enzyme systems required for biodegradation of wastes are usually best obtained from populations of indigenous microorganisms at a hazardous waste site where they have developed the ability to degrade particular kinds of molecules. Biological treatment offers a number of significant advantages and has considerable potential for the degradation of hazardous wastes, even *in situ*.

Under the label of **bioremediation**, the use of microbial processes to destroy hazardous wastes has been the subject of intense investigation in the waste treatment community for a number of years. Doubts still exist about claims for its effectiveness in a number of applications. It must be kept in mind, however, that there are many factors that can cause biodegradation to fail as a treatment process. Often physical conditions are such that mixing of wastes, nutrients, and electron acceptor species (such as oxygen) is too slow to permit biodegradation to occur at a useful rate. Low temperatures may make reactions too slow to be useful. Toxicants, such as heavy metals, may inhibit biological activity, and some metabolites produced by the microorganisms may be toxic to them.

Biodegradability

The **biodegradability** of a compound is influenced by its physical characteristics, such as solubility in water and vapor pressure, and by its chemical properties, including molar mass, molecular structure, and presence of various kinds of functional groups, some of which provide a "biochemical handle" for the initiation of biodegradation. With the appropriate organisms and under the right conditions, even substances such as phenol that are considered to be biocidal to most microorganisms can undergo biodegradation.

Recalcitrant or **biorefractory** substances are those that resist biodegradation and tend to persist and accumulate in the environment. Such materials are not necessarily toxic to organisms, but simply resist their metabolic attack. However, even some compounds regarded as biorefractory may be degraded by microorganisms that have had the opportunity to adapt to their biodegradation; for example, DDT is degraded by properly acclimated *Pseudomonas*. Chemical pretreatment, especially by partial oxidation, can make some kinds of recalcitrant wastes much more biodegradable.

Properties of hazardous wastes and their media can be changed to increase biodegradability. This can be accomplished by adjustment of conditions to optimum temperature, pH (usually in the range of 6-9), stirring, oxygen level, and material load. Biodegradation can be aided by removal of toxic organic and inorganic substances, such as heavy metal ions.

Aerobic Treatment

Aerobic waste treatment processes utilize aerobic bacteria and fungi that require molecular oxygen, O_2. These processes are often favored by microorganisms, in part because of the high energy yield obtained when molecular oxygen reacts with organic matter. Aerobic waste treatment is well adapted to the use of an activated sludge process. It can be applied to hazardous wastes such as chemical process wastes and landfill leachates. Some systems use powdered activated carbon as an additive to absorb organic wastes that are not biodegraded by microorganisms in the system.

Contaminated soils can be mixed with water and treated in a bioreactor to eliminate biodegradable contaminants in the soil. It is possible, in principle, to treat contaminated soils biologically in place by pumping oxygenated, nutrient-enriched water through the soil in a recirculating system.

Anaerobic Treatment

Anaerobic waste treatment in which microorganisms degrade wastes in the absence of oxygen can be practiced on a variety of organic hazardous wastes. Compared to the aerated activated sludge process, anaerobic digestion requires less energy; yields less sludge by-product; generates hydrogen sulfide (H_2S), which precipitates toxic heavy metal ions; and produces methane gas, CH_4, which can be used as an energy source.

The overall process for anaerobic digestion is a fermentation process in which organic matter is both oxidized and reduced. The simplified reaction for the anaerobic fermentation of a hypothetical organic substance, "$\{CH_2O\}$", is the following:

$$2\{CH_2O\} \rightarrow CO_2 + CH_4 \tag{20.8.1}$$

In practice, the microbial processes involved are quite complex. Most of the wastes for which anaerobic digestion is suitable consist of oxygenated compounds, such as acetaldehyde or methylethyl ketone.

Reductive Dehalogenation

Reductive dehalogenation is a mechanism by which halogen atoms are removed from organohalide compounds by anaerobic bacteria. It is an important means of detoxifying alkyl halides (particularly solvents), aryl halides, and organochlorine pesticides, all of which are important hazardous waste compounds, and which were discarded in large quantities in some of the older waste disposal dumps. Reductive dehlogenation is the only means by which some of the more highly halogenated waste compounds are biodegraded; such compounds include tetrachloroethene, hexachlorobenzene, pentachlorophenol, and the more highly chlorinated PCB congeners.[13]

The two general processes by which reductive dehalogenation occurs are **hydrogenolysis**, as shown by the example in Equation 20.8.2,

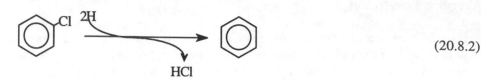

$$(20.8.2)$$

and **vicinal reduction**

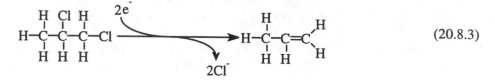

$$(20.8.3)$$

Vicinal reduction removes two adjacent halogen atoms, and works only on alkyl halides, not aryl halides. Both processes produce innocuous inorganic halide (Cl^-).

20.9. LAND TREATMENT AND COMPOSTING

Land Treatment

Soil may be viewed as a natural filter for wastes. Soil has physical, chemical, and biological characteristics that can enable waste detoxification, biodegradation, chemical decomposition, and physical and chemical fixation. Therefore, **land treatment** of wastes may be accomplished by mixing the wastes with soil under appropriate conditions.

Soil is a natural medium for a number of living organisms that may have an effect upon biodegradation of hazardous wastes. Of these, the most important are bacteria, including those from the genera *Agrobacterium*, *Arthrobacteri*, *Bacillus*, *Flavobacterium*, and *Pseudomonas*. *Actinomycetes* and fungi are important organisms in decay of vegetable matter and may be involved in biodegradation of wastes.

Microorganisms useful for land treatment are usually present in sufficient numbers to provide the inoculum required for their growth. The growth of these indigenous microorganisms may be stimulated by adding nutrients and an electron acceptor to act as an oxidant (for aerobic degradation) accompanied by mixing. The most commonly added nutrients are nitrogen and phosphorus. Oxygen can be added by pumping air underground or by treatment with hydrogen peroxide, H_2O_2. In some cases, such as for treatment of hydrocarbons on or near the soil surface, simple tillage provides both oxygen and the mixing required for optimum microbial growth.

Wastes that are amenable to land treatment are biodegradable organic substances. However, in soil contaminated with hazardous wastes, bacterial cultures may develop that are effective in degrading normally recalcitrant compounds through acclimation over a long period of time. Land treatment is most used for petroleum refining wastes and is applicable to the treatment of fuels and wastes from leaking underground storage tanks. It can also be applied to biodegradable organic chemical wastes, including some organohalide compounds. Land treatment is not suitable for the treatment of wastes containing acids, bases, toxic inorganic compounds, salts, heavy metals, and organic compounds that are excessively soluble, volatile, or flammable.

Composting

Composting of hazardous wastes is the biodegradation of solid or solidified materials in a medium other than soil. Bulking material, such as plant residue, paper, municipal refuse, or sawdust may be added to retain water and enable air to penetrate to the waste material. Successful composting of hazardous waste depends upon a number of factors, including those discussed above under land treatment. The first of these is the selection of the appropriate microorganism or **inoculum**. Once a successful composting operation is underway, a good inoculum is maintained by recirculating spent compost to each new batch. Other parameters that must be controlled include oxygen supply, moisture content (which should be maintained at a minimum of about 40%), pH (usually around neutral), and temperature. The composting process generates heat so, if the mass of the compost pile is sufficiently high, it can be self-heating under most conditions. Some wastes are deficient in nutrients, such as nitrogen, which must be supplied from commercial sources or from other wastes.

20.10. PREPARATION OF WASTES FOR DISPOSAL

Immobilization, stabilization, fixation, and solidification are terms that describe sometimes overlapping techniques whereby hazardous wastes are placed in a form suitable for long-term disposal. These aspects of hazardous waste management are addressed below.

Immobilization

Immobilization includes physical and chemical processes that reduce surface areas of wastes to minimize leaching. It isolates the wastes from their environment, especially groundwater, so that they have the least possible tendency to migrate. This is accomplished by physically isolating the waste, reducing its solubility, and decreasing its surface area. Immobilization usually improves the handling and physical characteristics of wastes.

Stabilization

Stabilization means the conversion of a waste from its original form to a physically and chemically more stable material that is less likely to cause problems during handling and disposal, and less likely to be mobile after disposal. Stabilization may include chemical reactions that generate products that are less volatile, soluble, and reactive. Solidification, which is discussed below, is one of the most common means of stabilization. Stabilization is required for land disposal of wastes. **Fixation** is a process that binds a hazardous waste in a less mobile and less toxic form; it means much the same thing as stabilization.

Solidification

Solidification may involve chemical reaction of the waste with the solidification agent, mechanical isolation in a protective binding matrix, or a combination of chemical and physical processes. It can be accomplished by evaporation of water from aqueous wastes or sludges, sorption onto solid material, reaction with cement, reaction with silicates, encapsulation, or imbedding in polymers or thermoplastic materials.

In many solidification processes, such as reaction with portland cement, water is an important ingredient of the hydrated solid matrix. Therefore, the solid should not be heated excessively or exposed to extremely dry conditions, which could result in diminished structural integrity from loss of water. In some cases, however, heating a solidified waste is an essential part of the overall solidification procedure. For example, an iron hydroxide matrix can be converted to highly insoluble, refractory iron oxide by heating. Organic constituents of solidified wastes may be converted to inert carbon by heating. Heating is an integral part of the process of vitrification (see below).

Sorption to a Solid Matrix Material

Hazardous waste liquids, emulsions, sludges, and free liquids in contact with sludges may be solidified and stabilized by fixing onto solid **sorbents**, including activated carbon (for organics), fly ash, kiln dust, clays, vermiculite, and various proprietary materials. Sorption may be done to convert liquids and semisolids to dry solids, improve waste handling, and reduce solubility of waste constituents. Sorption can also be used to improve waste compatibility with substances such as portland cement used for solidification and setting. Specific sorbents may also be used to stabilize pH and pE (a measure of the tendency of a medium to be oxidizing or reducing, see Chapter 4).

The action of sorbents can include simple mechanical retention of wastes, physical sorption, and chemical reactions. It is important to match the sorbent to the waste. A substance with a strong affinity for water should be employed for wastes containing excess water, and one with a strong affinity for organic materials should be used for wastes with excess organic solvents.

Thermoplastics and Organic Polymers

Thermoplastics are solids or semisolids that become liquified at elevated temperatures. Hazardous waste materials may be mixed with hot thermoplastic liquids and solidified in the cooled thermoplastic matrix, which is rigid but deformable. The thermoplastic material most used for this purpose is asphalt bitumen. Other thermoplastics, such as paraffin and polyethylene, have also been used to immobilize hazardous wastes.

Among the wastes that can be immobilized with thermoplastics are those containing heavy metals, such as electroplating wastes. Organic thermoplastics repel water and reduce the tendency toward leaching in contact with groundwater. Compared to cement, thermoplastics add relatively less material to the waste.

A technique similar to that described above uses **organic polymers** produced in contact with solid wastes to imbed the wastes in a polymer matrix. Three kinds of polymers that have been used for this purpose include polybutadiene, urea-formaldehyde, and vinyl ester-styrene polymers. This procedure is more complicated than is the use of thermoplastics but, in favorable cases, yields a product in which the waste is held more strongly.

Vitrification

Vitrification or **glassification** consists of imbedding wastes in a glass material. In this application, glass may be regarded as a high-melting-temperature inorganic thermoplastic. Molten glass can be used, or glass can be synthesized in contact with the waste by mixing and heating with glass constituents—silicon dioxide (SiO_2), sodium carbonate (Na_2CO_3), and calcium oxide (CaO). Other constituents may include boron oxide, B_2O_3, which yields a borosilicate glass that is especially resistant to changes in temperature and chemical attack. In some cases glass is used in conjunction with thermal waste destruction processes, serving to immobilize hazardous waste ash consituents. Some wastes are detrimental to the quality of the glass. Aluminum oxide, for example, may prevent glass from fusing.

Vitrification is relatively complicated and expensive, the latter because of the energy consumed in fusing glass. Despite these disadvantages, it is the favored immobilization technique for some special wastes and has been promoted for solidification of radionuclear wastes because glass is chemically inert and resistant to leaching. However, high levels of radioactivity can cause deterioration of glass and lower its resistance to leaching.

Solidification with Cement

Portland cement is widely used for solidification of hazardous wastes. In this application, Portland cement provides a solid matrix for isolation of the wastes, chemically binds water from sludge wastes, and may react chemically with wastes (for example, the calcium and base in portland cement react chemically with inorganic arsenic sulfide wastes to reduce their solubilities). However, most wastes are held physically in the rigid portland cement matrix and are subject to leaching.

As a solidification matrix, portland cement is most applicable to inorganic sludges containing heavy metal ions that form insoluble hydroxides and carbonates in the basic carbonate medium provided by the cement. The success of solidification with portland cement strongly depends upon whether or not the waste adversely affects the strength and stability of the concrete product. A number of substances—organic matter such as petroleum or coal; some silts and clays; sodium salts of arsenate, borate, phosphate, iodate, and sulfide; and salts of copper, lead, magnesium, tin, and zinc—are incompatible with portland cement because they interfere with its set and cure, producing a mechanically weak product and resulting in deterioration of the cement matrix with time. However, a reasonably good disposal form can be obtained by absorbing organic wastes with a solid material, which in turn is set in portland cement.

Solidification with Silicate Materials

Water-insoluble **silicates**, (pozzolanic substances) containing oxyanionic silicon such as SiO_3^{2-} are used for waste solidification. There are a number of such substances, some of which are waste products, including fly ash, flue dust, clay, calcium silicates, and ground-up slag from blast furnaces. Soluble silicates, such as sodium silicate, may also be used. Silicate solidification usually requires a setting agent, which may be portland cement (see above), gypsum (hydrated $CaSO_4$), lime, or compounds of aluminum, magnesium, or iron. The product may vary from a granular material to a concrete-like solid. In some cases the product is improved by additives, such as emulsifiers, surfactants, activators, calcium chloride, clays, carbon, zeolites, and various proprietary materials.

Success has been reported for the solidification of both inorganic wastes and organic wastes (including oily sludges) with silicates. The advantages and disadvantages of silicate solidification are similar to those of portland cement discussed above. One consideration that is especially applicable to fly ash is the presence in some silicate materials of leachable hazardous substances, which may include arsenic and selenium.

Encapsulation

As the name implies, **encapsulation** is used to coat wastes with an impervious material so that they do not contact their surroundings. For example, a water-soluble waste salt encapsulated in asphalt would not dissolve so long as the asphalt layer remains intact. A common means of encapsulation uses heated, molten thermoplastics, asphalt, and waxes that solidify when cooled. A more sophisticated approach to encapsulation is to form polymeric resins from monomeric substances in the presence of the waste.

Chemical Fixation

Chemical fixation is a process that binds a hazardous waste substance in a less mobile, less toxic form by a chemical reaction that alters the waste chemically. Physical and chemical fixation often occur together, and sometimes it is a little difficult to distinguish between them. Polymeric inorganic silicates containing some calcium and often some aluminum are the inorganic materials most widely used as a fixation matrix. Many kinds of heavy metals are chemically bound in such a matrix, as well as being held physically by it. Similarly, some organic wastes are bound by reactions with matrix constituents.

20.11. ULTIMATE DISPOSAL OF WASTES

Regardless of the destruction, treatment, and immobilization techniques used, there will always remain from hazardous wastes some material that has to be put somewhere. This section briefly addresses the ultimate disposal of ash, salts, liquids, solidified liquids, and other residues that must be placed where their potential to do harm is minimized.

Disposal Aboveground

In some important respects disposal aboveground, essentially in a pile designed to prevent erosion and water infiltration, is the best way to store solid wastes. Perhaps its most important advantage is that it avoids infiltration by groundwater that can result in leaching and groundwater contamination common to storage in pits and landfills. In a properly designed aboveground disposal facility, any leachate that is produced drains quickly by gravity to the leachate collection system where it can be detected and treated.

Aboveground disposal can be accomplished with a storage mound deposited on a layer of compacted clay covered with impermeable membrane liners laid somewhat above the original soil surface and shaped to allow leachate flow and collection. The slopes around the edges of the storage mound should be sufficiently great to allow good drainage of precipitation, but gentle enough to deter erosion.

Landfill

Landfill historically has been the most common way of disposing of solid hazardous wastes and some liquids, although it is being severely limited in many nations by new regulations and high land costs. Landfill involves disposal that is at least partially underground in excavated cells, quarries, or natural depressions. Usually fill is continued above ground to utilize space most efficiently and provide a grade for drainage of precipitation.

The greatest environmental concern with landfill of hazardous wastes is the generation of leachate from infiltrating surface water and groundwater with resultant contamination of groundwater supplies. Modern hazardous waste landfills provide elaborate systems to contain, collect, and control such leachate.

There are several components to a modern landfill. A landfill should be placed on a compacted, low-permeability medium, preferably clay, which is covered by a flexible-membrane liner consisting of watertight impermeable material. This liner is covered with granular material in which is installed a secondary drainage system. Next is another flexible-membrane liner, above which is installed a primary drainage system for the removal of leachate. This drainage system is covered with a layer of granular filter medium, upon which the wastes are placed. In the landfill, wastes of different kinds are separated by berms consisting of clay or soil covered with liner material. When the fill is complete, the waste is capped to prevent surface water infiltration and covered with compacted soil. In addition to leachate collection, provision may be made for a system to treat evolved gases, particularly when methane-generating biodegradable materials are disposed in the landfill.

The flexible-membrane liner made of rubber (including chlorosulfonated polyethylene) or plastic (including chlorinated polyethylene, high-density polyethylene, and polyvinylchloride) is a key component of state-of-the-art landfills. It controls seepage out of, and infiltration into, the landfill. Obviously, liners have to meet stringent standards to serve their intended purpose. In addition to being impermeable, the liner material must be strongly resistant to biodegradation, chemical attack, and tearing.

Capping is done to cover the wastes, prevent infiltration of excessive amounts of surface water, and prevent release of wastes to overlying soil and the atmosphere.

Caps come in a variety of forms and are often multilayered. Some of the problems that may occur with caps are settling, erosion, ponding, damage by rodents, and penetration by plant roots.

Surface Impoundment of Liquids

Many liquid hazardous wastes, slurries, and sludges are placed in **surface impoundments**, which usually serve for treatment and often are designed to be filled in eventually as landfill disposal sites. Most liquid hazardous wastes and a significant fraction of solids are placed in surface impoundments in some stage of treatment, storage, or disposal.

A surface impoundment may consist of an excavated "pit," a structure formed with dikes, or a combination thereof. The construction is similar to that discussed above for landfills in that the bottom and walls should be impermeable to liquids and provision must be made for leachate collection. The chemical and mechanical challenges to liner materials in surface impoundments are severe so that proper geological siting and construction with floors and walls composed of low-permeability soil and clay are important in preventing pollution from these installations.

Deep-Well Disposal of Liquids

Deep-well disposal of liquids consists of their injection under pressure to underground strata isolated by impermeable rock strata from aquifers. Early experience with this method was gained in the petroleum industry where disposal is required of large quantities of saline wastewater coproduced with crude oil. The method was later extended to the chemical industry for the disposal of brines, acids, heavy metal solutions, organic liquids, and other liquids.

A number of factors must be considered in deep-well disposal. Wastes are injected into a region of elevated temperature and pressure, which may cause chemical reactions to occur involving the waste constituents and the mineral strata. Oils, solids, and gases in the liquid wastes can cause problems such as clogging. Corrosion may be severe. Microorganisms may have some effects. Most problems from these causes can be mitigated by proper waste pretreatment.

The most serious consideration involving deep-well disposal is the potential contamination of groundwater. Although injection is made into permeable saltwater aquifers presumably isolated from aquifers that contain potable water, contamination may occur. Major routes of contamination include fractures, faults, and other wells. The disposal well itself can act as a route for contamination if it is not properly constructed and cased or if it is damaged.

20.12. LEACHATE AND GAS EMISSIONS

Leachate

The production of contaminated leachate is a possibility with most disposal sites. Therefore, new hazardous waste landfills require leachate collection/treatment systems and many older sites are required to have such systems retrofitted to them.

Modern hazardous waste landfills typically have dual leachate collection systems, one located between the two impermeable liners required for the bottom and sides of the landfill, and another just above the top liner of the double-liner system. The upper leachate collection system is called the primary leachate collection system, and the bottom is called the secondary leachate collection system. Leachate is collected in perforated pipes that are imbedded in granular drain material.

Chemical and biochemical processes have the potential to cause some problems for leachate collection systems. One such problem is clogging by insoluble manganese(IV) and iron(III) hydrated oxides upon exposure to air as described for water wells in Section 15.9.

Leachate consists of water that has become contaminated by wastes as it passes through a waste disposal site. It contains waste constituents that are soluble, not retained by soil, and not degraded chemically or biochemically. Some potentially harmful leachate constituents are products of chemical or biochemical transformations of wastes.

The best approach to leachate management is to prevent its production by limiting infiltration of water into the site. Rates of leachate production may be very low when sites are selected, designed, and constructed with minimal production of leachate as a major objective. A well-maintained, low-permeability cap over the landfill is very important for leachate minimization.

Hazardous Waste Leachate Treatment

The first step in treating leachate is to characterize it fully, particularly with a thorough chemical analysis of possible waste constituents and their chemical and metabolic products. The biodegradability of leachate constituents should also be determined.

The options available for the treatment of hazardous waste leachate are generally those that can be used for industrial wastewaters. These include biological treatment by an activated sludge or related process, and sorption by activated carbon, usually in columns of granular activated carbon. Hazardous waste leachate can be treated by a variety of chemical processes, including acid/base neutralization, precipitation, and oxidation/reduction. In some cases these treatment steps must precede biological treatment; for example, leachate exhibiting extremes of pH must be neutralized in order for microorganisms to thrive in it. Cyanide in the leachate may be oxidized with chlorine and organics with ozone, hydrogen peroxide promoted with ultraviolet radiation, or dissolved oxygen at high temperatures and pressures. Heavy metals may be precipitated with base, carbonate, or sulfide.

Leachate can be treated by a variety of physical processes. In some cases, simple density separation and sedimentation can be used to remove water-immiscible liquids and solids. Filtration is frequently required and flotation may be useful. Leachate solutes can be concentrated by evaporation, distillation, and membrane processes, including reverse osmosis, hyperfiltration, and ultrafiltration. Organic constituents can be removed from leachate by solvent extraction, air stripping, or steam stripping. In the case of volatile organic compounds in leachate (VOCs), care must be exercised to prevent excessive escape to the atmosphere, thus creating an air pollution problem as the result of leachate treatment.

Gas Emissions

In the presence of biodegradable wastes, methane and carbon dioxide gases are produced in landfills by anaerobic degradation (see Reaction 20.8.1). Gases may also be produced by chemical processes with improperly pretreated wastes, as would occur in the hydrolysis of calcium carbide to produce acetylene:

$$CaC_2 + 2H_2O \rightarrow C_2H_2 + Ca(OH)_2 \tag{20.12.1}$$

Odorous and toxic hydrogen sulfide, H_2S, may be generated by the chemical reaction of sulfides with acids or by the biochemical reduction of sulfate by anaerobic bacteria (*Desulfovibrio*) in the presence of biodegradable organic matter:

$$SO_4{}^{2-} + 2\{CH_2O\} + 2H^+ \xrightarrow[\text{bacteria}]{\text{Anaerobic}} H_2S + 2CO_2 + 2H_2O \tag{20.12.2}$$

Gases such as these may be toxic, combustible, or explosive. Furthermore, gases permeating through landfilled hazardous waste may carry along waste vapors, such as those of volatile aryl compounds and low-molar-mass halogenated hydrocarbons. Of these, the ones of most concern are benzene, 1,2-dibromoethane, 1,2-dichloroethane, carbon tetrachloride, chloroform, dichloromethane, tetrachloroethane, 1,1,1-trichloroethane, trichloroethylene, and vinyl chloride. Because of the hazards from these and other volatile species, it is important to minimize production of gases and, if significant amounts of gases are produced, they should be vented or treated by activated carbon sorption or flaring.

20.13. IN-SITU TREATMENT

In-situ treatment refers to waste treatment processes that can be applied to wastes in a disposal site by direct application of treatment processes and reagents to the wastes. Where possible, *in-situ* treatment is desirable for waste site remediation.

In-Situ Immobilization

In-situ immobilization is used to convert wastes to insoluble forms that will not leach from the disposal site. Heavy metal contaminants including lead, cadmium, zinc, and mercury, can be immobilized by chemical precipitation as the sulfides by treatment with gaseous H_2S or alkaline Na_2S solution. Disadvantages include the high toxicity of H_2S and the contamination potential of soluble sulfide. Although precipitated metal sulfides should remain as solids in the anaerobic conditions of a landfill, unintentional exposure to air can result in oxidation of the sulfide and remobilization of the metals as soluble sulfate salts.

Oxidation and reduction reactions can be used to immobilize heavy metals insitu. Oxidation of soluble Fe^{2+} and Mn^{2+} to their insoluble hydrous oxides, $Fe_2O_3 \cdot xH_2O$ and $MnO_2 \cdot xH_2O$, respectively, can precipitate these metal ions and coprecipitate other heavy metal ions. However, subsurface reducing conditions could later result in reformation of soluble reduced species. Reduction can be used *in-situ* to convert soluble, toxic chromate to insoluble chromium(III) compounds.

Chelation may convert metal ions to less mobile forms, although with most agents chelation has the opposite effect. A chelating agent called Tetran is supposed to form metal chelates that are strongly bound to clay minerals. The humin fraction of soil humic substances likewise immobilizes metal ions.

Vapor Extraction

Many important wastes have relatively high vapor pressures and can be removed by vapor extraction. This technique works for wastes in soil above the level of groundwater, that is, in the vadose zone. Simple in concept, vapor extraction involves pumping air into injection wells in soil and withdrawing it, along with volatile components that it has picked up, through extraction wells. The substances vaporized from the soil are removed by activated carbon or by other means. In some cases the air is pumped through an engine (which can be used to run the air pumps) and are destroyed by conditions in the engine's combustion chambers. It is relatively efficient compared to groundwater pumping because of the much higher flow rates of air through soil compared to water. Vapor extraction is most applicable to the removal of volatile organic compounds (VOCs) such as chloromethanes, chloroethanes, chloroethylenes (such as trichloroethylene), benzene, toluene, and xylene.

Solidification In-Situ

In situ solidification can be used as a remedial measure at hazardous waste sites. One approach is to inject soluble silicates followed by reagents that cause them to solidify. For example, injection of soluble sodium silicate followed by calcium chloride or lime forms solid calcium silicate.

Detoxification In-Situ

When only one or a limited number of harmful constituents is present in a waste disposal site, it may be practical to consider detoxification *in-situ*. This approach is most practical for organic contaminants including pesticides (organophosphate esters and carbamates), amides, and esters. Among the chemical and biochemical processes that can detoxify such materials are chemical and enzymatic oxidation, reduction, and hydrolysis. Chemical oxidants that have been proposed for this purpose include hydrogen peroxide, ozone, and hypochlorite.

Enzyme extracts collected from microbial cultures and purified have been considered for *in-situ* detoxification. One cell-free enzyme that has been used for detoxification of organophosphate insecticides is parathion hydrolase. The hostile environment of a chemical waste landfill, including the presence of enzyme-inhibiting heavy metal ions, is detrimental to many biochemical approaches to *in-situ* treatment. Furthermore, most sites contain a mixture of hazardous constituents, which might require several different enzymes for their detoxification.

Permeable Bed Treatment

Some groundwater plumes contaminated by dissolved wastes can be treated by a permeable bed of material placed in a trench through which the groundwater must flow. Limestone in a permeable bed neutralizes acid and precipitates some heavy metal hydroxides or carbonates. Synthetic ion exchange resins can be used in a permeable bed to retain heavy metals and even some anionic species, although competition with ionic species present naturally in the groundwater can cause some problems with their use. Activated carbon in a permeable bed will remove some organics, especially less soluble, higher molar mass organic compounds.

Permeable bed treatment requires relatively large quantities of reagent, which argues against the use of activated carbon and ion exchange resins. In such an application it is unlikely that either of these materials could be reclaimed and regenerated as is done when they are used in columns to treat wastewater. Furthermore, ions taken up by ion exchangers and organic species retained by activated carbon may be released at a later time, causing subsequent problems. Finally, a permeable bed that has been truly effective in collecting waste materials may, itself, be considered a hazardous waste requiring special treatment and disposal.

In-Situ Thermal Processes

Heating of wastes *in-situ* can be used to remove or destroy some kinds of hazardous substances. Steam injection and radio frequency and microwave heating have been proposed for this purpose. Volatile wastes brought to the surface by heating can be collected and held as condensed liquids or by activated carbon.

One approach to immobilizing wastes *in-situ* is high temperature vitrification using electrical heating. This process involves pouring conducting graphite on the surface between two electrodes and passing an electric current between the electrodes. In principle, the graphite becomes very hot and "melts" into the soil leaving a glassy slag in its path. Volatile species evolved are collected and, if the operation is successful, a nonleachable slag is left in place. It is easy to imagine problems that might occur, including difficulties in getting a uniform melt, problems from groundwater infiltration, and very high consumption of electricity.

Soil Washing and Flushing

Extraction with water containing various additives can be used to cleanse soil contaminated with hazardous wastes. When the soil is left in place and the water pumped into and out of it, the process is called **flushing**; when soil is removed and contacted with liquid the process is referred to as **washing**. Here, washing is used as a term applied to both processes.

The composition of the fluid used for soil washing depends upon the contaminants to be removed. The washing medium may consist of pure water or it may contain acids (to leach out metals or neutralize alkaline soil contaminants), bases (to neutralize contaminant acids), chelating agents (to solubilize heavy metals), surfactants (to enhance the removal of organic contaminants from soil and improve the ability of the water to emulsify insoluble organic species), or reducing agents (to reduce oxidized species). Soil contaminants may dissolve, form emulsions, or react chemically. Heavy metal salts; lighter aromatic hydrocarbons, such as toluene and xylenes; lighter organohalides, such as trichloro- or tetrachloroethylene; and light-to-medium molar mass aldehydes and ketones can be removed from soil by washing.

LITERATURE CITED

1. Hileman, Bette, "Another Stab at Superfund Reform," *Chemical and Engineering News*, June 28, 1999, pp. 20-21.

2. Jandrasi, Frank J., and Stephen Z. Masoomian, "Process Waste During Plant Design," *Environmental Engineering World*, **1**, 6-15 (1995).

3. Ray, Chittaranjan, Ravi K. Jain, Bernard A. Donahue, and E. Dean Smith, "Hazardous Waste Minimization Through Life Cycle Cost Analysis at Federal Facilities," *Journal of the Air and Waste Management Association*, **49**, 17-27 (1999).

4. Hall, Freddie E., Jr. , "OC-ALC Hazardous Waste Minimization Strategy: Reduction of Industrial Biological Sludge from Industrial Wastewater Treatment Facilities," *Proceedings of the Annual Meeting of the Air Waste Management Association*, **90**, 1052-6102 (1997).

5. Smith, Edward H. and Carlos Davis, "Hazardous Materials Balance Approach for Source Reduction and Waste Minimization," *Journal of Environmental Science and Health*, Part A., **32**, 171-193 (1997).

6. Richardson, Kelly E. and Terry Bursztynsky, "Refinery Waste Minimization," *Proceedings of HAZMACON 95*, 505-513 (1995).

7. Eckenfelder, W. Wesley, Jr. and A. H. Englande, Jr., "Chemical/Petrochemical Wastewater Management—Past, Present and Future," *Water Science Technology*, **34**, 1-7 (1996).

8. Ellis, Jeffrey I., "Waste Water Recycling and Pre-treatment Systems: an Alternative to Oil/Water Separators," *Proceedings of the Annual Meeting of the Air Waste Management Association*, **91**, 1052-6102 (1998).

9. Chang, Li-Yang, "A Waste Minimization Study of a Chelated Copper Complex in Wastewater—Treatability and Process Analysis," *Waste Management (New York)*, **15**, 209-20 (1995).

10. Macchi, G., M. Pagano, M. Santori, and G. Tiravanti, "Battery Industry Wastewater: Pb Removal and Produced Sludge," *Water Research*, **27**, 1511-1518 (1993).

11. Abda, Moshe and Yoram Oren, "Removal of Cadmium and Associated Contaminants from Aqueous Wastes by Fibrous Carbon Electrodes" *Water Research*, **27**, 1535-1544 (1993).

12. Paul, S. F., "Review of Thermal Plasma Research and Development for Hazardous Waste Remediation in the United States," *Thermal Plasmas for Hazardous Waste Treatment, Proceedings of the International Symposium on Plasma Physics "Piero Caldirola,"* Roberto Benocci, Giovanni Bonizzoni, and Elio Sindoni, Eds., World Scientific, Singapore, 1996, pp. 67-92.

13. Mohn, William W., and James M. Tiedje, "Microbial Reductive Dehalogenation," *Microbial Reviews*, **56**, 482-507 (1992).

SUPPLEMENTARY REFERENCES

ACS Task Force on Laboratory Waste Management, *Laboratory Waste Management: A Guidebook*, American Chemical Society, Washington, D.C., 1994.

Anderson, Todd A. and Joel R. Coats, Eds., *Bioremediation Through Rhizosphere Technology*, American Chemical Society, Washington, D.C., 1994.

Anderson, W. C., *Innovative Site Remediation Technology*, Springer Verlag, New York, 1995.

Armour, Margaret-Ann, *Hazardous Laboratory Chemicals Disposal Guide*, CRC Press/Lewis Publishers, Boca Raton, FL, 1996.

Barth, Edwin F., *Stabilization and Solidification of Hazardous Wastes* (Pollution Technology Review, No 186), Noyes Publications, Park Ridge, NJ, 1990.

Bentley, S. P., Ed., *Engineering Geology of Waste Disposal*, American Association of Petroleum Geologists, Houston, TX, 1995.

Bierma, Thomas J. and Frank L. Waterstraat, *Waste Minimization through Shared Savings: Chemical Supply Stategies*, John Wiley & Sons, New York, 1999.

Boardman, Gregory D., Ed., *Proceedings of the Twenty-Ninth Mid-Atlantic Industrial and Hazardous Waste Conference*, Technomic Publishing Co., Lancaster, PA, 1997.

Bodocsi, A., Michael E. Ryan, and Ralph R. Rumer, Eds., *Barrier Containment Technologies for Environmental Remediation Applications*, John Wiley & Sons, New York, 1995.

Cheremisinoff, Nicholas P., *Groundwater Remediation and Treatment Technologies*, Noyes Publications, Westwood, NJ, 1998.

Cheremisinoff, Nicholas P., *Biotechnology for Waste and Wastewater Treatment*, Noyes Publications, Park Ridge, NJ, 1996.

Cheremisinoff, Paul N., *Waste Minimization and Cost Reduction for the Process Industries*, Noyes Publications, Park Ridge, NJ, 1995.

Childers, Darin G., *Environmental Economics: Profiting from Waste Minimization: A Practical Guide to Achieving Improvements in Quality, Profitability, and Competitiveness through the Prevention of Pollution*, Water Environment Federation, Alexandria, VA, 1998.

Ciambrone, David F., *Waste Minimization As a Strategic Weapon*, CRC Press/Lewis Publishers, Boca Raton, FL, 1996.

Clark, J. H., Ed., *Chemistry of Waste Minimization*, Blackie Academic and Professional, New York, 1995.

Davis, John W., *Fast Track to Waste-Free Manufacturing: Straight Talk from a Plant Manager*, Productivity Press, Portland, OR, 1999.

Epps, John A. and Chin-Fu Tsang, Eds., *Industrial Waste: Scientific and Engineering Aspects*, Academic Press, San Diego, CA, 1996.

Freeman, Harry and Eugene F. Harris, Eds., *Hazardous Waste Remediation: Innovative Treatment Technologies*, Technomic Publishing Co., Lancaster, PA, 1995.

Frosch, Robert A., "Industrial Ecology: Adapting Technology for a Sustainable World," *Environment Magazine*, **37**, 16–37, 1995.

Gilliam, Michael T. and Carlton G. Wiles, Eds., *Stabilization and Solidification of Hazardous, Radioactive, and Mixed Wastes* (ASTM Special Technical Publication, 1123), American Society for Testing and Materials, Philadelphia, 1992.

Graedel, Thomas E. and B. R. Allenby, "Industrial Process Residues: Composition and Minimization," Chapter 15 in *Industrial Ecology*, Prentice Hall, Englewood Cliffs, NJ, 1995, pp. 204–230.

Haas, Charles N. and Richard J. Vamos, *Hazardous and Industrial Waste Treatment*, Prentice Hall Press, New York, 1995.

Hester, R. E., and R. M. Harrison, *Waste Treatment and Disposal* (Issues in Environmental Science and Technology, 3), Royal Society of Chemistry, London, 1995.

Hickey, Robert F. and Gretchen Smith, Eds., *Biotechnology in Industrial Waste Treatment and Bioremediation*, CRC Press/Lewis Publishers, Boca Raton, FL, 1996.

Hinchee, Robert E., Rodney S. Skeen, and Gregory D. Sayles, *Biological Unit Processes for Hazardous Waste Treatment*, Battelle Press, Columbus, OH 1995.

International Atomic Energy Agency Vienna, *Minimization of Radioactive Waste from Nuclear Power Plants and the Back End of the Nuclear Fuel Cycle*, International Atomic Energy Agency, Vienna, 1995.

Karnofsky, Brian, Ed., *Hazardous Waste Management Compliance Handbook*, 2nd ed., Van Nostrand Reinhold, New York, 1996.

LaGrega, Michael D., Phillip L. Buckingham, and Jeffrey C. Evans, *Hazardous Waste Management*, McGraw-Hill, New York, 1994.

Long, Robert B., *Separation Processes in Waste Minimization*, Marcel Dekker, New York, 1995.

Lewandowski, Gordon A. and Louis J. DeFilippi, *Biological Treatment of Hazardous Wastes*, Wiley, New York, 1997.

National Research Council, *Review and Evaluation of Alternative Chemical Disposal Technologies*, National Academy Press, Washington, D.C., 1997.

Nemorow, Nelson, *Zero Pollution for Industry : Waste Minimization Through Industrial Complexes*, John Wiley & Sons, New York, 1995.

Olsen, Scott, Barbara J. McKellar, and Kathy Kuntz, *Enhancing Industrial Competitiveness: First Steps to Recognizing the Potential of Energy Efficiency and Waste Minimization*, Wisconsin Demand-Side Demonstrations, Inc., Madison, WI, 1995.

Reed, Sherwood C., Ronald W. Crites, and E. Joe Middlebrooks, *Natural Systems for Waste Management and Treatment*, McGraw-Hill, New York, 1995.

Reinhardt, Peter A., K. Leigh Leonard, and Peter C. Ashbrook, Eds., *Pollution Prevention and Waste Minimization in Laboratories*, CRC Press/Lewis Publishers, Boca Raton, FL, 1996.

Roberts, Stephen M., Christopher M. Teaf, and Judy A. Bean, Eds., *Hazardous Waste Incineration: Evaluating the Human Health and Environmental Risks*, CRC Press/Lewis Publishers, Boca Raton, FL, 1999.

Rossiter, Alan P., Ed., *Waste Minimization Through Process Design*, McGraw-Hill, New York, 1995.

Sellers, Kathleen, *Fundamentals of Hazardous Waste Site Remediation*, CRC Press/Lewis Publishers, Boca Raton, FL, 1999.

Tedder, D. William and Frederick G. Pohland, Eds., *Emerging Technologies in Hazardous Waste Management 7*, Plenum, New York,1997.

Thomas, Suzanne T., *Facility Manager's Guide to Pollution Prevention and Waste Minimization*, BNA Books, Castro Valley, CA, 1995.

Waste Characterization and Treatment, Society for Mining Metallurgy & Exploration, Littleton, CO, 1998.

Water and Residue Treatment, Volume II, Hazardous Materials Control Research Institute, Silver Spring, MD, 1987.

Watts, Richard J., *Hazardous Wastes: Sources, Pathways, Receptors*, John Wiley & Sons, New York, 1997.

Williams, Paul T., *Waste Treatment and Disposal*, John Wiley & Sons, New York, 1998.

Winkler, M. A., *Biological Treatment of Waste-Water*, Halsted Press (John Wiley & Sons), New York, 1997.

Wise, Donald L., and Debra J. Trantolo, *Process Engineering for Pollution Control and Waste Minimization*, Marcel Dekker, Inc., New York, 1994.

QUESTIONS AND PROBLEMS

1. Place the following in descending order of desirability for dealing with wastes and discuss your rationale for doing so: (a) reducing the volume of remaining wastes by measures such as incineration, (b) placing the residual material in landfills, properly protected from leaching or release by other pathways, (c) treating residual material as much as possible to render it nonleachable and innocuous, (d) reduction of wastes at the source, (e) recycling as much waste as is practical.

2. Match the waste recycling process or industry from the column on the left with the kind of material that can be recycled from the list on the right, below:

1. Recycle as raw material to the generator	(a) Waste alkali
2. Utilization for pollution control or waste treatment	(b) Hydraulic and lubricating oils
	(c) Incinerable materials
3. Energy production	(d) Incompletely consumed feedstock material
4. Materials with agricultural uses	(e) Waste lime or phosphate-containing sludge
5. Organic substances	

3. What material is recycled using hydrofinishing, treatment with clay, and filtration?

4. What is the "most important operation in solvent purification and recycle" that is used to separate solvents from impurities, water, and other solvents?

5. Dissolved air flotation (DAF) is used in the secondary treatment of wastes. What is the principle of this technique? For what kinds of hazardous waste substances is it most applicable?

6. Match the process or industry from the column on the left with its "phase of waste treatment" from the list on the right, below:

1. Activated carbon sorption	(a) Primary treatment
2. Precipitation	(b) Secondary treatment
3. Reverse osmosis	(c) Polishing
4. Emulsion breaking	
5. Slurrying	

7. Distillation is used in treating and recycling a variety of wastes, including solvents, waste oil, aqueous phenolic wastes, and mixtures of ethylbenzene and styrene. What is the major hazardous waste problem that arises from the use of distillation for waste treatment?

8. Supercritical fluid technology has a great deal of potential for the treatment of hazardous wastes. What are the principles involved with the use of supercritical fluids for waste treatment? Why is this technique especially advantageous? Which substance is most likely to be used as a supercritical fluid in this application? For which kinds of wastes are supercritical fluids most useful?

9. What are some advantages of using acetic acid, compared, for example, to sulfuric acid, as a neutralizing agent for treating waste alkaline materials?

10. Which of the following would be **least likely** to be produced by, or used as a reagent for the removal of heavy metals by their precipitation from solution? (a) Na_2CO_3, (b) CdS, (e) $Cr(OH)_3$, (d) KNO_3, (e) $Ca(OH)_2$

11. Both $NaBH_4$ and Zn are used to remove metals from solution. How do these substances remove metals? What are the forms of the metal products?

12. Of the following, thermal treatment of wastes is **not** useful for (a) volume reduction, (b) destruction of heavy metals, (c) removal of volatile, combustible, mobile organic matter, (d) destruction of pathogenic materials, (e) destruction of toxic substances.

13. From the following, choose the waste liquid that is least amenable to incineration and explain why it is not readily incinerated: (a) methanol, (b) tetrachloroethylene, (c) acetonitrile, (d) toluene, (e) ethanol, (f) acetone.

14. Name and give the advantages of the process that is used to destroy more hazardous wastes by thermal means than are burned solely for the purpose of waste destruction.

15. What is the major advantage of fluidized-bed incinerators from the standpoint of controlling pollutant by-products?

16. What is the best way to obtain microorganisms to be used in the treatment of hazardous wastes by biodegradation?

17. What are the principles of composting? How is it used to treat hazardous wastes?

18. How is portland cement used in the treatment of hazardous wastes for disposal? What might be some disadvantages of such a use?

19. What are the advantages of aboveground disposal of hazardous wastes as opposed to burying wastes in landfills?

20. Describe and explain the best approach to managing leachate from hazardous waste disposal sites.

21 ENVIRONMENTAL BIOCHEMISTRY

21.1. BIOCHEMISTRY

The effects of pollutants and potentially harmful chemicals on living organisms are of particular importance in environmental chemistry. These effects are addressed under the topic of "Toxicological Chemistry" in Chapter 22, and for specific substances in Chapter 23. The current chapter, "Environmental Biochemistry," is designed to provide the fundamental background in biochemistry required to understand toxicological chemistry.

Most people have had the experience of looking through a microscope at a single cell. It may have been an ameba, alive and oozing about like a blob of jelly on the microscope slide, or a cell of bacteria, stained with a dye to make it show up more plainly. Or, it may have been a beautiful cell of algae with its bright green chlorophyll. Even the simplest of these cells is capable of carrying out a thousand or more chemical reactions. These life processes fall under the heading of **biochemistry**, that branch of chemistry that deals with the chemical properties, composition, and biologically-mediated processes of complex substances in living systems.

Biochemical phenomena that occur in living organisms are extremely sophisticated. In the human body complex metabolic processes break down a variety of food materials to simpler chemicals, yielding energy and the raw materials to build body constituents such as muscle, blood, and brain tissue. Impressive as this may be, consider a humble microscopic cell of photosynthetic cyanobacteria only about a micrometer in size, which requires only a few simple inorganic chemicals and sunlight for its existence. This cell uses sunlight energy to convert carbon from CO_2, hydrogen and oxygen from H_2O, nitrogen from NO_3^-, sulfur from SO_4^{2-}, and phosphorus from inorganic phosphate into all the proteins, nucleic acids, carbohydrates, and other materials that it requires to exist and reproduce. Such a simple cell accomplishes what could not be done by humans in even a vast chemical factory costing billions of dollars.

Ultimately, most environmental pollutants and hazardous substances are of concern because of their effects upon living organisms. The study of the adverse effects of substances on life processes requires some basic knowledge of biochemistry. Biochemistry is discussed in this chapter, with emphasis upon aspects that are especially pertinent to environmentally hazardous and toxic substances, including cell membranes, DNA, and enzymes.

Biochemical processes not only are profoundly influenced by chemical species in the environment, they largely determine the nature of these species, their degradation, and even their syntheses, particularly in the aquatic and soil environments. The study of such phenomena forms the basis of **environmental biochemistry.**[1]

Biomolecules

The biomolecules that constitute matter in living organisms are often polymers with molecular masses of the order of a million or even larger. As discussed later in this chapter, these biomolecules may be divided into the categories of carbohydrates, proteins, lipids, and nucleic acids. Proteins and nucleic acids consist of macromolecules, lipids are usually relatively small molecules, carbohydrates range from relatively small sugar molecules to high molar mass macromolecules such as those in cellulose.

The behavior of a substance in a biological system depends to a large extent upon whether the substance is hydrophilic ("water-loving") or hydrophobic ("water-hating"). Some important toxic substances are hydrophobic, a characteristic that enables them to traverse cell membranes readily. Part of the detoxification process carried on by living organisms is to render such molecules hydrophilic, therefore water-soluble and readily eliminated from the body.

21.2. BIOCHEMISTRY AND THE CELL

The focal point of biochemistry and biochemical aspects of toxicants is the **cell,** the basic building block of living systems where most life processes are carried out. Bacteria, yeasts, and some algae consist of single cells. However, most living things are made up of many cells. In a more complicated organism the cells have different functions. Liver cells, muscle cells, brain cells, and skin cells in the human body are quite different from each other and do different things. Cells are divided into two major categories depending upon whether or not they have a nucleus: **eukaryotic** cells have a nucleus and **prokaryotic** cells do not. Prokaryotic cells are found predominately in single-celled organisms such as bacteria.[2] Eukaryotic cells occur in multicellular plants and animals—higher life forms.

Major Cell Features

Figure 21.1 shows the major features of the **eukaryotic cell,** which is the basic structure in which biochemical processes occur in multicellular organisms. These features are the following:

- **Cell membrane,** which encloses the cell and regulates the passage of ions, nutrients, lipid-soluble ("fat-soluble") substances, metabolic products, toxicants, and toxicant metabolites into and out of the cell interior because of its varying **permeability** for different substances. The cell membrane protects the contents of the cell from undesirable outside influences. Cell membranes are composed in part of phospholipids that are arranged with their hydrophilic ("water-seeking") heads on the cell membrane surfaces and their hydrophobic ("water-repelling") tails inside the membrane. Cell membranes contain bodies of proteins that are involved in the transport of some substances through the membrane. One reason the cell membrane is very important in toxicology and environmental biochemistry is because it regulates the passage of toxicants and their products into and out of the cell interior. Furthermore, when its membrane is damaged by toxic sustances, a cell may not function properly and the organism may be harmed.

- **Cell nucleus,** which acts as a sort of "control center" of the cell. It contains the genetic directions the cell needs to reproduce itself. The key substance in the nucleus is **deoxyribonucleic** acid (**DNA**). **Chromosomes** in the cell nucleus are made up of combinations of DNA and proteins. Each chromosome stores a separate quantity of genetic information. Human cells contain 46 chromosomes. When DNA in the nucleus is damaged by foreign substances, various toxic effects, including mutations, cancer, birth defects, and defective immune system function may occur.

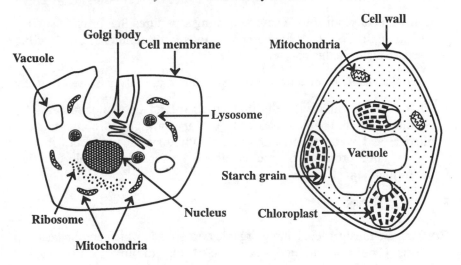

Figure 21.1. Some major features of the eukaryotic cell in animals (left) and plants (right).

- **Cytoplasm,** which fills the interior of the cell not occupied by the nucleus. Cytoplasm is further divided into a water-soluble proteinaceous filler called **cytosol,** in which are suspended bodies called **cellular organelles,** such as mitochondria or, in photosynthetic organisms, chloroplasts.

- **Mitochondria,** "powerhouses" which mediate energy conversion and utilization in the cell. Mitochondria are sites in which food materials— carbohydrates, proteins, and fats—are broken down to yield carbon dioxide, water, and energy, which is then used by the cell. The best example of this is the oxidation of the sugar glucose, $C_6H_{12}O_6$:

$$C_6H_{12}O_6 + 6O_2 \rightarrow 6CO_2 + 6H_2O + energy$$

This kind of process is called **cellular respiration.**

- **Ribosomes**, which participate in protein synthesis.

- **Endoplasmic reticulum**, which is involved in the metabolism of some toxicants by enzymatic processes.

- **Lysosome**, a type of organelle that contains potent substances capable of digesting liquid food material. Such material enters the cell through a "dent" in the cell wall, which eventually becomes surrounded by cell material. This surrounded material is called a **food vacuole**. The vacuole merges with a lysosome, and the substances in the lysosome bring about digestion of the food material. The digestion process consists largely of **hydrolysis reactions** in which large, complicated food molecules are broken down into smaller units by the addition of water.

- **Golgi bodies**, that occur in some types of cells. These are flattened bodies of material that serve to hold and release substances produced by the cells.

- **Cell walls** of plant cells. These are strong structures that provide stiffness and strength. Cell walls are composed mostly of cellulose, which will be discussed later in this chapter.

- **Vacuoles** inside plant cells that often contain materials dissolved in water.

- **Chloroplasts** in plant cells that are involved in photosynthesis (the chemical process which uses energy from sunlight to convert carbon dioxide and water to organic matter). Photosynthesis occurs in these bodies. Food produced by photosynthesis is stored in the chloroplasts in the form of **starch grains**.

21.3. PROTEINS

Proteins are nitrogen-containing organic compounds which are the basic units of live systems. Cytoplasm, the jelly-like liquid filling the interior of cells, consists largely of protein. Enzymes, which act as catalysts of life reactions, are proteins; they are discussed later in the chapter. Proteins are made up of **amino acids** joined together in huge chains. Amino acids are organic compounds which contain the carboxylic acid group, $-CO_2H$, and the amino group, $-NH_2$. They are sort of a hybrid of carboxylic acids and amines (See Chapter 29 if these terms are unfamiliar). Proteins are polymers or **macromolecules** of amino acids containing from approximately 40 to several thousand amino acid groups joined by peptide linkages. Smaller molecule amino acid polymers, containing only about 10 to about 40 amino acids per molecule, are called **polypeptides**. A portion of the amino acid left after the elimination of H_2O during polymerization is called a **residue**. The amino acid sequence of these residues is designated by a series of three-letter abbreviations for the amino acid.

Natural amino acids all have the following chemical group:

In this structure the $-NH_2$ group is always bonded to the carbon next to the $-CO_2H$ group. This is called the "alpha" location, so natural amino acids are alpha-amino acids. Other groups, designated as "R," are attached to the basic alpha-amino acid structure. The R groups may be as simple as an atom of H found in glycine,

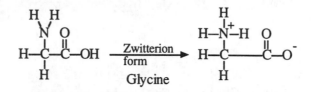

or, they may be as complicated as the structure,

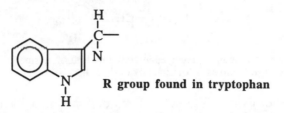

R group found in tryptophan

found in tryptophan. There are 20 common amino acids in proteins, examples of which are shown in Figure 21.2. The amino acids are shown with uncharged $-NH_2$ and $-CO_2H$ groups. Actually, these functional groups exist in the charged **zwitterion** form as shown for glycine, above.

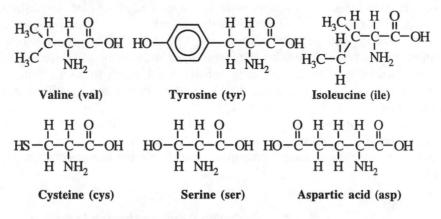

Figure 21.2. Examples of naturally occurring amino acids.

Amino acids in proteins are joined together in a specific way. These bonds are called the **peptide linkage**. The formation of peptide linkages is a condensation process involving the loss of water. Consider as an example the condensation of alanine, leucine, and tyrosine shown in Figure 21.3. When these three amino acids join together, two water molecules are eliminated. The product is a *tri*peptide since there are three amino acids involved. The amino acids in proteins are linked as shown for this tripeptide, except that many more monomeric amino acid groups are involved.

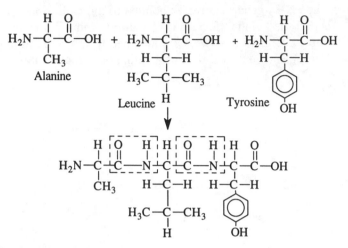

Figure 21.3. Condensation of alanine, leucine, and tyrosine to form a tripeptide consisting of three amino acids joined by peptide linkages (outlined by dashed lines).

Proteins may be divided into several major types that have widely varying functions. These are given in Table 21.1.

Protein Structure

The order of amino acids in protein molecules, and the resulting three-dimensional structures that form, provide an enormous variety of possibilites for **protein structure**. This is what makes life so diverse. Proteins have primary, secondary, tertiary, and quaternary structures. The structures of protein molecules determine the behavior of proteins in crucial areas such as the processes by which the body's immune system recognizes substances that are foreign to the body. Proteinaceous enzymes depend upon their structures for the very specific functions of the enzymes.

The order of amino acids in the protein molecule determines its primary structure. **Secondary protein structures** result from the folding of polypeptide protein chains to produce a maximum number of hydrogen bonds between peptide linkages:

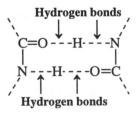

Illustration of hydrogen bonds between N and O atoms in peptide linkages, which constitute protein secondary structures

The nature of the R groups on the amino acids determines the secondary structure. Small R groups enable protein molecules to be hydrogen-bonded together in a parallel arrangement. With larger R groups the molecules tend to take a spiral form. Such a spiral is known as an **alpha-helix**.

Table 21.1. Major Types of Proteins

Type of protein	Example	Function and characteristics
Nutrient	Casein (milk protein)	Food source. People must have an adequate supply of nutrient protein with the right balance of amino acids for adequate nutrition.
Storage	Ferritin	Storage of iron in animal tissues
Structural	Collagen (tendons) keratin (hair)	Structural and protective components in organisms
Contractile	Actin, myosin in muscle tissue	Strong, fibrous proteins that can contract and cause movement to occur
Transport	Hemoglobin	Transport inorganic and organic species across cell membranes, in blood, between organs
Defense	- - -	Antibodies against foreign agents such as viruses produced by the immune system
Regulatory	Insulin, human growth hormone	Regulate biochemical processes such as sugar metabolism or growth by binding to sites inside cells or on cell membranes
Enzymes	Acetylcholinesterase	Catalysts of biochemical reactions (see Section 21.6.)

Tertiary structures are formed by the twisting of alpha-helices into specific shapes. They are produced and held in place by the interactions of amino side chains on the amino acid residues constituting the protein macromolecules. Tertiary protein structure is very important in the processes by which enzymes identify specific proteins and other molecules upon which they act. It is also involved with the action of antibodies in blood which recognize foreign proteins by their shape and react to them. This is basically what happens in the case of immunity to a disease where antibodies in blood recognize specific proteins from viruses or bacteria and reject them.

Two or more protein molecules consisting of separate polypeptide chains may be further attracted to each other to produce a **quaternary structure**.

Some proteins are **fibrous proteins**, which occur in skin, hair, wool, feathers, silk, and tendons. The molecules in these proteins are long and threadlike and are laid out parallel in bundles. Fibrous proteins are quite tough and they do not dissolve in water.

Aside from fibrous protein, the other major type of protein form is the **globular protein**. These proteins are in the shape of balls and oblongs. Globular proteins are relatively soluble in water. A typical globular protein is hemoglobin, the oxygen-carrying protein in red blood cells. Enzymes are generally globular proteins.

Denaturation of Proteins

Secondary, tertiary, and quaternary protein structures are easily changed by a process called **denaturation**. These changes can be quite damaging. Heating, exposure to acids or bases, and even violent physical action can cause denaturation to occur. The albumin protein in egg white is denatured by heating so that it forms a semisolid mass. Almost the same thing is accomplished by the violent physical action of an egg beater in the preparation of meringue. Heavy metal poisons such as lead and cadmium change the structures of proteins by binding to functional groups on the protein surface.

21.4. CARBOHYDRATES

Carbohydrates have the approximate simple formula CH_2O and include a diverse range of substances composed of simple sugars such as glucose:

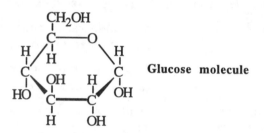

Glucose molecule

High-molar-mass **polysaccharides**, such as starch and glycogen ("animal starch"), are biopolymers of simple sugars.

When photosynthesis occurs in a plant cell, the energy from sunlight is converted to chemical energy in a carbohydrate. This carbohydrate may be transferred to some other part of the plant for use as an energy source. It may be converted to a water-insoluble carbohydrate for storage until it is needed for energy. Or it may be transformed to cell wall material and become part of the structure of the plant. If the plant is eaten by an animal, the carbohydrate is used for energy by the animal.

The simplest carbohydrates are the **monosaccharides**, also called **simple sugars**. Because they have six carbon atoms, simple sugars are sometimes called *hex*oses. Glucose (formula shown above) is the most common simple sugar involved in cell processes. Other simple sugars with the same formula but somewhat different structures are fructose, mannose, and galactose. These must be changed to glucose before they can be used in a cell. Because of its use for energy in body processes, glucose is found in the blood. Normal levels are from 65 to 110 mg glucose per 100 ml of blood. Higher levels may indicate diabetes.

Units of two monosaccharides make up several very important sugars known as **disaccharides**. When two molecules of monosaccharides join together to form a disaccharide,

$$C_6H_{12}O_6 + C_6H_{12}O_6 \rightarrow C_{12}H_{22}O_{11} + H_2O \qquad (21.4.1)$$

a molecule of water is lost. Recall that proteins are also formed from smaller amino acid molecules by condensation reactions involving the loss of water molecules. Disaccharides include sucrose (cane sugar used as a sweetener), lactose (milk sugar), and maltose (a product of the breakdown of starch).

Polysaccharides consist of many simple sugar units hooked together. One of the most important polysaccharides is **starch**, which is produced by plants for food storage. Animals produce a related material called **glycogen**. The chemical formula of starch is $(C_6H_{10}O_5)_n$, where n may represent a number as high as several hundreds. What this means is that the very large starch molecule consists of many units of $C_6H_{10}O_5$ joined together. For example, if n is 100, there are 6 times 100 carbon atoms, 10 times 100 hydrogen atoms, and 5 times 100 oxygen atoms in the molecule. Its chemical formula is $C_{600}H_{1000}O_{500}$. The atoms in a starch molecule are actually present as linked rings represented by the structure shown in Figure 21.4. Starch occurs in many foods, such as bread and cereals. It is readily digested by animals, including humans.

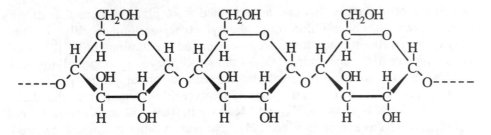

Figure 21.4. Part of a starch molecule showing units of $C_6H_{10}O_5$ condensed together.

Cellulose is a polysaccharide which is also made up of $C_6H_{10}O_5$ units. Molecules of cellulose are huge, with molecular weights of around 400,000. The cellulose structure (Figure 21.5) is similar to that of starch. Cellulose is produced by plants and forms the structural material of plant cell walls. Wood is about 60% cellulose, and cotton contains over 90% of this material. Fibers of cellulose are extracted from wood and pressed together to make paper.

Humans and most other animals cannot digest cellulose because they lack the enzyme needed to hydrolyze the oxygen linkages between the glucose molecules. Ruminant animals (cattle, sheep, goats, moose) have bacteria in their stomachs that break down cellulose into products which can be used by the animal. Chemical processes are available to convert cellulose to simple sugars by the reaction

$$(C_6H_{10}O_5)_n + nH_2O \rightarrow nC_6H_{12}O_6 \tag{21.4.2}$$
$$\text{cellulose} \qquad\qquad\qquad \text{glucose}$$

where n may be 2000-3000. This involves breaking the linkages between units of $C_6H_{10}O_5$ by adding a molecule of H_2O at each linkage, a hydrolysis reaction. Large amounts of cellulose from wood, sugar cane, and agricultural products go to waste each year. The hydrolysis of cellulose enables these products to be converted to sugars, which can be fed to animals.

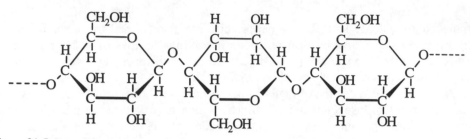

Figure 21.5. Part of the structure of cellulose.

Carbohydrate groups are attached to protein molecules in a special class of materials called **glycoproteins**. Collagen is a crucial glycoprotein that provides structural integrity to body parts. It is a major constituent of skin, bones, tendons, and cartilage.

21.5. LIPIDS

Lipids are substances that can be extracted from plant or animal matter by organic solvents, such as chloroform, diethyl ether, or toluene (Figure 21.6). Whereas carbohydrates and proteins are characterized predominately by the monomers (monosaccharides and amino acids) from which they are composed, lipids are defined essentially by their physical characteristic of organophilicity. The most common lipids are fats and oils composed of **triglycerides** formed from the alcohol glycerol, $CH_2(OH)CH(OH)CH_2OH$, and a long-chain fatty acid such as stearic acid, $CH_3(CH_2)_{16}COOH$ as shown in Figure 21.7. Numerous other biological materials, including waxes, cholesterol, and some vitamins and hormones, are classified as lipids. Common foods such as butter and salad oils are lipids. The longer chain fatty acids such as stearic acid are also organic-soluble and are classified as lipids.

Lipids are toxicologically important for several reasons. Some toxic substances interfere with lipid metabolism, leading to detrimental accumulation of lipids. Many toxic organic compounds are poorly soluble in water but are lipid-soluble, so that bodies of lipids in organisms serve to dissolve and store toxicants.

An important class of lipids consists of **phosphoglycerides** (glycerophosphatides), which may be regarded as triglyderides in which one of the acids bonded to glycerol is orthophosphoric acid. These lipids are especially important because they are essential constituents of cell membranes. These membranes consist of bilayers in which the hydrophilic phosphate ends of the molecules are on the outside of the membrane and the hydrophobic "tails" of the molecules are on the inside.

Waxes are also esters of fatty acids. However, the alcohol in a wax is not glycerol, but is often a very long chain alcohol. For example, one of the main compounds in beeswax is myricyl palmitate,

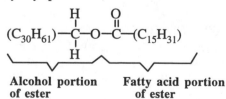

$$(C_{30}H_{61})-\overset{\overset{\displaystyle H}{|}}{\underset{\underset{\displaystyle H}{|}}{C}}-O-\overset{\overset{\displaystyle O}{\|}}{C}-(C_{15}H_{31})$$

Alcohol portion   **Fatty acid portion**
of ester   **of ester**

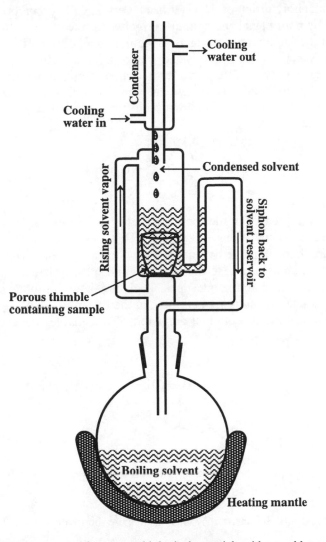

Figure 21.6. Lipids are extracted from some biological materials with a soxhlet extractor (above). The solvent is vaporized in the distillation flask by the heating mantle, rises through one of the exterior tubes to the condenser, and is cooled to form a liquid. The liquid drops onto the porous thimble containing the sample. Siphon action periodically drains the solvent back into the distillation flask. The extracted lipid collects as a solution in the solvent in the flask.

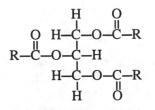

Figure 21.7. General formula of triglycerides, which make up fats and oils. The R group is from a fatty acid and is a hydrocarbon chain, such as $-(CH_2)_{16}CH_3$.

in which the alcohol portion of the ester has a very large hydrocarbon chain. Waxes are produced by both plants and animals, largely as protective coatings. Waxes are found in a number of common products. Lanolin is one of these. It is the "grease" in sheep's wool. When mixed with oils and water, it forms stable colloidal emulsions consisting of extremely small oil droplets suspended in water. This makes lanolin useful for skin creams and pharmaceutical ointments. Carnauba wax occurs as a coating on the leaves of some Brazilian palm trees. Spermaceti wax is composed largely of cetyl palmitate,

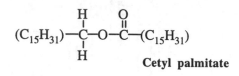

Cetyl palmitate

extracted from the blubber of the sperm whale. It is very useful in some cosmetics and pharmaceutical preparations.

Steroids are lipids found in living systems which all have the ring system shown in Figure 21.8 for cholesterol. Steroids occur in bile salts, which are produced by the liver and then secreted into the intestines. Their breakdown products give feces its characteristic color. Bile salts act upon fats in the intestine. They suspend very tiny fat droplets in the form of colloidal emulsions. This enables the fats to be broken down chemically and digested.

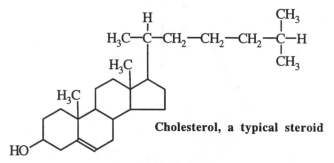

Cholesterol, a typical steroid

Figure 21.8. Steroids are characterized by the ring structure shown above for cholesterol.

Some steroids are **hormones**. Hormones act as "messengers" from one part of the body to another. As such, they start and stop a number of body functions. Male and female sex hormones (estrogens) are examples of steroid hormones. Hormones are given off by glands in the body called **endocrine glands**. The locations of important endocrine glands are shown in Figure 21.9.

21.6. ENZYMES

Catalysts are substances that speed up a chemical reaction without themselves being consumed in the reaction. The most sophisticated catalysts of all are those found in living systems. They bring about reactions that could not be performed at all, or only with great difficulty, outside a living organism. These catalysts are called **enzymes**. In addition to speeding up reactions by as much as ten- to a hundred million-fold, enzymes are extremely selective in the reactions they promote.

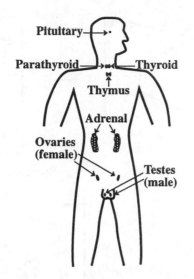

Figure 21.9. Locations of important endocrine glands.

Enzymes are proteinaceous substances with highly specific structures that inter-act with particular substances or classes of substances called **substrates**. Enzymes act as catalysts to enable biochemical reactions to occur, after which they are regenerated intact to take part in additional reactions. The extremely high specificity with which enzymes interact with substrates results from their "lock and key" action based upon the unique shapes of enzymes as illustrated in Figure 21.10. This illustration shows that an enzyme "recognizes" a particular substrate by its molecular structure and binds to it to produce an **enzyme-substrate complex**. This complex then breaks apart to form one or more products different from the original substrate, regenerating the unchanged enzyme, which is then available to catalyze additional reactions. The basic process for an enzyme reaction is, therefore,

$$\text{enzyme + substrate} \longleftrightarrow \text{enzyme-substrate complex} \longleftrightarrow$$
$$\text{enzyme + product} \qquad (21.6.1)$$

Several important things should be noted about this reaction. As shown in Figure 21.10, an enzyme acts on a specific substrate to form an enzyme-substrate complex because of the fit between their structures. As a result, something happens to the substrate molecule. For example, it might be split in two at a particular location. Then the enzyme-substrate complex comes apart, yielding the enzyme and products. The enzyme is not changed in the reaction and is now free to react again. Note that the arrows in the formula for enzyme reaction point both ways. This means that the reaction is **reversible**. An enzyme-substrate complex can simply go back to the enzyme and the substrate. The products of an enzymatic reaction can react with the enzyme to form the enzyme-substrate complex again. It, in turn, may again form the enzyme and the substrate. Therefore, the same enzyme may act to cause a reaction to go either way.

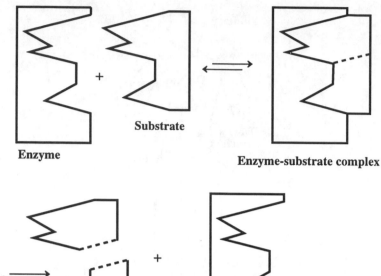

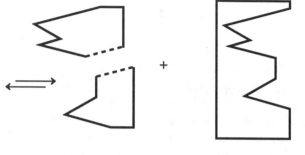

Figure 21.10. Representation of the "lock-and-key" mode of enzyme action which enables the very high specificity of enzyme-catalyzed reactions.

Some enzymes cannot function by themselves. In order to work, they must first be attached to **coenzymes**. Coenzymes normally are not protein materials. Some of the vitamins are important coenzymes.

Enzymes are named for what they do. For example, the enzyme given off by the stomach, which splits proteins as part of the digestion process, is called *gastric proteinase*. The "gastric" part of the name refers to the enzyme's origin in the stomach. The "proteinase" denotes that it splits up protein molecules. The common name for this enzyme is pepsin. Similarly, the enzyme produced by the pancreas that breaks down fats (lipids) is called *pancreatic lipase*. Its common name is steapsin. In general, lipase enzymes cause lipid triglycerides to dissociate and form glycerol and fatty acids.

The enzymes mentioned above are **hydrolyzing enzymes**, which bring about the breakdown of high-molecular-weight biological compounds by the addition of water. This is one of the most important types of the reactions involved in digestion. The three main classes of energy-yielding foods that animals eat are carbohydrates, proteins, and fats. Recall that the higher carbohydrates humans eat are largely disaccharides (sucrose, or table sugar) and polysaccharides (starch). These are formed by the joining together of units of simple sugars, $C_6H_{12}O_6$, with the elimination of an H_2O molecule at the linkage where they join. Proteins are formed by the condensation of amino acids, again with the elimination of a water molecule at each linkage. Fats are esters which are produced when glycerol and fatty acids link together. A water molecule is lost for each of these linkages when a protein, fat, or carbohydrate is synthesized. In order for these substances to be used as a food source, the reverse process must occur to break down large, complicated molecules of protein, fat, or carbohydrate to simple, soluble substances which can penetrate a cell membrane and take part in chemical processes in the cell. This reverse process is accomplished by hydrolyzing enzymes.

Biological compounds with long chains of carbon atoms are broken down into molecules with shorter chains by the breaking of carbon-carbon bonds. This commonly occurs by the elimination of CO_2 from carboxylic acids. For example, *pyruvate decarboxylase* enzyme acts upon pyruvic acid,

$$\underset{\textbf{Pyruvic acid}}{\overset{\displaystyle H-\overset{\displaystyle H}{\underset{\displaystyle H}{\overset{|}{\underset{|}{C}}}}-\overset{\displaystyle O}{\overset{\|}{C}}-\overset{\displaystyle O}{\overset{\|}{C}}-OH}{}} \xrightarrow[\text{decarboxylase}]{\text{Pyruvate}} \underset{\textbf{Acetaldehyde}}{\overset{\displaystyle H-\overset{\displaystyle H}{\underset{\displaystyle H}{\overset{|}{\underset{|}{C}}}}-\overset{\displaystyle O}{\overset{\|}{C}}-H}{}} \;+\; CO_2 \qquad (21.6.2)$$

to split off CO_2 and produce a compound with one less carbon. It is by such carbon-by-carbon breakdown reactions that long chain compounds are eventually degraded to CO_2 in the body, or that long-chain hydrocarbons undergo biodegradation by the action of microorganisms in the water and soil environments.

Oxidation and reduction are the major reactions for the exchange of energy in living systems. Cellular respiration is an oxidation reaction in which a carbohydrate, $C_6H_{12}O_6$, is broken down to carbon dioxide and water with the release of energy.

$$C_6H_{12}O_6 + 6O_2 \rightarrow 6CO_2 + 6H_2O + \text{energy} \qquad (21.6.3)$$

Actually, such an overall reaction occurs in living systems by a complicated series of individual steps. Some of these steps involve oxidation. The enzymes that bring about oxidation in the presence of free O_2 are called **oxidases**. In general, biological oxidation-reduction reactions are catalyzed by **oxidoreductase enzymes**.

In addition to the types of enzymes discussed above, there are many enzymes that perform miscellaneous duties in living systems. Typical of these are **isomerases**, which form isomers of particular compounds. For example, of the several simple sugars with the formula $C_6H_{12}O_6$, only glucose can be used directly for cell processes. The other isomers are converted to glucose by the action of isomerases. **Transferase enzymes** move chemical groups from one molecule to another, **lyase enzymes** remove chemical groups without hydrolysis and participate in the formation of C=C bonds or addition of species to such bonds, and **ligase enzymes** work in conjunction with ATP (adenosine triphosphate, a high-energy molecule that plays a crucial role in energy-yielding, glucose-oxidizing metabolic processes) to link molecules together with the formation of bonds such as carbon-carbon or carbon-sulfur bonds.

Enzyme action may be affected by many different things. Enzymes require a certain hydrogen ion concentration (pH) to function best. For example, gastric proteinase requires the acid environment of the stomach to work well. When it passes into the much less acidic intestines, it stops working. This prevents damage to the intestine walls, which would occur if the enzyme tried to digest them. Temperature is critical. Not surprisingly, the enzymes in the human body work best at around 98.6°F (37°C), which is the normal body temperature. Heating these enzymes to around 140°F permanently destroys them. Some bacteria that thrive in hot springs have enzymes that work best at temperatures as high as that of boiling water. Other "cold-seeking" bacteria have enzymes adapted to near the freezing point of water.

One of the greatest concerns regarding the effects of surroundings upon enzymes is the influence of toxic substances. A major mechanism of toxicity is the alteration or destruction of enzymes by toxic agents—cyanide, heavy metals, or organic compounds such as insecticidal parathion. An enzyme that has been destroyed obviously cannot perform its designated function, whereas one that has been altered may either not function at all or may act improperly. The detrimental effects of toxicants on enzymes are discussed in more detail in Chapter 22.

21.7. NUCLEIC ACIDS

The "essence of life" is contained in **deoxyribonucleic acid (DNA**, which stays in the cell nucleus) and **ribonucleic acid (RNA**, which functions in the cell cytoplasm). These substances, which are known collectively as **nucleic acids**, store and pass on essential genetic information that controls reproduction and protein synthesis.

The structural formulas of the monomeric constituents of nucleic acids are given in Figure 21.11. These are pyrimidine or purine nitrogen-containing bases, two sugars, and phosphate. DNA molecules are made up of the nitrogen-containing bases adenine, guanine, cytosine, and thymine; phosphoric acid (H_3PO_4); and the simple sugar 2-deoxy-β-D-ribofuranose (commonly called deoxyribose). RNA molecules are composed of the nitrogen-containing bases adenine, guanine, uracil, and cytosine; phosphoric acid (H_3PO_4) and the simple sugar β-D-ribofuranose (commonly called ribose).

The formation of nucleic acid polymers from their monomeric constituents may be viewed as the following steps.

- Monosaccharide (simple sugar) + cyclic nitrogenous base yields **nucleoside:**

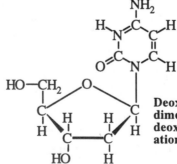

Deoxyctidine formed by the dimerization of cytosine and deoxyribose with the elimination of a molecule of H_2O.

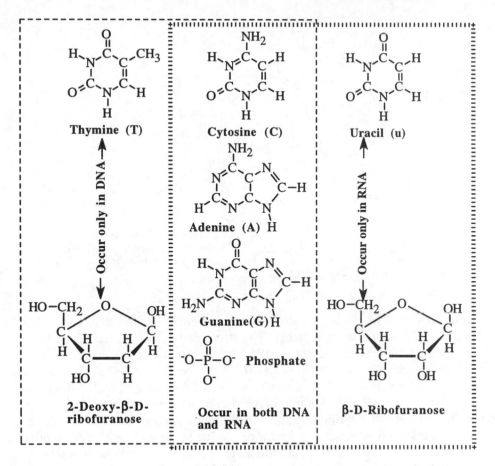

Figure 21.11. Constituents of DNA (enclosed by ----) and of RNA (enclosed by ⁗⁗).

- Nucleoside + phosphate yields **phosphate ester nucleotide**.

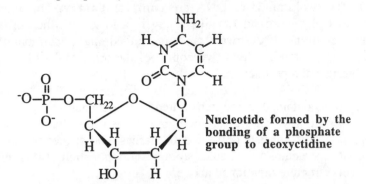

- Polymerized nucleotide yields **nucleic acid**. In the nucleic acid the phosphate negative charges are neutralized by metal cations (such as Mg^{2+}) or positively-charged proteins (histones).

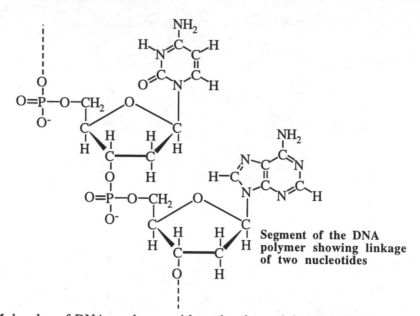

Segment of the DNA polymer showing linkage of two nucleotides

   Molecules of DNA are huge, with molecular weights greater than one billion. Molecules of RNA are also quite large. The structure of DNA is that of the famed "double helix." It was figured out in 1953 by an American scientist, James D. Watson, and Francis Crick, a British scientist. They received the Nobel prize for this scientific milestone in 1962. This model visualizes DNA as a so-called double α-helix structure of oppositely-wound polymeric strands held together by hydrogen bonds between opposing pyrimidine and purine groups. As a result, DNA has both a primary and a secondary structure; the former is due to the sequence of nucleotides in the individual strands of DNA, and the latter results from the α-helix interaction of the two strands. In the secondary structure of DNA, only cytosine can be opposite guanine and only thymine can be opposite adenine and vice versa. Basically, the structure of DNA is that of two spiral ribbons "counter-wound" around each other (Fig. 21.12). The two strands of DNA are **complementary**. This means that a particular portion of one strand fits like a key in a lock with the corresponding portion of another strand. If the two strands are pulled apart, each manufactures a new complementary strand, so that two copies of the original double helix result. This occurs during cell reproduction.

   The molecule of DNA is sort of like a coded message. This "message," the genetic information contained in and transmitted by nucleic acids, depends upon the sequence of bases from which they are composed. It is somewhat like the message sent by telegraph which consists only of dots, dashes, and spaces in between. The key aspect of DNA structure that enables storage and replication of this information is its double helix structure mentioned above.

Nucleic Acids in Protein Synthesis

   Whenever a new cell is formed, the DNA in its nucleus must be accurately reproduced from the parent cell. Life processes are absolutely dependent upon accurate protein synthesis as regulated by cell DNA. The DNA in a single cell must

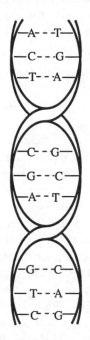

Figure 21.12. Representation of the double helix structure of DNA showing the allowed base pairs held together by hydrogen bonding between the phosphate/sugar polymer "backbones" of the two strands of DNA. The letters stand for adenine (A), cytosine (C), guanine (G), and thymine (T). The dashed lines, ---, represent hydrogen bonds.

be capable of directing the synthesis of up to 3000 or even more different proteins. The directions for the synthesis of a single protein are contained in a segment of DNA called a **gene**. The process of transmitting information from DNA to a newly-synthesized protein involves the following steps:

- The DNA undergoes **replication**. This process involves separation of a segment of the double helix into separate single strands which then replicate such that guanine is opposite cytosine (and vice versa) and adenine is opposite thymine (and vice versa). This process continues until a complete copy of the DNA molecule has been produced.

- The newly replicated DNA produces **messenger RNA (m-RNA)**, a complement of the single strand of DNA, by a process called **transcription**.

- A new protein is synthesized using m-RNA as a template to determine the order of amino acids in a process called **translation**.

Modified DNA

DNA molecules may be modified by the unintentional addition or deletion of nucleotides or by substituting one nucleotide for another. The result is a **mutation** that is transmittable to offspring. Mutations can be induced by chemical substances. This is a major concern from a toxicological viewpoint because of the detrimental effects of many mutations and because substances that cause mutations often cause cancer as well. DNA malfunction may also result in birth defects. The failure to control cell reproduction results in cancer. Radiation from X rays and radioactivity also disrupts DNA and may cause mutation.

21.8. RECOMBINANT DNA AND GENETIC ENGINEERING

As noted above, segments of DNA contain information for the specific syntheses of particular proteins. Within the last two decades it has become possible to transfer this information between organisms by means of **recombinant DNA technology**, which has resulted in a new industry based on **genetic engineering**. Most often the recipient organisms are bacteria, which can be reproduced (cloned) over many orders of magnitude from a cell that has acquired the desired qualities. Therefore, to synthesize a particular substance such as human insulin or growth hormone, the required genetic information can be transferred from a human source to bacterial cells, which then produce the substance as part of their metabolic processes.

The first step in recombinant DNA gene manipulation is to lyze, or "open up," a cell that has the genetic material needed and to remove this material from the cell. Through enzyme action the sought-after genes are cut from the donor DNA chain. These are next spliced into small DNA molecules. These molecules, called **cloning vehicles**, are capable of penetrating the host cell and becoming incorporated into its genetic material. The modified host cell is then reproduced many times and carries out the desired biosynthesis.

Early concerns about the potential of genetic engineering to produce "monster organisms" or new and horrible diseases have been largely allayed, although caution is still required with this technology. In the environmental area genetic engineering offers some hope for the production of bacteria engineered to safely destroy troublesome wastes and to produce biological substitutes for environmentally damaging synthetic pesticides.

Numerous possibilities exist for combining biology with chemistry to produce chemical feedstocks and products of various kinds. An example is production of polylactic acid using lactic acid produced enzymatically with corn and polymerized by standard chemical processes.[3] Much attention has been focused on the development of enzymes to perform a variety of chemical conversions. Another important area that uses transgenic organisms is the breeding of plants that produce natural insecticides, specifically the insecticide from *Bacillus thuringiensis*.

21.9. METABOLIC PROCESSES

Biochemical processes that involve the alteration of biomolecules fall under the category of **metabolism**. Metabolic processes may be divided into the two major categories of **anabolism** (synthesis) and **catabolism** (degradation of substances). An organism may use metabolic processes to yield energy or to modify the constituents of biomolecules.

Energy-Yielding Processes

Organisms can gain energy by one of three major processes, which are listed as follows:

- **Respiration,** in which organic compounds undergo catabolism that requires molecular oxygen (**aerobic respiration**) or that occurs in the absence of molecular oxygen (**anaerobic repiration**). Aerobic respiration uses the **Krebs cycle** to obtain energy from the following reaction:

$$C_6H_{12}O_6 + 6O_2 \rightarrow 6CO_2 + 6H_2O + energy$$

About half of the energy released is converted to short-term stored chemical energy, particularly through the synthesis of **adenosine triphosphate (ATP)** nucleotide. For longer-term energy storage, glycogen or starch polysaccharides are synthesized, and for still longer-term energy storage, lipids (fats) are generated and retained by the organism.

- **Fermentation,** which differs from respiration in not having an electron transport chain. Yeasts produce ethanol from sugars by fermentation:

$$C_6H_{12}O_6 \rightarrow 2CO_2 + 2C_2H_5OH$$

- **Photosynthesis,** in which light energy captured by plant and algal chloroplasts is used to synthesize sugars from carbon dioxide and water:

$$6CO_2 + 6H_2O + h\nu \rightarrow C_6H_{12}O_6 + 6O_2$$

Plants cannot always get the energy that they need from sunlight. During the dark they must use stored food. Plant cells, like animal cells, contain mitochondria in which stored food is converted to energy by cellular respiration.

Plant cells, which use sunlight as a source of energy and CO_2 as a source of carbon, are said to be **autotrophic**. In contrast, animal cells must depend upon organic material manufactured by plants for their food. These are called **heterotrophic** cells. They act as "middlemen" in the chemical reaction between oxygen and food material using the energy from the reaction to carry out their life processes.

21.10. METABOLISM OF XENOBIOTIC COMPOUNDS

When toxicants or their metabolic precursors (**protoxicants**) enter a living organism they may undergo several processes, including those that may make them more toxic or that detoxify them. Chapter 22 discusses the metabolic processes that toxicants undergo and the mechanisms by which they may cause damage to an organism. Emphasis is placed on **xenobiotic compounds,** which are those that are normally foreign to living organisms; on chemical aspects; and on processes that lead to products that can be eliminated from the organism. Of particular importance is **intermediary xenobiotic metabolism** which results in the formation of somewhat transient species that are different from both those ingested and the ultimate product that is excreted. These species may have significant toxicological effects. Xenobiotic compounds in general are acted upon by enzymes that function on a material that is in the body naturally—an **endogenous substrate**. For example, flavin-containing monooxygenase enzyme acts upon endogenous cysteamine to convert it to cystamine, but also functions to oxidize xenobiotic nitrogen and sulfur compounds.

Biotransformation refers to changes in xenobiotic compounds as a result of enzyme action. Reactions that are not mediated by enzymes may also be important in some cases. As examples of nonenzymatic transformations, some xenobiotic compounds bond with endogenous biochemical species without an enzyme catalyst, undergo hydrolysis in body fluid media, or undergo oxidation/reduction processes. The metabolic Phase I and Phase II reactions of xenobiotics discussed here and in Chapter 22 are enzymatic, however.

The likelihood that a xenobiotic species will undergo enzymatic metabolism in the body depends upon the chemical nature of the species. Compounds with a high degree of polarity, such as relatively ionizable carboxylic acids, are less likely to enter the body system and, when they do, tend to be quickly excreted. Therefore, such compounds are unavailable, or only available for a short time, for enzymatic metabolism. Volatile compounds, such as dichloromethane or diethyl ether, are expelled so quickly from the lungs that enzymatic metabolism is less likely. This leaves as the most likely candidates for enzymatic metabolic reactions **nonpolar lipophilic compounds**, those that are relatively less soluble in aqueous biological fluids and more attracted to lipid species. Of these, the ones that are resistant to enzymatic attack (PCBs, for example) tend to bioaccumulate in lipid tissue.

Xenobiotic species may be metabolized in many body tissues and organs. As part of the body's defense against the entry of xenobiotic species, the most prominent sites of xenobiotic metabolism are those associated with entry into the body, such as the skin and lungs. The gut wall through which xenobiotic species enter the body from the gastrointestinal tract is also a site of significant xenobiotic compound metabolism. The liver is of particular significance because materials entering systemic circulation from the gastrointestinal tract must first traverse the liver.

Phase I and Phase II Reactions

The processes that most xenobiotics undergo in the body can be divided into the two categories of Phase I reactions and Phase II reactions. A **Phase I reaction** introduces reactive, polar functional groups into lipophilic ("fat-seeking") toxicant molecules. In their unmodified forms, such toxicant molecules tend to pass through lipid-containing cell membranes and may be bound to lipoproteins, in which form they are transported through the body. Because of the functional group present, the product of a Phase I reaction is usually more water-soluble than the parent xenobiotic species, and more importantly, possesses a "chemical handle" to which a substrate material in the body may become attached so that the toxicant can be eliminated from the body. The binding of such a substrate is a **Phase II reaction**, and it produces a **conjugation product** that is amenable to excretion from the body.

In general, the changes in structure and properties of a compound that result from a Phase I reaction are relatively mild. Phase II processes, however, usually produce species that are much different from the parent compounds. It should be emphasized that not all xenobiotic compounds undergo both Phase I and Phase II reactions. Such a compound may undergo only a Phase I reaction and be excreted directly from the body. Or a compound that already possesses an appropriate functional group capable of conjugation may undergo a Phase II reaction without a preceding Phase I reaction. Phase I and Phase II reactions are discussed in more detail as they relate to toxicological chemistry in Chapter 22 and 23.

LITERATURE CITED

1. Stanley E. Manahan, *Toxicological Chemistry*, 2nd ed., Lewis Publishers/CRC Press, Inc., Boca Raton, FL, 1992.

2. White, David, *The Physiology and Biochemistry of Prokaryotes*, Oxford University Press, New York, 1999.

3. Thayer, Ann, "Best of All Industrial Worlds: Chemistry and Biology," *Chemical and Engineering News*, May 31, 1999, p. 18.

SUPPLEMENTARY REFERENCES

Bettelheim, Frederick A. and Jerry March, *Introduction to Organic and Biochemistry*, Saunders College Publishing, Fort Worth, TX, 1998.

Brownie, Alexander C. and John C. Kemohan, Illustrations by Jane Templeman and Chartwell Illustrators, *Biochemistry*, Churchill Livingstone, Edinburgh, Scotland, 1999.

Chesworth, J. M., T. Stuchbury, and J.R. Scaife, *An Introduction to Agricultural Biochemistry*, Chapman and Hall, London, 1998.

Garrett, Reginald H. and Charles M. Grisham, *Biochemistry*, Saunders College Publishing, Philadelphia, 1998.

Gilbert, Hiram F., Ed., *Basic Concepts in Biochemistry*, McGraw-Hill, Health Professions Division, New York, 2000.

Kuchel, Philip W., Ed., *Schaum's Outline of Theory and Problems of Biochemistry*, McGraw-Hill, New York, 1998.

Lea, Peter J. and Richard C. Leegood, Eds., *Plant Biochemistry and Molecular Biology*, 2nd ed., John Wiley & Sons, New York, 1999.

Marks, Dawn B., *Biochemistry*, Williams & Wilkins, Baltimore, 1999.

McKee, Trudy and James R. McKee, *Biochemistry: An Introduction*, WCB/McGraw-Hill, Boston, 1999.

Meisenberg,Gerhard and William H. Simmons, *Principles of Medical Biochemistry*, Mosby, St. Louis, 1998.

Switzer, Robert L. and Liam F. Garrity, *Experimental Biochemistry*, W. H. Freeman and Co., New York, 1999.

Voet, Donald, Judith G. Voet, and Charlotte Pratt, *Fundamentals of Biochemistry*, John Wiley & Sons, New York, 1998.

Vrana, Kent E., *Biochemistry*, Lippincott Williams & Wilkins, Philadelphia,, 1999.

Wilson, Keith and John M. Walker, *Principles and Techniques of Practical Biochemistry*, Cambridge University Press, New York, 1999.

QUESTIONS AND PROBLEMS

1. What is the toxicological importance of lipids? How are lipids related to hydrophobic pollutants and toxicants?

2. What is the function of a hydrolase enzyme?

3. Match the cell structure on the left with its function on the right, below:

   1. Mitochondria
   2. Endoplasmic reticulum
   3. Cell membrane
   4. Cytoplasm
   5. Cell nucleus

   (a) Toxicant metabolism
   (b) Fills the cell
   (c) Deoxyribonucleic acid
   (d) Mediate energy conversion and utilization
   (e) Encloses the cell and regulates the passage of materials into and out of the cell interior

4. The formula of simple sugars is $C_6H_{12}O_6$. The simple formula of higher carbohydrates is $C_6H_{10}O_5$. Of course, many of these units are required to make a molecule of starch or cellulose. If higher carbohydrates are formed by joining together molecules of simple sugars, why is there a difference in the ratios of C, H, and O atoms in the higher carbohydrates as compared to the simple sugars?

5. Why does wood contain so much cellulose?

6. What would be the chemical formula of a *tri*saccharide made by the bonding together of three simple sugar molecules?

7. The general formula of cellulose may be represented as $(C_6H_{10}O_5)_x$. If the molecular weight of a molecule of cellulose is 400,000, what is the estimated value of x?

8. During one month a factory for the production of simple sugars, $C_6H_{12}O_6$, by the hydrolysis of cellulose processes one million pounds of cellulose. The percentage of cellulose that undergoes the hydrolysis reaction is 40%. How many pounds of water are consumed in the hydrolysis of cellulose each month?

9. What is the structure of the largest group of atoms common to all amino acid molecules?

10. Glycine and phenylalanine can join together to form two different dipeptides. What are the structures of these two dipeptides?

11. One of the ways in which two parallel protein chains are joined together, or cross linked, is by way of an —S—S— link. What amino acid to you think might be most likely to be involved in such a link? Explain your choice.

12. Fungi, which break down wood, straw, and other plant material, have what are called "exoenzymes." Fungi have no teeth and cannot break up plant material physically by force. Knowing this, what do you suppose an exoenzyme is? Explain how you think it might operate in the process by which fungi break down something as tough as wood.

13. Many fatty acids of lower molecular weight have a bad odor. Speculate as to the reasons that rancid butter has a bad odor. What chemical compound is produced

that has a bad odor? What sort of chemical reaction is involved in its production?

14. The long-chain alcohol with 10 carbons is called decanol. What do you think would be the formula of decyl stearate? To what class of compounds would it belong?

15. Write an equation for the chemical reaction between sodium hydroxide and cetyl stearate. What are the products?

16. What type of endocrine gland is found only in females? What type of these glands is found only in males?

17. The action of bile salts is a little like that of soap. What function do bile salts perform in the intestine? Look up the action of soaps in Chapter 5, and explain how you think bile salts may function somewhat like soap.

18. If the structure of an enzyme is illustrated as,

how should the structure of its substrate be represented?

19. Look up the structures of ribose and deoxyribose. Explain where the "deoxy" came from in the name deoxyribose.

20. In what respect is an enzyme and its substrate like two opposite strands of DNA?

21. For what discovery are Watson and Crick noted?

22. Why does an enzyme no longer function if it is denatured?

22 TOXICOLOGICAL CHEMISTRY

22.1. INTRODUCTION TO TOXICOLOGY AND TOXICOLOGICAL CHEMISTRY

Ultimately, most pollutants and hazardous substances are of concern because of their toxic effects. The general aspects of these effects are addressed in this chapter under the heading of toxicological chemistry; the toxicological chemistry of specific classes of chemical substances is addressed in Chapter 23. In order to understand toxicological chemistry, it is essential to have some understanding of biochemistry, the science that deals with chemical processes and materials in living systems. Biochemistry was summarized in Chapter 21.

Toxicology

A **poison**, or **toxicant**, is a substance that is harmful to living organisms because of its detrimental effects on tissues, organs, or biological processes. **Toxicology** is the science of poisons. These definitions are subject to a number of qualifications. Whether a substance is poisonous depends upon the type of organism exposed, the amount of the substance, and the route of exposure. In the case of human exposure, the degree of harm done by a poison can depend strongly upon whether the exposure is to the skin, by inhalation, or through ingestion.

Toxicants to which subjects are exposed in the environment or occupationally may be in several different physical forms. This may be illustrated for toxicants that are inhaled. **Gases** are substances such as carbon monoxide in air that are normally in the gaseous state under ambient conditions of temperature and pressure. **Vapors** are gas-phase materials that have evaporated or sublimed from liquids or solids. **Dusts** are respirable solid particles produced by grinding bulk solids, whereas **fumes** are solid particles from the condensation of vapors, often metals or metal oxides. **Mists** are liquid droplets.

Often a toxic substance is in solution or mixed with other substances. A substance with which the toxicant is associated (the solvent in which it is dissolved or the solid medium in which it is dispersed) is called the **matrix**. The matrix may have a strong effect upon the toxicity of the toxicant.

There are numerous variables related to the ways in which organisms are exposed to toxic substances. One of the most crucial of these, **dose,** is discussed in Section 22.2. Another important factor is the **toxicant concentration**, which may range from the pure substance (100%) down to a very dilute solution of a highly potent poison. Both the **duration** of exposure per exposure incident and the **frequency** of exposure are important. The **rate** of exposure and the total time period over which the organism is exposed are both important situational variables. The exposure **site** and **route** also affect toxicity.

It is possible to classify exposures on the basis of acute vs. chronic and local *vs.* systemic exposure, giving four general categories. **Acute local** exposure occurs at a specific location over a time period of a few seconds to a few hours and may affect the exposure site, particularly the skin, eyes, or mucous membranes. The same parts of the body can be affected by **chronic local** exposure, for which the time span may be as long as several years. **Acute systemic** exposure is a brief exposure or exposure to a single dose and occurs with toxicants that can enter the body, such as by inhalation or ingestion, and affect organs, such as the liver, that are remote from the entry site. **Chronic systemic** exposure differs in that the exposure occurs over a prolonged time period.

In discussing exposure sites for toxicants it is useful to consider the major routes and sites of exposure, distribution, and elimination of toxicants in the body as shown in Figure 22.1. The major routes of accidental or intentional exposure to toxicants by humans and other animals are the skin (percutaneous route), the lungs (inhalation, respiration, pulmonary route), and the mouth (oral route); minor routes of exposure are rectal, vaginal, and parenteral (intravenous or intramuscular, a common means for the administration of drugs or toxic substances in test subjects). The way that a toxic substance is introduced into the complex system of an organism is strongly dependent upon the physical and chemical properties of the substance. The pulmonary system is most likely to take in toxic gases or very fine, respirable solid or liquid particles. In other than a respirable form, a solid usually enters the body orally. Absorption through the skin is most likely for liquids, solutes in solution, and semisolids, such as sludges.

The defensive barriers that a toxicant may encounter vary with the route of exposure. For example, toxic elemental mercury is absorbed through the alveoli in the lungs much more readily than through the skin or gastrointestinal tract. Most test exposures to animals are through ingestion or gavage (introduction into the stomach through a tube). Pulmonary exposure is often favored with subjects that may exhibit refractory behavior when noxious chemicals are administered by means requiring a degree of cooperation from the subject. Intravenous injection may be chosen for deliberate exposure when it is necessary to know the concentration and effect of a xenobiotic substance in the blood. However, pathways used experimentally that are almost certain not to be significant in accidental exposures can give misleading results when they avoid the body's natural defense mechanisms.

An interesting historical example of the importance of the route of exposure to toxicants is provided by cancer caused by contact of coal tar with skin. The major barrier to dermal absorption of toxicants is the **stratum corneum,** or horny layer.

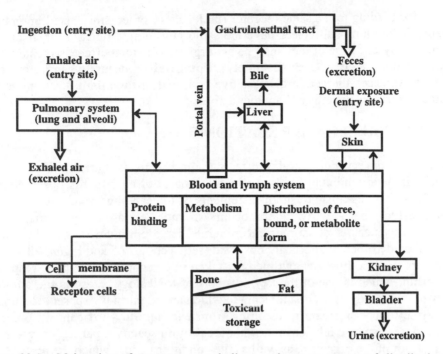

Figure 22.1.   Major sites of exposure, metabolism, and storage, routes of distribution and elimination of toxic substances in the body.

The permeability of skin is inversely proportional to the thickness of this layer, which varies by location on the body in the order soles and palms > abdomen, back, legs, arms > genital (perineal) area. Evidence of the susceptibility of the genital area to absorption of toxic substances is to be found in accounts of the high incidence of cancer of the scrotum among chimney sweeps in London described by Sir Percival Pott, Surgeon General of Britain during the reign of King George III. The cancer-causing agent was coal tar condensed in chimneys. This material was more readily absorbed through the skin in the genital areas than elsewhere, leading to a high incidence of scrotal cancer. (The chimney sweeps' conditions were aggravated by their lack of appreciation of basic hygienic practices, such as bathing and regular changes of underclothing.)

Organisms can serve as indicators of various kinds of pollutants. In this application, organisms are known as **biomonitors**. For example, higher plants, fungi, lichens, and mosses can be important biomonitors for heavy metal pollutants in the environment.

Synergism, Potentiation, and Antagonism

The biological effects of two or more toxic substances can be different in kind and degree from those of one of the substances alone. One of the ways in which this can occur is when one substance affects the way in which another undergoes any of the steps in the kinetic phase as discussed in Section 22.7 and illustrated in Figure 22.9. Chemical interaction between substances may affect their toxicities. Both substances may act upon the same physiologic function, or two substances may compete for binding to the same receptor (molecule or other entity acted upon by a toxicant). When both substances have the same physiologic function, their effects may be simply **additive** or they may be **synergistic** (the total effect is greater than the sum of the effects of each separately). **Potentiation** occurs when an inactive substance enhances the action of an active one, and **antagonism** when an active substance decreases the effect of another active one.

22.2. DOSE-RESPONSE RELATIONSHIPS

Toxicants have widely varying effects upon organisms. Quantitatively, these variations include minimum levels at which the onset of an effect is observed, the sensitivity of the organism to small increments of toxicant, and levels at which the ultimate effect (particularly death) occurs in most exposed organisms. Some essential substances, such as nutrient minerals, have optimum ranges above and below which detrimental effects are observed (see Section 22.5 and Figure 22.4).

Factors such as those just outlined are taken into account by the **dose-response** relationship, which is one of the key concepts of toxicology. **Dose** is the amount, usually per unit body mass, of a toxicant to which an organism is exposed. **Response** is the effect upon an organism resulting from exposure to a toxicant. In order to define a dose-response relationship, it is necessary to specify a particular response, such as death of the organism, as well as the conditions under which the response is obtained, such as the length of time from administration of the dose. Consider a specific response for a population of the same kinds of organisms. At relatively low doses, none of the organisms exhibits the response (for example, all live), whereas at higher doses all of the organisms exhibit the response (for example, all die).   In between, there is a range of doses over which some of the organisms respond in the specified manner and others do not, thereby defining a dose-response curve. Dose-response relationships differ among different kinds and strains of organisms, types of tissues, and populations of cells.

Figure 22.2 shows a generalized dose-response curve. Such a plot may be obtained, for example, by administering different doses of a poison in a uniform manner to a homogeneous population of test animals and plotting the cumulative percentage of deaths as a function of the log of the dose. The dose corresponding to the mid-point (inflection point) of the resulting S-shaped curve is the statistical estimate of the dose that would kill 50 percent of the subjects and is designated as LD_{50}. The estimated doses at which 5 percent (LD_5) and 95 percent (LD_{95}) of the test subjects die are obtained from the graph by reading the dose levels for 5 percent and 95 percent fatalities, respectively. A relatively small difference between LD_5 and LD_{95} is reflected by a steeper S-shaped curve, and vice versa. Statistically, 68 percent of all values on a dose-response curve fall within ± 1 standard deviation of the mean at LD_{50} and encompass the range from LD_{16} to LD_{84}.

22.3. RELATIVE TOXICITIES

Table 22.1 illustrates standard **toxicity ratings** that are used to describe estimated toxicities of various substances to humans. In terms of fatal doses to an adult human of average size, a "taste" of a supertoxic substances (just a few drops or less)

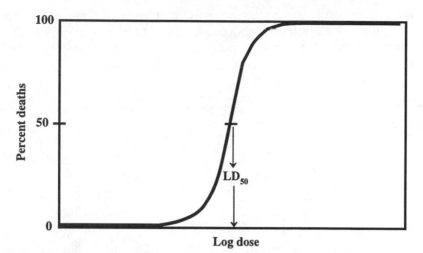

Figure 22.2.   Illustration of a dose-response curve in which the response is the death of the organism. The cumulative percentage of deaths of organisms is plotted on the Y axis.

is fatal. A teaspoonful of a very toxic substance could have the same effect. However, as much as a quart of a slightly toxic substance might be required to kill an adult human.

When there is a substantial difference between LD_{50} values of two different substances, the one with the lower value is said to be the more **potent**. Such a comparison must assume that the dose-response curves for the two substances being compared have similar slopes.

Nonlethal Effects

So far, toxicities have been described primarily in terms of the ultimate effect— deaths of organisms or lethality. This is obviously an irreversible consequence of exposure. In many, and perhaps most, cases, **sublethal** and **reversible** effects are of greater importance. This is obviously true of drugs, where death from exposure to a registered therapeutic agent is rare, but other effects, both detrimental and beneficial, are usually observed. By their very nature, drugs alter biological processes; therefore, the potential for harm is almost always present. The major consideration in establishing drug dose is to find a dose that has an adequate therapeutic effect without undesirable side effects. A dose-response curve can be established for a drug that progresses from noneffective levels through effective, harmful, and even lethal levels. A low slope for this curve indicates a wide range of effective dose and a wide **margin of safety** (see Figure 22.3). This term applies to other substances, such as pesticides, for which it is desirable to have a large difference between the dose that kills a target species and that which harms a desirable species.

22.4. REVERSIBILITY AND SENSITIVITY

Sublethal doses of most toxic substances are eventually eliminated from an organism's system. If there is no lasting effect from the exposure, it is said to be **reversible**. However, if the effect is permanent, it is termed **irreversible**. Irreversible effects of exposure remain after the toxic substance is eliminated from the organism. Figure 22.3 illustrates these two kinds of effects. For various chemicals and different subjects, toxic effects may range from the totally reversible to the totally irreversible.

Table 22.1. Toxicity Scale with Example Substances[1]

Substance	Approximate LD_{50}	Toxicity rating
	-10^5	1. Practically nontoxic $> 1.5 \times 10^4$ mg/kg
DEHP[2] $\longrightarrow$	$-$	
Ethanol $\longrightarrow$	-10^4	2. Slightly toxic, 5×10^3 to 1.5×10^4 mg/kg
Sodium chloride $\longrightarrow$	$-$	
Malathion $\longrightarrow$	-10^3	3. Moderately toxic, 500 to 5000 mg/kg
Chlordane $\longrightarrow$	$-$	
Heptachlor $\longrightarrow$	-10^2	4. Very toxic, 50 to 500 mg/kg
	$-$	
Parathion $\longrightarrow$	-10	
	$-$	5. Extremely toxic, 5 to 50 mg/kg
TEPP[3] $\longrightarrow$	-1	
	$-$	
Tetrodotoxin[4] $\longrightarrow$	-10^{-1}	
	$-$	
	-10^{-2}	6. Supertoxic, <5 mg/kg
	$-$	
TCDD[5] $\longrightarrow$	-10^{-3}	
	$-$	
	-10^{-4}	
	$-$	
Botulinus toxin $\longrightarrow$	-10^{-5}	

[1] Doses are in units of mg of toxicant per kg of body mass. Toxicity ratings on the right are given as numbers ranging from 1 (practically nontoxic) through 6 (supertoxic) along with estimated lethal oral doses for humans in mg/kg. Estimated LD_{50} values for substances on the left have been measured in test animals, usually rats, and apply to oral doses.

[2] Bis(2-ethylhexyl)phthalate

[3] Tetraethylpyrophosphate

[4] Toxin from pufferfish

[5] TCDD represents 2,3,7,8,-tetrachlorodibenzodioxin, commonly called "dioxin."

Hypersensitivity and Hyposensitivity

Examination of the dose-response curve shown in Figure 22.2 reveals that some subjects are very sensitive to a particular poison (for example, those killed at a dose corresponding to LD_5), whereas others are very resistant to the same substance (for example, those surviving a dose corresponding to LD_{95}). These two kinds of responses illustrate **hypersensitivity** and **hyposensitivity**, respectively; subjects in the mid-range of the dose-response curve are termed **normals**. These variations in response tend to complicate toxicology in that there is no specific dose guaranteed to yield a particular response, even in a homogeneous population.

In some cases hypersensitivity is induced. After one or more doses of a chemical, a subject may develop an extreme reaction to it. This occurs with penicillin, for example, in cases where people develop such a severe allergic response to the antibiotic that exposure is fatal if countermeasures are not taken.

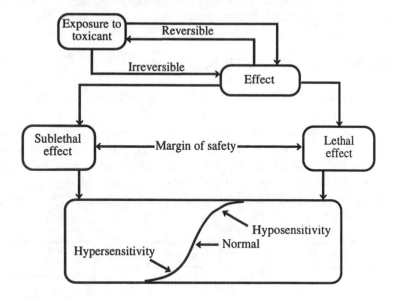

Figure 22.3. Effects of and responses to toxic substances.

22.5. XENOBIOTIC AND ENDOGENOUS SUBSTANCES

Xenobiotic substances are those that are foreign to a living system, whereas those that occur naturally in a biologic system are termed **endogenous**. The levels of endogenous substances must usually fall within a particular concentration range in order for metabolic processes to occur normally. Levels below a normal range may result in a deficiency response or even death, and the same effects may occur above the normal range. This kind of response is illustrated in Figure 22.4.

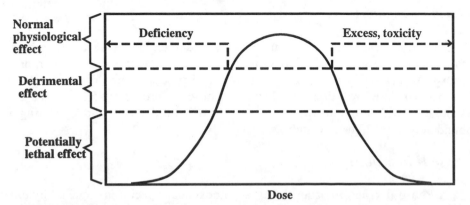

Figure 22.4. Biological effect of an endogenous substance in an organism showing optimum level, deficiency, and excess.

Examples of endogenous substances in organisms include various hormones, glucose (blood sugar), and some essential metal ions, including Ca^{2+}, K^+, and Na^+. The optimum level of calcium in human blood serum occurs over a rather narrow range of 9 – 9.5 milligrams per deciliter (mg/dL). Below these values a deficiency response known as hypocalcemia occurs, manifested by muscle cramping. At serum levels above about 10.5 mg/dL hypercalcemia occurs, the major effect of which is kidney malfunction.

22.6. TOXICOLOGICAL CHEMISTRY

Toxicological Chemistry

Toxicological chemistry is the science that deals with the chemical nature and reactions of toxic substances, including their origins, uses, and chemical aspects of exposure, fates, and disposal.[1] Toxicological chemistry addresses the relationships between the chemical properties and molecular structures of molecules and their toxicological effects. Figure 22.5 outlines the terms discussed above and the relationships among them.

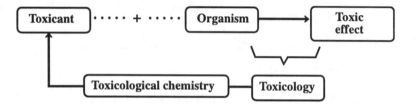
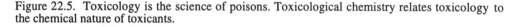

Figure 22.5. Toxicology is the science of poisons. Toxicological chemistry relates toxicology to the chemical nature of toxicants.

Toxicants in the Body

The processes by which organisms metabolize xenobiotic species are enzyme-catalyzed Phase I and Phase II reactions, which are described briefly here.[2,3]

Phase I Reactions

Lipophilic xenobiotic species in the body tend to undergo **Phase I reactions** that make them more water-soluble and reactive by the attachment of polar functional groups, such as –OH (Figure 22.6). Most Phase I processes are "microsomal mixed-function oxidase" reactions catalyzed by the cytochrome P-450 enzyme system associated with the **endoplasmic reticulum** of the cell and occurring most abundantly in the liver of vertebrates.[4]

Phase II Reactions

A **Phase II reaction** occurs when an endogenous species is attached by enzyme action to a polar functional group which often, though not always, is the result of a Phase I reaction on a xenobiotic species. Phase II reactions are called **conjugation**

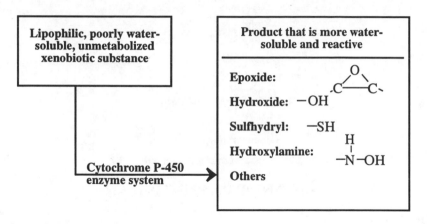

Figure 22.6. Illustration of Phase I reactions.

reactions in which enzymes attach **conjugating agents** to xenobiotics, their Phase I reaction products, and nonxenobiotic compounds (Figure 22.7). The **conjugation product** of such a reaction is usually less toxic than the original xenobiotic compound, less lipid-soluble, more water-soluble, and more readily eliminated from the body. The major conjugating agents and the enzymes that catalyze their Phase II reactions are glucuronide (UDP glucuronyltransferase enzyme), glutathione (glutathionetransferase enzyme), sulfate (sulfotransferase enzyme), and acetyl (acetylation by acetyltransferase enzymes). The most abundant conjugation products are glucuronides. A glucuronide conjugate is illustrated in Figure 22.8, where -X-R represents a xenobiotic species conjugated to glucuronide, and R is an organic moiety. For example, if the xenobiotic compound conjugated is phenol, HXR is HOC_6H_5, X is the O atom, and R represents the phenyl group, C_6H_5.

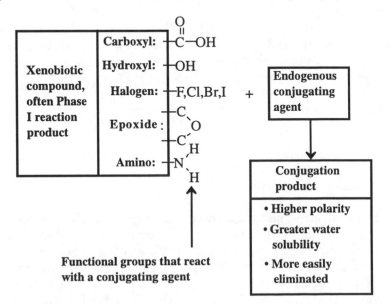

Figure 22.7. Illustration of Phase II reactions.

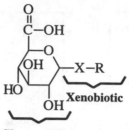

Figure 22.8. Glucuronide conjugate formed from a xenobiotic, HX-R.

22.7. KINETIC PHASE AND DYNAMIC PHASE

Kinetic Phase

The major routes and sites of absorption, metabolism, binding, and excretion of toxic substances in the body are illustrated in Figure 22.1. Toxicants in the body are metabolized, transported, and excreted; they have adverse biochemical effects; and they cause manifestations of poisoning. It is convenient to divide these processes into two major phases, a kinetic phase and a dynamic phase.

In the **kinetic phase**, a toxicant or the metabolic precursor of a toxic substance (**protoxicant**) may undergo absorption, metabolism, temporary storage, distribution, and excretion, as illustrated in Figure 22.9. A toxicant that is absorbed may be passed through the kinetic phase unchanged as an **active parent compound**, metabolized to a **detoxified metabolite** that is excreted, or converted to a toxic **active metabolite**. These processes occur through Phase I and Phase II reactions discussed above.

Dynamic Phase

In the **dynamic phase** (Figure 22.10) a toxicant or toxic metabolite interacts with cells, tissues, or organs in the body to cause some toxic response. The three major subdivisions of the dynamic phase are the following:

- **Primary reaction** with a receptor or target organ
- A **biochemical response**
- **Observable effects.**

Primary Reaction in the Dynamic Phase

A toxicant or an active metabolite reacts with a receptor. The process leading to a toxic response is initiated when such a reaction occurs. A typical example is when benzene epoxide, the initial product of the Phase I reaction of benzene (see Chapter 23, Figure 23.1), forms an adduct with a nucleic acid unit in DNA (receptor) resulting in alteration of the DNA. (Many species that cause a toxic response are reactive intermediates, such as benzene epoxide, which have a brief lifetime but a strong tendency to undergo reactions leading to a toxic response while they are around.[5])

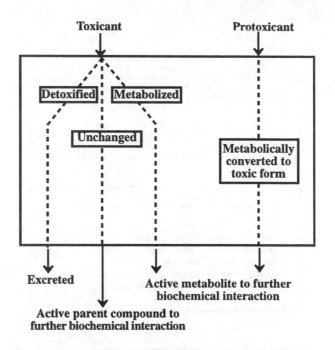

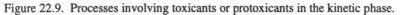

Figure 22.9. Processes involving toxicants or protoxicants in the kinetic phase.

This reaction is an **irreversible** reaction between a toxicant and a receptor. A **reversible** reaction that can result in a toxic response is illustrated by the binding between carbon monoxide and oxygen-transporting hemoglobin (Hb) in blood:

$$O_2Hb + CO \leftarrow\rightarrow COHb + O_2 \qquad\qquad (22.7.1)$$

Biochemical Effects in the Dynamic Phase

The binding of a toxicant to a receptor may result in some kind of biochemical effect. The major ones are the following:

- Impairment of enzyme function by binding to the enzyme, coenzymes, metal activators of enzymes, or enzyme substrates

- Alteration of cell membrane or carriers in cell membranes

- Interference with carbohydrate metabolism

- Interference with lipid metabolism resulting in excess lipid accumulation ("fatty liver")

- Interference with respiration, the overall process by which electrons are transferred to molecular oxygen in the biological oxidation of energy-yielding substrates

- Stopping or interfering with protein biosynthesis by their action on DNA

- Interference with regulatory processes mediated by hormones or enzymes.

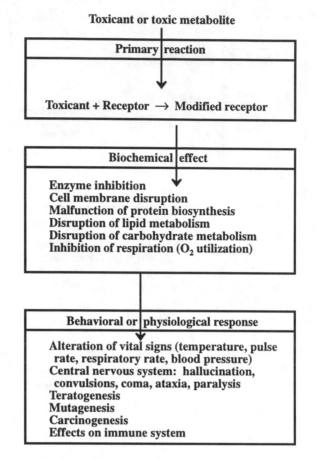

Figure 22.10. The dynamic phase of toxicant action.

Responses to Toxicants

Among the more immediate and readily observed manifestations of poisoning are alterations in the **vital signs** of **temperature, pulse rate, respiratory rate,** and **blood pressure**. Poisoning by some substances may cause an abnormal skin color (jaundiced, yellow skin from CCl_4 poisoning) or excessively moist or dry skin. Toxic levels of some materials or their metabolites cause the body to have unnatural **odors,** such as the bitter almond odor of HCN in tissues of victims of cyanide poisoning. Symptoms of poisoning manifested in the eye include **miosis** (excessive or prolonged contraction of the eye pupil), **mydriasis** (excessive pupil dilation), **conjunctivitis** (inflammation of the mucus membrane that covers the front part of the eyeball and the inner lining of the eyelids) and **nystagmus** (involuntary movement of the eyeballs). Some poisons cause a moist condition of the mouth, whereas others cause a dry mouth. Gastrointestinal tract effects including pain, vomiting, or paralytic ileus (stoppage of the normal peristalsis movement of the intestines) occur as a result of poisoning by a number of toxic substances.

Central nervous system poisoning may be manifested by **convulsions**, **paralysis**, **hallucinations**, and **ataxia** (lack of coordination of voluntary movements of the body), as well as abnormal behavior, including agitation, hyperactivity, disorientation, and delirium. Severe poisoning by some substances, including organophosphates and carbamates, causes **coma**, the term used to describe a lowered level of consciousness.

Prominent among the more chronic responses to toxicant exposure are mutations, cancer, and birth defects and effects on the immune system. Other observable effects, some of which may occur soon after exposure, include gastrointestinal illness, cardiovascular disease, hepatic (liver) disease, renal (kidney) malfunction, neurologic symptoms (central and peripheral nervous systems), and skin abnormalities (rash, dermatitis).

Often the effects of toxicant exposure are subclinical in nature. The most common of these are some kinds of damage to the immune system, chromosomal abnormalities, modification of functions of liver enzymes, and slowing of conduction of nerve impulses.

22.8. TERATOGENESIS, MUTAGENESIS, CARCINOGENESIS, AND EFFECTS ON THE IMMUNE AND REPRODUCTIVE SYSTEMS

Teratogenesis

Teratogens are chemical species that cause birth defects. These usually arise from damage to embryonic or fetal cells. However, mutations in germ cells (egg or sperm cells) may cause birth defects, such as Down's syndrome.

The biochemical mechanisms of teratogenesis are varied. These include enzyme inhibition by xenobiotics; deprivation of the fetus of essential substrates, such as vitamins; interference with energy supply; or alteration of the permeability of the placental membrane.

Mutagenesis

Mutagens alter DNA to produce inheritable traits. Although mutation is a natural process that occurs even in the absence of xenobiotic substances, most mutations are harmful. The mechanisms of mutagenicity are similar to those of carcinogenicity, and mutagens often cause birth defects as well. Therefore, mutagenic hazardous substances are of major toxicological concern.

Biochemistry of Mutagenesis

To understand the biochemistry of mutagenesis, it is important to recall from Chapter 21 that DNA contains the nitrogenous bases adenine, guanine, cytosine, and thymine. The order in which these bases occur in DNA determines the nature and structure of newly produced RNA, a substance generated as a step in the synthesis of new proteins and enzymes in cells. Exchange, addition, or deletion of any of the nitrogenous bases in DNA alters the nature of RNA produced and can change vital life processes, such as the synthesis of an important enzyme. This phenomenon, which can be caused by xenobiotic compounds, is a mutation that can be passed on to progeny, usually with detrimental results.

There are several ways in which xenobiotic species may cause mutations. It is beyond the scope of this work to discuss these mechanisms in detail. For the most part, however, mutations due to xenobiotic substances are the result of chemical alterations of DNA, such as those discussed in the two examples below.

Nitrous acid, HNO_2, is an example of a chemical mutagen that is often used to cause mutations in bacteria. To understand the mutagenic activity of nitrous acid it should be noted that three of the nitrogenous bases—adenine, guanine, and cytosine—contain the amino group, $-NH_2$. The action of nitrous acid is to replace amino groups with a hydroxy group. When this occurs, the DNA may not function in the intended manner, causing a mutation to occur.

Alkylation, the attachment of a small alkyl group, such as $-CH_3$ or $-C_2H_5$, to an N atom on one of the nitrogenous bases in DNA is one of the most common mechanisms leading to mutation. The methylation of "7" nitrogen in guanine in DNA to form N–Methylguanine is shown in Figure 22.11. O-alkylation may also occur by attachment of a methyl or other alkyl group to the oxygen atom in guanine.

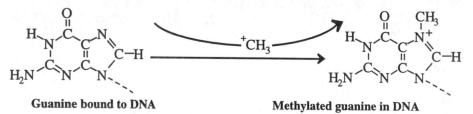

Guanine bound to DNA   **Methylated guanine in DNA**

Figure 22.11. Alkylation of guanine in DNA.

A number of mutagenic substances act as alkylating agents. Prominent among these are the compounds shown in Figure 22.12.

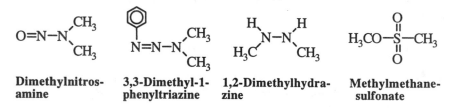

Dimethylnitros- **3,3-Dimethyl-1-** **1,2-Dimethylhydra-** **Methylmethane-**
amine   **phenyltriazine** **zine**   **sulfonate**

Figure 22.12. Examples of simple alkylating agents capable of causing mutations.

Alkylation occurs by way of generation of positively charged electrophilic species that bond to electron-rich nitrogen or oxygen atoms on the nitrogenous bases in DNA. The generation of such species usually occurs by way of biochemical and chemical processes. For example, dimethylnitrosamine (structure in Figure 22.12) is activated by oxidation through cellular NADPH to produce the following highly reactive intermediate:

This product undergoes several nonenzymatic transitions, losing formaldehyde and generating a methyl carbocation, $^+CH_3$, that can methylate nitrogenous bases on DNA

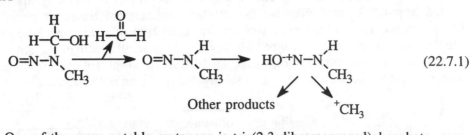

$$(22.7.1)$$

One of the more notable mutagens is tris(2,3-dibromopropyl)phosphate, commonly called "tris," that was used as a flame retardant in children's sleepwear. Tris was found to be mutagenic in experimental animals and metabolites of it were found in children wearing the treated sleepwear. This strongly suggested that tris is absorbed through the skin and its uses were discontinued.

Carcinogenesis

Cancer is a condition characterized by the uncontrolled replication and growth of the body's own (somatic) cells. **Carcinogenic agents** may be categorized as follows:

- Chemical agents, such as nitrosamines and polycyclic aromatic hydrocarbons

- Biological agents, such as hepadnaviruses or retroviruses

- Ionizing radiation, such as X-rays

- Genetic factors, such as selective breeding.

Clearly, in some cases, cancer is the result of the action of synthetic and naturally occurring chemicals. The role of xenobiotic chemicals in causing cancer is called **chemical carcinogenesis**. It is often regarded as the single most important facet of toxicology and is clearly the one that receives the most publicity.

Chemical carcinogenesis has a long history. As noted earlier in this chapter, in 1775 Sir Percival Pott, Surgeon General serving under King George III of England, observed that chimney sweeps in London had a very high incidence of cancer of the scrotum, which he related to their exposure to soot and tar from the burning of bituminous coal. Around 1900 a German surgeon, Ludwig Rehn, reported elevated incidences of bladder cancer in dye workers exposed to chemicals extracted from coal tar; 2-naphthylamine,

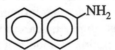

was shown to be largely responsible. Other historical examples of carcinogenesis include observations of cancer from tobacco juice (1915), oral exposure to radium from painting luminescent watch dials (1929), tobacco smoke (1939), and asbestos (1960).

Biochemistry of Carcinogenesis

Large expenditures of time and money on the subject in recent years have yielded a much better understanding of the biochemical bases of chemical carcinogenesis. The overall processes for the induction of cancer may be quite complex, involving numerous steps.[6] However, it is generally recognized that there are two major steps in carcinogenesis: an **initiation stage** followed by a **promotional stage**. These steps are further subdivided as shown in Figure 22.13.

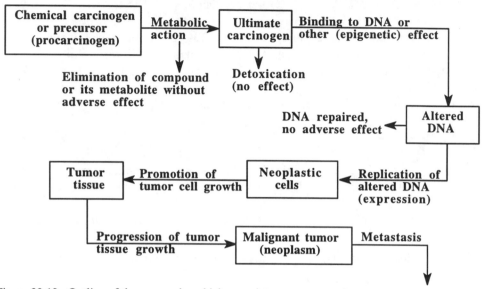

Figure 22.13. Outline of the process by which a carcinogen or procarcinogen may cause cancer.

Initiation of carcinogenesis may occur by reaction of a **DNA-reactive species** with DNA,[7] or by the action of an **epigenetic carcinogen** that does not react with DNA and is carcinogenic by some other mechanism.[8] Most DNA-reactive species are **genotoxic carcinogens** because they are also mutagens. These substances react irreversibly with DNA. They are either electrophilic or, more commonly, metabolically activated to form electrophilic species, as is the case with electrophilic $^+CH_3$ generated from dimethylnitrosamine, discussed under mutagenesis above. Cancer-causing substances that require metabolic activation are called **procarcinogens**. The metabolic species actually responsible for carcinogenesis is termed an **ultimate carcinogen**. Some species that are intermediate metabolites between precarcinogens and ultimate carcinogens are called **proximate carcinogens**. Carcinogens that do not require biochemical activation are categorized as **primary** or **direct-acting carcinogens**. Some procarcinogens and primary carcinogens are shown in Figure 22.14.

Most substances classified as epigenetic carcinogens are **promoters** that act after initiation. Manifestations of promotion include increased numbers of tumor cells and decreased length of time for tumors to develop (shortened latency period). Promoters do not initiate cancer, are not electrophilic, and do not bind with DNA. The classic example of a promoter is a substance known chemically as decanoyl phorbol acetate or phorbol myristate acetate, Which is extracted from croton oil.

Naturally occurring carcinogens that require bioactivation

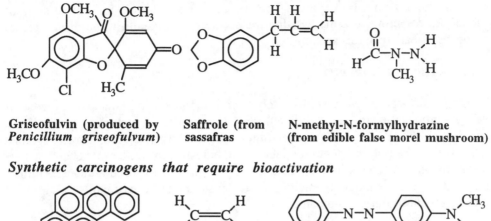

| Griseofulvin (produced by *Penicillium griseofulvum*) | Saffrole (from sassafras | N-methyl-N-formylhydrazine (from edible false morel mushroom) |

Synthetic carcinogens that require bioactivation

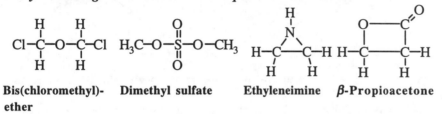

Benzo(a)pyrene   Vinyl chloride   4-Dimethylaminoazobenzene

Primary carcinogens that do not require bioactivation

Bis(chloromethyl)-   Dimethyl sulfate   Ethyleneimine   β-Propioacetone
ether

Figure 22.14. Examples of the major classes of naturally occurring and synthetic carcinogens, some of which require bioactivation, and others which act directly.

Alkylating Agents in Carcinogenesis

Chemical carcinogens usually have the ability to form covalent bonds with macromolecular life molecules.[9] Such covalent bonds can form with proteins, peptides, RNA, and DNA. Although most binding is with other kinds of molecules, which are more abundant, the DNA adducts are the significant ones in initiating cancer. Prominent among the species that bond to DNA in carcinogenesis are the alkylating agents which attach alkyl groups—methyl (CH_3) or ethyl (C_2H_5)—to DNA. A similar type of compound, **arylating agents**, act to attach aryl moieties, such as the phenyl group

 Phenyl group

to DNA. As shown by the examples in Figure 22.15, the alkyl and aryl groups become attached to N and O atoms in the nitrogenous bases that compose DNA. This alteration in DNA can trigger initiation of the sequence of events that results in the growth and replication of neoplastic (cancerous) cells. The reactive species that donate alkyl groups in alkylation are usually formed by metabolic activation through the action of enzymes. This process was shown for conversion of dimethylnitrosamine to a methylating metabolic intermediate in the discussion of mutagenesis earlier in this section.

Testing for Carcinogens

Only a few chemicals have definitely been established as human carcinogens. A well documented example is vinyl chloride, $CH_2=CHCl$, which is known to have caused a rare form of liver cancer (angiosarcoma) in individuals who cleaned autoclaves in the polyvinylchloride fabrication industry. In some cases chemicals are known to be carcinogens from epidemiological studies of exposed humans. Animals are used to test for carcinogenicity, and the results can be extrapolated, although with much uncertainty, to humans.

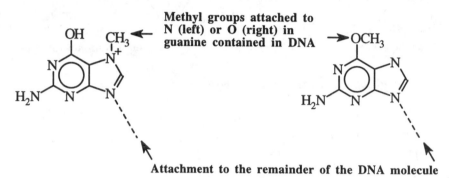

Figure 22.15. Alkylated (methylated) forms of the nitrogenous base guanine.

Bruce Ames Test

Mutagenicity used to infer carcinogenicity is the basis of the **Bruce Ames** test, in which observations are made of the reversion of mutant histidine-requiring *Salmonella* bacteria back to a form that can synthesize its own histidine.[10] The test makes use of enzymes in homogenized liver tissue to convert potential procarcinogens to ultimate carcinogens. Histidine-requiring *Salmonella* bacteria are inoculated onto a medium that does not contain histidine, and those that mutate back to a form that can synthesize histidine establish visible colonies that are assayed to indicate mutagenicity.

According to Bruce Ames, the pioneer developer of the test which bears his name, animal tests for carcinogens that make use of massive doses of chemicals have a misleading tendency to give results that cannot be accurately extrapolated to assess cancer risks from smaller doses of chemicals.[11] This is because the huge doses of chemicals used kill large numbers of cells, which the organism's body attempts to replace with new cells. Rapidly dividing cells greatly increase the likelihood of mutations that result in cancer simply as the result of rapid cell proliferation, not genotoxicity.

Immune System Response

The **immune system** acts as the body's natural defense system to protect it from xenobiotic chemicals; infectious agents, such as viruses or bacteria; and neoplastic cells, which give rise to cancerous tissue. Adverse effects on the body's immune system are being increasingly recognized as important consequences of exposure to hazardous substances.[12] Toxicants can cause **immunosuppression**, which is the impairment of the body's natural defense mechanisms. Xenobiotics can also cause the immune system to lose its ability to control cell proliferation, resulting in leukemia or lymphoma.

Another major toxic response of the immune system is **allergy** or **hypersensitivity**. This kind of condition results when the immune system overreacts to the presence of a foreign agent or its metabolites in a self-destructive manner. Among the xenobiotic materials that can cause such reactions are beryllium, chromium, nickel, formaldehyde, some kinds of pesticides, resins, and plasticizers.

Estrogenic Substances

A number of xenobiotic substances are thought to have adverse effects on animal and human reproductive systems by mimicking or interfering with the action of estrogens (see Section 21.5). Rodent experiments indicate that such substances, sometimes called exogenous estrogens or exoestrogens, may cause disorders of the reproductive tract and effects such as reduced sperm counts and semen production.[13] Another effect of some concern is the potential to cause hormone-dependent cancers. A wide variety of synthetic compounds, including phthalates, alkylphenols, organochlorine compounds, and polycyclic aromatic hydrocarbons, are suspected of being exoestrogens.

22.9. HEALTH HAZARDS

In recent years attention in toxicology has shifted away from readily recognized, usually severe, acute maladies that developed on a short time scale as a result of brief, intense exposure to toxicants, toward delayed, chronic, often less severe illnesses caused by long-term exposure to low levels of toxicants. Although the total impact of the latter kinds of health effects may be substantial, their assessment is very difficult because of factors such as uncertainties in exposure, low occurrence above background levels of disease, and long latency periods.

Assessment of Potential Exposure

A critical step in assessing exposure to toxic substances, such as those from hazardous waste sites is evaluation of potentially exposed populations. The most direct approach to this is to determine chemicals or their metabolic products in organisms. For inorganic species this is most readily done for heavy metals, radionuclides, and some minerals, such as asbestos. Symptoms associated with exposure to particular chemicals may also be evaluated. Examples of such effects include readily apparent effects, for example, skin rashes, or subclinical effects, such as chromosomal damage.

Epidemiological Evidence

Epidemiological studies applied to toxic environmental pollutants, such as those from hazardous wastes, attempt to correlate observations of particular illnesses with probable exposure to such wastes. There are two major approaches to such studies. One approach is to look for diseases known to be caused by particular agents in areas where exposure is likely from such agents in hazardous wastes. A second approach is to look for **clusters** consisting of an abnormally large number of cases of a particular disease in a limited geographic area, then attempt to locate sources of exposure to hazardous wastes that may be responsible. The most common types of maladies observed in clusters are spontaneous abortions, birth defects, and particular types of cancer.

Epidemiologic studies are complicated by long latency periods from exposure to onset of disease, lack of specificity in the correlation between exposure to a particular waste and the occurrence of a disease, and background levels of a disease in the absence of exposure to a hazardous waste capable of causing the disease.

Estimation of Health Effects Risks

An important part of estimating the risks of adverse health effects from exposure to toxicants involves extrapolation from experimentally observable data. Usually the end result needed is an estimate of a low occurrence of a disease in humans after a long latency period resulting from low-level exposure to a toxicant for a long period of time. The data available are almost always taken from animals exposed to high levels of the substance for a relatively short period of time. Extrapolation is then made using linear or curvilinear projections to estimate the risk to human populations. There are, of course, very substantial uncertainties in this kind of approach.

Risk Assessment

Toxicological considerations are very important in estimating potential dangers of pollutants and hazardous waste chemicals. One of the major ways in which toxicology interfaces with the area of hazardous wastes is in **health risk assessment**, providing guidance for risk management, cleanup, or regulation needed at a hazardous waste site based upon knowledge about the site and the chemical and toxicological properties of wastes in it. Risk assessment includes the factors of site characteristics; substances present, including indicator species; potential receptors; potential exposure pathways; and uncertainty analysis. It may be divided into the following major components:

- Identification of hazard
- Dose-response assessment
- Exposure assessment
- Risk characterization.

LITERATURE CITED

1. Manahan, Stanley E., *Toxicological Chemistry* , 2nd ed., Lewis Publishers/CRC Press, Boca Raton, FL, 1992.

2. Gonzalez, Frank J., "The Study of Xenobiotic-Metabolizing Enzymes and their Role in Toxicity In Vivo Using Targeted Gene Disruption," *Toxicology Letters*, **102**, 161-166 (1998).

3. Millburn, P., "The Fate of Xenobiotics in Mammals: Biochemical Processes," *Progress in Pesticide Biochemistry and Toxicology*, **8**, 1-86 (1995).

4. Ioannides, Costas, Ed., *Cytochromes P450: Metabolic and Toxicological Aspects*, CRC Press, Boca Raton, FL, 1996.

5. Snyder, Robert, Ed., *Biological Reactive Intermediates V: Basic Mechanistic Research in Toxicology and Human Risk Assessment*, Plenum Press, New York, 1996.

6. Arcos, Joseph C., Mary F. Argus, and Yin-tak Woo, Eds., *Chemical Induction of Factors Which Influence Carcinogenesis*, Birkhauser, Boston, 1995

7. Hemminki, K., Ed., *DNA Adducts: Identification and Biological Significance*, Oxford University Press, New York, 1994.

9. Pitot, Henry C., III, and Yvonne P. Dragan, "Chemical Carcinogenesis," Chapter 8 in *Casarett and Doull's Toxicology*, 5th ed., Curtis D. Klaassen, Mary O. Amdur, and John Doull, Eds., McGraw-Hill, New York, 1996, pp. 201-268.

10. Ames, Bruce N., "The Detection of Environmental Mutagens and Potential Carcinogens," *Cancer*, **53**, 1034-1040 (1984).

11. "Tests of Chemicals on Animals are Unreliable as Predictors of Cancer in Humans," *Environmental Science and Technology* **24**, 1990 (1990).

12. Griem, Peter, Marty Wulferink, Bernhardt Sachs, Jose B. Gonzalez, and Ernst Gleichmann, "Allergic and Autoimmune Reactions to Xenobiotics: How Do They Arise?," *Immunology Today*, **19**, 133-141 (1998).

13. Zacharewski, Tim, "*In Vitro* Bioassays for Assessing Estrogenic Substances," *Environmental Science and Technology*, **31**, 613-623 (1997).

SUPPLEMENTARY REFERENCES

Ballantyne, Bryan, Timothy Marrs, and Paul Turner, *General and Applied Toxicology: College Edition*, Stockton Press, New York, 1995.

Baselt, Randall C. and Robert H. Cravey, *Disposition of Toxic Drugs and Chemicals in Man*, Chemical Toxicology Institute, Foster City, CA, 1995.

Brandenberger, Hans and Robert A. Maes, Eds., *Analytical Toxicology: For Clinical, Forensic, and Pharmaceutical Chemists*, W. de Gruyter, Berlin, 1997.

Carey, John, Ed., *Ecotoxicological Risk Assessment of the Chlorinated Organic Chemicals*, SETAC Press, Pensacola, FL, 1998.

Cockerham, Lorris G., and Barbara S. Shane, *Basic Environmental Toxicology*, CRCPress/Lewis Publishers, Boca Raton, FL, 1994.

Cooper, Andre R., Leticia Overholt, Heidi Tillquist, and Douglas Jamison, *Cooper's Toxic Exposures Desk Reference with CD-ROM*, CRC Press/Lewis Publishers, Boca Raton, FL,1997.

Crompton, Thomas Roy, *Toxicants in The Aqueous Ecosystem*, John Wiley & Sons, New York, 1997.

Davidson, Victor L. and Donald B. Stittman, *Biochemistry*, 4th ed., Lippincott William & Wilkins, Philadelphia, 1999.

Draper, William M., Ed., *Environmental Epidemiology*, American Chemical Society, Washington, D.C., 1994.

Ellenhorn, Matthew J. and Sylvia Syma Ellenhorn, *Ellenhorn's Medical Toxicology: Diagnosis and Treatment of Human Poisoning*, 2nd ed., Williams & Wilkins, Baltimore, 1997.

Greenberg, Michael I., Richard J. Hamilton, and Scott D. Phillips, Eds., *Occupational, Industrial, and Environmental Toxicology*, Mosby, St. Louis,1997.

Hall, Stephen K., Joanna Chakraborty, and Randall J. Ruch, Eds., *Chemical Exposure and Toxic Responses*, CRC Press/Lewis Publishers, Boca Raton, FL, 1997.

Hodgson, Ernest, and Patricia E. Levi, *Introduction to Biochemical Toxicology*, 2nd Ed., Appleton & Lange, Norwalk, CT., 1994.

Johnson, Barry L., *Impact of Hazardous Waste on Human Health: Hazard, Health Effects, Equity and Communication Issues*, Ann Arbor Press, Chelsea, MI, 1999.

Jorgensen, Sven Erik, B. Halling-Sorensen, and Henrick Mahler, *Handbook of Estimation Methods in Ecotoxicology and Environmental Chemistry*, CRC Press/Lewis Publishers, Boca Raton, FL, 1998.

Keith, Lawrence H., *Environmental Endocrine Disruptors: A Handbook of Property Data*, John Wiley & Sons, New York,1997.

Kent, Chris, *Basics of Toxicology*, John Wiley & Sons, New York, 1998.

Klaassen, Curtis D., Mary O. Amdur, and John Doull, Eds., *Casarett and Doull's Toxicology: The Basic Science of Poisons Companion*, 5th ed., McGraw-Hill, Health Professions Division, New York, 1996.

Klaassen, Curtis D. and John B. Watkins, III, Eds., *Casarett and Doull's Toxicology: The Basic Science of Poisons Companion Handbook*, 5th ed., McGraw-Hill, Health Professions Division, New York, 1999.

Kneip, Theodore J., and John V. Crable, Eds., *Methods for Biological Monitoring: A Manual for Assessing Human Exposure to Hazardous Substances*, American Public Health Association, Washington, D.C., 1988.

Landis, Wayne G. and Ming-Ho Yu, *Introduction to Environmental Toxicology: Impacts of Chemicals upon Ecological Systems*, 2nd ed., CRC Press/Lewis Publishers, Boca Raton, FL, 1999.

Lippmann, Morton, Ed., *Environmental Toxicants: Human Exposures and their Health Effects*, 2nd ed., John Wiley & Sons, New York, 1999.

Linardakis, Nikos M. and Christopher R. Wilson, *Biochemistry*, 2nd ed., McGraw-Hill, New York, 1998.

Liverman, Catharyn T., Ed., *Toxicology and Environmental Health Information Resources: The Role of the National Library of Medicine*, National Academy Press, Washington, D.C., 1997.

Lu, Frank C., *Basic Toxicology: Fundamentals, Target Organs, and Risk Assessment*, 3rd ed., Taylor & Francis, London, U.K., 1996.

Malachowski, M. J., *Health Effects of Toxic Substances*, Government Institutes, Rockville, MD, 1995.

Marks, Dawn B., *Biochemistry*, 3rd ed., Williams & Wilkins, Baltimore, 1999.

Milman, Harry A. and Elizabeth K. Weisburger, *Handbook of Carcinogen Testing*, 2nd ed., Noyes Publications, Park Ridge, NJ, 1994.

Nichol, John, *Bites and Stings. The World of Venomous Animals*, Facts on File, New York, 1989.

Ostler, Neal K., Thomas E. Byrne, and Michael J. Malachowski, *Health Effects of Hazardous Materials*, Neal K. Ostler, Prentice Hall, Upper Saddle River, NJ,1996.

Ostrander, Gary K., Ed., *Techniques in Aquatic Toxicology*, CRC Press/Lewis Publishers, Boca Raton, FL, 1996.

Patnaik, Pradyot, *A Comprehensive Guide to the Hazardous Properties of Chemical Substances*, 2nd ed., John Wiley & Sons, Inc., New York, 1999.

Richardson, Mervyn, *Environmental Xenobiotics*, Taylor and Francis, London, 1996.

Richardson, Mervyn, *Environmental Toxicology Assessment*, Taylor & Francis, London, 1995.

Rom, William N., Ed., *Environmental & Occupational Medicine*, 3rd ed., Lippincott-Raven Publishers, Philadelphia, 1998.

Shaw, Ian C. and John Chadwick, *Principles of Environmental Toxicology*, Taylor & Francis, London, 1998.

Shibamoto, Takayuki, *Chromatographic Analysis of Environmental and Food Toxicants*, Marcel Dekker, New York, 1998.

Stine, Karen E. and Thomas M. Brown, *Principles of Toxicology*, CRC Press/Lewis Publishers, Boca Raton, FL, 1996.

Sipes, I. Glenn, Charlene A. McQueen, and A. Jay Gandolfi, *Comprehensive Toxicology*, Pergamon, New York, 1997.

Timbrell, John A., *Introduction to Toxicology*, 2nd ed., Taylor & Francis, London, 1995.

Vallejo Rosero, Maria del Carmen, *Toxicologia Ambiental: Fuentes, Cinetica y Efectos de los Contaminantes*, Santafe de Bogota: Fondo Nacional Universitario, Bogota, Columbia, 1997.

Wexler, Philip, Ed., *Encyclopedia of Toxicology*, Academic Press, San Diego, CA, 1998.

Williamson, John A., Peter J. Fenner, Joseph W. Burnett, and Jacqueline F. Rifkin, Eds, *Venomous and Poisonous Marine Animals: A Medical and Biological Handbook*, University of New South Wales Press, Sydney, Australia , 1996.

Zelikoff, Judith, and Peter L. Thomas, Eds., *Immunotoxicology of Environmental and Occupational Metals*, Taylor and Francis, London, 1998.

QUESTIONS AND PROBLEMS

1. How are conjugating agents and Phase II reactions involved with some toxicants?

2. What is the toxicological importance of proteins, particularly as related to protein structure?

3. What is the toxicological importance of lipids? How are lipids related to hydrophobic pollutants and toxicants?

4. What are Phase I reactions? What enzyme system carries them out? Where is this enzyme system located in the cell?

5. Name and describe the science that deals with the chemical nature and reactions of toxic substances, including their origins, uses, and chemical aspects of exposure, fates, and disposal.

6. What is a dose-response curve?

7. What is meant by a toxicity rating of 6?

8. What are the three major subdivisions of the *dynamic phase* of toxicity, and what happens in each?

9. Characterize the toxic effect of carbon monoxide in the body. Is its effect reversible or irreversible? Does it act on an enzyme system?

10. Of the following, choose the one that is **not** a biochemical effect of a toxic substance: (a) impairment of enzyme function by binding to the enzyme, (b) alteration of cell membrane or carriers in cell membranes, (c) change in vital signs, (d) interference with lipid metabolism, (e) interference with respiration.

11. Distinguish among teratogenesis, mutagenesis, carcinogenesis, and immune system effects. Are there ways in which they are related?

12. As far as environmental toxicants are concerned, compare the relative importance of acute and chronic toxic effects and discuss the difficulties and uncertainties involved in studying each.

13. What are some of the factors that complicate epidemiologic studies of toxicants?

23 TOXICOLOGICAL CHEMISTRY OF CHEMICAL SUBSTANCES

23.1. INTRODUCTION

Toxicological chemistry, defined and discussed in Chapter 22, centers on the relationship between the chemical nature of toxicants and their toxicological effects. This chapter discusses this relationship with regard to some of the major pollutants and hazardous substances. The first section deals with toxicological aspects of elements (particularly heavy metals) the presence of which in a compound frequently means that the compound is toxic. It also discusses the toxicities of some commonly used elemental forms, such as the chemically uncombined elemental halogens. The following section discusses the toxicological chemistry of inorganic compounds, many of which are produced from industrial processes. It also provides a brief discussion of organometallic compounds.The last section deals with the toxicology of organic compounds. The toxicological properties of hydrocarbons and oxygen-containing organic compounds are discussed as well as other organic substances containing functional groups, such as alcohols and ketones. This section also discusses the toxicities of organic nitrogen, halide, sulfur, and phosphorous compounds, some of which are used as pesticides or military poisons.

ATSDR Toxicological Profiles

A very useful source of information about the toxicological chemistry of various kinds of toxic substances is published by the U. S. Department of Health and Human Services, Public Health Service Agency for Toxic Substances and Disease Registry ASTDR's Toxicological Profiles. These detailed documents are available on CD-ROM.[1] The substances covered are given in Table 23.1.

Table 23.1. Materials Listed by ATSDR[1]

Acetone	1,2-Dibromoethane	Naphthalene
Acrolein	1,4-Dichlorobenzene	Nickel
Acrylonitrile	3,3'-Dichlorobenzidine	Nitrobenzene
Aldrin/Dieldrin	1,1-Dichloroethane	2-Nitrophenol/
Alpha-,Beta-,Gamma-	1,2-Dichloroethane	4-Nitrophenol
and Delta-Hexachloro-	1,1-Dichloroethene	Otto Fuels
cyclohexane	1,2-Dichloroethene	Pentachlorophenol
Aluminum	1,3-Dichloropropene	Phenol
Ammonia	Diethyl Phthalate	Plutonium
Arsenic	1,3-Dinitrobenzene/	Polybrominated
Asbestos	1,3,5-Trinitrobenzene	Biphenyls
Automotive Gasoline	Dinitrocresols	Polychlorinated
Barium	Dinitrophenols	Biphenyls
Benzene	2,4-Dinitrotoluene/	Polycyclic Aromatic
Benzidine	2,6-Dinitrotoluene	Hydrocarbons (PAH's)
Beryllium	1,2-Diphenylhydrazine	Radon
Bis(2-Chloroethyl) Ether	Disulfoton	RDX
Boron	Endosulfan	Selenium
Bromomethane	Endrin	Silver
1,3-Butadiene	Ethylbenzene	Stoddard Solvent
2-Butanone	Ethylene Glycol and	1,1,2,2-Tetrachloroethane
Cadmium	Propylene Glycol	Tetrachloroethylene
Carbon Disulfide	Fluorides, Hydrogen	Tetryl
Carbon Tetrachloride	Fluoride, and Fluorine	Thallium
Chlordane	Fuel Oils	Thorium
Chlorobenzene	Heptachlor/Heptachlor	Tin
Chlorodibenzofurans	Epoxide	Titanium Tetrachloride
Chloroethane	Hexachlorobenzene	Toluene
Chloroform	Hexachlorobutadiene	Toxaphene
Chloromethane	2-Hexanone	1,1,1-Trichloroethane
Chlorpyrifos	Hydraulic Fluids	1,1,2-Trichloroethane
Chromium	Isophorone	Trichloroethylene
Coal Tar Pitch, and	Jet Fuels (Jp4 And Jp7)	2,4,6-Trichlorophenol
Coal Tar Pitch Volatiles	Lead	2,4,6-Trinitrotoluene
Cobalt	Manganese	Uranium
Copper	Mercury	Used Mineral-Based
Cresols: *o*-Cresol, *p*-	Methoxychlor	Crankcase Oil
Cresol, *m*-Cresol	Methyl Parathion	Vanadium
Cyanide	Methyl Tert-Butyl Ether	Vinyl Acetate
4,4'-Ddt,4,4'-Dde,4,4'-Ddd	4, 4'-Methylenebis-(2-	Vinyl Chloride
Di (2-Ethylhexyl) Phthalate	Chloroaniline) (MBOCA)	White Phosphorus
Di-N-Butylphthalate	Methylene Chloride	Wood Creosote, Coal Tar
Diazinon	Mirex And chlordecone	Creosote, Coal Tar
1,2-Dibromo-	N-Nitrosodi-N-Propylamine	Xylenes
3-Chloropropane	N-Nitrosodiphenylamine	Zinc

23.2. TOXIC ELEMENTS AND ELEMENTAL FORMS

Ozone

Ozone (O_3, see Chapters 9, 13, and 14) has several toxic effects. Air containing 1 ppm by volume ozone has a distinct odor. Inhalation of ozone at this level causes severe irritation and headache. Ozone irritates the eyes, upper respiratory system, and lungs. Inhalation of ozone can sometimes cause fatal pulmonary edema. *Pulmonary* refers to lungs and *edema* to an accumulation of fluid in tissue spaces; therefore, *pulmonary edema* is an abnormal accumulation of fluid in lung tissue. Chromosomal damage has been observed in subjects exposed to ozone.

Ozone generates free radicals in tissue. These reactive species can cause lipid peroxidation, oxidation of sulfhydryl (–SH) groups, and other destructive oxidation processes. Compounds that protect organisms from the effects of ozone include radical scavengers, antioxidants, and compounds containing sulfhydryl groups.

White Phosphorus

Elemental white phosphorus can enter the body by inhalation, by skin contact, or orally. It is a systemic poison, that is, one that is transported through the body to sites remote from its entry site. White phosphorus causes anemia, gastrointestinal system dysfunction, bone brittleness, and eye damage. Exposure also causes **phossy jaw**, a condition in which the jawbone deteriorates and becomes fractured.

Elemental Halogens

Elemental **fluorine** (F_2) is a pale yellow, highly reactive gas that is a strong oxidant. It is a toxic irritant and attacks skin, eye tissue, and the mucous membranes of the nose and respiratory tract. **Chlorine** (Cl_2) gas reacts in water to produce a strongly oxidizing solution. This reaction is responsible for some of the damage caused to the moist tissue lining the respiratory tract when the tissue is exposed to chlorine. The respiratory tract is rapidly irritated by exposure to 10-20 ppm of chlorine gas in air, causing acute discomfort that warns of the presence of the toxicant. Even brief exposure to 1,000 ppm of Cl_2 can be fatal.

Bromine (Br_2) is a volatile, dark red liquid that is toxic when inhaled or ingested. Like chlorine and fluorine, it is strongly irritating to the mucous tissue of the respiratory tract and eyes and may cause pulmonary edema. The toxicological hazard of bromine is limited somewhat because its irritating odor elicits a withdrawal response.

Elemental solid **iodine** (I_2) is irritating to the lungs much like bromine or chlorine. However, the relatively low vapor pressure of iodine limits exposure to I_2 vapor.

Heavy Metals

Heavy metals (Section 7.3) are toxic in their chemically combined forms and some, notably mercury, are toxic in the elemental form. The toxic properties of some of the most hazardous heavy metals and metalloids are discussed here.

Although not truly a *heavy* metal, **beryllium** (atomic mass 9.01) is one of the more hazardous toxic elements. Its most serious toxic effect is berylliosis, a condition manifested by lung fibrosis and pneumonitis, which may develop after a latency period of 5-20 years. Berylliumis is a hypersensitizing agent and exposure to it causes skin granulomas and ulcerated skin. Beryllium was used in the nuclear weapons program in the U. S., and it is believed that 500 to 1000 cases of beryllium poisoning have occurred or will occur in the future as a result of exposure to workers. In July 1999, the U. S. Department of Energy acknowledged these cases of beryllium poisoning and announced proposed legislation to compensate the victims in a program expected to cost up to $15 million at its peak.[2]

Cadmium adversely affects several important enzymes; it can also cause painful osteomalacia (bone disease) and kidney damage. Inhalation of cadmium oxide dusts and fumes results in cadmium pneumonitis characterized by edema and pulmonary epithelium necrosis (death of tissue lining lungs).

Lead, widely distributed as metallic lead, inorganic compounds, and organometallic compounds, has a number of toxic effects, including inhibition of the synthesis of hemoglobin. It also adversely affects the central and peripheral nervous systems and the kidneys. Its toxicological effects have been widely studied.

Arsenic is a metalloid which forms a number of toxic compounds. The toxic +3 oxide, As_2O_3, is absorbed through the lungs and intestines. Biochemically, arsenic acts to coagulate proteins, forms complexes with coenzymes, and inhibits the production of adenosine triphosphate (ATP) in essential metabolic processes involving the utilization of energy.

Arsenic is the toxic agent in one of the great environmental catastrophes of the last century, the result of its ingestion through well water in Bangladesh.[3] Several million of the wells in question were installed in Bangladesh starting in the 1970s using funds provided by the United Nations Children's Fund (UNICEF). Providing an abundant source of microbiologically safe drinking water, they were very successful in reducing water-borne diseases, especially cholera and dysentery. In 1992 a problem with arsenic contamination of many of the wells was shown to exist, and since that time tens of thousands of people have exhibited symptoms of arsenicosis, manifested by skin discoloration and other symptoms. It is likely that many more people in Bangladesh will become ill and die prematurely from arsenic poisoning.

Elemental **mercury** vapor can enter the body through inhalation and be carried by the bloodstream to the brain where it penetrates the blood-brain barrier. It disrupts metabolic processes in the brain causing tremor and psychopathological symptoms such as shyness, insomnia, depression, and irritability. Divalent ionic mercury, Hg^{2+}, damages the kidney. Organometallic mercury compounds such as dimethylmercury, $Hg(CH_3)_2$, are also very toxic.

23.3. TOXIC INORGANIC COMPOUNDS

Cyanide

Both **hydrogen cyanide** (HCN) and **cyanide salts** (which contain CN^- ion) are rapidly acting poisons; a dose of only 60–90 mg is sufficient to kill a human. Metabolically, cyanide bonds to iron(III) in iron-containing ferricytochrome oxidase enzyme (see enzymes, Section 21.6), preventing its reduction to iron(II) in the oxidative phosphorylation process by which the body utilizes O_2. This prevents utilization of oxygen in cells, so that metabolic processes cease.

Carbon Monoxide

Carbon monoxide, CO, is a common cause of accidental poisonings. At CO levels in air of 10 parts per million (ppm) impairment of judgment and visual perception occur; exposure to 100 ppm causes dizziness, headache, and weariness; loss of consciousness occurs at 250 ppm; and inhalation of 1,000 ppm results in rapid death. Chronic long-term exposures to low levels of carbon monoxide are suspected of causing disorders of the respiratory system and the heart.

After entering the blood stream through the lungs, carbon monoxide reacts with hemoglobin (Hb) to convert oxyhemoglobin (O_2Hb) to carboxyhemoglobin (COHb):

$$O_2Hb + CO \rightarrow COHb + O_2 \qquad\qquad (23.3.1)$$

In this case, hemoglobin is the receptor (Section 22.7) acted on by the carbon monoxide toxicant. Carboxyhemoglobin is much more stable than oxyhemoglobin so that its formation prevents hemoglobin from carrying oxygen to body tissues.

Nitrogen Oxides

The two most common toxic oxides of nitrogen are NO and NO_2, of which the latter is regarded as the more toxic. Nitrogen dioxide causes severe irritation of the innermost parts of the lungs resulting in pulmonary edema. In cases of severe exposures, fatal bronchiolitis fibrosa obliterans may develop approximately three weeks after exposure to NO_2. Fatalities may result from even brief periods of inhalation of air containing 200–700 ppm of NO_2. Biochemically, NO_2 disrupts lactic dehydrogenase and some other enzyme systems, possibly acting much like ozone, a stronger oxidant mentioned in Section 23.2. Free radicals, particularly HO·, are likely formed in the body by the action of nitrogen dioxide and the compound probably causes **lipid peroxidation** in which the C=C double bonds in unsaturated body lipids are attacked by free radicals and undergo chain reactions in the presence of O_2, resulting in their oxidative destruction.

Nitrous oxide, N_2O is used as an oxidant gas and in dental surgery as a general anesthetic. This gas was once known as "laughing gas," and was used in the late 1800s as a "recreational gas" at parties held by some of our not-so-staid Victorian ancestors. Nitrous oxide is a central nervous system depressant and can act as an asphyxiant.

Hydrogen Halides

Hydrogen halides (general formula HX, where X is F, Cl, Br, or I) are relatively toxic gases. The most widely used of these gases are HF and HCl; their toxicities are discussed here.

Hydrogen Fluoride

Hydrogen fluoride, (HF, mp -83.1°C, bp 19.5°C) is used as a clear, colorless liquid or gas or as a 30–60% aqueous solution of **hydrofluoric acid**, both referred to here as HF. Both are extreme irritants to any part of the body that they contact, causing ulcers in affected areas of the upper respiratory tract. Lesions caused by contact with HF heal poorly, and tend to develop gangrene.

Fluoride ion, F^-, is toxic in soluble fluoride salts, such as NaF, causing **fluorosis**, a condition characterized by bone abnormalities and mottled, soft teeth. Livestock are especially susceptible to poisoning from fluoride fallout on grazing land; severely afflicted animals become lame and even die. Industrial pollution has been a common source of toxic levels of fluoride. However, about 1 ppm of fluoride used in some drinking water supplies prevents tooth decay.

Hydrogen Chloride

Gaseous **hydrogen chloride** and its aqueous solution, called **hydrochloric acid**, both denoted as HCl, are much less toxic than HF. Hydrochloric acid is a natural physiological fluid present as a dilute solution in the stomachs of humans and other animals. However, inhalation of HCl vapor can cause spasms of the larynx as well as pulmonary edema and even death at high levels. The high affinity of hydrogen chloride vapor for water tends to dehydrate eye and respiratory tract tissue.

Interhalogen Compounds and Halogen Oxides

Interhalogen compounds, including ClF, BrCl, and BrF_3, are extremely reactive and are potent oxidants. They react with water to produce hydrohalic acid solutions (HF, HCl) and nascent oxygen {O}. Too reactive to enter biological systems in their original chemical state, interhalogen compounds tend to be powerful corrosive irritants that acidify, oxidize, and dehydrate tissue, much like those of the elemental forms of the elements from which they are composed. Because of these effects, skin, eyes, and mucous membranes of the mouth, throat, and pulmonary systems are especially susceptible to attack.

Major halogen oxides, including fluorine monoxide (OF_2), chlorine monoxide (Cl_2O), chlorine dioxide (ClO_2), chlorine heptoxide (Cl_2O_7), and bromine monoxide (Br_2O), tend to be unstable, highly reactive, and toxic compounds that pose hazards similar to those of the interhalogen compounds discussed above. Chlorine dioxide, the most commonly used halogen oxide, is employed for odor control and bleaching wood pulp. As a substitute for chlorine in water disinfection, it produces fewer undesirable chemical by-products, particularly trihalomethanes.

The most important of the oxyacids and their salts formed by halogens are hypochlorous acid, HOCl, and hypochlorites, such as NaOCl, used for bleaching and disinfection. The hypochlorites irritate eye, skin, and mucous membrane tissue because they react to produce active (nascent) oxygen ({O}) and acid as shown by the reaction below:

$$HClO \rightarrow H^+ + Cl^- + \{O\} \tag{23.3.2}$$

Inorganic Compounds of Silicon

Silica (SiO_2, quartz) occurs in a variety of types of rocks such as sand, sandstone, and diatomaceous earth. **Silicosis** resulting from human exposure to silica dust from construction materials, sandblasting, and other sources has been a common occupational disease. A type of pulmonary fibrosis that causes lung nodules and makes victims more susceptible to pneumonia and other lung diseases, silicosis is one of the most common disabling conditions resulting from industrial exposure to hazardous substances. It can cause death from insufficient oxygen, or from heart failure in severe cases.

Silane, SiH_4, and disilane, H_3SiSiH_3, are examples of inorganic **silanes**, which have H-Si bonds. Numerous organic ("organometallic") silanes exist in which alkyl moieties are substituted for H. Little information is available regarding the toxicities of silanes.

Silicon tetrachloride, $SiCl_4$, is the only industrially significant compound of the **silicon tetrahalides**, a group of compounds with the general formula SiX_4, where X is a halogen. The two commercially produced **silicon halohydrides**, general formula $H_{4-x}SiX_x$, are dichlorosilane (SiH_2Cl_2) and trichlorosilane, ($SiHCl_3$). These compounds are used as intermediates in the synthesis of organosilicon compounds and in the production of high-purity silicon for semiconductors. Silicon tetrachloride and trichlorosilane, fuming liquids which react with water to give off HCl vapor, have suffocating odors and are irritants to eye, nasal, and lung tissue.

Asbestos

Asbestos is the name given to a group of fibrous silicate minerals, typically those of the serpentine group, for which the approximate chemical formula is $Mg_3(Si_2O_5)(OH)_4$. Asbestos has been widely used in structural materials, brake linings, insulation, and pipe manufacture. Inhalation of asbestos may cause asbestosis (a pneumonia condition), mesothelioma (tumor of the mesothelial tissue lining the chest cavity adjacent to the lungs), and bronchogenic carcinoma (cancer originating with the air passages in the lungs) so that uses of asbestos have been severely curtailed and widespread programs have been undertaken to remove the material from buildings.

Inorganic Phosphorus Compounds

Phosphine (PH_3), a colorless gas that undergoes autoignition at 100°C, is a potential hazard in industrial processes and in the laboratory. Symptoms of poisoning from potentially fatal phosphine gas include pulmonary tract irritation, depression of the central nervous system, fatigue, vomiting, and difficult, painful breathing.

Tetraphosphorus decoxide, P_4O_{10}, is produced as a fluffy white powder from the combustion of elemental phosphorus and reacts with water from air to form syrupy orthophosphoric acid. Because of the formation of acid by this reaction and its dehydrating action, P_4O_{10} is a corrosive irritant to skin, eyes and mucous membranes.

The most important of the **phosphorus halides**, general formulas PX_3 and PX_5, is phosphorus pentachloride used as a catalyst in organic synthesis, as a chlorinating agent, and as a raw material to make phosphorus oxychloride ($POCl_3$). Because they react violently with water to produce the corresponding hydrogen halides and oxo phosphorus acids,

$$PCl_5 + 4H_2O \rightarrow H_3PO_4 + 5HCl \qquad (23.3.3)$$

the phosphorus halides are strong irritants to eyes, skin, and mucous membranes.

The major **phosphorus oxyhalide** in commercial use is phosphorus oxychloride ($POCl_3$), a faintly yellow fuming liquid. Reacting with water to form toxic vapors of hydrochloric acid and phosphorous acid (H_3PO_3), phosphorus oxyhalide is a strong irritant to the eyes, skin, and mucous membranes.

Inorganic Compounds of Sulfur

A colorless gas with a foul, rotten-egg odor, **hydrogen sulfide** is very toxic. In some cases inhalation of H_2S kills faster than even hydrogen cyanide; rapid death ensues from exposure to air containing more than about 1000 ppm H_2S due to asphyxiation from respiratory system paralysis. Lower doses cause symptoms that include headache, dizziness, and excitement due to damage to the central nervous system. General debility is one of the numerous effects of chronic H_2S poisoning.

Sulfur dioxide, SO_2, dissolves in water, to produce sulfurous acid, H_2SO_3; hydrogen sulfite ion, HSO_3^-; and sulfite ion, SO_3^{2-}. Because of its water solubility, sulfur dioxide is largely removed in the upper respiratory tract. It is an irritant to the eyes, skin, mucous membranes, and respiratory tract. Some individuals are hypersensitive to sodium sulfite (Na_2SO_3), which has been used as a chemical food preservative. Because of threats to hypersensitive individuals, these uses were severely restricted in the U.S. in early 1990.

Number-one in synthetic chemical production, **sulfuric acid** (H_2SO_4) is a severely corrosive poison and dehydrating agent in the concentrated liquid form; it readily penetrates skin to reach subcutaneous tissue causing tissue necrosis with effects resembling those of severe thermal burns. Sulfuric acid fumes and mists irritate eye and respiratory tract tissue, and industrial exposure has even caused tooth erosion in workers.

The more important halides, oxides, and oxyhalides of sulfur are listed in Table 23.2. The major toxic effects of these compounds are given in the table.

Organometallic Compounds

The toxicological properties of some organometallic compounds—pharmaceutical organoarsenicals, organomercury fungicides, and tetraethyllead antiknock gasoline additives—that have been used for many years are well known. However, toxicological experience is lacking for many relatively new organometallic compounds that are now being used in semiconductors, as catalysts, and for chemical synthesis, so they should be treated with great caution until proven safe.

Table 23.2. Inorganic Sulfur Compounds

Compound name	Formula	Properties
Sulfur		
Monofluoride	S_2F_2	Colorless gas, mp -104°C, bp 99°C, toxicity similar to HF
Tetrafluoride	SF_4	Gas, bp -40°C, mp -124°C, powerful irritant
Hexafluoride	SF_6	Colorless gas, mp -51°C, remarkably low toxicity when pure, but often contaminated with toxic lower fluorides
Monochloride	S_2Cl_2	Oily, fuming orange liquid, mp -80°C, bp 138°C, strong irritant to eyes, skin, and lungs
Tetrachloride	SCl_4	Brownish/yellow liquid/gas, mp -30°C, Decom. below 0°C, irritant
Trioxide	SO_3	Solid anhydride of sulfuric acid reacts with moisture or steam to produce sulfuric acid
Sulfuryl chloride	SO_2Cl_2	Colorless liquid, mp -54°C, bp 69°C, used for organic synthesis, corrosive toxic irritant
Thionyl chloride	$SOCl_2$	Colorless-to-orange fuming liquid, mp -105°C, bp 79°C, toxic corrosive irritant
Carbon oxysulfide	COS	Volatile liquid by-product of natural gas or petroleum refining, toxic narcotic
Carbon disulfide	CS_2	Colorless liquid, industrial chemical, narcotic and central nervous system anesthetic

Organometallic compounds often behave in the body in ways totally unlike the inorganic forms of the metals that they contain. This is due in large part to the fact that, compared to inorganic forms, organometallic compounds have an organic nature and higher lipid solubility.

Organolead Compounds

Perhaps the most notable toxic organometallic compound is tetraethyllead, $Pb(C_2H_5)_4$, a colorless, oily liquid that was widely used as a gasoline additive to boost octane rating. Tetraethyllead has a strong affinity for lipids and can enter the body by all three common routes of inhalation, ingestion, and absorption through the skin. Acting differently from inorganic compounds in the body, it affects the central nervous system with symptoms such as fatigue, weakness, restlessness, ataxia, psychosis, and convulsions. Recovery from severe lead poisoning tends to be slow. In cases of fatal tetraethyllead poisoning, death has occurred as soon as one or two days after exposure.

Organotin Compounds

The greatest number of organometallic compounds in commercial use are those of tin—tributyltin chloride and related tributyltin (TBT) compounds. These compounds have bactericidal, fungicidal, and insecticidal properties. They have particular environmental significance because of their widespread applications as industrial biocides, now increasingly limited because of their environmental and toxicological effects.[4] Organotin compounds are readily absorbed through the skin, sometimes causing a skin rash. They probably bind with sulfur groups on proteins and appear to interfere with mitochondrial function.

Carbonyls

Metal carbonyls, regarded as extremely hazardous because of their toxicities include nickel tetracarbonyl ($Ni(CO)_4$), cobalt carbonyl, and iron pentacarbonyl. Some of the hazardous carbonyls are volatile and readily taken into the body through the respiratory tract or through the skin. The carbonyls affect tissue directly and they break down to toxic carbon monoxide and products of the metal, which have additional toxic effects.

Reaction Products of Organometallic Compounds

An example of the production of a toxic substance from the burning of an organometallic compound is provided by the combustion of diethylzinc:

$$Zn(C_2H_5)_2 + 7O_2 \rightarrow ZnO(s) + 5H_2O(g) + 4CO_2(g) \qquad (23.3.4)$$

Zinc oxide is used as a healing agent and food additive. However, inhalation of zinc oxide fume particles produced by the combustion of zinc organometallic compounds causes zinc **metal fume fever**. This is an uncomfortable condition characterized by elevated temperature and "chills."

23.4. TOXICOLOGY OF ORGANIC COMPOUNDS

Alkane Hydrocarbons

Gaseous methane, ethane, propane, *n*-butane, and isobutane (both C_4H_{10}) are regarded as **simple asphyxiants** that form mixtures with air containing insufficient oxygen to support respiration. The most common toxicological occupational problem associated with the use of hydrocarbon liquids in the workplace is dermatitis, caused by dissolution of the fat portions of the skin and characterized by inflamed, dry, scaly skin. Inhalation of volatile liquid 5–8 carbon *n*-alkanes and branched-chain alkanes may cause central nervous system depression manifested by dizziness and loss of coordination. Exposure to *n*-hexane and cyclohexane results in loss of myelin (a fatty substance constituting a sheath around certain nerve fibers) and degeneration of axons (part of a nerve cell through which nerve impulses are transferred out of the cell). This has resulted in multiple disorders of the nervous system (**polyneuropathy**) including muscle weakness and impaired sensory function of the hands and feet. In the body, *n*-hexane is metabolized to 2,5-hexanedione:

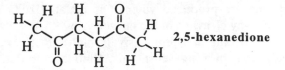

2,5-hexanedione

This Phase I oxidation product can be observed in urine of exposed individuals and is used as a biological monitor of exposure to *n*-hexane.[5]

Alkene and Alkyne Hydrocarbons

Ethylene, a widely used colorless gas with a somewhat sweet odor, acts as a simple asphyxiant and anesthetic to animals and is phytotoxic (toxic to plants). The toxicological properties of propylene (C_3H_6) are very similar to those of ethylene. Colorless, odorless, gaseous 1,3-butadiene is an irritant to eyes and respiratory system mucous membranes; at higher levels it can cause unconsciousness and even death. Acetylene, H-C≡C-H, is a colorless gas with a garlic odor. It acts as an asphyxiant and narcotic, causing headache, dizziness, and gastric disturbances. Some of these effects may be due to the presence of impurities in the commercial product.

Benzene and Aromatic Hydrocarbons

Inhaled benzene is readily absorbed by blood, from which it is strongly taken up by fatty tissues. For the non-metabolized compound, the process is reversible and benzene is excreted through the lungs. As shown in Figure 23.1, benzene is converted to phenol by a Phase I oxidation reaction (see Section 22.6) in the liver.

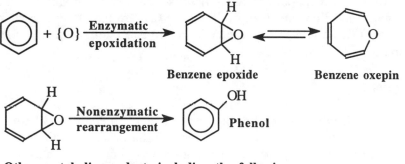

Other metabolic products including the following:

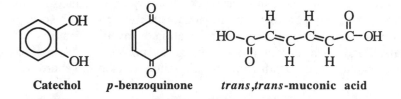

Catechol   *p*-benzoquinone   *trans,trans*-muconic acid

Figure 23.1. Conversion of benzene to phenol in the body.

The reactive and short-lived benzene epoxide intermediate known to occur in this reaction is probably responsible for much of the unique toxicity of benzene, which involves damage to bone marrow. In addition to phenol, several other oxygenated derivatives of benzene are produced when it is metabolized, as is *trans,trans-muconic acid*, produced by cleavage of the benzene ring.[6]

Benzene is a skin irritant, and progressively higher local exposures can cause skin redness (erythema), burning sensations, fluid accumulation (edema), and blistering. Inhalation of air containing about 7 g/m^3 of benzene causes acute poisoning within an hour because of a narcotic effect upon the central nervous system manifested progressively by excitation, depression, respiratory system failure, and death. Inhalation of air containing more than about 60 g/m^3 of benzene can be fatal within a few minutes.

Long-term exposures to lower levels of benzene cause nonspecific symptoms, including fatigue, headache, and appetite loss. Chronic benzene poisoning causes blood abnormalities, including a lowered white cell count, an abnormal increase in blood lymphocytes (colorless corpuscles introduced to the blood from the lymph glands), anemia, a decrease in the number of blood platelets required for clotting (thrombocytopenia), and damage to bone marrow. It is thought that preleukemia, leukemia, or cancer may result.

Toluene

Toluene, a colorless liquid boiling at 101.4°C, is classified as moderately toxic through inhalation or ingestion; it has a low toxicity by dermal exposure. Toluene can be tolerated without noticeable ill effects in ambient air up to 200 ppm. Exposure to 500 ppm may cause headache, nausea, lassitude, and impaired coordination without detectable physiological effects. Massive exposure to toluene has a narcotic effect, which can lead to coma. Because it possesses an aliphatic side chain that can be oxidized enzymatically to products that are readily excreted from the body (see the metabolic reaction scheme in Figure 23.2), toluene is much less toxic than benzene.

Naphthalene

As is the case with benzene, **naphthalene** undergoes a Phase I oxidation reaction that places an epoxide group on the aromatic ring. This process is followed by Phase II conjugation reactions to yield products that can be eliminated from the body in the urine.

Exposure to naphthalene can cause anemia and marked reductions in red cell count, hemoglobin, and hematocrit in individuals exhibiting genetic susceptibility to these conditions. Naphthalene causes skin irritation or severe dermatitis in sensitized individuals. Headaches, confusion, and vomiting may result from inhalation or ingestion of naphthalene. Death from kidney failure occurs in severe instances of poisoning.

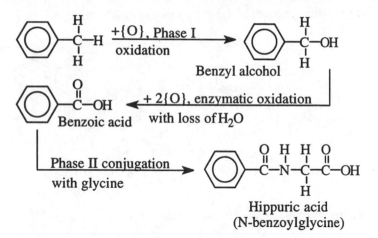

Figure 23.2. Metabolic oxidation of toluene with conjugation to hippuric acid, which is excreted with urine.

Polycyclic Aromatic Hydrocarbons

Benzo[a]pyrene (see Section 10.4) is the most studied of the polycyclic aromatic hydrocarbons (PAHs). Some metabolites of PAH compounds, particularly the 7,8-diol-9,10-epoxide of benzo[a]pyrene shown in Figure 23.3, are known to cause cancer. There are two stereoisomers of this metabolite, both of which are known to be potent mutagens and presumably can cause cancer.

Oxygen-Containing Organic Compounds

Oxides

Hydrocarbon **oxides** such as ethylene oxide and propylene oxide,

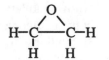

Ethylene oxide   **Propylene oxide**

which are characterized by an **epoxide** functional group bridging oxygen between two adjacent C atoms, are significant for both their uses and their toxic effects. Ethylene oxide, a gaseous colorless, sweet-smelling, flammable, explosive gas used as a chemical intermediate, sterilant, and fumigant, has a moderate to high toxicity, is a mutagen, and is carcinogenic to experimental animals. Inhalation of relatively low levels of this gas results in respiratory tract irritation, headache, drowsiness, and dyspnea, whereas exposure to higher levels causes cyanosis, pulmonary edema, kidney damage, peripheral nerve damage, and even death. Propylene oxide is a colorless, reactive, volatile liquid (bp 34°C) with uses similar to those of ethylene oxide and similar, though less severe, toxic effects. The toxicity of 1,2,3,4-butadiene epoxide, the oxidation product of 1,3-butadiene, is notable in that it is a direct–acting (primary) carcinogen.

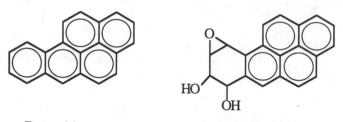

Benzo(a)pyrene   7,8-Diol-9,10-epoxide
                                         of benzo(a)pyrene

Figure 23.3. Benzo[a]pyrene and its carcinogenic metabolic product.

Alcohols

Human exposure to the three light alcohols shown in Figure 23.4. is common because they are widely used industrially and in consumer products.

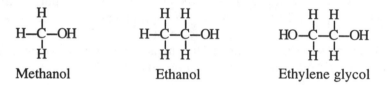

Methanol   Ethanol   Ethylene glycol

Figure 23.4. **Alcohols** such as these three compounds are oxygenated compounds in which the hydroxyl functional group is attached to an alkyl or alkenyl hydrocarbon skeleton.

Methanol, which has caused many fatalities when ingested accidentally or consumed as a substitute for beverage ethanol, is metabolically oxidized to formaldehyde and formic acid. In addition to causing acidosis, these products affect the central nervous system and the optic nerve. Acute exposure to lethal doses causes an initially mild inebriation, followed in about 10–20 hours by unconsciousness, cardiac depression, and death. Sublethal exposures can cause blindness from deterioration of the optic nerve and retinal ganglion cells. Inhalation of methanol fumes may result in chronic, low-level exposure.

Ethanol is usually ingested through the gastrointestinal tract, but can be absorbed as vapor by the alveoli of the lungs. Ethanol is oxidized metabolically more rapidly than methanol, first to acetaldehyde (discussed later in this section), then to CO_2. Ethanol has numerous acute effects resulting from central nervous system depression. These range from decreased inhibitions and slowed reaction times at 0.05% blood ethanol, through intoxication, stupor, and—at more than 0.5% blood ethanol—death. Ethanol also has a number of chronic effects, of which the addictive condition of alcoholism and cirrhosis of the liver are the most prominent.[7]

Despite its widespread use in automobile cooling systems, exposure to ethylene glycol is limited by its low vapor pressure. However, inhalation of droplets of ethylene glycol can be very dangerous. In the body, ethylene glycol initially stimulates the central nervous system, then depresses it. Glycolic acid, chemical formula $HOCH_2CO_2H$, formed as an intermediate metabolite in the metabolism of ethylene glycol, may cause acidemia, and oxalic acid produced by further oxidation may precipitate in the kidneys as solid calcium oxalate, CaC_2O_4, causing clogging.

Of the higher alcohols, 1-butanol is an irritant, but its toxicity is limited by its low vapor pressure. Unsaturated (alkenyl) allyl alcohol, $CH_2=CHCH_2OH$, has a pungent odor and is strongly irritating to eyes, mouth, and lungs.

Phenols

Figure 23.5 shows some of the more important phenolic compounds, aryl analogs of alcohols which have properties much different from those of the aliphatic and olefinic alcohols. Nitro groups ($-NO_2$) and halogen atoms (particularly Cl) bonded to the aromatic rings strongly affect the chemical and toxicological behavior of phenolic compounds.

Although the first antiseptic used on wounds and in surgery, phenol is a protoplasmic poison that damages all kinds of cells and is alleged to have caused "an astonishing number of poisonings" since it came into general use.[8] The acute toxicological effects of phenol are largely upon the central nervous system and death can occur as soon as one-half hour after exposure. Acute poisoning by phenol can cause severe gastrointestinal disturbances, kidney malfunction, circulatory system failure, lung edema, and convulsions. Fatal doses of phenol may be absorbed through the skin. Key organs damaged by chronic phenol exposure include the spleen, pancreas, and kidneys. The toxic effects of other phenols resemble those of phenol.

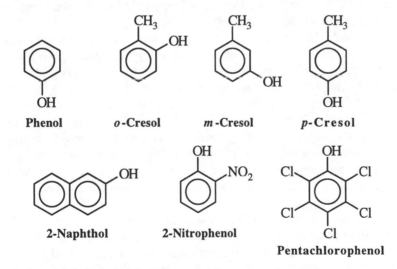

Figure 23.5. Some phenols and phenolic compounds.

Aldehydes and Ketones

Aldehydes and ketones are compounds that contain the carbonyl (C=O) group, as shown by the examples in Figure 23.6.

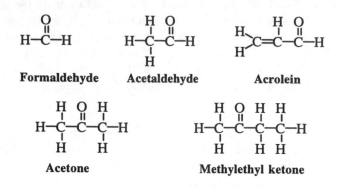

Figure 23.6. Commercially and toxicologically significant aldehydes and ketones.

Formaldehyde is uniquely important because of its widespread use and toxicity. In the pure form, formaldehyde is a colorless gas with a pungent, suffocating odor. It is commonly encountered as **formalin,** a 37–50% aqueous solution of formaldehyde containing some methanol. Exposure to inhaled formaldehyde via the respiratory tract is usually due to molecular formaldehyde vapor, whereas exposure by other routes is usually due to formalin. Prolonged, continuous exposure to formaldehyde can cause hypersensitivity. A severe irritant to the mucous membrane linings of both the respiratory and alimentary tracts, formaldehyde reacts strongly with functional groups in molecules. Formaldehyde has been shown to be a lung carcinogen in experimental animals. The toxicity of formaldehyde is largely due to its metabolic oxidation product, formic acid (see below).

The lower aldehydes are relatively water-soluble and intensely irritating. These compounds attack exposed moist tissue, particularly the eyes and mucous membranes of the upper respiratory tract. (Some of the irritating properties of photochemical smog, Chapter 13, are due to the presence of aldehydes.) However, aldehydes that are relatively less soluble can penetrate further into the respiratory tract and affect the lungs. Colorless, liquid acetaldehyde is relatively less toxic than acrolein and acts as an irritant and systemically as a narcotic to the central nervous system. Extremely irritating, lachrimating acrolein vapor has a choking odor and inhalation of it can cause severe damage to respiratory tract membranes. Tissue exposed to acrolein may undergo severe necrosis, and direct contact with the eye can be especially hazardous.

The ketones shown in Figure 23.6. are relatively less toxic than the aldehydes. Pleasant-smelling acetone can act as a narcotic; it causes dermatitis by dissolving fats from skin. Not many toxic effects have been attributed to methyl ethyl ketone. It is suspected of having caused neuropathic disorders in shoe factory workers.

Carboxylic Acids

Formic acid, HCO_2H, is a relatively strong acid that is corrosive to tissue. In Europe, decalcifier formulations for removing mineral scale that contain about 75% formic acid are sold; and children ingesting these solutions have suffered corrosive lesions to mouth and esophageal tissue. Although acetic acid as a 4–6% solution in vinegar is an ingredient of many foods, pure acetic acid (glacial acetic acid) is extremely corrosive to tissue that it contacts. Ingestion of, or skin contact with acrylic acid can cause severe damage to tissues.

Ethers

The common ethers have relatively low toxicities because of the low reactivity of the C–O–C functional group which has very strong carbon-oxygen bonds. Exposure to volatile diethyl ether is usually by inhalation and about 80% of this compound that gets into the body is eliminated unmetabolized as the vapor through the lungs. Diethyl ether depresses the central nervous system and is a depressant widely used as an anesthetic for surgery. Low doses of diethyl ether cause drowsiness, intoxication, and stupor, whereas higher exposures cause unconsciousness and even death.

Acid Anhydrides

Strong-smelling, intensely lachrimating **acetic anhydride,**

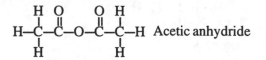

is a systemic poison. It is very corrosive to the skin, eyes, and upper respiratory tract, causing blisters and burns that heal only slowly. Levels in the air should not exceed 0.04 mg/m^3, and adverse effects to the eyes have been observed at about 0.4 mg/m^3.

Esters

Many esters (Figure 23.7) have relatively high volatilities so that the pulmonary system is a major route of exposure. Because of their generally good solvent properties, esters penetrate tissues and tend to dissolve body lipids. For example, vinyl acetate acts as a skin defatting agent. Because they hydrolyze in water, ester toxicities tend to be the same as the toxicities of the acids and alcohols from which they were formed. Many volatile esters exhibit asphyxiant and narcotic action. Whereas many of the naturally occurring esters have insignificant toxicities at low doses, allyl acetate and some of the other synthetic esters are relatively toxic.

Insofar as potential health effects are concerned, di-(2-ethylhexyl) phthalate (DEHP) is arguably the ester of most concern. This is because of the use of this ester at levels of around 30% as a plasticizer to impart flexibility to poly(vinyl chloride) (PVC) plastic. As a consequence of the widespread use of DEHP-containing PVC plastics, DEHP has become a ubiquitous contaminant found in water, sediment, food, and biological samples. The most acute concern arises from its use in medical applications, particularly bags used to hold intravenous solutions administered to medical patients.[9] As a result of medical use, DEHP enters the blood of hemophiliacs, kidney dialysis patients, and premature and high-risk infants.[10] Although the acute toxic effects of DEHP are low, such widespread direct exposure of humans is worrisome.

Organonitrogen Compounds

Organonitrogen compounds constitute a large group of compounds with diverse toxicities. Examples of several of the kinds of organonitrogen compounds discussed here are given in Figure 23.8.

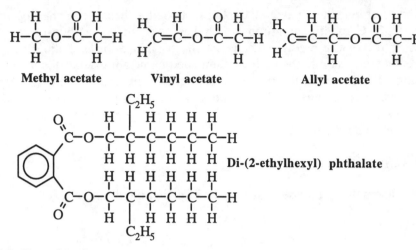

Methyl acetate   Vinyl acetate   Allyl acetate

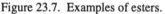

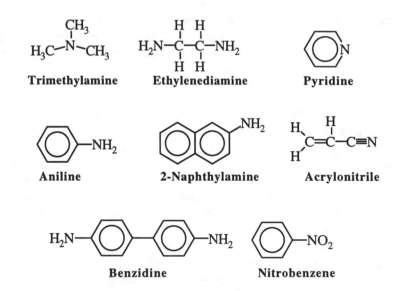

Figure 23.7. Examples of esters.

Trimethylamine   Ethylenediamine   Pyridine

Aniline   2-Naphthylamine   Acrylonitrile

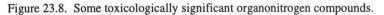

Benzidine   Nitrobenzene

Figure 23.8. Some toxicologically significant organonitrogen compounds.

Aliphatic Amines

The lower amines, such as the methylamines, are rapidly and easily taken into the body by all common exposure routes. They are basic and react with water in tissue,

$$R_3N + H_2O \rightarrow R_3NH^+ + OH^- \qquad\qquad (23.4.1)$$

raising the pH of the tissue to harmful levels, acting as corrosive poisons (especially to sensitive eye tissue), and causing tissue necrosis at the point of contact. Among the systemic effects of amines are necrosis of the liver and kidneys, lung hemorrhage and edema, and sensitization of the immune system. The lower amines are among the more toxic substances in routine, large-scale use.

Ethylenediamine is the most common of the **alkyl polyamines**, compounds in which two or more amino groups are bonded to alkane moieties. Its toxicity rating is only 3, but it is a strong skin sensitizer and can damage eye tissue.

Carbocyclic Aromatic Amines

Aniline is a widely used industrial chemical and is the simplest of the **carbocyclic aromatic amines**, a class of compounds in which at least one substituent group is an aromatic hydrocarbon ring bonded directly to the amino group. There are numerous compounds with many industrial uses in this class of amines. Some of the carbocyclic aromatic amines have been shown to cause cancer in the human bladder, ureter, and pelvis, and are suspected of being lung, liver, and prostate carcinogens. A very toxic colorless liquid with an oily consistency and distinct odor, aniline readily enters the body by inhalation, ingestion, and through the skin. Metabolically, aniline converts iron(II) in hemoglobin to iron(III). This causes a condition called **methemoglobinemia**, characterized by cyanosis and a brown-black color of the blood, in which the hemoglobin can no longer transport oxygen in the body. This condition is not reversed by oxygen therapy.

Both **1-naphthylamine** (α-naphthylamine) and **2-naphthylamine** (β-naphthylamine) are proven human bladder carcinogens. In addition to being a proven human carcinogen, **benzidine**, 4,4'-diaminobiphenyl, is highly toxic and has systemic effects that include blood hemolysis, bone marrow depression, and kidney and liver damage. It can be taken into the body orally, by inhalation into the lungs, and by skin sorption.

Pyridine

Pyridine, a colorless liquid with a sharp, penetrating, "terrible"odor, is an aromatic amine in which an N atom is part of a 6-membered ring. This widely used industrial chemical is only moderately toxic with a toxicity rating of 3. Symptoms of pyridine poisoning include anorexia, nausea, fatigue, and, in cases of chronic poisoning, mental depression. In a few rare cases pyridine poisoning has been fatal.

Nitriles

Nitriles contain the $-C\equiv N$ functional group. Colorless, liquid **acetonitrile**, CH_3CN, is widely used in the chemical industry. With a toxicity rating of 3–4, acetonitrile is considered relatively safe, although it has caused human deaths, perhaps by metabolic release of cyanide. **Acrylonitrile**, a colorless liquid with a peach-seed (cyanide) odor, is highly reactive because it contains both nitrile and

C=C groups. Ingested, absorbed through the skin, or inhaled as vapor, acrylonitrile metabolizes to release deadly HCN, which it resembles toxicologically.

Nitro Compounds

The simplest of the **nitro compounds, nitromethane** H_3CNO_2, is an oily liquid that causes anorexia, diarrhea, nausea, and vomiting, and damages the kidneys and liver. **Nitrobenzene**, a pale yellow, oily liquid with an odor of bitter almonds or shoe polish, can enter the body by all routes. It has a toxic action much like that of aniline, converting hemoglobin to methemoglobin, which cannot carry oxygen to body tissue. Nitrobenzene poisoning is manifested by cyanosis.

Nitrosamines

N-nitroso compounds (**nitrosamines**) contain the N–N=O functional group and have been found in a variety of materials to which humans may be exposed, including beer, whiskey, and cutting oils used in machining. Cancer may result from exposure to a single large dose or from chronic exposure to relatively small doses of some nitrosamines. Once widely used as an industrial solvent and known to cause liver damage and jaundice in exposed workers, dimethylnitrosamine was shown to be carcinogenic from studies starting in the 1950s.

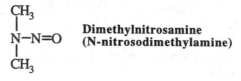

Dimethylnitrosamine
(N-nitrosodimethylamine)

Isocyanates and Methyl Isocyanate

Compounds with the general formula R–N=C=O, **isocyanates** are widely used industrial chemicals noted for the high chemical and metabolic reactivity of their characteristic functional group. **Methyl isocyanate**, $H_3C–N=C=O$, was the toxic agent involved in the catastrophic industrial poisoning in Bhopal, India, on December 2, 1984, the worst industrial accident in history. In this incident several tons of methyl isocyanate were released, killing 2000 people and affecting about 100,000. The lungs of victims were attacked; survivors suffered long-term shortness of breath and weakness from lung damage, as well as numerous other toxic effects including nausea and bodily pain.

Organonitrogen Pesticides

Pesticidal *carbamates* are characterized by the structural skeleton of carbamic acid outlined by the dashed box in the structural formula of carbaryl in Figure 23.9. Widely used on lawns and gardens, insecticidal **carbaryl** has a low toxicity to mammals. Highly water-soluble **carbofuran** is a systemic insecticide in that it is taken up by the roots and leaves of plants; insects that feed on the leaves are poisoned. The toxic effects to animals of carbamates are due to the fact that they inhibit acetylcholinesterase directly without the need to first undergo biotransformation. This effect is relatively reversible because of metabolic hydrolysis of the carbamate ester.

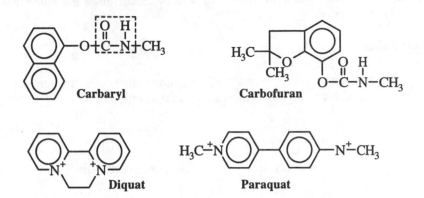

Carbaryl   Carbofuran

Diquat   Paraquat

Figure 23.9. Examples of organonitrogen pesticides.

Reputed to have "been responsible for hundreds of human deaths,"[11] herbicidal **paraquat** has a toxicity rating of 5. Dangerous or even fatal acute exposures can occur by inhalation of spray, skin contact, and ingestion. Paraquat is a systemic poison that affects enzyme activity and is devastating to a number of organs. Pulmonary fibrosis results in animals that have inhaled paraquat aerosols, and the lungs are also adversely affected by nonpulmonary exposure. Acute exposure may cause variations in the levels of catecholamine, glucose, and insulin. The most prominent initial symptom of poisoning is vomiting, followed within a few days by dyspnea, cyanosis, and evidence of impairment of the kidneys, liver, and heart. Pulmonary fibrosis, often accompanied by pulmonary edema and hemorrhaging, is observed in fatal cases.

Organohalide Compounds

Alkyl Halides

The toxicities of alkyl halides, such as carbon tetrachloride, CCl_4, vary a great deal with the compound. Most of these compounds cause depression of the central nervous system, and individual compounds exhibit specific toxic effects.

During its many years of use as a consumer product, carbon tetrachloride compiled a grim record of toxic effects which led the U. S. Food and Drug Administration (FDA) to prohibit its household use in 1970. It is a systemic poison that affects the nervous system when inhaled, and the gastrointestinal tract, liver, and kidneys when ingested. The biochemical mechanism of carbon tetrachloride toxicity involves reactive radical species, including

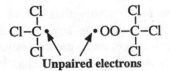

Unpaired electrons

that react with biomolecules, such as proteins and DNA. The most damaging such reaction occurs in the liver as **lipid peroxidation**, consisting of the attack of free radicals on unsaturated lipid molecules, followed by oxidation of the lipids through a free radical mechanism.

Alkenyl Halides

The most significant **alkenyl** or **olefinic organohalides** are the lighter chlorinated compounds, such as vinyl chloride and tetrachloroethylene:

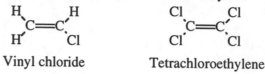

Vinyl chloride   Tetrachloroethylene

Because of their widespread use and disposal in the environment, the numerous acute and chronic toxic effects of the alkenyl halides are of considerable concern.

The central nervous system, respiratory system, liver, and blood and lymph systems are all affected by vinyl chloride exposure, which has been widespread because of this compound's use in poly(vinyl chloride) manufacture. Most notably, vinyl chloride is carcinogenic, causing a rare angiosarcoma of the liver. This deadly form of cancer has been observed in workers chronically exposed to vinyl chloride while cleaning autoclaves in the poly(vinyl chloride) fabrication industry. The alkenyl organohalide, 1,1-dichloroethylene, is a suspect human carcinogen based upon animal studies and its structural similarity to vinyl chloride The toxicities of both 1,2-dichloroethylene isomers are relatively low. These compounds act in different ways in that the *cis* isomer is an irritant and narcotic, whereas the *trans* isomer affects both the central nervous system and the gastrointestinal tract, causing weakness, tremors, cramps, and nausea. A suspect human carcinogen, trichloroethylene has caused liver carcinoma in experimental animals and is known to affect numerous body organs. Like other organohalide solvents, trichloroethylene causes skin dermatitis from dissolution of skin lipids, and it can affect the central nervous and respiratory systems, liver, kidneys, and heart. Symptoms of exposure include disturbed vision, headaches, nausea, cardiac arrhythmias, and burning/tingling sensations in the nerves (paresthesia). Tetrachloroethylene damages the liver, kidneys, and central nervous system. It is a suspect human carcinogen.

Aryl Halides

Individuals exposed to irritant monochlorobenzene by inhalation or skin contact suffer symptoms to the respiratory system, liver, skin, and eyes. Ingestion of this compound causes effects similar to those of toxic aniline, including incoordination, pallor, cyanosis, and eventual collapse.

The dichlorobenzenes are irritants that affect the same organs as monochlorobenzene. Some animal tests have suggested that 1,2-dichlorobenzene is a potential cancer-causing substance. *Para*-dichlorobenzene (1,4-dichlorobenzene), a chemical used in air fresheners and mothballs, has been known to cause profuse rhinitis (running nose), nausea, jaundice, liver cirrhosis, and weight loss associated with anorexia. It is not known to be a carcinogen. Its major urinary metabolite is 2,5-dichlorophenol, which is eliminated principally as the glucuronide or sulfate.[12]

Because of their once widespread use in electrical equipment, as hydraulic fluids, and in many other applications, polychlorinated biphenyls (PCBs, see Section 7.12) became widespread, extremely persistent environmental pollutants. PCBs have a strong tendency to undergo bioaccumulation in lipid tissue. Polybrominated biphenyl analogs (PBBs) were much less widely used and distributed. However, PBBs were involved in one major incident that resulted in catastrophic agricultural losses when livestock feed contaminated with PBB flame retardant caused massive livestock poisoning in Michigan in 1973.

Organohalide Pesticides

Exhibiting a wide range of kind and degree of toxic effects, many organohalide insecticides (see Section 7.11) affect the central nervous system, causing tremor, irregular eye jerking, changes in personality, and loss of memory. Such symptoms are characteristic of acute DDT poisoning. However, the acute toxicity of DDT to humans is very low and, when used for the control of typhus and malaria in World War II, it was applied directly to people. The chlorinated cyclodiene insecticides— aldrin, dieldrin, endrin, chlordane, heptachlor, endosulfan, and isodrin—act on the brain, releasing betaine esters and causing headaches, dizziness, nausea, vomiting, jerking muscles, and convulsions. Dieldrin, chlordane, and heptachlor have caused liver cancer in test animals, and some chlorinated cyclodiene insecticides are teratogenic or fetotoxic. Because of these effects, aldrin, dieldrin, heptachlor, and— more recently—chlordane have been prohibited from use in the U. S.

The major **chlorophenoxy** herbicides are 2,4-dichlorophenoxyacetic acid (2,4-D), 2,4,5-trichlorophenoxyacetic acid (2,4,5-T or Agent Orange), and Silvex. Large doses of 2,4-dichlorophenoxyacetic acid have been shown to cause nerve damage, (peripheral neuropathy), convulsions, and brain damage. According to a National Cancer Institute study,[13] Kansas farmers who had handled 2,4-D extensively have suffered 6 to 8 times the incidence of non-Hodgkins lymphoma as comparable unexposed populations. With a toxicity somewhat less than that of 2,4-D, Silvex is largely excreted unchanged in the urine. The toxic effects of 2,4,5-T (used as a herbicidal warfare chemical called "Agent Orange") have resulted from the presence of 2,3,7,8-tetrachloro-*p*-dioxin (TCDD, commonly known as "dioxin", discussed below), a manufacturing by-product. Autopsied carcasses of sheep poisoned by this herbicide have exhibited nephritis, hepatitis, and enteritis.

TCDD

Polychlorinated dibenzodioxins are compounds which have the same basic structure as that of TCDD (2,3,7,8-tetrachlorodibenzo-*p*-dioxin),

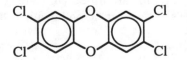

TCDD (2,3,7,8-tetrachloro-dibenzo-p-dioxin)

but have different numbers and locations of chlorine atoms on the ring structure. Extremely toxic to some animals, the toxicity of TCDD to humans is rather uncertain; it is known to cause a skin condition called chloracne. TCDD has been a manufacturing by-product of some commercial products (see the discussion of 2,4,5-T, above), a contaminant identified in some municipal incineration emissions, and a widespread environmental pollutant from improper waste disposal. This compound has been released in a number of industrial accidents, the most massive of which exposed several tens of thousands of people to a cloud of chemical emissions spread over an approximately 3-square-mile area at the Givaudan-La Roche Icmesa manufacturing plant near Seveso, Italy, in 1976. On an encouraging note from a toxicological perspective, no abnormal occurrences of major malformations were found in a study of 15,291 children born in the area within 6 years after the release.[14]

Chlorinated Phenols

The chlorinated phenols used in largest quantities have been **pentachlorophenol** (Chapter 7) and the trichlorophenol isomers used as wood preservatives. Although exposure to these compounds has been correlated with liver malfunction and dermatitis, contaminant polychlorinated dibenzodioxins may have caused some of the observed effects.

Organosulfur compounds

Despite the high toxicity of H_2S, not all organosulfur compounds are particularly toxic. Their hazards are often reduced by their strong, offensive odors that warn of their presence.

Inhalation of even very low concentrations of the alkyl **thiols**, such as methanethiol, H_3CSH, can cause nausea and headaches; higher levels can cause increased pulse rate, cold hands and feet, and cyanosis. In extreme cases, unconsciousness, coma, and death occur. Like H_2S, the alkyl thiols are precursors to cytochrome oxidase poisons.

An oily, water-soluble liquid, **methylsulfuric acid** is a strong irritant to skin, eyes, and mucous tissue. Colorless, odorless **dimethyl sulfate** is highly toxic and is a

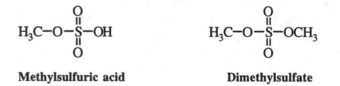

Methylsulfuric acid   **Dimethylsulfate**

primary carcinogen that does not require bioactivation to cause cancer. Skin or mucous membranes exposed to dimethyl sulfate develop conjunctivitis and inflammation of nasal tissue and respiratory tract mucous membranes following an initial latent period during which few symptoms are observed. Damage to the liver and kidneys, pulmonary edema, cloudiness of the cornea, and death within 3–4 days can result from heavier exposures.

Sulfur Mustards

A typical example of deadly **sulfur mustards**, compounds used as military poisons, or "poison gases," is mustard oil (bis(2-chloroethyl)sulfide), the structure of which is shown at the top of the next page. An experimental mutagen and primary carcinogen, mustard oil produces vapors that penetrate deep within tissue, resulting

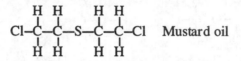

in destruction and damage at some depth from the point of contact; penetration is very rapid, so that efforts to remove the toxic agent from the exposed area are ineffective after 30 minutes. This military "blistering gas" poison causes tissue to become severely inflamed with lesions that often become infected. These lesions in the lung can cause death.

Organophosphorus Compounds

Organophosphorus compounds have varying degrees of toxicity. Some of these compounds, such as the "nerve gases" produced as industrial poisons, are deadly in minute quantities. The toxicities of major classes of organophosphate compounds are discussed in this section.

Organophosphate Esters

Some organophosphate esters are shown in Figure 23.10. **Trimethyl phosphate** is probably moderately toxic when ingested or absorbed through the skin, whereas moderately toxic **triethyl phosphate**, $(C_2H_5O)_3PO$, damages nerves and inhibits acetylcholinesterase. Notoriously toxic **tri-*o*-cresyl phosphate, TOCP,** apparently is metabolized to products that inhibit acetylcholinesterase. Exposure to TOCP causes degeneration of the neurons in the body's central and peripheral nervous systems with early symptoms of nausea, vomiting, and diarrhea accompanied by severe abdominal pain. About 1–3 weeks after these symptoms have subsided, peripheral paralysis develops manifested by "wrist drop" and "foot drop," followed by slow recovery, which may be complete or leave a permanent partial paralysis.

Briefly used in Germany as a substitute for insecticidal nicotine, **tetraethyl pyrophosphate, TEPP,** is a very potent acetylcholinesterase inhibitor. With a toxicity rating of 6 (supertoxic), TEPP is deadly to humans and other mammals.

Phosphorothionate and Phosphorodithioate Ester Insecticides

Because esters containing the P=S (thiono) group are resistant to nonenzymatic hydrolysis and are not as effective as P=O compounds in inhibiting acetylcholinesterase, they exhibit higher insect:mammal toxicity ratios than their nonsulfur analogs. Therefore, **phosphorothionate** and **phosphorodithioate** esters (Figure 23.11.) have been widely used as insecticides. The insecticidal activity of these compounds requires metabolic conversion of P=S to P=O (oxidative desulfuration). Environmentally, organophosphate insecticides are superior to many of the organochlorine insecticides because the organophosphates readily undergo biodegradation and do not bioaccumulate.

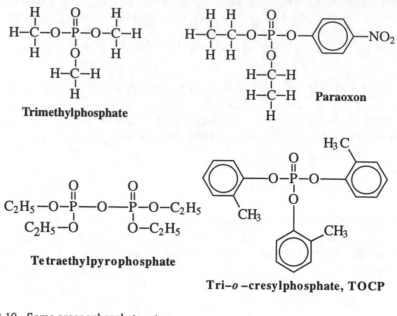

Figure 23.10. Some organophosphate esters.

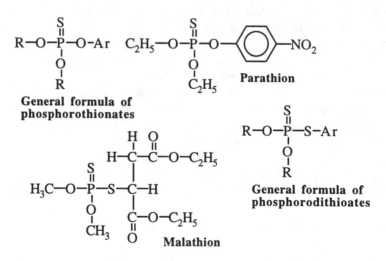

Figure 23.11. Phosphorothionate and phosphorodithioate ester insecticides. Malathion contains hydrolyzable carboxyester linkages.

The first commercially successful phosphorothionate/phosphorodithioate ester insecticide was **parathion**, O,O-diethyl-O-p-nitrophenylphosphorothionate, first licensed for use in 1944. This insecticide has a toxicity rating of 6 (supertoxic). Since its use began, several hundred people have been killed by parathion, including 17 of 79 people exposed to contaminated flour in Jamaica in 1976. As little as 120 mg of parathion has been known to kill an adult human, and a dose of 2 mg has been fatal to a child. Most accidental poisonings have occurred by absorption through the skin. Methylparathion (a closely related compound with methyl groups instead of ethyl groups) is regarded as extremely toxic, and in August 1999 the U. S. Environmental Protection Agency proposed severely curtailing its use. In order for parathion to have a toxic effect, it must be converted metabolically to paraoxon (Figure 23.10), which is a potent inhibitor of acetylcholinesterase. Because of the time required for this conversion, symptoms develop several hours after exposure, whereas the toxic effects of TEPP or paraoxon develop much more rapidly. Humans poisoned by parathion exhibit skin twitching and respiratory distress. In fatal cases, respiratory failure occurs due to central nervous system paralysis.

Malathion is the best known of the phosphorodithioate insecticides. It has a relatively high insect:mammal toxicity ratio because of its two carboxyester linkages which are hydrolyzable by carboxylase enzymes (possessed by mammals, but not insects) to relatively nontoxic products. For example, although malathion is a very effective insecticide, its LD_{50} for adult male rats is about 100 times that of parathion.

Organophosphorus Military Poisons

Powerful inhibitors of acetylcholinesterase enzyme, organophosphorus "nerve gas" military poisons include **Sarin** and **VX**, for which structural formulas are shown below. (The possibility that military poisons such as these might be used in war was a major concern during the 1991 Mid-East conflict, which, fortunately, ended without their being employed.) A systemic poison to the central nervous system that is readily absorbed as a liquid through the skin, Sarin may be lethal at doses as low as about 0.01 mg/kg; a single drop can kill a human.

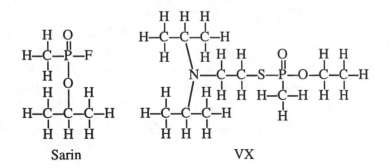

Sarin   VX

LITERATURE CITED

1. U.S. Department of Health and Human Services, Public Health Service Agency for Toxic Substances and Disease Registry, *ATSDR's Toxicological Profiles on CD-ROM*, CRC Press, Boca Raton, FL, 1999.

2. Wald, Matthew L., "Government Admits Work on Weapons Had Role in Illness," *New York Times*, July 15, 1999, p. 1.

3. Lepkowski, Wil, "Arsenic Crisis Spurs Scientists," *Chemical and Engineering News*, May 17, 1999, pp. 45-49.

4. De Mora, Stephen J., *Tributyltin: Case Study of an Environmental Contaminant*, Cambridge University Press, Cambridge, U.K., 1996.

5. Ichihara, Gaku, Isao Saito, Michihiro Kamijima, Xiaozhong Yu, Eiji Shibata, Machiko Toida, and Yasuhiro Takeuchi, "Urinary 2,5-Hexanedione Increases with Potentiation of Neurotoxicity in Chronic Coexposure to *n*-Hexane and Methyl Ethyl Ketone," *International Archives of Occupational and Environmental Health*, **72**, 100-104 (1998).

6. Scherer, Gerhard, Thomas Renner, and Michael Meger, "Analysis and Evaluation of *Trans,trans*-muconic Acid as a Biomarker for Benzene Exposure," *Journal of Chromatography, B: Biomedical Science Applications*, **717**, 179-199 (1998).

7. Lieber, Charles S., "Ethanol Metabolism, Cirrhosis and Alcoholism," *Clinical Chimica Acta*, **257**, 59-84 (1997).

8. Gosselin, Robert E., Roger P. Smith, and Harold C. Hodge, "Phenol," in *Clinical Toxicology of Commercial Products*, 5th ed., Williams and Wilkins, Baltimore/London, 1984, pp. III-344–III-348.

9. Ember, Lois, "Baxter Agrees to Study Alternatives to PVC," *Chemical and Engineering News*, April 12, 1999, pp. 12-13.

10. "New Study Suggests Risks from Plasticizer," *Chemical and Engineering News*, June 21, 1999, p. 24.

11. Gosselin, Robert E., Roger P. Smith, and Harold C. Hodge, "Paraquat," in *Clinical Toxicology of Commercial Products,* 5th ed., Williams and Wilkins, Baltimore/London, 1984, pp. III-328–III-336.

12. U.S. Department of Health and Human Services, Public Health Service Agency for Toxic Substances and Disease Registry, "1,4-Dichlorobenzene," *ATSDR's Toxicological Profiles on CD-ROM*, CRC Press, Boca Raton, FL, 1999.

13. Silberner, J., "Common Herbicide Linked to Cancer," *Science News*, **130**(11), 167–174 (1986).

14. "Dioxin Is Found Not To Increase Birth Defects," *New York Times*, March 18, 1988, p. 12.

SUPPLEMENTARY REFERENCES

Baselt, Randall C. and Robert H. Cravey, *Disposition of Toxic Drugs and Chemicals in Man*, Chemical Toxicology Institute, Foster City, CA, 1995.

Carey, John, Ed., *Ecotoxicological Risk Assessment of the Chlorinated Organic Chemicals*, SETAC Press, Pensacola, FL, 1998.

Cockerham, Lorris G., and Barbara S. Shane, *Basic Environmental Toxicology* , CRCPress/Lewis Publishers, Boca Raton, FL, 1994.

Cooper, Andre R., Leticia Overholt, Heidi Tillquist, and Douglas Jamison, *Cooper's Toxic Exposures Desk Reference with CD-ROM*, CRC Press/Lewis Publishers, Boca Raton, FL,1997.

Ellenhorn, Matthew J. and Sylvia Syma Ellenhorn, *Ellenhorn's Medical Toxicology: Diagnosis and Treatment of Human Poisoning*, 2nd ed., Williams & Wilkins, Baltimore, 1997.

Greenberg, Michael I., Richard J. Hamilton, and Scott D. Phillips, Eds., *Occupational, Industrial, and Environmental Toxicology*, Mosby, St. Louis, 1997.

Hall, Stephen K., Joanna Chakraborty, and Randall J. Ruch, Eds., *Chemical Exposure and Toxic Responses*, CRC Press/Lewis Publishers, Boca Raton, FL, 1997.

Hodgson, Ernest, and Patricia E. Levi, *Introduction to Biochemical Toxicology*, 2nd Ed., Appleton & Lange, Norwalk, CT., 1994.

Johnson, Barry L., *Impact of Hazardous Waste on Human Health: Hazard, Health Effects, Equity and Communication Issues*, Ann Arbor Press, Chelsea, MI, 1999.

Klaassen, Curtis D., Mary O. Amdur, and John Doull, Eds., *Casarett and Doull's Toxicology: The Basic Science of Poisons Companion*, 5th ed., McGraw-Hill, Health Professions Division, New York, 1996.

Klaassen, Curtis D. and John B. Watkins, III, Eds., *Casarett and Doull's Toxicology: The Basic Science of Poisons Companion Handbook*, 5th ed., McGraw-Hill, Health Professions Division, New York, 1999.

Kneip, Theodore J., and John V. Crable, Eds., *Methods for Biological Monitoring: A Manual for Assessing Human Exposure to Hazardous Substances*, American Public Health Association, Washington, D.C., 1988.

Lippmann, Morton, Ed., *Environmental Toxicants: Human Exposures and their Health Effects*, 2nd ed., John Wiley & Sons, New York, 1999.

Liverman, Catharyn T., Ed., *Toxicology and Environmental Health Information Resources: The Role of the National Library of Medicine*, National Academy Press, Washington, D.C., 1997.

Malachowski, M. J., *Health Effects of Toxic Substances*, Government Institutes, Rockville, MD, 1995.

Ostler, Neal K., Thomas E. Byrne, and Michael J. Malachowski, *Health Effects of Hazardous Materials*, Neal K. Ostler, Prentice Hall, Upper Saddle River, NJ,1996.

Ostrander, Gary K., Ed., *Techniques in Aquatic Toxicology*, CRC Press/Lewis Publishers, Boca Raton, FL, 1996.

Patnaik, Pradyot, *A Comprehensive Guide to the Hazardous Properties of Chemical Substances*, 2nd ed., John Wiley & Sons, Inc., New York, 1999.

Richardson, Mervyn, *Environmental Xenobiotics*, Taylor and Francis, London, 1996.

Richardson, Mervyn, *Environmental Toxicology Assessment*, Taylor & Francis, London, 1995.

Rom, William N., Ed., *Environmental & Occupational Medicine*, 3rd ed., Lippincott-Raven Publishers, Philadelphia, 1998.

Vallejo Rosero, Maria del Carmen, *Toxicologia Ambiental: Fuentes, Cinetica y Efectos de los Contaminantes*, Santafe de Bogota: Fondo Nacional Universitario, Bogota, Colombia, 1997.

Zelikoff, Judith, and Peter L. Thomas, Eds., *Immunotoxicology of Environmental and Occupational Metals*, Taylor and Francis, London, 1998.

QUESTIONS AND PROBLEMS

1. List and discuss two elements that are invariably toxic in their elemental forms. For another element, list and discuss two elemental forms, one of which is quite toxic and the other of which is essential for the body. In what sense is even the toxic form of this element "essential for life?"

2. What is a toxic substance that bonds to iron(III) in iron-containing ferricytochrome oxidase enzyme, preventing its reduction to iron(II) in the oxidative phosphorylation process by which the body utilizes O_2?

3. What are interhalogen compounds, and which elemental forms do their toxic effects most closely resemble?

4. Name and describe the three health conditions that may be caused by inhalation of asbestos.

5. Why might tetraethyllead be classified as "the most notable toxic organometallic compound"?

6. What is the most common toxic effect commonly attributed to low-molar-mass alkanes?

7. Information about the toxicities of many substances to humans is lacking because of limited data on direct human exposure. (Volunteers to study human health effects of toxicants are in notably short supply.) However, there is a great deal of information available about human exposure to phenol and the adverse effects of such exposure. Explain.

8. Comment on the toxicity of the compound below:

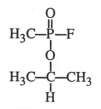

9. What are neuropathic disorders? Why are organic solvents frequently the cause of such disorders?

10. What is a major metabolic effect of aniline? What is this effect called? How is it manifested?

11. What are the organic compounds characterized by the N–N=O functional group? What is their major effect on health?

12. What structural group is characteristic of carbamates? For what purpose are these compounds commonly used? What are their major advantages in such an application?

13. What is lipid peroxidation? Which common toxic substance is known to cause lipid peroxidation?

14. Biochemically, what do organophosphate esters such as parathion do that could classify them as "nerve poisons"?

15. Although benzene and toluene have a number of chemical similarities, their metabolisms and toxic effects are quite different. Explain.

24 CHEMICAL ANALYSIS OF WATER AND WASTEWATER

24.1. GENERAL ASPECTS OF ENVIRONMENTAL CHEMICAL ANALYSIS

Scientists' understanding of the environment can only be as good as their knowledge of the identities and quantities of pollutants and other chemical species in water, air, soil, and biological systems. Therefore, proven, state-of-the-art techniques of chemical analysis, properly employed, are essential to environmental chemistry. Now is a very exciting period in the evolution of analytical chemistry, characterized by the development of new and improved analysis techniques that enable detection of much lower levels of chemical species and a vastly increased data throughput. These developments pose some challenges. Because of the lower detection limits of some instruments, it is now possible to see quantities of pollutants that would have escaped detection previously, resulting in difficult questions regarding the setting of maximum allowable limits of various pollutants. The increased output of data from automated instruments has in many cases overwhelmed human capacity to assimilate and understand it.

Challenging problems still remain in developing and utilizing techniques of environmental chemical analysis. Not the least of these problems is knowing which species should be measured, or even whether or not an analysis should be performed at all. The quality and choice of analyses is much more important than the number of analyses performed. Indeed, a persuasive argument can be made that, given modern capabilities in analytical chemistry, too many analyses of environmental samples are performed, whereas fewer, more carefully planned analyses would yield more useful information.

In addition to a discussion of water analysis, this chapter covers some of the general aspects of environmental chemical analysis and the major techiques that are used to determine a wide range of analytes (species measured). Many techniques are common to water, air, soil, and biological sample analyses and reference is made to them in chapters that follow.

Error and Quality Control

A crucial aspect of any chemical analysis is the validity and quality of the data that it produces. All measurements are subject to error, which may be **systematic** (of the same magnitude and same direction) or **random** (varying in both magnitude and direction). Systematic errors cause the measured values to vary consistently from the true values, this variation is known as the **bias**. The degree to which a measured value comes close to the actual value of an analytical measurement is called the **accuracy** of the measurement, reflecting both systematic and random errors. It is essential for the analyst to determine these error components in the measurement of environmental samples, including water samples. The identification and control of systematic and random errors falls in the category of **quality control (Q C)** procedures. It is beyond the scope of this chapter to go into any detail on these crucial procedures for which the reader is referred to a work on standard methods for the analysis of water.[1]

In order for results from a laboratory to be meaningful, the laboratory needs a quality assurance plan specifying measures taken to produce data of known quality. An important aspect of such a plan is the use of laboratory control standards consisting of samples with very accurately known analyte levels in a carefully controlled matrix. Such standard reference materials are available in the U. S. for many kinds of samples from the National Institute of Standards and Technology (NIST).

Many environmental analytes are present at very low levels which challenge the ability of the method used to detect and accurately quantify them. Therefore, the **detection limit** of a method of analysis is quite important. Defining detection limit has long been a controversial topic in chemical analysis. Every analytical method has a certain degree of noise. The detection limit is an expression of the lowest concentration of analyte that can be measured above the noise level with a specified degree of confidence in an analytical procedure. In the detection of analyte, two kinds of errors can be defined. A Type I error occurs when the measurement finds an analyte present when it actually is absent. A Type II error occurs when the measurement finds an analyte absent when it is actually present.

Detection limits can be further categorized into several different subcategories. The **instrument detection limit (IDL)** is the analyte concentration capable of producing a signal three times the standard deviation of the noise. The **lower level of detection (LLD)** is the quantity of analyte that will produce a measurable signal 99 percent of the time; it is about 2 times the IDL. The **method detection limit (MDL)** is measured like the LLD except that the analyte is taken through the whole analytical procedure, including steps such as extraction and sample cleanup; it is about 4 times the IDL . Finally, the **practical quantitation limit (PQL)**, which is about 20 times the IDL, is the lowest level achievable among laboratories in routine analysis.

24.2. CLASSICAL METHODS

Before sophisticated instrumentation became available, most important water quality parameters and some air pollutant analyses were done by **classical methods**, which require only chemicals, balances to measure masses, burets, volumetric flasks and pipets to measure volumes, and other simple laboratory glassware. The two major classical methods are **volumetric analysis**, in which volumes of reagents are measured, and **gravimetric analysis**, in which masses are measured. Some of these methods are still used today, and many have been adapted to instrumental and automated procedures.

The most common classical methods for pollutant analysis are titrations, largely used for water analysis. Some of the titration procedures used are discussed in this section.

Acidity (see Section 3.7) is determined simply by titrating hydrogen ion with base. Titration to the methyl orange endpoint (pH 4.5) yields the "free acidity" due to strong acids (HCl, H_2SO_4). Carbon dioxide does not, of course, appear in this category. Titration to the phenolphthalein endpoint, pH 8.3, yields total acidity and accounts for all acids except those weaker than HCO_3^-.

Alkalinity may be determined by titration with H_2SO_4 to pH 8.3 to neutralize bases as strong as, or stronger than, carbonate ion,

$$CO_3^{2-} + H^+ \rightarrow HCO_3^- \tag{24.2.1}$$

or by titration to pH 4.5 to neutralize bases weaker than CO_3^{2-}, but as strong as, or stronger than, HCO_3^-:

$$HCO_3^- + H^+ \rightarrow H_2O + CO_2(g) \tag{24.2.2}$$

Titration to the lower pH yields total alkalinity.

The ions involved in water hardness, a measure of the total concentration of calcium and magnesium in water, are readily titrated at pH 10 with a solution of EDTA, a chelating agent discussed in Sections 3.10 and 3.13. The titration reaction is

$$Ca^{2+}(\text{or } Mg^{2+}) + H_2Y^{2-} \rightarrow CaY^{2-}(\text{or } MgY^{2-}) + 2H^+ \tag{24.2.3}$$

where H_2Y^{2-} is the partially ionized EDTA chelating agent. Eriochrome Black T is used as an indicator, and it requires the presence of magnesium, with which it forms a wine red complex.

Several oxidation-reduction titrations can be used for environmental chemical analysis. Oxygen is determined in water by the Winkler titration. The first reaction in the Winkler method is the oxidation of manganese(II) to manganese(IV) by the oxygen analyte in a basic medium; this reaction is followed by acidification of the brown hydrated MnO_2 in the presence of I^- ion to release free I_2, then titration of the liberated iodine with standard thiosulfate, using starch as an endpoint indicator:

$$Mn^{2+} + 2OH^- + \tfrac{1}{2}O_2 \rightarrow MnO_2(s) + H_2O \tag{24.2.4}$$

$$MnO_2(s) + 2I^- + 4H^+ \rightarrow Mn^{2+} + I_2 + 2H_2O \tag{24.2.5}$$

$$I_2 + 2S_2O_3^{2-} \rightarrow S_4O_6^{2-} + 2I^- \tag{24.2.6}$$

A back calculation from the amount of thiosulfate required yields the original quantity of dissolved oxygen (DO) present. Biochemical oxygen demand, BOD (see

Section 7.9), is determined by adding a microbial "seed" to the diluted sample, saturating with air, incubating for five days, and determining the oxygen remaining. The results are calculated to show BOD as mg/L of O_2. A BOD of 80 mg/L, for example, means that biodegradation of the organic matter in a liter of the sample would consume 80 mg of oxygen.

24.3. SPECTROPHOTOMETRIC METHODS

Absorption Spectrophotometry

Absorption spectrophotometry of light-absorbing species in solution, historically called colorimetry when visible light is absorbed, is still used for the analysis of many water and some air pollutants. Basically, absorption spectrophotometry consists of measuring the percent transmittance (%T) of monochromatic light passing through a light-absorbing solution as compared to the amount passing through a blank solution containing everything in the medium but the sought-for constituent (100%). The absorbance (A) is defined as the following:

$$A = \log \frac{100}{\%T} \qquad (24.3.1)$$

The relationship between A and the concentration (C) of the absorbing substance is given by Beer's law:

$$A = abC \qquad (24.3.2)$$

where a is the absorptivity, a wavelength-dependent parameter characteristic of the absorbing substance; b is the path length of the light through the absorbing solution; and C is the concentration of the absorbing substance. A linear relationship between A and C at constant path length indicates adherence to Beer's law. In many cases, analyses may be performed even when Beer's law is not obeyed, if a suitable calibration curve is prepared. A color-developing step usually is required in which the sought-for substance reacts to form a colored species, and in some cases a colored species is extracted into a nonaqueous solvent to provide a more intense color and a more concentrated solution.

A number of solution spectrophotometric methods have been used for the determination of water and air pollutants. Some of these are summarized in Table 24.1.

Atomic Absorption and Emission Analyses

Atomic absorption analysis is commonly used for the determination of metals in environmental samples. This technique is based upon the absorption of monochromatic light by a cloud of atoms of the analyte metal. The monochromatic light can be produced by a source composed of the same atoms as those being analyzed. The source produces intense electromagnetic radiation with a wavelength exactly the same as that absorbed by the atoms, resulting in extremely high selectivity. The basic components of an atomic absorption instrument are shown in Figure 24.1. The

Table 24.1. Solution Spectrophotometric (Colorimetric) Methods of Analysis

Analyte	Reagent and Method
Ammonia	Alkaline mercury(II) iodide reacts with ammonia, producing colloidal orange-brown $NH_2Hg_2I_3$, which absorbs light between 400 and 500 nanometers (nm)
Arsenic	Reaction of arsine, AsH_3, with silver diethylthiocarbamate in pyridine, forming a red complex
Boron	Reaction with curcumin, forming red rosocyanine
Bromide	Reaction of hypobromite with phenol red to form bromphenol blue-type indicator
Chlorine	Development of color with orthotolidine
Cyanide	Formation of a blue dye from reaction of cyanogen chloride, CNCl, with pyridine-pyrazolone reagent, measured at 620 nm
Fluoride	Decolorization of a zirconium-dye colloidal precipitate ("lake") by formation of colorless zirconium fluoride and free dye
Nitrate and nitrite	Nitrate is reduced to nitrite, which is diazotized with sulfanilamide and coupled with N-(l-naphthyl)-ethylenediamine dihydrochloride to produce a highly colored azo dye measured at 540 nm
Nitrogen, Kjeldahl-phenate method	Digestion in sulfuric acid to NH_4^+ followed by treatment with alkaline phenol reagent and sodium hypochlorite to form blue indophenol measured at 630 nm
Phenols	Reaction with 4-aminoantipyrine at pH 10 in the presence of potassium ferricyanide, forming an antipyrine dye which is extracted into pyridine and measured at 460 nm
Phosphate	Reaction with molybdate ion to form a phosphomolybdate which is selectively reduced to intensely colored molybdenum blue
Selenium	Reaction with diaminobenzidine, forming colored species absorbing at 420 nm
Silica	Formation of molybdosilicic acid with molybdate, followed by reduction to a heteropoly blue measured at 650 nm or 815 nm
Sulfide	Formation of methylene blue
Surfactants	Reaction with methylene blue to form blue salt
Tannin and lignin	Blue color from tungstophosphoric and molybdophosphoric acids

key element is the hollow cathode lamp in which atoms of the analyte metal are energized such that they become electronically excited and emit radiation with a very narrow wavelength band characteristic of the metal. This radiation is guided by the appropriate optics through a flame into which the sample is aspirated. In the flame, most metallic compounds are decomposed, and the metal is reduced to the elemental state, forming a cloud of atoms. These atoms absorb a fraction of radiation in the flame. The fraction of radiation absorbed increases with the concentration of the sought-for element in the sample according to the Beer's law relationship (Eq. 24.3.2). The attenuated light beam next goes to a monochromator to eliminate extraneous light resulting from the flame, and then to a detector.

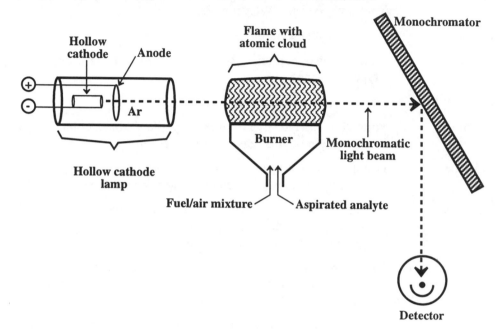

Figure 24.1. The basic components of a flame atomic absorption spectrophotometer.

Atomizers other than a flame can be used. The most common of these is the graphite furnace, an electrothermal atomization device which consists of a hollow graphite cylinder placed so that the light beam passes through it. A small sample of up to 100 μL is inserted in the tube through a hole in the top. An electric current is passed through the tube to heat it—gently at first to dry the sample, then rapidly to vaporize and excite the metal analyte. The absorption of metal atoms in the hollow portion of the tube is measured and recorded as a spike-shaped signal. A diagram of a graphite furnace with a typical output signal is shown in Figure 24.2. The major advantage of the graphite furnace is that it gives detection limits up to 1000 times lower than those of conventional flame devices.

A special technique for the flameless atomic absorption analysis of mercury involves room-temperature reduction of mercury to the elemental state by tin(II) chloride in solution, followed by sweeping the mercury into an absorption cell with air. Nanogram (10^{-9}g) quantities of mercury can be determined by measuring mercury absorption at 253.7 nm.

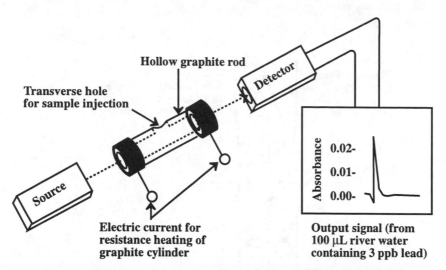

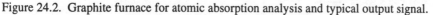

Figure 24.2. Graphite furnace for atomic absorption analysis and typical output signal.

Atomic Emission Techniques

Metals may be determined in water, atmospheric particulate matter, and biolog-ical samples very well by observing the spectral lines emitted when they are heated to a very high temperature. An especially useful atomic emission technique is inductively coupled plasma atomic emission spectroscopy (ICP/AES). The "flame" in which analyte atoms are excited in plasma emission consists of an incandescent plasma (ionized gas) of argon heated inductively by radiofrequency energy at 4-50 MHz and 2-5 kW (Figure 24.3). The energy is transferred to a stream of argon through an induction coil, producing temperatures up to 10,000 K. The sample atoms are subjected to temperatures around 7000 K, twice those of the hottest conventional flames (for example, nitrous oxide-acetylene operates at 3   s200 K). Since emission of light increases exponentially with temperature, lower detection limits are obtained. Furthermore, the technique enables emission analysis of some of the environmentally important metalloids such as arsenic, boron, and selenium. Interfering chemical reactions and interactions in the plasma are minimized as compared to flames. Of greatest significance, however, is the capability of analyzing as many as 30 elements simultaneously, enabling a true multielement analysis technique. Plasma atomization combined with mass spectrometric measurement of analyte elements is a relatively new technique that is an especially powerful means for multielement analysis.

24.4. ELECTROCHEMICAL METHODS OF ANALYSIS

Several useful techniques for water analysis utilize electrochemical sensors. These techniques may be potentiometric, voltammetric, or amperometric. Potenti-ometry is based upon the general principle that the relationship between the potential of a measuring electrode and that of a reference electrode is a function of the log of the activity of an ion in solution. For a measuring electrode responding selectively to a particular ion, this relationship is given by the Nernst equation,

$$E = E^o + \frac{2.303RT}{zF} \log(a_z) \tag{24.4.1}$$

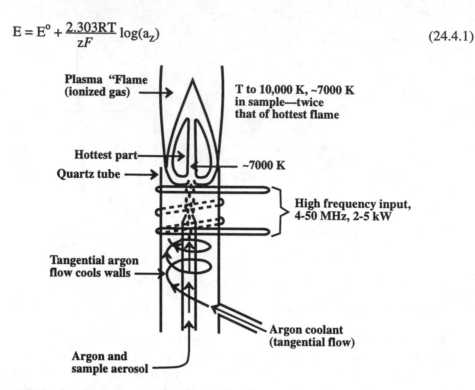

Plasma "Flame (ionized gas) → **T to 10,000 K, ~7000 K in sample—twice that of hottest flame**

Hottest part —
Quartz tube → **~7000 K**

High frequency input, 4-50 MHz, 2-5 kW

Tangential argon flow cools walls →

Argon coolant (tangential flow)

Argon and sample aerosol —

Figure 24.3. Schematic diagram showing inductively coupled plasma used for optical emission spectroscopy.

where E is the measured potential; E^o is the standard electrode potential; R is the gas constant; T is the absolute temperature; z is the signed charge (+ for cations, - for anions); F is the Faraday constant; and a is the activity of the ion being measured. At a given temperature, the quantity $2.303RT/F$ has a constant value; at 25°C it is 0.0592 volt (59.2 mv). At constant ionic strength, the activity, a, is directly proportional to concentration, and the Nernst equation may be written as the following for electrodes responding to Cd^{2+} and F^-, respectively:

$$E \text{ (in mv)} = E^o + \frac{59.2}{2} \log [Cd^{2+}] \tag{24.4.2}$$

$$E = E^o - 59.2 \log [F^-] \tag{24.4.3}$$

Electrodes that respond more or less selectively to various ions are called **ion-selective electrodes**. Generally, the potential-developing component is a membrane of some kind that allows for selective exchange of the sought-for ion. The glass electrode used for the measurement of hydrogen-ion activity and pH is the oldest and most widely used ion-selective electrode. The potential is developed at a glass membrane that selectively exchanges hydrogen ion in preference to other cations, giving a Nernstian response to hydrogen ion activity, a_{H^+}:

$$E = E^o + 59.2 \log(a_{H^+}) \tag{24.4.4}$$

Of the ion-selective electrodes other than glass electrodes, the fluoride electrode is the most successful. It is well-behaved, relatively free of interferences, and has an adequately low detection limit and a long range of linear response. Like all ion-selective electrodes, its electrical output is in the form of a potential signal that is proportional to log of concentration. A small error in E leads to a variation in log of concentration, which leads to relatively high concentration errors.

Voltammetric techniques, the measurement of current resulting from potential applied to a microelectrode, have found some applications in water analysis. One such technique is differential-pulse polarography, in which the potential is applied to the microelectrode in the form of small pulses superimposed on a linearly increasing potential. The current is read near the end of the voltage pulse and compared to the current just before the pulse was applied. It has the advantage of minimizing the capacitive current from charging the microelectrode surface, which sometimes obscures the current due to the reduction or oxidation of the species being analyzed. Anodic-stripping voltammetry involves deposition of metals on an electrode surface over a period of several minutes followed by stripping them off very rapidly using a linear anodic sweep. The electrodeposition concentrates the metals on the electrode surface, and increased sensitivity results. An even better technique is to strip the metals off using a differential pulse signal. A differential-pulse anodic-stripping voltammogram of copper, lead, cadmium, and zinc in tap water is shown in Figure 24.4.

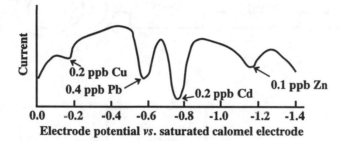

Figure 24.4. Differential-pulse anodic-stripping voltammogram of tap water at a mercury-plated, wax-impregnated graphite electrode.

24.5. CHROMATOGRAPHY

First described in the literature in the early 1950s, gas chromatography has played an essential role in the analysis of organic materials. Gas chromatography is both a qualitative and quantitative technique; for some analytical applications of environmental importance, it is remarkably sensitive and selective. Gas chromatography is based upon the principle that when a mixture of volatile materials transported by a carrier gas is passed through a column containing an adsorbent solid phase or, more commonly, an absorbing liquid phase coated on a solid material, each volatile component will be partitioned between the carrier gas and the solid or liquid. The length of time required for the volatile component to traverse the column is proportional to the degree to which it is retained by the nongaseous phase. Since different components may be retained to different degrees, they will emerge from the end of the column at different times. If a suitable detector is available, the time at which the component emerges from the column and the quantity of the component are both measured. A recorder trace of the detector response appears as peaks of different sizes, depending upon the quantity of material producing the detector response. Both quantitative and (within limits) qualitative analyses of the sought-for substances are obtained.

The essential features of a gas chromatograph are shown schematically in Figure 24.5. The carrier gas generally is argon, helium, hydrogen, or nitrogen. The sample is injected as a single compact plug into the carrier gas stream immediately ahead of the column entrance. If the sample is liquid, the injection chamber is heated to vaporize the liquid rapidly. The separation column may consist of a metal or glass tube packed with an inert solid of high surface area covered with a liquid phase, or it may consist of an active solid, which enables the separation to occur. More commonly, capillary columns are now employed which consist of very small diameter, very long tubes in which the liquid phase is coated on the inside of the column.

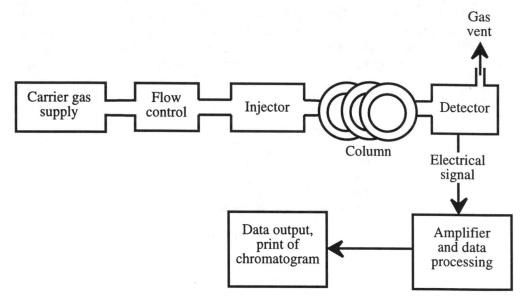

Figure 24.5. Schematic diagram of the essential features of a gas chromatograph.

The component that primarily determines the sensitivity of gas chromatographic analysis and, for some classes of compounds, the selectivity as well, is the detector. One such device is the thermal conductivity detector, which responds to changes in the thermal conductivity of gases passing over it. The electron-capture detector, which is especially useful for halogenated hydrocarbons and phosphorus compounds, operates through the capture of electrons emitted by a beta-particle source. The flame-ionization gas chromatographic detector is very sensitive for the detection of organic compounds. It is based upon the phenomenon by which organic compounds form highly conducting fragments, such as C^+, in a flame. Application of a potential gradient across the flame results in a small current that may be readily measured. The mass spectrometer, described in Section 24.6, may be used as a detector for a gas chromatograph. A combined gas chromatograph/mass spectrometer (GC/MS) instrument is an especially powerful analytical tool for organic compounds.

Chromatographic analysis requires that a compound exhibit at least a few mm of vapor pressure at the highest temperature at which it is stable. In many cases, organic compounds that cannot be chromatographed directly may be converted to derivatives that are amenable to gas chromatographic analysis. It is seldom possible to analyze organic compounds in water by direct injection of the water into the gas chromatograph; higher concentration is usually required. Two techniques commonly employed to remove volatile compounds from water and concentrate them are extraction with solvents and purging volatile compounds with a gas, such as helium; concentrating the purged gases on a short column; and driving them off by heat into the chromatograph.

High-Performance Liquid Chromatography

A liquid mobile phase used with very small column-packing particles enables high-resolution chromatographic separation of materials in the liquid phase. Very high pressures up to several thousand psi are required to get a reasonable flow rate in such systems. Analysis using such devices is called **high-performance liquid chromatography** (HPLC) and offers an enormous advantage in that the materials analyzed need not be changed to the vapor phase, a step that often requires preparation of a volatile derivative or results in decomposition of the sample. The basic features of a high-performance liquid chromatograph are the same as those of a gas chromatograph, shown in Figure 24.5, except that a solvent reservoir and high-pressure pump are substituted for the carrier gas source and regulator. A hypothetical HPLC chromatogram is shown in Figure 24.6. Refractive index and ultraviolet detectors are both used for the detection of peaks coming from the liquid chromatograph column. Fluorescence detection can be especially sensitive for some classes of compounds. Mass spectrometric detection of HPLC effluents has lead to the development of LC/MS analysis. Somewhat difficult in practice, this technique can be a powerful tool for the determination of analytes that cannot be subjected to gas chromatography. High-performance liquid chromatography has emerged as a very useful technique for the analysis of a number of water pollutants.

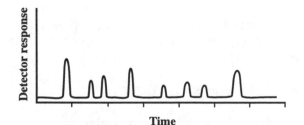

Figure 24.6. Hypothetical HPLC chromatogram.

Chromatographic Analysis of Water Pollutants

The U. S. Environmental Protection Agency has developed a number of chromatography-based standard methods for determining water pollutants.[2] Some of these methods use the purge-and-trap technique, bubbling gas through a column of water to purge volatile organics from the water followed by trapping the organics on solid sorbents, whereas others use solvent extraction to isolate and concentrate the organics. These methods are summarized in Table 24.2.

Ion Chromatography

The liquid chromatographic determination of ions, particularly anions, has enabled the measurement of species that used to be very troublesome for water chemists. This technique is called **ion chromatography**, and its development has been facilitated by special detection techniques using so-called suppressors to enable detection of analyte ions in the chromatographic effluent. Ion chromatography has been developed for the determination of most of the common anions, including arsenate, arsenite, borate, carbonate, chlorate, chlorite, cyanide, the halides, hypochlorite, hypophosphite, nitrate, nitrite, phosphate, phosphite, pyrophosphate, selenate, selenite, sulfate, sulfite, sulfide, trimetaphosphate, and tripolyphosphate. Cations, including the common metal ions, can also be determined by ion chromatography.

24.6. MASS SPECTROMETRY

Mass spectrometry is particularly useful for the identification of specific organic pollutants. It depends upon the production of ions by an electrical discharge or chemical process, followed by separation based on the charge-to-mass ratio and measurement of the ions produced. The output of a mass spectrometer is a mass spectrum, such as the one shown in Figure 24.8. A mass spectrum is characteristic of a compound and serves to identify it. Computerized data banks for mass spectra have been established and are stored in computers interfaced with mass spectrometers. Identification of a mass spectrum depends upon the purity of the compound from which the spectrum is taken. Prior separation by gas chromatography with continual sampling of the column effluent by a mass spectrometer, commonly called gas chromatography-mass spectrometry (GC/MS), is particularly effective in the analysis of organic pollutants.

24.7. ANALYSIS OF WATER SAMPLES

The preceding sections of this chapter have covered the major kinds of analysis techniques that are used on water. In this section several specific aspects of water analysis are addressed.

Physical Properties Measured in Water

The commonly determined physical properties of water are color, residue (solids), odor, temperature, specific conductance, and turbidity. Most of these terms are self-explanatory and will not be discussed in detail. All of these properties either

Table 24.2. Chromatography-based EPA Methods for Organic Compounds in Water

Class of compounds	GC	GC/MS	HPLC	Example analytes
		Method Number		
Purgeable halocarbons	601			Carbon tetrachloride
Purgeable aromatics	602			Toluene
Acrolein and acrylo-nitrile	603			Acrolein
Phenols	604			Phenol and chlorophenols
Benzidines			605	Benzidine
Phthalate esters	606			Bis(2-ethylhexylphthalate)
Nitrosamines	607			N-nitroso-N-dimethylamine
Organochlorine pesticides and PCB's	608			Heptachlor, PCB 1016
Nitroaromatics and isophorone	609			Nitrobenzene
Polycyclic aromatic hydrocarbons	610		610	Benzo[a]pyrene
Haloethers	611			Bis(2-chloroethyl) ether
Chlorinated hydrocarbons	612			1,3-Dichlorobenzene
2,3,7,8-Tetrachloro-dibenzo-p-dioxin		613		2,3,7,8-TCDD
Organophosphorus pesticides	614			Malathion
Chlorinated Herbicides	615			Dinoseb
Triazine Pesticides	619			Atrazine
Purgeable organics		624		Ethylbenzene
Base/neutrals and acids		625		More than 70 organic compounds
Dinitro aromatic pesticides		646		Basalin (Fluchloralin)
Volatile organic compounds		1624		Vinyl chloride

influence or reflect the chemistry of the water. Solids, for example, arise from chemical substances either suspended or dissolved in the water and are classified physically as total, filterable, nonfilterable, or volatile. Specific conductance is a measure of the degree to which water conducts alternating current and reflects, therefore, the total concentration of dissolved ionic material. By necessity, some physical properties must be measured in the water without sampling (see discussion of water sampling below).

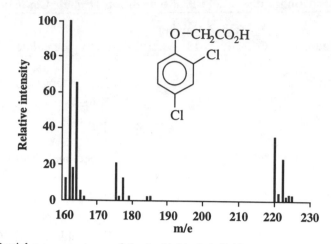

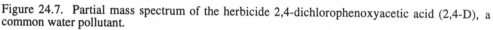

Figure 24.7. Partial mass spectrum of the herbicide 2,4-dichlorophenoxyacetic acid (2,4-D), a common water pollutant.

Water Sampling

It is beyond the scope of this text to describe water sampling procedures in detail. It must be emphasized, however, that the acquisition of meaningful data demands that correct sampling and storage procedures be used. These procedures may be quite different for various species in water. In general, separate samples must be collected for chemical and biological analysis because the sampling and preservation techniques differ significantly. Usually, the shorter the time interval between sample collection and analysis, the more accurate the analysis will be. Indeed, some analyses must be performed in the field within minutes of sample collection. Others, such as the determination of temperature, must be done on the body of water itself. Within a few minutes after collection, water pH may change, dissolved gases (oxygen, carbon dioxide, hydrogen sulfide, chlorine) may be lost, or other gases (oxygen, carbon dioxide) may be absorbed from the atmosphere. Therefore, analyses of temperature, pH, and dissolved gases should always be performed in the field. Furthermore, precipitation of calcium carbonate accompanies changes in the pH-alkalinity-calcium carbonate relationship following sample collection. Analysis of a sample after standing may thus give erroneously low values for calcium and total hardness.

Oxidation-reduction reactions may cause substantial errors in analysis. For example, soluble iron(II) and manganese(II) are oxidized to insoluble iron(III) and manganese(IV) compounds when an anaerobic water sample is exposed to atmospheric oxygen. Microbial activity may decrease phenol or biological oxygen demand (BOD) values, change the nitrate-nitrite-ammonia balance, or alter the relative proportions of sulfate and sulfide. Iodide and cyanide frequently are oxidized. Chromium(VI) in solution may be reduced to insoluble chromium(III). Sodium, silicate, and boron are leached from glass container walls.

Samples can be divided into two major categories. **Grab samples** are taken at a single time and in a single place. Therefore, they are very specific with respect to time and location. **Composite samples** are collected over an extended time and may encompass different locations as well. In principle, the average results from a large number of grab samples give the same information as a composite sample. A composite sample has the advantage of providing an overall picture from only one analysis. On the other hand, it may miss extreme concentrations and important variations that occur over time and space.

Solid-Phase Extractors

The ease and effectiveness of various kinds of solid-phase devices for water sampling is steadily increasing their use in water analysis. Based upon size and physical configuration, at least three categories of such devices are available. One of these is the conventional solid-phase extractor (SPE) containing an extracting solid in a column. Activated carbon has been used for decades for this purpose, but synthetic materials, such as those composed of long hydrocarbon chains (C18) bound to solids have been found to be quite useful. A typical procedure uses a polymer-divinylstyrene extraction column to remove pesticides from water.[3] The pesticide analytes are eluted from the SPE with ethyl acetate and measured by gas chromatography. A mean recovery of 85% has been reported.

A clever approach to sulfide analysis using SPE has been described.[4] The water sample is sucked into an airtight syringe to prevent exposure to sulfide-oxidizing atmospheric oxygen and is immediately reacted with N,N-dimethyl-*p*-phenylenediamine sulfate and $FeCl_3$, which produces methylene blue, a colored compound used as an indicator. The resulting solution is forced through a Sep-Pak C18 solid phase extractor to remove the methylene blue, which is stable for at least 30 days on the solid phase. After elution with a mixture of methanol and 0.01 M HCl, the absorbance of the methylene blue is measured at 659 nm to quantitate the sulfide.

Solid-phase microextraction (SPME) devices constitute a second kind of solid-phase extractor. These make use of very small diameter devices in which analytes are bonded directly to the extractor walls, then eluted directly into a chromatograph. The use of SPME devices for the determination of haloethers in water has been described.[5]

A third kind of device, disks composed of substances that bind with and remove analytes from water when the water is filtered through them, are available for a number of classes of substances and are gaining in popularity because of their simplicity and convenience. As an example, solid phase extraction disks can be used to remove and concentrate radionuclides from water, including ^{99}Tc, ^{137}Cs, ^{90}Sr, ^{238}Pu.[6] Organic materials sampled from water with such disks include haloacetic acids[7] and acidic and neutral herbicides.[8]

Water Sample Preservation

It is not possible to completely protect a water sample from changes in composition. However, various additives and treatment techniques can be employed to minimize sample deterioration. These methods are summarized in Table 24.3.

The most general method of sample preservation is refrigeration to 4°C. Freezing normally should be avoided because of physical changes—formation of precipitates and loss of gas—which may adversely affect sample composition. Acidification is commonly applied to metal samples to prevent their precipitation, and it also slows microbial action. In the case of metals, the samples should be filtered before adding acid to enable determination of dissolved metals. Sample holding times vary, from zero for parameters such as temperature or dissolved oxygen measured by a probe, to 6 months for metals. Many different kinds of samples, including those to be analyzed for acidity, alkalinity, and various forms of nitrogen or phosphorus, should not be held for more than 24 hours. Details on water sample preservation are to be found in standard references on water analysis.[9] Instructions should be followed for each kind of sample in order to ensure meaningful results.

Table 24.3. Preservatives and Preservation Methods Used with Water Samples

Preservative or technique used	Effect on sample	Type of samples for which the method is employed
Nitric acid	Keeps metals in solution	Metal-containing samples
Sulfuric acid	Bactericide	Biodegradable samples containing organic carbon, oil, or grease
	Formation of sulfates with volatile bases	Samples containing amines or ammonia
Sodium hydroxide	Formation of sodium salts from volatile acids	Samples containing volatile organic acids or cyanides
Chemical reaction	Fix a particular constituent	Samples to be analyzed for dissolved oxygen using the Winkler method

Total Organic Carbon in Water

The importance and possible detrimental effects of dissolved organic compounds in water were discussed in Chapter 7. Dissolved organic carbon exerts an oxygen demand in water, often is in the form of toxic substances, and is a general indicator of water pollution. Therefore, its measurement is quite important. The measurement of total organic carbon, TOC, is now recognized as the best means of assessing the organic content of a water sample. The measurement of this parameter has been facilitated by the development of methods which, for the most part, totally oxidize the dissolved organic material to produce carbon doxide. The amount of carbon dioxide evolved is taken as a measure of TOC.

TOC can be determined by a technique that uses a dissolved oxidizing agent promoted by ultraviolet light. Potassium peroxydisulfate, $K_2S_2O_8$, can be used as an oxidizing agent to be added to the sample. Phosphoric acid is also added to the sample, which is sparged with air or nitrogen to drive off CO_2 formed from HCO_3^- and CO_3^{2-} in solution. After sparging, the sample is pumped to a chamber containing a lamp emitting ultraviolet radiation of 184.9 nm. This radiation produces reactive free radical species such as the hydroxyl radical, $HO\cdot$, discussed extensively as a photochemical reaction intermediate in Chapters 9, 12, and 13. These active species bring about the rapid oxidation of dissolved organic compounds as shown in the following general reaction:

$$\text{Organics} + HO\cdot \rightarrow CO_2 + H_2O \tag{24.7.1}$$

After oxidation is complete, the CO_2 is sparged from the system and measured with a gas chromatographic detector or by absorption in ultrapure water followed by a conductivity measurement. Figure 24.8 is a schematic of a TOC analyzer.

Measurement of Radioactivity in Water

There are several potential sources of radioactive materials that may contaminate water (see Section 7.13). Radioactive contamination of water is normally detected by measurements of gross beta and gross alpha activity, a procedure that is simpler than detecting individual isotopes. The measurement is made from a sample formed by evaporating water to a very thin layer on a small pan, which is then inserted inside an internal proportional counter. This setup is necessary because beta particles can penetrate only very thin detector windows, and alpha particles have essentially no

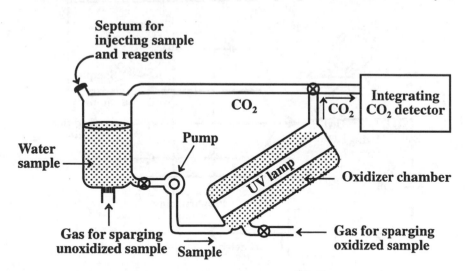

Figure 24.8. TOC analyzer employing UV-promoted sample oxidation.

penetrating power. More detailed information can be obtained for radionuclides that emit gamma rays by the use of gamma spectrum analysis. This technique employs solid state detectors to resolve rather closely spaced gamma peaks in the sample's spectra. In conjunction with multichannel spectrometric data analysis, it is possible to determine a number of radionuclides in the same sample without chemical separation. This method requires minimal sample preparation.

Biological Toxins

Toxic substances produced by microorganisms are of some concern in water. Photosynthetic cyanobacteria and some kinds of algae growing in water produce potentially troublesome toxic substances. An immunoassay method of analysis (see Chapter 25, Section 25.5) for such toxins has been described.[10]

Summary of Water Analysis Procedures

The main chemical parameters commonly determined in water are summarized in Table 24.4. In addition to these, a number of other solutes, especially specific organic pollutants, may be determined in connection with specific health hazards or incidents of pollution.

24.8. AUTOMATED WATER ANALYSES

Huge numbers of water analyses must often be performed in order to get meaningful results and for reasons of economics. This has resulted in the development of a number of automated procedures in which the samples are introduced through a sampler and the analyses performed and results posted without manual manipulation of reagents and apparatus. Such procedures have been developed and instruments marketed for the determination of a number of analytes, including alkalinity, sulfate,

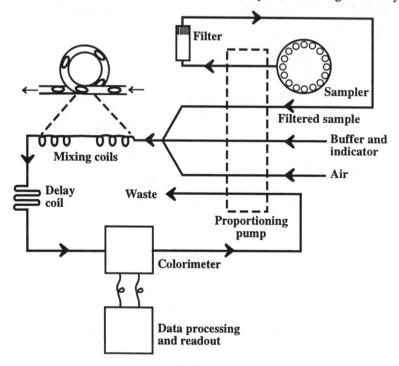

Figure 24.9. Automated analyzer system for the determination of total alkalinity in water. Addition of a water sample to a methyl orange solution buffered to pH 3.1 causes a loss of color in proportion to the alkalinity in the sample.

Table 24.4. Chemical Parameters Commonly Determined in Water

Chemical species	Significance in water	Methods of analysis
Acidity	Indicative of industrial pollution or acid mine drainage	Titration
Alkalinity	Water treatment, buffering, algal productivity	Titration
Aluminum	Water treatment, buffering	AA,[1] ICP[2]
Ammonia	Algal productivity, pollutant	Spectrophotometry
Arsenic	Toxic pollutant	Spectrophotometry, AA, ICP
Barium	Toxic pollutant	AA, ICP
Beryllium	Toxic pollutant	AA, ICP, fluorimetry
Boron	Toxic to plants	Spectrophotometry, ICP
Bromide	Seawater intrusion, industrial waste	Spectrophotometry, potentiometry, ion chromatography
Cadmium	Toxic pollutant	AA, ICP
Calcium	Hardness, productivity, treatment	AA, ICP, titration
Carbon dioxide	Bacterial action, corrosion	Titration, calculation
Chloride	Saline water contamination	Titration, electrochemical, ion chromatography
Chlorine	Water treatment	Spectrophotometry
Chromium	Toxic pollutant (hexavalent Cr)	AA, ICP, colorimetry
Copper	Plant growth	AA, ICP
Cyanide	Toxic pollutant	Spectrophotometry, potentiometry, ion chromatography
Fluoride	Water treatment, toxic at high levels	Spectrophotometry, potentiometry, ion chromatography
Hardness	Water quality, water treatment	AA, titration
Iodide	Seawater intrusion, industrial waste	Catalytic effect, potentiometry, ion chromatography
Iron	Water quality, water treatment	AA, ICP, colorimetry
Lead	Toxic pollutant	AA, ICP, voltammetry

Table 24.4 (Cont.)

Lithium	May indicate some pollution	AA, ICP, flame photometry
Magnesium	Hardness	AA, ICP
Manganese	Water quality (staining)	AA, ICP
Mercury	Toxic pollutant	Flameless atomic absorption
Methane	Anaerobic bacterial action	Combustible-gas indicator
Nitrate	Algal productivity, toxicity	Spectrophotometry, ion chromatography
Nitrite	Toxic pollutant	Spectrophotometry, ion chromatography
Nitrogen (albuminoid)	Proteinaceous material	Spectrophotometry
(organic)	Organic pollution indicator	Spectrophotometry
Oil and grease	Industrial pollution	Gravimetry
Organic carbon	Organic pollution indicator	Oxidation-CO_2 measurement
Organic contaminants	Organic pollution indicator	Activated carbon adsorption
Oxygen	Water quality	Titration, electrochemical
Oxygen demand (biochemical)	Water quality and pollution	Microbiological-titration
(chemical)	Water quality and pollution	Chemical oxidation-titration
Ozone	Water treatment	Titration
Pesticides	Water pollution	Gas chromatography
pH	Water quality and pollution	Potentiometry
Phenols	Water pollution	Distillation-colorimetry
Phosphate	Productivity, pollution	Spectrophotometry
Phosphorus (hydrolyzable)	Water quality and pollution	Spectrophotometry
Potassium	Productivity, pollution	AA, ICP, flame photometry
Selenium	Toxic pollutant	Spectrophotometry, ICP, neutron activation
Silica	Water quality	Spectrophotometry, ICP
Silver	Water pollution	AA, ICP

Table 24.4 (Cont.)

Sodium	Water quality, saltwater intrusion	AA, ICP, flame photometry
Strontium	Water quality	AA, ICP, flame photometry
Sulfate	Water quality, water pollution	Ion chromatography
Sulfide	Water quality, water pollution	Spectrophotometry, titra-ion, chromatography
Sulfite	Water pollution, oxygen scavenger	Titration, ion chromatography
Surfactants	Water pollution	Spectrophotometry
Tannin, Lignin	Water quality, water pollution	Spectrophotometry
Vanadium	Water quality, water pollution	ICP
Zinc	Water quality, water pollution	AA, ICP

[1] AA denotes atomic absorption
[2] ICP stands for inductively coupled plasma techniques, including atomic emission and detection of plasma-atomized atoms by mass spectrometry.

ammonia, nitrate/nitrite, and metals. Colorimetric procedures are popular for such automated analytical instruments, using simple, rugged colorimeters for absorbance measurements. Figure 24.9 shows an automated analytical system for the determination of alkalinity. The reagents and sample liquids are transported through the analyzer by a peristaltic pump consisting basically of rollers moving over flexible tubing. By using different sizes of tubing, the flow rates of the reagents are proportioned. Air bubbles are introduced into the liquid stream to aid mixing and to separate one sample from another. Mixing of the sample and various reagents is accomplished in mixing coils. Since many color-developing reactions are not rapid, a delay coil is provided that allows the color to develop before reaching the colorimeter. Bubbles are removed from the liquid stream by a debubbler pror to introduction into the flowcell for colorimetric analysis.

LITERATURE CITED

1. "Data Quality," Section 1030 in *Standard Methods for the Examination of Water and Wastewater*, 20th ed., Clesceri, Lenore, S., Arnold E. Greenberg, Andrew D. Eaton, and Mary Ann H. Franson, Eds., American Public Health Association, Washington, D.C., 1998, pp. 1-13–1-22.

2. *Understanding Environmental Methods* (CD/ROM version), Genium Publishing Corporation, Schenectady, NY, 1998.

3. Pihlstrom, Tuija, Anna Hellstrom, and Victoria Axelsson, "Gas Chromatographic Analysis of Pesticides in Water with Off-Line Solid Phase Extraction," *Analytica Chimica Acta*, **356**, 155-163 (1997).

4. Okumura, Minoru, Naoaki Yano, Kaoru Fujinaga, Yasushi Seike, and Shuji Matsuo, "In Situ Preconcentration Method for Trace Dissolved Sulfide in

Environmental Water Samples Using Solid-Phase Extraction Followed by Spectrophotometric Determination," *Analytical Science*, **15**, 427-431 (1999).

5. Wennrich, Luise, Werner Engewald, and Peter Popp, "GC Trace Analysis of Haloethers in Water. Comparison of Different Extraction Techniques," *International Journal of Environmental Analytical Chemistry*, **73**, 31-41 (1999).

6. Beals, D. M., W. G. Britt, J. P. Bibler, and D. A. Brooks, "Radionuclide Analysis Using Solid Phase Extraction Disks," *Journal of Radioanalytical and Nuclear Chemistry*, **236**, 187-191 (1998).

7. Martinez, D., F. Borrull, M. Calull, and J. Ruana; Colom, "Application of Solid-Phase Extraction Membrane Disks in the Determination of Haloacetic Acids in Water by Gas Chromatography-Mass Spectrometry," *Chromatographia*, **48**, 811-816, (1998).

8. Thompson, T. S. and B. D. Miller, "Use of Solid Phase Extraction Disks for the GC-MS Analysis of Acidic and Neutral Herbicides in Drinking Water," *Chemosphere*, **36**, 2867-2878, (1998).

9. Clesceri, Lenore, S., Arnold E. Greenberg, Andrew D. Eaton, and Mary Ann H. Franson, Eds., *Standard Methods for the Examination of Water and Wastewater*, 20th ed., American Public Health Association, Washington, D.C., 1998.

10. Rivasseau, Corinne, Pascale Racaud, Alain Deguin, and Marie Claire Hennion, "Evaluation of an ELISA Kit for Monitoring Microcystins (Cyanobacterial toxins) in Water and Algae Environmental Samples," *Environmental Science and Technology*, **33**, 1520-1527 (1999).

SUPPLEMENTARY REFERENCES

Dieken, Fred P., *Methods Manual for Chemical Analysis of Water and Wastes*, Alberta Environmental Centre, Vergeville, Alberta, Canada (1996).

Garbarino, John R. and Tedmund M. Struzeski, *Methods Of Analysis By The U.S. Geological Survey National Water Quality Laboratory—Determination Of Elements In Whole-Water Digests Using Inductively Coupled Plasma-Optical Emission Spectrometry And Inductively Coupled Plasma-Mass Spectrometry*, U. S. Department of the Interior U. S. Geological Survey, Denver, 1998.

Keith, Lawrence H., *Environmental Sampling and Analysis: A Practical Guide*, Lewis Publishers, Boca Raton, FL, 1991.

Meyers, R. A., Ed., *The Encyclopedia of Environmental Analysis and Remediation*, John Wiley and Sons, New York, 1998.

Patnaik, Pradyot, *Handbook of Environmental Analysis: Chemical Pollutants in Air, Water, Soil, and Solid Wastes*, CRC Press/Lewis Publishers, Boca Raton, FL, 1997.

Richardson, Susan D., "Water Analysis," *Analytical Chemistry*, **71**, 281R-215R (1999).

QUESTIONS AND PROBLEMS

1. A soluble water pollutant forms ions in solution and absorbs light at 535 nm. What are two physical properties of water influenced by the presence of this pollutant?

2. A sample was taken from the bottom of a deep, stagnant lake. Upon standing, bubbles were evolved from the sample; the pH went up; and a white precipitate formed. From these observations, what may be said about the dissolved CO_2 and hardness in the water?

3. For which of the following analytes may nitric acid be used as a water sample preservative: H_2S; CO_2; metals; coliform bacteria; cyanide?

4. In the form of what compound is oxygen fixed in the Winkler analysis of O_2?

5. Of the following analytical techniques, the water analysis technique that would best distinguish between the hydrated $Ag(H_2O)_6^+$ ion and the complex $Ag(NH_3)_2^+$ ion by direct measurement of the uncomplexed ion is: (a) neutron-activation analysis, (b) atomic absorption, (c) inductively coupled plasma atomic emission spectroscopy, (d) potentiometry, (e) flame emission.

6. A water sample was run through the colorimetric procedure for the analysis of nitrate, giving 55.0% transmittance. A sample containing 1.00 ppm nitrate run through the exactly identical procedure gave 24.6% transmittance. What was the concentration of nitrate in the first sample?

7. What is the molar concentration of HCl in a water sample containing HCl as the only contaminant and having a pH of 3.80?

8. A 200-mL sample of water required 25.12 mL of 0.0200N standard H_2SO_4 for titration to the methyl orange endpoint, pH 4.5. What was the total alkalinity of the original sample?

9. Analysis of a lead-containing sample by graphite-furnace atomic absorption analysis gave a peak of 0.075 absorbance units when 50 microliters of pure sample was injected. Lead was added to the sample such that the added concentration of lead was 6.0 micrograms per liter. Injection of 50 microliters of "spiked" sample gave an absorbance of 0.115 absorbance units. What was the concentration of lead in the original sample?

10. In a 2.63×10^{-4} M standard fluoride solution, a fluoride electrode read - 0.100 volts versus a reference electrode, and it read -0.118 volts in an appropriately processed fluoride sample. What was the concentration of fluoride in the sample?

11. The activity of iodine-131 ($t_{1/2}$ = 8 days) in a water sample 24 days after collection was 520 pCi/liter. What was the activity on the day of collection?

12. Neutron irradiation of exactly 2.00 mL of a standard solution containing 1.00 mg/L of unknown heavy metal "X" for exactly 30 seconds gave an activity of 1,257 counts per minute, when counted exactly 33.5 minutes after the irradiation, measured for a radionuclide product of "X" having a half-life of 33.5

minutes. Irradiation of an unknown water sample under identical conditions (2.00 mL, 30.0 seconds, same neutron flux) gave 1,813 counts per minute when counted 67.0 minutes after irradiation. What was the concentration of "X" in the unknown sample?

13. Why is magnesium-EDTA chelate added to a magnesium-free water sample before it is to be titrated with EDTA for Ca^{2+} ?

14. For what type of sample is the flame-ionization detector most useful?

15. Manganese from a standard solution was oxidized to MnO_4^- and diluted such that the final solution contained 1.00 mg/L of Mn. This solution had an absorbance of 0.316. A 10.00 mL wastewater sample was treated to develop the MnO_4 color and diluted to 250.0 mL. The diluted sample had an absorbance of 0.296. What was the concentration of Mn in the original wastewater sample?

25 ANALYSIS OF WASTES AND SOLIDS

25.1. INTRODUCTION

The analysis of hazardous wastes of various kinds for a variety of potentially dangerous substances is one of the most important aspects of hazardous waste management.[1] These analyses are performed for a number of reasons including tracing the sources of wastes, assessing the hazards posed by the wastes to surroundings and to waste remediation personnel, and determining the best means of waste treatment. This chapter is a brief overview of several of the main considerations applied to the analysis of wastes. Here, wastes are broadly defined to include all kinds of solids, semisolids, sludges, liquids, contaminated soils, sediments, and other kinds of materials that are either wastes themselves or are contaminated by wastes.

For the most part, the substances determined as part of waste analysis, the *analytes*, are measured by techniques that are used for the determination of the same analytes in water (see methods described in Chapter 24) and, to a lesser extent, in air. However, the preparation techniques that must be employed for waste analysis are usually more complex than those used for the same analytes in water. That is because the matrices in which the waste analytes are contained are usually relatively complicated, which makes it difficult to recover all the analytes from the waste and which introduces interfering substances. As a result, the lower limits at which substances can be measured in wastes (the practical quantitation limit, see Section 24.1) are usually much higher than in water.

There are several distinct steps in the analysis of a waste. Compared to water, wastes are often highly heterogeneous, which makes the first step, collection of representative samples, difficult. Whereas water samples can often be introduced into an analytical instrument with minimal preparation, the processing of hazardous wastes to get a sample that can be introduced into an instrument is often relatively complicated. Such processing can consist of dilution of oily samples with an organic solvent, extraction of organic analytes into an organic solvent, evolution and collection of volatile organic carbon analytes, or digestion of solids with strong acids and oxidants to extract metals for atomic spectrometric analysis. The products of these processes must often be subjected to relatively complicated sample cleanup procedures to remove contaminants that might interfere with the analysis or damage the analytical instrument.

Over a number of years, the U. S. Environmental Protection Agency has developed specialized methods for the characterization of wastes. These methods are given in the publication entitled *Test Methods for Evaluating Solid Waste, Physical/Chemical Methods*, which is periodically updated to keep it current.[2] Because of the difficult and exacting nature of many of the procedures in this work and because of the hazards associated with the use of reagents such as strong acids and oxidants used for sample digestion and solvents used to extract organic analytes, anyone attempting analyses of hazardous waste materials should use this resource and follow procedures carefully with special attention to precautions. "SW-846," as it is commonly known, is available in a convenient CD/ROM form as part of a comprehensive summary of environmental analytical methods.[3]

25.2. SAMPLE DIGESTION

In order to analyze a solid waste sample by flame atomic absorption spectroscopy, graphite furnace absorption spectroscopy, inductively coupled argon plasma spectroscopy, or inductively coupled argon plasma mass spectrometry, the sample must first be digested to get the analyte metals in solution. Digestion dissolves only those fractions of metals that can be put into solution under relatively extreme conditions and therefore enables measurement of available metals. It should be noted that sample digestion procedures generally use highly corrosive, dangerous reagents which are strong acids and strong oxidants. Therefore, digestion should be carried out only by carefully trained personnel using the proper equipment, including fume hoods and adequate personnel protection.

EPA Method 3050 is a procedure for acid digestion of sediments, sludges, and soils. A sample of up to 2 g is treated with a mixture of nitric acid and hydrogen peroxide; the sample is then refluxed with either con. HNO_3 or con. HCl, then refluxed with dilute HCl, filtered, and the filtrate analyzed for metals.

Microwave heating can be used to assist the digestion of samples. The procedure for the digestion of aqueous liquids consists of mixing a 45 mL sample with 5 mL of concentrated nitric acid, placing it in a fluorocarbon (Teflon) digestion vessel, and heating for 20 minutes. After digestion is complete, the sample is cooled, solids are separated by filtration or centrifugation, and the liquid remaining is analyzed by the appropriate atomic spectrometric technique.

Method 3052 is a procedure for microwave assisted acid digestion of siliceous and organically based matrices. It can be used on a variety of kinds of samples including biological tissues, oils, oil-contaminated soils, sediments, sludges, and soil. This method is not appropriate for analyses of leachable metals, but is used for measurement of total metals. A sample of up to 0.5 g is digested with microwave heating for 15 minutes in an appropriate fluorocarbon polymer container in an appropriate acid mixture. Commonly, the reagents employed are a mixture of 9 mL of con. nitric acid and 3 mL hydrofluoric acid, although other acid mixtures employing reagents such as con. HCl and hydrogen peroxide may be used. The sample is heated in the microwave oven to 180°C and held at that temperature for at least 9.5 minutes. After heating, the residual solids are filtered off and the filtrate analyzed for metals.

Many kinds of hazardous waste samples contain metals dissolved or suspended in viscous petroleum products, including oils, oil sludges, tars, waxes, paints, paint sludges, and other hydrocarbon materials. Method 3031 can be used to dissolve these metals—including antimony, arsenic, barium, beryllium cadmium, chromium, cobalt, copper, lead molybdenum, nickel, selenium, silver, thallium, vanadium, and zinc—in a form suitable for atomic spectrometric analysis. The procedure involves mixing 0.5 g of sample with 0.5 g of finely ground $KMnO_4$ and 1.0 mL of con. H_2SO_4, which causes a strongly exothermic reaction to occur as the hydrocarbon matrix is oxidized. After the reaction has subsided, 2 mL of con. HNO_3 and 2 mL of con. HCl are added, the sample is filtered with filter paper, the filter paper is digested with con. HCl, and the sample is diluted and analyzed for metals.

25.3. ANALYTE ISOLATION FOR ORGANICS ANALYSIS

The determination of organic analytes requires that they be isolated from the sample matrix. Since organic analytes are generally soluble in organic solvents, they can usually be extracted from samples with a suitable solvent. Although extraction works well for nonvolatile and semivolatile analytes, it is not so suitable for volatile organic compounds, which are readily vaporized during sample processing. The volatile materials are commonly isolated by techniques that take advantage of their high vapor pressures.

SOLVENT EXTRACTION

Method 3500 is a procedure for extracting nonvolatile or semivolatile compounds from a liquid or solid sample. The sample is extracted with an appropriate solvent, dried, and concentrated in a Kuderna-Danish apparatus prior to further processing for analysis.

A number of methods more complicated than Method 3500 have been devised for extracting nonvolatile and semivolatile analytes from waste samples. Method 3540 uses extraction with a Soxhlet extractor. This device, illustrated in Chapter 21, Figure 21.7, for the extraction of lipids from biological tissue, provides for recirculation of continuously redistilled fresh solvent over a sample of soils, sludges, and wastes. The sample is first mixed with anhydrous Na_2SO_4 to dry it, then placed inside an extraction thimble in the Soxhlet apparatus, which redistills a relatively small volume of extraction solvent over the sample. After extraction, the sample may be dried, concentrated, and exchanged into another solvent prior to analysis.

Method 3545 uses pressurized fluid extraction at 100°C and a pressure up to 2000 psi to remove organophilic analyte species from solid samples including soils, clays, sediments, sludges, and waste solids. Used for the extraction of semivolatile organic compounds, organophosphorus pesticides, organochlorine pesticides, chlorinated herbicides, and PCBs, it requires less solvent and takes less time than the Soxhlet extraction described above. Before extraction, the finely ground 10–30 g sample should be dried to prevent residual water from interfering with the extraction. It may be air-dried or dried with anhydrous Na_2SO_4 or pelletized diatomaceous earth. An extraction time of 5 minutes is typically used.

Method 3550 uses sonication with ultrasound to expedite the extraction of nonvolatile and semivolatile organic compounds from solids including soils, sludges, and wastes. The procedure calls for subjecting the finely divided dried sample mixed with solvent to ultrasound for a brief period of time. Low concentration samples may be subjected to multiple extractions with additional fresh solvent.

Although the requirement for specialized high pressure equipment has limited its application, extraction with supercritical carbon dioxide maintained at temperatures and pressures above the critical point where separate liquid and vapor phases do not exist is a very effective means of extracting some organic analytes. Method 3561 is used to extract polycyclic aromatic hydrocarbons such as acenaphthene, benzo(a)pyrene, fluorene, and pyrene from solid samples. The actual procedure is relatively complicated and is divided into three steps. The first step extracts the more volatile compounds with pure supercritical carbon dioxide at moderately low density and temperature. Less volatile PAHs are removed in the second step using supercritical carbon dioxide and methanol modifier. A third step using pure carbon dioxide is employed to purge the modifier from the system. The theory of supercritical fluid extraction of PCBs from sediments has been described.[4]

A comparison of several means of extracting PCBs from sewage treatment plant sludge (Soxhlet extraction, ultrasound, and shaking) has shown differences in the efficacy of extraction among these techniques depending upon the congeners extracted. The authors concluded that no single extraction method was optimum for all congeners.[5]

Sample Preparation for Volatile Organic Compounds

Many different volatile organic compounds are found at waste sites and in various kinds of waste solid and sludge samples. These include benzene, bromomethane, chloroform, 1,4-dichlorobenzene, dichloromethane, styrene, toluene, vinyl chloride, and the xylene isomers. Various methods are employed to isolate and concentrate volatile organic compounds from the solid, sludge, or liquid matrices in which they are contained.

Method 5021, "Volatile Organic Compounds in Soils and Other Solid Matrices Using Equilibrium Headspace Analysis" is used to isolate volatile organic compounds from soil, sediment, or solid waste samples for determination by gas chromatography or gas chromatography/mass spectrometry. This procedure makes use of a special glass headspace vial that contains at least 2 g of sample. A matrix modifying preservative solution, internal standards, and surrogate compounds may be added to the sample and mixed thoroughly by rotating the vial. The sample is heated to 85°C for approximately one hour prior to injection into the gas chromatograph. For injection, helium is forced under pressure into the vial and a sample of the gas in the headspace above the solid sample is forced out of the vial and into the gas chromatograph system for quantitation.

Method 5030 is a purge-and-trap procedure that can be used to collect for gas chromatographic analysis poorly water soluble compounds that have boiling points below 200°C, which includes a wide variety of compounds commonly occurring in hazardous wastes. For samples in water, helium is bubbled through the sample and the volatile analyte compounds are absorbed on a sorbent column. Solid samples can be dispersed in methanol and the methanol added to water for the purging step. After purging is complete, the sorbent column is heated and flushed with carrier gas to sweep the sample compounds into the gas chromatograph for qualitative and quantitative analysis of the volatile organic compounds present.

Method 5035 is a closed-system, purge-and-trap and extraction method applicable to the determination of volatile organic compounds in soil, sediment, and waste samples. For the determination of very low levels of volatile organic compounds in soil, a sealed sample vial is used that remains sealed throughout the sample processing operations. An approximately 5-gram sample is weighed into a sample vial containing a stirring bar and sodium bisulfate preservative solution, and the vial is sealed and transported to the laboratory as soon as possible. For analysis, reagent water, surrogates, and internal standards are injected into the vial without opening it. The contents of the vial are purged with helium gas into an appropriate sampling trap, then flushed into the gas chromatograph for measurement.

Method 5031 is used to isolate volatile, nonpurgeable, water-soluble compounds by azeotropic distillation. The analyte compounds for which it is suitable include acetone, acetonitrile, acrylonitrile, allyl alcohol, 1-butanol, t-butyl alcohol, crotonaldehyde, 1,4-dioxane, ethanol, ethyl acetate, ethylene oxide, isobutyl alcohol, methanol, methylethyl ketone, methylisobutyl ketone, n-nitroso-di-n-butylamine, paraldehyde, 2-pentanone, 2-picoline, 1-propanol, 2-propanol, propionitrile, pyridine, and, o-toluidine. The technique takes advantage of the formation of an azeotrope liquid mixture of water and analytes which boil at a constant temperature and give off vapors of a constant composition. For separation of the analyte species, a 1-liter sample is adjusted to pH 7 with a buffer, and a small sample of condensate enriched in analyte species is collected. The organics are measured in the azeotrope solution distilled from the sample.

Method 5032 employs vacuum distillation to isolate volatile organic analytes from liquid, solid, and oily waste matrices, and even animal tissues. Examples of compounds isolated for analysis by this procedure include acetone, benzene, carbon disulfide, chloroform, ethanol, styrene, tetrachloroethene, vinyl chloride, and the o,m,p-xylene isomers. Such compounds should be no more than minimally soluble in water and should boil below 180°C. The sample in water is distilled under a vacuum and the water condensed in a chilled condenser. Volatile analyte constituents not condensed with the water vapor are swept into a cryogenic trap maintained at liquid nitrogen temperature of -196°C for collection. For analysis, the contents of the trap are evaporated at an elevated temperature and swept into the chromatograph.

Method 3585, Waste Dilution for Volatile Organics, is employed to place a non-aqueous waste sample of volatile organics into the appropriate form for injection into a gas chromatograph. It is applicable to samples containing analytes at levels of 1 mg/kg or higher. The procedure calls for placing a 1 g oil-phase sample into a vial marked for a volume of 10 mL, sealing the vial, diluting with n-hexadecane or other appropriate solvent, and mixing to dissolve the sample. Because the diluted sample usually contains residual materials from the sample with a tendency to foul the gas chromatograph, the sample is injected through a replaceable direct injection liner containing Pyrex glass wool.

25.4. SAMPLE CLEANUP

Most waste, soil, and sediment samples result in extraction of extraneous substances that can result in the observation of extraneous peaks, be detrimental to peak resolution and column efficiency, and be damaging to expensive columns and detectors. **Sample cleanup** refers to a number of measures that can be taken to remove these constituents from sample extracts by a number of procedures including distillation, partitioning with immiscible solvents, adsorption chromatography, gel permeation chromatography, or chemical destruction of interfering substances with acid, alkali, or oxidizing agents; two or more of these techniques may be used in combination. The most widely applicable cleanup technique is gel permeation chromatography, which can be used to separate substances with high molecular weights from the analytes of interest. Treatment by adsorption column chromatography with alumina, Florisil, or silica gel can be used to isolate a relatively narrow polarity range of analytes away from interfering substances. Acid-base partitioning can be used in the determination of materials such as chlorophenoxy herbicides and phenols to separate acidic, basic, and neutral organics. Table 25.1 shows the uses of the main sample cleanup techniques.

Table 25.1. Sample Cleanup Techniques and Their Applications

Number	Type	Applications
3610	Alumina column	Phthalate esters, nitrosamines
3611	Alumina column cleanup and separation of petroleum wastes	Polycyclic aromatic hydrocarbons, petroleum wastes
3620	Florisil column	Phthalate esters, nitrosamines, organochlorine pesticides, PCBs, chlorinated hydrocarbons, organophosphorus pesticides
3630	Silica gel	Polycyclic aromatic hydrocarbons
3630(b)	Silica gel	Phenols
3640	Gel permeation chromatography	Phenols, phthalate esters, nitrosamines, organochlorine pesticides, PCBs, nitroaromatics, cyclic ketones, polycyclic aromatic hydrocarbons, chlorinated hydrocarbons, organophosphorus pesticides, priority pollutant semivolatiles
3650	Acid-base liquid/liquid partition	Phenols, priority pollutant semivolatiles
3660	Sulfur cleanup	Organochlorine pesticides, PCBs, Priority pollutant semivolatiles

Alumina column cleanup makes use of highly porous granular aluminum oxide. Available in acidic, neutral, and basic pH ranges, this solid is packed into a column topped with a water-absorbing substance over which the sample is eluted with a suitable solvent, which leaves interferences on the column. After elution, the sample is concentrated, exchanged with another solvent if necessary, then analyzed. Florisil is an acidic magnesium silicate and a registered trade name of Floridin Co. It is used in a column cleanup procedure in a manner similar to alumina. Silica gel is a weakly acidic amorphous silicon oxide. It can be activated by heating for several hours at 150-160°C and used for the separation of hydrocarbons. Deactivated silica gel containing 10-20% water acts as an adsorbent for compounds with ionic and nonionic functionalities such as dyes, alkali metal cations, terpenoids, and plasticizers. It is used in a column as described for alumina above. Gel-permeation chromatography separates solutes by size carried over a hydrophobic gel by organic solvents. A gel must be chosen that will separate the appropriate size range of analytes and interferences. The gel is preswelled before loading onto a column and flushed extensively with solvent before the sample is introduced for separation.

25.5 IMMUNOASSAY SCREENING OF WASTES

Immunoassay has emerged as a useful technique for screening wastes for specific kinds of pollutants. Commercial immunoassay techniques have been developed that permit very rapid analyses of large numbers of samples. A variety of immunoassay techniques have been developed. These techniques all use biologically produced antibodies that bind specifically to analytes or classes of analytes. This binding is combined with chemical processes that enable detection through a signal-producing species (reporter reagent) such as enzymes, chromophores, fluorophores, and luminescent compounds. The reporter reagent binds with the antibody. When an analyte is added to the antibody to displace the reagent, the concentration of displaced reagent is proportional to the level of analyte displacing it from the antibody. Detection of the displaced reporter reagent enables quantification of the analyte.

Immunoassay techniques are divided into the two major categories of heterogeneous and homogeneous; the former requires a separation (washing) step whereas the latter does not require such a step. Typically, when heterogeneous procedures are used, the antibody is immobilized on a solid support on the inner surface of a disposable test tube. The sample is contacted with the antibody displacing reporter reagent, which is removed by washing. The amount of reagent displaced, commonly measured spectrophotometrically, is proportional to the amount of analyte added. Very widely used enzyme immunoassays make use of reporter reagent molecules bound with enzymes, and kits are available for enzyme-linked immunosorbent assays (ELISA) of a number of organic species likely to be found in hazardous wastes.

Immunoassay techniques have been approved for the determination of numerous analytes commonly found in hazardous wastes. Where the EPA method numbers are given in parentheses, these include pentachlorophenol (4010), 2,4-dichlorophenoxyacetic acid (4015), polychlorinated biphenyls (4020), petroleum hydrocarbons (4030), polycyclic aromatic hydrocarbons (4035), toxaphene (4040), chlordane (4041), DDT (4042), TNT explosives in soil (4050), and hexahydro-1,3,5-trinitro-

1,3,5-triazine (RDX) in soil (4051). Enzyme-linked immunosorbent assays have been reported for monitoring pentachlorophenol, BTEX (benzene, toluene, ethylbenzene, and *o*-, *m*-, and *p*-xylene) in industrial effluents.[6]

25.6. DETERMINATION OF CHELATING AGENTS

Strong chelating agents in wastes have been found to play an important role in the mobility of heavy metals and metal radionuclides at waste disposal sites, with their potential to contaminate groundwater. Therefore, the determination of chelating agents, such as ethylenediaminetetraacetic acid (EDTA) and N-(2-hydroxyethyl)-ethylenediaminetriacetic acid (HEDTA), is an important analytical procedure for wastes. Most chelating agents of concern at waste sites are polar and nonvolatile, which prevents their determination by direct gas chromatographic methods. One of the more satisfactory methods for their determination makes use of derivatization to produce volatile species suitable for gas chromatographic analysis.

A method has been described for the determination of chelating agents in hazardous wastes from radioactive wastes using derivatization followed by gas chromatographic/mass spectrometric analysis (GC/MS).[7] The study involved wastes contained in a potentially leaking double-shell storage tank at the U. S. Department of Energy Hanford site. Treatment with BF_3 and methanol produced volatile derivatives that were measured by GC/MS. In addition to EDTA and HEDTA, the study showed the presence of nitrilotriacetate (NTA) and citrate chelating agents, and chelating nitrosoiminiodiacetate was produced as an artifact of the analytical procedure.

25.7. TOXICITY CHARACTERISTIC LEACHING PROCEDURE

The **Toxicity Characteristic Leaching Procedure** (TCLP) is specified to determine the potential toxicity hazard of various kinds of wastes.[8] The test was designed to estimate the availability to organisms of both inorganic and organic species in hazardous materials present as liquids, solids, or multiple phase mixtures by producing a leachate, the TCLP extract, which is analyzed for the specific toxicants listed in Table 25.2.

The procedure for conducting the TCLP is rather involved. The procedure need not be run at all if a total analysis of the sample reveals that none of the pollutants specified in the procedure could exceed regulatory levels. At the opposite end of the scale, analysis of any of the liquid fractions of the sample showing that any regulated species would exceed regulatory levels even after the dilutions involved in the TCLP measurement have been carried designate the sample as hazardous, and the TCLP measurement is not required.

In conducting the TCLP test, if the waste is a liquid containing less than 0.5% solids, it is filtered through a 0.6–0.8 µm glass fiber filter and the filtrate is designated as the TCLP extract. At solids levels exceeding 0.5%, any liquid present is filtered off for separate analysis and the solid is extracted to provide a TCLP extract (after size reduction, if the particles exceed certain size limitations). The choice of the extraction fluid is determined by the pH of the aqueous solution produced from shaking a mixture of 5 g of solids and 96.5 mL of water. If the pH is less than 5.0, a pH 4.93 acetic acid/sodium acetate buffer is used for extraction; otherwise, the extraction fluid used is a pH of 2.88±0.05 solution of dilute acetic acid. Extractions

Table 25.2. Contaminants Determined in TCLP Procedure

EPA hazard-ous waste number	Contaminant	Regulatory level, mg/L	EPA hazard-ous waste number	Contaminant	Regulatory level, mg/L
Heavy metals (metalloids)					
D004	Arsenic	5.0	D033	Hexachloro-butadiene	0.5
D005	Barium	100.0			
D006	Cadmium	1.0	D034	Hexachloro-ethane	3.0
D007	Chromium	5.0			
D008	Lead	5.0	D035	Methylethyl ketone	
D009	Mercury	0.2			200.0
D010	Selenium	1.0	D036	Nitrobenzene	2.0^2
D011	Silver	5.0	D037	Pentachloro-phenol	
					100.0
Organics			D038	Pyridine	5.0^2
			D039	Tetrachloro-ethylene	0.7
D018	Benzene	0.5			
D019	Carbon tetrachloride	0.5	D040	Trichloroethylene	0.5
			D041	2,4,5-Trichloro-phenol	
D021	Chloro-benzene	100.0			400.0
			D042	2,4,6-Trichloro-phenol	2.0
D022	Chloroform	6.0			
D023	*o*-Cresol	200.0^1	D043	Vinyl chloride	0.2
D024	*m*-Cresol	200.0^1			
D025	*p*-Cresol	200.0^1	*Pesticides*		
D026	Cresol	200.0^1			
D027	1,4-Dichlor-obenzene	7.5	D012	Endrin	0.02
			D013	Lindane	0.4
D028	1,2-Dichlor-oethane	0.5	D014	Methoxychlor	10.0
			D015	Toxaphene	0.5
D029	1,1-Dichlor-oethylene	0.7	D016	2,4-D	10.0
			D017	2,4,5-TP (Silvex)	1.0
D030	2,4-Dinitro-toluene	0.13^2	D020	Chlordane	0.03
			D031	Heptachlor (and its epoxide)	0.008
D032	Hexachloro-benzene	0.13^2			

[1] If *o*-, *m*-, and *p*-Cresol concentrations cannot be differentiated, the total cresol (D026) concentration is used. The regulatory level of total cresol is 200 mg/L.

[2] Quantitation limit is greater than the calculated regulatory level. The quantitation limit therefore becomes the regulatory level.

are carried out in a sealed container rotated end-over-end for 18 hours. The liquid portion is then separated and analyzed for the specific substances given in Table 25.2. If values exceed the regulatory limits, the waste is designated as "toxic."

LITERATURE CITED

1. Quevauviller, P. H., Ed, "The Role of Analysis in Solid Waste Management," *Trends in Analytical Chemistry*, **17**, 1-71 (1998).

2. *Test Methods for Evaluating Solid Waste, Physical/Chemical Methods*, EPA Publication SW-846, 3rd ed., (1986), as amended by Updates I (1992), II (1993, 1994, 1995) and III (1996), U.S. Government Printing Office, Washington, D.C.

3. *Understanding Analytical Methods*, CD/ROM Version 2.0, Genium Publishing Corporation, Schenectady, NY, 1998.

4. Pilorz, Karol, Erland Bjoerklund, Soren Bowadt, Lennart Mathiasson, and Steven Hawthorne, "Determining PCB Sorption/Desorption Behavior on Sediments Using Selective Supercritical Fluid Extraction. 2. Describing PCB Extraction with Simple Diffusion Models," *Environmental Science and Technology*, **33**, 2204-2212 (1999).

5. Sulkowski, W. and A. Rosinska, "Comparison of the Efficiency of Extraction Methods for Polychlorinated Biphenyls from Environmental Wastes," *Journal of Chromatography*, **845**, 349-355 (1999).

6. Castillo, M., A. Oubiña, and D. Barcelo, "Evaluation of ELISA Kits Followed by Liquid Chromatography-Atmospheric Pressure Chemical Ionization-Mass Spectrometry for the Determination of Organic Pollutants in Industrial Effluents," *Environmental Science and Technology*, **32**, 2180-2184 (1998).

7. Grant, K.E., G. M. Mong, R. B. Lucke, and J. A. Campbell, "Quantitative Determination of Chelators and their Degradation Products in Mixed Hazardous Wastes from Tank 241-SY-101 Using Derivatization GC/MS," *Journal of Radioanalytical and Nuclear Chemistry*, **221**, 383-402 (1996).

8. "Toxicity Characteristic Leaching Procedure," Test Method 1311 in *Test Methods for Evaluating Solid Waste, Physical/Chemical Methods*, EPA Publication SW-846, 3rd ed., (November, 1986), as amended by Updates I, II, IIA, U.S. Government Printing Office, Washington, D.C.

SUPPLEMENTARY REFERENCES

Gavasci, R., F. Lombardi, A Polettini, and P. Sirini, "Leaching Tests on Solidified Products," *Journal of Solid Waste Technology Management*, **25**, 14-20 (1998).

Williford, Clint W., Jr. and R. Mark Bricka, "Extraction of TNT from Aggregate Soil Fractions, *Journal of Hazardous Materials*, **66**, 1-13 (1999).

Pradyot, Patnaik, Ed., *Handbook of Environmental Analysis: Chemical Pollutants in Air, Water, Soil, and Solid Wastes*, CRC Press, Boca Raton, FL (1997).

QUESTIONS AND PROBLEMS

1. Explain the uses of microwave in hazardous waste analysis. How is ultrasound employed in hazardous waste analysis?

2. Does sample digestion necessarily give an analysis leading to total metals? Why might it not be advantageous to measure total metals in a sample?

3. What is the distinction between a Kuderna-Danish apparatus and a Soxhlet apparatus?

4. How is anhydrous Na_2SO_4 used in organics analysis?

5. How does the purge-and-trap procedure differ from azeotropic distillation? For what kinds of compounds would the two procedures be employed?

6. What is the purpose of sample cleanup? Why is cleanup more commonly applied to samples to be analyzed for organic contaminants than for metals?

7. For what purpose is a cryogenic trap used in organics analysis? What advantage might it have over the solid sorbent traps used in conventional purge-and-trap analysis?

8. What do benzene, bromomethane, chloroform, 1,4-dichlorobenzene, dichloromethane, styrene, toluene, vinyl chloride, and o-xylene have in common?

9. What is the principle of immunoassay? What makes it specific for compounds or narrow classes of compounds? Why might it be especially suitable as a survey technique for hazardous waste sites? What is ELISA?

10. In what sense is the TCLP a measure of available toxicants?

11. Under what circumstances is it unnecessary to run the TCLP in evaluating the toxicity hazard of a waste material?

12. Which is the most widely applicable sample cleanup technique? What are three other kinds of materials used in column cleanup of samples?

26 AIR AND GAS ANALYSIS

26.1. ATMOSPHERIC MONITORING

Good analytical methodology, particularly that applicable to automated analysis and continuous monitoring, is essential to the study and alleviation of air pollution. The atmosphere is a particularly difficult analytical system because of the very low levels of substances to be analyzed; sharp variations in pollutant level with time and location; differences in temperature and humidity; and difficulties encountered in reaching desired sampling points, particularly those substantially above the earth's surface. Furthermore, although improved techniques for the analysis of air pollutants are continually being developed, a need still exists for new analytical methodology and the improvement of existing methodology.

Much of the earlier data on air pollutant levels were unreliable as a result of inadequate analysis and sampling methods. An atmospheric pollutant analysis method does not have to give the actual value to be useful. One which gives a relative value may still be helpful in establishing trends in pollutants levels, determining pollutant effects, and locating pollution sources. Such methods may continue to be used while others are being developed.

Air Pollutants Measured

The air pollutants generally measured may be placed in several different categories. In the U.S., one such category contains materials for which ambient (surrounding atmosphere) standards have been set by the Environmental Protection Agency. These are sulfur dioxide, carbon monoxide, nitrogen dioxide, nonmethane hydrocarbons, and particulate matter. The standards are categorized as primary and secondary. Primary standards are those defining the level of air quality necessary to protect public health. Secondary standards are designed to provide protection against known or expected adverse effects of air pollutants, particularly upon materials, vegetation, and animals. Another group of air pollutants to be measured consists of those known to be specifically hazardous to human health, such as asbestos, beryllium, and mercury. Still another category of air pollutants contains those regulated in new installations of selected stationary sources, such as coal-cleaning plants, cotton gins, lime plants, and paper mills. Some pollutants in this category are visible emissions, acid (H_2SO_4) mist, particulate matter, nitrogen oxides, and sulfur oxides. These substances often must be monitored in the stack to ensure that emissions standards are being met. A fourth category consists of the emissions of mobile sources (motor vehicles)—hydrocarbons, CO, and NO_x. A fifth group consists of miscellaneous elements and compounds, such as certain heavy metals, fluoride, chlorine, phosphorus, polycyclic aromatic hydrocarbons (PAH), polychlorinated biphenyls, odorous compounds, reactive organic compounds, and radionuclides.

Much of the remainder of this chapter is devoted to a discussion of the analytical methods for most of the species mentioned above. For some species, an analytical method is well developed and reasonably satisfactory. For others, improved methods would be useful. The development of analytical techniques for air pollutants is a very active area of research and development in analytical chemistry.

The units in which air pollutants and air-quality parameters are expressed include for gases and vapors, $\mu g/m^3$ (alternatively, ppm by volume); for weight of particulate matter, $\mu g/m^3$; for particulate matter count, number per cubic meter; for visibility, kilometers; for instantaneous light transmission, percentage of light transmitted; for emission and sampling rates, m^3/min; for pressure, mm Hg; and for temperature, degrees Celsius. Air volumes should be converted to conditions of 10°C and 760 mm Hg (1 atm), assuming ideal gas behavior.

26.2. SAMPLING

The ideal atmospheric analysis techniques are those that work successfully without sampling, such as long-path laser resonance absorption monitoring. For most analyses, however, various types of sampling are required. In some very sophisticated monitoring systems, samples are collected and analyzed automatically and the results are transmitted to a central receiving station. Often, however, a batch sample is collected for later chemical analysis.

The analytical result from a sample can only be as good as the method employed to obtain that sample. A number of factors enter into obtaining a good sample. The size of the sample required (total volume of air sampled) decreases with increasing concentration of pollutant and increasing sensitivity of the analytical method. Often a sample of ten or more cubic meters is required. The sampling rate is determined by the equipment used and generally ranges from approximately 0.003 m^3/min to 3.0 m^3/min. The duration of sampling time influences the result obtained, as shown in Figure 26.1. The actual concentration of the pollutant is shown by the solid line. A sample collected over an eight-hour period has the concentration shown in the dashed line, whereas samples taken over one-hour intervals exhibit the concentration levels shown by the dotted line. Sampling techniques are discussed briefly for specific kinds of analytes later in this chapter.

The most straightforward technique for the collection of particles is sedimentation. A sedimentation collector may be as simple as a glass jar equipped with a funnel. Liquid is sometimes added to the collector to prevent the solids from being blown out. Filtration is the most common technique for sampling particulate matter.

Filters composed of fritted (porous) glass, porous ceramics, paper fibers, cellulose fibers, fiberglass, asbestos, mineral wool, or plastic may be used. A special type of filter is the membrane filter, which yields high flow rates with small, moderately uniform pores. Impingers, as the name implies, collect particles from a relatively high-velocity air stream directed at a surface.

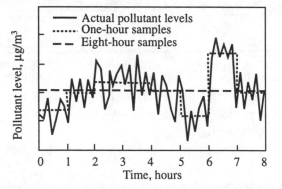

Figure 26.1. Effect of duration of sampling upon observed values of air pollutant levels.

Sampling for vapors and gases may range from methods designed to collect only one specific pollutant to those designed to collect all pollutants. Essentially all pollutants may be removed from an air sample cryogenically by freezing or by liquifying the air in collectors maintained at a low temperature. Absorption in a solvent, such as by bubbling the gas through a liquid, can be used for the collection of gaseous pollutants. Adsorption, in which a gas collects on the surface of a solid, is particularly useful for the collection of samples to be analyzed by gas chromatography.

Denuders are among the more useful sampling devices for certain kinds of air pollutants. Denuders solve a major sampling problem by enabling collection of gas-phase pollutants free of contamination by particles. Otherwise, for some pollutants such as acids, it is not possible to distinguish relative amounts of analytes in the gas phase from those in particulate form.

Diffusion denuders run a laminar flow airstream through a tube, the walls of which are covered with a sorptive or reactive collecting medium for the analytes of interest.[1] The diffusion coefficents of small particles are only about 10^{-4} those of gases, so that the particles go through the tube, and the gases diffuse to the walls and are collected. Open-bore denuders consist of tubes with coated walls. A more efficient device is the **annular denuder** composed of concentric tubes separated by 1-2 mm of annular space. **Thermodenuders** use heat to drive off analytes collected by the device. They can be used for semicontinuous analysis by employing a collection/analysis cycle involving alternate collection and thermal desorption. **Diffusion scrubbers** are denuders in which walls are composed of membranes such that gas from the sample goes through the wall and is collected in a collecting liquid.

26.3. METHODS OF ANALYSIS

A very large number of different analytical techniques are used for atmospheric pollutant analysis. Some of these, the uses of which are not confined to atmospheric analysis, were discussed in Chapters 24 and 25. Techniques confined largely to atmospheric analysis are discussed in the remainder of this chapter. A summary of some of the main instrumental techniques for air monitoring is presented in Table 26.1.

Table 26.1. The Main Techniques Used for Air Pollutant Analysis

Pollutant	Method	Potential interferences
SO_2 (total S)	Flame photometric (FPD)	H_2S, CO
SO_2	Gas chromatography (FPD)	H_2S, CO
SO_2	Spectrophotometric (pararosaniline wet chemical)	H_2S, HCl, NH_3, NO_2, O_3
SO_2	Electrochemical	H_2S, HCl, NH_3, NO, NO_2, O_3, C_2H_4,
SO_2	Conductivity	HCl, NH_3, NO_2
SO_2	Gas-phase spectrophotometric	NO, NO_2, O_3,
O_3	Chemiluminescent	H_2S
O_3	Electrochemical	NH_3, NO_2, SO_2
O_3	Spectrophotometric (potassium iodide reaction, wet chemical)	NH_3, NO_2, NO, SO_2
O_3	Gas-phase spectrophotometric	NO_2, NO, SO_2
CO	Infrared	CO_2 (at high levels)
CO	Gas chromatography	- - -
CO	Electrochemical	NO, C_2H_4
CO	Catalytic combustion-thermal detection	NH_3
CO	Infrared fluorescence	- - -
CO	Mercury replacement ultraviolet photometric	C_2H_4
NO_2	Chemiluminescent	NH_3, NO, NO_2, SO_2
NO_2	Spectrophotometric (azo-dye reaction, wet chemical)	NO, SO_2, NO_2, O_3
NO_2	Electrochemical	HCl, NH_3, NO, NO_2, SO_2, O_3, CO
NO_2	Gas-phase spectrophotometric	NH_3, NO, NO_2, SO_2, CO
NO_2	Conductivity	HCl, NH_3, NO, NO_2, SO_2

    The U. S. Environmental Protection Agency specifies reference methods of analysis for selected air pollutants to determine compliance with the primary and secondary national ambient air quality standards for those pollutants. These methods are published annually in the *Code of Federal Regulations*.[2] These methods are not necessarily state-of-the art, and in some cases are outdated and cumbersome. However, for regulatory and legal purposes they provide reliable measurements proven to be valid.

26.4. DETERMINATION OF SULFUR DIOXIDE

The reference method for the analysis of sulfur dioxide is the spectrophotometric pararosaniline method first described by West and Gaeke[3] and subsequently optimized.[4] It is applicable to the analysis of 0.005-5 ppm SO_2 in ambient air. The method makes use of a collecting solution of 0.04 M potassium tetrachloromercurate to collect sulfur dioxide according to the following reaction:

$$HgCl_4^{2-} + SO_2 + H_2O \rightarrow HgCl_2SO_3^{2-} + 2H^+ + 2Cl^- \tag{26.4.1}$$

Typically, samplings involves scrubbing 30 liters of air through 10 mL of scrubbing solution with a collection efficiency of around 95%. The $HgCl_2SO_3^{2-}$ complex stabilizes readily oxidized sulfur dioxide against the oxidizing agents of oxidants such as ozone and nitrogen oxides. For analysis, sulfur dioxide in the scrubbing medium is reacted with formaldehyde:

$$HCHO + SO_2 + 2H_2O \rightarrow HOCH_2SO_3H \tag{26.4.2}$$

The adduct formed is then reacted with uncolored organic pararosaniline hydrochloride to produce a red-violet dye. Although NO_2 at levels above about 2 ppm interferes, the interference may be eliminated by reducing the NO_2 to N_2 gas with sulfamic acid, H_2NSO_3H. The level of SO_2 measured spectrophotometrically is corrected to standard ambient conditions of 25°C and atmospheric pressure (101 kiloPascals). The lower limit of detection of SO_2 in 10 mL of the collecting solution is 0.75 µg. This represents a concentration of 25 µg SO_2/m^3 (0.01 ppm) in an air sample of 30 standard liters (short-term sampling), and a concentration of 13 µg SO_2/m^3 (0.005 ppm) in an air sample of 288 standard liters (long-term sampling).

Performed manually, the West-Gaeke method for sulfur dioxide analysis is cumbersome and complicated. However, the method has been refined to the point that it can be done automatically with continuous-monitoring equipment. A block diagram of such an analyzer is shown in Figure 26.2.

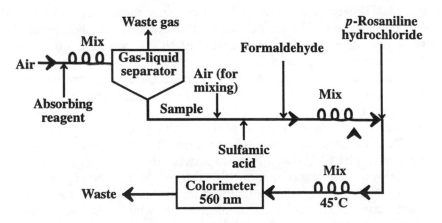

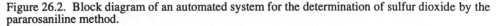

Figure 26.2. Block diagram of an automated system for the determination of sulfur dioxide by the pararosaniline method.

Conductimetry was used as the basis for a commercial continuous sulfur dioxide analyzer as early as 1929. Generally, the sulfur dioxide is collected in a hydrogen peroxide solution and increased conductance of the sulfuric acid solution is measured. Several types of sulfur dioxide monitors are based on amperometry, in which an electrical current is measured that is proportional to the SO_2 in a collecting solution. Sulfur dioxide can be determined by ion chromatography by bubbling SO_2 through hydrogen peroxide solution to produce SO_4^{2-}, followed by analysis of the sulfate by ion chromatography, a method that separates ions on a chromatography column and detects them very sensitively by conductivity measurement. Flame photometry, sometimes in combination with gas chromatography, is used for the detection of sulfur dioxide and other gaseous sulfur compounds. The gas is burned in a hydrogen flame, and the sulfur emission line at 394 nm is measured.

Several direct spectrophotometric methods are used for sulfur dioxide measurement, including nondispersive infrared absorption, Fourier transform infrared analysis (FTIR), ultraviolet absorption, molecular resonance fluorescence, and second-derivative spectrophotometry. The principles of these methods are the same for any gas measured.

26.5. NITROGEN OXIDES

Several methods have been used to determine nitrogen oxides. As noted in Table 26.1, these methods include electrochemical methods, direct measurements of nitrogen oxides spectrophotometrically in the gas phase, and a wet chemical method based on formation of an azo dye. However, gas-phase chemiluminescence is the favored method of NO_x analysis.[5] The general phenomenon of chemiluminescence was defined in Section 9.7. It results from the emission of light from electronically excited species formed by a chemical reaction. In the case of NO, ozone is used to bring about the reaction, producing electronically excited nitrogen dioxide:

$$NO + O_3 \rightarrow NO_2{}^* + O_2 \qquad\qquad (26.5.1)$$

The species loses energy and returns to the ground state through emission of light in the 600-3000 nm range. The emitted light is measured by a photomultiplier; its intensity is proportional to the concentration of NO. A schematic diagram of the device used is shown in Figure 26.3.

Since the chemiluminescence detector system depends upon the reaction of O_3 with NO, it is necessary to convert NO_2 to NO in the sample prior to analysis. This is accomplished by passing the air sample over a thermal converter, which brings about the desired conversion. Analysis of such a sample gives NO_x, the sum of NO and NO_2. Chemiluminescence analysis of a sample that has not been passed over the thermal converter gives NO. The difference between these two results is NO_2.

Other nitrogen compounds besides NO and NO_2 undergo chemiluminescence by reacting with O_3, and these may interfere with the analysis if present at an excessive level. Particulate matter also causes interference which may be overcome by employing a membrane filter on the air inlet.

This analysis technique is illustrative of chemiluminescence analysis in general. Chemiluminescence is an inherently desirable technique for the analysis of atmospheric pollutants because it avoids wet chemistry, is basically simple, and lends itself well to continuous monitoring and instrumental methods. Another chemiluminescence method, that employed for the analysis of ozone, is described in the next section.

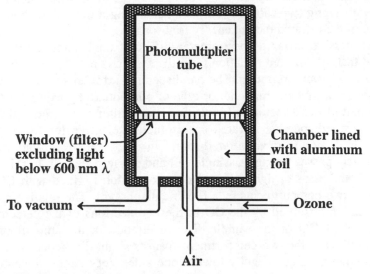

Figure 26.3. Chemiluminescence detector for NO_x.

26.6. ANALYSIS OF OXIDANTS

The atmospheric oxidants that are commonly analyzed include ozone, hydrogen peroxide, organic peroxides, and chlorine. The classic manual method for the analysis of oxidants is based upon their oxidation of I^- ion followed by spectrophotometric measurement of the product. The sample is collected in 1% KI buffered at pH 6.8. Oxidants react with iodide ion as shown by the following reaction of ozone:

$$O_3 + 2H^+ + 3I^- \rightarrow I_3^- + O_2 + H_2O \qquad (26.6.1)$$

The absorbance of the colored I_3^- product is measured spectrophotometrically at 352 nm. Generally, the level of oxidant is expressed in terms of ozone, although it should be noted that not all oxidants—PAN, for example—react with the same efficiency as O_3. Oxidation of I^- may be used to determine oxidants in a concentration range of several hundredths of a part per million to approximately 10 ppm. Nitrogen dioxide gives a limited response to the method, and reducing substances interfere seriously.

Now the favored method for oxidant analysis uses chemiluminescence.[6] The chemiluminescent reaction is that between ozone and ethylene. Chemiluminescence from this reaction occurs over a range of 300-6000 nm, with a maximum at 435 nm. The intensity of emitted light is directly proportional to the level of ozone. Ozone concentrations ranging from 0.003–30 ppm may be measured. Ozone for calibrating the instrument is generated photochemically from the absorption of ultraviolet radiation by oxygen.

26.7. ANALYSIS OF CARBON MONOXIDE

Carbon monoxide is analyzed in the atmosphere by nondispersive infrared spectrometry.[7] This technique depends upon the fact that carbon monoxide absorbs infrared radiation strongly at certain wavelengths. Therefore, when such radiation is passed through a long (typically 100 cm) cell containing a trace of carbon monoxide levels, more of the infrared radiant energy is absorbed.

A nondispersive infrared spectrometer differs from standard infrared spectrometers in that the infrared radiation from the source is not dispersed according to wavelength by a prism or grating. The nondispersive infrared spectrometer is made very specific for a given compound, or type of compound, by using the sought-for material as part of the detector, or by placing it in a filter cell in the optical path. A diagram of a nondispersive infrared spectrometer selective for CO is shown in Figure 26.4. Radiation from an infrared source in "chopped" by a rotating device so that it alternately passes through a sample cell and a reference cell. In this particular instrument, both beams of light fall on a detector which is filled with CO gas and separated into two compartments by a flexible diaphragm. The relative amounts of infrared radiation absorbed by the CO in the two sections of the detector depend upon the level of CO in the sample. The difference in the amount of infrared radiation absorbed in the two-compartments causes slight differences in heating, so that the diaphragm bulges slightly toward one side. Very slight movement of the diaphragm can be detected and recorded. By means of this device, carbon monoxide can be measured from 0-150 ppm, with a relative accuracy of ±5% in the optimum concentration range.

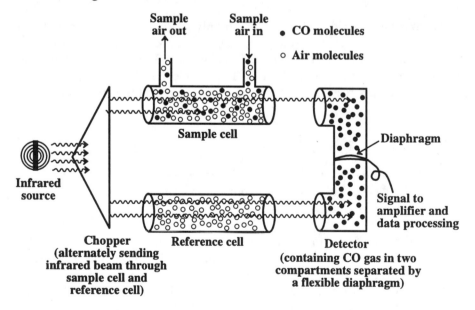

Figure 26.4. Nondispersive infrared spectrometer for the determination of carbon monoxide in the atmosphere.

Flame-ionization gas chromatography detection can also be used for the analysis of carbon monoxide. This detector system was described in detail in Chapter 24,

Section 24.5. It is selective for hydrocarbons, and conversion of CO to methane in the sample is required. This is accomplished by reaction with hydrogen over a nickel catalyst at 360 °C:

$$CO + 3H_2 \rightarrow CH_4 + H_2O \qquad (26.7.1)$$

A major advantage of this approach is that the same basic instrumentation may be used to measure hydrocarbons.

Carbon monoxide may also be analyzed by measuring the heat produced by its catalytic oxidation to CO_2 over a catalyst consisting of a mixture of MnO_2 and CuO. Differences in temperature between a cell in which the oxidation is occurring and a reference cell through which part of the sample is flowing are measured by thermistors. A vanadium oxide catalyst can be used for the oxidation of hydrocarbons, enabling their simultaneous analysis.

26.8. DETERMINATION OF HYDROCARBONS AND ORGANICS

Monitoring of hydrocarbons in atmospheric samples takes advantage of the very high sensitivity of the hydrogen flame ionization detector to measure this class of compounds.[8] Known quantities of air are run through the flame ionization detector 4 to 12 times per hour to provide a measure of total hydrocarbon content. A separate portion of each sample goes into a stripper column to remove water, carbon dioxide, and nonmethane hydrocarbons. Methane and carbon monoxide, which are not retained by the stripper column, are separated by a chromatographic column, passed through a catalytic reduction tube, and then to a flame ionization detector. Eluting first, methane is not changed by the reduction tube and is detected as methane by the detector. The carbon monoxide is reduced to methane, as shown by Reaction 26.7.1 in the preceding section, then detected as the methane product by the flame ionization detector. Concentrations of nonmethane hydrocarbons are given by subtracting the methane concentrations from the total hydrocarbons.

Using the method described above, total hydrocarbons can be determined in a range of 0-13 mg/m^3, corresponding to 0-10 ppm. Methane can be measured over a range of 0-6.5 mg/m^3 (0-10 ppm).

Determination of Specific Organics in the Atmosphere

The method for hydrocarbons described above gives total hydrocarbons and methanes. In some cases it is important to have a method to determine individual organics because of their toxicities, ability to form photochemical smog, as indicators of photochemical smog, and as means of tracing sources of pollution. Numerous techniques have been published for the determination of organic compounds in the atmosphere. For example, whole air samples can be collected in Tedlar bags, concentrated cryogenically at -180°C, then thermally desorbed and measured with high-resolution capillary column gas chromatography.[9]

The U. S. Environmental Protection Agency has published several procedures for determining the identities and concentrations of key organic air pollutants classified as air toxins.[10] These are listed in Table 26.2.

26.9. ANALYSIS OF PARTICULATE MATTER

Particles are almost always removed from air or gas (such as exhaust flue gas) prior to analysis. The two main approaches to particle isolation are filtration and removal by methods that cause the gas stream to undergo a sharp bend such that particles are collected on a surface.[11]

Filtration

The method commonly used for determining the quantity of total suspended particulate matter in the atmosphere draws air over filters that remove the particles.[12] This device, called a **Hi-Vol sampler**, is essentially a glorified vacuum cleaner that draws air through a filter. The samplers are usually placed under a shelter which excludes precipitation and particles larger than about 0.1 mm in diameter, favoring collection of particles up to 25-50 μm diameter. These devices efficiently collect particles from a large volume of air, typically 2000 m³, and typically over a 24-hour period.

The filters used in a Hi-Vol sampler are usually composed of glass fibers and have a collection efficiency of at least 99% for particles with 0.3 μm diameter. Particles with diameters exceeding 100 μm remain on the filter surface, whereas particles with diameters down to approximately 0.1 μm are collected on the glass fibers in the filters. Efficient collection is achieved by using very small diameter fibers (less than 1 μm) for the filter material.

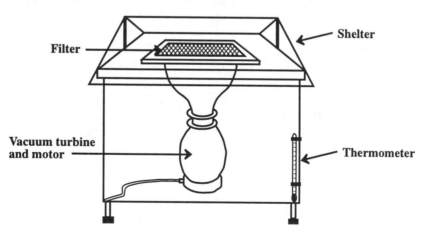

Figure 26.5. Hi-Vol sampler for the collection of particulate matter from the atmosphere for analysis.

The technique described here is most useful for determining total levels of particulate matter. Prior to taking the sample, the filter is maintained at 15-35°C at 50% relative humidity for 24 hours, then weighed. After sampling for 24 hours, the filter is removed and equilibrated for 24 hours under the same conditions used prior to its installation on the sampler. The filter is then weighed and the quantity of particulate matter per unit volume of air is calculated.

Table 26.2. Methods for the Determination of Air Toxins in Ambient Air

Method	Title (abbreviated)	Examples of analytes measured
TO1	Volatile organic compounds[1]	Benzene, toluene, chloroform, trichloro-ethylene, *n*-heptane, chloroform
TO2	Volatile organic compounds[2]	Vinyl chloride, acrylonitrile, chloroform, carbon tetrachloride, toluene
TO3	Volatile organic compounds[3]	Chloroform, 1,2-dichloroethane, benzene, trichloroethylene, chlorobenzene
TO4	Organochlorine pesticides and polychlorinated biphenyls[4]	Aldrin, chlordane, 4,4'-DDT, 2,4,5-trichlorobiphenyl, 2,2',4,4',5,5'-hexachlorobiphenyl
TO5	Aldehydes and ketones[5]	Formaldehyde, acetaldehyde, acetone, methylethyl ketone, benzaldehyde
TO6	Phosgene[6]	Phosgene
TO7	Determination of N-nitrosodimethylamine[7]	N-nitrosodimethylamine
TO8	Phenol and methylphenols (cresols)[8]	Phenol, *o*-, *m*-, *p*-cresol
TO9	Polychlorinated dibenzo-*p*-dioxins[9]	1,2,3,4-Tetrachlorodibenzo-*p*-dioxin, 1,2,3,4,7,8-hexachlorodibenzo-*p*-dioxin
TO10	Organochlorine pesticides[10]	Lindane, chlordane, DDT, dichlorvos, heptachlor, methoxychlor
TO11	Formaldehyde[11]	Formaldehyde
TO12	Nonmethane organic compounds[12]	Organic compounds
TO13	Polynuclear (polycyclic) aromatic hydrocarbons[13]	Acenaphthene, anthracene, benzo(a)pyrene, chrysene, fluorene, naphthalene, pyrene
TO14	Volatile organic compounds (VOCs)[14]	Dichlorodifluoromethane, vinyl chloride, dichloromethane, benzene, trichloroethene, ethylene dibromide, styrene, trans-1,3-dichloropropene, hexachlorobutadiene

[1] Adsorption on Tenax, gas chromatography/mass spectrometry (GC/MS) analysis
[2] Carbon molecular sieve adsorption, GC/MS
[3] Cryogenic preconcentration, gas chromatography/flame ionization detector (GC/FID)
[4] Polyurethane foam collection, gas chromatography/electron capture detector (GC/ECD)

5   Air drawn into impinger containing HCl/2,4-dinitrophenylhydrazine/isooctane to form dinitrophenylhydrazones measured by high performance liquid chromatography (HPLC)

6   Air passed through a solution of aniline, forming carbanilide, HPLC

7   Collection on adsorbent cartridge, GC/MS

8   Collection in NaOH, HPLC

9   Collection on polyurethane foam, gas chromatography/ high-resolution mass spectrometry (HRGC/HRMS).

10 Polyurethane foam sampling, GC/ECD

11 Cartridge sampling, HPLC

12 Cryogenic preconcentration, direct flame ionization detection (FID)

13 Collection on filters and adsorbent cartridges, HPLC

14 Passivated canister sampling, gas chromatography (GC)

The range over which particulate matter can be measured is approximately 2–750 $\mu g/m^3$, where volume is expressed at 25°C and 1 atm (760 mm Hg, 101 kPa) pressure. The lower limit is determined by limitations in measuring mass, and the upper limit by limited flow rate when the filter becomes clogged.

Size separation of particles can be achieved by filtration through successively smaller filters in a **stacked filter unit**. Another approach uses the **virtual impactor**, a combination of an air filter and an impactor (discussed below). In the virtual impactor, the gas stream being sampled is forced to make a sharp bend. Particles larger than about 2.5 μm do not make the bend and are collected on a filter. The remaining gas stream is then filtered to remove smaller particles. A similar approach is the basis for a reference method for the collection of particulate matter with an aerodynamic diameter less than or equal to a nominal 10 micrometers, PM_{10}.[13] As discussed in Chapter 10, small particulate matter is of special concern because of its respirability and other properties. This has led to the definition of another category of particulate matter, $PM_{2.5}$, consisting of particles having an aerodynamic diameter less than or equal to a nominal 2.5 μm. To determine $PM_{2.5}$, ambient air is drawn into an inertial particle size separator and the $PM_{2.5}$ particles are collected on a polytetrafluoroethylene (PTFE) filter for weighing over a period of about 24 hours. The lower detection limit of the method is about 2 $\mu g/m^3$.

Results obtained by the analysis of particulate matter collected by filters should be treated with some caution. A number of reactions may occur on the filter and during the process of removing the sample from the filter. This can cause serious misinterpretation of data. For example, volatile particulate matter may be lost from the filter. Furthermore, because of chemical reactions on the filter, the material analyzed may not be the material that was collected. *Artifact particulate matter* forms from the oxidation of acid gases on alkaline glass fibers. This is especially a problem with sulfur dioxide retained and oxidized to sulfate, especially on alkaline filters. Artifact particulate matter gives an exaggerated value of particulate matter concentration.

One of the major difficulties in particle analysis is the lack of suitable filter material. Different filter materials serve very well for specific applications, but none is satisfactory for all applications. Fiber filters composed of polystyrene are very good for elemental analysis because of low background levels of inorganic materials. However, they are not useful for organic analysis. Glass-fiber filters have good weighing qualities and are therefore very useful for determining total particle concentration; however, metals, silicates, sulfates, and other species are readily leached from the fine glass fibers, introducing error into analysis for inorganic pollutant analysis.

Collection by Impactors

Impactors cause a relatively high velocity gas stream to undergo a sharp bend such that particles are collected on a surface impacted by the stream. The device may be called a dry or wet impactor depending upon whether collecting surface is dry or wet; wet surfaces aid particle retention. Size segregation can be achieved with an impactor because larger particles are preferentially impacted, and smaller particles continue in the gas stream. The cascade impactor, illustrated in Figure 26.6, accomplishes size separation by directing the gas stream onto a series of collection slides through successively smaller orifices, which yield successively higher gas velocities. Particles may break up into smaller pieces from the impact of impingement; therefore, in some cases impingers yield erroneously high values for levels of smaller particles.

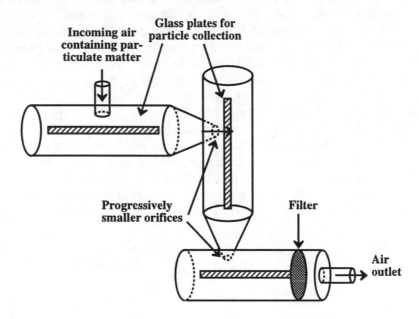

Figure 26.6. Schematic representation of a cascade impactor for the collection of progressively smaller particles.

Particle Analysis

A number of chemical analysis techniques can be used to characterize atmospheric pollutants. These include atomic absorption, inductively coupled plasma techniques, X-ray fluorescence, neutron activation analysis, and ion-selective electrodes for fluoride analysis. Chemical microscopy is an extremely useful technique for the characterization of atmospheric particles. Either visible or electron microscopy may be employed. Particle morphology and shape tell the experienced microscopist a great deal about the material being examined. Reflection, refraction, microchemical tests, and other techniques may be employed to further characterize the materials being examined. Microscopy may be used for determining levels of specific kinds of particles and for determining particle size.

X-Ray Fluorescence

X-Ray fluorescence is another multielement analysis technique that can be applied to a wide variety of environmental samples. It is especially useful for the characterization of atmospheric particulate matter, but it can be applied to some water and soil samples as well. This technique is based upon measurement of X-rays emitted when electrons fall back into inner shell vacancies created by bombardment with energetic X-rays, gamma radiation, or protons. The emitted X-rays have an energy characteristic of the particular atom. The wavelength (energy) of the emitted radiation yields a qualitative analysis of the elements, and the intensity of radiation from a particular element provides a quantitative analysis. A schematic diagram of a wavelength-dispersive X-ray fluorescence spectrophotometer is shown in Figure 26.7. An excitation source, normally an X-ray tube emitting "white" X-rays (a continuum), produces a primary beam of energetic radiation which excites fluorescent X-rays in in the sample. A radioactive source emitting gamma rays or protons from an accelerator may also be used for excitation. For best results, the sample should be mounted as a thin layer, which means that segments of air filters containing fine particulate matter make ideal samples. The fluorescent X-rays are passed through a collimator to select a parallel secondary beam, which is dispersed according to wavelength by diffraction with a crystal monochromator. The monochromatic X-rays in the secondary beam are counted by a detector which rotates at a degree twice that of the crystal to scan the spectrum of emitted radiation. Energy-selective detectors of the Si(Li) semiconductor type enable measurement of fluorescent X-rays of different energies without the need for wavelength dispersion. Instead, the energies of a number of lines falling on a detector simultaneously are distinguished.

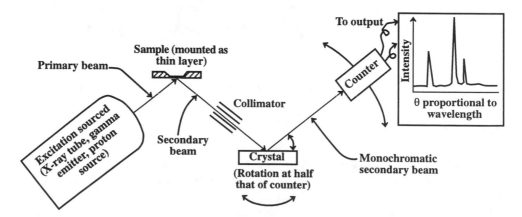

Figure 26.7. Wavelength-dispersive X-ray fluorescence spectrophotometer.

electronically. An energy-dispersive X-ray fluorescence spectrum from an atmospheric particulate sample is shown in Figure 26.8. A significant advantage of X-ray fluorescence multielement analysis is that sensitivities and detection limits do not vary greatly across the periodic table as they do with methods such as neutron activation analysis or atomic absorption. Proton-excited X-ray emission is particularly sensitive.

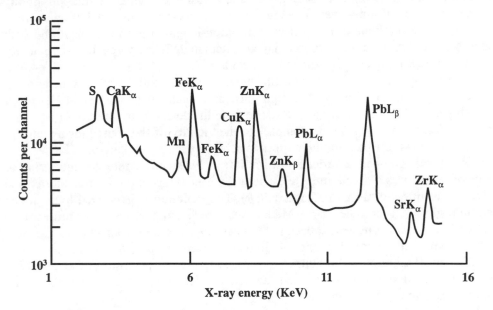

Fig 26.8. Energy-dispersive X-ray fluorescence spectrum from an atmospheric particle sample.

Determination of Lead in Particulate Matter

Because of its toxicity, widespread industrial use, and former use in gasoline, lead is one of the most important contaminants in atmospheric particulate matter. The U. S. Environmental Protection Agency specifies a method for the determination of lead in atmospheric particulate matter.[14] Lead collected from the atmosphere is extracted from particulate matter with heated nitric acid or by a mixture of nitric and hydrochloric acid facilitated by ultrasonication. The lead in the extract is then measured by flame atomic absorption spectrometry. The lower detection limit is typically 0.07 µg Pb/m^3, using a collection volume of 2,400 m^3 of air.

26.10. DIRECT SPECTROPHOTOMETRIC ANALYSIS OF GASEOUS AIR POLLUTANTS

From the foregoing discussion, it is obvious that measurement techniques that depend upon the use of chemical reagents, particularly liquids, are cumbersome and complicated, It is a tribute to the ingenuity of instrument designers that such techniques are being applied successfully to atmospheric pollutant monitoring. Direct spectrophotometric techniques are much more desirable when they are available and when they are capable of accurate analysis at the low levels required. [15] One such technique, nondispersive infrared spectrophotometry, was described in

Section 26.7 for the analysis of carbon monoxide. Three other direct spectrophoto-metric methods are Fourier transform infrared spectroscopy,[16] tunable diode laser spectroscopy, and, the most important of all, differential optical absorption spectroscopy.[17] These techniques may be used for point air monitoring, in which a sample is monitored at a given point, generally by measurement in a long absorption cell. In-stack monitoring may be performed to measure effluents. A final possibility is the collection of long-line data (sometimes using sunlight as a radiation source), yielding concentrations in units of concentration-length (ppm-meters). If the path length is known, the concentration may be calculated. This approach is particularly useful for measuring concentrations in stack plumes.

The low levels of typical air constituents require long path lengths, sometimes up to several kilometers, for spectroscopic measurements. These may be achieved by locating the radiation source some distance from the detector, by the use of a distant retroreflector to reflect the radiation back to the vicinity of the source, or by cells in which a beam is reflected multiple times to achieve a long path length.

A typical open-path Fourier transform infrared system for remote monitoring of air pollutants uses a single unit (telescope) that functions as both a transmitter and receiver of infrared radiation (Figure 26.9). The radiation is generated by a silicon carbide glower, modulated by a Michaelson inteferometer, and transmitted to a retroreflector, which reflects it back to the telescope where its intensity is measured. The modulated infrared signal, called an inteferogram, is processed by a mathematical algorithm, the Fourier transform, to give a spectrum of the absorbing substances. This spectrum is fitted mathematically to spectra of the absorbing species to give their concentrations.

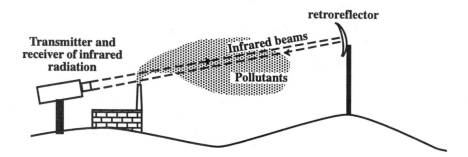

Figure 26.9. FTIR system for remote sensing of air pollutants.

Dispersive absorption spectrometers are basically standard spectrometers with a monochromator for selection of the wavelength to be measured. They are used to measure air pollutants by determining absorption at a specified part of the spectrum of the sought-for material. Of course, other gases or particulate matter that absorb or scatter light at the chosen wavelength interfere. These instruments are generally applied to in-stack monitoring. Sensitivity is increased by using long path lengths or by pressurizing the cell.

Second-derivative spectroscopy is a useful technique for trace gas analysis. Basically, this technique varies the wavelength by a small value around a specified nominal wavelength. The second derivative of light intensity versus wavelength is obtained. In conventional absorption spectrophotometry, a decrease in light intensity as the light passes through a sample indicates the presence of at least one substance—and possibly many—absorbing at that wavelength. Second-derivative spectroscopy, however, provides information regarding the change in intensity with wavelength, thereby indicating the presence of specific absorption lines or bands which may be superimposed on a relatively high background of absorption. Much higher specificity is obtained. The spectra obtained by second-derivative spectrometry in the ultraviolet region show a great deal of structure and are quite characteristic of the compounds being observed.

Lidar, which stands for *light detection and ranging* (analogous to radar, *radio detection and ranging*), is finding numerous applications in atmospheric monitoring.[18] Lidar systems send short pulses of light or infrared radiation into the atmosphere and collect radiation scattered back from molecules or particles in the atmosphere. Computer analysis of the signal enables analysis of species in the atmosphere.

LITERATURE CITED

1. Appell, B. R., "Sampling of Selected Labile Atmospheric Pollutants," Chapter 1 in *Measurement Challenges in Atmospheric Chemistry*, Advances in Chemistry Series **233**, American Chemical Society, Washington, D.C., 1993, pp. 1-40.

2. 40 *Code of Federal Regulations*, Part 50, Office of the Federal Register, National Archives and Records Administration, Washington, D.C., July 1, annually.

3. West, P. W., and G. C. Gaeke, "Fixation of Sulfur Dioxide as Disulfito-mercurate(II) and Subsequent Colorimetric Estimation," *Analytical Chemistry*, **28**, 1816-1810 (1956).

4. "Reference Method for the Determination of Sulfur Dioxide in the Atmosphere (Pararosaniline Method)," 40 *Code of Federal Regulations*, Part 50, Appendix A, 1999.

5. "Measurement Principle and Calibration Procedure for the Measurement of Nitrogen Dioxide in the Atmosphere (Gas-Phase Chemiluminescence)," 40 *Code of Federal Regulations*, Part 50, Appendix F, 1999.

6. "Measurement Principle and Calibration Procedure for the Measurement of Ozone in the Atmosphere," 40 *Code of Federal Regulations*, Part 50, Appendix D, 1999.

7. "Measurement Principle and Calibration Procedure for the Measurement of Carbon Monoxide in the Atmosphere (Nondispersive Infrared Photometry)" 40 *Code of Federal Regulations*, Part 50, Appendix C, 1999.

8. "Reference Method for the Determination of Hydrocarbons Corrected for Methane," 40 *Code of Federal Regulations*, Part 50, Appendix E, 1999.

9. Fukui, Yoshiko and Paul Doskey, "Measurement Technique for Organic Nitrates and Halocarbons in Ambient Air," *Journal of High Resolution Chromatography*, **21**, 201-208 (1998).

10. "Air Toxins," *Understanding Environmental Methods*, CD/ROM Version 2.0, Genium Publishing Corp., Schenectady, NY, 1998

11. Cahill, Thomas A., and Paul Wakabayashi, "Compositional Analysis of Size-Segregated Aerosol Samples," Chapter 7 in *Measurement Challenges in Atmospheric Chemistry*, Advances in Chemistry Series **233**, American Chemical Society, Washington, D.C., 1993, pp. 211-228.

12. "Reference Method for the Determination of Suspended Particulate Matter in the Atmosphere (High-Volume Method)" 40 *Code of Federal Regulations*, Part 50, Appendix B, 1999.

13. "Reference Method for the Determination of Particulate Matter as PM_{10} in the Atmosphere (High-Volume Method)" 40 *Code of Federal Regulations*, Part 50, Appendix J, 1999.

14. "Reference Method for the Determination of Lead in Suspended Particulate Matter Collecteld from Ambient Air," 40 *Code of Federal Regulations*, Part 50, Appendix G, 1999.

15. Roscoe, Howard K. and Kevin C. Clemitshaw, "Measurement Techniques in Gas-Phase Tropospheric Chemistry: A Selective View of the Past, Present, and Future, *Science*, 1065-1072 (1997).

16. Kagann, Robert H. and Robert J. Kricks, "Improvements in Open-Path FTIR for Measurement of Ambient Air at Industrial Facilities," *Proceedings of the Annual ISA Analytical Division Symposium*, **32**, 69-73 (1999).

17. Dasgupta, Purnendu K., "Automated Measurement of Atmospheric Trace Gases," Chapter 2 in *Measurement Challenges in Atmospheric Chemistry*, Advances in Chemistry Series **233**, American Chemical Society, Washington, DC, 1993, pp. 41-90.

18. Holmén, Britt A., William E. Eichinger, and Robert G. Flocchini, *Environmental Science and Technology*, **32**, 3968-3076 (1998).

SUPPLEMENTAL REFERENCES

Bucholtz, Frank, Ed., *Environmental Monitoring and Instrumentation*, Optical Society of America, Washington, D.C., 1997.

Jacob, Daniel J., *Introduction to Atmospheric Chemistry*, Princeton University Press, Princeton, NJ, 1999.

Keith Lawrence H. and Mary M. WalkerEds., *Handbook of Air Toxics: Sampling, Analysis, and Properties*, CRC Press/Lewis Publishers, Boca Raton, FL, 1995.

Matson, P. A. and R. C. Harriss, Eds., *Biogenic Trace Gases: Measuring Emissions from Soil and Water*, Blackwell Science, Cambridge, MA, 1995.

Meier, Arndt, *Determination of Atmospheric Trace Gas Amounts and Corresponding Natural Isotopic Ratios by Means of Ground-Based FTIR Spectroscopy in the High Arctic*, Alfred-Wegener-Institut für Polar und Meeresforschung ; Bremen, Germany, 1997.

Optical Society of America, *Laser Applications to Chemical and Environmental Analysis*, Optical Society of America, Washington, D.C., 1998.

Seinfeld, John H., *Atmospheric Chemistry and Physics*, John Wiley & Sons, Inc., New York, NY, 1998.

Wight, Gregory D., *Fundamentals of Air Sampling*, CRC Press/Lewis Publishers, Boca Raton, FL, 1994.

Willeke, Klaus, and Paul A. Baron, Eds., *Aerosol Measurement: Principles, Techniques, and Applications*, Van Nostrand Reinhold, New York, 1993.

Winegar, Eric D. and Lawrence H. Keith, Eds., *Sampling and Analysis of Airborne Pollutants*, CRC Press/Lewis Publishers, Boca Raton, FL, 1993.

QUESTIONS AND PROBLEMS

1. What device is employed to make a nondispersive infrared analyzer selective for the compound being determined?

2. Suggest how mass spectrometry would be most useful in air pollutant analysis.

3. What is a required characteristic of the absorption cell used for the direct spectrophotometric measurement of gaseous pollutants in the atmosphere, and how may this characteristic be achieved in a cell of manageable dimensions?

4. If 0.250 g of particulate matter is the minimum quantity required for accurate weighing on a Hi-Vol sampler filter, how long must such a sampler be operated at a flow rate of 2.00 m^3/min to collect a sufficiently large sample in an atmosphere containing 5 $\mu g/m^3$ of particulate matter?

5. The atmosphere around a chemical plant is suspected of containing a number of heavy metals in the form of particulate matter. After a review of Chapter 10, suggest several methods that would be useful for a qualitative and roughly quantitative analysis of the metals in the particulate matter.

6. Assume that the signal from a chemiluminescence analyzer for NO is proportional to NO concentration. For the same rate of air flow, an instrument gave a signal of 135 microamp for an air sample that had been passed over a thermal converter and 49 μamp with the converter out of the stream. A standard sample containing 0.233 ppm NO gave a signal of 80 μamp. What was the level of NO_2 in the atmospheric sample?

7. A permeation tube containing NO_2 lost 208 mg of the gas in 124 min at 20°C. What flow rate of air at 20°C should be used with the tube to prepare a standard atmospheric sample containing exactly 1.00 ppm by volume of NO_2?

8. What solution should be used in the "wet impinger" designed to collect samples for the determination of metals in the atmosphere?

9. An atmosphere contains 0.10 ppm by volume of SO_2 at 25°C and 1.00 atm pressure. What volume of air would have to be sampled to collect 1.00 mg of SO_2 in tetrachloromercurate solution?

10. Assume that 20% of the surface of a membrane filter used to collect particulate matter consists of circular openings with a uniform diameter of 0.45 μm. How many openings are on the surface of a filter with a diameter of 5.0 cm?

11. Some atmospheric pollutant analysis methods have been used in the past that later have been shown not to give the "true" value. In what respects may such methods still be useful?

12. How may ion chromatography be used for the analysis of nonionic gases?

27 ANALYSIS OF BIOLOGICAL MATERIALS AND XENOBIOTICS

27.1. INTRODUCTION

As defined in Chapter 22 a xenobiotic species is one that is foreign to living systems. Common examples include heavy metals, such as lead, which serve no physiologic function, and synthetic organic compounds that are not made in nature. Exposure of organisms to xenobiotic materials is a very important consideration in environmental and toxicological chemistry. Therefore, the determination of exposure by various analytical techniques is one of the more crucial aspects of environmental chemistry.

This chapter deals with the determination of xenobiotic substances in biological materials. Although such substances can be measured in a variety of tissues, the greatest concern is their presence in human tissues and other samples of human origin. Therefore, the methods described in this chapter apply primarily to exposed human subjects. They are essentially identical to methods used on other animals and, in fact, most were developed through animal studies. Significantly different techniques may be required for plant or microbiological samples.

The measurement of xenobiotic substances and their metabolites in blood, urine, breath, and other samples of biological origin to determine exposure to toxic substances is called **biological monitoring**. Comparison of the levels of analytes measured with the degree and type of exposure to foreign substances is a crucial aspect of toxicological chemistry. It is an area in which rapid advances are being made. For current information regarding this area, the reader is referred to excellent reviews of the topic,[1,2] and several books on biological monitoring such as those by Angerer, Draper, Baselt, and Kneip and coauthors as listed in the back of this chapter under "Supplementary References," are available as well.

The two main approaches to workplace monitoring of toxic chemicals are workplace monitoring, using samplers that sample xenobiotic substances from workplace air, and biological monitoring. Although the analyses are generally much more difficult, biological monitoring is a much better indicator of exposure because it measures exposure to all routes—oral and dermal as well as inhalation—and it gives an integrated value of exposure. Furthermore, biological monitoring is very useful in determining the effectiveness of measures taken to prevent exposure, such as protective clothing and hygienic measures.

27.2. INDICATORS OF EXPOSURE TO XENOBIOTICS

The two major considerations in determining exposure to xenobiotics are the type of sample and the type of analyte. Both of these are influenced by what happens to a xenobiotic material when it gets into the body. For some exposures, the entry site composes the sample. This is the case, for example, in exposure to asbestos fibers in the air, which is manifested by lesions to the lung. More commonly, the analyte may appear at some distance from the site of exposure, such as lead in bone that was originally taken in by the respiratory route. In other cases the original xenobiotic is not even present in the analyte. An example of this is methemoglobin in blood, the result of exposure to aniline absorbed through the skin.

The two major kinds of samples analyzed for xenobiotics exposure are blood and urine. Both of these kinds of samples are analyzed for systemic xenobiotics, which are those that are transported in the body and metabolized in various tissues. Xenobiotic substances, their metabolites, and their adducts are absorbed into the body and transported through it in the bloodstream. Therefore, blood is of unique importance as a sample for biological monitoring. Blood is not a simple sample to process, and subjects often object to the process of taking it. Upon collection, blood may be treated with an anticoagulant, usually a salt of ethylenediaminetetraacetic acid (EDTA), and processed for analysis as whole blood. It may also be allowed to clot and be centrifuged to remove solids; the liquid remaining is blood serum.

Recall from Chapter 22 that as the result of Phase 1 and Phase 2 reactions, xenobiotics tend to be converted to more polar and water soluble metabolites. These are eliminated with the urine, making urine a good sample to analyze as evidence of exposure to xenobiotic substances. Urine has the advantage of being a simpler matrix than blood and one that subjects more readily give for analysis. Other kinds of samples that may be analyzed include breath (for volatile xenobiotics and volatile metabolites), air or nails (for trace elements, such as selenium), adipose tissue (fat), and milk (obviously limited to lactating females). Various kinds of organ tissue can be analyzed in cadavers, which can be useful in trying to determine cause of death by poisoning.

The choice of the analyte actually measured varies with the xenobiotic substance to which the subject has been exposed. Therefore, it is convenient to divide xenobiotic analysis on the basis of the type of chemical species determined. The most straightforward analyte is, of course, the xenobiotic itself. This applies to elemental xenobiotics, especially metals, which are almost always determined in the elemental form. In a few cases organic xenobiotics can also be determined as the parent compound. However, organic xenobiotics are commonly metabolized to other products by Phase 1 and Phase 2 reactions. Commonly, the Phase 1 reaction product is measured, often after it is hydrolyzed from the Phase 2 conjugate, using enzymes or acid hydrolysis procedures. Thus, for example, *trans,trans*-muconic acid can be measured as evidence of exposure to the parent compound benzene. In other cases a

Phase 2 reaction product is measured, for example, hippuric acid determined as evidence of exposure to toluene. Some xenobiotics or their metabolites form adducts with endogenous materials in the body, which are then measured as evidence of exposure. A simple example is the adduct formed between carbon monoxide and hemoglobin, carboxyhemoglobin. More complicated examples are the adducts formed by the carcinogenic Phase 1 reaction products of polycyclic aromatic hydrocarbons with DNA or hemoglobin. Another class of analytes consists of endogenous substances produced upon exposure to a xenobiotic material. Methemoglobin formed as a result of exposure to nitrobenzene, aniline, and related compounds is an example of such a substance which does not contain any of the original xenobiotic material. Another class of substance causes measurable alterations in enzyme activity. The most common example of this is the inhibition of acetylcholinesterase enzyme by organophosphates or carbamate insecticides.

27.3. DETERMINATION OF METALS

Direct Analysis of Metals

Several biologically important metals can be determined directly in body fluids, especially urine, by atomic absorption. In the simplest cases the urine is diluted with water or with acid and a portion analyzed directly by graphite furnace atomic absorption, taking advantage of the very high sensitivity of that technique for some metals. Metals that can be determined directly in urine by this approach include chromium, copper, lead, lithium, and zinc. Very low levels of metals can be measured using a graphite furnace atomic absorption technique, and Zeeman background correction with a graphite furnace enables measurement of metals in samples that contain enough biological material to cause significant amounts of "smoke" during the atomization process, so that ashing the samples is less necessary.

A method has been published for the determination of a variety of metals in diluted blood and serum using inductively coupled plasma atomization with mass spectrometric detection.[3] Blood was diluted 10-fold and serum 5-fold with a solution containing ammonia, Triton X-100 surfactant, and EDTA. Detection limits adequate for measurement in blood or serum were found for cadmium, cobalt, copper, lead, rubidium, and zinc.

Metals in Wet-Ashed Blood and Urine

Several toxicologically important metals are readily determined from wet-ashed blood or urine using atomic spectroscopic techniques. The ashing procedure may vary, but always entails heating the sample with strong acid and oxidant to dryness and redissolving the residue in acid. A typical procedure is digestion of blood or urine for cadmium analysis, which consists of mixing the sample with a comparable volume of concentrated nitric acid, heating to a reduced volume, adding 30% hydrogen peroxide oxidant, heating to dryness, and dissolving in nitric acid prior to measurement by atomic absorption or emission. Mixtures of nitric, sulfuric, and perchloric acid are effective though somewhat hazardous media for digesting blood, urine, or tissue. Wet ashing followed by atomic absorption analysis can be used for the determination in blood or urine of cadmium, chromium, copper, lead, manganese, and zinc, among other metals. Although atomic absorption, especially highly sensitive graphite furnace atomic absorption, has long been favored for measuring metals in biological samples, the multielement capability and other advantages of inductively coupled plasma atomic spectroscopy has led to its use for determining metals in blood and urine samples.[4]

Extraction of Metals for Atomic Absorption Analysis

A number of procedures for the determination of metals and biological samples call for the extraction of the metal with an organic chelating agent in order to remove interferences and concentrate the metal to enable detection of low levels. The urine or blood sample may be first subjected to wet ashing to enable extraction of the metal. Beryllium from an acid-digested blood or urine sample may be extracted by acetylacetone into methylisobutyl ketone prior to atomic absorption analysis. Virtually all of the common metals can be determined by this approach using appropriate extractants.

The availability of strongly chelating extracts for a number of metals has lead to the development of procedures in which the metal is extracted from minimally treated blood or urine, then quantified by atomic absorption analysis. The metals for which such extractions can be used include cobalt, lead, and thallium extracted into organic solvent as the dithiocarbamate chelate, and nickel extracted into methyl-isobutyl ketone as a chelate formed with ammonium pyrrolidinedithiocarbamate.

Methods for several metals or metalloids involve conversion to a volatile form. Arsenic, antimony, and selenium can be reduced to their volatile hydrides, AsH_3, SbH_3, and H_2Se, repectively, which can be determined by atomic absorption or other means. Mercury is reduced to volatile mercury metal, which is evolved from solution and measured by cold vapor atomic absorption.

27.4. DETERMINATION OF NONMETALS AND INORGANIC COMPOUNDS

Relatively few nonmetals require determination in biological samples. One important example is fluoride, which occurs in biological fluids as the fluoride ion, F^-. In some cases of occupational exposure or exposure through food or drinking water, excessive levels of fluoride in the body can be a health concern. Fluoride is readily determined potentiometrically with a fluoride ion-selective electrode. The sample is diluted with an appropriate buffer and the potential of the fluoride electrode measured very accurately vs. a reference electrode, with the concentration calculated from a calibration plot. Even more accurate values can be obtained by the use of standard addition in which the potential of the electrode system in a known volume of sample is read, a measured amount of standard fluoride is added, and the shift in potential is used to calculate the unknown concentration of fluoride.

Another nometal for which a method of determining biological exposure would be useful is white phosphorus, the most common and relatively toxic elemental form. Unfortunately, there is not a chemical method suitable for the determination of exposure to white phosphorus that would distinguish such exposure from relatively high background levels of organic and inorganic phosphorus in body fluids and tissues.

Toxic cyanide can be isolated in a special device called a Conway microdiffusion cell by treatment with acid, followed by collection of the weakly acidic HCN gas that is evolved in a base solution. The cyanide released can be measured spectrophotometrically by formation of a colored species.

Carbon monoxide is readily determined in blood by virtue of the colored carboxyhemoglogin that it forms with hemoglobin. The procedure consists of measuring the absorbances at wavelengths of 414, 421, and 428 nm of the blood sample, a sample through which oxygen has been bubbled to change all the hemoglobin to the oxyhemoglobin form, and a sample through which carbon monoxide has been bubbled to change all the hemoglobin to carboxyhemoglobin. With the appropriate calculations, a percentage conversion to carboxyhemoglobin can be obtained.

27.5. DETERMINATION OF PARENT ORGANIC COMPOUNDS

A number of organic compounds can be measured as the unmetabolized compound in blood, urine, and breath. In some cases the sample can be injected along with its water content directly into a gas chromatograph. Direct injection is used for the measurement of acetone, n-butanol, dimethylformamide, cyclopropane, halothane, methoxyflurane, diethyl ether, isopropanol, methanol, methyln-butyl ketone, methyl chloride, methylethyl ketone, toluene, trichloroethane, and trichloroethylene.

For the determination of volatile compounds in blood or urine, a straightforward approach is to liberate the analyte at an elevated temperature allowing the volatile compound to accumulate in headspace above the sample followed by direct injection of headspace gas into a gas chromatograph. A reagent such as perchloric acid may be added to deproteinize the blood or urine sample and facilitate release of the volatile xenobiotic compound. Among the compounds determined by this approach are acetaldehyde, dichloromethane, chloroform, carbon tetrachloride, benzene, trichloroethylene, toluene, cyclohexane, and ethylene oxide. The use of multiple detectors for the gas chromatographic determination of analytes in headspace increases the versatility of this technique and enables the determination of a variety of physiologically important volatile organic compounds.[5]

Purge-and-trap techniques in which volatile analytes are evolved from blood or urine in a gas stream and collected on a trap for subsequent chromatographic analysis have been developed. Such a technique employing gas chromatographic separation and Fourier transform infrared detection has been described for a number of volatile organic compounds in blood.[6]

27.6. MEASUREMENT OF PHASE 1 AND PHASE 2 REACTION PRODUCTS

Phase 1 Reaction Products

For a number of organic compounds the most accurate indication of exposure is to be obtained by determining their Phase 1 reaction products. This is because many compounds are metabolized in the body and don't show up as the parent compound.

And those fractions of volatile organic compounds that are not metabolized may be readily eliminated with expired air from the lungs and may thus be missed. In cases where a significant fraction of the xenobiotic compound has undergone a Phase 2 reaction, the Phase 1 product may be regenerated by acid hydrolysis.

One of the compounds commonly determined as its Phase 1 metabolite is benzene,[7] which undergoes the following reactions in the body (see Chapter 23, Section 23.4):

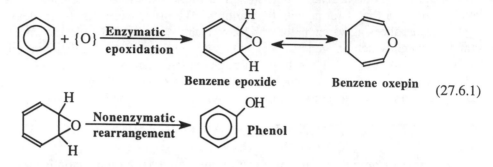

Benzene epoxide   Benzene oxepin   (27.6.1)

Phenol

Therefore, exposure to benzene can be determined by analysis of urine for phenol. Although a very sensitive colorimetric method for phenol involving diazotized *p*-nitroaniline has long been available, gas chromatographic analysis is now favored. The urine sample is treated with perchloric acid to hydrolyze phenol conjugates and the phenol is extracted into diisopropyl ether for chromatographic analysis. Two other metabolic products of benzene, *trans,trans*-muconic acid[8] and S-phenyl mercapturic acid.[9] are now commonly measured as more specific biomarkers of benzene exposure.

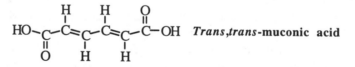

Trans,trans-muconic acid

Insecticidal carbaryl undergoes the following metabolic reaction:

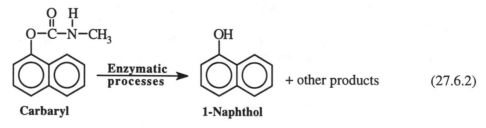

Carbaryl   1-Naphthol   + other products   (27.6.2)

Therefore, the analysis of 1-naphthol in urine indicates exposure to carbaryl. The 1-naphthol that is conjugated by a Phase 2 reaction is liberated by acid hydrolysis, then determined spectrophotometrically or by chromatography

In addition to the examples discussed above, a number of other xenobiotics are measured by their phase one reaction products. These compounds and their metabolites are listed in Table 27.1. These methods are for metabolites in urine. Normally the urine sample is acidified to release the Phase 1 metabolites from Phase 2 conjugates that they might have formed and, except where direct sample injection is employed, the analyte is collected as vapor or extracted into an organic solvent. In some cases the analyte is reacted with a reagent that produces a volatile derivative that is readily separated and detected by gas chromatography.

Phase 2 Reaction Products

Hippuric acids, which are formed as Phase 2 metabolic products from toluene, the xylenes, benzoic acid, ethylbenzene, and closely related compounds, can be determined as biological markers of exposure. The formation of hippuric acid from toluene is shown in Chapter 23, Figure 23.2, and the formation of 4-methylhippuric acid from *p*-xylene is shown below:

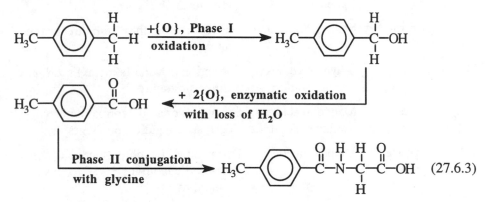

(27.6.3)

Other metabolites that may be formed from aryl solvent precursors include mandelic acid and phenylgloxylic acid.

Exposure to toluene can be detected by extracting hippuric acid from acidified urine into diethyl ether/isopropanol and direct ultraviolet absorbance measurement of the extracted acid at 230 nm. When the analysis is designed to detect the xylenes, ethylbenzene, and related compounds, several metabolites related to hippuric acid may be formed and the ultraviolet spectrometric method does not give the required specificity. However, the various acids produced from these compounds can be extracted from acidified urine into ethyl acetate, derivatized to produce volatile species, and quantified by gas chromatography.

A disadvantage to measuring toluene exposure by hippuric acid is the production of this metabolite from natural sources, and the determination of tolulylmercapturic acid is now favored as a biomarker of toluene exposure.[10] An interesting sidelight is that dietary habits can cause uncertainties in the measurement of xenobiotic metabolites. An example of this is the measurement of worker exposure to 3-chloropropene by the production of allylmercapturic acid.[11] This metabolite is also produced by garlic, and garlic consumption by workers was found to be a confounding factor in the method. Thiocyanate monitored as evidence of exposure to cyanide is increased markedly by the consumption of cooked cassava!

Mercapturates

Mercapturates are proving to be very useful Phase 2 reaction products for measuring exposure to xenobiotics, especially because of the sensitive determination of

Table 27.1. Phase 1 Reaction Products of Xenobiotics Determined

Parent compound	Metabolite	Method of analysis
Cyclohexane	Cyclohexanol	Extraction of acidified, hydrolyzed urine with dichloromethane followed by gas chromatography
Diazinone	Organic phosphates	Colorimetric determination of phosphates
p-Dichlorobenzene	2,5-Dichlorophenol	Extraction into benzene, gas chromatographic analysis
Dimethylformamide	Methylformamide	Gas chromatography with direct sample introduction
Dioxane	β-hydroxyethoxyacetic acid	Formation of volatile methyl ester, gas chromatography
Ethybenzene	Mandelic acid and related aryl acids	Extraction of acids, formation of volatile derivatives, gas chromatography
Ethylene glycol monomethyl ether	Methoxyacetic acid	Extracted with dichloromethane, converted to volatile methyl derivative, gas chromatography
Formaldehyde	Formic acid	Gas chromatography of volatile formic acid deriviative
Hexane	2,5-Hexanedione	Gas chromatography after extraction with dichloromethane
n-heptane	2-Heptanone, valerolactone, 2,5-heptanedione	Measurement in urine by GC/MS
Isopropanol	Acetone	Gas chromatography following extraction with methylethyl ketone
Malathion	Organic phosphates	Colorimetric determination of phosphates
Methanol	Formic acid	Gas chromatography of volatile formic acid derivative
Methyl bromide	Bromide ion	Formation of volatile organobromine compounds, gas chromatography
Nitrobenzene	p-Nitrophenol	Gas chromatography of volatile derivative
Parathion	p-Nitrophenol	Gas chromatography of volatile derivative
Polycyclic aryl hydrocarbons	1-Hydroxypyrene	HPLC of urine

Table 27.1. (Cont.)

Styrene	Mandelic acid	Extraction of acids, formation of volatile derivatives, gas chromatography
Tetrachloro-ethylene	Trichloroacetic acid	Extracted into Pyridine and measured colorimetrically
Trichloroethane	Trichloroacetic acid	Extracted into Pyridine and measured colorimetrically
Trichloro-ethylene	Trichloroacetic acid	Extracted into Pyridine and measured colorimetrically

these substances by HPLC separation, and fluorescence detection of their *o*-phthaldialdehyde derivatives. In addition to toluene mentioned above, the xenobiotics for which mercapturates may be monitored include styrene, structurally similar to toluene; acrylonitrile; allyl chloride; atrazine; butadiene; and epichlorohydrin.

The formation of mercapturates or mercapturic acid derivatives by metabolism of xenobiotics is the result of a Phase 2 conjugation by glutathione. **Glutathione** (commonly abbreviated GSH) is a crucial conjugating agent in the body. This compound is a tripeptide, meaning that it is composed of three amino acids linked together. These amino acids and their abbreviations are glutamic acid (Glu), cysteine (Cys), and glycine (Gly). The formula of glutathione may be represented as illustrated in Figure 27.1, where the SH is shown specifically because of its crucial

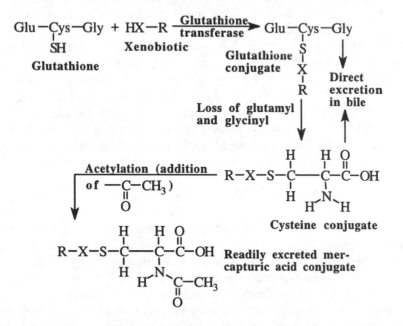

Figure 27.1. Glutathione conjugate of a xenobiotic species (HX-R) followed by formation of glutathione and cysteine conjugate intermediates (which may be excreted in bile) and acetylation to form readily excreted mercapturic acid conjugate.

role in forming the covalent link to a xenobiotic compound. Glutathione conjugate may be excreted directly, although this is rare. More commonly, the GSH conjugate undergoes further biochemical reactions that produce mercapturic acids (compounds with N-acetylcysteine attached) or other species. The specific mercapturic acids can be monitored as biological markers of exposure to the xenobiotic species that result in their formation. The overall process for the production of mercapturic acids as applied to a generic xenobiotic species, HX-R (see previous discussion), is illustrated in Figure 27.1.

27.7. DETERMINATION OF ADDUCTS

Determination of adducts is often a useful and elegant means of measuring exposure to xenobiotics. Adducts, as the name implies, are substances produced when xenobiotic substances add to endogenous chemical species. The measurement of carbon monoxide from its hemoglobin adduct was discussed in Section 27.4. In general, adducts are produced when a relatively simple xenobiotic molecule adds to a large macromolecular biomolecule that is naturally present in the body. The fact that adduct formation is a mode of toxic action, such as occurs in the methylation of DNA during carcinogenesis (Chapter 22, Section 22.8), makes adduct measurement as a means of biological monitoring even more pertinent.

Adducts to hemoglobin are perhaps the most useful means of biological monitoring by adduct formation. Hemoglobin is, of course, present in blood, which is the most accurate type of sample for biological monitoring. Adducts to blood plasma albumin are also useful monitors and have been applied to the determination of exposure to toluene diisocyanate, benzo(a)pyrene, styrene, styrene oxide, and aflatoxin B1. The DNA adduct of styrene oxide has been measured to indicate exposure to carcinogenic styrene oxide.[13]

One disadvantage of biological monitoring by adduct formation can be the relatively complicated prodedures and expensive, specialized instruments requried. Lysing red blood cells may be required to release the hemoglobin adducts, derivatization may be necessary, and the measurements of the final analyte species can require relatively sophisticated instrumental techniques. Despite these complexities, the measurement of hemoglobin adducts is emerging as a method of choice for a number of xenobiotics including acrylamide, acrylonitrile, 1,3-butadiene, 3,3' dichlorobenzidine, ethylene oxide, and hexahydrophthalic anhydride.

27.8. THE PROMISE OF IMMUNOLOGICAL METHODS

As discussed in Chapter 25, Section 25.5, immunoassay methods offer distinct advantages in specificity, selectivity, simplicity, and costs. Although used in simple test kits for blood glucose and pregnancy testing, immunoassay methods have been limited in biological monitoring of xenobiotics, in part because of interferences in complex biological systems. Because of their inherent advantages, however, it can be anticipated that immunoassays will grow in importance for biological monitoring of xenobiotics.[14] As an example of such an application, polychlorinated biphenyls (PCBs) have been measured in blood plasma by immunoassay.[15]

In addition to immunoassay measurement of xenobiotics and their metabolites, immunological techniques can be used for the separation of analytes from complex biological samples employing immobilized antibodies. This approach has been used to isolate aflatoxicol from urine and enable its determination along with aflatoxins B1, B2, G1, G2, M1, and Q1 using high-performance liquid chromatography and post-column derivatization/fluorescence detection.[16] A monoclonal antibody reactive with S-phenylmercapturic acid, an important Phase 2 reaction product of benzene resulting from glutathione conjugation, has been generated from an appropriate hapten-protein conjugate. The immobilized antibody has been used in a column to enrich S-phenylmercapturic acid from the urine of workers exposed to benzene.[17] Many more such applications can be anticipated in future years.

LITERATURE CITED

1. Draper, William M., Kevin Ashley, Clifford R. Glowacki, and Paul R. Michael, "Industrial Hygiene Chemistry: Keeping Pace with Rapid Change in the Workplace," *Analytical Chemistry*, **71**, 33R-60R (1999). A comprehensive review of this topic is published every two years in *Analytical Chemistry*.

2. Atio, A., "Special Issue: Biological Monitoring in Occupational and Environmental Health," *Science of the Total Environment*, **199**, 1-226 (1997).

3. Barany, Ebba, Ingvar A. Bergdahl, Andrejs Schutz, Staffan Skerfving, and Agneta Oskarsson, "Inductively Coupled Plasma Mass Spectrometry for Direct Multielement Analysis of Diluted Human Blood and Serum," *Journal of Analytical Atomic Spectroscopy*, **12**, 1005-1009 (1997).

4. Paschal, Daniel C., Bill G. Ting, John C. Morrow, James L. Pirkle, Richard J. Jackson, Eric J. Sampson, Dayton T. Miller, and Kathleen L. Caldwell, "Trace Metals in Urine of United States Residents: Reference Range Concentrations," *Environmental Research*, **76**, 53-59 (1998).

5. Schroers, H.-J. and E. Jermann, "Determination of Physiological Levels of Volatile Organic Compounds in Blood Using Static Headspace Capillary Gas Chromatography with Serial Triple Detection, *Analyst*, **123**, 715-720 (1998).

6. Ojanpera, Ilkka, Katja Pihlainen, and Erkki Vuori, "Identification Limits for Volatile Organic Compounds in the Blood by Purge-and-Trap GC-FTIR," *Journal of Analytical Toxicology*, **22**, 290-295 (1998).

7. Agency for Toxic Substances and Disease Registry, U. S. Department of Health and Human Services, *Toxicological Profile for Benzene*, CD/ROM Version, CRC Press/Lewis Publishers, Boca Raton, FL, 1999.

8. Scherer, Gerhard, Thomas Renner, and Michael Meger, "Analysis and Evaluation of *Trans,trans*-muconic Acid as a Biomarker for Benzene Exposure," *Journal of Chromatography, B: Biomedical Science Applications*, **717**, 179-199 (1998).

9. Boogaard, Pieter J. and Nico J. Van Sittert, "Suitability of S-phenyl Mercapturic Acid and *Trans-trans*-muconic acid as Biomarkers for Exposure to Low Concentrations of Benzene," *Environmental Health Perspectives Supplement*, **104**, 1151-1157 (1996).

10. Angerer, J., M. Schildbach, and A. Kramer, "S-Toluylmercapturic Acid in the Urine of Workers Exposed to Toluene: A New Biomarker for Toluene Exposure," *Archives of Toxicology*, **72**, 119-123 (1998).

11. De Ruij, Ben M., Pieter J. Boogard, Jan N. M. Commandeur, Nico J. van Sittert, and Nico P. E. Verneulen, "Allylmercapturic Acid as Urinary Biomarker of Human Exposure to Allyl Chloride," *Occupational and Environmental Medicine*, **54**, 653-661 (1997).

12. Brashear, Wayne T., Christel T. Bishop, and Richat Abbas, "Electrospray Analysis of Biological Samples for Trace Amounts of Trichloroacetic Acid, Dichloroacetic acid, and Monochloroacetic Acid" *Journal of Analytical Toxicology*, **21**, 330-334 (1997).

13. Rappaport, Stephen M., Karen Yeowell-O'Connell, William Bodell, Janice W. Yager, and Elaine Symanski, "An Investigation of Multiple Biomarkers Among Workers Exposed to Styrene and Styrene-7,8-oxide," *Cancer Research*, **56**, 5410-5416 (1996).

14. Wengatz, Ingrid, Adam S. Harris, S. Douglass Gilman, Monika Wortberg, Horacio Kido, Ferenc Szurdoki, Marvin H. Goodrow, Lynn L. Jaeger, and Donald W. Stoutamire, "Recent Developments in Immunoassays and Related Methods for the Detection of Xenobiotics," *ACS Symposium Series 646 (Environmental Immunochemical Methods)*, American Chemical Society, Washington, DC, 1996, pp. 110-126.

15. Griffin, P., K. Jones, and J. Cocker, "Biological Monitoring of Polychlorinated Biphenyls in Plasma: A Comparison of Enzyme-linked Immunosorbent Assay and Gas Chromatography Detection Methods," *Biomarkers*, **2**, 193-195 (1997).

16. Kussak, Anders, Barbro Andersson, Kurt Andersson, and Carl-Axel Nilsson, "Determination of Aflatoxicol in Human Urine by Immunoaffinity Column Clean-up and Liquid Chromatography," *Chemosphere*, **36**, 1841-1848 (1998).

17. Ball, Lathan, Alan S. Wright, Nico J. Van Sittert, and Paul Aston, "Immuno-enrichment of Urinary S-phenylmercapturic Acid," *Biomarkers*, **2**, 29-33 (1997).

SUPPLEMENTARY REFERENCES

Angerer, J. K., and K. H. Schaller, *Analyses of Hazardous Substances in Biological Materials*, Vol. 1, VCH, Weinheim, Germany, 1985.

Angerer, J. K., and K. H. Schaller, *Analyses of Hazardous Substances in Biological Materials*, Vol. 2, VCH, Weinheim, Germany, 1988.

Angerer, J. K., and K. H. Schaller, *Analyses of Hazardous Substances in Biological Materials*, Vol. 3, VCH, Weinheim, Germany, 1991.

Angerer, J. K., and K. H. Schaller, *Analyses of Hazardous Substances in Biological Materials*, Vol. 4, VCH, Weinheim, Germany, 1994.

Angerer, J. K., and K. H. Schaller, *Analyses of Hazardous Substances in Biological Materials*, Vol. 5, John Wiley & Sons, New York, 1996.

Angerer, J. K., and K. H. Schaller, *Analyses of Hazardous Substances in Biological Materials*, John Wiley & Sons, New York, 1999.

Baselt, Randall C., *Biological Monitoring Methods for Industrial Chemicals*, 2nd ed., PSG Publishing Company, Inc., Littleton, MA, 1988.

Committee on National Monitoring of Human Tissues, Board on Environmental Studies and Toxoicology, Commission on Life Sciences, *Monitoring Human Tissues for Toxic Substances*, National Academy Press, Washington, D.C., 1991.

Draper, William M., Kevin Ashley, Clifford R. Glowacki, and Paul R. Michael, "Industrial Hygiene Chemistry: Keeping Pace with Rapid Change in the Workplace," *Analytical Chemistry*, **71**, 33R-60R (1999). A comprehensive review of this topic is published every two years in *Analytical Chemistry*.

Ellenberg, Hermann, *Biological Monitoring: Signals from the Environment*, Braunschweig, Vieweg, Germany, 1991.

Hee, Shane Que, *Biological Monitoring: An Introduction*, Van Nostrand Reinhold, New York, 1993.

Ioannides, Costas, Ed., *Cytochromes P450: Metabolic and Toxicological Aspects*, CRC Press, Boca Raton, FL, 1996.

Kneip, Theodore J. and John V. Crable, *Methods for Biological Monitoring*, American Public Health Association, Washington, DC, 1988.

Lauwerys, Robert R. and Perrine Hoet, *Industrial Chemical Exposure: Guidelines for Biological Monitoring*, 2nd ed., CRC Press/Lewis Publishers, Boca Raton, FL, 1993.

Mendelsohn, Mortimer L., John P. Peeters, and Mary Janet Normandy, Eds., *Biomarkers and Occupational Health: Progress and Perspectives*, Joseph Henry Press, Washington, D.C., 1995.

Minear, Roger A., Allan M. Ford, Lawrence L. Needham, and Nathan J. Karch, Eds., *Applications of Molecular Biology in Environmental Chemistry*, CRC Press/Lewis Publishers, Boca Raton, FL, 1995.

Richardson, Mervyn, Ed., *Environmental Xenobiotics*, Taylor & Francis, London, 1996.

Saleh, Mahmoud A., Jerry N. Blancato, and Charles H. Nauman, *Biomarkers of Human Exposure to Pesticides*, American Chemical Society, Washington, D.C., 1994.

Singh,Ved Pal, Ed., *Biotransformations: Microbial Degradation of Health-Risk Compounds*, Elsevier, Amsterdam, 1995.

Travis, Curtis C., Ed., *Use of Biomarkers in Assessing Health and Environmental Impacts of Chemical Pollutants*, Plenum Press, New York, 1993.

822   **Environmental Chemistry**

Williams, W. P., *Human Exposure to Pollutants: Report on the Pilot Phase of the Human Exposure Assessment Locations Programme*, United Nations Environment Programme, New York, 1992.

World Health Organization, *Biological Monitoring of Chemical Exposure in the Workplace*, World Health Organization, Geneva, Switzerland, 1996.

QUESTIONS AND PROBLEMS

1. Personnel monitoring in the workplace is commonly practiced with vapor samplers that workers carry around. How does this differ from biological monitoring? In what respects is biological monitoring superior?

2. Why is blood arguably the best kind of sample for biological monitoring? What are some of the disadvantages of blood in terms of sampling and sample processing? What are some disadvantages of blood as a matrix for analysis? What are the advantages of urine? Discuss why urine might be the kind of sample most likely to show metabolites and least likely to show parent species.

3. Distinguish among the following kinds of analytes measured for biological monitoring: parent compound, Phase 1 reaction product, Phase 2 reaction product, adducts.

4. What is wet ashing? For what kinds of analytes is wet ashing of blood commonly performed? What kinds of reagents are used for wet ashing, and what are some of the special safety precautions that should be taken with the use of these kinds of reagents for wet ashing?

5. What species is commonly measured potentiometrically in biological monitoring?

6. Compare the analysis of Phase 1 and Phase 2 metabolic products for biological monitoring. How are Phase 2 products converted back to Phase 1 metabolites for analysis?

7. Which biomolecule is most commonly involved in the formation of adducts for biological monitoring? What is a problem with measuring adducts for biological monitoring?

8. What are two general uses of immunology in biological monitoring? What is a disadvantage of immunological techniques? Discuss the likelihood that immunological techniques will find increasing use in the future as a means of biological monitoring.

9. The determination of DNA adducts is a favored means of measuring exposure to carcinogens. Based upon what is known about the mechanism of carcinogenicity, why would this method be favored? What might be some limitations of measuring DNA adducts as evidence of exposure to carcinogens?

10. How are mercapturic acid conjugates formed? What special role do they play in biological monitoring? What advantage do they afford in terms of measurement?

11. For what kinds of xenobiotics is trichloroacetic acid measured? Suggest the

pathways by which these compounds might form trichloroacetic acid metabolically.

12. Match each xenobiotic species from the column on the left below with the analyte that is measured in its biological monitoring from the column on the right.

1. Methanol	(a)	Mandelic acid
2. Malathion	(b)	A diketone
3. Styrene	(c)	Organic phosphates
4. Nitrobenzene	(d)	Formic acid
5 *n*-Heptane	(e)	*p*-Nitrophenol

28 FUNDAMENTALS OF CHEMISTRY

28.1. INTRODUCTION

This chapter is designed to give those readers who have had little exposure to chemistry the basic knowledge needed to understand the material in the rest of the book. Although it is helpful for the reader to have had several courses in chemistry, including organic chemistry and quantitative chemical analysis, most of the material in this book can be understood with less. Indeed, a reader willing to do some independent study on the fundamentals of chemistry can understand much of the material in this book without ever having had any formal chemistry course work.

Chapter 28, "Fundamentals of Chemistry," can serve two purposes. For the reader who has had no chemistry, it provides the concepts and terms basic to general chemistry. A larger category of reader consists of those who have had at least one chemistry course, but whose chemistry background, for various reasons, is inadequate. By learning the material in this chapter, plus the subject matter of Chapter 29, "Fundamentals of Organic Chemistry," these readers can comprehend the rest of the material in the book. For a more complete coverage of basic chemistry readers should consult one of a number of basic chemistry books, such as *Fundamentals of Environmental Chemistry*[1] and other supplementary references listed at the end of the chapter.

Chemistry is the science of matter. Therefore, it deals with all of the things that surround humankind, and with all aspects of the environment. Chemical properties and processes are central to environmental science. A vast variety of chemical reactions occur in water, for example, including acid-base reactions, precipitation reactions, and oxidation-reduction reactions largely mediated by microorganisms. Atmospheric chemical phenomena are largely determined by photochemical processes and chain reactions. A large number of organic chemical processes occur in the atmosphere. The geosphere, including soil, is the site of many chemical processes, particularly those that involve solids. The biosphere obviously is where the many biochemical processes crucial to the environment and to the toxic effects of chemicals occur.

This chapter emphasizes several aspects of chemistry. It begins with a discussion of the fundamental subatomic particles that make up all matter, and explains how these are assembled to produce atoms. In turn, atoms join together to make compounds. Chemical reactions and chemical equations that represent them are discussed. Solution chemistry is especially important to aquatic chemistry and is addressed in a separate section. The important, vast discipline of organic chemistry is crucial to all parts of the environment and is addressed in Chapter 29.

28.2. ELEMENTS

All substances are composed of only about a hundred fundamental kinds of matter called **elements**. Elements, themselves, may be of environmental concern. The "heavy metals," including lead, cadmium, and mercury, are well recognized as toxic substances in the environment. Elemental forms of otherwise essential elements may be very toxic or cause environmental damage. Oxygen in the form of ozone, O_3, is the agent most commonly associated with atmospheric smog pollution and is very toxic to plants and animals. Elemental white phosphorus is highly flammable and toxic.

Each element is made up of very small entities called **atoms**; all atoms of the same element behave identically chemically. The study of chemistry, therefore, can logically begin with elements and the atoms of which they are composed. Each element is designated by an **atomic number**, a name, and a **chemical symbol**, such as carbon, C; potassium, K (for its Latin name kalium); or cadmium, Cd. Each element has a characteristic **atomic mass** (atomic weight), which is the average mass of all atoms of the element. Atomic numbers of the elements are integrals ranging from 1 for hydrogen, H, to somewhat more than 100 for some of the transuranic elements (those beyond uranium). Atomic number is a unique, important way of designating each element, and it is equal to the number of protons in the nuclei of each atom of the element (see discussion of subatomic particles and atoms, below).

Subatomic Particles and Atoms

Figure 28.1 represents an atom of deuterium, a form of hydrogen. It is seen that such an atom is made up of even smaller **subatomic particles**—positively charged **protons**, negatively charged **electrons**, and uncharged (neutral) **neutrons**.

Subatomic Particles

The subatomic particles differ in mass and charge. Their masses are expressed by the **atomic mass unit, u** (also called the **dalton**), which is also used to express the masses of individual atoms and molecules (aggregates of atoms). The atomic mass unit is defined as a mass equal to exactly 1/12 that of an atom of carbon-12, the isotope of carbon that contains 6 protons and 6 neutrons in its nucleus.

The proton, p, has a mass of 1.007277 u and a unit charge of +1. This charge is equal to 1.6022×10^{-19} coulombs, where a coulomb is the amount of electrical charge involved in a flow of electrical current of 1 ampere for 1 second. The neutron, n, has no electrical charge and a mass of 1.009665 u. The proton and neu-

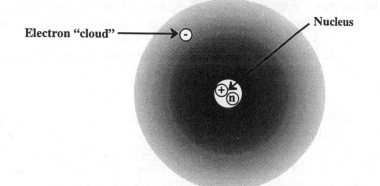

Figure 28.1. Representation of a deuterium atom. The nucleus contains one proton (+) and one neutron (n). The electron (-) is in constant, rapid motion around the nucleus forming a cloud of negative electrical charge, the density of which drops off with increasing distance from the nucleus.

tron each have a mass of essentially 1 u and are said to have a *mass number* of 1. (Mass number is a useful concept expressing the total number of protons and neutrons, as well as the approximate mass, of a nucleus or subatomic particle.) The electron, *e*, has a unit electrical charge of -1. It is very light, however, with a mass of only 0.00054859 u, about 1/1840 that of the proton or neutron. Its mass number is 0. The properties of protons, neutrons, and electrons are summarized in Table 28.1.

Table 28.1. Properties of Protons, Neutrons, and Electrons

Subatomic particle	Symbol	Unit charge	Mass number	Mass in u	Mass in grams
Proton[1]	p	+1	1	1.007277	1.6726×10^{-24}
Neutron[1]	n	0	1	1.008665	1.6749×10^{-24}
Electron[1]	e	-1	0	0.000549	9.1096×10^{-28}

[1] The mass number and charge of each of these kinds of particles may be indicated by a superscipt and subscript, respectively, as in the symbols ^1_1p, ^1_0n, and $^0_{-1}e$.

Although it is convenient to think of the proton and neutron as having the same mass, and each is assigned a mass number of 1, it is seen in Table 28.1 that their exact masses differ slightly from each other. Furthermore, the mass of an atom is not exactly equal to the sum of the masses of subatomic particles composing the atom. This is because of the energy relationships involved in holding the subatomic particles together in atoms so that the masses of the atom's constituent subatomic particles do not add up to exactly the mass of the atom.

Atom Nucleus and Electron Cloud

Protons and neutrons, which have relatively high masses compared to electrons, are contained in the positively charged **nucleus** of the atom. The nucleus has essentially all of the mass, but occupies virtually none of the volume, of the atom. An uncharged atom has the same number of electrons as protons. The electrons in an atom are contained in a cloud of negative charge around the nucleus that occupies most of the volume of the atom. These concepts are emphasized in Figure 28.2.

Isotopes

Atoms with the *same* number of protons, but *different* numbers of neutrons in their nuclei are called **isotopes**. They are chemically identical atoms of the same element, but have different masses and may differ in their nuclear properties. Some isotopes are **radioactive isotopes** or **radionuclides**, which have unstable nuclei that give off charged particles and gamma rays in the form of **radioactivity**. Radioactivity may have detrimental, or even fatal, health effects; a number of hazardous substances are radioactive and they can cause major environmental problems. The most striking example of such contamination resulted from a massive explosion and fire at a power reactor in the Ukrainian city of Chernobyl in 1986.

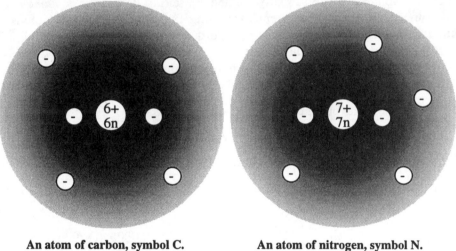

An atom of carbon, symbol C. Each C atom has 6 protons (+) in its nucleus, so the atomic number of C is 6. The atomic mass of C is 12.

An atom of nitrogen, symbol N. Each N atom has 7 protons (+) in its nucleus, so the atomic number of N is 7. The atomic mass of N is 14.

Figure 28.2. Atoms of carbon and nitrogen

Important Elements

An abbreviated list of a few of the most important elements that the reader should learn at this point is given in Table 28.2. A complete list of elements is given on the inside back cover of the book.

The Periodic Table

When elements are considered in order of increasing atomic number, it is observed that their properties are repeated in a periodic manner. For example, ele-

Table 28.2. List of Some of the More Important Common Elements

Element	Symbol	Atomic number	Atomic mass	Significance
Aluminum	Al	13	26.9815	Abundant in Earth's crust
Argon	Ar	18	39.948	Noble gas
Arsenic	As	33	74.9216	Toxic metalloid
Bromine	Br	35	79.904	Toxic halogen
Cadmium	Cd	48	112.40	Toxic heavy metal
Calcium	Ca	20	40.08	Abundant essential element
Carbon	C	6	12.011	"Life element"
Chlorine	Cl	17	35.453	Halogen
Copper	Cu	29	63.54	Useful metal
Fluorine	F	9	18.998	Halogen
Helium	He	2	4.00260	Lightest noble gas
Hydrogen	H	1	1.008	Lightest element
Iodine	I	53	126.904	Halogen
Iron	Fe	26	55.847	Important metal
Lead	Pb	82	207.19	Toxic heavy metal
Magnesium	Mg	12	24.305	Light metal
Mercury	Hg	80	200.59	Toxic heavy metal
Neon	Ne	10	20.179	Noble gas
Nitrogen	N	7	14.0067	Important nonmetal
Oxygen	O	8	15.9994	Abundant, essential nonmetal
Phosphorus	P	15	30.9738	Essential nonmetal
Potassium	K	19	39.0983	Alkali metal
Silicon	Si	14	28.0855	Abundant metalloid
Silver	Ag	47	107.87	Valuable, reaction-resistant metal
Sulfur	S	16	32.064	Essential element, occurs in air pollutant SO_2
Sodium	Na	11	22.9898	Essential, abundant alkali metal
Tin	Sn	50	118.69	Useful metal
Uranium	U	92	238.03	Fissionable metal used for nuclear fuel
Zinc	Zn	30	65.37	Useful metal

ments with atomic numbers 2, 10, and 18 are gases that do not undergo chemical reactions and consist of individual atoms, whereas those with atomic numbers larger by 1—elements with atomic numbers 3, 11, and 19—are unstable, highly reactive metals. An arrangement of the elements in a manner that reflects this recurring behavior is known as the **periodic table** (Figure 28.3). The periodic table is extremely useful in understanding chemistry and predicting chemical behavior. As shown in Figure 28.3, the entry for each element in the periodic table gives the element's atomic number, name, symbol, and atomic mass. More detailed versions of the table include other information as well.

Features of the Periodic Table

Groups of elements having similar chemical behavior are contained in vertical columns in the periodic table. **Main group** elements may be designated as A groups (1A and 2A on the left, 3A through 8A on the right). **Transition elements** are those between main groups 2A and 3A. **Noble gases** (group 8A), a group of gaseous elements that are virtually chemically unreactive, are in the far right column. The chemical similarities of elements in the same group are especially pronounced for groups 1A, 2A, 7A, and 8A.

Horizontal rows of elements in the periodic table are called **periods**, the first of which consists of only hydrogen (H) and helium (He). The second period begins with atomic number 3 (lithium) and terminates with atomic number 10 (neon), whereas the third goes from atomic number 11 (sodium) through 18 (argon). The fourth period includes the first row of transition elements, whereas lanthanides and actinides are listed separately at the bottom of the table.

Electrons in Atoms

Although a detailed discussion of the placement of electrons in atoms determines how the atoms behave chemically and, therefore, the chemical properties of each element, it is beyond the scope of this chapter to discuss electronic structure in detail. Several key points pertaining to this subject are mentioned here.

Electrons in atoms are present in **orbitals** in which the electrons have different energies, orientations in space, and average distances from the nucleus. Each orbital may contain a maximum of 2 electrons. The placement of electrons in their orbitals determines the chemical behavior of an atom; in this respect the outermost orbitals and the electrons contained in them are the most important. These **outer electrons** are the ones beyond those of the immediately preceding noble gas in the periodic table. They are of particular importance because they become involved in the sharing and transfer of electrons through which chemical bonding occurs that results in the formation of huge numbers of different substances from only a few elements.

Much of environmental chemistry is concerned with electrons in atoms. In Chapters 9 and 13 are discussed examples in which the absorption of electromagnetic radiation promotes electrons to higher energy levels, forming reactive excited species and reactive free radicals with unpaired electrons. Atomic absorption and emission methods of elemental analysis involve transitions of electrons between energy levels.

Figure — The periodic table of the elements. (rotated on page)

Active metals: groups 1A, 2A
Nonmetals: right portion
Transition metals

1A 1	2A 2	3B 3	4B 4	5B 5	6B 6	7B 7	8B 8	8B 9	8B 10	1B 11	2B 12	3A 13	4A 14	5A 15	6A 16	7A 17	8A 18
1 H 1.0079																	2 He 4.00260
3 Li 6.941	4 Be 9.01218											5 B 10.81	6 C 12.011	7 N 14.0067	8 O 15.9994	9 F 18.998403	10 Ne 20.179
11 Na 22.98977	12 Mg 24.305											13 Al 26.98154	14 Si 28.0855	15 P 30.97376	16 S 32.06	17 Cl 35.453	18 Ar 39.948
19 K 39.0983	20 Ca 40.078	21 Sc 44.9559	22 Ti 47.88	23 V 50.9415	24 Cr 51.996	25 Mn 54.9380	26 Fe 55.847	27 Co 58.9332	28 Ni 58.69	29 Cu 63.546	30 Zn 65.38	31 Ga 69.72	32 Ge 72.61	33 As 74.9216	34 Se 78.96	35 Br 79.904	36 Kr 83.80
37 Rb 85.4678	38 Sr 87.62	39 Y 88.9059	40 Zr 91.22	41 Nb 92.9064	42 Mo 95.94	43 Tc (98)	44 Ru 101.07	45 Rh 102.9055	46 Pd 106.42	47 Ag 107.8682	48 Cd 112.41	49 In 114.82	50 Sn 118.69	51 Sb 121.75	52 Te 127.60	53 I 126.9045	54 Xe 131.29
55 Cs 132.9054	56 Ba 137.33	57 *La 138.9055	72 Hf 178.49	73 Ta 180.9479	74 W 183.85	75 Re 186.207	76 Os 190.2	77 Ir 192.22	78 Pt 195.08	79 Au 196.9665	80 Hg 200.59	81 Tl 204.383	82 Pb 207.2	83 Bi 208.9804	84 Po (209)	85 At (210)	86 Rn (222)
87 Fr (223)	88 Ra 226.0254	89 +Ac 227.0278	104 Rf (261)	105 Ha (262)	106 Unh (263)	107 Uns (262)		109 Une (266)									

*Lanthanide series

58 Ce 140.12	59 Pr 140.9077	60 Nd 144.24	61 Pm (145)	62 Sm 150.36	63 Eu 151.96	64 Gd 157.25	65 Tb 158.9254	66 Dy 162.50	67 Ho 164.9304	68 Er 167.26	69 Tm 168.9342	70 Yb 173.04	71 Lu 174.967

+Actinide series

90 Th 232.0381	91 Pa 231.0359	92 U 238.0289	93 Np (237)	94 Pu (244)	95 Am (243)	96 Cm (247)	97 Bk (247)	98 Cf (251)	99 Es (252)	100 Fm (257)	101 Md (258)	102 No (259)	103 Lr (260)

The larger and smaller labels reflect two different numbering schemes in common usage.

Figure 28.3. The periodic table of the elements.

Lewis Structures and Symbols of Atoms

Outer electrons are called **valence electrons** and are represented by dots in **Lewis symbols**, as shown for carbon and argon in Figure 28.4, below:

$$\cdot \overset{\displaystyle \cdot}{\underset{\displaystyle \cdot}{C}} \cdot \qquad\qquad\qquad \overset{\displaystyle \cdot\cdot}{\underset{\displaystyle \cdot\cdot}{:Ar:}}$$

Lewis symbol of carbon   **Lewis symbol of argon**

Figure 28.4. Lewis symbols of carbon and argon.

The four electrons shown for the carbon atom are those added beyond the electrons possessed by the noble gas that immediately precedes carbon in the periodic table (helium, atomic number 2). Eight electrons are shown around the symbol of argon. This is an especially stable electron configuration for noble gases known as an **octet**. (Helium is the exception among noble gases in that it has a stable shell of only two electrons.) When atoms interact through the sharing, loss, or gain of electrons to form molecules and chemical compounds (see Section 28.3) many attain an octet of outer shell electrons. This tendency is the basis of the **octet rule** of chemical bonding. (Two or three of the lightest elements, most notably hydrogen, attain stable helium-like electron configurations containing two electrons when they become chemically bonded.)

Metals, Nonmetals, and Metalloids

Elements are divided between metals and nonmetals; a few elements with intermediate character are called metalloids. **Metals** are elements that are generally solid, shiny in appearance, electrically conducting, and malleable—that is, they can be pounded into flat sheets without disintegrating. They tend to have only 1–3 outer electrons, which they may lose in forming chemical compounds. Examples of metals are iron, copper, and silver. Most metallic objects that are commonly encountered are not composed of just one kind of elemental metal, but are alloys consisting of homogeneous mixtures of two or more metals. **Nonmetals** often have a dull appearance, are not at all malleable, and frequently occur as gases or liquids. Color-less oxygen gas, green chlorine gas (transported and stored as a liquid under pressure), and brown bromine liquid are common nonmetals. Nonmetals tend to have close to a full octet of outer-shell electrons, and in forming chemical compounds they gain or share electrons. **Metalloids**, such as silicon or arsenic, are elements with properties intermediate between those of metals and nonmetals. Under some conditions, a metalloid may exhibit properties of metals, and under other conditions, properties of nonmetals.

28.3. CHEMICAL BONDING

Only a few elements, particularly the noble gases, exist as individual atoms; most atoms are joined by chemical bonds to other atoms. For example, elemental hydrogen exists as **molecules,** each consisting of 2 H atoms linked by a **chemical bond** as shown in Figure 28.5. Because hydrogen molecules contain 2 H atoms, they are said to be diatomic and are denoted by the **chemical formula,** H_2. The H atoms in the H_2 molecule are held together by a **covalent bond** made up of 2 electrons, each contributed by one of the H atoms, and shared between the atoms. (Bonds formed by transferring electrons between atoms are described later in this section.) The shared electrons in the covalent bonds holding the H_2 molecule together are represented by two dots between the H atoms in Figure 28.5. By analogy with Lewis symbols defined in the preceding section, such a representation of molecules showing outer-shell and bonding electrons as dots is called a **Lewis formula.**

H_2

The H atoms in ele-
mental hydrogen

are held together by chemical bonds in molecules

That have the chemical
formula H_2.

$$H\cdot + \cdot H \;\longrightarrow\; H\!:\!H$$
Lewis structure of H_2

Figure 28.5. Molecule and Lewis formula of H_2.

Chemical Compounds

Most substances consist of two or more elements joined by chemical bonds. As an example consider the chemical combination of hydrogen and oxygen shown in Figure 28.6. Oxygen, chemical symbol O, has an atomic number of 8 and an atomic mass of 16.00, and it exists in the elemental form as diatomic molecules of O_2. Hydrogen atoms combine with oxygen atoms to form molecules in which 2 H atoms are bonded to 1 O atom in a substance with a chemical formula of H_2O (water). A substance such as H_2O that consists of a chemically bonded combination of two or more elements is called a **chemical compound.** In the chemical formula for water the letters H and O are the chemical symbols of the two elements in the compound and the subscript 2 indicates that there are 2 H atoms per O atom. (The absence of a subscript after the O denotes the presence of just 1 O atom in the molecule.).

As shown in Figure 28.6, each of the hydrogen atoms in the water molecule is connected to the oxygen atom by a chemical bond composed of two electrons shared between the hydrogen and oxygen atoms. For each bond one electron is contributed by the hydrogen and one by oxygen. The two dots located between each H and O in the Lewis formula of H_2O represent the two electrons in the covalent bond joining these atoms. Four of the electrons in the octet of electrons surrounding O are involved in H-O bonds and are called bonding electrons. The other four electrons shown around the oxygen that are not shared with H are nonbonding outer electrons.

Molecular Structure

As implied by the representations of the water molecule in Figure 28.6, the atoms and bonds in H_2O form an angle somewhat greater than 90 degrees. The shapes of molecules are referred to as their **molecular geometry**, which is crucial in determining the chemical and toxicological activity of a compound and structure-activity relationships.

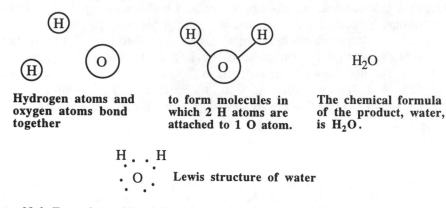

Hydrogen atoms and oxygen atoms bond together   **to form molecules in which 2 H atoms are attached to 1 O atom.**   **The chemical formula of the product, water, is H₂O.**

H_2O

Lewis structure of water

Figure 28.6. Formation and Lewis formula of a chemical compound, water.

Ionic Bonds

As shown in Figure 28.7, the transfer of electrons from one atom to another produces charged species called **ions**. Positively charged ions are called **cations** and negatively charged ions are called **anions**. Ions that make up a solid compound are held together by **ionic bonds** in a **crystalline lattice** consisting of an ordered arrangement of the ions in which each cation is largely surrounded by anions and each anion by cations. The attracting forces of the oppositely charged ions in the crystalline lattice constitute ionic bonds in the compound.

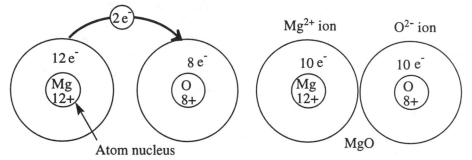

Atom nucleus

The transfer of 2 electrons from a magnesium atom to an oxygen atom   **yields Mg²⁺ and O²⁻ ions that are bonded together by ionic bonds in the compound MgO.**

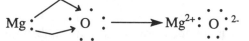

Formation of ionic MgO as shown by Lewis structures and symbols. In MgO, Mg has lost 2 electrons and is in the +2 oxidation state {Mg(II)} and O has gained 2 electrons and is in the -2 oxidation state.

Figure 28.7. Ionic bonds are formed by the transfer of electrons and the mutual attraction of oppositely charged ions in a crystalline lattice.

The formation of magnesium oxide is shown in Figure 28.7. In naming this compound, the cation is simply given the name of the element from which it was formed, magnesium. However, the ending of the name of the anion, ox*ide*, is different from that of the element from which it was formed, ox*ygen*.

Rather than individual atoms that have lost or gained electrons, many ions are groups of atoms bonded together covalently and having a net charge. A common example of such an ion is the ammonium ion, NH_4^+,

$$
\begin{array}{c}
\text{H} \\
\overset{\cdot\cdot}{} \quad + \\
\text{H}\overset{\cdot\cdot}{\underset{\cdot\cdot}{\text{N}}}\text{H} \\
\text{H}
\end{array}
\quad \text{Lewis formula of the ammonium ion}
$$

which consists of 4 hydrogen atoms covalently bonded to a single nitrogen (N) atom and having a net electrical charge of +1 for the whole cation, as shown by its Lewis formula above.

Summary of Chemical Compounds and the Ionic Bond

The preceding several pages have just covered some material on chemical compounds and bonds that are essential to understand chemistry. To summarize, these are the following:

- Atoms of two or more different elements can form *chemical bonds* with each other to yield a product that is entirely different from the elements.

- Such a substance is called a *chemical compound.*

- The *formula* of a chemical compound gives the symbols of the elements and uses subscripts to show the relative numbers of atoms of each element in the compound.

- *Molecules* of some compounds are held together by *covalent bonds* consisting of shared electrons.

- Another kind of compound consists of *ions* composed of electrically charged atoms or groups of atoms held together by *ionic bonds* that exist because of the mutual attraction of oppositely charged ions.

Molecular Mass

The average mass of all molecules of a compound is its **molecular mass** (formerly called molecular weight). The molecular mass of a compound is calculated by multiplying the atomic mass of each element by the relative number of atoms of the element, then adding all the values obtained for each element in the compound. For example, the molecular mass of NH_3 is $14.0 + 3 \times 1.0 = 17.0$. As another example consider the following calculation of the molecular mass of ethylene, C_2H_4.

1. The chemical formula of the compound is C_2H_4.

2. Each molecule of C_2H_4 consists of 2 C atoms and 4 H atoms.

3. From the periodic table or Table 28.2, the atomic mass of C is 12.0 and that of H is 1.0.

4. Therefore, the molecular mass of C_2H_4 is

$$\underbrace{12.0 + 12.0}_{\text{From 2 C atoms}} + \underbrace{1.0 + 1.0 + 1.0 + 1.0}_{\text{From 4 H atoms}} = 28.0.$$

From 2 C atoms   From 4 H atoms

Oxidation State

The loss of two electrons from the magnesium atom as shown in Figure 28.7 is an example of **oxidation**, and the Mg^{2+} ion product is said to be in the +2 **oxidation state**. (A positive oxidation state or oxidation number is conventionally denoted by a Roman numeral in parentheses following the name or symbol of an element as in magnesium(II) and Mg(II)). In gaining 2 negatively charged electrons in the reaction that produces magnesium oxide, the oxygen atom is **reduced** and is in the -2 oxidation state. (Unlike positive oxidation numbers, negative ones are not conventionally shown by Roman numerals in parentheses.) In chemical terms an **oxidizer** is a species that takes electrons from a reducing agent in a chemical reaction. Many hazardous waste substances are oxidizers or strong reducers and oxidation/reduction is the driving force behind many dangerous chemical reactions. For example, the reducing tendencies of the carbon and hydrogen atoms in propane cause it to burn violently or explode in the presence of oxidizing oxygen in air. The oxidizing ability of concentrated nitric acid, HNO_3, enables it to react destructively with organic matter, such as cellulose or skin.

Covalently bonded atoms that have not actually lost or gained electrons to produce ions are also assigned oxidation states. This can be done because in covalent compounds electrons are not shared equally. Therefore, an atom of an element with a greater tendency to attract electrons is assigned a negative oxidation number compared to the positive oxidation number assigned to an element with a lesser tendency to attract electrons. For example, Cl atoms attract electrons more strongly than do H atoms so that in hydrogen chloride gas, HCl, the Cl atom is in the -1 oxidation state and the H atoms are in the +1 oxidation state. **Electronegativity** values are assigned to elements on the basis of their tendencies to attract electrons.

The oxidation state (oxidation number) of an element in a compound may have a strong influence on the hazards posed by the compound. For example, chromium from which each atom has lost 3 electrons to form a chemical compound, designated as chromium(III) or Cr(III), is not toxic, whereas chromium in the +6 oxidation state (CrO_4^{2-}, chromate) is regarded as a cancer-causing chemical when inhaled.

28.4. CHEMICAL REACTIONS AND EQUATIONS

Chemical reactions occur when substances are changed to other substances through the breaking and formation of chemical bonds. For example, water is produced by the chemical reaction of hydrogen and oxygen:

Hydrogen plus oxygen yields water

Chemical reactions are written as **chemical equations**. The chemical reaction between hydrogen and water is written as the **balanced chemical equation**

$$2H_2 + O_2 \rightarrow 2H_2O \tag{28.4.1}$$

in which the arrow is read as "yields" and separates the hydrogen and oxygen **reactants** from the water **product**. Note that because elemental hydrogen and elemental oxygen occur as *diatomic molecules* of H_2 and O_2, respectively, it is necessary to write the equation in a way that reflects these correct chemical formulas of the elemental form. All correctly written chemical equations are **balanced** in that *the same number of each kind of atom must be shown on both sides of the equation.* The equation above is balanced because of the following:

On the left

- There are 2 H_2 *molecules* each containing 2 H *atoms* for a total of 4 H atoms on the left.

- There is 1 O_2 *molecule* containing 2 O *atoms* for a total of 2 O atoms on the left.

On the right

- There are 2 H_2O *molecules* each containing 2 H *atoms* and 1 O *atom* for a total of 4 H atoms and 2 O atoms on the right.

Reaction Rates

Most chemical reactions give off heat and are classified as exothermic reactions. The rate of a reaction may be calculated by the Arrhenius equation, which contains absolute temperature ($K = °C + 273$) in an exponential term. As a general rule the speed of a reaction doubles for each 10°C increase in temperature. Reaction rate factors are important factors in fires or explosions involving hazardous chemicals.

28.5. SOLUTIONS

A liquid **solution** is formed when a substance in contact with a liquid becomes dispersed homogeneously throughout the liquid in a molecular form. The substance, called a **solute**, is said to **dissolve**. The liquid is called a **solvent**. There may be no readily visible evidence that a solute is present in the solvent; for example, a deadly poisonous solution of sodium cyanide in water looks like pure water. The solution may have a strong color, as is the case for intensely purple solutions of potassium permanganate, $KMnO_4$. It may have a strong odor, such as that of ammonia, NH_3, dissolved in water. Solutions may consist of solids, liquids, or gases dissolved in a solvent. Technically, it is even possible to have solutions in which a solid is a solvent, although such solutions are not discussed in this book.

Solution Concentration

The quantity of solute relative to that of solvent or solution is called the **solution concentration**. Concentrations are expressed in numerous ways. Very high concentrations are often given as percent by weight. For example, commercial concentrated hydrochloric acid is 36% HCl, meaning that 36% of the weight has come from dissolved HCl and 64% from water solvent. Concentrations of very dilute solutions, such as those of hazardous waste leachate containing low levels of contaminants, are expressed as weight of solute per unit volume of solution. Common units are milligrams per liter (mg/L) or micrograms per liter (μg/L). Since a liter of water weighs essentially 1,000 grams, a concentration of 1 mg/L is equal to 1 part per million (ppm) and a concentration of 1 μg/L is equal to 1 part per billion (ppb).

Chemists often express concentrations in moles per liter, or **molarity**, M. Molarity is given by the relationship

$$M = \frac{\text{Number of moles of solute}}{\text{Number of liters of solution}} \tag{28.5.1}$$

The number of moles of a substance is its mass in grams divided by its molar mass. For example, the molecular mass of ammonia, NH_3, is 14 + 1 + 1 +1, so a mole of ammonia has a mass of 17 g. Therefore, 17 g of NH_3 in 1 L of solution has a value of M equal to 1 mole/L.

Water as a Solvent

Most liquid wastes are solutions or suspensions of waste materials in water. Water has some unique properties as a solvent which arise from its molecular structure as represented by the Lewis structure of water below:

$$\begin{matrix} \text{H} \cdot \cdot \\ (+) \ \vdots \ \text{O} \ \vdots \ (-) \\ \text{H} \cdot \cdot \end{matrix}$$

The H atoms are not on opposite sides of the O atom and the two H–O bonds form an angle of 105°. Furthermore, the O atom (-2 oxidation state) is able to attract electrons more strongly than the 2 H atoms (each in the +1 oxidation state) so that the molecule is **polar**, with the O atom having a partial negative charge and the end of the molecule with the 2 H atoms having a partial positive charge. This means that water molecules can cluster around ions with the positive ends of the water molecules attracted to negatively charged anions and the negative end to positively charged cations. This kind of interaction is part of the general phenomenon of **solvation**. It is specifically called **hydration** when water is the solvent and is partially responsible for water's excellent ability to dissolve ionic compounds including acids, bases, and salts.

Water molecules form a special kind of bond called a **hydrogen bond** with each other and with solute molecules that contain O, N, or S atoms. As its name implies, a hydrogen bond involves a hydrogen atom held between two other atoms of O, N, or S. Hydrogen bonding is partly responsible for water's ability to solvate and dissolve chemical compounds capable of hydrogen bonding.

As noted above, the water molecule is a polar species, which affects its ability to act as a solvent. Solutes may likewise have polar character. In general, solutes with polar molecules are more soluble in water than nonpolar ones. The polarity of an impurity solute in wastewater is a factor in determining how it may be removed from water. Nonpolar organic solutes are easier to take out of water by an adsorbent species such as activated carbon than are more polar solutes.

Solutions of Acids and Bases

Acid-Base Reactions

The reaction between H^+ ion from an acid and OH^- ion from a base is a **neutralization** reaction. As a specific example consider the reaction of H^+ from a solution of sulfuric acid, H_2SO_4, and OH^- from a solution of calcium hydroxide:

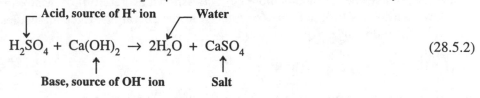

$$H_2SO_4 + Ca(OH)_2 \rightarrow 2H_2O + CaSO_4 \tag{28.5.2}$$

In addition to water, which is always the product of a neutralization reaction, the other product is calcium sulfate, $CaSO_4$. This compound is a **salt** composed of Ca^{2+} ions and SO_4^{2-} ions held together by ionic bonds. A salt, consisting of a cation other than H^+ and an anion other than OH^-, is the other product in addition to water produced when an acid and base react. Some salts are hazardous substances and environmental pollutants because of their dangerous or harmful properties. Some examples are the following:

- Ammonium perchlorate, NH_4ClO_4, (reactive oxidant)
- Barium cyanide, $Ba(CN)_2$ (toxic)
- Lead acetate, $Pb(C_2H_3O_2)_2$ (toxic)
- Thallium(I) carbonate, Tl_2CO_3 (toxic)

Concentration of H^+ Ion and pH

Acids such as HCl and sulfuric acid (H_2SO_4) produce H^+ ion, whereas bases, such as sodium hydroxide and calcium hydroxide (NaOH and $Ca(OH)_2$, respectively), produce hydroxide ion, OH^-. Molar concentrations of hydrogen ion, $[H^+]$, range over many orders of magnitude and are conveniently expressed by pH defined as

$$pH = -\log [H^+] \tag{28.5.3}$$

In absolutely pure water at 25°C, the value of $[H^+]$ is exactly 1×10^{-7} mole/L, the pH is 7.00, and the solution is **neutral** (neither acidic nor basic). **Acidic** solutions have pH values of less than 7, and **basic** solutions have pH values of greater than 7.

Strong acids and strong bases are **corrosive** substances that exhibit extremes of pH. They are destructive to materials and flesh. Strong acids can react with cyanide and sulfide compounds to release highly toxic hydrogen cyanide (HCN) or hydrogen sulfide (H_2S) gases, respectively. Bases liberate noxious ammonia gas (NH_3) from solid ammonium compounds.

Metal Ions Dissolved in Water

Metal ions dissolved in water have some unique characteristics that influence their properties as natural water constituents and heavy metal pollutants, and in biological systems. The formulas of metal ions are usually represented by the symbol for the ion followed by its charge. For example, iron(II) ion (from a compound such as iron(II) sulfate, $FeSO_4$) dissolved in water is represented as Fe^{2+}. Actually, in water solution each iron(II) ion is strongly solvated and bonded to water molecules, so that the formula is more correctly shown as $Fe(H_2O)_6{}^{2+}$. Many metal ions have a tendency to lose hydrogen ions from the solvating water molecules as shown by the following:

$$Fe(H_2O)_6{}^{2+} \rightarrow Fe(OH)(H_2O)_5 + H^+ \tag{28.5.4}$$

Ions of the next higher oxidation state, iron(III), have such a tendency to lose H^+ ion in aqueous solution that, except in rather highly acidic solutions, they precipitate out as solid hydroxides, such as iron(III) hydroxide, $Fe(OH)_3$:

$$Fe(H_2O)_6{}^{3+} \rightarrow Fe(OH)_3(s) + 3H_2O + 3\,H^+ \tag{28.5.5}$$

Complex Ions Dissolved in Water

It was noted above that metal ions are solvated (hydrated) by binding to water molecules in aqueous solution. Some species in solution have a stronger tendency than water to bond to metal ions. An example of such a species is cyanide ion, CN^-, which displaces water molecules from some metal ions in solution as shown below:

$$Ni(H_2O)_4{}^{2+} + 4CN^- \xleftarrow{\quad} \rightarrow Ni(CN)_4{}^{2-} + 4H_2O \tag{28.5.6}$$

The species that bonds to the metal ion, cyanide in this case, is called a **ligand**, and the product of the reaction is a **complex ion** or metal complex. The overall process is called **complexation**.

Colloidal Suspensions

Very small particles on the order of 1 micrometer or less in size, called **colloidal particles**, may stay suspended in a liquid for an indefinite period of time. Such a mixture is a **colloidal suspension** and it behaves in many respects like a solution. Colloidal suspensions are used in many industrial applications. Many waste materials are colloidal and are often emulsions consisting of colloidal liquid droplets suspended in another liquid, usually wastewater. One of the challenges in dealing with colloidal wastes is to remove a relatively small quantity of colloidal material from a large quantity of water by precipitating the colloid. This process is called **coagulation** or **flocculation** and is often brought about by the addition of chemical agents.

Solution Equilibria

Many of the phenomena in aquatic chemistry (Chapters 3–8) and geochemistry (Chapters 15 and 16) involve solution equilibrium. In a general sense, solution equilibrium deals with the extent to which **reversible** acid-base, solubilization (precipitation), complexation, or oxidation-reduction reactions proceed in a forward or backward direction. This is expressed for a generalized equilibrium reaction,

$$aA + bB \longleftrightarrow cC + dD \qquad (28.5.7)$$

by the following **equilibrium constant expression**:

$$\frac{[C]^c[D]^d}{[A]^a[B]^b} = K \qquad (28.5.8)$$

where K is the **equilibrium constant**.

A reversible reaction may approach equilibrium from either direction. In the example above, if A were mixed with B, or C were mixed with D, the reaction would proceed in a forward or reverse direction such that the concentrations of species—[A], [B], [C], and [D]—substituted into the equilibrium expression gave a value equal to K.

As expressed by **Le Châtelier's principle**, a stress placed upon a system in equilibrium will shift the equilibrium to relieve the stress. For example, adding product "D" to a system in equilibrium will cause Reaction 28.5.7 to shift to the left, consuming "C" and producing "A" and "B," until the equilibrium constant expression is again satisfied. This **mass action effect** is the driving force behind many environmental chemical phenomena.

In most cases this book uses concentrations and pressures in equilibrium constant expression calculations. When this is done, K is not exactly constant with varying concentrations and pressures; it is an *approximate equilibrium constant* that applies only to limited conditions. **Thermodynamic equilibrium constants** are more exact forms derived from thermodynamic data that make use of *activities* in place of concentrations. At a specified temperature, the value of a thermodynamic equilibrium constant is applicable over a wide concentration range. The activity of a species, commonly denoted as a_X for species "X," expresses how effectively it interacts with its surroundings, such as other solutes or electrodes in solution. (The analogy may be drawn of an environmental chemistry class with a "concentration" of 20 students per classroom. Their "activity" in relating to the subject is likely to be much higher on a cold, rainy day than on a balmy, sunny day in springtime.) Activities approach concentrations at low values of concentration. The thermodynamic equilibrium constant expression for Reaction 28.5.7 is expressed as the following in terms of activities:

$$\frac{a_C{}^c\, a_D{}^d}{a_A{}^a\, a_B{}^b} = K \tag{28.5.9}$$

There are several major kinds of equilibria in aqueous solution. One of these is acid-base equilibrium (see Chapter 3) as exemplified by the ionization of acetic acid, HAc,

$$HAc \xleftarrow{\;\;}{\longrightarrow} H^+ + Ac^- \tag{28.5.10}$$

for which the acid dissociation constant is

$$\frac{[H^+][Ac^-]}{[HAc]} = K = 1.75 \times 10^{-5} \quad \text{(at 25°C)} \tag{28.5.11}$$

Very similar expressions are obtained for the formation and dissociation of metal **complexes** or **complex ions** (Chapter 3) formed by the reaction of a metal ion in solution with a **complexing agent** or **ligand**, both of which are capable of independent existence in solution. This can be shown by the reaction of iron(III) ion and thiocyanate ligand

$$Fe^{3+} + SCN^- \xleftarrow{\;\;}{\longrightarrow} FeSCN^{2+} \tag{28.5.12}$$

for which the **formation constant expression** is:

$$\frac{[FeSCN^{2+}]}{[Fe^{3+}][SCN^-]} = K_f = 1.07 \times 10^3 \quad \text{(at 25°C)} \tag{28.5.13}$$

The bright red color of the $FeSCN^{2+}$ complex formed is used to test for the presence of iron(III) in acid mine water (Chapter 7).

An example of an **oxidation-reduction reaction**, which involves the transfer of electrons between species, is

$$MnO_4^- + 5Fe^{2+} + 8H^+ \xleftarrow{\;\;}{\longrightarrow} Mn^{2+} + 5Fe^{3+} + 4H_2O \tag{28.5.14}$$

for which the equilibrium expression is:

$$\frac{[Mn^{2+}][Fe^{3+}]^5}{[MnO_4^-][Fe^{2+}]^5[H^+]^8} = K = 3 \times 10^{62} \quad \text{(at 25°C)} \tag{28.5.15}$$

The value of K is calculated from the Nernst equation, as explained in Chapter 4.

Distribution between Phases

As discussed in Chapter 5, many important environmental chemical phenomena involve distribution of species between phases. This most commonly involves the equilibria between species in solution and in a solid phase. **Solubility equilibria** (see Chapter 5) deal with reactions such as,

$$AgCl(s) \rightarrow Ag^+ + Cl^- \tag{28.5.16}$$

in which one of the participants is a solid that is slightly soluble (virtually insoluble). For the example shown above, the equilibrium constant is,

$$[Ag^+][Cl^-] = K_{sp} = 1.82 \times 10^{-10} \quad (\text{at } 25°C) \tag{28.5.17}$$

a **solubility product**. Note that in the equilibrium constant expression there is not a value given for the solid AgCl. This is because the activity of a solid is constant at a specific temperature and is contained in the value of K_{sp}.

An important example of distribution between phases is that of a hazardous waste species partitioned between water and a body of immiscible organic liquid in a hazardous waste site. The equilibrium for such a reaction,

$$X(aq) \leftrightarrow X(org) \tag{28.5.18}$$

is described by the **distribution law** expressed by a **distribution coefficient** or **partition coefficient** in the following form:

$$\frac{[X(org)]}{[X(aq)]} = K_d \tag{28.5.19}$$

LITERATURE CITED

1. Manahan, Stanley E., *Fundamentals of Environmental Chemistry*, CRC Press, Boca Raton, FL, 1993.

SUPPLEMENTARY REFERENCES

Brown, Theodore L., H. Eugene Lemay, and Bruce Edward Bursten, *Chemistry: The Central Science*, 8th ed., Prentice Hall, Upper Saddle River, NJ, 1999.

Burns, Ralph A., *Fundamentals of Chemistry*, 3rd ed., Prentice Hall, Upper Saddle River, NJ, 1999.

Dickson, T. R., *Introduction to Chemistry*, 8th ed, John Wiley & Sons, New York, 1999.

Ebbing, Darrell D. and Steven D. Gammon, *General Chemistry*, 6th ed., Houghton Mifflin, Boston, 1999.

Hein, Morris and Susan Arena, *Foundations of College Chemistry/With Infotrak*, 10th ed, Brooks/Cole Publishing Co., Pacific Grove, CA, 1999.

Hill, John W., and Doris K. Kolb, *Chemistry for Changing Times*, 8th ed., Prentice Hall, Upper Saddle River, NJ, 1998.

Kotz, John C. and Paul Treichel, *Chemistry and Chemical Reactivity*, 4th ed, Saunders College Publishing, Philadelphia, 1998.

McMurry, John and Mary E. Castellion, *Fundamentals of General, Organic and Biological Chemistry*, Prentice Hall, Upper Saddle River, NJ, 1999.

Rosenberg, Jerome L. and Lawrence M. Epstein, *Schaum's Outline of Theory and Problems of College Chemistry*, 8th edition, McGraw-Hill, New York, 1997.

QUESTIONS AND PROBLEMS

1. What distinguishes a radioactive isotope from a "normal" stable isotope?

2. Why is the periodic table so named?

3. After examining Figure 28.7, consider what might happen when an atom of sodium (Na), atomic number 11, loses an electron to an atom of fluorine, (F), atomic number 9. What kinds of particles are formed by this transfer of a negatively charged electron? Is a chemical compound formed? If so, what is it called?

4. Match the following:

   1. O_2 (a) Element consisting of individual atoms

   2. NH_3 (b) Element consisting of chemically bonded atoms

   3. Ar   (c) Ionic compound

   4. NaCl   (d) Covalently bound compound

5. Consider the following atom:

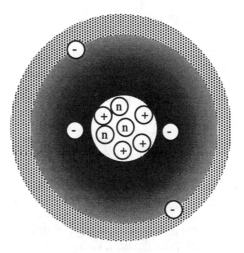

How many electrons, protons, and neutrons does it have? What is its atomic number? Give the name and chemical symbol of the element of which it is composed.

6. Give the chemical formula and molecular mass of the molecule represented below:

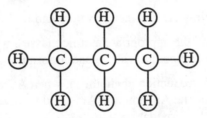

7. Calculate the molecular masses of (a) C_2H_2, (b) N_2H_4, (c) Na_2O, (d) O_3 (ozone), (e) PH_3, (f) CO_2.

8. Is the equation, $H_2 + O_2 \rightarrow H_2O$, a balanced chemical equation? Explain. Point out the reactants and products in the equation.

9. Define and distinguish the differences between environmental chemistry, environmental biochemistry, and toxicological chemistry.

10. An uncharged atom has the same number of _____ as protons. The electrons in an atom are contained in a cloud of _____ around the nucleus that occupies most of the _____ of the atom.

11. Match:

   1. Argon   (a) A halogen

   2. Hydrogen   (b) Fissionable element

   3. Uranium   (c) Noble gas

   4. Chlorine   (d) Has an isotope with a mass number of 2

   5. Mercury   (e) Toxic heavy metal

12. The entry for each element in the periodic table gives the element's _____ _____ and the periodic table is arranged horizontally in _____ and vertically in _____.

13. Electrons in atoms occupy _____ in which electrons have different _____. Each orbital may contain a maximum of _____ electrons.

14. The Lewis symbol of carbon is _____ in which each dot represents _____.

15. Elements that are generally solid, shiny in appearance, electrically conducting, and malleable are _____ whereas elements that tend to have a dull appearance, are not at all malleable, and frequently occur as gases or liquids are _____. Elements with intermediate properties are called _____.

16. In the Lewis formula,

$$H : H$$

the two dots represent _____.

17. Explain why H_2 is not a chemical compound whereas H_2O is a chemical compound.

18. Using examples, distinguish between covalent and ionic bonds.

19. In terms of c, h, o, and the appropriate atomic masses, write a formula for the molecular mass of a compound with a general formula of $C_cH_hO_o$.

20. Considering oxidation/reduction phenomena, when Al reacts with O_2 to produce Al_2O_3, which contains the Al^{3+} ion, the Al is said to have been _____ and is in the _____ oxidation state.

21. Calculate the concentration in moles per liter of (a) a solution that is 27.0% H_2SO_4 by mass and that has a density of 1,198 g/L, and (b) of a solution that is 1 mg/L H_2SO_4 having a density of 1,000 g/L.

22. Calculate the pH of the second solution described in the preceding problem, keeping in mind that each H_2SO_4 molecule yields two H^+ ions.

23. Write a balanced neutralization reaction between NaOH and H_2SO_4.

24. Distinguish between solutions and colloidal suspensions.

25. What is the nature of the Fe^{3+} ion? Why are solutions containing this ion acidic?

26. What kind of species is $Ni(CN)_4^{2-}$?

27. What is the solubility product expression for lead sulfate, $PbSO_4$, in terms of concentrations of Pb^{2+} ion and SO_4^{2-} ion?

29 ORGANIC CHEMISTRY

29.1. ORGANIC CHEMISTRY

Most carbon-containing compounds are **organic chemicals** and are addressed by the subject of **organic chemistry**. Organic chemistry is a vast, diverse discipline because of the enormous number of organic compounds that exist as a consequence of the versatile bonding capabilities of carbon. Such diversity is due to the ability of carbon atoms to bond to each other through single (2 shared electrons) bonds, double (4 shared electrons) bonds, and triple (6 shared electrons) bonds in a limitless variety of straight chains, branched chains, and rings.

Among organic chemicals are included the majority of important industrial compounds, synthetic polymers, agricultural chemicals, biological materials, and most substances that are of concern because of their toxicities and other hazards. Pollution of the water, air, and soil environments by organic chemicals is an area of significant concern.

Chemically, most organic compounds can be divided among hydrocarbons, oxygen-containing compounds, nitrogen-containing compounds, sulfur-containing compounds, organohalides, phosphorus-containing compounds, or combinations of these kinds of compounds. Each of these classes of organic compounds is discussed briefly here.

All organic compounds, of course, contain carbon. Virtually all also contain hydrogen and have at least one C–H bond. The simplest organic compounds, and those easiest to understand, are those that contain only hydrogen and carbon. These compounds are called **hydrocarbons** and are addressed first among the organic compounds discussed in this chapter. Hydrocarbons are used here to illustrate some of the most fundamental points of organic chemistry, including organic formulas, structures, and names.

Molecular Geometry in Organic Chemistry

The three-dimensional shape of a molecule, that is, its molecular geometry, is particularly important in organic chemistry. This is because its molecular geometry determines in part the properties of an organic molecule, particularly its interactions with biological systems and how it is metabolized by organisms. Shapes of molecules are represented in drawings by lines of normal, uniform thickness for bonds in the plane of the paper; broken lines for bonds extending away from the viewer; and heavy lines for bonds extending toward the viewer. These conventions are shown by the example of dichloromethane, CH_2Cl_2, an important organochloride solvent and extractant illustrated in Figure 29.1.

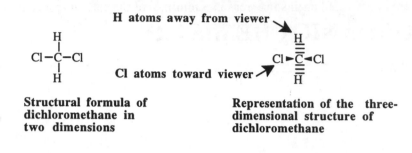

Structural formula of
dichloromethane in
two dimensions

Representation of the three-
dimensional structure of
dichloromethane

Figure 29.1. Structural formulas of dichloromethane, CH_2Cl_2; the formula on the right provides a three-dimensional representation.

29.2. HYDROCARBONS

As noted above, hydrocarbon compounds contain only carbon and hydrogen. The major types of hydrocarbons are alkanes, alkenes, alkynes, and aryl compounds. Examples of each are shown in Figure 29.2.

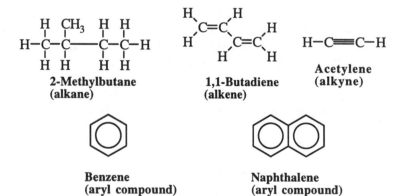

2-Methylbutane
(alkane)

1,1-Butadiene
(alkene)

Acetylene
(alkyne)

Benzene
(aryl compound)

Naphthalene
(aryl compound)

Figure 29.2. Examples of major types of hydrocarbons.

Alkanes

Alkanes, also called **paraffins** or **aliphatic hydrocarbons**, are hydrocarbons in which the C atoms are joined by single covalent bonds (sigma bonds) consisting of two shared electrons (see Section 28.3). Some examples of alkanes are shown in Figure 29.2. As with other organic compounds, the carbon atoms in alkanes may form straight chains, branched chains, or rings. These three kinds of alkanes are, respectively, **straight-chain alkanes**, **branched-chain alkanes**, and **cycloalkanes**. As shown in Figure 29.2, a typical branched chain alkane is 2-methylbutane, a volatile, highly flammable liquid. It is a component of gasoline, which may explain why it is commonly found as an air pollutant in urban air. The general molecular formula for straight- and branched-chain alkanes is C_nH_{2n+2}, and that of cyclic alkanes is C_nH_{2n}. The four hydrocarbon molecules in Figure 29.3 contain 8 carbon atoms each. In one of the molecules, all of the carbon atoms are in a straight chain and in two they are in branched chains, whereas in a fourth, 6 of the carbon atoms are in a ring.

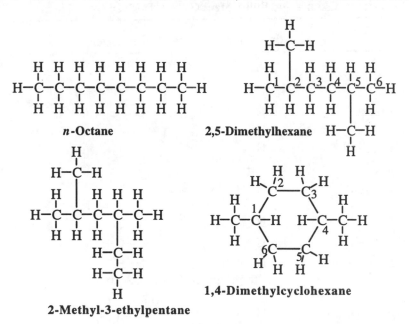

Figure 29.3. Structural formulas of four hydrocarbons, each containing 8 carbon atoms, that illustrate the structural diversity possible with organic compounds. Numbers used to denote locations of atoms for purposes of naming are shown on two of the compounds.

Formulas of Alkanes

Formulas of organic compounds present information at several different levels of sophistication. **Molecular formulas**, such as that of octane (C_8H_{18}), give the number of each kind of atom in a molecule of a compound. As shown in Figure 29.3, however, the molecular formula of C_8H_{18} may apply to several alkanes, each one of which has unique chemical, physical, and toxicological properties. These different compounds are designated by **structural formulas** showing the order in which the atoms in a molecule are arranged. Compounds that have the same molecular but different structural formulas are called **structural isomers**. Of the compounds shown in Figure 29.3, n-octane, 2,5-dimethylhexane, and 2-methyl-3-ethylpentane are structural isomers, all having the formula C_8H_{18}, whereas 1,4-dimethylcyclohexane is not a structural isomer of the other three compounds because its molecular formula is C_8H_{16}.

Alkanes and Alkyl Groups

Most organic compounds can be derived from alkanes. In addition, many important parts of organic molecules contain one or more alkane groups minus a hydrogen atom bonded as substituents onto the basic organic molecule. As a consequence of these factors, the names of many organic compounds are based upon alkanes, and it is useful to know the names of some of the more common alkanes and substituent groups derived from them as shown in Table 29.1.

Table 29.1. Some Alkanes and Substituent Groups Derived from Them.

Alkane	Substituent groups derived from alkane

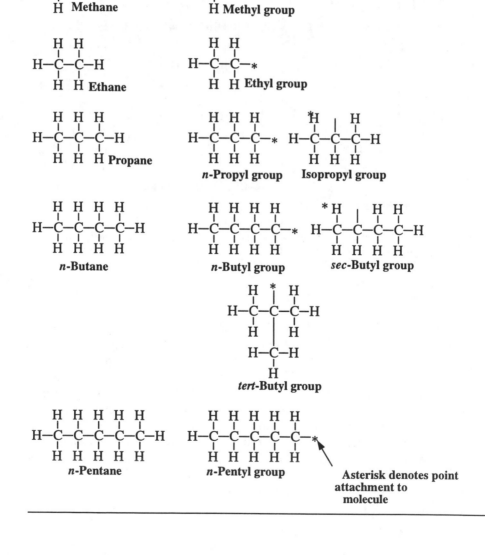

Names of Alkanes and Organic Nomenclature

Systematic names, from which the structures of organic molecules can be deduced, have been assigned to all known organic compounds. The more common organic compounds, including many toxic and hazardous organic sustances, likewise have **common names** that have no structural implications. Although it is not possible to cover organic nomenclature in any detail in this chapter, the basic approach to nomenclature (naming) is presented in the chapter along with some pertinent examples. The simplest approach is to begin with names of alkane hydro-carbons.

Consider the alkanes shown in Figure 29.3. The fact that *n*-octane has no side chains is denoted by "*n*", that it has 8 carbon atoms is denoted by "oct," and that it is an alkane is indicated by "ane." The names of compounds with branched chains or atoms other than H or C attached make use of numbers that stand for positions on the longest continuous chain of carbon atoms in the molecule. This convention is illus-trated by the second compound in Figure 29.3. It gets the hexane part of the name from the fact that it is an alkane with 6 carbon atoms in its longest continuous chain ("hex" stands for 6). However, it has a methyl group (CH_3) attached on the second carbon atom of the chain and another on the fifth. Hence the full systematic name of the compound is 2,5-dimethylhexane, where "di" indicates two methyl groups. In the case of 2-methyl-3-ethylpentane, the longest continuous chain of carbon atoms contains 5 carbon atoms, denoted by *pent*ane, a methyl group is attached to the second carbon atom, and an ethyl group, C_2H_5, on the third carbon atom The last compound shown in the figure has 6 carbon atoms in a ring, indicated by the prefix "cyclo," so it is a cyclo*hex*ane compound. Furthermore, the carbon in the ring to which one of the methyl groups is attached is designated by "1" and another methyl group is attached to the fourth carbon atom around the ring. Therefore, the full name of the compound is 1,4-dimethylcyclohexane.

Summary of Organic Nomenclature as Applied to Alkanes

Naming relatively simple alkanes is a straightforward process. The basic rules to be followed are the following:

1. The name of the compound is based upon the longest continuous chain of carbon atoms. (The structural formula may be drawn such that this chain is not immediately obvious.)

2. The carbon atoms in the longest continous chain are numbered sequentially from one end. The end of the chain from which the numbering is started is chosen to give the lower numbers for substituent groups in the final name. For example, the compound,

$$H-\overset{\overset{\displaystyle H}{|}}{\underset{\underset{\displaystyle H}{|}}{C}}-\overset{\overset{\displaystyle H}{|}}{\underset{\underset{\displaystyle H}{|}}{C}}-\overset{\overset{\displaystyle H}{|}}{\underset{\underset{\displaystyle H}{|}}{C}}-\overset{\overset{\displaystyle H}{|}}{\underset{\underset{\displaystyle CH_3}{|}}{C}}-\overset{\overset{\displaystyle H}{|}}{\underset{\underset{\displaystyle H}{|}}{C}}-H$$

could be named 4-methylpentane (numbering the 5-carbon chain from the left), but should be named 2-methylpentane (numbering the 5-carbon chain from the right).

3. All groups attached to the longest continuous chain are designated by the number of the carbon atoms to which they are attached and by the name of the substituent group ("2-methyl" in the example cited in Step 2, above).

4. A prefix is used to denote multiple substitutions by the same kind of group. This is illustrated by 2,2,3-trimethylpentane

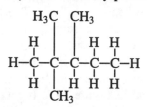

in which the prefix *tri* is used to show that *three* methyl groups are attached to the pentane chain.

5. The complete name is assigned such that it denotes the longest continuous chain of carbon atoms and the name and location on this chain of each substituent group.

Reactions of Alkanes

Alkanes contain only C-C and C-H bonds, both of which are relatively strong. For that reason they have little tendency to undergo many kinds of reactions common to some other organic chemicals, such as acid-base reactions or low-temperature oxidation-reduction reactions. However, at elevated temperatures alkanes readily undergo oxidation, more specifically combustion, with molecular oxygen in air as shown by the following reaction of propane:

$$C_3H_8 + 5O_2 \rightarrow 3CO_2 + 4H_2O + heat \qquad (29.7.1)$$

Common alkanes are highly flammable and the more volatile lower molecular mass alkanes form explosive mixtures with air. Furthermore, combustion of alkanes in an oxygen-deficient atmosphere or in an automobile engine produces significant quantities of carbon monoxide, CO, the toxic properties of which are discussed in Chapter 23.

In addition to combustion, alkanes undergo **substitution reactions** in which one or more H atoms on an alkane are replaced by atoms of another element. The most common such reaction is the replacement of H by chlorine, to yield **organochlorine** compounds. For example, methane reacts with chlorine to give chloromethane. This reaction begins with the dissociation of molecular chlorine, usually initiated by ultraviolet electromagnetic radiation:

$$Cl_2 + UV\ energy \rightarrow Cl\cdot + Cl\cdot \qquad (29.7.2)$$

The Cl· product is a **free radical** species in which the chlorine atom has only 7 outer shell electrons as shown by the Lewis symbol,

$$:\overset{..}{\underset{..}{Cl}}\cdot$$

instead of the favored octet of 8 outer-shell electrons. In gaining the octet required for chemical stability, the chlorine atom is very reactive. It abstracts a hydrogen from methane,

$$Cl\cdot + CH_4 \rightarrow HCl + CH_3\cdot \qquad (29.7.3)$$

to yield HCl gas and another reactive species with an unpaired electron, $CH_3\cdot$, called methyl radical. The methyl radical attacks molecular chlorine,

$$CH_3\cdot + Cl_2 \rightarrow CH_3Cl + Cl\cdot \qquad (29.7.4)$$

to give the chloromethane (CH_3Cl) product and regenerate Cl·, which can attack additional methane as shown in Reaction 29.7.3. The reactive Cl· and $CH_3\cdot$ species continue to cycle through the two preceding reactions.

The reaction sequence shown above illustrates three important aspects of chemistry that are shown to be very important in the discussion of atmospheric chemistry in Chapters 9-14. The first of these is that a reaction may be initiated by a **photochemical process** in which a photon of "light" (electromagnetic radiation) energy produces a reactive species, in this case the Cl· atom. The second point illustrated is the high chemical reactivity of **free radical species** with unpaired electrons and incomplete octets of valence electrons. The third point illustrated is that of **chain reactions**, which can multiply manyfold the effects of a single reaction-initiating event, such as the photochemical dissociation of Cl_2.

Alkenes and Alkynes

Alkenes or **olefins** are hydrocarbons that have double bonds consisting of 4 shared electrons. The simplest and most widely manufactured alkene is ethylene,

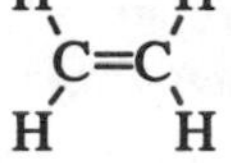

$$\underset{H}{\overset{H}{\diagdown}}C=C\underset{H}{\overset{H}{\diagup}} \quad \textbf{Ethylene (ethene)}$$

used for the production of polyethylene polymer. Another example of an important alkene is 1,3-butadiene (Figure 29.3), widely used in the manufacture of polymers, particularly synthetic rubber. The lighter alkenes, including ethylene and 1,3-butadiene, are highly flammable. Like other gaseous hydrocarbons, they form explosive mixtures with air.

Acetylene (Figure 29.3) is an **alkyne**, a class of hydrocarbons characterized by carbon-carbon triple bonds consisting of 6 shared electrons. Acetylene is a highly flammable gas that forms dangerously explosive mixtures with air. It is used in large quantities as a chemical raw material. Acetylene is the fuel in oxyacetylene torches used for cutting steel and for various kinds of welding applications.

Addition Reactions

The double and triple bonds in alkenes and alkynes have "extra" electrons capable of forming additional bonds. So the carbon atoms attached to these bonds can add atoms without losing any atoms already bonded to them, and the multiple bonds are said to be **unsaturated**. Therefore, alkenes and alkynes both undergo **addition reactions** in which pairs of atoms are added across unsaturated bonds as shown in the reaction of ethylene with hydrogen to give ethane:

$$
\begin{array}{c}
\text{H} \qquad \text{H} \\
\diagdown \quad \diagup \\
\text{C=C} \qquad + \quad \text{H–H} \quad \longrightarrow \quad \text{H–C–C–H} \\
\diagup \quad \diagdown \\
\text{H} \qquad \text{H}
\end{array}
\qquad (29.7.5)
$$

This is an example of a **hydrogenation reaction**, a very common reaction in organic synthesis, food processing (manufacture of hydrogenated oils), and petroleum refining. Another example of an addition reaction is that of HCl gas with acetylene to give vinyl chloride:

$$
\begin{array}{c}
\text{H} \qquad \text{H} \\
\diagdown \quad \diagup \\
\text{C=C} \qquad + \quad \text{H–Cl} \quad \longrightarrow \quad \text{H–C–C–Cl} \\
\diagup \quad \diagdown \\
\text{H} \qquad \text{H}
\end{array}
\qquad (29.7.6)
$$

This kind of reaction, which is not possible with alkanes, adds to the chemical and metabolic versatility of compounds containing unsaturated bonds and is a factor contributing to their generally higher toxicities. It makes unsaturated compounds much more chemically reactive, more hazardous to handle in industrial processes, and more active in atmospheric chemical processes such as smog formation (see Chapter 13).

Alkenes and *Cis-trans* Isomerism

As shown by the two simple compounds in Figure 29.4, the two carbon atoms connected by a double bond in alkenes cannot rotate relative to each other. For this reason, another kind of isomerism, known as *cis-trans* isomerism, is possible for alkenes. *Cis-trans* isomers have different parts of the molecule oriented differently in space, although these parts occur in the same order. Both alkenes illustrated in Fig-

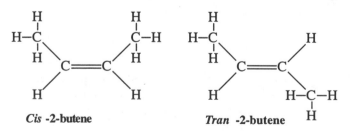

Figure 29.4. *Cis* and *trans* isomers of the alkene, 2-butene.

ure 29.4 have a molecular formula of C_4H_8. In the case of *cis*-2-butene, the two CH_3 (methyl) groups attached to the C=C carbon atoms are on the same side of the double bond, whereas in *trans*-2-butene they are on opposite sides.

Condensed Structural Formulas

To save space, structural formulas are conveniently abbreviated as **condensed structural formulas** such as $CH_3CH(CH_3)CH(C_2H_5)CH_2CH_3$ for 2-methyl-3-ethylpentane, where the CH_3 (methyl) and C_2H_5 (ethyl) groups are placed in parentheses to show that they are branches attached to the longest continuous chain of carbon atoms, which contains 5 carbon atoms. It is understood that each of the methyl and ethyl groups is attached to the carbon immediately preceding it in the condensed structural formula (methyl attached to the second carbon atom, ethyl to the third).

As illustrated by the examples in Figure 29.5, the structural formulas of organic molecules may be represented in a very compact form by lines and by figures such as hexagons. The ends and intersections of straight line segments in these formulas

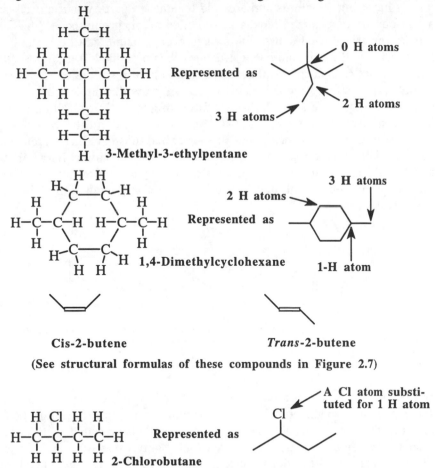

Figure 29.5. Representation of structural formulas with lines. A carbon atom is understood to be at each corner and at the end of each line. The numbers of hydrogen atoms attached to carbons at several specific locations are shown with arrows.

indicate the locations of carbon atoms. Carbon atoms at the terminal ends of lines are understood to have three H atoms attached, C atoms at the intersections of two lines are understood to have *two* H atoms attached to each, *one* H atom is attached to a carbon represented by the intersection of three lines, and *no* hydrogen atoms are bonded to C atoms where four lines intersect. Other atoms or groups of atoms, such as the Cl atom or OH group, that are substituted for H atoms are shown by their symbols attached to a C atom with a line.

Aryl Hydrocarbons

Benzene (Figure 29.6) is the simplest of a large class of **aryl** or **aromatic** hydrocarbons. Many important aryl compounds have substituent groups containing atoms of elements other than hydrogen and carbon and are called **aryl compounds** or **aromatic compounds**. Most aromatic compounds discussed in this book contain 6-carbon-atom benzene rings as shown for benzene, C_6H_6, in Figure 29.13. Aromatic compounds have ring structures and are held together in part by particularly stable bonds that contain delocalized clouds of so-called π (pi, pronounced "pie") electrons. In an oversimplified sense, the structure of benzene can be visualized as resonating between the two equivalent structures shown on the left in Figure 29.6 by the shifting of electrons in chemical bonds to form a hybrid structure. This structure can be shown more simply and accurately by a hexagon with a circle in it.

Aryl compounds have special characteristics of **aromaticity**, which include a low hydrogen:carbon atomic ratio; C–C bonds that are quite strong and of intermediate length between such bonds in alkanes and those in alkenes; tendency to undergo substitution reactions rather than the addition reactions characteristic of alkenes; and delocalization of π electrons over several carbon atoms. The last phenomenon adds substantial stability to aromatic compounds and is known as **resonance stabilization**.

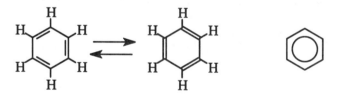

Figure 29.6. Representation of the aromatic benzene molecule with two resonance structures (left) and, more accurately, as a hexagon with a circle in it (right). Unless shown by symbols of other atoms, it is understood that a C atom is at each corner and that one H atom is bonded to each C atom.

First isolated from hydrocarbon liquids produced in the coking of coal (heating in absence of air to produce a carbonaceous residue used in steelmaking), benzene, toluene, and other low-molecular-mass aryl compounds have found widespread use as solvents, in chemical synthesis, and as a fuel. Many toxic substances, environmental pollutants, and hazardous waste compounds, such as benzene, toluene, naphthalene, and chlorinated phenols, are aryl compounds (see Figure 29.7). As shown in Figure 29.7, some arenes, such as naphthalene and the polycyclic aromatic compound, benzo(a)pyrene, contain fused rings.

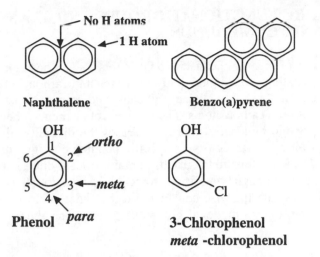

Naphthalene

Benzo(a)pyrene

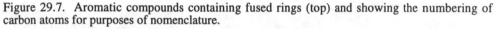

Phenol *para*

3-Chlorophenol
meta -chlorophenol

Figure 29.7. Aromatic compounds containing fused rings (top) and showing the numbering of carbon atoms for purposes of nomenclature.

Benzene and Naphthalene

Benzene is a volatile, colorless, highly flammable liquid that is consumed as a raw material for the manufacture of phenolic and polyester resins, polystyrene plastics, alkylbenzene surfactants, chlorobenzenes, insecticides, and dyes. It is hazardous both for its ignitability and toxicity (exposure to benzene causes blood abnormalities that may develop into leukemia). Naphthalene is the simplest member of a large number of polycyclic (multicyclic) aromatic hydrocarbons having two or more fused rings. It is a volatile white crystalline solid with a characteristic odor and has been used to make mothballs. The most important of the many chemical derivatives made from naphthalene is phthalic anhydride, from which phthalate ester plasticizers are synthesized.

Polycyclic Aromatic Hydrocarbons

Benzo(a)pyrene (Figure 29.7) is the most studied of the polycyclic aromatic hydrocarbons (PAHs), which are characterized by condensed ring systems ("chicken wire" structures). These compounds are formed by the incomplete combustion of other hydrocarbons, a process that consumes hydrogen in preference to carbon. The carbon residue is left in the thermodynamically favored condensed aromatic ring system of the PAH compounds.

Because there are so many partial combustion and pyrolysis processes that favor production of PAHs, these compounds are encountered abundantly in the atmosphere, soil, and elsewhere in the environment from sources that include engine exhausts, wood stove smoke, cigarette smoke, and char-broiled food. Coal tars and petroleum residues such as road and roofing asphalt have high levels of PAHs. Some PAH compounds, including benzo(a)pyrene, are of toxicological concern because they are precursors to cancer-causing metabolites.

29.3. ORGANIC FUNCTIONAL GROUPS AND CLASSES OF ORGANIC COMPOUNDS

The discussion of organic chemistry so far in this chapter has emphasized hydrocarbon compounds, those that contain only hydrogen and carbon. It has been shown that hydrocarbons may exist as alkanes, alkenes, and arenes, depending upon the kinds of bonds between carbon atoms. The presence of elements other than hydrogen and carbon in organic molecules greatly increases the diversity of their chemical behavior. **Functional groups** consist of specific bonding configurations of atoms in organic molecules. Most functional groups contain at least one element other than carbon or hydrogen, although two carbon atoms joined by a double bond (alkenes) or triple bond (alkynes) are likewise considered to be functional groups. Table 29.2 shows some of the major functional groups that determine the nature of organic compounds.

Organooxygen Compounds

The most common types of compounds with oxygen-containing functional groups are epoxides, alcohols, phenols, ethers, aldehydes, ketones, and carboxylic acids. The functional groups characteristic of these compounds are illustrated by the examples of oxygen-containing compounds shown in Figure 29.8.

Ethylene oxide is a moderately to highly toxic, sweet-smelling, colorless, flammable, explosive gas used as a chemical intermediate, sterilant, and fumigant. It is a mutagen and a carcinogen to experimental animals. It is classified as hazardous for both its toxicity and ignitability. **Methanol** is a clear, volatile, flammable liquid alcohol used for chemical synthesis, as a solvent, and as a fuel. It is used as a gasoline additive to reduce emissions of carbon monoxide and other air pollutants. Ingestion of methanol can be fatal, and blindness can result from sublethal doses. **Phenol** is a dangerously toxic aryl alcohol widely used for chemical synthesis and polymer manufacture. **Methyltertiarybutyl ether**, MTBE, is an ether that has become the octane booster of choice to replace tetraethyllead in gasoline. **Acrolein** is an alkenic aldehyde and a volatile, flammable, highly reactive chemical. It forms explosive peroxides upon prolonged contact with O_2. An extreme lachrimator and strong irritant, acrolein is quite toxic by all routes of exposure. **Acetone** is the lightest of the ketones. Like all ketones, acetone has a carbonyl (C=O) group that is bonded to *two* carbon atoms (that is, it is somewhere in the middle of a carbon atom chain). Acetone is a good solvent and is chemically less reactive than the aldehydes which all have the functional group,

$$-\overset{\overset{\displaystyle O}{\|}}{C}-H$$

in which binding of the C=O to H makes the molecule significantly more reactive. **Propionic acid** is a typical organic carboxylic acid. The -CO_2H group of carboxylic acids may be viewed as the most oxidized functional group (other than peroxides) on an oxygenated organic compound, and carboxylic acids may be synthesized by oxidizing alcohols or aldehydes that have an -OH group or C=O group on an end carbon atom.

Table 29.2. Examples of Some Important Functional Groups

Type of functional group	Example compound	Structural formula of group[1]
Alkene (olefin)	Propene (propylene)	
Alkyne	Acetylene	
Alcohol (-OH attached to alkyl group)	2-Propanol	
Phenol (-OH attached to aryl group)	Phenol	
Ketone	Acetone	

$$\text{(When } -\overset{\overset{\text{O}}{\|}}{\text{C}}-\text{H group is on end carbon,}$$
compound is an aldehyde)

Amine	Methylamine	
Nitro compounds	Nitromethane	
Sulfonic acids	Benzenesulfonic acid	
Organohalides	1,1–Dichloroethane	

[1] Functional group outlined by dashed line

Organonitrogen Compounds

Figure 29.9 shows examples of three classes of the many kinds of compounds that contain N (amines, nitrosamines, and nitro compounds). Nitrogen occurs in many functional groups in organic compounds, some of which contain nitrogen in ring structures, or along with oxygen.

Methylamine is a colorless, highly flammable gas with a strong odor. It is a severe irritant affecting eyes, skin, and mucous membranes. Methylamine is the simplest of the **amine** compounds, which have the general formula, where the R's are hydrogen or hydrocarbon groups, at least one of which is the latter.

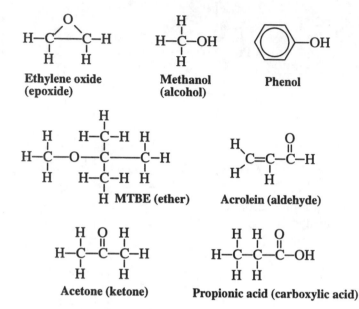

Figure 29.8. Examples of oxygen-containing organic compounds that may be significant as wastes, toxic substances, or environmental pollutants.

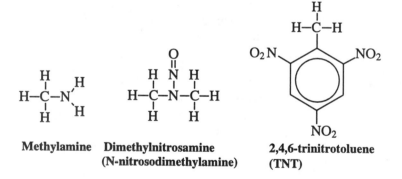

Figure 29.9. Examples of organonitrogen that may be significant as wastes, toxic substances, or environmental pollutants.

Dimethylnitrosamine is an N-nitroso compound, all of which are characterized by the N-N=O functional group. It was once widely used as an industrial solvent, but caused liver damage and jaundice in exposed workers. Subsequently, numerous other N-nitroso compounds, many produced as by-products of industrial operations and food and alcoholic beverage processing, were found to be carcinogenic.

Solid **2,4,6-trinitrotoluene** (TNT) has been widely used as a military explosive. TNT is moderately to very toxic and has caused toxic hepatitis or aplastic anemia in exposed individuals, a few of whom have died from its toxic effects. It belongs to the general class of nitro compounds characterized by the presence of -NO_2 groups bonded to a hydrocarbon structure.

Some organonitrogen compounds are chelating agents that bind strongly to metal ions and play a role in the solubilization and transport of heavy metal wastes. Prominent among these are salts of the aminocarboxylic acids which, in the acid form, have -CH_2CO_2H groups bonded to nitrogen atoms. An important example of such a compound is the monohydrate of trisodium nitrilotriacetate (NTA):

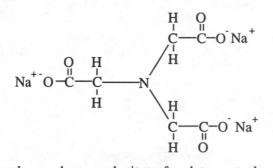

This compound can be used as a substitute for detergent phosphates to bind to calcium ion and make the detergent solution basic. NTA is used in metal plating formulations. It is highly water soluble and quickly eliminated with urine when ingested. It has a low acute toxicity and no chronic effects have been shown for plausible doses. However, concern does exist over its interaction with heavy metals in waste treatment processes and in the environment.

Organohalide Compounds

Organohalides (Figure 29.10) exhibit a wide range of physical and chemical properties. These compounds consist of halogen-substituted hydrocarbon molecules, each of which contains at least one atom of F, Cl, Br, or I. They may be saturated (**alkyl halides**), unsaturated (**alkenyl halides**), or aromatic (**aryl halides**). The most widely manufactured organohalide compounds are chlorinated hydrocarbons, many of which are regarded as environmental pollutants or as hazardous wastes.

Alkyl Halides

Substitution of halogen atoms for one or more hydrogen atoms on alkanes gives **alkyl halides**, example structural formulas of which are given in Figure 29.10. Most of the commercially important alkyl halides are derivatives of alkanes of low molecular mass. A brief discussion of the uses of the compounds listed in Figure 29.10 is given here to provide an idea of the versatility of the alkyl halides.

Dichloromethane is a volatile liquid with excellent solvent properties for nonpolar organic solutes. It has been used as a solvent for the decaffeination of coffee, in paint strippers, as a blowing agent in urethane polymer manufacture, and to depress vapor pressure in aerosol formulations. Once commonly sold as a solvent and stain remover, highly toxic **carbon tetrachloride** is now largely restricted to uses

862   **Environmental Chemistry**

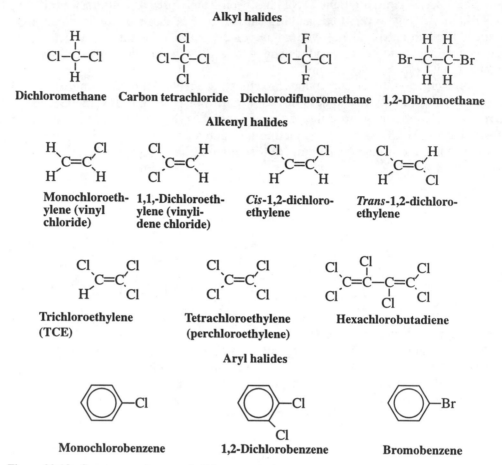

Figure 29.10. Some example organohalide compounds.

as a chemical intermediate under controlled conditions, primarily to manufacture chlorofluorocarbon refrigerant fluid compounds, which are also discussed in this section. Insecticidal **1,2-dibromoethane** has been consumed in large quantities as a lead scavenger in leaded gasoline and to fumigate soil, grain, and fruit (Fumigation with this compound has been discontinued because of toxicological concerns). An effective solvent for resins, gums, and waxes, it serves as a chemical intermediate in the syntheses of some pharmaceutical compounds and dyes.

Alkenyl Halides

Viewed as hydrocarbon-substituted derivatives of alkenes, the **alkenyl** or **olefinic organohalides** contain at least one halogen atom and at least one carbon-carbon double bond. The most significant of these are the lighter chlorinated compounds, such as those illustrated in Figure 29.17.

Vinyl chloride is consumed in large quantities as a raw material to manufacture pipe, hose, wrapping, and other products fabricated from polyvinylchloride plastic. This highly flammable, volatile, sweet-smelling gas is a known human carcinogen.

As shown in Figure 29.10, there are three possible dichloroethylene compounds, all clear, colorless liquids. Vinylidene chloride forms a copolymer with vinyl chloride used in some kinds of coating materials. The geometrically isomeric 1,2-dichloroethylenes are used as organic synthesis intermediates and as solvents. **Trichloroethylene** is a clear, colorless, nonflammable, volatile liquid. It is an excellent degreasing and drycleaning solvent and has been used as a household solvent and for food extraction (for example, in decaffeination of coffee). Colorless, nonflammable liquid **tetrachloroethylene** has properties and uses similar to those of trichloroethylene. **Hexachlorobutadiene**, a colorless liquid with an odor somewhat like that of turpentine, is used as a solvent for higher hydrocarbons and elastomers, as a hydraulic fluid, in transformers, and for heat transfer.

Aryl Halides

Aryl halide derivatives of benzene and toluene have many uses in chemical synthesis, as pesticides and raw materials for pesticides manufacture, as solvents, and a diverse variety of other applications. These widespread uses over many decades have resulted in substantial human exposure and environmental contamination. Three example aryl halides are shown in Figure 29.17. Monochlorobenzene is a flammable liquid boiling at 132°C. It is used as a solvent, heat transfer fluid, and synthetic reagent. Used as a solvent, 1,2-dichlorobenzene is employed for degreasing hides and wool. It also serves as a synthetic reagent for dye manufacture. Bromobenzene is a liquid boiling at 156°C that is used as a solvent, motor oil additive, and intermediate for organic synthesis.

Halogenated Naphthalene and Biphenyl

Two major classes of halogenated aryl compounds containing two benzene rings are made by the chlorination of naphthalene and biphenyl and have been sold as mixtures with varying degrees of chlorine content. Examples of chlorinated naphthalenes, and polychlorinated biphenyls (PCBs discussed later), are shown in Figure 29.11. The less highly chlorinated of these compounds are liquids, and those with higher chlorine contents are solids. Because of their physical and chemical stabilities and other desirable qualities, these compounds have had many uses, including heat transfer fluids, hydraulic fluids, and dielectrics. Polybrominated biphenyls (PBBs) have served as flame retardants. However, because chlorinated naphthalenes, PCBs, and PBBs are environmentally extremely persistent, their uses have been severely curtailed.

Chlorofluorocarbons, Halons, and Hydrogen-Containing Chlorofluorocarbons

Chlorofluorocarbons (**CFCs**) are volatile 1- and 2-carbon compounds that contain Cl and F bonded to carbon. These extremely stable and nontoxic compounds are discussed in some detail in Section 12.7. They were once widely used in the fabrication of flexible and rigid foams, and as fluids for refrigeration and air conditi-

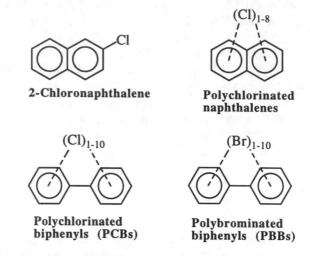

Figure 29.11. Halogenated naphthalenes and biphenyls.

tioning, but have now been essentially phased out because of their potential to cause harm to the stratospheric ozone layer. The most widely manufactured of these compounds in the past were CCl_3F (CFC-11), CCl_2F_2 (CFC-12), $C_2Cl_3F_3$ (CFC-113), $C_2Cl_2F_4$ (CFC-114), and C_2ClF_5 (CFC-115). **Halons** are related compounds that contain bromine and are used in fire extinguisher systems. The most commonly produced commercial halons were $CBrClF_2$ (Halon-1211), $CBrF_3$ (Halon-1301), and $C_2Br_2F_4$ (Halon-2402), where the sequence of numbers denotes the number of carbon, fluorine, chlorine, and bromine atoms, respectively, per molecule. Halons have also been implicated as ozone-destroying gases in the stratosphere and are being phased out, although finding suitable replacements has been difficult.

Hydrohalocarbons are hydrogen-containing chlorofluorocarbons (HCFCs) and hydrogen-containing fluorocarbons (HFCs) that are now produced as substitutes for chlorofluorocarbons. These compounds include CH_2FCF_3 (HFC-134a, a substitute for CFC-12 in automobile air conditioners and refrigeration equipment), $CHCl_2CF_3$ (HCFC-123, substitute for CFC-11 in plastic foam-blowing), CH_3CCl_2F (HCFC-141b, substitute for CFC-11 in plastic foam-blowing), and $CHClF_2$ (HCFC-22, air conditioners and manufacture of plastic foam food containers). Because each molecule of these compounds has at least one H-C bond, which is much more readily broken than C-Cl or C-F bonds, the HCFCs do not persist in the atmosphere and pose essentially no threat to the stratospheric ozone layer.

Chlorinated Phenols

The chlorinated phenols, particularly **pentachlorophenol** and the trichlorophenol isomers, are significant hazardous wastes. These compounds are biocides that are used to treat wood to prevent rot by fungi and to prevent termite infestation. They are toxic, causing liver malfunction and dermatitis. However, contaminant polychlorinated dibenzodioxins ("dioxin") may be responsible for some of the observed effects. Pentachlorophenol and other aryl halides and aryl hydrocarbons used as wood preservatives are encountered at many hazardous waste sites in waste-waters and sludges.

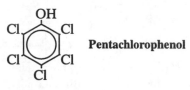

Pentachlorophenol

Organosulfur Compounds

The chemistry of sulfur is similar to but perhaps more diverse than that of oxygen. Whereas, with the exception of peroxides, most chemically combined organic oxygen is in the -2 oxidation state, sulfur occurs in the -2, +4, and +6 oxidation states. Many organosulfur compounds are noted for their foul, "rotten egg" or garlic odors. A number of example organosulfur compounds are shown in Figure 29.12.

Thiols and Thioethers

Substitution of alkyl or aryl hydrocarbon groups such as phenyl and methyl for H on hydrogen sulfide, H_2S, leads to a number of different organosulfur **thiols** (mercaptans, R–SH) and **sulfides**, also called thioethers (R–S–R). Structural formulas of examples of these compounds are shown in Figure 29.12.

Methanethiol and other lighter alkyl thiols are fairly common air pollutants that have "ultragarlic" odors; both 1- and 2-butanethiol are associated with skunk odor. Gaseous methanethiol is used as an odorant leak-detecting additive for natural gas, propane, and butane; it is also employed as an intermediate in pesticide synthesis. A toxic, irritating volatile liquid with a strong garlic odor, 2-propene-1-thiol (allyl mercaptan) is a typical alkenyl mercaptan. Benzenethiol (phenyl mercaptan) is the simplest of the aryl thiols. It is a toxic liquid with a severely "repulsive" odor.

Alkyl sulfides or thioethers contain the C-S-C functional group. The lightest of these compounds is dimethyl sulfide, a volatile liquid (bp 38°C) that is moderately toxic by ingestion. It was mentioned in Chapter 11, Section 11.4, as a major source of gaseous sulfur entering the atmosphere over the oceans due to its production by marine organisms. Cyclic sulfides contain the C-S-C group in a ring structure. The most common of these compounds is thiophene, a heat-stable liquid (bp 84°C) with a solvent action much like that of benzene, that is used to make pharmaceuticals, dyes, and resins. Its saturated analog is tetrahydrothiophene, or thiophane.

Nitrogen-Containing Organosulfur Compounds

Many important organosulfur compounds also contain nitrogen. One such compound is **thiourea**, the sulfur analog of urea. Its structural formula is shown in Figure 29.12. Thiourea and **phenylthiourea** have been used as rodenticides. Commonly called ANTU, **1-naphthylthiourea** is an excellent rodenticide that is virtually tasteless and has a very high rodent:human toxicity ratio.

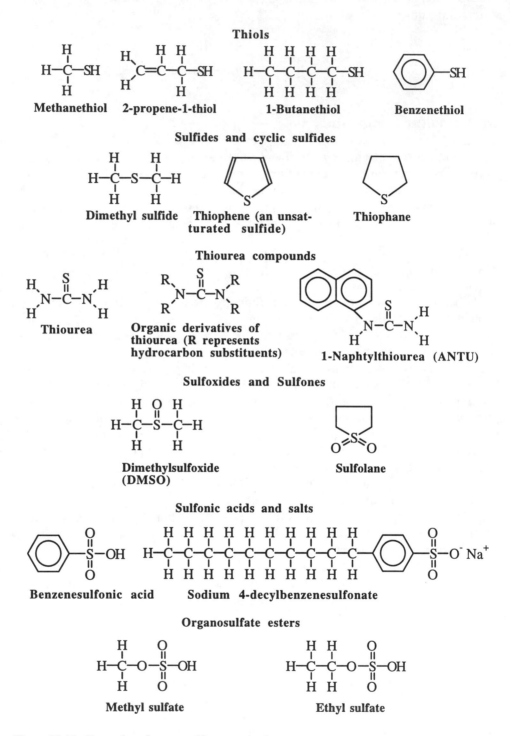

Figure 29.12. Examples of organosulfur compounds.

Sulfoxides and Sulfones

Sulfoxides and **sulfones** (Figure 29.12) contain both sulfur and oxygen. **Dimethylsulfoxide** (DMSO) is a liquid with numerous uses and some very interesting properties. It is used to remove paint and varnish, as a hydraulic fluid, mixed with water as an antifreeze solution, and in pharmaceutical applications as an anti-inflammatory and bacteriostatic agent. A polar aprotic (no ionizable H) solvent with a relatively high dielectric constant, **sulfolane** dissolves both organic and inorganic solutes. It is the most widely produced sulfone because of its use in an industrial process called BTX processing in which it selectively extracts benzene, toluene, and xylene from aliphatic hydrocarbons; as the solvent in the Sulfinol process by which thiols and acidic compounds are removed from natural gas; as a solvent for polymerization reactions; and as a polymer plasticizer.

Sulfonic Acids, Salts, and Esters

Sulfonic acids and sulfonate salts contain the $-SO_3H$ and $-SO_3^-$ groups, respectively, attached to a hydrocarbon moiety. The structural formula of benzenesulfonic acids and of sodium 1-(p-sulfophenyl)decane, a biodegradable detergent surfactant, are shown in Figure 29.19. The common sulfonic acids are water-soluble strong acids that lose virtually all ionizable H^+ in aqueous solution. They are used commercially to hydrolyze fat and oil esters to fatty acids and glycerol.

Organic Esters of Sulfuric Acid

Replacement of 1 H on sulfuric acid, H_2SO_4, with a hydrocarbon group yields an acid ester, and replacement of both yields an ester. Examples of these esters are shown in Figure 29.10. Sulfuric acid esters are used as alkylating agents, which act to attach alkyl groups (such as methyl) to organic molecules in the manufacture of agricultural chemicals, dyes, and drugs. **Methylsulfuric acid** and **ethylsulfuric acid** are oily, water-soluble liquids that are strong irritants to skin, eyes, and mucous tissue.

Organophosphorus Compounds

Alkyl and Aryl Phosphines

The first two examples in Figure 29.11, illustrate that the structural formulas of alkyl and aryl phosphine compounds may be derived by substituting organic groups for the H atoms in phosphine (PH_3), the hydride of phosphorus discussed as a toxic inorganic compound in Chapter 23, Section 23.3. **Methylphosphine** is a colorless, reactive gas. Crystalline, solid **triphenylphosphine** has a low reactivity and moderate toxicity when inhaled or ingested.

As shown by the reaction,

$$4C_3H_9P + 26O_2 \rightarrow 12O_2 + 18H_2O + P_4O_{10} \qquad (29.8.4)$$

combustion of aryl and alkyl phosphines produces P_4O_{10}, a corrosive, irritant, toxic substance that reacts with moisture in the air to produce droplets of corrosive orthophosphoric acid, H_3PO_4.

Organophosphate Esters

The structural formulas of three esters of orthophosphoric acid (H_3PO_4) and an ester of pyrophosphoric acid ($H_4P_2O_6$) are shown in Figure 29.13. Although **trimethylphosphate** is considered to be only moderately toxic, **tri-*o*-cresylphosphate, TOCP,** has a notorious record of poisonings. **Tetraethylpyrophosphate, TEPP,** was developed in Germany during World War II as a substitute for insecticidal nicotine. Although it is a very effective insecticide, its use in that application was of very short duration because it kills almost everything else, too.

Phosphorothionate Esters

Parathion, shown in Figure 29.13, is an example of **phosphorothionate** esters. These compounds are used as insecticidal acetylcholinesterase inhibitors. They contain the P=S (thiono) group, which increases their insect:mammal toxicity ratios. Since the first organophosphate insecticides were developed in Germany during the 1930s and 1940s, many insecticidal organophosphate compounds have been synthesized. One of the earliest and most successful of these is **parathion,** *O,O*-diethyl-*O-p*-nitrophenylphosphorothionate (banned from use in the U.S. in 1991 because of its acute toxicity to humans). From a long-term environmental standpoint, organophosphate insecticides are superior to the organohalide insecticides that they largely displaced because the organophosphates readily undergo biodegradation and do not bioaccumulate.

29.4. SYNTHETIC POLYMERS

A large fraction of the chemical industry worldwide is devoted to polymer manufacture, which is very important in the area of hazardous wastes, as a source of environmental pollutants, and in the manufacture of materials used to alleviate environmental and waste problems. Synthetic **polymers** are produced when small molecules called **monomers** bond together to form a much smaller number of very large molecules. Many natural products are polymers; for example, cellulose produced by trees and other plants and found in wood, paper, and many other materials, is a polymer of the sugar glucose. Synthetic polymers form the basis of many industries, such as rubber, plastics, and textiles manufacture.

An important example of a polymer is that of polyvinylchloride, shown in Figure 29.14. This polymer is synthesized in large quantities for the manufacture of water and sewer pipe, water-repellant liners, and other plastic materials. Other major polymers include polyethylene (plastic bags, milk cartons), polypropylene, (impact-resistant plastics, indoor-outdoor carpeting), polyacrylonitrile (Orlon, carpets), polystyrene (foam insulation), and polytetrafluoroethylene (Teflon coatings, bearings); the monomers from which these substances are made are shown in Figure 29.15.

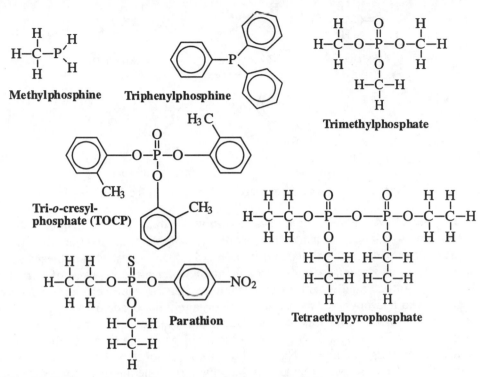

Figure 29.13. Some representative organophosphorus compounds.

Many of the hazards from the polymer industry arise from the monomers used as raw materials. Many monomers are reactive and flammable, with a tendency to form explosive vapor mixtures with air. All have a certain degree of toxicity; vinyl chloride is a known human carcinogen. The combustion of many polymers may result in the evolution of toxic gases, such as hydrogen cyanide (HCN) from polyacrylonitrile, or hydrogen chloride (HCl) from polyvinylchloride. Another hazard presented by plastics results from the presence of **plasticizers** added to provide essential properties such as flexibility. The most widely used plasticizers are phthalates, which are environmentally persistent, resistant to treatment processes, and prone to undergo bioaccumulation.

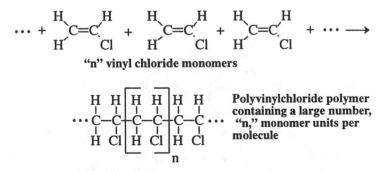

Figure 29.14. Polyvinylchloride polymer.

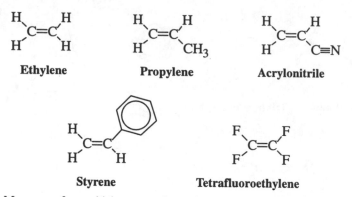

Figure 29.15. Monomers from which commonly used polymers are synthesized.

Polymers have a number of applications in waste treatment and disposal. Waste disposal landfill liners are made from synthetic polymers, as are the fiber filters which remove particulate pollutants from flue gas in baghouses. Membranes used for ultrafiltration and reverse osmosis treatment of water are composed of very thin sheets of synthetic polymers. Organic solutes can be removed from water by sorption onto hydrophobic (water-repelling) organophilic beads of Amberlite XAD resin. Heavy metal pollutants are removed from wastewater by cation exchange resins made of polymers with anionic functional groups. Typically, these resins exchange harmless sodium ion, Na^+, on the solid resin for toxic heavy metal ions in water. Figure 29.16 shows a segment of the polymeric structure of a cation exchange resin in the sodium form. In the treatment of heavy-metal-containing waste solutions, these resins can exchange toxic heavy metal ions in solution, such as Cd^{2+}, for nontoxic Na^+ ions. Ion exchange resins are used in nuclear reactors to remove traces of metals, some of which may be radioactive, from the water used in the reactor for heat exchange. Ion exchange resins have also been developed in which the ion-exchanging functional group is an iminiodiacetate $\{-N(CH_2CO_2^-)_2\}$ group that has a particularly strong affinity for heavy metals.

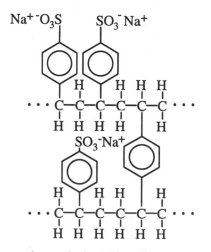

Figure 29.16. Polymeric cation exchanger in the sodium form.

SUPPLEMENTARY REFERENCES

Atkins, Robert C. and Francis A. Carey, *Organic Chemistry: A Brief Course*, 2nd ed., McGraw-Hill, New York, 1997.

Brown, William H. and Christopher S. Foote, *Organic Chemistry*, 2nd ed., Saunders College Publishing, Fort Worth, TX, 1998.

Bruice, Paula Yurkanis, *Organic Chemistry*, 2nd ed., Prentice Hall, Upper Saddle River, NJ, 1998

Ege, Seyhan N., Organic Chemistry: Structure and Reactivity, 4th ed., Houghton Mifflin, Boston, 1999.

Faber, Kurt, *Biotransformations in Organic Chemistry: A Textbook*, Springer Verlag, Berlin, 1997.

Fessenden, Ralph J., Joan S. Fessenden, and Marshall Logue, *Organic Chemistry*, 6th ed., Brooks/Cole Pub Co; Pacific Grove, CA, 1998.

Hornback, Joseph M., *Organic Chemistry*, Brooks/Cole Pub Co; Pacific Grove, CA, 1998.

Johnson, A. William, Invitation to Organic Chemistry, and Bartlett Publishers, Sudbury, MA, 1999.

McMurry, John, *Organic Chemistry*, 5th ed., Brooks/Cole/Thomson Learning, Pacific Grove, CA, 1999.

McMurry, John and Mary E. Castellion, *Fundamentals of General, Organic and Biological Chemistry*, Prentice Hall, Upper Saddle River, NJ, 1999.

Ouellette, Robert J., *Organic Chemistry: A Brief Introduction*, 2nd ed., Prentice Hall, Upper Saddle River, NJ, 1998.

Schwarzenbach, Rene P., Phillip M. Gschwend, and Dieter M. Imboden, *Environmental Organic Chemistry*, John Wiley & Sons, New York, 1993.

Simpson, Peter, *Basic Concepts in Organic Chemistry—A Programmed Learning Approach*, Chapman and Hall, London, 1994.

Solomons, T. W. Graham, *Organic Chemistry*, 6th ed., John Wiley & Sons, New York, 1998.

Sorrell, Thomas N., *Organic Chemistry*, University Science Books, Sausalito, CA, 1999.

Timberlake,Karen C., *Chemistry: An Introduction to General, Organic, and Biological Chemistry*, Benjamin/Cummings, Menlo Park, CA, 1999.

Vollhardt, K. Peter C. and Neil E. Schore, *Organic Chemistry: Structure and Function*, 3rd ed., W.H. Freeman, New York, 1999.

Wade, L. G., Jr., *Organic Chemistry* , 4th ed., Prentice Hall, Upper Saddle River, NJ, 1999.

QUESTIONS AND PROBLEMS

1. Explain the bonding properties of carbon that makes organic chemistry so diverse.

2. Distinguish among alkanes, alkenes, alkynes, and aryl compounds. To which general class of organic compounds do all belong?

3. In what sense are alkanes saturated? Why are alkenes more reactive than alkanes?

4. Name the compound below:

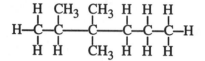

5. What is indicated by "*n*" in a hydrocarbon name?

6. Discuss the chemical reactivity of alkanes. Why are they chemically reactive or unreactive?

7. Discuss the chemical reactivity of alkenes. Why are they chemically reactive or unreactive?

8. What are the characteristics of aromaticity? What are the chemical reactivity characteristics of aromatic compounds?

9. Describe chain reactions, discussing what is meant by free radicals and photochemical processes.

10. Define, with examples, what is meant by isomerism.

11. Describe how the two forms of 1,2-dichloroethylene can be used to illustrate *cis-trans* isomerism.

12. Give the structural formula corresponding to the condensed structural formula of $CH_3CH(C_2H_5)CH(C_2H_5)CH_2CH_3$.

13. Discuss how organic functional groups are used to define classes of organic compounds.

14. Give the functional groups corresponding to (a) alcohols, (b) aldehydes, (c) carboxylic acids, (d) ketones, (e) amines, (f) thiol compounds, and (g) nitro compounds.

15. Give an example compound of each of the following: epoxides, alcohols, phenols, ethers, aldehydes, ketones, and carboxylic acids.

16. Which functional group is characteristic of N-nitroso compounds, and why are these compounds toxicologically significant?

17. Give an example of each of the following: Alkyl halides, alkenyl halides, aryl halides.

18. Give an example compound of a chlorinated naphthalene and of a PCB.

19. What explains the tremendous chemical stability of CFCs? What kinds of compounds are replacing CFCs? Why?

20. How does a thio differ from a thioether?

21. How do sulfoxides differ from sulfones?

22. Which inorganic compound is regarded as the parent compound of alkyl and aryl phosphines? Give an example of each of these.

23. What are organophosphate esters and what is their toxicological significance?

24. Define what is meant by a polymer and give an example of one.

INDEX